IMMUNOGL(

FactsBook

Other books in the FactsBook Series:

Robin Callard and Andy Gearing
The Cytokine FactsBook

Steve Watson and Steve Arkinstall
The G-Protein Linked Receptor FactsBook

Shirley Ayad, Ray Boot-Handford, Martin J. Humphries, Karl E. Kadler and C. Adrian Shuttleworth
The Extracellular Matrix FactsBook, 2nd edn

Grahame Hardie and Steven Hanks
The Protein Kinase FactsBook
The Protein Kinase FactsBook CD-Rom

Edward C. Conley
The Ion Channel FactsBook
I: Extracellular Ligand-Gated Channels

Edward C. Conley
The Ion Channel FactsBook
II: Intracellular Ligand-Gated Channels

Edward C. Conley and William J. Brammar
The Ion Channel FactsBook
IV: Voltage-Gated Channels

Kris Vaddi, Margaret Keller and Robert Newton
The Chemokine FactsBook

Marion E. Reid and Christine Lomas-Francis
The Blood Group Antigen FactsBook

A. Neil Barclay, Marion H. Brown, S.K. Alex Law, Andrew J. McKnight, Michael G. Tomlinson and P. Anton van der Merwe
The Leucocyte Antigen FactsBook, 2nd edn

Robin Hesketh
The Oncogene and Tumour Suppressor Gene FactsBook, 2nd edn

Jeffrey K. Griffith and Clare E. Sansom
The Transporter FactsBook

Tak W. Mak, Josef Penninger, John Rader, Janet Rossant and Mary Saunders
The Gene Knockout FactsBook

Bernard J. Morley and Mark J. Walport
The Complement FactsBook

Steven G.E. Marsh, Peter Parham and Linda Barber
The HLA FactsBook

Hans G. Drexler
The Leukemia-Lymphoma Cell Line FactsBook

Clare M. Isacke and Michael A. Horton
The Adhesion Molecule FactsBook, 2nd edn

Marie-Paule Lefrance and Gerard Lefranc
The T Cell ReceptorFactsBook (not yet published)

THE IMMUNOGLOBULIN *FactsBook*

Marie-Paule Lefranc
Gérard Lefranc

IMGT, the international ImMunoGeneTics database
Laboratoire d'ImmunoGénétique Moléculaire
Université Montpellier II,
Institut de Génétique Humaine CNRS,
Montpellier, France

ACADEMIC PRESS
A Harcourt Science and Technology Company
San Diego San Francisco New York Boston
London Sydney Tokyo

This book is printed on acid-free paper.

Academic Press
A Harcourt Science and Technology Company
Harcourt Place, 32 Jamestown Road, London NW1 7BY, UK
http://www.academicpress.com

Academic Press
A Harcourt Science and Technology Company
525 B Street, Suite 1900, San Diego, California 92101-4495, USA
http://www.academicpress.com

ISBN 0-12-441351-X

A catalogue record for this book is available from the British Library

Typeset by Mackreth Media Services, Hemel Hempstead, UK

Printed and bound in the United Kingdom

Transfered to Digital Printing, 2011

Contents

Section III THE HUMAN IMMUNOGLOBULIN IGK GENES

Part 1 IGKC

Part 2 IGKJ

Part 3 IGKV

Section IV THE HUMAN IMMUNOGLOBULIN IGL GENES

Preface

The authors wish to acknowledge the IMGT team that contributed to the completion of this book. In particular, we would like to thank Nathalie Bosc, Valérie Contet, Géraldine Folch, Christèle Jean and Dominique Scaviner, the motivated and enthusiastic IMGT annotators for their invaluable contribution and expertise. Sandrine Beranger, Olivier Elemento, Françoise Marlhens, Pauline Rodrigues and Manuel Ruiz helped with figures for the introductory chapters. We are very grateful to Gérard Mennessier for the IMGT/Collier de Perles tool development, Véronique Giudicelli and Denys Chaume for the bioinformatics and computer management of IMGT, the international ImMunoGeneTics database (http://imgt.cines.fr).

The authors wish to acknowledge the funding by the Ministère de l'Education Nationale, the Ministère de la Recherche, the Université Montpellier II, the CNRS, the European Community and the Région Languedoc-Roussillon.

The authors hope that there are a minimum of omissions and inaccuracies and that these can be rectified in later editions. It would be appreciated if such points were forwarded to the Editor, Human Immunoglobulin FactsBook, Academic Press Ltd, Harcourt Place, 32 Jamestown Road, London NW1 7BY, UK.

Back from left: *Nathalie Bosc, Françoise Marlhens, Olivier Elemento, Denys Chaume, Valérie Contet,* ***Gérard Lefranc****;*
front from left: ***Marie-Paule Lefranc****, Pauline Rodrigues, Dominique Scaviner, Christèle Jean, Géraldine Folch, Véronique Giudicelli.*

Abbreviations

a or A	Adenine (purine base of DNA and RNA)
c or C	Cytosine (pyrimidine base of DNA and RNA)
C	Constant
CDR	Complementarity Determining Region
CH	Immunoglobulin heavy constant exon or domain
CNRS	Centre National de la Recherche Scientifique
D	Diversity
DDBJ	DNA DataBank of Japan
E	Enhancer
EMBL	European Molecular Biology Laboratory Nucleotide Sequence Database
FR	Framework
g or G	Guanine (purine base of DNA and RNA)
GDB	Genome Database
GenBank	US Nucleotide Sequence Database
HUGO	HUman Genome Organization
Ig	Immunoglobulin
IMGT	The international ImMunoGeneTics database
J	Joining
kb	Kilobase
kDa	Kilodalton
LIGM-DB	Laboratoire d'ImmunoGénétique Moléculaire DataBase (Immunoglobulin and T cell receptor database), part of IMGT
nt	Nucleotide
OMIM	Online Mendelian Inheritance in Man (MIM)
ORF	Open Reading Frame
RS	Recombination signal
S	Switch
SRS	Sequence Retrieval System, a database query system developed by EMBL
t or T	Thymine (pyrimidine base of DNA)
V	Variable

Links to database or molecular biology server web sites quoted in this book are available from IMGT Bloc-notes, http://imgt.cines.fr

IMGT standardized labels used in this book

Label name	Definition
1st-CYS	codon (3 nucleotides) for Cysteine in conserved position in FR1
2nd-CYS	codon (3 nucleotides) for Cysteine in conserved position in FR3
3'D-HEPTAMER	7 nucleotide recombination site like CACAGTG, part of a 3'D-RS
3'D-NONAMER	9 nucleotide recombination site like ACAAAAACC, part of a 3'D-RS
3'D-RS	recombination signal including the 3'D-HEPTAMER, 3'D-SPACER and 3'D-NONAMER in 3' of the D-REGION of a D-SEGMENT
3'UTR	3' untranslated sequence, EMBL feature Key signification
5'D-HEPTAMER	7 nucleotide recombination site like CACTGTG, part of a 5'D-RS
5'D-NONAMER	9 nucleotide recombination site like GGTTTTTGT, part of a 5'D-RS
5'D-RS	recombination signal including the 5'D-NONAMER, 5'D-SPACER and 5'D-HEPTAMER in 5' of the D-REGION of a D-SEGMENT or in 5' of the D-REGION of D-J-SEGMENT
5'UTR	5' untranslated sequence, EMBL feature Key signification
ACCEPTOR_ SPLICE	splicing site in 5' of coding region (nagnn), with splicing occurring after g
C-GENE	genomic DNA including C-REGION (and INTRONs if present) with 5' UTR and 3' UTR
C-REGION	coding region of C-GENE or corresponding region in cDNA
CDR1-IMGT	first complementarity determining region according to the IMGT unique numbering
CDR2-IMGT	second complementarity determining region according to the IMGT unique numbering
CDR3-IMGT	third complementarity determining region according to the IMGT unique numbering
CH-S	3' end of CH3 or CH4 exon or independent exon which encodes the hydrophilic C-terminal end of soluble Ig or corresponding region in cDNA
CH1	first exon of Ig heavy C-GENE or corresponding coding region in cDNA
CH2	second exon of Ig heavy C-GENE (or part of the second exon when hinge sequence belongs to the same exon) or corresponding coding region in cDNA
CH3	third exon of Ig heavy C-GENE or corresponding coding region in cDNA
CH4	fourth exon of Ig heavy C-GENE or corresponding coding region in cDNA
CONSERVED-TRP	codon (3 nucleotides) for Tryptophan in conserved position in FR2-IMGT

Label name	Definition
D-GENE	see D-SEGMENT
D-REGION	coding region of D-SEGMENT (plus 1 or 2 nucleotide(s) after the 5'D-HEPTAMER and/or before the 3'D-HEPTAMER, if present), or corresponding region in cDNA
D-SEGMENT	germline genomic DNA including D-REGION with 5' UTR and 3' UTR (also designated as D-GENE)
DECAMER	10 nucleotide regulation site or decanucleotide, includes OCTAMER, in the 5'UTR of a V-, V-D- or V-D-J-GENE
DELETION	point out a deletion compared to other sequences
DONOR_SPLICE	splicing site in 3' of coding region (ngt), with splicing occurring before g
FR1-IMGT	first framework according to the IMGT unique numbering
FR2-IMGT	second framework according to the IMGT unique numbering
FR3-IMGT	third framework according to the IMGT unique numbering
H	hinge exon of Ig heavy C-GENE or corresponding region in cDNA
H1	first hinge exon of Ig heavy C-GENE or corresponding region in cDNA
H2	second hinge exon of Ig heavy C-GENE or corresponding region in cDNA
H3	third hinge exon of Ig heavy C-GENE or corresponding region in cDNA
H4	fourth hinge exon of Ig heavy C-GENE or corresponding region in cDNA
H5	fifth hinge exon of Ig heavy C-GENE or corresponding region in cDNA
INSERTION	point out an insertion of 1 or more nucleotide(s) compared with old release of the sequence or with a similar sequence
INT_DONOR_SPLICE	alternative donor splice site located in a coding region
J-C-CLUSTER	genomic DNA in germline configuration including at least one J-SEGMENT and one C-GENE
J-GENE	see J-SEGMENT
J-HEPTAMER	7 nucleotide recombination site, like CACAGTG, part of a J-RS
J-NONAMER	9 nucleotide recombination site, like GGTTTTTGT, part of a J-RS
J-REGION	coding region of J-SEGMENT (plus 1 or 2 nucleotide(s) after J-HEPTAMER, if present) or corresponding region in cDNA
J-RS	recombination signal including J-HEPTAMER, J-SPACER and J-NONAMER in 5' of J-REGION of a J-SEGMENT or J-SEQUENCE
J-SEGMENT	germline genomic DNA including J-REGION with 5'UTR and 3'UTR (also designated as J-GENE)
JUNCTION	coding region encompassing the V-J or V-D-J junction from 2nd CYS to the J-PHE or J-TRP of the J-REGION
L-INTRON-L	sequence including L-PART1, V-INTRON and L-PART2 in genomic DNA or corresponding sequence in unspliced cDNA
L-PART1	exon encoding the first part of the leader peptide of a V-, V-D-, V-D-J- or V-J-GENE or corresponding region in unspliced cDNA

Label name	Definition
L-PART2	5′ region of V-EXON encoding the second part of leader peptide of a V-, V-D-, V-D-J- or V-J-GENE or corresponding region in unspliced cDNA
M	membrane exon of genomic C-GENE or corresponding region in cDNA
M1	1st membrane exon of genomic C-GENE or corresponding region in cDNA
M2	2nd membrane exon of genomic C-GENE or corresponding region in cDNA
N-AND-D-REGION	coding region encompassing the N diversity sequence(s) and coding region of D-SEGMENT(s) when limits between N- and D-REGIONS are unknown or corresponding region in cDNA
N-REGION	coding region encompassing the N diversity sequence
OCTAMER	8 nucleotide regulation site or octanucleotide, in the 5′UTR of a V-, V-D-, V-D-J- or V-J-GENE
STOP-CODON	codon which stops gene translation
TATA_BOX	TATA signal in eukaryotic promoters
V-CLUSTER	genomic DNA in germline configuration including more than one V-GENE
V-D-J-EXON	rearranged genomic DNA including L-PART2, V-, any D- and N-REGION, and J-REGION
V-D-J-GENE	rearranged genomic DNA including L-PART1, V-INTRON and V-D-J-EXON, with the 5′UTR and 3′UTR
V-D-J-REGION	coding region including V-, any D- and N-REGION, and J-REGION in rearranged genomic DNA or corresponding region in cDNA
V-EXON	germline genomic DNA including L-PART2 and V-REGION
V-HEPTAMER	7 nucleotide recombination site, like CACAGTG, part of V-RS
V-INTRON	non-coding sequence between L-PART1 and V-EXON in genomic DNA or corresponding sequence in unspliced cDNA
V-J-EXON	rearranged genomic DNA including L-PART2, V- and J- REGION
V-J-GENE	rearranged genomic DNA including L-PART1, V-INTRON and V-J-EXON, with the 5′UTR and 3′UTR
V-J-REGION	coding region including V- and J-REGION, in rearranged genomic DNA or corresponding region in cDNA
V-NONAMER	9 nucleotide recombination site, like ACAAAAACC, part of V-RS
V-REGION	coding region of V-GENE without the leader peptide (plus 1 or 2 nucleotide(s) before the V-HEPTAMER, if present) or corresponding region in cDNA
V-RS	recombination signal including V-HEPTAMER, V-SPACER and V-NONAMER in 3′ of V-REGION of a V-GENE or V-SEQUENCE
V-SPACER	12 or 23 nucleotide spacer between the V-HEPTAMER and the V-NONAMER of a V-RS

Aide-Mémoire

Useful restriction sites

*Bam*HI	G⬇GATCC
*Eco*RI	G⬇AATTC
*Hin*dIII	A⬇AGCTT
*Kpn*I	GGTAC⬇C
*Pst*I	CTGCA⬇G
*Pvu*II	CAG⬇CTG
*Sac*I (*Sst*I)	GAGCT⬇C
*Taq*I	T⬇CGA
*Xba*I	T⬇CTAGA
*Xho*I	C⬇TCGAG

Amino Acid Abbreviations

Amino acid	Abbreviations	
Alanine	Ala	A
Arginine	Arg	R
Asparagine	Asn	N
Aspartic acid	Asp	D
Asparagine or Aspartic acid	Asx	B
Cysteine	Cys	C
Glutamine	Gln	Q
Glutamic acid	Glu	E
Glutamine or Glutamic acid	Glx	Z
Glycine	Gly	G
Histidine	His	H
Isoleucine	Ile	I
Leucine	Leu	L
Lysine	Lys	K
Methionine	Met	M
Phenylalanine	Phe	F
Proline	Pro	P
Serine	Ser	S
Threonine	Thr	T
Tryptophan	Trp	W
Tyrosine	Tyr	Y
Valine	Val	V

Genetic code

Nucleotide position in codon					
first	second				third
	U	C	A	G	
U	UUU Phe	UCU Ser	UAU Tyr	UGU Cys	U
	UUC Phe	UCC Ser	UAC Tyr	UGC Cys	C
	UUA Leu	UCA Ser	UAA Stop	UGA Stop	A
	UUG Leu	UCG Ser	UAG Stop	UGG Trp	G
C	CUU Leu	CCU Pro	CAU His	CGU Arg	U
	CUC Leu	CCC Pro	CAC His	CGC Arg	C
	CUA Leu	CCA Pro	CAA Gln	CGA Arg	A
	CUG Leu	CCG Pro	CAG Gln	CGG Arg	G
A	AUU Ile	ACU Thr	AAU Asn	AGU Ser	U
	AUC Ile	ACC Thr	AAC Asn	AGC Ser	C
	AUA Ile	ACA Thr	AAA Lys	AGA Arg	A
	AUG Met	ACG Thr	AAG Lys	AGG Arg	G
G	GUU Val	GCU Ala	GAU Asp	GGU Gly	U
	GUC Val	GCC Ala	GAC Asp	GGC Gly	C
	GUA Val	GCA Ala	GAA Glu	GGA Gly	A
	GUG Val	GCG Ala	GAG Glu	GGG Gly	G

Section I

THE INTRODUCTORY CHAPTERS

1 Introduction

SCOPE OF THE BOOK

The primary aim of this book is to provide a compendium of the human germline immunoglobulin genes which are used to create the human antibody repertoire. The book includes entries for 203 genes and for 459 alleles, with a total of 837 sequences displayed (Section II). Prior to the entries there are four introductory chapters (Section I). This first chapter defines the data content and data selection criteria based on The international ImMunoGeneTics database (IMGT) Scientific chart[1] and IMGT-ONTOLOGY concepts[2]. Chapter 2 is a short overview on the structural and biological properties of the human immunoglobulins. Chapter 3 provides a summary of the molecular mechanisms of the synthesis of the human immunoglobulin chains. Chapter 4 represents a major IMGT contribution by providing, in a unique document, the first complete description of the immunoglobulin germline repertoire in humans.

SELECTION OF THE DATA

The individual entries comprise all the human immunoglobulin constant genes, and germline variable, diversity, and joining genes which have at least one functional or open reading frame (ORF) allele, and which are localized in the three major loci. Selected data are from IMGT[1,3–5] (http://imgt.cines.fr), created in Montpellier in 1989 by M.-P. Lefranc (Université Montpellier II, CNRS), and more particularly from the IMGT/LIGM-DB database, and from the IMGT Repertoire[6]. The selection criteria of the individual entries are defined in the IMGT Scientific chart[1] (http://imgt.cines.fr) and in the IMGT-ONTOLOGY "IDENTIFICATION" and "CLASSIFICATION" concepts[2], some of which are briefly summarized in the following paragraphs.

The "IDENTIFICATION" concept

The "IDENTIFICATION" concept allows scientists to identify immunoglobulin sequences according to fundamental biological and immunogenetic characteristics[2]. These are as follows.

"Molecule type"

Three instances are considered: genomic deoxyribonucleic acid (DNA), complementary DNA (cDNA), and protein.

"Gene type"

Four types of genes are involved in immunoglobulin synthesis, the variable (V), diversity (D) and joining (J) genes which encode the antigen binding sites, and the constant (C) genes which encode the part of the polypeptide chains which has effector properties.

"Configuration"

The configuration defines the status of the genes: "germline" or "rearranged" for the V, D and J genes. This concept is particularly important because it is unique to the immunoglobulin and T cell receptor V, D and J genes. Note that the C genes do not rearrange directly and therefore their configuration is not defined.

"Chain type"

The chain type identifies the nature of the peptidic chain potentially encoded by the immunoglobulin genes. There are three main instances which are defined by the C gene sequence characteristics: Ig-Heavy, Ig-Light-Kappa and Ig-Light-Lambda.

"Functionality"

The definition of functionality is based on the sequence analysis. As examples, the instances functional (for germline V, D, J and for C sequences) and productive (for rearranged V-J-C and V-D-J-C sequences) mean that the coding regions have an ORF without a stop codon, and that there is no described defect in the splicing sites, and/or recombination signals and/or regulatory elements. According to the gravity of the identified defects, the functionality can be defined as ORF, pseudogene or vestigial (for germline V, D, J and for C genes)[7]. Complete definitions are available in the IMGT Scientific chart.

The "CLASSIFICATION" concept

The "CLASSIFICATION" concept (Fig. 1) organizes the immunogenetic knowledge useful to name and classify the immunoglobulin genes[2].

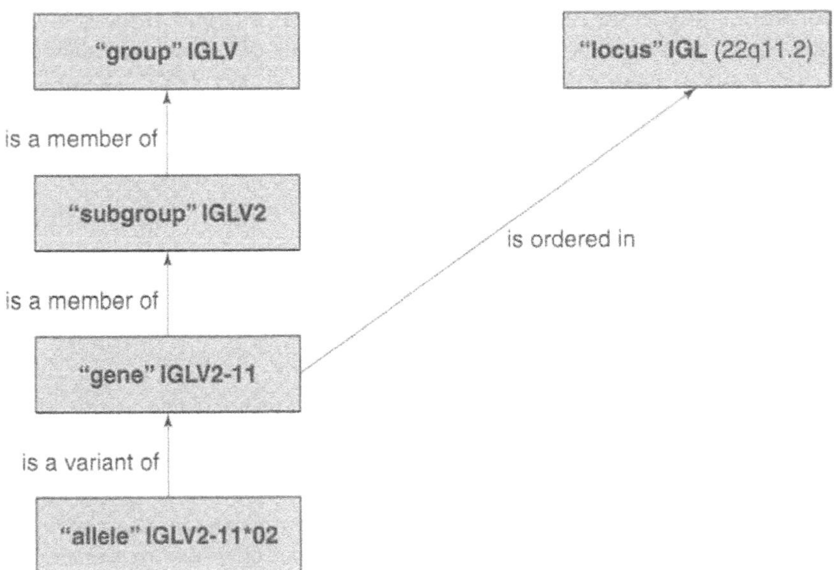

Figure 1 *The "CLASSIFICATION" concept in the IMGT-ONTOLOGY.*

"Locus"

A locus is a group of immunoglobulin genes that are ordered and are localized in the same chromosomal location in a given species. The human genome includes three main immunoglobulin loci: IGH (14q32.33), IGK (2p11.2) and IGL (22q11.2). Immunoglobulin genes have also been identified in other chromosomal locations outside the main loci which represent new instances of the concept locus. However, the genes they contain, designated as orphons, are not functional.

"Group"

A group is a set of genes which share the same "gene type" (V, D, J or C) and participate potentially in the synthesis of a polypeptide of the same "chain type". By extension, a group includes the related pseudogenes and orphons.

"Subgroup"

A subgroup is a set of genes which belong to the same group, in a given species, and which share at least 75% identity at the nucleotide level (in the germline configuration for V, D and J).

"Gene"

A gene is defined as a DNA sequence that can be potentially transcribed and/or translated (this definition includes the regulatory elements in 5′ and 3′, and the introns, if present). Instances of the "gene" concept are gene names. By extension, orphons and pseudogenes are also instances of the "gene" concept.

For each gene, IMGT has defined a reference sequence[1]. For the V, D and J genes, the reference sequence corresponds to a germline entity. The rules for the choice of the reference sequences are described in the IMGT Scientific chart.

"Allele"

An allele is a polymorphic variant of a gene. Alleles are described, exhaustively and in a standardized way, for the four "core" coding regions, that is for the germline V-REGIONs, D-REGIONs, J-REGIONs and for the C-REGIONs, from immunoglobulin genes. These alleles refer to sequence polymorphisms, with mutations described at the sequence level[4,7]. Their sequences are compared to the reference sequence designated as *01 (see IMGT Scientific chart for IMGT description of mutations and IMGT allele nomenclature for sequence polymorphisms).

DESCRIPTION OF THE DATA

The description of the individual gene entries is based on the "DESCRIPTION" concept of the IMGT-ONTOLOGY[2], and, for the V-REGIONs, on the setting up of the IMGT unique numbering[7–9].

The "DESCRIPTION" concept

The "DESCRIPTION" concept provides a standardized description of the organization and of the components of the immunoglobulin sequences, and a characterization of their specific and conserved motifs. A list of the IMGT labels used in this book is provided. Prototypes have been set up to graphically represent the description and configuration of an immunoglobulin gene[3] (Fig. 2). For example, the prototype V-GENE represents a genomic V gene in the germline configuration, whereas V-J-GENE represents genomic V and J genes in the rearranged configuration for a light chain, and V-D-J-GENE represents genomic V, D and J genes in the rearranged configuration for a heavy chain (Fig. 2).

Germline genomic DNA

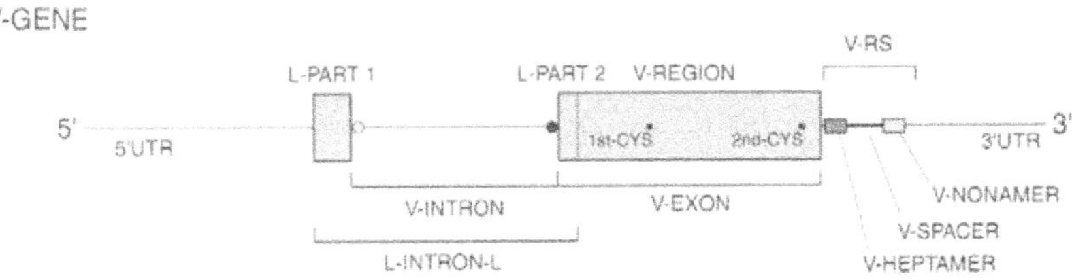

Rearranged genomic DNA

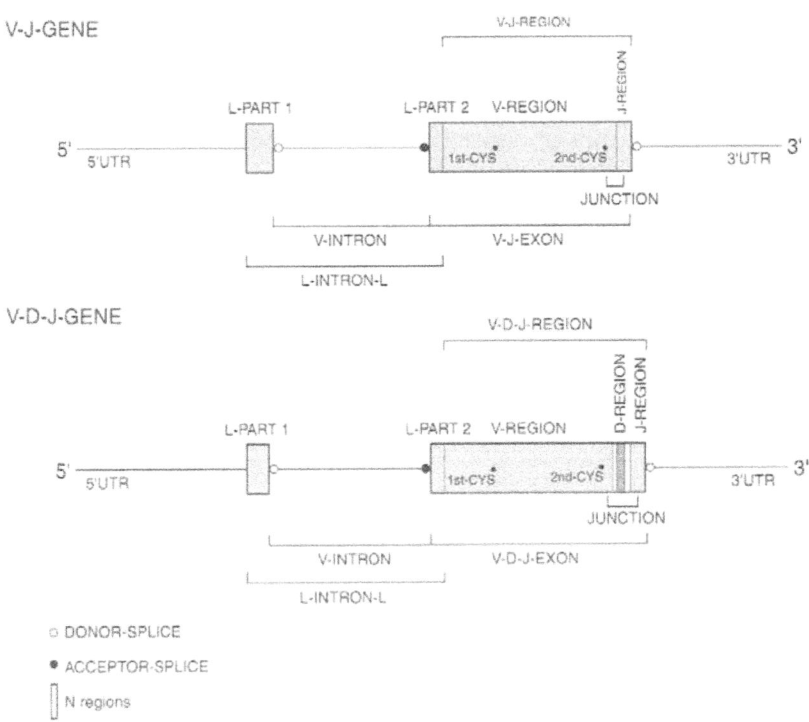

Figure 2 *Prototypes of a variable gene in the germline (V-GENE) or rearranged (V-J-GENE for a light chain, V-D-J-GENE for a heavy chain) configuration. Labels (in capital letters) are those used for the sequence description in IMGT (http://imgt.cines.fr).*

The IMGT unique numbering for the V-REGIONs

The IMGT unique numbering[7-9] relies on the high conservation of the structure of the variable region. This numbering, set up after aligning more than 5000 sequences, takes into account and combines the definition of the framework (FR) and complementarity determining regions (CDR)[10], structural data from X-ray diffraction studies[11], and the characterization of the hypervariable loops[12]. The delimitations of the FR-IMGT and CDR-IMGT regions have been defined, and correspondence between the IMGT numbering and the other numberings has been established[9].

The IMGT unique numbering has many advantages:

- It allows an easy comparison between sequences coding the variable regions, whatever the antigen receptor (immunoglobulins or T cell receptors), the chain type (heavy or light chains for immunoglobulins) or the species.
- In the IMGT unique numbering, the conserved amino acids always have the same position, for instance Cysteine 23, Tryptophan 41, Leucine 89 and Cysteine 104. The hydrophobic amino acids of the framework regions are also found in conserved positions.
- This unique numbering has allowed the redefinition of the limits of the FR and CDR. The FR-IMGT and CDR-IMGT lengths themselves become crucial information, characterizing the variable regions belonging to a group, a subgroup and/or a gene.
- Framework amino acids (and codons) located at the same position in different sequences can be compared without requiring sequence alignments. This also holds for amino acids belonging to CDR-IMGT of the same length.
- The IMGT unique numbering has allowed a standardized IMGT description of mutations for the IMGT description of allele polymorphisms and somatic hypermutations of the variable regions[4,7].
- The unique numbering is used as the output of the IMGT/V-QUEST alignment tool (http://imgt.cines.fr) which analyses the immunoglobulin variable (germline or rearranged) sequences according to IMGT criteria[1]. In IMGT/V-QUEST, a variable rearranged sequence is compared to the appropriate sets of V-REGION, D-REGION and J-REGION alleles from the IMGT reference directory. The results show, aligned with the input sequence, the sequences of the most homologous V-REGION alleles and, if appropriate, D-REGION (for heavy chains) and J-REGION alleles. The aligned V-REGION sequences are displayed according to the IMGT unique numbering and with the FR-IMGT and CDR-IMGT delimitations.

By facilitating comparisons between the sequences and the descriptions of alleles and mutations, the IMGT unique numbering represents a big step forward in the analysis of the immunoglobulin sequences of all species. Moreover, it gives insight into the structural configuration of the variable domain and opens interesting views on the evolution of the sequences of the V-set, since this numbering has been applied with success to all the sequences belonging to the V-set of the immunoglobulin superfamily, including non-rearranging sequences in vertebrates (CD4, *Xenopus* CTX,...) and in invertebrates (*Drosophila* Amalgam, *Drosophila* Fasciclin II, etc.)[7-9].

Correspondence between numberings

Table 1 gives the correspondence between the IMGT unique Lefranc numbering[7-9] and the different Kabat numberings[13] for the immunoglobulin variable regions.

Table 2 gives the correspondence between the IMGT CH (immunoglobulin heavy constant exon or domain) numbering, used in the IGHG individual entries in Section II and the protein EU heavy chain numbering[14].

Table 1. *Correspondence between the V-REGION numberings.*

IGHV Human IGHV6-1					IGKV Human IGKV1-5					IGLV Human IGLV2-23				
FR1-IMGT														
1	1	cag	GLN	Q	1	1	gac	ASP	D	1	1	cag	GLN	Q
2	2	gta	VAL	V	2	2	atc	ILE	I	2	2	tct	SER	S
3	3	cag	GLN	Q	3	3	cag	GLN	Q	3	3	gcc	ALA	A
4	4	ctg	LEU	L	4	4	atg	MET	M	4	4	ctg	LEU	L
5	5	cag	GLN	Q	5	5	acc	THR	T	5	5	act	THR	T
6	6	cag	GLN	Q	6	6	cag	GLN	Q	6	6	cag	GLN	Q
7	7	tca	SER	S	7	7	tct	SER	S	7	7	cct	PRO	P
8	8	ggt	GLY	G	8	8	cct	PRO	P	8	8	gcc	ALA	A
9	9	cca	PRO	P	9	9	tcc	SER	S	9	9	tcc	SER	S
10		---	---	-	10	10	acc	THR	T	10	10	---	---	-
11	10	gga	GLY	G	11	11	ctg	LEU	L	11	11	gtg	VAL	V
12	11	ctg	LEU	L	12	12	tct	SER	S	12	12	tct	SER	S
13	12	gtg	VAL	V	13	13	gca	ALA	A	13	13	ggg	GLY	G
14	13	aag	LYS	K	14	14	tct	SER	S	14	14	tct	SER	S
15	14	ccc	PRO	P	15	15	gta	VAL	V	15	15	cct	PRO	P
16	15	tcg	SER	S	16	16	gga	GLY	G	16	16	gga	GLY	G
17	16	cag	GLN	Q	17	17	gac	ASP	D	17	17	cag	GLN	Q
18	17	acc	THR	T	18	18	aga	ARG	R	18	18	tcg	SER	S
19	18	ctc	LEU	L	19	19	gtc	VAL	V	19	19	atc	ILE	I
20	19	tca	SER	S	20	20	acc	THR	T	20	20	acc	THR	T
21	20	ctc	LEU	L	21	21	atc	ILE	I	21	21	atc	ILE	I
22	21	acc	THR	T	22	22	act	THR	T	22	22	tcc	SER	S
23	22	tgt	**CYS**	**C**	23	23	tgc	**CYS**	**C**	23	23	tgc	**CYS**	**C**
24	23	gcc	ALA	A	24	24	cgg	ARG	R	24	24	act	THR	T
25	24	atc	ILE	I	25	25	gcc	ALA	A	25	25	gga	GLY	G
26	25	tcc	SER	S	26	26	agt	SER	S	26	26	acc	THR	T
CDR1-IMGT														
27	26	ggg	GLY	G	27	27	cag	GLN	Q	27	27	agc	SER	S
28	27	gac	ASP	D	28	*28	agt	SER	S	28	*27A	agt	SER	S
29	28	agt	SER	S	29	*29	att	ILE	I	29	*27B	gat	ASP	D
30	29	gtc	VAL	V	30	*30	agt	SER	S	30	*27C	gtt	VAL	V
31	30	tct	SER	S	31	*31	agc	SER	S	31	*28	ggg	GLY	G
32	31	agc	SER	S	32	*32	tgg	TRP	W	32	*29	agt	SER	S
33	32	aac	ASN	N	33	*	---	---	-	33	*30	tat	TYR	Y
34	33	agt	SER	S	34	*	---	---	-	34	*31	aac	ASN	N
35	*34	gct	ALA	A	35	*	---	---	-	35	*32	ctt	LEU	L
36	*35	gct	ALA	A	36	*	---	---	-	36		---	---	-
37		---	---	-	37	*	---	---	-	37		---	---	-
38		---	---	-	38	*	---	---	-	38		---	---	-

Table 1. *Continued.*

IGHV Human IGHV6-1					IGKV Human IGKV1-5					IGLV Human IGLV2-23				
FR2-IMGT														
39	*35A	tgg	TRP	W	**39**	33	ttg	LEU	L	**39**	33	gtc	VAL	V
40	*35B	aac	ASN	N	**40**	34	gcc	ALA	A	**40**	34	tcc	SER	S
41	36	tgg	**TRP**	**W**	**41**	35	tgg	**TRP**	**W**	**41**	35	tgg	**TRP**	**W**
42	37	atc	ILE	I	**42**	36	tat	TYR	Y	**42**	36	tac	TYR	Y
43	38	agg	ARG	R	**43**	37	cag	GLN	Q	**43**	37	caa	GLN	Q
44	39	cag	GLN	Q	**44**	38	cag	GLN	Q	**44**	38	cag	GLN	Q
45	40	tcc	SER	S	**45**	39	aaa	LYS	K	**45**	39	cac	HIS	H
46	41	cca	PRO	P	**46**	40	cca	PRO	P	**46**	40	cca	PRO	P
47	42	tcg	SER	S	**47**	41	ggg	GLY	G	**47**	41	ggc	GLY	G
48	43	aga	ARG	R	**48**	42	aaa	LYS	K	**48**	42	aaa	LYS	K
49	44	ggc	GLY	G	**49**	43	gcc	ALA	A	**49**	43	gcc	ALA	A
50	45	ctt	LEU	L	**50**	44	cct	PRO	P	**50**	44	ccc	PRO	P
51	46	gag	GLU	E	**51**	45	aag	LYS	K	**51**	45	aaa	LYS	K
52	47	tgg	TRP	W	**52**	46	ctc	LEU	L	**52**	46	ctc	LEU	L
53	48	ctg	LEU	L	**53**	47	ctg	LEU	L	**53**	47	atg	MET	M
54	49	gga	GLY	G	**54**	48	atc	ILE	I	**54**	48	att	ILE	I
55	50	agg	ARG	R	**55**	49	tat	TYR	Y	**55**	49	tat	TYR	Y
CDR2-IMGT														
56	51	aca	THR	T	**56**	50	gat	ASP	D	**56**	50	gag	GLU	E
57	52	tac	TYR	Y	**57**	51	gcc	ALA	A	**57**	51	ggc	GLY	G
58	*52A	tac	TYR	Y	**58**	52	tcc	SER	S	**58**	52	agt	SER	S
59	*52B	agg	ARG	R	**59**		---	---	-	**59**		---	---	-
60	*53	tcc	SER	S	**60**		---	---	-	**60**		---	---	-
61	*54	aag	LYS	K	**61**		---	---	-	**61**		---	---	-
62	*55	tgg	TRP	W	**62**		---	---	-	**62**		---	---	-
63	*56	tat	TYR	Y	**63**		---	---	-	**63**		---	---	-
64	*57	aat	ASN	N	**64**		---	---	-	**64**		---	---	-
65	*	---	---	-	**65**		---	---	-	**65**		---	---	-
FR3-IMGT														
66	58	gat	ASP	D	**66**	53	agt	SER	S	**66**	53	aag	LYS	K
67	59	tat	TYR	Y	**67**	54	ttg	LEU	L	**67**	54	cgg	ARG	R
68	60	gca	ALA	A	**68**	55	gaa	GLU	E	**68**	55	ccc	PRO	P
69	61	gta	VAL	V	**69**	56	agt	SER	S	**69**	56	tca	SER	S
70	62	tct	SER	S	**70**	57	ggg	GLY	G	**70**	57	ggg	GLY	G
71	63	gtg	VAL	V	**71**	58	gtc	VAL	V	**71**	58	gtt	VAL	V
72	64	aaa	LYS	K	**72**	59	cca	PRO	P	**72**	59	tct	SER	S
73		---	---	-	**73**		---	---	-	**73**		---	---	-
74	65	agt	SER	S	**74**	60	tca	SER	S	**74**	60	aat	ASN	N
75	66	cga	ARG	R	**75**	61	agg	ARG	R	**75**	61	cgc	ARG	R
76	67	ata	ILE	I	**76**	62	ttc	PHE	F	**76**	62	ttc	PHE	F
77	68	acc	THR	T	**77**	63	agc	SER	S	**77**	63	tct	SER	S
78	69	atc	ILE	I	**78**	64	ggc	GLY	G	**78**	64	ggc	GLY	G
79	70	aac	ASN	N	**79**	65	agt	SER	S	**79**	65	tcc	SER	S
80	71	cca	PRO	P	**80**	66	gga	GLY	G	**80**	66	aag	LYS	K
81	72	gac	ASP	D	**81**		---	---	-	**81**		---	---	-
82	73	aca	THR	T	**82**		---	---	-	**82**		---	---	-

Table 1. *Continued.*

IGHV Human IGHV6-1					IGKV Human IGKV1-5					IGLV Human IGLV2-23				
83	74	tcc	SER	S	**83**	67	tct	SER	S	**83**	67	tct	SER	S
84	75	aag	LYS	K	**84**	68	ggg	GLY	G	**84**	68	ggc	GLY	G
85	76	aac	ASN	N	**85**	69	aca	THR	T	**85**	69	aac	ASN	N
86	77	cag	GLN	Q	**86**	70	gaa	GLU	E	**86**	70	acg	THR	T
87	78	ttc	PHE	F	**87**	71	ttc	PHE	F	**87**	71	gcc	ALA	A
88	79	tcc	SER	S	**88**	72	act	THR	T	**88**	72	tcc	SER	S
89	80	ctg	LEU	L	**89**	73	ctc	LEU	L	**89**	73	ctg	LEU	L
90	81	cag	GLN	Q	**90**	74	acc	THR	T	**90**	74	aca	THR	T
91	82	ctg	LEU	L	**91**	75	atc	ILE	I	**91**	75	atc	ILE	I
92	82A	aac	ASN	N	**92**	76	agc	SER	S	**92**	76	tct	SER	S
93	82B	tct	SER	S	**93**	77	agc	SER	S	**93**	77	ggg	GLY	G
94	82C	gtg	VAL	V	**94**	78	ctg	LEU	L	**94**	78	ctc	LEU	L
95	83	act	THR	T	**95**	79	cag	GLN	Q	**95**	79	cag	GLN	Q
96	84	ccc	PRO	P	**96**	80	cct	PRO	P	**96**	80	gct	ALA	A
97	85	gag	GLU	E	**97**	81	gat	ASP	D	**97**	81	gag	GLU	E
98	86	gac	ASP	D	**98**	82	gat	ASP	D	**98**	82	gac	ASP	D
99	87	acg	THR	T	**99**	83	ttt	PHE	F	**99**	83	gag	GLU	E
100	88	gct	ALA	A	**100**	84	gca	ALA	A	**100**	84	gct	ALA	A
101	89	gtg	VAL	V	**101**	85	act	THR	T	**101**	85	gat	ASP	D
102	90	tat	TYR	Y	**102**	86	tat	TYR	Y	**102**	86	tat	TYR	Y
103	91	tac	TYR	Y	**103**	87	tac	TYR	Y	**103**	87	tac	TYR	Y
CDR3-IMGT														
104	92	tgt	**CYS**	**C**	**104**	88	tgc	**CYS**	**C**	**104**	88	tgc	**CYS**	**C**
105	93	gca	ALA	A	**105**	89	caa	GLN	Q	**105**	89	tgc	CYS	C
106	94	aga	ARG	R	**106**	90	cag	GLN	Q	**106**	90	tca	SER	S
107	95	---	---	-	**107**	91	tat	TYR	Y	**107**	91	tat	TYR	Y
108					**108**	92	aat	ASN	N	**108**	92	gca	ALA	A
109					**109**	93	agt	SER	S	**109**	93	ggt	GLY	G
110					**110**	94	tat	TYR	Y	**110**	94	agt	SER	S
111					**111**	95	tct	SER	S	**111**	95	agc	SER	S
112										**112**	95A	act	THR	T
113										**113**	95B	tta	LEU	L
114										**114**	95C	---	---	-
115										**115**	95D	---	---	-

For each V-REGION group, one germline sequence is shown with, on the left-hand side of each column, the IMGT unique Lefranc numbering (in bold)[7-9] and the corresponding Kabat numbering[13]. Positions of missing amino acids (shown with dashes) are reported to the 3′ end of the CDR-IMGT.
* Positions for which it is not possible to make changes from one numbering to the other, automatically. 1st-CYS 23, CONSERVED-TRP 41 and 2nd-CYS 104 are in bold.

Table 2. *Correspondence between the C-REGION numberings.*

CH1 (98 aa)		CH2 (110 aa)(1)		CH3 (107 aa)(2)	
IMGT numbering 1–98	EU numbering 118–215	IMGT numbering 1–110	EU numbering 231–340	IMGT numbering 1–107	EU numbering 341–446(3)
1	118	1	231	1	341
2	119	2	232	2	342
3	120	3	233	3	343
4	121	4	234	4	344
5	122	5	235	5	345
6	123	6	236	6	346
7	124	7	237	7	347
8	125	8	238	8	348
9	126	9	239	9	349
10	127	10	240	10	350
11	128	11	241	11	351
12	129	12	242	12	352
13	130	13	243	13	353
14	131	14	244	14	354
15	132	15	245	15	355
16	133	16	246	16	356
17	134	17	247	17	357
18	135	18	248	18	358
19	136	19	249	19	359
20	137	20	250	20	360
21	138	21	251	21	361
22	139	22	252	22	362
23	140	23	253	23	363
24	141	24	254	24	364
25	142	25	255	25	365
26	143	26	256	26	366
Cys 27	Cys 144	27	257	Cys 27	Cys 367
28	145	28	258	28	368
29	146	29	259	29	369
30	147	30	260	30	370
31	148	Cys 31	Cys 261	31	371
32	149	32	262	32	372
33	150	33	263	33	373
34	151	34	264	34	374
35	152	35	265	35	375
36	153	36	266	36	376
37	154	37	267	37	377
38	155	38	268	38	378
39	156	39	269	39	379
40	157	40	270	40	380
41	158	41	271	41	381

Table 2. *Continued.*

CH1 (98 aa)		CH2 (110 aa)(1)		CH3 (107 aa)(2)	
IMGT numbering 1–98	EU numbering 118–215	IMGT numbering 1–110	EU numbering 231–340	IMGT numbering 1–107	EU numbering 341–446(3)
42	159	42	272	42	382
43	160	43	273	43	383
44	161	44	274	44	384
45	162	45	275	45	385
46	163	46	276	46	386
47	164	47	277	47	387
48	165	48	278	48	388
49	166	49	279	49	389
50	167	50	280	50	390
51	168	51	281	51	391
52	169	52	282	52	392
53	170	53	283	53	393
54	171	54	284	54	394
55	172	55	285	55	395
56	173	56	286	56	396
57	174	57	287	57	397
58	175	58	288	58	398
59	176	59	289	59	399
60	177	60	290	60	400
61	178	61	291	61	401
62	179	62	292	62	402
63	180	63	293	63	403
64	181	64	294	64	404
65	182	65	295	65	405
66	183	66	296	66	406
67	184	67	297	67	407
68	185	68	298	68	408
69	186	69	299	69	409
70	187	70	300	70	410
71	188	71	301	71	411
72	189	72	302	72	412
73	190	73	303	73	413
74	191	74	304	74	414
75	192	75	305	75	415
76	193	76	306	76	416
77	194	77	307	77	417
78	195	78	308	78	418
79	196	79	309	79	419
80	197	80	310	80	420
81	198	81	311	81	421
82	199	82	312	82	422

Table 2. *Continued.*

CH1 (98 aa)		CH2 (110 aa)(1)		CH3 (107 aa)(2)	
IMGT numbering 1–98	EU numbering 118–215	IMGT numbering 1–110	EU numbering 231–340	IMGT numbering 1–107	EU numbering 341–446(3)
Cys 83	Cys 200	83	313	83	423
84	201	84	314	84	424
85	202	85	315	Cys 85	Cys 425
86	203	86	316	86	426
87	204	87	317	87	427
88	205	88	318	88	428
89	206	89	319	89	429
90	207	90	320	90	430
91	208	Cys 91	Cys 321	91	431
92	209	92	322	92	432
93	210	93	323	93	433
94	211	94	324	94	434
95	212	95	325	95	435
96	213	96	326	96	436
97	214	97	327	97	437
98	215	98	328	98	438
–	–	99	329	99	439
–	–	100	330	100	440
–	–	101	331	101	441
–	–	102	332	102	442
–	–	103	333	103	443
–	–	104	334	104	444
–	–	105	335	105	445
–	–	106	336	CH-S 106	446
–	–	107	337	CH-S 107	–
–	–	108	338	–	–
–	–	109	339	–	–
–	–	110	340	–	–

(1) The human IGHG1, IGHG3 and IGHG4 CH2 exons encode 110 amino acids. The IGHG2 CH2 exon encodes 109 amino acids, due to a three nucleotide (nt) deletion corresponding to codon 3 (see individual gene entries in Section II).
(2) The last two amino acids of the IGHG CH3 exons belong to the CH-S which encode the heavy chain C-terminus of the secreted immunoglobulins (see Chapter 3).
(3) The EU last amino acid Lysine 447 has not been characterized.
The variable region length is different from one immunoglobulin chain to another, moreover the hinge length varies between gamma chains of different IgG subclasses, the EU heavy chain gamma1 numbering (EU is an IgG1)[14] is therefore arbitrary but, in the absence of a unique numbering for the constant region, is frequently used in the literature for the gamma chains. An IMGT numbering based on the CH three-dimensional (3D)-structure is currently being developed. This numbering will also be applicable to the immunoglobulin chains (mu, delta, alpha, epsilon) of other classes. Hinge amino acids are not shown.

ORGANIZATION OF THE DATA

Nomenclature

The IMGT gene name (gene symbol) and the IMGT full name are given. The concepts of classification (Fig. 1) have been used to set up a unique nomenclature of the immunoglobulin genes[15–21]. A four-letter root designates the "group": IGHV, IGHD, IGHJ and IGHC for the immunoglobulin heavy genes; IGKV, IGKJ and IGKC for the immunoglobulin kappa genes; IGLV, IGLJ and IGLC for the immunoglobulin lambda genes.

Gene names are derived from the four-letter root, by adding, if necessary, number(s) and/or letter(s) to allow unambigous identification of the gene, a single number or letter being used whenever it is possible. IMGT nomenclature has been approved by the HUGO (HUman Genome Organization) Nomenclature Committee (http://www.gene.ucl.ac.uk/nomenclature) in 1999.

Definition and functionality

The definition includes information on the functionality of a gene, and if necessary, that of its alleles. It also comprises eventual structural or biological particularities.

Gene location

The chromosomal location of the gene is given.

Nucleotide and amino acid sequences

Alignments of all known germline sequences assigned to the different alleles, by comparison to the allele*01 are displayed. The translation of the allele*01, and the nucleotide mutations and corresponding amino acid changes of the other alleles are shown. Dashes indicate identical nucleotides. Dots indicate gaps according to the IMGT numbering.

Allele names of the V-REGIONs, D-REGIONs, J-REGIONs and C-REGIONs comprise the IMGT gene name followed by an asterisk and a two-figure number. The V-REGIONs, D-REGIONs, J-REGIONs and C-REGIONs selected as references for the allele polymorphism description have the number *01; other alleles are designated by increasing numbers (*02, *03, ...) based, if possible, on the chronological order of their publication, and/or confirmation of data by different authors. Note that the number *01 does not mean necessarily that other alleles are already known, but it signifies that any new polymorphic sequence will be described by comparison to that allele *01. IMGT accession numbers are indicated for each allele. Although the IMGT accession numbers are the same as those from the EMBL/GenBank/DDBJ generalist databases, the content of the IMGT/LIGM-DB flat files differs by the expertised annotations added by IMGT. IMGT data are available from IMGT/LIGM-DB, IMGT Repertoire, and from Sequence Retrieval System (SRS) sites (available from the IMGT Home page, http://imgt.cines.fr). References indicated by a number between brackets are listed at the end of the group entries.

Framework and complementarity determining regions

For the V-GENE entries, the length (in number of amino acids) of the FR and CDR are indicated. The limits of the FR and CDR are based on the IMGT unique numbering.

Collier de Perles

The IMGT Collier de Perles[1,4,5,9] are two-dimensional (2D) graphical representations of the immunoglobulin variable regions, with FR-IMGT and CDR-IMGT delimitations. Collier de Perles 2D representations provide information on the amino acid positions in the beta-strands and loops of the variable domain and allows quick visualization of amino acids which are important for the structural configuration of the V-REGION.

Amino acids are shown in the one letter abbreviation. Hydrophobic amino acids (hydropathy index with positive value) and Tryptophan (W) found at a given position in more than 50% of analysed immunoglobulin sequences are shown in dark grey. All Proline (P) are shown in pale grey. The CDR-IMGT are limited by amino acids shown in squares, which belong to the neighbouring FR-IMGT. Hatched circles or squares correspond to missing positions according to the IMGT unique numbering.

Arrows indicate the direction of the beta-strands and their different designations in 3D structure. This information has to be used carefully if not supported by experimental data.

For a given germline V-GENE, the length of the three CDR-IMGT is shown in brackets after the gene name, and separated by dots. For example, IGKV1-6 [6.3.7] means that in the germline IGKV1-6 gene, the CDR1-IMGT, CDR2-IMGT and CDR3-IMGT regions are six, three and seven amino acids long, respectively.

Genome database accession numbers

All IMGT human immunoglobulin genes[15–21] have been entered into GDB, Genome Database, Toronto, Canada (http://www.gdb.org), and into LocusLink at NCBI (National Center for Biotechnology Information), Bethesda, USA (http://www.ncbi.nlm.nih.gov/LocusLink), and entry accession numbers are provided. Links to the individual IMGT, GDB and LocusLink gene entries are available from http://imgt.cines.fr from IMGT Repertoire. Links to OMIM (Online Mendelian Inheritance in Man, MIM) (http://www.ncbi.nlm.nih.gov/Omim) are cited when there are existing entries in OMIM.

References

[1] Lefranc, M.-P. et al. (1999) Nucleic Acids Res. 27, 209–212.
[2] Giudicelli, V. et al. (1999) Bioinformatics 15, 1047–1054.
[3] Giudicelli, V. et al. (1997) Nucleic Acids Res. 25, 206–211.
[4] Lefranc, M.-P. et al. (1998) Nucleic Acids Res. 26, 297–303.
[5] Ruiz, M. et al. (2000) Nucleic Acids Res. 28, 219–221.
[6] Lefranc, M.-P. (2000) BIOforum international 4, 98–100.
[7] Lefranc, M.-P. (1998) Exp. Clin. Immunogenet. 15, 1–7.
[8] Lefranc, M.-P. (1997) Immunol. Today 18, 509.

[9] Lefranc, M.-P. (1999) The Immunologist 7, 132–136.
[10] Kabat, E.A. et al. (1987) In Sequences of Proteins of Immunological Interest. Public Health Service, NIH, Washington DC.
[11] Satow, Y. et al. (1986) J. Mol. Biol. 190, 593–604.
[12] Chothia, C. and Lesk, A.M. (1987) J. Mol. Biol. 196, 901–917.
[13] Kabat, E.A. et al. (1991) In Sequences of Proteins of Immunological Interest. NIH Publication 91–3242, Washington DC.
[14] Edelman, G.M.et al. (1969) Proc. Natl Acad. Sci. USA 63, 78–85.
[15] Lefranc, M.-P. (2000) InCurrent Protocols in Immunology A.1P.1-A.1p.37, John Wiley and Sons, New York, USA.
[16] Pallarès, N. et al. (1998) Exp. Clin. Immunogenet. 15, 8–18.
[17] Barbié, V. and Lefranc, M.-P. (1998) Exp. Clin. Immunogenet. 15, 171–183.
[18] Pallarès, N. et al. (1999) Exp. Clin. Immunogenet. 16, 36–60.
[19] Ruiz, M. et al. (1999) Exp. Clin. Immunogenet. 16, 173–184.
[20] Scaviner, D. et al. (1999) Exp. Clin. Immunogenet. 16, 234–240.
[21] Lefranc, M.-P. (2000) The Immunologist 8, 80–87.

2 Immunoglobulin structural and biological properties

THE IMMUNOGLOBULIN DUAL ROLE: ANTIGEN BINDING AND EFFECTOR FUNCTION

Immunoglobulins or antibodies serve a dual role in immunity: they both recognize antigens on the surface of foreign bodies such as bacteria and viruses, and trigger elimination mechanisms such as cell lysis and phagocytosis, to rid the body of these invading cells and particles.

Basic immunoglobulin structure and dual function

The immunoglobulins share a basic four-chain structure composed of two identical heavy chains of 50–70 kDa, and two identical light chains of 25 kDa[1–2] (Fig. 1). The discovery of the action of proteolytic enzymes on IgG1 by Porter in 1959[3] provided the first great insight into antibody structure. The general Y shaped configuration of an immunoglobulin corresponds to two Fab (fragment antigen binding) arms and an Fc (fragment crystallizable) region (Fig. 1). These fragments are obtained by papain-digestion of IgG1. The two identical Fab arms, each comprising the N-terminal half of the heavy chain associated with the light chain, are involved in the antigen recognition and binding, whilst the Fc region, formed from the C-terminal half of the two heavy chains, interacts with effector molecules such as the complement component C1q and the Fc receptors. These effector molecules bind to the Fc of the antibodies, coated at the surface of the foreign antigens, and trigger elimination processes. Activation of the classical complement cascade generates a variety of potent biological molecules, which promote phagocytosis, chemotaxis and formation of the membrane attack complex, resulting in cell lysis. The pathway is triggered by the interaction of Cl, a protein complex of C1q, Clr and Cls, with antigen–antibody complexes. It is the C1q head region which interacts directly with the immunoglobulin Fc. Binding of antigen–antibody complexes or aggregated immunoglobulins to the Fc receptors triggers cell functions which serve important roles against pathogenic agents as well as in the regulation of antibody protection.

Heavy and light chain types

Human immunoglobulins are divided into five classes or isotypes, IgM, IgD, IgG, IgA and IgE – each with distinct heavy chains (termed mu, delta, gamma, alpha and epsilon, respectively) which differ by their physico-chemical properties, their isotypic antigenic determinants and in their biological function[4–8]. The IgG and IgA classes in humans are further subdivided into four subclasses (IgG1, IgG2, IgG3 and IgG4) and two subclasses (IgA1 and IgA2), respectively, on their heavy chain characteristics (gamma1, gamma2, gamma3 and gamma4, alpha1 and alpha2). Further allotypic variants of certain subclasses or class have also been identified which represent genetic markers for the gamma1, gamma2, gamma3, alpha2 and epsilon heavy chains, and are, consequently called G1m, G2m, G3m, A2m and Em allotypes.

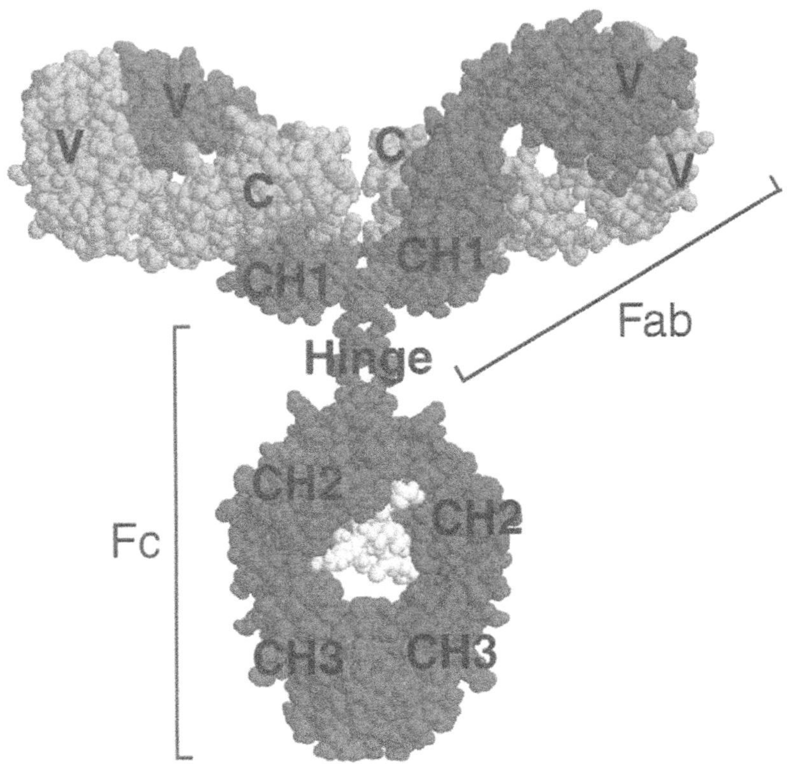

Figure 1 *3D representation of an IgG immunoglobulin. The corresponding PDB (Protein Data Bank) format file is from the Mike's Immunoglobulin Structure/ Function site* (http://www.umass.edu/microbio/rasmol/padlan.htm). *The light chains are in pale grey. The heavy chains are in dark grey. Carbohydrate chains between the CH2 domains are in white. V = variable domain; C = constant domain of the light chain; CH1, CH2, CH3 = constant domains of the heavy chain.*

The allotypes are polymorphic markers detected by serological methods, and are present in different individuals of the same species[9]. The Gm and Am allotypic determinants, inherited in fixed combinations or Gm-haplotypes are useful tools in the characterization of populations[9–11] and genetics of immunoglobulins[9,12–14].

The two light chain types, kappa and lambda, are common to all five classes[6–8]. Either light chain type can associate with any of the heavy chain types, but in any particular immunoglobulin, both light and both heavy chains are identical. The kappa to lambda ratio in the serum of healthy individuals is approximately 2 to 1. Four lambda isotypes have been identified by the presence or absence of serological markers (Mcg, Kern and Oz)[15]. Three Km allotypes have been characterized[9].

Variable and constant domains

Each chain folds up into domains of approximately 100 to 110 amino acids (Fig. 2). There are two domains for a light chain, and four or five domains for a heavy chain. The N-terminal domain of the light and heavy chains is the variable (V) domain which exhibits an enormous diversity between different immunoglobulins. Each variable domain comprises a beta-sheet "framework" supporting three hypervariable loops or CDRs, which are spatially close to each other and constitute the antigen binding site[16,17] (Fig. 2). The other domains, designated as constant domains, are identical between chains from the same class, subclass and with the same allotypes. There is one constant (C) domain in the kappa and lambda light chain, three (CH1, CH2 and CH3) in the delta, gamma, or alpha heavy chains with a flexible hinge region between CH1 and CH2 (Fig. 1), and four (CH1–CH4) in the mu and epsilon heavy chains. In IgM and IgE, the CH2 domain replaces the hinge, and the CH3 and CH4 domains correspond to the CH2 and CH3 in IgG, IgD and IgA. The modular structure of the immunoglobulin chains in domains (12 for IgG, IgD and IgA, 14 for IgM and IgE) is extensively used for the construction and expression of engineered antibodies[18,19]. Some examples are shown in Fig. 3.

In intact immunoglobulins the domains usually associate into pairs through multiple non-covalent lateral interactions. However, the CH2 domains of the IgG, IgD and IgA, and the "equivalent" CH3 domains of the IgM and IgE are unpaired but stabilized with interposed N-linked, branched carbohydrate chains (Fig. 1). All human immunoglobulin heavy chains are glycosylated[20] (Table 1). The number of potential N-glycosylation sites per heavy chain is reported in Table 2, and their positions are indicated in Section II (IGHC protein display). O-glycosylation characterizes the hinge of the delta and alpha1 chains.

Membrane immunoglobulins

Membrane immunoglobulins (or mIg) occur as monomers (two light chains and two heavy chains), anchored by their heavy chains in the surface of the B cells. They are associated with CD79a (Igα, mb-1) and CD79b (Igβ, B29)[21]. CD79a and CD79b exist at the surface as a disulphide-linked heterodimer. Both are composed of a single Ig-like domain. One CD79a-CD79b heterodimer and one immunoglobulin monomer constitute the B cell receptor (BcR). mIgM and mIgD are widely found on the cell surfaces of B cells where they function as a specific antigen receptor. mIgG, mIgA and mIgE are at the surface of memory B cells (see Chapter 3). Cross-linking of the BcR induces the tyrosylphosphorylation of ITAM (immunoreceptor Tyrosine activation motif) on the intracytoplasmic region of CD79a and CD79b, and the signalling cascade leading to B cell activation.

Secreted immunoglobulins

Secreted IgG, IgD and IgE are monomeric, whereas IgM occurs as a pentamer. IgA occurs predominantly as a monomer in the serum and as a dimer in seromucous secretions. In the following paragraphs, we will give a short overview of the principal structural and biological properties of the secreted immunoglobulins belonging to the five classes, in human.

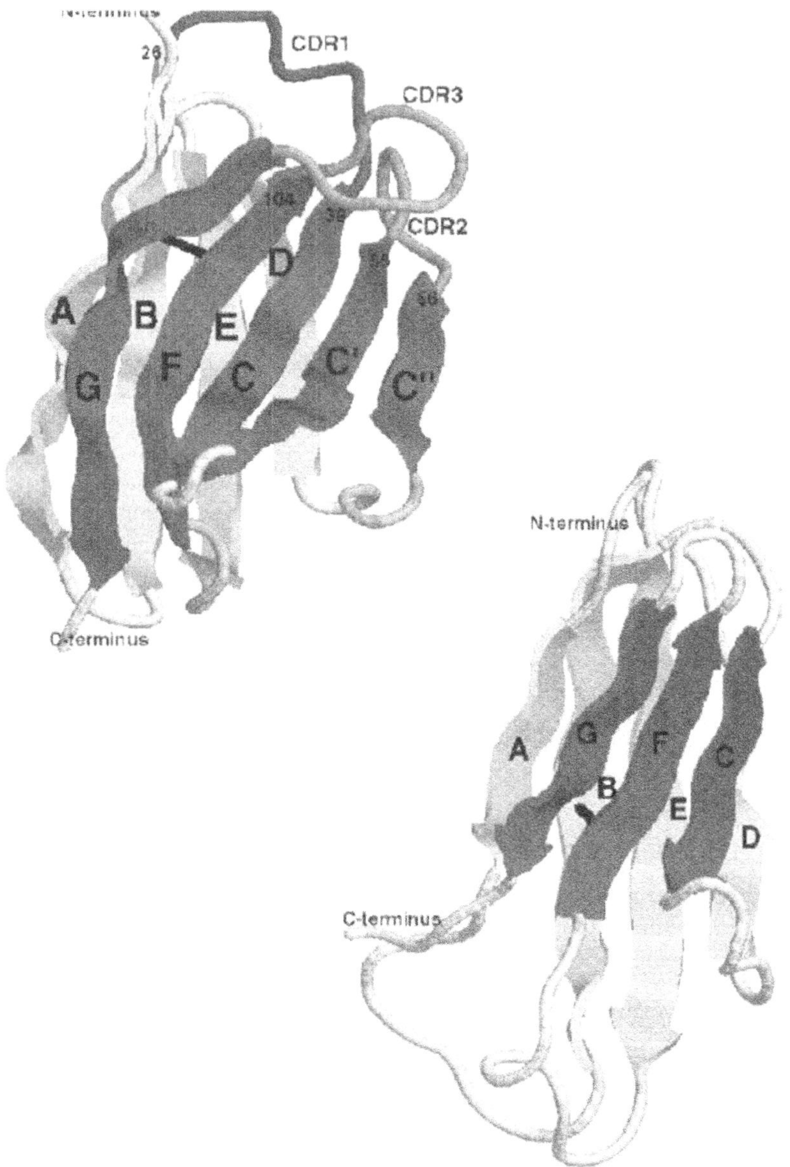

Figure 2 *3D representations (cartoon model) of the variable domain (A) and of the CH1 constant domain (B) of an immunoglobulin heavy chain. These figures show the spreading of the strands labelled A to G, in the two antiparallel sheets (the beta-sheet in contact with the domain of the other chain is in dark grey, and the external beta-sheet is in light grey). The disulphide bridges of the domains are shown as black cylinders. The FR-IMGT amino acid positions which limit the CDR-IMGT of the variable domain[17] (A) are indicated (See Section II, Collier de Perles).*

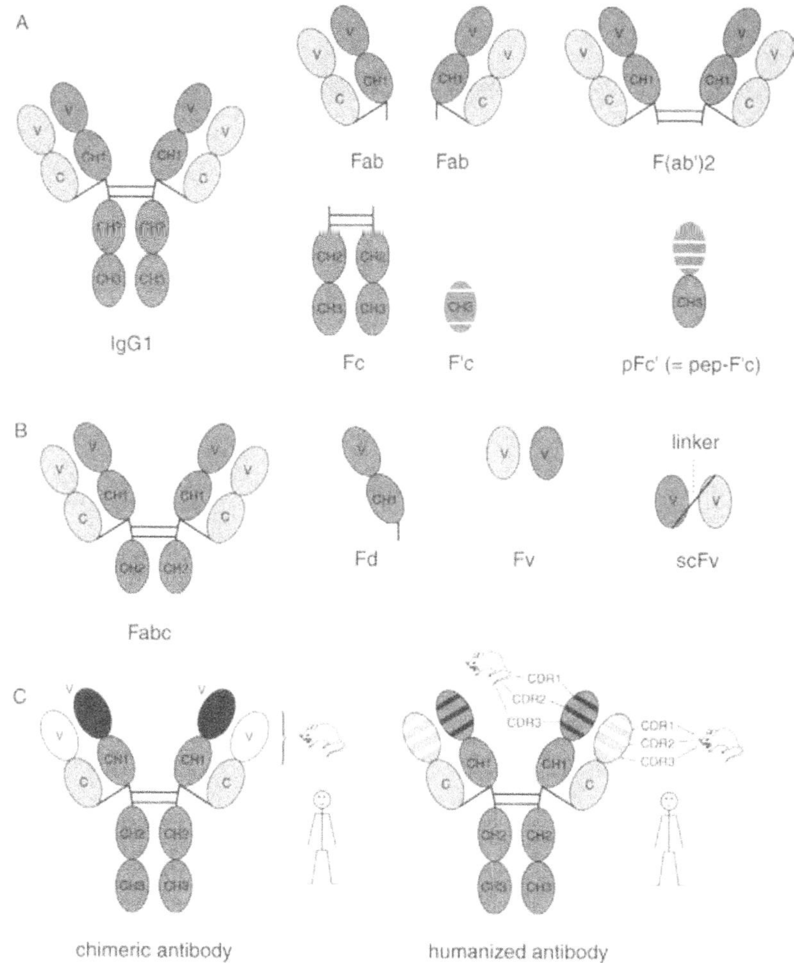

Figure 3 *Modular organization of the immunoglobulin IgG domains following: (A) IgG1 proteolytic cleavage by the papain (Fab, Fc) or the pepsin (F(ab')2, several polypeptides, the bigger being pFc' (also designated as pep-F'c)). F'c results from a papain digestion in presence of Cysteine. Due to the domain organization of the immunoglobulins, Fab and Fc of all Ig classes or subclasses are frequently constructed by recombinant engineering; (B) antibody engineering: Fabc, Fd, Fv, scFv (single chain Fv). Combinational library scFv are frequently displayed at the surface of filamentous phages[18]; (C) antibody "humanization": chimeric antibody and humanized antibody[19]. The murine (or other species) sequences which confer the antibody specificity correspond to the variable domains (V) in a chimeric antibody, or to the CDRs in a humanized antibody. Murine (or other species) sequences are shown in white (light chains) and black (heavy chains). The human light and heavy chain domains are shown in pale and dark grey, respectively. V = variable domain; C = constant domain of the light chain; CH1, CH2, CH3 = constant domains of the heavy chain.*

Table 1. *Structural and biological properties of human immunoglobulins.*

Classes and subclasses / Properties	IgM	IgD	IgG				IgA		IgE
			IgG1	IgG2	IgG3	IgG4	IgA1	IgA2	
Molecular weight of secreted form (kDa) (1)	950 (p)	170–180	150	150	155–165	150	160 (m) 300 (d)	160 (m) 350 (d)	190
Chain composition	$(\kappa_2\mu_2)5$ or $(\lambda_2\mu_2)5$	$\kappa_2\delta_2$ or $\lambda_2\delta_2$	$\kappa_2\gamma 1_2$ or $\lambda_2\gamma 1_2$	$\kappa_2\gamma 2_2$ or $\lambda_2\gamma 2_2$	$\kappa_2\gamma 3_2$ or $\lambda_2\gamma 3_2$	$\kappa_2\gamma 4_2$ or $\lambda_2\gamma 4_2$	$(\kappa_2\alpha 1_2)1–2$ or $(\lambda_2\alpha 1_2)1–2$	$(\kappa_2\alpha 2_2)1–2$ or $(\lambda_2\alpha 2_2)1–2$	$\kappa_2\varepsilon_2$ or $\lambda_2\varepsilon_2$
Functional valency	5 or 10	2	2	2	2	2	2 or 4	2 or 4	2
Structure (2)	monomer (mb) pentamer (s)	monomer	monomer				monomer (mb,s) dimer (sec)		monomer
Interheavy disulphide bonds per monomer	1	1	2	4	11	2	2	2	1
Other chain	J chain (16 kDa)	–	–				J chain (16 kDa) secretory component (70 kDa)		–
Sedimentation coefficient in Svedberg unit (S)	18–20	7	6.5–7.0				7, 10, 13, 15, 17		7.9
Carbohydrate average (%)	10–12	9–14	2–3				7–11		12–13
Adult level range (age 16–60) in serum (g/L) (3)	0.25–3.1	0.03–0.4	5–12	2–6 8.0–16.8	0.5–1	0.2–1	1.4–4.2	0.2–0.5	0.0001–0.0002
Approximate % total Ig in adult serum	10	0.2	45–53	11–15	3–6	1–4	11–14	1–4	0.004
Synthetic rate (mg/kg weight/day)	3.3	0.2	33	33	33	33	19–29	3.3–5.3	0.002

Continued

Table 1. *Continued.*

Properties \ Classes and subclasses	IgM	IgD	IgG				IgA		IgE
			IgG1	IgG2	IgG3	IgG4	IgA1	IgA2	
Biological half-life (day)	5–10	2–8	21–24	21–24	7–8	21–24	5–7	4–5	1–5
Transplacental transfer	0	0	++	+	++	++	0	0	0
Complement activation classical pathway (C1q)	+++	0	++	+	+++	0	0	0	0
Complement activation alternative pathway	0	0	0	0	0	0	+	0	0
Binding macrophages and other phagocytic cells (FcγR) (4)	0	+	+++	+/−	+++	+/−	0	0	0
Binding to mast cells and basophils (FcεR)	0	0	0	0	0	0	0	0	+++
Binding to epithelial poly-Ig receptor	+	0	0	0	0	0	+++	+++	0
Reactivity with *Staphylococcus* protein A	0	0	++	++	(0)[a]	++	0	0	0

[a]IgG3 allotype dependent.
(1) Approximate molecular weight (m = monomer, d = dimer, p = pentamer).
(2) mb = membrane, s = secreted, sec = secretory (see text).
(3) Total Ig adult level range (age 16–60) in serum: 9.5–21.7 g/L.
(4) +/−: Binding depends on the FcγR isoform and on the type of cell (see text).
0: null, +: low, ++: mean, +++: high for the biological properties.
Information from this table has been compiled from references 7 and 8 and from IMGT, http://imgt.cines.fr.

Table 2. *Human immunoglobulin chain characteristics.*

Chain types / Chain characteristics	Heavy chain									Light chain	
	μ	δ	γ1	γ2	γ3	γ4	α1	α2	ε	κ	λ
Molecular weight (kDa)	70	60	50	50	53–57	50	55	55	70	25	25
Number of variable domains	1	1	1	1	1	1	1	1	1	1	1
Number of constant domains	4	3	3	3	3	3	3	3	4	1	1
Length in number of amino acids											
Hinge	0	58	15	12	32, 47 or 62	12	19	6	0	–	–
Membrane Ig transmembrane region	25	25	25	25	25	25	25	25	25	–	–
Membrane Ig intracytoplasmic region	3	3	28	28	28	28	14	14	28	–	–
Secreted Ig tailpiece (1)	20	9	2	2	2	2	20	20	2	–	–
N-glycosylation sites (2)	5	3	1	1	2	1	2	4	7	–	–
O-glycosylation sites (3)	–	7	–	–	–	–	8	–	–	–	–
Allotypes	none	none	G1m	G2m	G3m	none	none	A2m	Em	Km	none

(1) Encoded by CH-S (see Chapter 3).
(2) Only indicated for the C-REGION.
(3) O-glycosylation sites are in the delta and alpha1 hinge regions.
Data from this table are from IMGT Repertoire (http://imgt.cines.fr), and from Section II (individual gene entries and IGHC protein display).

STRUCTURAL AND BIOLOGICAL PROPERTIES OF THE SECRETED IMMUNOGLOBULINS

IgM

IgM represents about 10% of total serum immunoglobulins in human and is largely confined to the intravascular pool. It exists almost exclusively as a polymeric form (pentamer) made up of five monomer units associated with a small polypeptide designated as J (joining) chain[22]. IgM is the predominant antibody produced early in the immune response. Pentameric IgM is decavalent with small antigens but only pentavalent with larger antigens, presumably due to steric hindrance (Table 1). A disulphide bridge connects the mu heavy chains between CH2 and CH3. Disulphide bridges between the CH3 and the tailpieces of the different monomers are involved in the IgM polymerization. A single J chain is present per Ig. This amounts to 1.5% of pentameric IgM. The conserved features of the J chain (16 kDa) is the presence of a carbohydrate acceptor sequence Asn-Ile-Ser, and of eight Cysteines. Six of the Cysteines form three intradisulphide bridges, and two are linked to the penultimate Cysteine of the mu chain.

The five Fc domains (CH3 and CH4) are arranged into a planar pentamer (Fc5). Electron microscope studies have revealed that uncomplexed IgM has a "star" conformation with the 10 Fab arms protruding out from the Fc5. On binding to an antigenic surface, the F(ab')2 dislocate out of the plane of the central Fc5 disc, giving a "staple" or "crab-like" conformation[6]. In this latter conformation, IgM is a very efficient activator of the classical complement pathway. C1q interacts directly with the CH3 domain of IgM.

IgD

IgD represents less than 1% of total serum immunoglobulins. IgD has a long, probably extended, hinge region of 58 amino acids which would allow great flexibility in the relative position of the two Fab arms. The hinge N-terminus half is heavily O-glycosylated (four–five oligosaccharides in one myeloma protein and seven in another, although the hinge region sequences are identical). The hinge C-terminus half is rich in charged amino acids (3 Arg, 6 Lys, 10 Glu) and is very susceptible to proteolytic attack, which makes serum IgD unstable (Table 1).

IgG

IgG is the major antibody class in normal human serum forming ~70% of the total immunoglobulins. IgG is a monomer and is evenly distributed between intravascular and extravascular pools. IgG is the predominant antibody of the secondary immune responses. There are four subclasses in humans.

The effector molecules binding IgG Fc are C1q, the Fcgamma receptors (FcγR) present on the surface of many cells of the immune system, the neonatal Fc receptor (FcRn) which transports maternal IgG to the fœtus, and the bacterial Fc receptors, protein A and protein G, which are believed to mask the bacteria through immobilized immunoglobulins on their surface.

C1q interacts directly with the CH2 domain of IgG. Binding to monomeric IgG is weak, but when several IgGs bind to, and effectively aggregate at an antigenic surface, two or more C1q heads may bind simultaneously leading to tighter binding and activation of the complement cascade. There are marked differences in ability to activate complement: IgG1 and IgG3 activate well, IgG2 less well, IgG4 not at all (Table 1).

Three classes of human FcγR have been described: FcγRI (CD64) are receptors with high affinity for IgG Fc and possess three Ig-like extracellular domains, FcγRII (CD32), and FcγRIII (CD16) are receptors with lower affinity and possess two Ig-like domains[23]. FcγRI, FcγRII and FcγRIII on macrophages and NK (Natural Killer) cells mediate antibody-dependent cell-mediated cytotoxicity (ADCC) and phagocytosis, whilst FcγRI, FcγRII and possibly FcγRIII on neutrophils are able to trigger release of activated oxygen species. The cellular responses also comprise endocytosis, enhanced antigen presentation and regulation of the antibody production, depending on the particular FcγR isoform and the type of cell[23]. FcγRI (CD64) displays high affinity for monomeric human IgG1 and IgG3, whilst the affinity for human IgG4 is about 10-fold lower and human IgG2 does not bind. The human FcγRII (CD32) binds IgG1 and IgG3. IgG4 does not bind, whilst the binding of IgG2 is controlled by an allotypic determinant in certain forms of the receptor. FcγRI and FcγRII appear to recognize overlapping but non-identical sites in the lower hinge region of IgG.

The crystal structures of the Fc fragment of human IgG1 in complex with *Staphylococcus aureus* protein A[24], or streptococcal protein G[25], and that of the Fc of rat with FcRn[26] revealed binding sites at the interface between the CH2 and CH3 Fc domains. The crystal structure of the human IgG1 Fc fragment–FcγRIII complex shows that FcγRIII binds to the two CH2 domains and lower hinge of the Fc[27].

IgA

IgA forms about 15–20% of total serum immunoglobulins where it occurs as a monomer. IgA is the predominant immunoglobulin in seromucous secretions such as saliva, tracheobronchial secretions, colostrum, milk and genitourinary secretions, where it is found in a dimeric form known as secretory IgA (sIgA). There are two subclasses of IgA, with IgA1 being the predominant (80–90%) subclass in serum. In contrast to serum IgA, secretory IgA shows roughly equal proportions of the two subclasses. The two IgA subclasses differ in the hinge. IgA1 has an effective structural hinge of 19 amino acids containing eight potential O-glycosylation sites. IgA2 has a structural hinge of six amino acids including five Proline which is likely to be relatively short, and by its nature, resistant to proteolysis. A further peculiarity of IgA2 is that for most molecules (allotype A2m(1)), the light chain is disulphide bridged, not to the heavy chain but to the light chain of the other Fab unit. The CH2 domain of both IgA subclasses has seven Cysteines. Two are involved in the usual intradomain disulphide bridge, another two in a second intradomain bridge and one is thought to be free, possibly for interaction with secretory component (see below). The remaining two form interheavy disulphide bridges. There is a further intradomain disulphide linkage in CH1 in addition to the conserved domain disulphide.

Secretory IgA

The dimer involves J chain (16 kDa) and another polypeptide known as secretory component (SC)(70 kDa). Selective transport of polymeric IgA through epithelial cells depends on the incorporation of the J chain into the polymers. Two of the J chain Cysteines are linked to the penultimate Cysteine of the alpha chains. The J chain, which was identified initially in human IgA[28], amounts to 4% of dimeric human IgA. The SC, unlike Ig and J chain which are produced by plasma cells, is synthesized in epithelial cells. With extra segments to attach it to the epithelial cell membrane, SC serves as a receptor for polymeric Ig (poly-Ig) containing J chain, i.e. IgA (or IgM). After endocytosis and transport, cleavage of the poly-Ig/poly-Ig receptor complex releases poly-Ig (poly-Ig with the J chain) associated with SC[7]. This process is particularly important for secretory IgA release. The poly-Ig receptor is composed, in its poly-Ig binding portion (i.e. SC) of five highly conserved Ig-like domains of approximately 100 amino acids. SC (70 kDa) probably interacts non-covalently with the Fc and J chain and forms a single disulphide bridge to one of the monomers of dimeric IgA.

Effector function

IgA can activate the alternative complement pathway and bind to specific FcαR. FcαR is present on monocytes, macrophages, neutrophils and eosinophils and can mediate ADCC, phagocytosis and degranulation. FcαR has two extracellular Ig-like domains and spans the membrane once. FcαR binds at a site in the IgA CH2 domain.

IgE

IgE, though a trace Ig in serum, is found bound through specific receptors on the cell surface of mast cells and basophils in all individuals. It is involved in protection against helminthic parasites but is most commonly associated with atopic allergies. IgE binds with high affinity to FcϵRI on the surface of mast cells and basophils. Aggregation of the receptor by binding of multivalent antigens, such as pollen, to prebound IgE results in cell degranulation and release of mediators of the allergic response. The FcϵRI binding site on IgE involves CH3 (next to the interface between CH2 and CH3). The crystal structure of the Fc fragment of human IgE and its high-affinity receptor FcϵRIα reveals that one receptor binds one Fc asymmetrically: CH3 domain of each epsilon chain of the Fc is bound to two different sites on the FcϵRI receptor[29]. IgE binds with lower affinity to a second receptor, FcϵRII, present on monocytes, B cells and platelets, which plays a role in cytotoxicity against parasites such as schistosomes. The interaction between IgE and FcϵRII appears to require the presence of the IgE CH2, CH3 and CH4 domains, the latter serving to promote the dimerization of the two epsilon chains, necessary for receptor binding.

References

[1] Fleischmann, J.B. et al. (1962) Arch. Biochem. Biophys. Suppl. 1, 174–190.
[2] Edelman, G.M. et al. (1969) Proc. Natl Acad. Sci. USA 63, 78–85.
[3] Porter, R.R. (1959) Biochem. J. 73, 119–126.
[4] Strosberg, D. (1986) In Immunologie, Ed. J.F. Bach. Flammarion Médecine Sciences, Paris.
[5] Janeway, C.A. and Travers, P. (1997) In Immunobiologie, Eds C.A. Janeway and P. Travers, De Boeck and Larder, Paris, pp. 111–151.
[6] Burton, D.R. (1987) In Molecular Genetics of Immunoglobulin, Eds. F. Calabi and M.S. Neuberger. Elsevier, Oxford, pp. 1–50.

[7] Lefranc, M.-P. and Lefranc, G. (1992) In Hématologie de Bernard Dreyfus, Eds J. Breton-Gorius et al. Flammarion Médecine-Sciences, Paris, pp. 197–254.
[8] Hamilton, R.G. (1997) In Handbook of Human Immunology, Eds M.S. Leffell et al. CRC Press, New York, pp. 65–109.
[9] Lefranc, M.-P. and Lefranc, G. (1990) In The Human IgG Subclasses, Ed. F. Shakibs. Pergamon Press, Oxford, pp. 43–78.
[10] Lefranc, G. et al. (1978) Hum. Genet. 41, 197–209.
[11] Osipova, L.P. et al. (1999) Hum. Genet. 105, 530–541.
[12] Lefranc, G. et al. (1976) Amer. J. Hum. Genet. 28, 51–61.
[13] Lefranc, G. et al. (1983) Eur. J. Immunol. 13, 240–244.
[14] Wiebe, V. et al. (1994) Hum. Genet. 93, 520–528.
[15] Dariavach, P. (1987) Proc. Natl Acad. Sci. USA 84, 9074–9078.
[16] Kabat, E.A. et al. (1987) In Sequences of Proteins of Immunological Interest, Public Health Service, NIH, Washington DC.
[17] Lefranc, M.-P. (1999) The Immunologist 7, 132–136.
[18] Lefranc, M.-P. and Lefranc, G. (1997) Ingénierie des anticorps, Les Editions INSERM, Paris, pp. 1–103.
[19] Jones, P.T. et al. (1986) Nature 321, 522–525.
[20] Jefferis, R. et al. (1998) Immunol. Rev. 163, 59–76.
[21] Reth, M. (1992) Annu. Rev. Immunol. 10, 97–121.
[22] Koshland, M.E. (1985) Annu. Rev. Immunol. 3, 425–453.
[23] Gessner, J.E. et al. (1998) Ann. Hematol. 76, 231–248.
[24] Deisenhofer, J. et al. (1981) Biochemistry 20, 2361–2370.
[25] Sauer-Eriksson, A.E. et al. (1995) Structure 3, 265–278.
[26] Burmeister, W.P. et al. (1994) Nature 372, 379–383.
[27] Sonderman, P. et al. (2000) Nature 406, 267–273.
[28] Halpern, M.S. and Koshland, M.E. (1970) Nature 228, 1276–1278.
[29] Garman, S.C. et al. (2000) Nature 406, 259–266.

3 Synthesis of the immunoglobulin chains

The immunoglobulins comprise two identical heavy chains, associated with two identical light chains, kappa or lambda (Fig. 1). In humans, the genes encoding the heavy chains, the kappa light chains and the lambda light chains, are located in the IGH, IGK and IGL loci on chromosomes 14 (14q32.33), 2 (2p11.2) and 22 (22q11.2), respectively (see Chapter 4). The synthesis of the immunoglobulin heavy and light chains requires gene rearrangements, at the DNA level, in the IGH, IGK and IGL loci during the B cell differentiation[1–3]. Chronologically, the synthesis of the mu heavy chains precedes that of the light chains. The stages of differentiation, from the hematopoietic stem cell to the mature B cell which expresses IgM and IgD, occur in the bone marrow, and are antigen independent (Fig. 2). The final differentiation stages, from the mature B cell to the plasma cell which express secreted immunoglobulin of diverse classes or subclasses occur in the germinal centres of the secondary lymphoid organs, and are antigen dependent (Fig. 2).

SYNTHESIS OF THE MU HEAVY CHAINS

D-J and V-D-J rearrangements in the IGH locus

The heavy (IGH) locus comprises variable (V), diversity (D), joining (J) and constant (C) genes. The variable domain of a heavy chain, or V-D-J-REGION, is generated by the junction at the DNA level of three genes: a variable gene IGHV, a diversity gene IGHD and a joining gene IGHJ. The synthesis requires two successive rearrangements. First, one of the D genes is joined to one of the J genes with deletion of the intermediary DNA as an excision loop, then one of the IGHV genes is joined to the partially rearranged D-J gene to generate a completely rearranged IGHV-D-J gene (Fig. 3). This second rearrangement is also accompanied by the formation of an excision loop which is cleaved off. The rearranged IGHV-D-J gene is transcribed with the IGHM gene, the most 5′ IGHC gene in the locus, into a IGHV-D-J-M (or IGHV-D-J-Cmu) pre-messenger RNA. The IGHM gene encodes the four domains (CH1 to CH4) of the mu heavy chain constant region. After splicing of the pre-messenger RNA, translation of the messenger RNA, and elimination of the signal peptide by a peptidase in the endoplasmic reticulum, a mature mu chain is produced.

SYNTHESIS OF THE KAPPA AND LAMBDA LIGHT CHAINS

V-J rearrangements in the IGK and IGL loci

The kappa (IGK) locus and the lambda (IGL) locus comprise variable (V), joining (J) and constant (C) genes. The variable domain of a kappa chain or a lambda chain, or V-J-REGION, is generated by the junction, at the DNA level, of two genes: a variable and a joining genes, with deletion of the intermediary DNA to create a rearranged IGKV-J gene (Fig. 4), or a IGLV-J gene (not shown). The rearranged

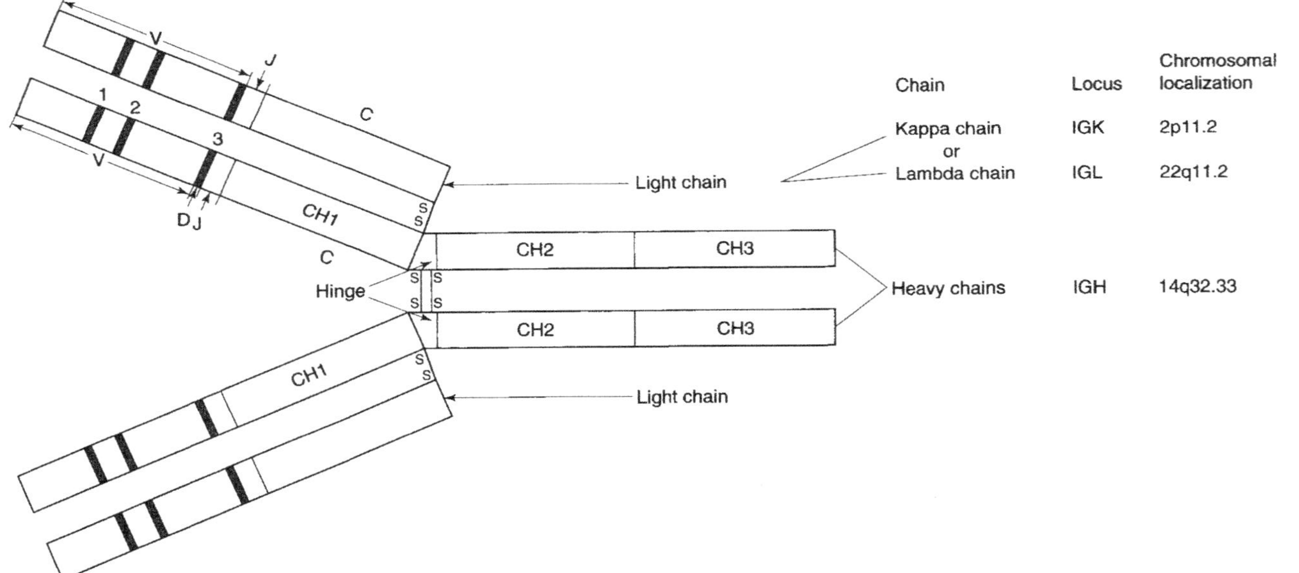

Figure 1. *Schematic representation of an immunoglobulin. The variable domain of a light chain, or V-J-REGION, is encoded by two rearranged genes (one IGKV rearranged to one IGKJ for a kappa chain, one IGLV rearranged to one IGLJ for a lambda chain). The variable domain of a heavy chain, or V-D-J-REGION, is encoded by three rearranged genes (IGHV, IGHD and IGHJ). The three hypervariable regions (or CDRs) 1, 2 and 3 (hatched in the figure) determine the recognition and binding site to the antigen, in the 3D structure. The constant region, or C-REGION, of the light chain is encoded by the IGKC gene (for the kappa chains), or one of the IGLC genes (for the lambda chains), and comprises a unique constant domain. The constant region of the heavy chain is encoded by one of the IGHC genes, and comprises three–four constant domains. CH1, CH2 and CH3 represent the constant domains of a heavy gamma chain. V = V-REGION, J = J-REGION, D = D-REGION (or more exactly N-AND-D-REGION to take into account the N-diversity, see Chapter 3), C = C-REGION. Intrachain disulphide bridge of each of the 12 domains are not show. Interchain disulphide bridges correspond to those of an IgG1.*

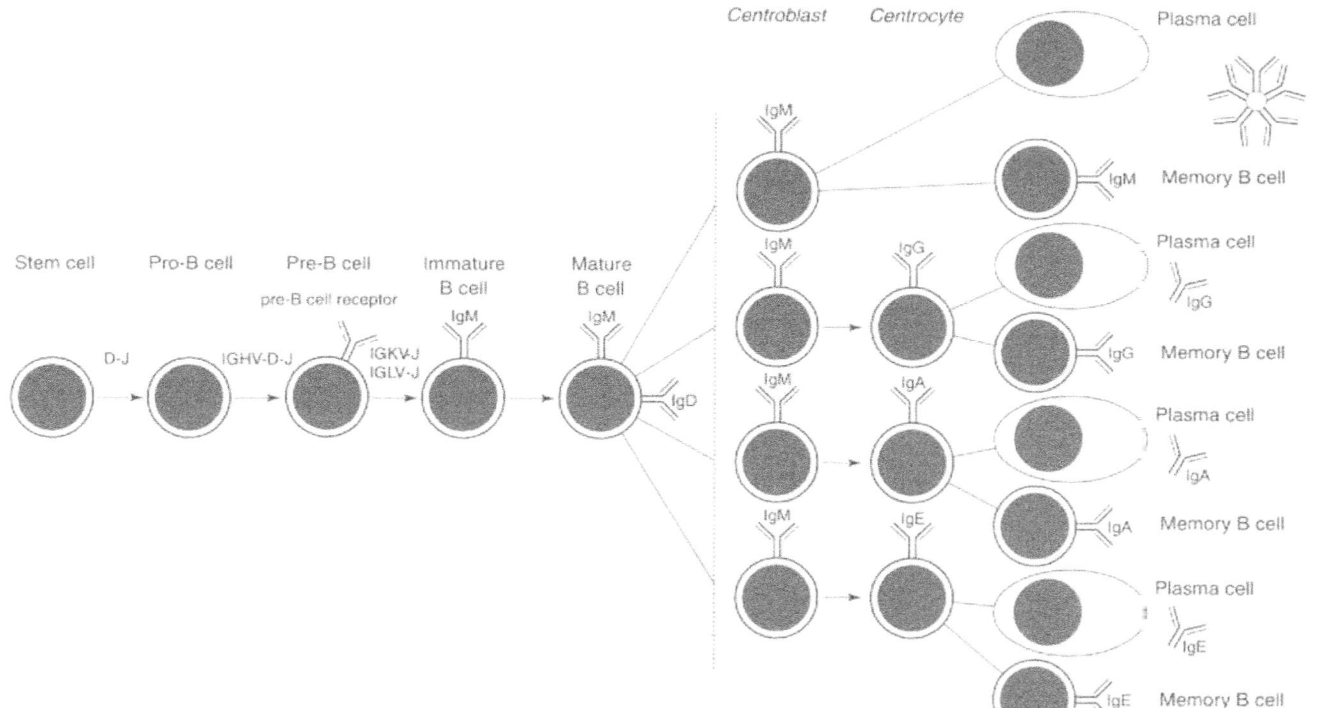

Figure 2 *B cell differentiation. The B cell differentiation from the hematopoietic stem cell to the mature B cell, in the bone marrow, is antigen independent. The final differentiation stages, from the mature B cell to the plasma cell and memory B cell, in the germinal centres of the secondary lymphoid organs, is antigen dependent, and generally requires cooperation between B and T cells.*

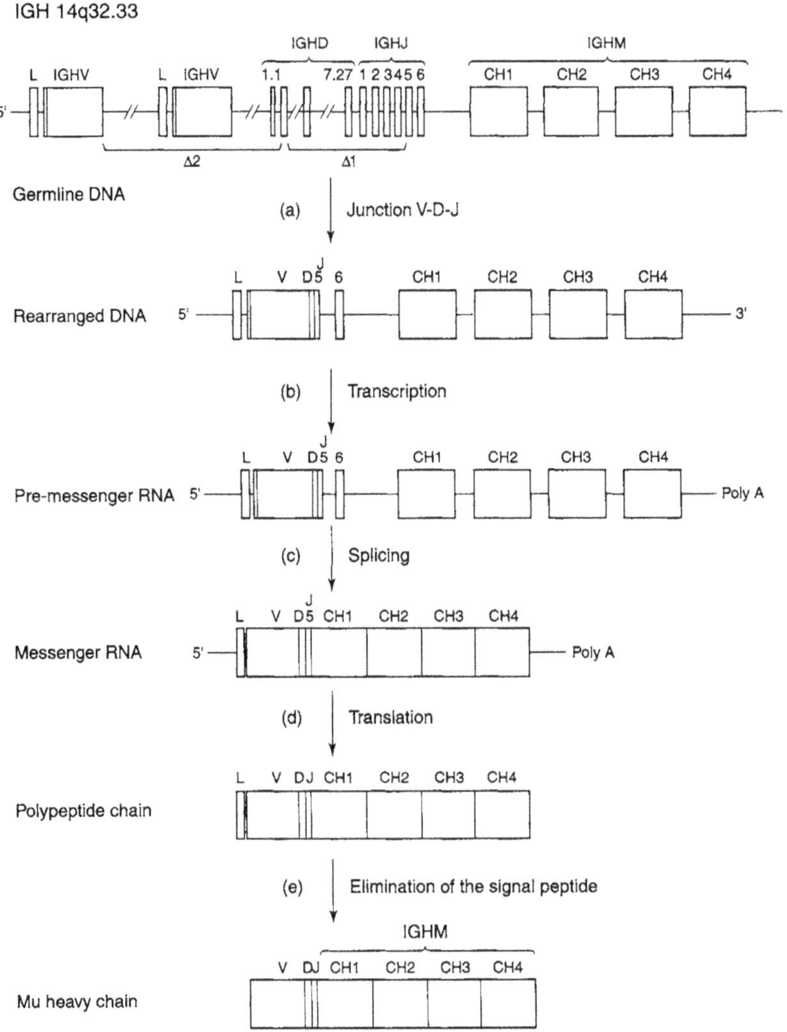

Figure 3 *Synthesis of a mu heavy chain. (a) At the DNA level, in a first step, one of the IGHD genes is joined to one of the IGHJ genes, with deletion of the intermediary DNA, to create a partially rearranged D-J gene. In a second step, one of the IGHV genes is joined to D-J, with deletion of the intermediary DNA, to generate a completely rearranged IGHV-D-J gene. (b) The rearranged IGHV-D-J gene is transcribed with the IGHM gene into a IGHV-D-J-M (or IGHV-D-J-Cmu) pre-messenger RNA. (c) The RNA sequences corresponding to the introns and to the non-used IGHJ genes are excised by splicing, and a mature messenger which comprises the spliced coding regions and the 5′ and 3′ untranslated sequences, is obtained. (d) The messenger RNA is translated into a polypeptide chain by the ribosomes. (e) The signal peptide is cleaved off by a peptidase following the entry of the polypeptide chain in the endoplasmic reticulum, and a mature mu heavy chain is produced. L = L-PART1 (L: for Leader).*

IGKV-J (or IGLV-J) gene is transcribed with the IGKC gene (or one of the IGLC genes) into a IGKV-J-C (or IGLV-J-C) pre-messenger RNA. The unique IGKC gene, or one of the functional IGLC genes, with their single exon, encodes the single domain of the constant region of the kappa or lambda chains, respectively. After splicing of the pre-messenger RNA, translation of the messenger RNA and elimination of the signal peptide from the polypeptide chain in the endoplasmic reticulum, a mature kappa (or lambda) light chain is produced.

ORIGIN OF THE VARIABLE DOMAIN DIVERSITY OF THE IMMUNOGLOBULINS

The diversity of the variable domain of the immunoglobulin chains arises mainly from combinatorial diversity, junctional diversity and somatic hypermutations.

Combinatorial diversity

The combinatorial diversity is created by the somatic V-D-J and V-J rearrangements, and the association of different variable domains of heavy and light chains to form the antibody recognition sites. The somatic IGKV-J, IGLV-J and IGHV-D-J rearrangements require the presence of recombination signal (RS) sequences which are located in 3′ of the V genes, 5′ of the J genes, and on both sides of the D genes[4] (Fig. 5). RS which are recognized by the recombinase enzyme comprise conserved nucleotide sequences (heptamer and nonamer) separated by a spacer of 12 ± 1 or 23 ± 1 nucleotides. In contrast to heptamer sequences which are palindromic and well conserved, and to nonamer sequences which are rich in A or T, spacer sequences are not conserved. However, spacer lengths are important, and efficient rearrangements occur between RS of different lengths, that is one with a 12 ± 1 spacer, and another one with a 23 ± 1 spacer (12/23 joining rule)[5]. The potential repertoire resulting from the combinatorial diversity depends on the number of V, D and J genes, and on their functionality, as this will be described in Chapter 4.

Junctional diversity

The junctional diversity is mostly represented by the N-diversity. The N-diversity (N, for nucleotides) is frequently observed at the V-D-J junctions of the immunoglobulin heavy chains, and represents the major source of the CDR3 diversity[6]. It results from the addition of nucleotides at random by the terminal deoxynucleotidyltransferase TdT[7]. This nucleotide addition preferentially involves G, is template independent, and is frequently preceded by the deletion of nucleotides at the ends of the DNA fragments which recombine (3′ end of the V-REGION, 5′ end of the J-REGION and both ends of the D-REGION). If there is no nucleotide deletion, P-nucleotides may be observed[8]. P-nucleotides are short inverted-repeat sequences identified at the V-(D)-J junctions. These nucleotides are adjacent to and complementary of the intact coding ends of the rearranged V-REGION, D-REGION or J-REGION. The P-nucleotides result from the dissymmetric opening of the hairpin formed at the extremities of the coding regions during the V-J or V-D-J rearrangements[9].

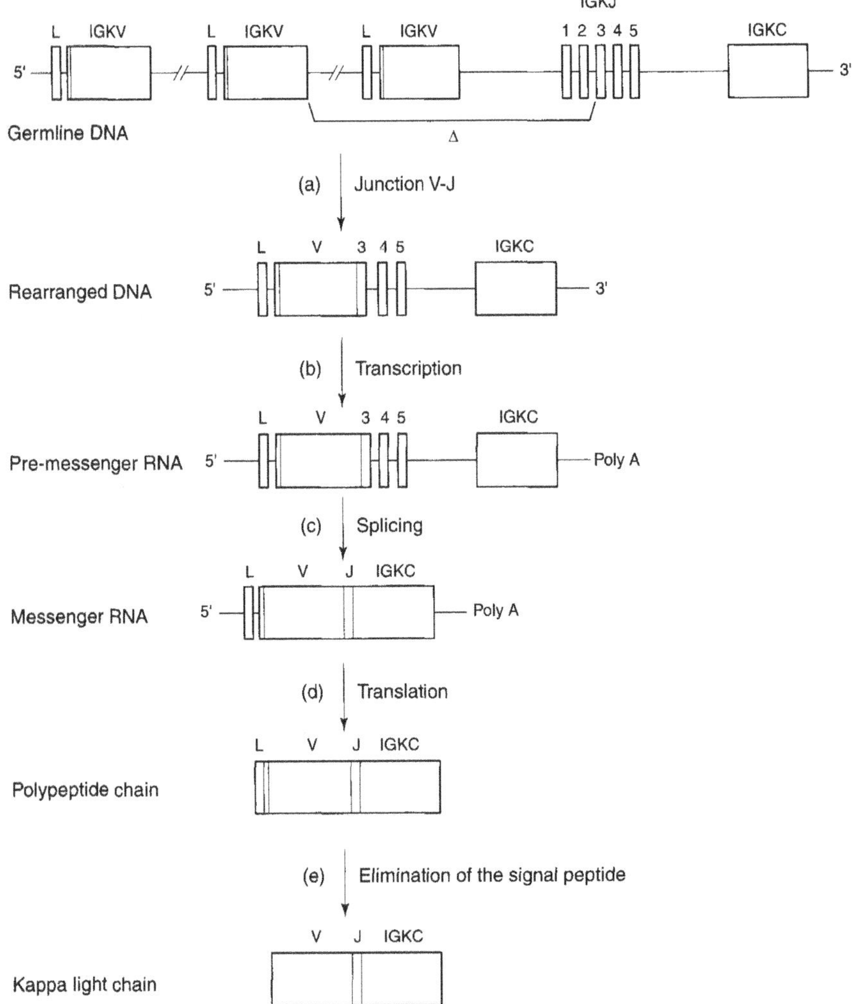

Figure 4 *Synthesis of a kappa light chain. (a) At the DNA level, one of the IGKV gene is joined to one of the five IGKJ genes, with deletion of the intermediary DNA, to create a rearranged IGKV-J gene. (b) The rearranged IGKV-J gene is transcribed with the IGKC gene into a IGKV-J-C pre-messenger RNA. (c) The RNA sequences corresponding to the introns and to the non-used IGKJ genes are excised by splicing, and a mature messenger which comprises the spliced coding regions, and the 5′ and 3′ untranslated sequences, is obtained. (d) The messenger RNA is translated into a polypeptide chain by the ribosomes. (e) The signal peptide is cleaved off by a peptidase following the entry of the polypeptide chain in the endoplasmic reticulum, and a mature kappa light chain is produced. L = L-PART1 (L: for Leader).*

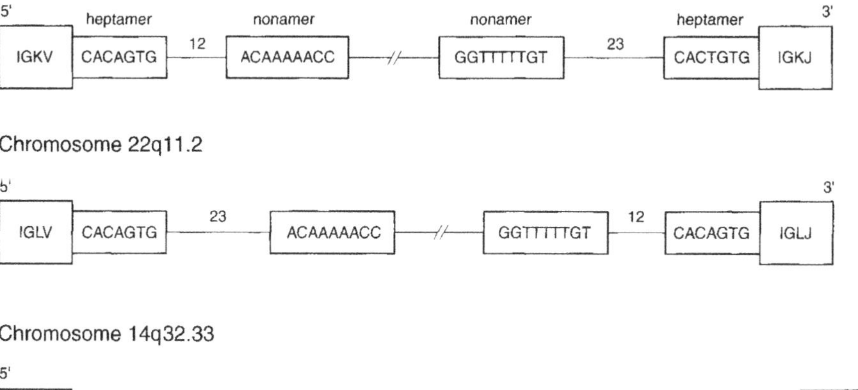

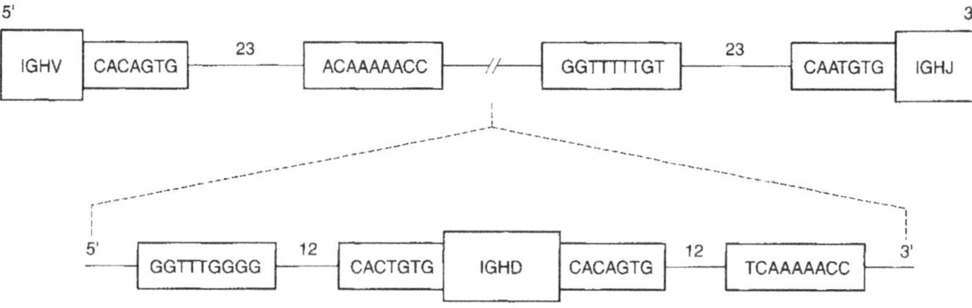

Figure 5 *Examples of human immunoglobulin V, D and J gene recombination signals. Although the heptamer and the nonamer sequences are well conserved, there are sequence differences between individual genes as shown in the RS tables in Section II. Spacer lengths are 12 ± 1 or 23 ± 1 bp. This is known as the 12/23 rule.*

The N-diversity mechanism creates at the D-J and V-D-J junctions nucleotide sequences which do not exist in the germline DNA, and which therefore, characterize a given B cell. However, two-thirds of the resulting junctions are out of frame, and only the sequences which have an in frame junction can be found expressed in productive transcripts. In 1982, Alt and Baltimore proposed a model to explain the mechanism of N-diversity[6]. This model has been confirmed by experimental data. This mechanism of diversity is particularly important for the antibody specificity, since the V-D-J junction encodes the heavy chain CDR3, in contact with the antigen.

Somatic hypermutations

Somatic hypermutations appear during the B cell maturation in the germinal centres of the secondary lymphoid organs (spleen and lymph nodes). They specifically affect the rearranged V-J and V-D-J genes during the antigen dependent stages of differentiation and represent a major mechanism for generating antibody diversity[10].

CO-EXPRESSION OF THE MEMBRANE MU AND DELTA HEAVY CHAINS

Alternative splicing of a IGHV-D-J-Cmu-Cdelta pre-messenger

During its differentiation, an immature B cell becomes a mature B cell which expresses simultaneously membrane IgM and IgD. The variable domains of the mu and delta heavy chains are identical, and are encoded by the same rearranged IGHV-D-J gene. This can be explained by the synthesis of a long IGHV-D-J-Cmu-Cdelta pre-messenger of about 20 kilobases (kb) (the distance separating IGHM and IGHD being 6 kb in the human IGH locus). Mature IGHV-D-J-Cmu and IGHV-D-J-Cdelta messengers are obtained by alternative splicing of the pre-messenger, and are then translated into membrane mu or delta heavy chains[11-13].

EXPRESSION OF GAMMA, EPSILON AND ALPHA CHAINS

Switch recombination

The mature B cell which enters the lymph nodes expresses IgM and IgD. In the lymph nodes, as a result of a B and T cell cooperation, there is interaction of the CD40 on B cell with its ligand, CD40L, on T cell, and in the presence of interleukines, the B cell starts to express IgG (IgG1, IgG2, IgG3 or IgG4) or IgA (IgA1 or IgA2) or IgE. This switch results in a change of the constant region of the heavy chain, while maintaining the expression of the same antibody specificity. In the switch recombination, the rearranged IGHV-D-J gene, which was previously associated with the IGHM gene, is brought into the proximity of one of the other IGHC genes. The recombination occurs between switch sequences located at about 2 kb in 5′ of each IGHC gene (except IGHD)(Fig. 6). The switch sequences, of about 2 kb, are composed of 20–80 nucleotide motifs repeated in tandem. These motifs contain short GGGCT and GAGCT repeats and, near the recombination site, TGGG or TGAG. Sequences chi GCTGGTGG, promoters of the lambda phage generalized recombination, have also been described in murine switch sequences, although their functional significance in the IGH locus is not known. The switch leads to the deletion of the IGHC genes located between the rearranged IGHV-D-J and the IGHC to be used, and to the formation of excision loops which are cleaved off[14-18].

EXPRESSION OF DELTA CHAINS IN PLASMA CELLS

Only a minority of normal plasma cells and rare B cell malignancies express exclusively IgD (IgM^- IgD^+ B cells). The low frequency has been explained by the lack of a recognizable switch sequence between Cmu and Cdelta. However, a region, designated as sigma delta, contains a relatively high content of pentameric repeats with an extremely G-rich area and appears to function as a vestigial switch recombination site leading to the expression of delta chains in germinal centre B cells and plasma cells[19-21].

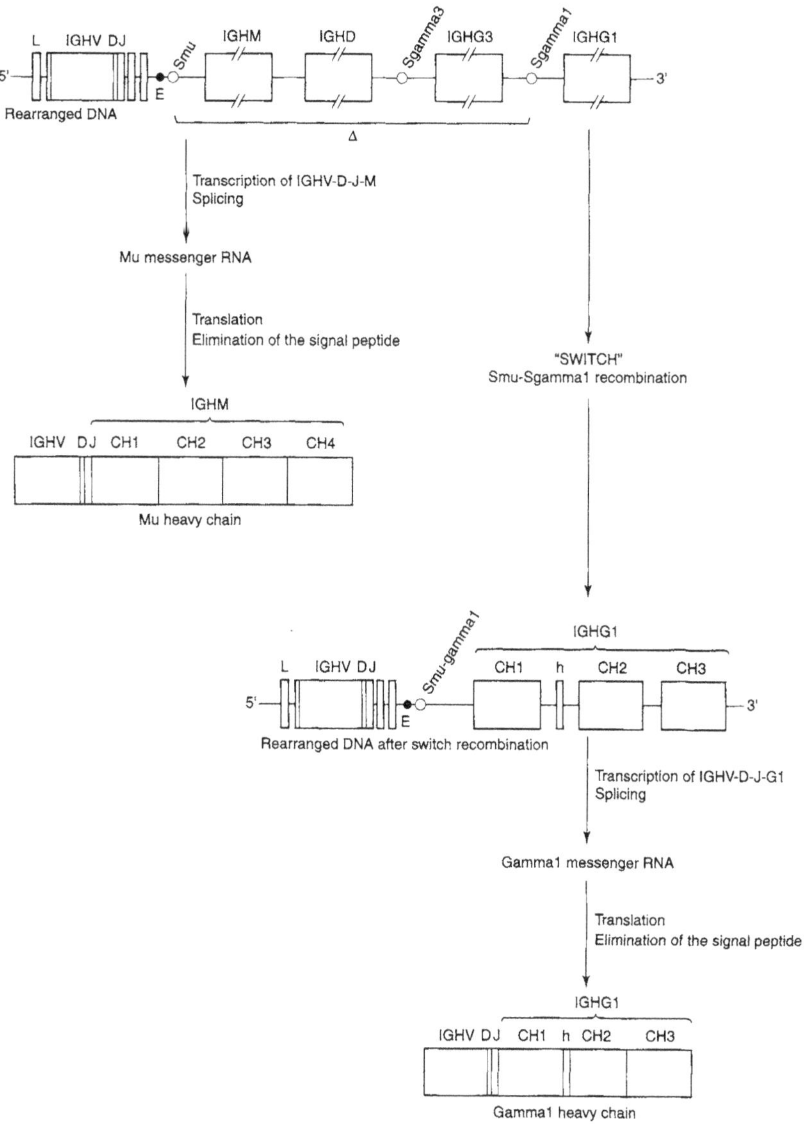

Figure 6 *Class switch IgM-IgG1: Smu-Sgamma1 recombination. In a B cell which expresses IgM on the cell surface, a productive rearranged IGHV-D-J gene on one chromosome 14 is transcribed with the IGHM gene. Before the switch recombination, all the IGHC genes are present in the IGH locus. During the switch recombination, a novel DNA rearrangement occurs in the IGH locus, between the Smu (Switch mu) sequence and another S sequence located in 5' of a more downstream IGHC gene (for example, Sgamma1 upstream of IGHG1). This leads to the deletion of the intermediary DNA and to the loss of the IGHC genes located between the two S sequences which recombine. The enhancer (E) located between the most 3' IGHJ and Smu is retained during the switch recombination.*

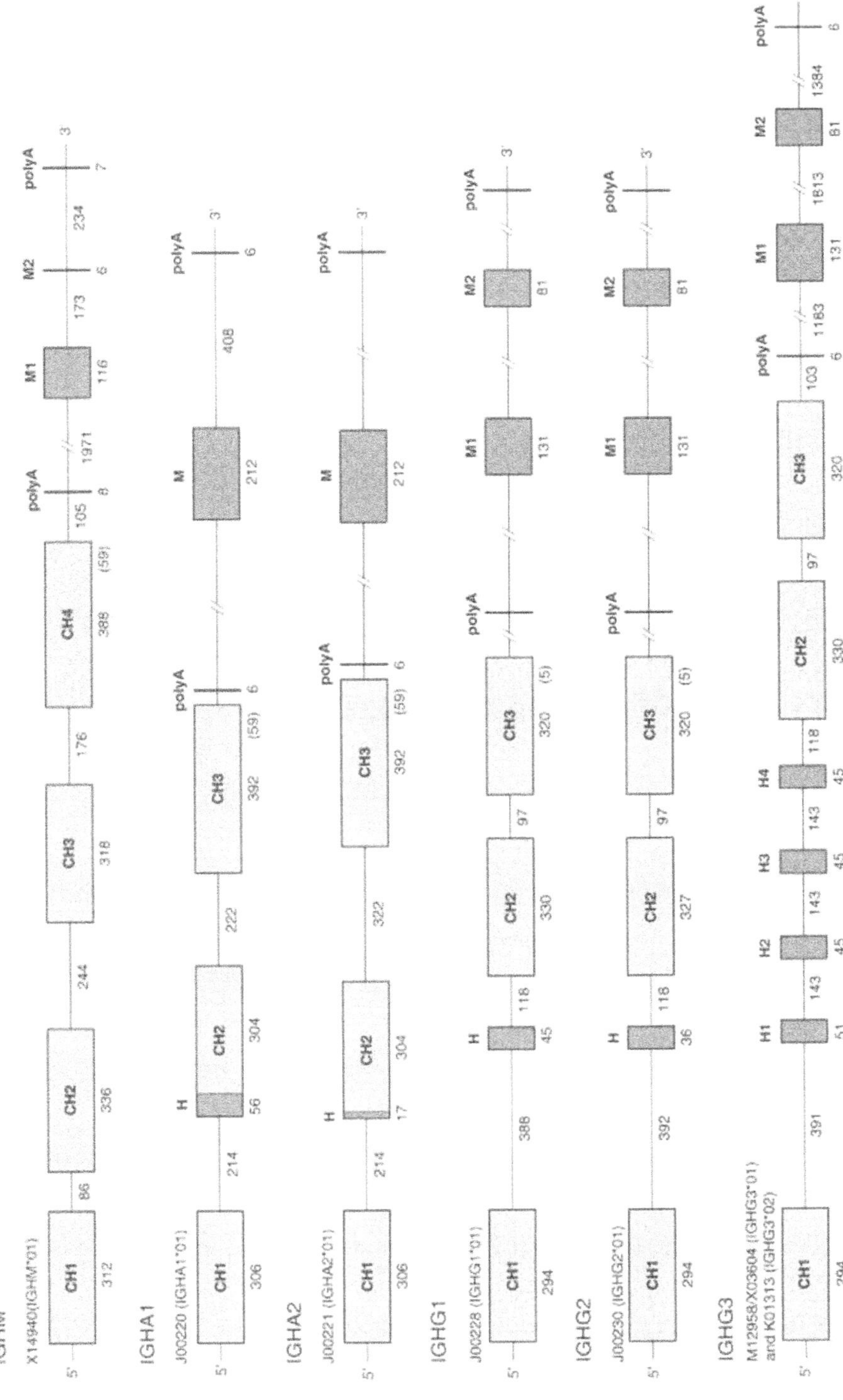
IGHM
X14940 (IGHM*01)
CH1
CH2
CH3
CH4
M1
M2
polyA
IGHA1
J00220 (IGHA1*01)
H
M
IGHA2
J00221 (IGHA2*01)
IGHG1
J00228 (IGHG1*01)
IGHG2
J00230 (IGHG2*01)
IGHG3
M12958/X03604 (IGHG3*01)
and K01313 (IGHG3*02)
H1
H2
H3
H4
5'
3'

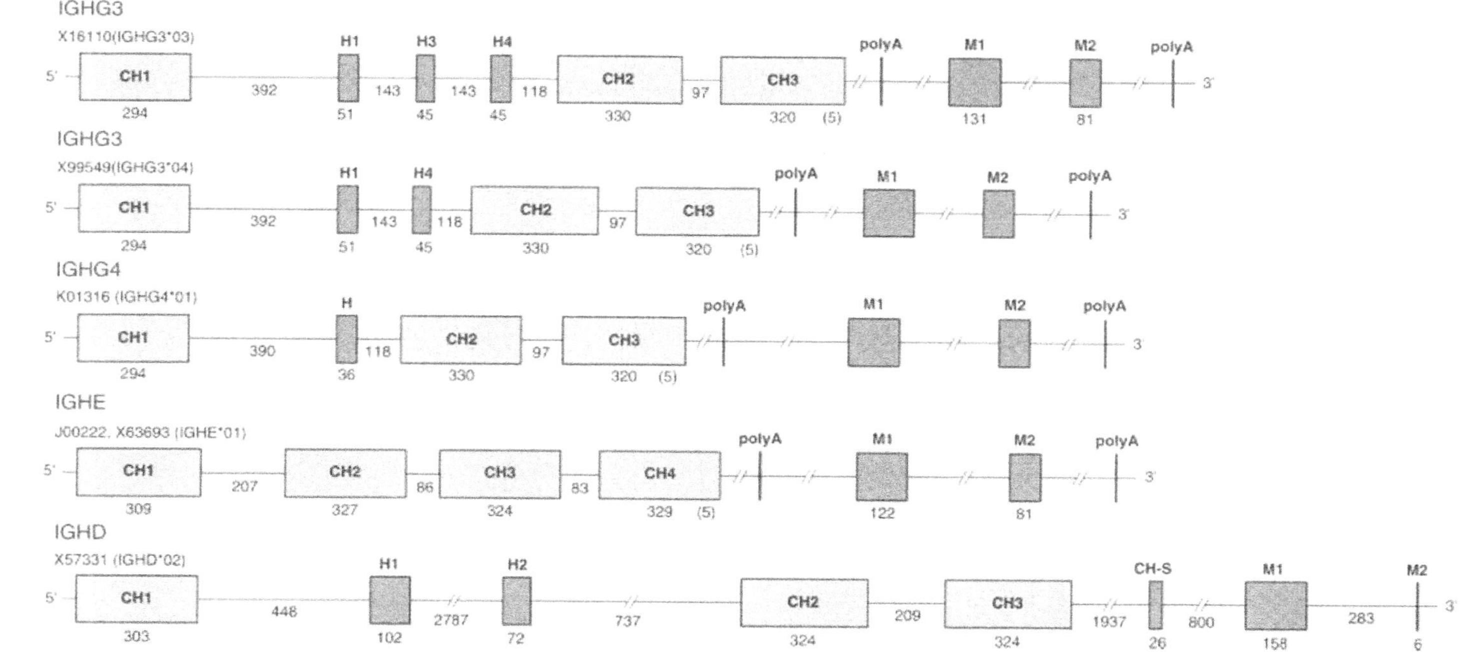

Figure 7 *Structure of the human immunoglobulin constant genes. Sequences and references for the individual human IGHC genes and alleles are given in Section II. Between parentheses are the number of nucleotides corresponding to the CH-S region.*

MEMBRANE AND SECRETED IMMUNOGLOBULINS

B cell membrane immunoglobulin heavy chains have a hydrophobic C-terminal end which holds them anchored in the plasma membrane, whereas plasma cell secreted immunoglobulins have an hydrophilic end[22]. Membrane and secreted immunoglobulins result from an alternative splicing of heavy chain transcripts.

Membrane and secreted mu heavy chains

The C-terminal region of the membrane mu chain is encoded by two small exons, M1 and M2 located at about 2 kb in 3′ of the CH4 exon[23], M1 encodes 39 amino acids, whereas M2 only encodes two amino acids (Fig. 7). These 41 amino acids represent the anchor region of the membrane mu chain which comprises an extracellular region of 13 amino acids between the CH4 domain and the membrane, a hydrophobic transmembrane region of 25 amino acids and a short cytoplasmic region of three amino acids. The C-terminal region of the secreted mu chain comprises 20 amino acids encoded by the 3′ end of the CH4 exon (designated as CH-S).

For the synthesis of a membrane mu chain, it is the poly A site located in 3′ of the M2 exon, and the splicing site located in the CH4 exon, at the 5′ limit of CH-S which are used (Fig. 8). This splicing deletes the CH-S sequence and its stop codon, as well as the sequence between CH4 and M1, and between M1 and M2. For the synthesis of a secreted mu chain, it is the poly A site located 103 bp from the 3′ end of the CH4 exon, and the stop codon at the 3′ end of CH4 which are used (Fig. 8). The expression of the membrane and secreted mu chain depends on the selection of the poly A used[24].

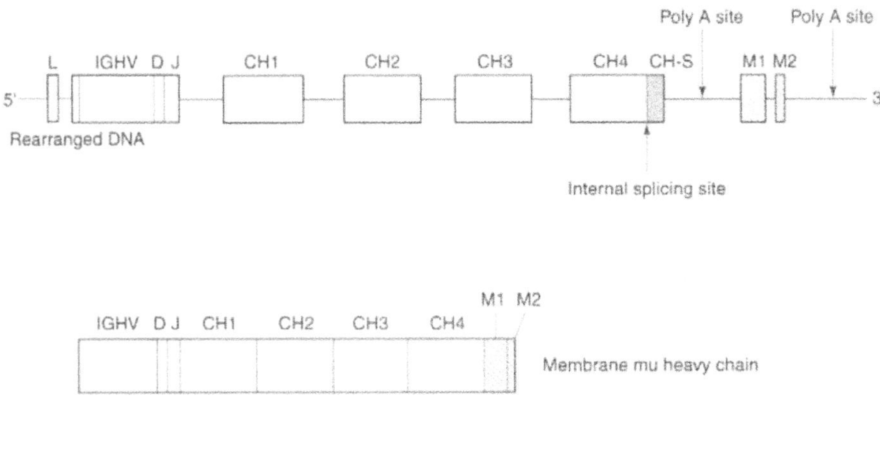

Figure 8 *Synthesis of a membrane and of a secreted mu heavy chain.*

Membrane and secreted delta chains

The organization of the 3′ region of the IGHD gene differs due to the presence of a small independent CH-S exon, located at 1.9 kb in 3′ of the CH3 exon, and which encodes the nine last amino acids of the secreted delta chains. The M1 and M2 exons, located at 0.8 kb and 1.1 kb in 3′ of CH-S, respectively, encode the transmembrane and the cytoplasmic region; M1 encodes 53 amino acids whereas M2 encodes two amino acids (Fig. 7). The expression of the membrane and secreted delta chain depends on the selection of the poly A used : poly A in 3′ of the IGHD exon M2, for the synthesis of the membrane delta chain, or poly A in 3′ of CH-S, for the synthesis of the secreted delta chain[25].

Membrane and secreted gamma, alpha and epsilon chains

The expression of the membrane and secreted gamma, alpha and epsilon chains follows the same mechanisms as those described for the mu chain, the CH-S being part of a domain exon (CH3 or CH4, depending on the IGHC gene)(Fig. 7).

ALLELIC AND ISOTYPIC EXCLUSION – REARRANGEMENT CHRONOLOGY

The B cells, and the plasma cells which derive from them, display:

- allelic exclusion: in most cases, the only productive genes are either those of the paternal chromosome, or those of the maternal chromosome, but usually never the two together (functional haploidy).
- isotypic exclusion: a single type of light chain, kappa or lambda, and usually a single type of heavy chain belonging to a given subclass, are synthesised.

The mechanisms of the allelic and isotypic exclusion are still poorly understood. However, molecular analysis has shown that the excluded allele is usually either non-rearranged, or unproductively rearranged (IGK locus in B cells synthesising a lambda chain)[26,27].

During the B cell differentiation, the IGH locus on one chromosome 14 undergoes first a D-J, then a V-D-J rearrangement. A productive rearrangement allows the synthesis of a mu heavy chain in the cytoplasm of the pre-B cells. The mu chain is expressed at the surface of the pre-B cells in association with a lambda-like chain and a V pre-B chain, which constitute together the pre-B cell receptor[28,29] (Figs. 2 and 9). The expression of the surface IgM at a later stage requires the synthesis of kappa or lambda light chains, that is a productive V-J rearrangement of the IGK or IGL loci. It is the expression of the pre-B cell receptor at the surface of a pre-B cell which gives the signal which inhibits further IGHV-D-J rearrangements on chromosome 14 (Fig. 9), and the signal which starts the light chain V-J rearrangements. Chronologically the V-J rearrangements of the IGK locus usually precede those of the IGL locus[26,27]. A chromosome 2 will be rearranged first. If the resulting rearrangement is productive, kappa light chains will be synthesized, which will allow the production of functional IgM, the other chromosome 2 and both chromosomes 22 remaining non-rearranged. If the first rearrangement is non-productive, the other chromosome 2, then the chromosome

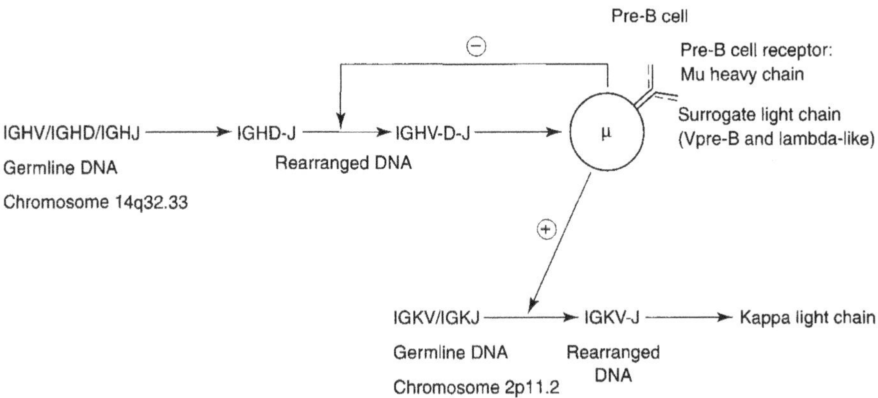

Figure 9 *Regulation of the immunoglobulin gene rearrangements in the pre-B cell. The expression of the pre-B cell receptor gives the signal which inhibits further IGHV-D-J rearrangements on chromosome 14, and the signal which starts the IGKV-J rearrangements on chromosome 2.*

22 will be rearranged until a productive rearrangement allows the synthesis of a light chain. Thus, in a B cell which produces antibodies, generally only one chromosome 14 is productive (that is expressing a functional mu chain) and only one chromosome 2 or 22 is productive (that is expressing a functional kappa or lambda chain). Ig genes on the other chromosomes are either non-rearranged, or rearranged but non-productive, or deleted.

REGULATION OF THE IG GENE EXPRESSION: ENHANCERS

In order to synthesize complete Ig heavy and light chains, the rearranged IGHV-D-J and IGKV-J (or IGLV-J) genes are transcribed with the Cmu and with the Ckappa (or one Clambda) genes, respectively. The transcription level, low in the first B cell development stages, becomes very high in the plasma cells which result from the clonal proliferation[30]. The V genes possess a promoter sequence in 5′ of the signal peptide[31] and can be transcribed before they rearrange. These transcripts designated as "germline transcripts" or "sterile transcripts" probably correspond to an opening, and therefore to better accessibility of the chromatin before the rearrangements. However, that transcription remains low. The C genes can also be transcribed from promoter sequences located upstream of the J genes but again the level of transcription is low and the transcripts are degraded in the nucleus[32–34].

Murine and human immunoglobulin transcription enhancers are the first enhancers described in the DNA of eucaryote cells[35–38]. By different approaches, several groups have simultaneously shown that a DNA segment, located between the most 3′ IGHJ and the switch Smu sequence is not only able to increase the transcription, but also possesses the properties of the enhancers previously described in the viruses. These enhancers (i) activate the transcription whatever their orientation and their position (in 5′ or 3′) relative to the gene promoter, (ii) only activate the promoters, located in *cis*, that is on the same chromosome, and (iii) activate *in vitro* genes others than those to which they are normally associated *in vivo* and increase their transcription, even when located at a distance of several kb[39].

The presence of an enhancer in the human immunoglobulin loci has been demonstrated between the most 3′ IGHJ and the IGHM gene[40], and between the most 3′ IGKJ and the IGKC gene[41,42]. When a IGHV-D-J or IGKV-J rearrangement occur, the promoter sequence in 5′ of the IGHV or IGKV genes is not modified[32,43], but this promoter is now closer to the enhancer sequences located in 3′ of the IGHJ or IGKJ genes. By decreasing the distance between the V gene promoter and the enhancer, the IGHV-D-J and IGKV-J rearrangements allow the interaction of factors binding to these sequences, and consequently an increased transcription of the IGHV-D-J-Cmu and IGKV-J-Ckappa transcripts. During the switch recombination, the IGH enhancer, being localized at more than 1 kb upstream from the Smu sequence, is retained in the locus and can therefore be used for the expression of all the heavy chain classes and subclasses. A second enhancer has been described in 3′ of the IGKC gene[44]. Two 3′ enhancers have been characterized in the human IGH locus, one downstream of IGHA1, and another downstream of IGHA2, within 25 kb of each gene, respectively[45]. These enhancers were duplicated along with part of the IGH locus[46–49], which occurred between about 30 and 60 million years ago[50]. An enhancer has also been localized in the IGL locus in 3′ of IGLC7, the most 3′ IGLC gene[51]. This enhancer consists of three modules located 6, 9.8 and 11.7 kb downstream of IGLC7[52].

References

1 Brack, C. et al. (1978) Cell 15, 1–14.
2 Tonegawa, S. (1983) Nature 302, 575–581.
3 Weigert, M. et al. (1980) Nature 283, 497–499.
4 Sakano, H. et al. (1979) Nature 280, 288–294.
5 Early, P. et al. (1980) Cell 19, 981–992.
6 Alt, F.W. and Baltimore, D. (1982) Proc. Natl Acad. Sci. USA 79, 4118–4122.
7 Landau, N.R. et al. (1984) Proc. Natl Acad. Sci. USA 81, 5836–5840.
8 Lafaille, J.J. et al. (1989) Cell 59, 859–870.
9 Lewis, S.M. (1994) Proc. Natl Acad. Sci. USA 91, 1332–1336.
10 Gearhart, P.J. et al. (1981) Nature 291, 29–34.
11 Knapp, M.R. et al. (1982) Proc. Natl Acad. Sci. USA 79, 2996–3000.
12 Maki, R. et al. (1981) Cell 24, 353–365.
13 Kerr, W.G. et al. (1991) J. Immunol. 146, 3314–3321.
14 Cory, S. et al. (1980) Cell 19, 37–51.
15 Honjo, T. and Kataoka, T. (1978) Proc. Natl Acad. Sci. USA 75, 2140–2144.
16 Rabbitts, T.H. et al. (1980) Nature 283, 351–356.
17 Iwasato, T. et al. (1990) Cell 62, 143–149.
18 Matsuoka, M. et al. (1990) Cell 62, 135–142.
19 Kluin, P.M. et al. (1995) Eur. J. Immunol. 25, 3504–3508.
20 Arpin, C. et al. (1998) J. Exp. Med. 187, 1169–1178.
21 Liu, Y.J. et al. (1996) Immunity 4, 603–613.
22 Kehry, M. et al. 1980) Cell 21, 393–406.
23 Rabbitts, T.H. et al. (1981) Nucleic Acids Res. 9, 4509–4524.
24 Nelson, K.J. et al. (1983) Mol. Cell. Biol. 3, 1317–1332.

[25] Blattner, F.R. and Tucker, P.W. (1984) Nature 307, 417–422.
[26] Hieter, P.A. et al. (1981) Nature 290, 368–372.
[27] Korsmeyer, S.J. et al. (1981) Proc. Natl Acad. Sci. USA 78, 7096–7100.
[28] Hollis, G.F et al. (1989) Proc. Natl Acad. Sci. USA 86, 5552–5556.
[29] Schiff, C. et al. (1990) Int. Immunol. 2, 201–207.
[30] Mather, E.L. and Perry, R.E. (1981) Nucleic Acids Res. 9, 6855–6867.
[31] Bentley, D.L. et al. (1982) Nucleic Acids Res. 10, 1841–1856.
[32] Perry, R.B. et al. (1980) Proc. Natl Acad. Sci. USA 77, 1937–1941.
[33] Kemp, D.J. et al. (1980) Proc. Natl Acad. Sci. USA 77, 2876–2880.
[34] Van Ness, B.G. et al. (1981) Cell 27, 593–602.
[35] Banerji, J. et al. (1983) Cell 33, 729–740.
[36] Gillies, S.D. et al. (1983) Cell 33, 717–728.
[37] Mercola, M. et al. (1983) Science 221, 663–665.
[38] Neuberger, M.S. (1983) EMBO J. 2, 1373–1378.
[39] Lefranc, G. and Lefranc, M.-P. (1990) Biochimie 72, 7–17.
[40] Hayday, A.C. et al. (1984) Nature 307, 334–340.
[41] Emorine, L. et al. (1983) Nature 304, 447–449.
[42] Gimble, J.M. and Max, E.E (1987) Mol. Cell. Biol. 7, 15–25.
[43] Clarke, C. et al. (1982) Nucleic Acids Res. 10, 7731–7749.
[44] Judde, J.G. and Max, E.E. (1992) Mol. Cell. Biol. 12, 5206–5216.
[45] Mills, F.C. et al. (1997) J. Exp. Med. 186, 845–858.
[46] Flanagan, J.G. and Rabbitts T.H. (1982) Nature 300, 709–713.
[47] Lefranc, M.-P. et al. (1982) Nature 300, 760–762.
[48] Rabbitts, T.H. et al. (1983) In Genetic Rearrangement, Eds K.F. Chater et al. Croom Helm, London and Canberra, pp. 143–154.
[49] Keyeux, G. et al. (1989) Genomics 5, 431–441.
[50] Harindranath, N. et al. (1998) Gene 221, 215–224.
[51] Blomberg, B.B. et al. (1991) J. Immunol. 147, 2354–2358.
[52] Asenbauer, H. et al. (1999) Eur. J. Immunol. 29, 713–724.

4 Chromosomal localization, organization of the loci and potential repertoire

THE HUMAN IGH LOCUS

Chromosomal localization of the human IGH locus

The human IGH locus is located on chromosome 14[1], at band 14q32.33, at the telomeric extremity of the long arm[2,3] (Fig. 1). The orientation of the locus has been determined by the analysis of translocations, involving the IGH locus, in leukemia and lymphoma.

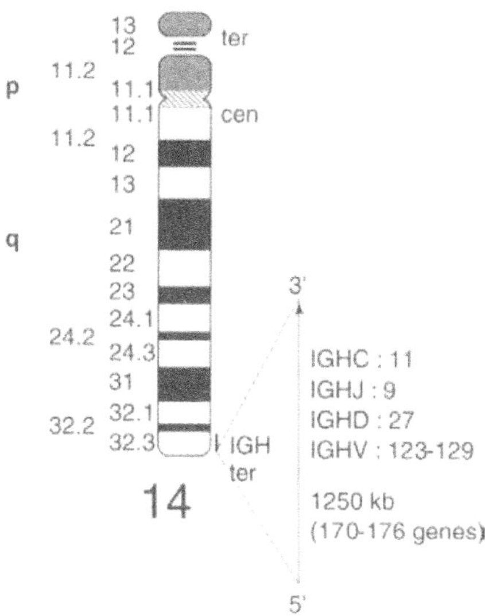

Figure 1 *Chromosomal localization of the human IGH locus at 14q32.33. A vertical line indicates the localization of the IGH locus at 14q32.33. The arrow indicates the orientation 5′ → 3′ of the locus, and the gene group order in the locus. The arrow is proportional to the size of the locus, indicated in kb. The total number of genes in the locus is shown between parentheses. Seven non-mapped IGHV genes which have a provisional designation are not included in this figure. The number of functional genes defines the potential IGH repertoire which comprises 76–84 genes (38–46 IGHV, 23 IGHD, 6 IGHJ and 9 IGHC) per haploid genome.*

Organization of the human IGH locus

The human IGH locus at 14q32.33 spans 1250 kb (Fig. 2). It consists of 123 to 129 IGHV genes[4-10], depending on the haplotypes, 27 IGHD genes[11-15] belonging to seven subgroups, nine IGHJ genes[15-16] and, in the most frequent haplotype, 11 IGHC genes[17-31]. Eighty-two to 88 IGHV genes belong to seven subgroups, whereas 41 pseudogenes, which are too divergent to be assigned to subgroups, have been assigned to the clans. Seven non-mapped IGHV genes have been described as corresponding to insertion/deletion polymorphisms but have not yet been precisely located. The most 5′ IGHV genes occupy a position very close to the chromosome 14q telomere whereas the IGHC genes are in a more centromeric position.

The potential genomic IGH repertoire per haploid genome comprises 38 to 46 functional IGHV genes belonging to six or seven subgroups depending on the haplotypes[32], 23 IGHD, 6 IGHJ and, in the most frequent haplotype, 9 IGHC genes. Thirty-five IGH genes have been found outside the main locus in other chromosomal localizations. These genes designated as orphons cannot contribute to the synthesis of the immunoglobulin chains, even if they have an ORF. Nine IGHV orphons and 10 IGHD orphons have been described on chromosome 15 (15q11.2), and 16 IGHV orphons on chromosome 16 (16p11.2)[10]. In addition, one IGHC processed gene, IGHEP2 is localized on chromosome 9 (9p24.2–p24.1)[33]. This is, so far, the only processed Ig gene described.

The total number of human IGH genes per haploid genome is 170–176 (206 to 212 genes, if the orphons and the processed gene are included) of which 76–84 genes are functional (see summary in pages 65 and 66, Tables 8 and 9).

List of the IGH genes on chromosome 14 at 14q32.33 and correspondence between nomenclatures

A list of the functional or ORF human IGH genes is displayed in Table 1.

IGHD gene nomenclature

IGHD genes are designated by a number for the subgroup followed by a hyphen and a number for the localization from 5′ to 3′ in the locus[14,15].

Figure 2 *Representation of the human IGH locus at 14q32.33. The boxes representing the genes are not to scale. Exons are not shown. SWITCH sequences are represented by a filled circle upstream of the IGHC genes. The seven non-mapped IGHV genes which correspond to insertion/deletion polymorphisms but which have not yet been precisely located are not shown. Pseudogenes which could not be assigned to subgroups with functional genes are designated by a Roman numeral between parentheses, corresponding to the clans, followed by a hyphen, and a number for the localization from 3′ to 5′ in the locus. All these pseudogenes have truncations. Horizontal arrows indicate the extent of the allelic IGHC multigene deletions[34-41] or duplications[42-45] described in healthy individuals. I to VI refer to the type of deletion (for review, see ref. 41) duplication or triplication, defined by the deleted, duplicated or triplicated genes. Ex: deletion I (del G1–EP1–A1–GP–G2–G4)[34-36].*

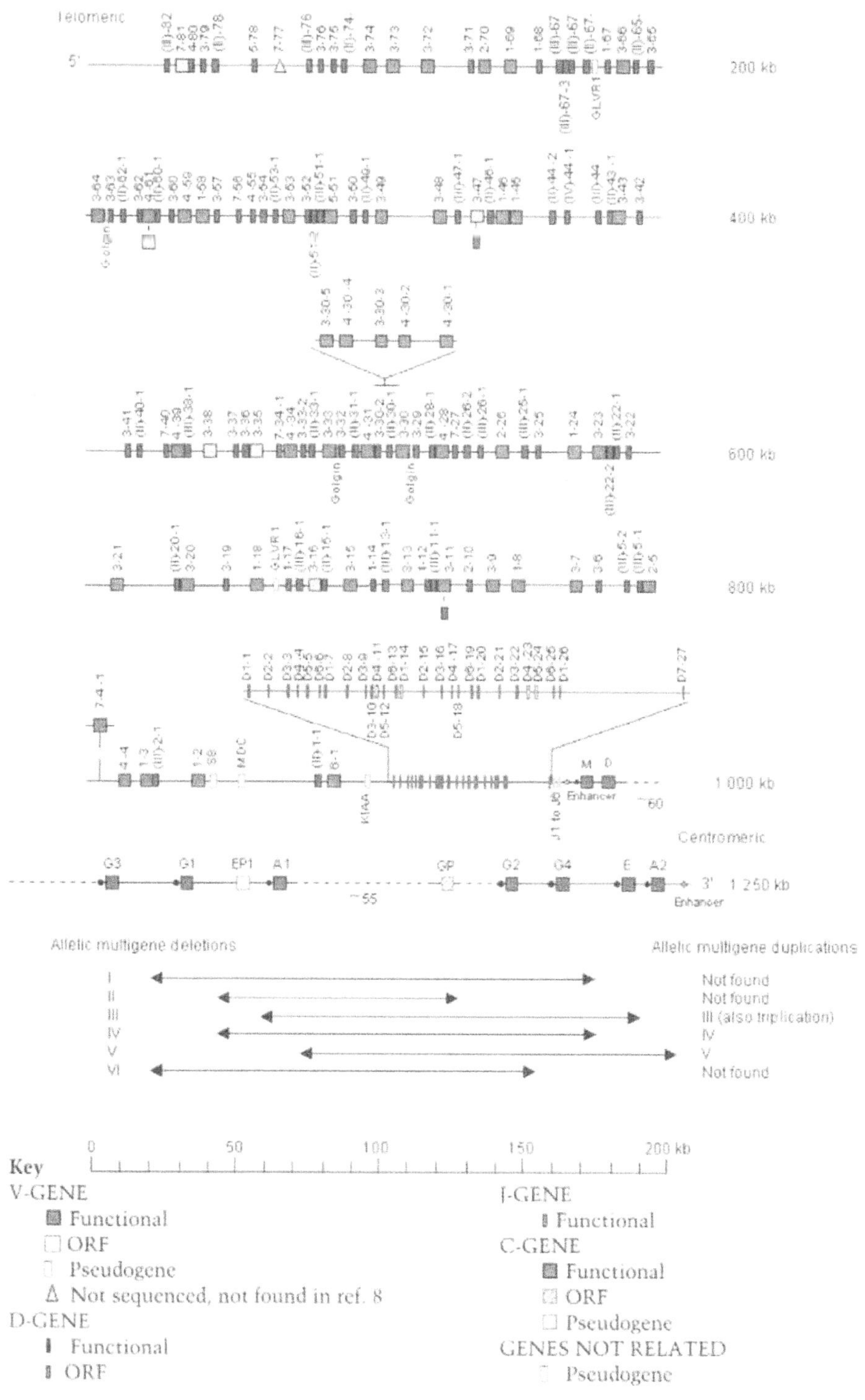
Telomeric
5'
200 kb
400 kb
600 kb
800 kb
1 000 kb
Centromeric
3'
1 250 kb
Enhancer
Allelic multigene deletions
Allelic multigene duplications
Not found
Not found
III (also triplication)
IV
V
Not found
0
50
100
150
200 kb
Key
V-GENE
Functional
ORF
Pseudogene
Not sequenced, not found in ref. 8
D-GENE
Functional
ORF
J-GENE
Functional
C-GENE
Functional
ORF
Pseudogene
GENES NOT RELATED
Pseudogene

IGHV gene nomenclature

IGHV genes are designated by a number for the subgroup, followed by a hyphen and a number for the localization from 3′ to 5′ in the locus[10,32]. Seven genes which have been described as insertion/deletion polymorphisms but which have not yet been precisely located have a provisional designation and are designated by a number for the subgroup, followed by a hyphen and a small letter: five of these genes (IGHV1-c, IGHV1-f, IGHV3-d, IGHV4-b and IGHV5-a) have at least an ORF or a functional allele and therefore have been included in Table 1.

Table 1. *List of functional or ORF human IGH genes.*

IMGT gene groups	IMGT gene names	IMGT functionality	IMGT number of alleles	Other nomenclatures IGHV: ref. [8]	
IGHC	IGHA1	F	1	–	–
	IGHA2	F	3	–	–
	IGHD	F	2	–	–
	IGHE	F	3	–	–
	IGHG1	F	2	–	–
	IGHG2	F	2	–	–
	IGHG3	F	6	–	–
	IGHG4	F	3	–	–
	IGHGP	ORF	1	–	–
	IGHM	F	3	–	–
IGHD	IGHD1-1	F	1	–	M4[8]
	IGHD1-7	F	1	–	M1[13]
	IGHD1-14	ORF	1	–	M2[13]
	IGHD1-20	F	1	–	M3[8]
	IGHD1-26	F	1	–	M′3[8]
	IGHD2-2	F	3	D4[11]	LR4[8]
	IGHD2-8	F	2	D1[11]	LR1[13]
	IGHD2-15	F	1	D2[11]	LR2[8]
	IGHD2-21	F	2	D3[11]	LR3[8]
	IGHD3-3	F	2	23/7[12]	XP4[13]
	IGHD3-9	F	1	21/0,5[12]	XP1[13]
	IGHD3-10	F	2	21/7[12]	XP′1[13]
	IGHD3-16	F	1	21/10[12]	XP2[8]
	IGHD3-22	F	1	21/9[12]	XP3[8]
	IGHD4-4	F	1	–	A4[13]
	IGHD4-11	ORF	1	–	A1[13]
	IGHD4-17	F	1	–	A2[8]
	IGHD4-23	ORF	1	–	A3[8]
	IGHD5-5	F	1	–	K4[13]
	IGHD5-12	F	1	–	K1[13]
	IGHD5-18	F	1	–	K2[8]

Continued

Table 1. *Continued.*

IMGT gene groups	IMGT gene names	IMGT functionality	IMGT number of alleles	Other nomenclatures IGHV: ref. [8]	
	IGHD5-24	ORF	1	–	K3[8]
	IGHD6-6	F	1	–	N4[13]
	IGHD6-13	F	1	–	N1[13]
	IGHD6-19	F	1	–	N2[8]
	IGHD6-25	F	1	–	N3[8]
	IGHD7-27	F	1	DHQ52[16]	–
IGHJ	IGHJ1	F	1	–	–
	IGHJ2	F	1	–	–
	IGHJ3	F	2	–	–
	IGHJ4	F	3	–	–
	IGHJ5	F	2	–	–
	IGHJ6	F	3	–	–
IGHV	IGHV1-2	F	4	1-2	–
	IGHV1-3	F	2	1-3	–
	IGHV1-8	F	1	1-8	–
	IGHV1-18	F	2	1-18	–
	IGHV1-24	F	1	1-24P	–
	IGHV1-45	F	3	1-45	–
	IGHV1-46	F	3	1-46	–
	IGHV1-58	F	1	1-58	–
	IGHV1-69	F	7	1-69	–
	IGHV1-c	ORF	1	–	–
	IGHV1-f	F	2	–	–
	IGHV2-5	F	9	2-5	–
	IGHV2-26	F	1	2-26	–
	IGHV2-70	F	12	2-70	–
	IGHV3-7	F	2	3-7	–
	IGHV3-9	F	1	3-9	–
	IGHV3-11	F, P	3	3-11	–
	IGHV3-13	F	2	3-13	–
	IGHV3-15	F	8	3-15	–
	IGHV3-16	ORF	1	3-16P	–
	IGHV3-20	F	1	3-20	–
	IGHV3-21	F	2	3-21	–
	IGHV3-23	F	3	3-23	–
	IGHV3-30	F	19 (a)	3-30	–
	(*)IGHV3-30-3	F	2	–	–
	(*)IGHV3-30-5	F	(a)	–	–
	IGHV3-33	F	5	3-33	–
	IGHV3-35	ORF	1	3-35	–
	IGHV3-38	ORF	2	3-38P	–

Continued

Table 1. *Continued.*

IMGT gene groups	IMGT gene names	IMGT functionality	IMGT number of alleles	Other nomenclatures IGHV: ref. [8]	
	IGHV3-43	F	2	3-43	–
	IGHV3-47	ORF, P	3	3-47P	–
	IGHV3-48	F	3	3-48	–
	IGHV3-49	F	3	3-49	–
	IGHV3-53	F	2	3-53	–
	IGHV3-64	F	5	3-64	–
	IGHV3-66	F	3	3-66	–
	IGHV3-72	F	2	3-72	–
	IGHV3-73	F	1	3-73	–
	IGHV3-74	F	3	3-74	–
	IGHV3-d	F	1	–	–
	IGHV4-4	F	8	4-4	–
	IGHV4-28	F	5	4-28	–
	(*)IGHV4-30-1	F	(b)	–	–
	(*)IGHV4-30-2	F	4	–	–
	(*)IGHV4-30-4	F	6	–	–
	IGHV4-31	F	10 (b)	4-31	–
	IGHV4-34	F	13	4-34	–
	IGHV4-39	F	6	4-39	–
	IGHV4-59	F	9	4-59	–
	IGHV4-61	F, ORF	7	4-61	–
	IGHV4-b	F	2	–	–
	IGHV5-51	F	5	5-51	–
	IGHV5-a	F, P	4	–	–
	IGHV6-1	F	2	6-1	–
	(*)IGHV7-4-1	F	3	–	–
	IGHV7-81	ORF	1	7-81	–

(a) Sequences of the polymorphic IGHV3-30-5 gene cannot be differentiated from those of the IGHV3-30 gene. It is therefore not excluded that some of these "alleles" belong exclusively to one or the other gene.
(b) Sequences of the polymorphic IGHV4-30-1 gene cannot be differentiated from those of the IGHV4-31 gene. It is therefore not excluded that some of these "alleles" belong exclusively to one or the other gene.
* Allelic polymorphisms by insertion/deletion which concern: a 50 kb insertion of five genes (3-30-5, 4-30-4, 3-30-3, 4-30-2 and 4-30-1) observed in 45% Caucasoids; the IGHV7-4-1 gene.
Only the genes with at least one functional or ORF allele are shown. For a complete listing of the human IGHV genes including the pseudogenes, see ref. 10, 15 and IMGT gene data: http://imgt.cines.fr/cgi-bin/IMGTlect.jv?query=203.
For information on individual IGH genes:
http://imgt.cines.fr/cgi-bin/IMGTlect.jv?query=202+genename.
Example: IMGT Repertoire for IGHA1:
http://imgt.cines.fr/cgi-bin/IMGTlect.jv?query=202+IGHA1.

Number of the human IGHV germline variable genes at 14q32.33 and potential repertoire

The repertoire of the human IGHV germline variable genes at 14q32.33 is shown in Table 2.

Overview

123–129 IGHV genes on 900 kb: 82–88 IGHV genes belonging to 7 subgroups, 41 pseudogenes assigned to the clans and / non-mapped genes[8,10].

38–44 FUNCTIONAL
4 ORF
78 PSEUDOGENE
1 FUNCTIONAL or PSEUDOGENE
1 ORF or PSEUDOGENE
1 FUNCTIONAL or ORF.

Potential repertoire

38–46 FUNCTIONAL IGHV genes belonging to six or seven subgroups.

Table 2. *Repertoire of the human IGHV germline variable genes at 14q32.33.*

Subgroup	Functional	ORF	Pseudogene	Total
IGHV1	9	–	5	14
IGHV2	3	–	1	4
IGHV3	18–20**(+1)*	3(+1)*	24(+2)*	47–49**
IGHV4	6–9**(+1)*	(+1)*	2	9–12**
IGHV5	1	–	1	2
IGHV6	1	–	–	1
IGHV7	0–1**	1	4	5–6**
IGHV(II)	–	–	22	22
IGHV(III)	–	–	18	18
IGHV(IV)	–	–	1	1
Total	38–44(+2)*	4(+2)*	78(+2)*	123–129**

* The genes have alleles with different functionality: FUNCTIONAL or PSEUDOGENE (IGHV3-11); ORF or PSEUDOGENE (IGHV3-47); FUNCTIONAL or ORF (IGHV4-61).
** Allelic polymorphisms by insertion/deletion which concern: a 50 kb insertion of five genes (3-30-5, 4-30-4, 3-30-3, 4-30-2, 4-30-1) in 45% Caucasoids; the IGHV7-4-1 gene.
(II), (III), (IV) Clans for the pseudogenes which could not be assigned to subgroups with functional genes. All these pseudogenes have truncations.
Clans comprise, respectively:
clan I: IGHV1, IGHV5 and IGHV7 subgroup genes;
clan II: IGHV2, IGHV4 and IGHV6 subgroup genes and pseudogenes IGHV(II);
clan III: IGHV3 subgroup genes and pseudogenes IGHV(III);
clan IV: one pseudogene IGHV(IV)-44.
The seven non-mapped genes are not included in this table, three of them are functional (1-f, 3-d, 4-b), one is functional or pseudogene (5-a), one is ORF (1-c), one is a pseudogene (3-g), and the last one (3-h) is too partial to determine its functionality.

THE HUMAN IGK LOCUS

Chromosomal localization of the human IGK locus

The human IGK locus is located on chromosome 2[46], on the short arm, at band 2p11.2[47] (Fig. 3). The orientation of the locus has been determined by the analysis of translocations, involving the IGK locus, in leukemia and lymphoma.

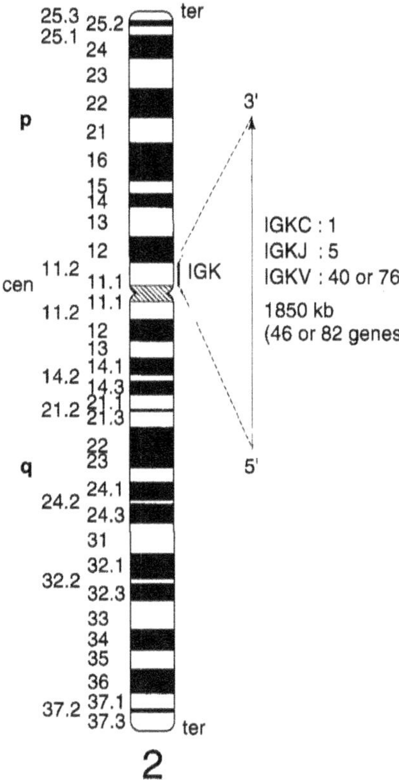

Figure 3 *Chromosomal localization of the human IGK locus at 2p11.2. A vertical line indicates the localization of the IGK locus at 2p11.2. The arrow indicates the orientation 5′ → 3′ of the locus, and the gene group order in the locus. The arrow is proportional to the size of the locus, indicated in kb. The total number of genes in the locus is shown between parentheses; the lower number (46 genes) corresponds to the rare haplotype which only comprises the proximal V-CLUSTER of 40 IGKV genes, whereas the higher number (82 genes) corresponds to the common haplotype which comprises both the distal and the proximal V-CLUSTERs and 76 IGKV genes. The number of functional genes defines the potential IGK repertoire which comprises, if both the proximal and distal clusters are present, 37–41 genes (31–35 IGKV, 5 IGKJ and 1 IGKC) per haploid genome.*

Organization of the human IGK locus

The human IGK locus at 2p12 spans 1820 kb (Fig. 4). It consists of 76 IGKV genes[32,48–54] belonging to seven subgroups, five IGKJ genes[32,54–55] and a unique IGKC gene[56]. The 76 IGKV genes are organized in two clusters separated by 800 kb (Fig. 5). The IGKV distal cluster in 5′ of the locus and in the most centromeric position) spans 400 kb and comprises 36 genes. The IGKV proximal cluster (in 3′ of the locus, closer to IGKC, and in the most telomeric position) spans 600 kb and comprises 40 genes.

The potential genomic IGK repertoire per haploid genome comprises 31–35 functional IGKV genes belonging to five subgroups, the five IGKJ genes and the unique IGKC gene. One rare IGKV haplotype has been described which contains only the proximal cluster. This haplotype comprises the 40 proximal IGKV genes belonging to seven subgroups, of which 17–19 are functional and belong to five subgroups.

Twenty-five IGKV orphons have been identified and sequenced: two on the short arm of chromosome two but outside of the main IGK locus, 12 on the long arm of chromosome 2, five on chromosome 22, one on chromosome 1, one on chromosome 15, and four outside of chromosome 2[54]. If both the proximal and distal IGKV clusters are present, the total number of human IGK genes per haploid genome is 82 (107, if the orphons are included) of which 37 to 41 are functional. If only the proximal IGKV cluster is present, the total number of genes per haploid genome is 46 (71 genes, if the orphons are included) of which 23–25 genes are functional (see summary in pages 65 and 66, Tables 8 and 9).

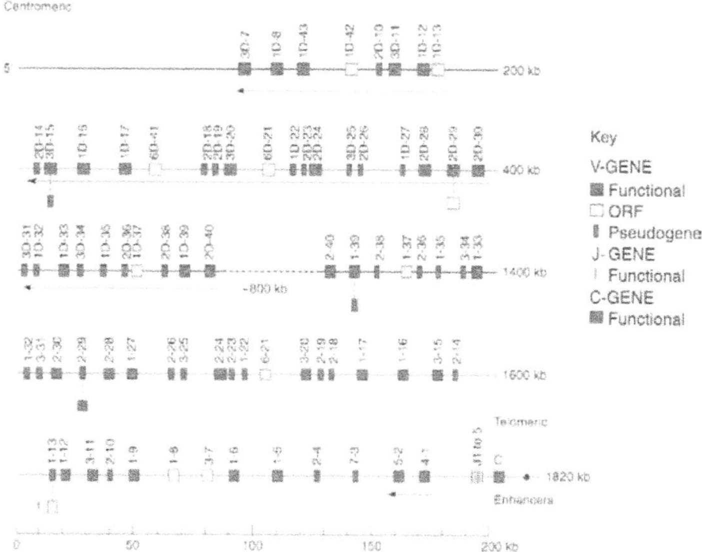

Figure 4 *Representation of the human IGK locus at 2p12. The boxes representing the genes are not to scale. Exons are not shown. The IGKV genes of the proximal V-CLUSTER are designated by a number for the subgroup, followed by a hyphen and a number for the localization from 3′ to 5′ in the locus. The IGKV genes of the distal duplicated V-CLUSTER are designated by the same numbers as the corresponding genes in the proximal V-CLUSTER, with the letter D added. Arrows show the IGKV genes polarity of which is opposite to that of the J-C-CLUSTER.*

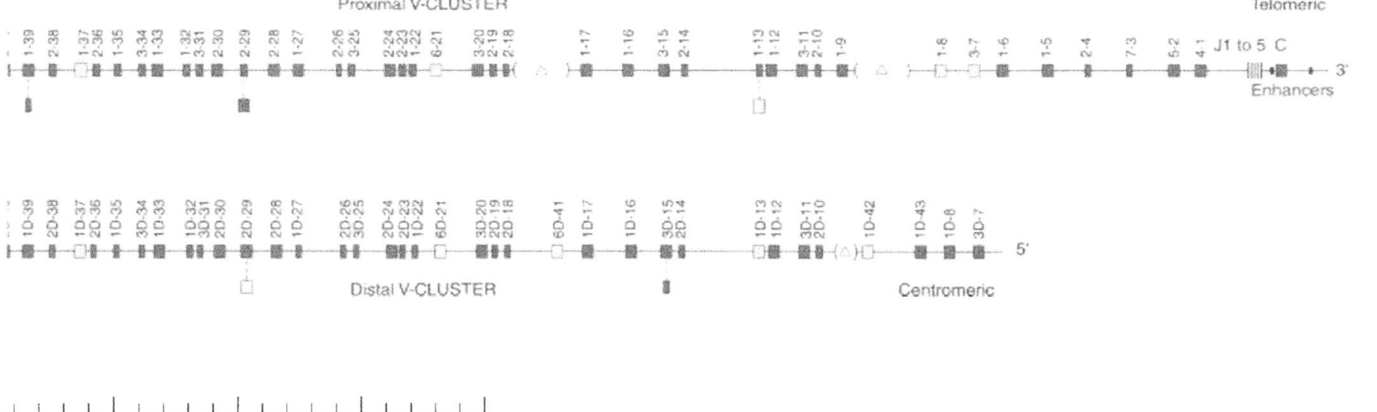

Figure 5 *Alternative representation of the human IGK locus, taking into account the polymorphic duplication in the locus[48–49]. Small triangles indicate parts which are absent in the proximal or distal V-CLUSTER when considering the duplication in the human IGK locus.*

List of the IGK genes on chromosome 2 at 2p11.2 and correspondence between nomenclatures

A list of the functional or ORF human IGK genes is displayed in Table 3.

IGKV gene nomenclature

IGKV genes are designated by a number for the subgroup, followed by a hyphen and a number for the localization from 3′ to 5′ in the locus[32,54]. The IGKV genes of the distal duplicated V-CLUSTER are designated by the same number as the corresponding genes in the proximal V-CLUSTER, with the letter D added.

Table 3. *List of functional or ORF human IGK genes.*

IMGT gene groups	IMGT gene names	IMGT functionality	IMGT number of alleles	Other nomenclature Ref.[49]
Proximal cluster				
IGKC	IGKC	F	4	–
IGKJ	IGKJ1	F	1	–
	IGKJ2	F	1	–
	IGKJ3	F	1	–
	IGKJ4	F	1	–
	IGKJ5	F	1	–
IGKV	IGKV1-5	F	3	L12, L12a
	IGKV1-6	F	1	L11
	IGKV1-8	ORF	1	L9
	IGKV1-9	F	1	L8
	IGKV1-12	F	1 (+1?)*	L5
	IGKV1-13	P, ORF?	1 (+1?)*	L4
	IGKV1-16	F	1	L1
	IGKV1-17	F	1	A30
	IGKV1-27	F	1	A20
	IGKV1-33	F	1	O18, O18a
	IGKV1-37	ORF	1	O14
	IGKV1-39	F, P	2	O12, O12a
	IGKV2-24	F	1	A23
	IGKV2-28	F	1	A19
	IGKV2-29	P, F	2	A18a, A18b
	IGKV2-30	F	1	A17
	IGKV2-40	F	2	O11, O11a
	IGKV3-7	ORF	3	L10, L10a
	IGKV3-11	F	2	L6, L6a
	IGKV3-15	F	1	L2
	IGKV3-20	F	2	A27, A27a
	IGKV4-1	F	1	B3
	IGKV5-2	F	1	B2
	IGKV6-21	ORF	1	A26

Continued

Table 3. *Continued.*

IMGT gene groups	IMGT gene names	IMGT functionality	IMGT number of alleles	Other nomenclature Ref.[49]
Distal cluster				
IGKV	IGKV1D-8	F	1	L24, L24a
	IGKV1D-12	F	1 (+1?)*	L19
	IGKV1D-13	ORF	1 (+1?)*	L18
	IGKV1D-16	F	2	L15, L15a
	IGKV1D-17	F	1	L14
	IGKV1D-33	F	1	O8
	IGKV1D-37	ORF	1	O4
	IGKV1D-39	F	1	O2
	IGKV1D-42	ORF	1	L22
	IGKV1D-43	F	1	L23, L23a
	IGKV2D-24	F	1	A7
	IGKV2D-28	F	1	A3
	IGKV2D-29	F, ORF	2	A2a, A2c
	IGKV2D-30	F	1	A1
	IGKV2D-40	F	1	O1
	IGKV3D-7	F	1	L25
	IGKV3D-11	F	1	L20
	IGKV3D-15	F, P	2	L16, L16a, L16b, L16c
	IGKV3D-20	F	1	A11, A11a
	IGKV6D-21	ORF	1	A10
	IGKV6D-41	ORF	1	A14

*Between parentheses are alleles which could not be assigned to the proximal or distal cluster gene.

Only the genes with at least one functional or ORF alleles are shown. For a complete listing of the human IGKV genes including the pseudogenes, see ref. 54, and IMGT gene data: http://imgt.cines.fr/cgi-bin/IMGTlect.jv?query=203.

For information on individual IGK genes:
http://imgt.cines.fr/cgi-bin/IMGTlect.jv?query=202+genename.

Example: IMGT Repertoire for IGKV1–39:
http://imgt.cines.fr/cgi-bin/IMGTlect.jv?query=202+IGKV1-39.

Number of the human IGKV germline variable genes at 2p11.2 and potential repertoire

The repertoire of the human IGKV germline variable genes at 2p11.2, for the haplotypes with both distal and proximal V-CLUSTERs is shown in Table 4. The repertoire for the rare haplotype with only the proximal V-CLUSTER is shown in Table 5.

Distal and proximal V-CLUSTERs (for the haplotypes with both clusters)

Overview

76 IGKV genes belonging to seven subgroups, on 1800 kb:
- 31 FUNCTIONAL
- 9 ORF
- 32 PSEUDOGENE
- 1 FUNCTIONAL or ORF
- 3 FUNCTIONAL or PSEUDOGENE.

Potential repertoire

31–35 FUNCTIONAL IGKV genes belonging to five subgroups

Table 4. *Repertoire of the human IGKV germline variable genes at 2p11.2, for the haplotypes with both the distal and proximal V-CLUSTERs.*

Subgroup	Functional	ORF	Pseudogene	Total
IGKV1	15(+1)*	5	8(+1)*	29
IGKV2	8(+2)*	(+1)*	17(+1)*	27
IGKV3	6(+1)*	1	6(+1)*	14
IGKV4	1	–	–	1
IGKV5	1	–	–	1
IGKV6	–	3	–	3
IGKV7	–	–	1	1
Total	31(+4)*	9(+1)*	32(+3)*	76

* The following genes have alleles with different functionality: FUNCTIONAL or ORF (IGKV2D-29) FUNCTIONAL or PSEUDOGENE (IGKV1-39, IGKV2-29, IGKV3D-15).

Proximal V-CLUSTER (for the haplotype without the distal V-CLUSTER)

Table 5 displays the repertoire of the human IGKV germline variable genes at 2p11.2 for the rare haplotype with only the proximal V-CLUSTER.

Overview

40 IGKV genes belonging to seven subgroups, on 600 kb:
17 FUNCTIONAL
4 ORF
17 PSEUDOGENE
2 FUNCTIONAL or PSEUDOGENE.

Potential repertoire

17–19 FUNCTIONAL IGKV genes belonging to five subgroups.

Table 5. *Repertoire of the human IGKV germline variable genes at 2p11.2 for the rare haplotypes with only the proximal V-CLUSTER.*

Subgroup	Functional	ORF	Pseudogene	Total
IGKV1	8(+1)*	2	4(+1)*	15
IGKV2	4(+1)*	–	9(+1)*	14
IGKV3	3	1	3	7
IGKV4	1	–	–	1
IGKV5	1	–	–	1
IGKV6	–	1	–	1
IGKV7	–	–	1	1
Total	17(+2)*	4	17(+2)*	40

* The following genes have alleles with different functionality: FUNCTIONAL or PSEUDOGENE (IGKV1-39, IGKV2-29).

THE HUMAN IGL LOCUS

Chromosomal localization of the human IGL locus

The human IGL locus is located on chromosome 22[57], on the long arm, at band 22q11.2[58] (Fig. 6). The orientation of the locus has been determined by the analysis of translocations, involving the IGL locus, in leukemia and lymphoma. Sequencing of the long arm of chromosome 22 showed that it encompasses about 35 megabases of DNA and that the IGL locus is localized at six megabases from the centromere[59]. Although the correlation between DNA sequences and chromosomal bands has not yet been made, the localization of the IGL locus can be refined to 22q11.2.

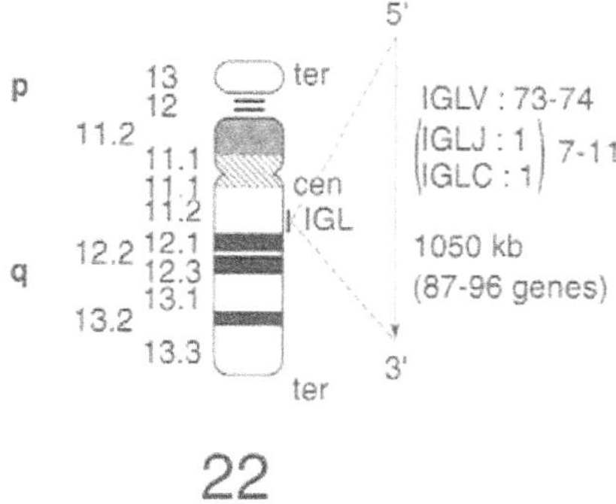

Figure 6 *Chromosomal localization of the human IGL locus at 22q11.2. A vertical line indicates the localization of the IGL locus at 22q11.2. The arrow indicates the orientation 5′ → 3′ of the locus, and the gene group order in the locus. The arrow is proportional to the size of the locus, indicated in kb. The total number of genes in the locus is shown between parentheses. Depending on the haplotype, there are 7 to 11 IGLC genes. In the 7-IGLC gene haplotype, each IGLC gene is preceded, in 5′, by one IGLJ gene. Although the additional IGLC genes, in the 8-, 9-, 10- and 11-IGLC gene haplotypes have not yet been sequenced, they are also probably preceded by one IGLJ gene. The number of functional genes define the potential IGL repertoire which comprises, in the 7-IGLC gene haplotype, 37–43 genes (29–33 IGLV, 4–5 IGLJ and 4–5 IGLC) per haploid genome.*

Organization of the human IGL locus

The human IGL locus at 22q11.2 spans 1050 kb (Fig. 7). It consists of 73–74 IGLV genes[32,60–64], localized on 900 kb, seven to 11 IGLJ and seven to 11 IGLC genes depending on the haplotypes, each IGLC gene being preceded by one IGLJ gene[65–68]. Fifty-six–57 genes belong to 11 subgroups, whereas 17 pseudogenes which are too divergent to be assigned to subgroups, have been assigned to the clans. The most 5′ IGLV genes occupy the more centromeric position, whereas the IGLC genes, in 3′ of the locus, are the most telomeric genes in the IGL locus. The potential genomic IGL repertoire per haploid genome comprises 29–33 functional IGLV genes belonging to 10 subgroups, four–five IGLJ, and four–five IGLC functional genes in the 7-IGLC gene haplotype. One, two, three or four additional IGLC genes, each one probably preceded by one IGLJ, have been shown to characterize IGLC haplotypes with eight, nine, 10 or 11 genes[69–72], but these genes have not yet been sequenced. Two IGLV orphons have been identified on chromosome 8 at 8q11.2 and one of them belonging to subgroup 8 has been sequenced. The recent sequencing of the chromosome 22q showed that the IGL locus is localized at six megabases from the centromere[59]. Two IGLC orphons and two IGLV orphons have also been characterized on 22q outside of the major IGL locus[59] (See also IMGT Repertoire, http://imgt.cines.fr). The total number of human IGL genes per haploid genome is 87–96 (93–102, if the orphons are included) of which 37–43 genes are functional (see summary in Tables 8 and 9, pages 65 and 66).

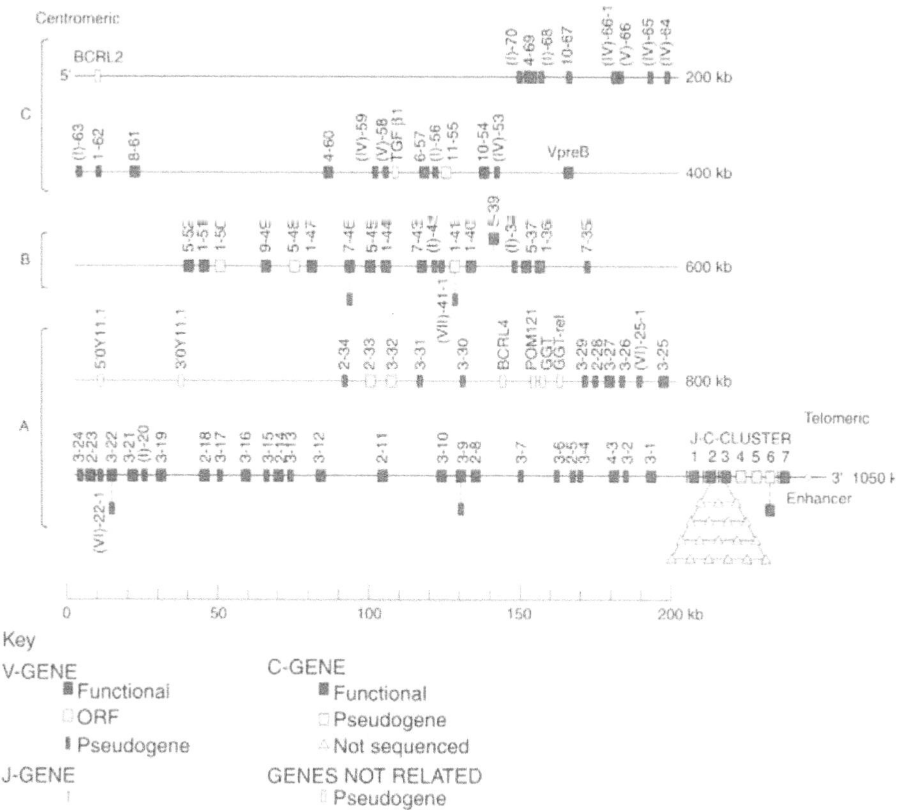

Figure 7 *Representation of the human IGL locus at 22q11.2. The boxes representing the genes are not to scale. Exons are not shown. A, B, C refer to three distinct V-CLUSTERs based on the IGLV gene subgroup content*[62]*. Pseudogenes which could not be assigned to subgroups with functional genes are designated by a Roman numeral between parentheses, corresponding to the clans (see Table 7), followed by a hyphen and a number for the localization from 3' to 5' in the locus. IGLV(IV)-66-1 has been identified in D87004 by IMGT curators (G. Folch, V. Giudicelli, M.-P. Lefranc imgt@ligm.igh.cnrs.fr). The vestigial sequences have been attributed to the clans VI and VII as IGLV(VI)-22-1 (λvg2), IGLV(VII)-41-1 (λvg1). IGLV(VI)-25-1 (λvg3) has been identified in D86994 by IGMT curators (N. Bosc, M.-P. Lefranc) between IGLV3-25 and IGLV3-26.*

List of the IGL genes on chromosome 22 at 22q11.2 and correspondence between nomenclatures

A list of the functional or ORF human IGL genes is displayed in Table 6.

IGLV gene nomenclature

IGLV genes are designated by a number for the subgroup[60,62] followed by a hyphen and a number for the localization from 3′ to 5′ in the locus. In the IGLV gene name column, the IGLV genes are listed, for each subgroup, according to their position from 3′ to 5′ in the locus.

Table 6. *List of functional or ORF human IGL genes.*

IMGT gene groups	IMGT gene names	IMGT functionality	IMGT reference sequence accession numbers	IMGT number of alleles	Other nomenclatures	
					Ref.[60,62]	Ref.[63]
IGLC	IGLC1	F	J00252	3	–	–
	IGLC2	F	J00253	2	–	–
	IGLC3	F	J00254	4	–	–
	IGLC4	P	J03009	2	–	–
	IGLC5	P	J03010	2	–	–
	IGLC6	F, P	J03011	5	–	–
	IGLC7	F	X51755	2	–	–
IGLJ	IGLJ1	F	X04457	1	–	–
	IGLJ2	F	M15641	1	–	–
	IGLJ3	F	M15642	2	–	–
	IGLJ4	ORF	X51755	1	–	–
	IGLJ5	ORF	X51755	2	–	–
	IGLJ6	ORF	M18338	1	–	–
	IGLJ7	F	X51755	2	–	–
IGLV	IGLV1-36	F	Z73653	1	1a	1-11
	IGLV1-40	F	M94116	3	1e	1-13
	IGLV1-41	ORF, P	M94118	2	1d	1-14P
	IGLV1-44	F	Z73654	1	1c	1-16
	IGLV1-47	F	Z73663	2	1g	1-17
	IGLV1-50	ORF	M94112	1	1f	1-18
	IGLV1-51	F	Z73661	2	1b	1-19
	IGLV2-8	F	X97462	3	2c	1-2
	IGLV2-11	F	Z73657	3	2e	1-3
	IGLV2-14	F	Z73664	4	2a2	1-4
	IGLV2-18	F	Z73642	4	2d	1-5
	IGLV2-23	F	X14616	3	2b2	1-7
	IGLV2-33	ORF	Z73643	3	2f	1-9
	IGLV3-1	F	X57826	1	3r	2-1
	IGLV3-9	F, P	X97473	3	3j	2-6

Continued

Table 6. *Continued.*

IMGT gene groups	IMGT gene names	IMGT functionality	IMGT reference sequence accession numbers	IMGT number of alleles	Other nomenclatures Ref.[60,62]	Ref.[63]
	IGLV3-10	F	X97464	2	3p	2-7
	IGLV3-12	F	Z73658	2	3i	2-8
	IGLV3-16	F	X97471	1	3a	2-11
	IGLV3-19	F	X56178	1	3l	2-13
	IGLV3-21	F	X71966	3	3h	2-14
	IGLV3-22	F, P	Z73666	2	3e	2-15
	IGLV3-25	F	X97474	3	3m	2-17
	IGLV3-27	F	D86994	1	–	2-19
	IGLV3-32	ORF	Z73645	1	3i1	2-23P
	IGLV4-3	F	X57828	1	4c	5-1
	IGLV4-60	F	Z73667	2	4a	5-4
	IGLV4-69	F	Z73648	2	4b	5-6
	IGLV5-37	F	Z73672	1	5e	4-1
	(*)IGLV5-39	F	Z73668	1	5a	–
	IGLV5-45	F	Z73670	3	5c	4-2
	IGLV5-48	ORF	Z73649	1	5d	4-3
	IGLV5-52	F	Z73669	1	5b	4-4
	IGLV6-57	F	Z73673	1	6a	1-22
	IGLV7-43	F	X14614	1	7a	3-2
	IGLV7-46	F, P	Z73674	3	7b	3-3
	IGLV8-61	F	Z73650	2	8a	3-4
	IGLV9-49	F	Z73675	3	9a	5-2
	IGLV10-54	F	Z73676	3	10a	1-20
	IGLV11-55	ORF	D86996	1	–	4-6

* An allelic polymorphism by insertion/deletion which concerns the IGLV5-39 gene. One, two, three or four additional IGLC genes, each one probably preceded by one IGLJ, have been shown to characterize IGLC haplotypes with 8, 9, 10 or 11 genes[66,70,71], but these genes have not yet been sequenced and are not shown in this table.

Only the genes with at least one functional or ORF allele are shown. For a complete listing of the human IGLV genes including the pseudogenes, see ref. 64 and IMGT gene data: http://imgt.cines.fr/cgi-bin/IMGTlect.jv?query=203.

For information on individual IGL genes:
http://imgt.cines.fr/cgi-bin/IMGTlect.jv?query=202+genename.

Example: IMGT Repertoire for IGLV3-30:
http://imgt.cines.fr/cgi-bin/IMGTlect.jv?query=202+IGLV3-30.

Number of the human IGLV germline variable genes at 22q11.2 and potential repertoire

The repertoire of the human IGLV germline variable genes at 22q11.2 is shown in Table 7.

Overview

73–74 IGLV genes[62–64] on 900 kb: 56–57 IGLV genes belonging to 11 subgroups and 17 pseudogenes assigned to the clans.

29 FUNCTIONAL
5–6 ORF
35 PSEUDOGENE
3 FUNCTIONAL or PSEUDOGENE
1 ORF or PSEUDOGENE.

Potential repertoire

29–33 FUNCTIONAL IGLV genes belonging to 10 subgroups.

Table 7. *Repertoire of the human IGLV germline variable genes at 22q11.2.*

Subgroup		Functional	ORF	Pseudogene	Total
IGLV1	(B)	5	1(+1)*	(+1)*	7
	(C)	–	–	1	1
IGLV2	(A)	5	1	3	9
IGLV3	(A)	8(+2)*	1	12(+2)*	23
IGLV4	(A)	1	–	–	1
	(C)	2	–	–	2
IGLV5	(B)	3–4	1	–	4–5**
IGLV6	(C)	1	–	–	1
IGLV7	(B)	1(+1)*	–	1(+1)*	3
IGLV8	(C)	1	–	–	1
IGLV9	(B)	1	–	–	1
IGLV10	(C)	1	–	1	2
IGLV11	(C)	–	1	–	1
IGLV(I)	(A)	–	–	1	1
	(B)	–	–	2	2
	(C)	–	–	4	4
IGLV(IV)	(C)	–	–	5	5
IGLV(V)	(C)	–	–	2	2
IGLV(VI)	(A)	–	–	2	2
IGLV(VII)	(B)	–	–	1	1
Total		29–30(+3)*	5(+1)*	35(+4)*	73–74**

* The following genes have alleles with different functionality: ORF or PSEUDOGENE (IGLV1-41); FUNCTIONAL or PSEUDOGENE (IGLV3-9, IGLV3-22, IGLV7-46).
** An allelic polymorphism by insertion/deletion which concerns IGLV5-39[60].
(A), (B), (C) Three distinct V-CLUSTERs based on the IGLV gene subgroup content[62].
(I), (IV), (V) Clans for the pseudogenes which could not be assigned to subgroups with functional genes. Clans comprise, respectively:
clan I: IGLV1, IGLV2, IGLV6 and IGLV10 subgroup genes and pseudogenes IGLV(I)-20, -38, -42, -56, -63, -68 and -70;
clan II: IGLV3 subgroup genes; clan III: IGLV7 and IGLV8 subgroup genes;
clan IV: IGLV5 and IGLV11 subgroup genes and pseudogenes IGLV(IV)-53, -59, -64, -65 and -66-1;
clan V: IGLV4 and IGLV9 subgroup genes and pseudogenes IGLV(V)-58 and -66;
clan VI: pseudogenes IGLV(VI)-22-1 and -25-1; clan VII: pseudogene IGLV(VII)-41-1.

Table 8. *Total number of human immunoglobulin genes per haploid genome.*

Locus	Major loci						Number of orphons	Total number of genes (including orphons)
	Chromosomal localization	V	D	J	C	Total number of genes in the major locus		
IGH	14q32.33	123–129	27	9	11[a]	170–176[b]	36[c]	206–212[b,c]
IGK	2p11.2	(40[d] or) 76	0	5	1	(46[d] or) 82	25	(71[d] or)–107
IGL	22q11.2	73–74	0	7–11	7–11	87–96	6	93–102

[a] Allelic IGHC multigene deletions[34–41], duplications and triplications[42–45] have been described in healthy individuals. The number of IGHC genes may vary from five (deletion I, in Fig. 2) to probably 19 (triplication III, in Fig. 2) per haploid genome.
[b] Not included the seven non-mapped IGHV genes.
[c] Included the IGHC processed gene, IGHEP2, localized on chromosome 9 (9p24.2-p24.1).
[d] Number of genes in the rare IGKV haplotype without the distal V-CLUSTER.
Major loci refer to the IGH locus at 14q32.33, the IGK locus at 2p11.2 and the IGL locus at 22q11.2, genes of which are involved in the immunoglobulin chain synthesis. Orphons are genes located outside of the main loci, which cannot contribute to the Ig chain synthesis.
25 IGHV, 10 IGHD, 25 IGKV, 4 IGLV, 2 IGLC orphons have been identified.
The only immunoglobulin processed gene so far described, IGHEP2, has been included with the orphons in this Table.

Table 9. *Number of functional human immunoglobulin genes per haploid genome.*

Locus	Chromosomal localization	Locus size (kb)	V	D	J	C	Number of functional genes	Combinatorial diversity (range per locus)
IGH	14q32.33	1250	38–46	23	6	9[a]	76–84	$38 \times 23 \times 6 = 5244$ (m) $46 \times 23 \times 6 = 6348$ (M)
IGK	2p11.2	1820	31–35	0	5	1	37–41	$31 \times 5 = 155$ (m) $35 \times 5 = 175$ (M)
		500[b]	17–19[b]	0	5	1	23–25[b]	$17 \times 5 = 85$ (m)[b] $19 \times 5 = 95$ (M)[b]
IGL	22q11.2	1050	29–33	0	4-5	4-5	37–43	$29 \times 4 = 116$ (m) $33 \times 5 = 165$ (M)

[a] In haplotypes with multigene deletion (see Fig. 2 and reference 41 for review), the number of functional IGHC genes is five (deletions I, III, and V), six (deletions IV and VI), or eight (deletion II), per haploid genome. In haplotypes with multigene duplication or triplication, the exact number of functional IGHC genes per haploid genome is not known.
[b] In the rare IGKV haplotype without the distal V-CLUSTER.
The range of the theoretical combinatorial diversity indicated takes into account the minimum (m) and the maximum (M) number of functional V, D and J genes in each of the major IGH, IGK and IGL loci.

References

[1] Croce, C.M. et al. (1979) Proc. Natl Acad. Sci. USA 76, 3416–3419..
[2] Kirsch, I.R. et al. (1982) Science 216, 301–303.
[3] McBride, O.W. et al. (1982) Nucleic Acids Res. 10, 8155–8170.
[4] Shin, E.K. et al. (1991) EMBO J. 10, 3641–3645.
[5] Matsuda, F. et al. (1993) Nature Genetics 3, 88–94.
[6] Cook, G.P. et al. (1994) Nature Genetics 7, 162–168.
[7] Cook, G.P. and Tomlinson, I.M. (1995) Immunol. Today 16, 237–242.
[8] Matsuda, F. et al. (1998) J. Exp. Med. 188, 2151–2162.
[9] Matsuda, F. and Honjo T. (1999) The Immunologist 7, 171–176.
[10] Pallarès, N. et al. (1999) Exp. Clin. Immunogenet. 16, 36–60.
[11] Siebenlist, U. et al. (1981) Nature 294, 632–635.
[12] Buluwela, L. et al. (1988) EMBO J. 7, 2003–2010.
[13] Ichahara, Y. et al. (1988) EMBO J. 13, 4141–4150.
[14] Corbett, S. et al. (1997) J. Mol. Biol. 270, 587–597.
[15] Ruiz, M. et al. (1999) Exp. Clin. Immunogenet. 16, 173–184.
[16] Ravetch, J.V. et al. (1981) Cell 27, 583–591.
[17] Flanagan, J.G. and Rabbitts, T.H. (1982) Nature 300, 709–713.
[18] Lefranc, M.-P. et al. (1982) Nature 300, 760–762.
[19] Lefranc, M.-P. et al. (1983) Mol. Biol. Med. 1, 207–217.
[20] Rabbitts, T.H. et al. (1981) Nucleic Acids Res. 9, 4509–4524.
[21] White, M.B. et al. (1985) Science 228, 733–737.
[22] Huck, S. et al. (1986) Nucleic Acids Res. 14, 1779–1789.
[23] Huck, S. et al. (1989) Immunogenetics 30, 250–257.
[24] Ellison, J.W. et al. (1982) Nucleic Acids Res. 10, 4071–4079.
[25] Max, E.E. et al. (1982) Cell 29, 691–699.
[26] Bensmana, M. et al. (1988) Nucleic Acids Res. 16, 3108.
[27] Ellison, J.W. and Hood, L.E. (1982) Proc. Natl Acad. Sci. USA 79, 1984–1988.
[28] Ellison, J.W. et al. (1981) DNA 1, 11–18.
[29] Flanagan, J.G. and Rabbitts, T.H. (1982) EMBO J. 1, 655–660.
[30] Flanagan, J.G. et al. (1984) Cell 36, 681–688.
[31] Bensmana, M. et al. (1991) Cytogenet. Cell Genet. 56, 128.
[32] Scaviner, D. et al. (1999) Exp. Clin. Immunogenet. 16, 234–240.
[33] Battey, J. et al. (1982) Proc. Natl Acad. Sci. USA 79, 5956–5960.
[34] Lefranc, M.-P. et al. (1982) Nature 300, 760–762.
[35] Lefranc, G. et al. (1983) Eur. J. Immunol. 13, 240–244.
[36] Lefranc, M.-P. et al. (1983) Mol. Biol. Med. 1, 207–217.
[37] Migone, N. et al. (1984) Proc. Natl Acad. Sci. USA 81, 5811–5815.
[38] Lefranc, M.-P. and Lefranc, G. (1987) FEBS Letters 213, 231–237.
[39] Lefranc, M.-P. et al. (1991) Immunodeficiency Rev. 2, 265–281.
[40] Keyeux, G. et al. (1989) Genomics 5, 431–441.
[41] Wiebe, V. et al. (1994) Hum. Genet. 93, 520–528.
[42] Bottaro, A. et al. (1991) Am. J. Hum. Genet. 48, 745–756.
[43] Bottaro, A. et al. (1993) Immunogenetics 37, 356–363.
[44] Brusco, A. et al. (1993) Immunodeficiency 4, 243–244.
[45] Brusco, A. et al..(1994) J. Immunol. 152, 129–135.
[46] Malcolm, S. et al. (1982) Proc. Natl Acad. Sci. USA, 79, 4957–4961.
[47] McBride, O.W. et al. (1982) J. Exp. Med. 155, 1480–1490.

[48] Zachau, H.G. (1993) Gene 135, 167–173.
[49] Zachau, H.G. (1996) The Immunologist 4, 49–54.
[50] Hüber, C. et al. (1993) Eur. J. Immunol. 23, 2868–2875.
[51] Schäble, K.F. and Zachau, H.G. (1993) Biol. Chem. Hoppe-Seyler 374, 1001–1022
[52] Cox, J.P. et al. (1994) Eur. J. Immunol. 24, 827–836.
[53] Schäble, K.F. et al. (1994) Biol. Chem. Hoppe-Seyler 375, 189–199.
[54] Barbié, V. and Lefranc, M.-P. (1998) Exp. Clin. Immunogenet. 15, 171–183.
[55] Hieter, P.A. et al. (1982) J. Biol. Chem. 257, 1516–1522.
[56] Hieter, P.A. et al. (1980) Cell 22,197–207.
[57] Erikson, J. et al. (1981) Nature 294, 173–175.
[58] Emanuel, B.S. et al. (1985) Nucleic Acids Res. 13, 381–387.
[59] Dunham, I. et al. (1999) Nature, 402, 489–495.
[60] Frippiat, J.-P. et al. (1995) Hum. Mol. Genet. 4, 983–991.
[61] Kawasaki, K. et al. (1995) Genome Res. 5, 125–135.
[62] Williams, S.C. et al. (1996) J. Mol. Biol. 264, 220–232.
[63] Kawasaki, K. et al. (1997) Genome Res. 7, 250–261.
[64] Pallarès, N. et al. (1998) Exp. Clin. Imm. 15, 8–18.
[65] Hieter, P.A. et al. (1981) Nature 294, 536–540.
[66] Taub, R.A. et al. (1983) Nature 304, 172–174.
[67] Dariavach, P. et al. (1987) Proc. Natl Acad. Sci. 84, 9074–9078.
[68] Vasicek, T.J. and Leder, P. (1990) J. Exp. Med. 1990, 609–620.
[69] Taub, R.A. et al. (1983) Nature 304, 172–174.
[70] Ghanem, N. et al. (1988) Exp. Clin. Immunogenet. 5, 186–195.
[71] Kay, P.H. et al. (1992) Immunogenetics 35, 341–343.
[72] Lefranc, M.-P. et al. (1999) Hum. Genet. 104, 361–369.

Section II

THE HUMAN IMMUNOGLOBULIN IGH GENES

Part 1

IGHC

Nomenclature

IGHA1: Immunoglobulin heavy constant alpha 1.

Definition and functionality

IGHA1 is one of the nine functional genes of the human IGHC group which comprises 11 mapped genes.

Gene location

IGHA1 is in the IGH locus on chromosome 14 at 14q32.33.

Gene structure

The IGHA1 gene comprises three exons, with the hinge region being encoded at the beginning of the second exon. This organization is similar to that of the IGHA2 gene but different from the IGHD and IGHG genes which have separate hinge exons.

Nucleotide and amino acid sequences for human IGHA1

The nucleotide between parentheses at the beginning of exons comes from a DONOR_SPLICE (n from ngt). The first nucleotide from an INT_DONOR_SPLICE is underlined (n from ngt).

The Cysteines involved in the CH1, CH2 and CH3 intrachain disulphide bridges are shown with their number and letter **C** in bold. The Cysteines involved in the H-L or H-H interchain disulphide bridges are shown with only the letter **C** in bold. References are shown in square brackets.

```
                                                1   2   3   4   5   6   7   8   9  10  11  12  13  14  15  16  17  18  19  20
                                                A   S   P   T   S   P   K   V   F   P   L   S   L   C   S   T   Q   P   D   G
J00220   ,IGHA1*01,alpha1 (CH1)[13]   (G)CA TCC CCG ACC AGC CCC AAG GTC TTC CCG CTG AGC CTC TGC AGC ACC CAG CCA GAT GGG
AF067420,IGHA1*01 (cDNA)(S)     [12]   (-)-- --- --- --- --- --- --- --- --- --- --- --- --- --- --- --- --- --- --- ---
                                       21  22  23  24  25  26  27  28  29  30  31  32  33  34  35  36  37  38  39  40
                                        N   V   V   I   A   C   L   V   Q   G   F   F   P   Q   E   P   L   S   V   T
J00220   ,IGHA1*01,alpha1 (CH1)        AAC GTG GTC ATC GCC TGC CTG GTC CAG GGC TTC TTC CCC CAG GAG CCA CTC AGT GTG ACC
AF067420,IGHA1*01 (cDNA)(S)            --- --- --- --- --- --- --- --- --- --- --- --- --- --- --- --- --- --- --- ---
                                       41  42  43  44  45  46  47  48  49  50  51  52  53  54  55  56  57  58  59  60
                                        W   S   E   S   G   Q   G   V   T   A   R   N   F   P   P   S   Q   D   A   S
J00220   ,IGHA1*01,alpha1 (CH1)        TGG AGC GAA AGC GGA CAG GGC GTG ACC GCC AGA AAC TTC CCA CCC AGC CAG GAT GCC TCC
AF067420,IGHA1*01 (cDNA)(S)            --- --- --- --- --- --- --- --- --- --- --- --- --- --- --- --- --- --- --- ---
                                       61  62  63  64  65  66  67  68  69  70  71  72  73  74  75  76  77  78  79  80
                                        G   D   L   Y   T   T   S   S   Q   L   T   L   P   A   T   Q   C   L   A   G
J00220   ,IGHA1*01,alpha1 (CH1)        GGG GAC CTG TAC ACC ACG AGC AGC CAG CTG ACC CTG CCG GCC ACA CAG TGC CTA GCC GGC
AF067420,IGHA1*01 (cDNA)(S)            --- --- --- --- --- --- --- --- --- --- --- --- --- --- --- --- --- --- --- ---
                                       81  82  83  84  85  86  87  88  89  90  91  92  93  94  95  96  97  98  99 100
                                        K   S   V   T   C   H   V   K   H   Y   T   N   P   S   Q   D   V   T   V   P
J00220   ,IGHA1*01,alpha1 (CH1)        AAG TCC GTG ACA TGC CAC GTG AAG CAC TAC ACG AAT CCC AGC CAG GAT GTG ACT GTG CCC
AF067420,IGHA1*01 (cDNA)(S)            --- --- --- --- --- --- --- --- --- --- --- --- --- --- --- --- --- --- --- ---
                                      101 102
                                        C   P
J00220   ,IGHA1*01,alpha1 (CH1)        TGC CCA G
AF067420,IGHA1*01 (cDNA)(S)            --- --- -

                                        _________________________________ H ______________________________
                                        1   2   3   4   5   6   7   8   9  10  11  12  13  14  15  16  17  18  19  20
                                        V   P   S   T   P   P   T   P   S   P   S   T   P   P   T   P   S   P   S   C
J00220   ,IGHA1*01,alpha1 (H+CH2)      TT CCC TCA ACT CCA CCT ACC CCA TCT CCC TCA ACT CCA CCT ACC CCA TCT CCC TCA TGC
AF067420,IGHA1*01 (cDNA)(S)            -- --- --- --- --- --- --- --- --- --- --- --- --- --- --- --- --- --- --- ---
                                       21  22  23  24  25  26  27  28  29  30  31  32  33  34  35  36  37  38  39  40
                                        C   H   P   R   L   S   L   H   R   P   A   L   E   D   L   L   L   G   S   E
J00220   ,IGHA1*01,alpha1 (H+CH2)      TGC CAC CCC CGA CTG TCA CTG CAC CGA CCG GCC CTC GAG GAC CTG CTC TTA GGT TCA GAA
AF067420,IGHA1*01 (cDNA)(S)            --- --- --- --- --- --- --- --- --- --- --- --- --- --- --- --- --- --- --- ---
                                       41  42  43  44  45  46  47  48  49  50  51  52  53  54  55  56  57  58  59  60
                                        A   N   L   T   C   T   L   T   G   L   R   D   A   S   G   V   T   F   T   W
J00220   ,IGHA1*01,alpha1 (H+CH2)      GCG AAC CTC ACG TGC ACA CTG ACC GGC CTG AGA GAT GCC TCA GGT GTC ACC TTC ACC TGG
AF067420,IGHA1*01 (cDNA)(S)            --- --- --- --- --- --- --- --- --- --- --- --- --- --- --- --- --- --- --- ---
```

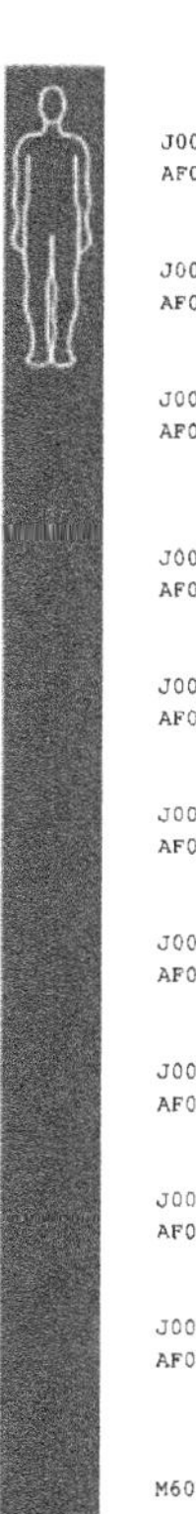

```
                                   61  62  63  64  65  66  67  68  69  70  71  72  73  74  75  76  77  78  79  80
                                    T   P   S   S   G   K   S   A   V   Q   G   P   P   E   R   D   L   C   G   C
J00220   ,IGHA1*01,alpha1 (H+CH2)  ACG CCC TCA AGT GGG AAG AGC GCT GTT CAA GGA CCA CCT GAG CGT GAC CTC TGT GGC TGC
AF067420,IGHA1*01 (cDNA)(S)        --- --- --- --- --- --- --- --- --- --- --- --- --- --- --- --- --- --- --- ---
                                   81  82  83  84  85  86  87  88  89  90  91  92  93  94  95  96  97  98  99 100
                                    Y   S   V   S   S   V   L   P   G   C   A   E   P   W   N   H   G   K   T   F
J00220   ,IGHA1*01,alpha1 (H+CH2)  TAC AGC GTG TCC AGT GTC CTG CCG GGC TGT GCC GAG CCA TGG AAC CAT GGG AAG ACC TTC
AF067420,IGHA1*01 (cDNA)(S)        --- --- --- --- --- --- --- --- --- --- --- --- --- --- --- --- --- --- --- ---
                                  101 102 103 104 105 106 107 108 109 110 111 112 113 114 115 116 117 118 119 120
                                    T   C   T   A   A   Y   P   E   S   K   T   P   L   T   A   T   L   S   K   S
J00220   ,IGHA1*01,alpha1 (H+CH2)  ACT TGC ACT GCT GCC TAC CCC GAG TCC AAG ACC CCG CTA ACC GCC ACC CTC TCA AAA TCC G
AF067420,IGHA1*01 (cDNA)(S)        --- --- --- --- --- --- --- --- --- --- --- --- --- --- --- --- --- --- --- --- -

                                    1   2   3   4   5   6   7   8   9  10  11  12  13  14  15  16  17  18  19  20
                                    G   N   T   F   R   P   E   V   H   L   L   P   P   P   S   E   E   L   A   L
J00220   ,IGHA1*01,alpha1 (CH3)    GA AAC ACA TTC CGG CCC GAG GTC CAC CTG CTG CCG CCG CCG TCG GAG GAG CTG GCC CTG
AF067420,IGHA1*01 (cDNA)(S)        -- --- --- --- --- --- --- --- --- --- --- --- --- --- --- --- --- --- --- ---
                                   21  22  23  24  25  26  27  28  29  30  31  32  33  34  35  36  37  38  39  40
                                    N   E   L   V   T   L   T   C   L   A   R   G   F   S   P   K   D   V   L   V
J00220   ,IGHA1*01,alpha1 (CH3)    AAC GAG CTG GTG ACG CTG ACG TGC CTG GCA CGC GGC TTC AGC CCC AAG GAC GTG CTG GTT
AF067420,IGHA1*01 (cDNA)(S)        --- --- --- --- --- --- --- --- --- --- --- --- --- --- --- --- --- --- --- ---
                                   41  42  43  44  45  46  47  48  49  50  51  52  53  54  55  56  57  58  59  60
                                    R   W   L   Q   G   S   Q   E   L   P   R   E   K   Y   L   T   W   A   S   R
J00220   ,IGHA1*01,alpha1 (CH3)    CGC TGG CTG CAG GGG TCA CAG GAG CTG CCC CGC GAG AAG TAC CTG ACT TGG GCA TCC CGG
AF067420,IGHA1*01 (cDNA)(S)        --- --- --- --- --- --- --- --- --- --- --- --- --- --- --- --- --- --- --- ---
                                   61  62  63  64  65  66  67  68  69  70  71  72  73  74  75  76  77  78  79  80
                                    Q   E   P   S   Q   G   T   T   T   F   A   V   T   S   I   L   R   V   A   A
J00220   ,IGHA1*01,alpha1 (CH3)    CAG GAG CCC AGC CAG GGC ACC ACC ACC TTC GCT GTG ACC AGC ATA CTG CGC GTG GCA GCC
AF067420,IGHA1*01 (cDNA)(S)        --- --- --- --- --- --- --- --- --- --- --- --- --- --- --- --- --- --- --- ---
                                   81  82  83  84  85  86  87  88  89  90  91  92  93  94  95  96  97  98  99 100
                                    E   D   W   K   K   G   D   T   F   S   C   M   V   G   H   E   A   L   P   L
J00220   ,IGHA1*01,alpha1 (CH3)    GAG GAC TGG AAG AAG GGG GAC ACC TTC TCC TGC ATG GTG GGC CAC GAG GCC CTG CCG CTG
AF067420,IGHA1*01 (cDNA)(S)        --- --- --- --- --- --- --- --- --- --- --- --- --- --- --- --- --- --- --- ---
                                  101 102 103 104 105 106 107 108 109 110 111 112 113 114 115 116 117 118 119 120
                                    A   F   T   Q   K   T   I   D   R   L   A   G   K   P   T   H   V   N   V   S
J00220   ,IGHA1*01,alpha1 (CH3)    GCC TTC ACA CAG AAG ACC ATC GAC CGC TTG GCG GGT AAA CCC ACC CAT GTC AAT GTG TCT
AF067420,IGHA1*01 (cDNA)(S)        --- --- --- --- --- --- --- --- --- --- --- --- --- --- --- --- --- --- --- ---
                                  121 122 123 124 125 126 127 128 129 130 131
                                    V   V   M   A   E   V   D   G   T   C   Y   *
J00220   ,IGHA1*01,alpha1 (CH3)    GTT GTC ATG GCG GAG GTG GAC GGC ACC TGC TAC
AF067420,IGHA1*01 (cDNA)(S)        --- --- --- --- --- --- --- --- --- --- ---

                                    1   2   3   4   5   6   7   8   9  10  11  12  13  14  15  16  17  18  19  20
                                    G   S   C   S   V   A   D   W   Q   M   P   P   P   Y   V   V   L   D   L   P
M60193   ,IGHA1*01,alpha1 (M)  [21] (G)GC TCT TGC TCT GTT GCA GAT TGG CAG ATG CCG CCT CCC TAT GTG GTG CTG GAC TTG CCG
                                   21  22  23  24  25  26  27  28  29  30  31  32  33  34  35  36  37  38  39  40
                                    Q   E   T   L   E   E   E   T   P   G   A   N   L   W   P   T   T   I   T   F
M60193   ,IGHA1*01,alpha1 (M)      CAG GAG ACC CTG GAG GAG GAG ACC CCC GGC GCC AAC CTG TGG CCC ACC ACC ATC ACC TTC
                                   41  42  43  44  45  46  47  48  49  50  51  52  53  54  55  56  57  58  59  60
                                    L   T   L   F   L   L   S   L   F   Y   S   T   A   L   T   V   T   S   V   R
M60193   ,IGHA1*01,alpha1 (M)      CTC ACC CTC TTC CTG CTG AGC CTG TTC TAT AGC ACA GCA CTG ACC GTG ACC AGC GTC CGG
                                   61  62  63  64  65  66  67  68  69  70  71
                                    G   P   S   G   N   R   E   G   P   Q   Y   *
M60193   ,IGHA1*01,alpha1 (M)      GGC CCA TCT GGC AAC AGG GAG GGC CCC CAG TAC
```

(s) Transcript of a secreted chain.

Genome database accession numbers

GDB:119332 LocusLink: 3493 OMIM: 146900

Nomenclature

IGHA2: Immunoglobulin heavy constant alpha 2.

Definition and functionality

IGHA2 is one of the nine functional genes of the human IGHC group which comprises 11 mapped genes.

Gene location

IGHA2 is in the IGH locus on chromosome 14 at 14q32.33.

Gene structure

The IGHA2 gene comprises three exons, with the hinge region being encoded at the beginning of the second exon. This organization is similar to that of the IGHA2 gene but different from the IGHD and IGHG genes which have separate hinge exons.

Nucleotide and amino acid sequences for human IGHA2

The nucleotide between parentheses at the beginning of exons comes from a DONOR_SPLICE (n from ngt). The first nucleotide from an INT_DONOR_SPLICE is underlined (n from ngt).

The Cysteines involved in the CH1, CH2 and CH3 intrachain disulphide bridges are shown with their number and letter **C** in bold. The Cysteines involved in the H-L or H-H interchain disulphide bridges are shown with only the letter **C** in bold.

```
                                          1   2   3   4   5   6   7   8   9  10  11  12  13  14  15  16  17  18  19  20
                                          A   S   P   T   S   P   K   V   F   P   L   S   L   D   S   T   P   Q   D   G
J00221  ,IGHA2*01,alpha2 (CH1) [13]  (G)CA TCC CCG ACC AGC CCC AAG GTC TTC CCG CTG AGC CTC GAC AGC ACC CCC CAA GAT GGG
M60192  ,IGHA2*02,alpha2       [21]  (-)-- --- --- --- --- --- --- --- --- --- --- --- --- --- --- --- --- --- --- ---
                 TOU II-5
S71043  ,IGHA2*03,alpha2       [26]  (-)-- --- --- --- --- --- --- --- --- --- --- --- --- --- --- --- --- --- --- ---
                                         21  22  23  24  25  26  27  28  29  30  31  32  33  34  35  36  37  38  39  40
                                          N   V   V   V   A   C   L   V   Q   G   F   F   P   Q   E   P   L   S   V   T
J00221  ,IGHA2*01,alpha2 (CH1)           AAC GTG GTC GTC GCA TGC CTG GTC CAG GGC TTC TTC CCC CAG GAG CCA CTC AGT GTG ACC
M60192  ,IGHA2*02,alpha2                 --- --- --- --- --- --- --- --- --- --- --- --- --- --- --- --- --- --- --- ---
                 TOU II-5
S71043  ,IGHA2*03,alpha2                 --- --- --- --- --- --- --- --- --- --- --- --- --- --- --- --- --- --- --- ---
                                         41  42  43  44  45  46  47  48  49  50  51  52  53  54  55  56  57  58  59  60
                                          W   S   E   S   G   Q   N   V   T   A   R   N   F   P   P   S   Q   D   A   S
J00221  ,IGHA2*01,alpha2 (CH1)           TGG AGC GAA AGC GGA CAG AAC GTG ACC GCC AGA AAC TTC CCA CCT AGC CAG GAT GCC TCC
M60192  ,IGHA2*02,alpha2                 --- --- --- --- --- --- --- --- --- --- --- --- --- --- --- --- --- --- --- ---
                 TOU II-5
S71043  ,IGHA2*03,alpha2                 --- --- --- --- --- --- --- --- --- --- --- --- --- --- --- --- --- --- --- ---
                                         61  62  63  64  65  66  67  68  69  70  71  72  73  74  75  76  77  78  79  80
                                          G   D   L   Y   T   T   S   S   Q   L   T   L   P   A   T   Q   C   P   D   G
J00221  ,IGHA2*01,alpha2 (CH1)           GGG GAC CTG TAC ACC ACG AGC AGC CAG CTG ACC CTG CCG GCC ACA CAG TGC CCA GAC GGC
M60192  ,IGHA2*02,alpha2                 --- --- --- --- --- --- --- --- --- --- --- --- --- --- --- --- --- --- --- ---
                 TOU II-5
S71043  ,IGHA2*03,alpha2                 --- --- --- --- --- --- --- --- --- --- --- --- --- --- --- --- --- --- --- ---
                                         81  82  83  84  85  86  87  88  89  90  91  92  93  94  95  96  97  98  99 100
                                          K   S   V   T   C   H   V   K   H   Y   T   N   P   S   Q   D   V   T   V   P
J00221  ,IGHA2*01,alpha2 (CH1)           AAG TCC GTG ACA TGC CAC GTG AAG CAC TAC ACG AAT CCC AGC CAG GAT GTG ACT GTG CCC
                                                                                          S
M60192  ,IGHA2*02,alpha2                 --- --- --- --- --- --- --- --- --- --- --- --- T-- --- --- --- --- --- --- ---
                 TOU II-5                                                                 S
S71043  ,IGHA2*03,alpha2                 --- --- --- --- --- --- --- --- --- --- --- --- T-- --- --- --- --- --- --- ---
```

```
                                               101 102
                                                C   P
J00221  ,IGHA2*01,alpha2 (CH1)  (1)            TGC CCA G
                                                    R
M60192  ,IGHA2*02,alpha2        (1)            --- -G- -
                  TOU II-5                          R
S71043  ,IGHA2*03,alpha2                       --- -G- -

                                                _________ H _________
                                                 1   2   3   4   5   6   7   8   9  10  11  12  13  14  15  16  17  18  19  20
                                                 V   P   P   P   P   P   C   C   H   P   R   L   S   L   H   R   P   A   L   E
J00221  ,IGHA2*01,alpha2 (H+CH2)                 TT CCC CCA CCT CCC CCA TGC TGC CAC CCC CGA CTG TCG CTG CAC CGA CCG GCC CTC GAG
M60192  ,IGHA2*02,alpha2                         -- --- --- --- --- --- --- --- --- --- --- --- --- --- --- --- --- --- --- ---
                  TOU II-5
AJ012264,IGHA2*02,alpha2        [13]                                                                                    --- ---
                  TOU II-5
S71043  ,IGHA2*03,alpha2                         -- --- --- --- --- --- --- --- --- --- --- --- --- --- --- --- --- --- --- ---
                                                21  22  23  24  25  26  27  28  29  30  31  32  33  34  35  36  37  38  39  40
                                                 D   L   L   L   G   S   E   A   N   L   T   C   T   L   T   G   L   R   D   A
J00221  ,IGHA2*01,alpha2 (H+CH2)                GAC CTG CTC TTA GGT TCA GAA GCG AAC CTC ACG TGC ACA CTG ACC GGC CTG AGA GAT GCC
M60192  ,IGHA2*02,alpha2                        --- --- --- --- --- --- --- --- --- --- --- --- --- --- --- --- --- --- --- ---
                  TOU II-5
AJ012264,IGHA2*02,alpha2                        --- --- --- --- --- --- --- --- --- --- --- --- --- --- --- --- --- --- --- ---
                  TOU II-5
S71043  ,IGHA2*03,alpha2                        --- --- --- --- --- --- --- --- --- --- --- --- --- --- --- --- --- --- --- ---
                                                41  42  43  44  45  46  47  48  49  50  51  52  53  54  55  56  57  58  59  60
                                                 S   G   A   T   F   T   W   T   P   S   S   G   K   S   A   V   Q   G   P   P
J00221  ,IGHA2*01,alpha2 (H+CH2)                TCT GGT GCC ACC TTC ACC TGG ACG CCC TCA AGT GGG AAG AGC GCT GTT CAA GGA CCA CCT
M60192  ,IGHA2*02,alpha2                        --- --- --- -
                  TOU II-5
AJ012264,IGHA2*02,alpha2                        --- --- --- --- --- --- --- --- --- --- --- --- --- --- --- --- --- --- --- ---
                  TOU II-5
S71043  ,IGHA2*03,alpha2                        --- --- --- --- --- --- --- --- --- --- --- --- --- --- --- --- --- --- --- ---
                                                61  62  63  64  65  66  67  68  69  70  71  72  73  74  75  76  77  78  79  80
                                                 E   R   D   L   C   G   C   Y   S   V   S   S   V   L   P   G   C   A   Q   P
J00221  ,IGHA2*01,alpha2 (H+CH2)                GAG CGT GAC CTC TGT GGC TGC TAC AGC GTG TCC AGT GTC CTG CCT GGC TGT GCC CAG CCA
AJ012264,IGHA2*02,alpha2                        --- --- --- --- --- --- --- --- --- --- --- --- --- --- --- --- --- --- --- ---
                  TOU II-5
S71043  ,IGHA2*03,alpha2                        --- --- --- --- --- --- --- --- --- --- --- --- --- --- --- --- --- --- --- ---
                                                81  82  83  84  85  86  87  88  89  90  91  92  93  94  95  96  97  98  99 100
                                                 W   N   H   G   E   T   F   T   C   T   A   A   H   P   E   L   K   T   P   L
J00221  ,IGHA2*01,alpha2 (H+CH2)                TGG AAC CAT GGG GAG ACC TTC ACC TGC ACT GCT GCC CAC CCC GAG TTG AAG ACC CCA CTA
AJ012264,IGHA2*02,alpha2                        --- --- --- --- --- --- --- --- --- --- --- --- --- --- --- --- --- --- --- ---
                  TOU II-5
S71043  ,IGHA2*03,alpha2                        --- --- --- --- --- --- --- --- --- --- --- --- --- --- --- --- --- --- --- ---
                                               101 102 103 104 105 106 107
                                                 T   A   N   I   T   K   S
J00221  ,IGHA2*01,alpha2 (H+CH2)                ACC GCC AAC ATC ACA AAA TCC G
AJ012264,IGHA2*02,alpha2                        --- --- --- --- --- --- --- -
                  TOU II-5
S71043  ,IGHA2*03,alpha2                        --- --- --- --- --- --- --- -

                                                 1   2   3   4   5   6   7   8   9  10  11  12  13  14  15  16  17  18  19  20
                                                 G   N   T   F   R   P   E   V   H   L   L   P   P   P   S   E   E   L   A   L
J00221  ,IGHA2*01,alpha2 (CH3)                   GA AAC ACA TTC CGG CCC GAG GTC CAC CTG CTG CCG CCG CCG TCG GAG GAG CTG GCC CTG
AJ012264,IGHA2*02,alpha2                         -- --- --- --- --- --- --- --- --- --- --- --- --- --- --- --- --- --- --- ---
                  TOU II-5
S71043  ,IGHA2*03,alpha2                         -- --- --- --- --- --- --- --- --- --- --- --- --- --- --- --- --- --- --- ---
                                                21  22  23  24  25  26  27  28  29  30  31  32  33  34  35  36  37  38  39  40
                                                 N   E   L   V   T   L   T   C   L   A   R   G   F   S   P   K   D   V   L   V
J00221  ,IGHA2*01,alpha2 (CH3)                  AAC GAG CTG GTG ACG CTG ACG TGC CTG GCA CGT GGC TTC AGC CCC AAG GAT GTG CTG GTT
AJ012264,IGHA2*02,alpha2                        --- --- --- --- --- --- --- --- --- --- --- --- --- --- --- --- --- --- --- ---
                  TOU II-5
S71043  ,IGHA2*03,alpha2                        --- --- --- --- --- --- --- --- --- --- --- --- --- --- --- --- --- --- --- ---
                                                41  42  43  44  45  46  47  48  49  50  51  52  53  54  55  56  57  58  59  60
                                                 R   W   L   Q   G   S   Q   E   L   P   R   E   K   Y   L   T   W   A   S   R
J00221  ,IGHA2*01,alpha2 (CH3)                  CGC TGG CTG CAG GGG TCA CAG GAG CTG CCC CGC GAG AAG TAC CTG ACT TGG GCA TCC CGG
AJ012264,IGHA2*02,alpha2                        --- --- --- --- --- --- --- --- --- --- --- --- --- --- --- --- --- --- --- ---
                  TOU II-5
S71043  ,IGHA2*03,alpha2                        --- --- --- --- --- --- --- --- --- --- --- --- --- --- --- --- --- --- --- ---
                                                61  62  63  64  65  66  67  68  69  70  71  72  73  74  75  76  77  78  79  80
                                                 Q   E   P   S   Q   G   T   T   T   F   A   V   T   S   I   L   R   V   A   A
J00221  ,IGHA2*01,alpha2 (CH3)                  CAG GAG CCC AGC CAG GGC ACC ACC ACC TTC GCT GTG ACC AGC ATA CTG CGC GTG GCA GCC
                                                                                     Y
AJ012264,IGHA2*02,alpha2                        --- --- --- --- --- --- --- --- --- -A- --- --A --- --- --- --- --- --- --- --T
                  TOU II-5
S71043  ,IGHA2*03,alpha2                        --- --- --- --- --- --- --- --- --- --- --- --- --- --- --- --- --- --- --- ---
                                                81  82  83  84  85  86  87  88  89  90  91  92  93  94  95  96  97  98  99 100
                                                 E   D   W   K   K   G   D   T   F   S   C   M   V   G   H   E   A   L   P   L
J00221  ,IGHA2*01,alpha2 (CH3)                  GAG GAC TGG AAG AAG GGG GAC ACC TTC TCC TGC ATG GTG GGC CAC GAG GCC CTG CCG CTG
                                                                         E
AJ012264,IGHA2*02,alpha2                        --- --- --- --- --- --- --G --- --- --- --- --- --- --- --- --- --- --- --- ---
                  TOU II-5
S71043  ,IGHA2*03,alpha2                        --- --- --- --- --- --- --- --- --- --- --- --- --- --- --- --- --- --- --- ---
                                               101 102 103 104 105 106 107 108 109 110 111 112 113 114 115 116 117 118 119 120
                                                 A   F   T   Q   K   T   I   D   R   L   A   G   K   P   T   H   V   N   V   S
J00221  ,IGHA2*01,alpha2 (CH3)                  GCC TTC ACA CAG AAG ACC ATC GAC CGC TTG GCG GGT AAA CCC ACC CAT GTC AAT GTG TCT
                                                                                     M                           I
AJ012264,IGHA2*02,alpha2                        --- --- --- --- --- --- --- --- --- A-- --- --- --- --- --- --C A-- --- --- ---
                  TOU II-5
```

```
S71043  ,IGHA2*03,alpha2               --- --- --- --- --- --- --- --- --- --- --- --- --- --- --- --- --- --- --- ---
                                       121 122 123 124 125 126 127 128 129 130 131
                                        V   V   M   A   E   V   D   G   T   C   Y   *
J00221  ,IGHA2*01,alpha2 (CH3)         GTT GTC ATG GCG GAG GTG GAC GGC ACC TGC TAC
                                                            A
AJ012264,IGHA2*02,alpha2               --- --- --- --- --- -C- --T --- --- --- ---
              TOU II-5
S71043  ,IGHA2*03,alpha2               --- --- --- --- --- --- --- --- --- --- ---

                                         1   2   3   4   5   6   7   8   9  10  11  12  13  14  15  16  17  18  19  20
                                        G   S   C   C   V   A   D   W   Q   M   P   P   P   Y   V   V   L   D   L   P
M60194  ,IGHA2*02,alpha2 (M)     [21] (G)GC TCT TGC TGT GTT GCA GAT TGG CAG ATG CCG CCT CCC TAT GTG GTG CTG GAC TTG CCG
              TOU II-5
                                        21  22  23  24  25  26  27  28  29  30  31  32  33  34  35  36  37  38  39  40
                                        Q   E   T   L   E   E   E   T   P   G   A   N   L   W   P   T   T   I   T   F
M60194  ,IGHA2*02,alpha2 (M)           CAG GAG ACC CTG GAG GAG GAG ACC CCC GGC GCC AAC CTG TGG CCC ACC ACC ATC ACC TTC
              TOU II-5
                                        41  42  43  44  45  46  47  48  49  50  51  52  53  54  55  56  57  58  59  60
                                        L   T   L   F   L   L   S   L   F   Y   S   T   A   L   T   V   T   S   V   R
M60194  ,IGHA2*02,alpha2 (M)           CTC ACC CTC TTC CTG CTG AGC CTG TTC TAT AGC ACA GCA CTG ACC GTG ACC AGC GTC CGG
              TOU II-5
                                        61  62  63  64  65  66  67  68  69  70  71
                                        G   P   S   G   K   R   E   G   P   Q   Y   *
M60194  ,IGHA2*02,alpha2 (M)           GGC CCA TCT GGC AAG AGG GAG GGC CCC CAG TAC
              TOU II-5
```

Note:

(1) There is no H-L interchain disulphide bridge in IgA2 with A2m(1) alpha 2 chains, whereas a disulphide bridge is present in IgA2 with A2m(2) alpha 2 chains. The disulphide bridge involves Cys 101 in A2m(2) alpha 2 chains, and the presence of Pro 102 in A2m(1) alpha 2 chains hinders this disulphide bridge formation.

The A2m(1) and A2m(2) alpha chains are encoded by the IGHA2*01 and IGHA2*02 alleles, respectively.

Genome database accession numbers

GDB:119333 LocusLink: 3494 OMIM: 147000

IGHD

Nomenclature

IGHD: Immunoglobulin heavy constant delta.

Definition and functionality

IGHD is one of the nine functional genes of the human IGHC group which comprises 11 mapped genes.

Gene location

IGHD is in the IGH locus on chromosome 14 at 14q32.33.

Gene structure

The IGHD gene comprises three main exons encoding the CH1, CH2 and CH3 domains, two hinge (H) exons, one CH-S exon (which encodes the hydrophilic C-terminal end of soluble immunoglobulin) and two transmembrane exons (M1 and M2). The separate CH-S exon is a characteristic of the IGHD gene.

Nucleotide and amino acid sequences for human IGHD

The nucleotide between parentheses at the beginning of exons comes from a DONOR_SPLICE (n from ngt).

The Cysteines involved in the CH1, CH2 and CH3 intrachain disulphide bridges are shown with their number and letter **C** in bold. The Cysteines involved in the H-L or H-H interchain disulphide bridges are shown with only the letter **C** in bold.

```
                                         1   2   3   4   5   6   7   8   9  10  11  12  13  14  15  16  17  18  19  20
                                         A   P   T   K   A   P   D   V   F   P   I   I   S   G   C   R   H   P   K   D
K02875  ,IGHD*01 (CH1)        [15]  (G)CA CCC ACC AAG GCT CCG GAT GTG TTC CCC ATC ATA TCA GGG TGC AGA CAC CCA AAG GAT
X57331  ,IGHD*02              [22]  (-)-- --- --- --- --- --- --- --- --- --- --- --- --- --- --- --- --- --- --- ---
                                        21  22  23  24  25  26  27  28  29  30  31  32  33  34  35  36  37  38  39  40
                                         N   S   P   V   V   L   A   C   L   I   T   G   Y   H   P   T   S   V   T   V
K02875  ,IGHD*01 (CH1)                 AAC AGC CCT GTG GTC CTG GCA TGC TTG ATA ACT GGG TAC CAC CCA ACG TCC GTG ACT GTC
X57331  ,IGHD*02                       --- --- --- --- --- --- --- --- --- --- --- --- --- --- --- --- --- --- --- ---
                                        41  42  43  44  45  46  47  48  49  50  51  52  53  54  55  56  57  58  59  60
                                         T   W   Y   M   G   T   Q   S   Q   P   Q   R   T   F   P   E   I   Q   R   R
K02875  ,IGHD*01 (CH1)                 ACC TGG TAC ATG GGG ACA CAG AGC CAG CCC CAG AGA ACC TTC CCT GAG ATA CAA AGA CGG
X57331  ,IGHD*02                       --- --- --- --- --- --- --- --- --- --- --- --- --- --- --- --- --- --- --- ---
                                        61  62  63  64  65  66  67  68  69  70  71  72  73  74  75  76  77  78  79  80
                                         D   S   Y   Y   M   T   S   S   Q   L   S   T   P   L   Q   Q   W   R   Q   G
K02875  ,IGHD*01 (CH1)                 GAC AGC TAC TAC ATG ACA AGC AGC CAG CTC TCC ACC CCC CTC CAG CAG TGG CGC CAA GGC
X57331  ,IGHD*02                       --- --- --- --- --- --- --- --- --- --- --- --- --- --- --- --- --- --- --- ---
                                        81  82  83  84  85  86  87  88  89  90  91  92  93  94  95  96  97  98  99 100
                                         E   Y   K   C   V   V   Q   H   T   A   S   K   S   K   K   E   I   F   R   W
K02875  ,IGHD*01 (CH1)                 GAG TAC AAA TGC GTG GTC CAG CAC ACC GCC AGC AAG AGT AAG AAG GAG ATC TTC CGC TGG
X57331  ,IGHD*02                       --- --- --- --- --- --- --- --- --- --- --- --- --- --- --- --- --- --- --- ---
                                       101
                                         P
K02875  ,IGHD*01 (CH1)                 CCA G
X57331  ,IGHD*02                       --- -

                                         1   2   3   4   5   6   7   8   9  10  11  12  13  14  15  16  17  18  19  20
                                         E   S   P   K   A   Q   A   S   S   V   P   T   A   Q   P   Q   A   E   G   S
K02876  ,IGHD*01 (H1)         [15]      AG TCT CCA AAG GCA CAG GCC TCC TCC GTG CCC ACT GCA CAA CCC CAA GCA GAG GGC AGC
X57331  ,IGHD*02                        -- --- --- --- --- --- --- --- --A --- --- --- --- --- --- --- --- --- --- ---
                                        21  22  23  24  25  26  27  28  29  30  31  32  33  34
                                         L   A   K   A   T   T   A   P   A   T   T   R   N   T
K02876  ,IGHD*01 (H1)                  CTC GCC AAG GCA ACC ACA GCC CCA GCC ACC ACC CGT AAC ACA G
X57331  ,IGHD*02                       --- --- --- --- --- --- --- --- --- --- --- --- --- --- -

                                         1   2   3   4   5   6   7   8   9  10  11  12  13  14  15  16  17  18  19  20
                                         G   R   G   G   E   E   K   K   K   E   K   E   K   E   E   Q   E   E   R   E
K02877  ,IGHD*01 (H2)         [15]      GA AGA GGA GGA GAA GAG AAG AAG AAG GAG AAG GAG AAA GAG GAA CAA GAA GAG AGA GAG
X57331  ,IGHD*02                        -- --- --- --- --- --- --- --- --- --- --- --- --- --- --- --- --- --- --- ---
```

```
                                21  22  23  24
                                T   K   T   P
K02877  ,IGHD*01 (H2)           ACA AAG ACA CCA G
X57331  ,IGHD*02                --- --- --- --- -

                                 1   2   3   4   5   6   7   8   9  10  11  12  13  14  15  16  17  18  19  20
                                 E   C   P   S   H   T   Q   P   L   G   V   Y   L   L   T   P   A   V   Q   D
K02878  ,IGHD*01 (CH2)     [15]  AG TGT CCG AGC CAC ACC CAG CCT CTT GGC GTC TAC CTG CTA ACC CCT GCA GTG CAG GAC
X57331  ,IGHD*02                 -- --- --- --- --- --- --- --- --- --- --- --- --- --- --- --- --- --- --- ---
                                21  22  23  24  25  26  27  28  29  30  31  32  33  34  35  36  37  38  39  40
                                 L   W   L   R   D   K   A   T   F   T   C   F   V   V   G   S   D   L   K   D
K02878  ,IGHD*01 (CH2)          CTG TGG CTC CGG GAC AAA GCC ACC TTC ACC TGC TTC GTG GTG GGC AGT GAC CTG AAG GAT
X57331  ,IGHD*02                --- --- --- --- --- --- --- --- --- --- --- --- --- --- --- --- --- --- --- ---
                                41  42  43  44  45  46  47  48  49  50  51  52  53  54  55  56  57  58  59  60
                                 A   H   L   T   W   E   V   A   G   K   V   P   T   G   G   V   E   E   G   L
K02878  ,IGHD*01 (CH2)          GCT CAC CTG ACC TGG GAG GTG GCT GGG AAG GTC CCC ACA GGG GGC GTG GAG GAA GGG CTG
X57331  ,IGHD*02                --- --- --- --- --- --- --- --- --- --- --- --- --- --- --- --- --- --- --- ---
                                61  62  63  64  65  66  67  68  69  70  71  72  73  74  75  76  77  78  79  80
                                 L   E   R   H   S   N   G   S   Q   S   Q   H   S   R   L   T   L   P   R   S
K02878  ,IGHD*01 (CH2)          CTG GAG CGG CAC AGC AAC GGC TCC CAG AGC CAG CAC AGC CGT CTG ACC CTG CCC AGG TCC
X57331  ,IGHD*02                --- --- --- --- --- --- --- --- --- --- --- --- --- --- --- --- --- --- --- ---
                                81  82  83  84  85  86  87  88  89  90  91  92  93  94  95  96  97  98  99 100
                                 L   W   N   A   G   T   S   V   T   C   T   L   N   H   P   S   L   P   P   Q
K02878  ,IGHD*01 (CH2)          TTG TGG AAC GCG GGG ACC TCC GTC ACC TGC ACA CTG AAC CAT CCC AGC CTC CCA CCC CAG
X57331  ,IGHD*02                --- --- --- --- --- --- --- --- --- --- --- --- --- --- --- --- --- --- --- ---
                               101 102 103 104 105 106 107 108
                                 R   L   M   A   L   R   E   P
K02878  ,IGHD*01 (CH2)          AGG TTG ATG GCG CTG AGA GAA CCC G
X57331  ,IGHD*02                --- --- --- --- --- --- --- --- -

                                 1   2   3   4   5   6   7   8   9  10  11  12  13  14  15  16  17  18  19  20
                                 A   A   Q   A   P   V   K   L   S   L   N   L   L   A   S   S   D   P   P   E
K02879  ,IGHD*01 (CH3)     [15]  CT GCG CAG GCA CCC GTC AAG CTT TCT CTG AAC CTG CTG GCC TCG TCT GAC CCT CCC GAG
X57331  ,IGHD*02                 -- --- --- --- --- --- --- --- --C --- --- --- --- --- --- --- --- --- --- ---
                                21  22  23  24  25  26  27  28  29  30  31  32  33  34  35  36  37  38  39  40
                                 A   A   S   W   L   L   C   E   V   S   G   F   S   P   P   N   I   L   L   M
K02879  ,IGHD*01 (CH3)          GCG GCC TCG TGG CTC CTG TGT GAG GTG TCT GGC TTC TCG CCC CCC AAC ATC CTC CTG ATG
X57331  ,IGHD*02                --- --- --- --- --- --- --- --- --- --- --- --- --- --- --- --- --- --- --- ---
                                41  42  43  44  45  46  47  48  49  50  51  52  53  54  55  56  57  58  59  60
                                 W   L   E   D   Q   R   E   V   N   T   S   G   F   A   P   A   R   P   P   P
K02879  ,IGHD*01 (CH3)          TGG CTG GAG GAC CAG CGT GAG GTG AAC ACT TCT GGG TTT GCC CCC GCA CGC CCC CCT CCA
X57331  ,IGHD*02                --- --- --- --- --- --- --- --- --- --- --- --- --- --- --- --- --- --- --- ---
                                61  62  63  64  65  66  67  68  69  70  71  72  73  74  75  76  77  78  79  80
                                 Q   P   R   S   T   T   F   W   A   W   S   V   L   R   V   P   A   P   P   S
K02879  ,IGHD*01 (CH3)          CAG CCC AGG AGC ACC ACG TTC TGG GCC TGG AGT GTG CTG CGT GTC CCA GCC CCG CCC AGC
X57331  ,IGHD*02                --- --- --- --- --- --- --- --- --- --- --- --- --- --- --- --- --- --- --- ---
                                81  82  83  84  85  86  87  88  89  90  91  92  93  94  95  96  97  98  99 100
                                 P   Q   P   A   T   Y   T   C   V   V   S   H   E   D   S   R   T   L   L   N
K02879  ,IGHD*01 (CH3)          CCT CAG CCA GCC ACC TAC ACG TGT GTG GTC AGC CAC GAG GAC TCC CGG ACT CTG CTC AAC
X57331  ,IGHD*02                --- --- --- --- --- --- --- --- --- --- --- --- --- --- --- --- --- --- --- ---
                               101 102 103 104 105 106 107 108
                                 A   S   R   S   L   E   V   S
K02879  ,IGHD*01 (CH3)          GCC AGC CGG AGC CTA GAA GTC AGC T
X57331  ,IGHD*02                --- --- --- --- --- --- --- --- -

                                 1   2   3   4   5   6   7   8   9
                                 Y   V   T   D   H   G   P   M   K   *
K02880  ,IGHD*01 (CH-S)    [15]  AT GTA ACA GAC CAT GGC CCC ATG AAA
X57331  ,IGHD*02                 -- --- --- --- --- --- --- --- ---

                                 1   2   3   4   5   6   7   8   9  10  11  12  13  14  15  16  17  18  19  20
                                 Y   L   A   M   T   P   L   I   P   Q   S   K   D   E   N   S   D   D   Y   T
K02881  ,IGHD*01 (M1)      [15] (T)AC CTG GCC ATG ACC CCC CTG ATC CCT CAG AGC AAG GAT GAG AAC AGC GAT GAC TAC ACG
X57331  ,IGHD*02                (-)-- --- --- --- --- --- --- --- --- --- --- --- --- --- --- --- --- --- --- ---
                                21  22  23  24  25  26  27  28  29  30  31  32  33  34  35  36  37  38  39  40
                                 T   F   D   D   V   G   S   L   W   T   T   L   S   T   F   V   A   L   F   I
K02881  ,IGHD*01 (M1)           ACC TTT GAT GAT GTG GGC AGC CTG TGG ACC ACC CTG TCC ACG TTT GTG GCC CTC TTC ATC
X57331  ,IGHD*02                --- --- --- --- --- --- --- --- --- --- --- --- --- --- --- --- --- --- --- ---
                                41  42  43  44  45  46  47  48  49  50  51  52  53
                                 L   T   L   L   Y   S   G   I   V   T   F   I   K
K02881  ,IGHD*01 (M1)           CTC ACC CTC CTC TAC AGC GGC ATT GTC ACT TTC ATC AAG
X57331  ,IGHD*02                --- --- --- --- --- --- --- --- --- --- --- --- ---

                                 1   2
                                 V   K   *
K02882  ,IGHD*01 (M2)      [15] GTG AAG
X57331  ,IGHD*02                --- ---
```

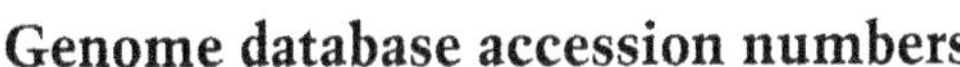

Genome database accession numbers

GDB:120084 LocusLink: 3495 OMIM: 147170

Nomenclature

IGHE: Immunoglobulin heavy constant epsilon.

Definition and functionality

IGHE is one of the nine functional genes of the human IGHC group which comprises 11 mapped genes.

Gene location

IGHE is in the IGH locus on chromosome 14 at 14q32.33.

Gene structure

The IGHE gene comprises four exons encoding the CH1, CH2, CH3 and CH4 domains, and two transmembrane exons (M1 and M2).

Nucleotide and amino acid sequences for human IGHE

The nucleotide between parentheses at the beginning of exons comes from a DONOR_SPLICE (n from ngt). The first nucleotide from an INT_DONOR_SPLICE is underlined (n from ngt).

The Cysteines involved in the CH1, CH2, CH3 and CH4 intrachain disulphide bridges are shown with their number and letter **C** in bold. The Cysteines involved in the H-L or H-H interchain disulphide bridges are shown with only the letter **C** in bold.

```
                                                   1   2   3   4   5   6   7   8   9  10  11  12  13  14  15  16  17  18  19  20
                                                   A   S   T   Q   S   P   S   V   F   P   L   T   R   C   C   K   N   I   P   S
J00222  ,IGHE*01,Cepsilon1 (CH1)[11]  (G)CC TCC ACA CAG AGC CCA TCC GTC TTC CCC TTG ACC CGC TGC TGC AAA AAC ATT CCC TCC

V00555  ,IGHE*02,Cepsilon-1        [4]  (-)-- --- --- --- --- --- --- --- --- --- --- --- --- --- --- --- --- --- --- ---

[9](1)  ,IGHE*03,epsilon-1              ( )-- --- --- --- --- --- --- --- --- --- --- --- --- --- --- --- --- --- --- ---
                                       21  22  23  24  25  26  27  28  29  30  31  32  33  34  35  36  37  38  39  40
                                        N   A   T   S   V   T   L   G   C   L   A   T   G   Y   F   P   E   P   V   M
J00222  ,IGHE*01,Cepsilon1 (CH1)       AAT GCC ACC TCC GTG ACT CTG GGC TGC CTG GCC ACG GGC TAC TTC CCG GAG CCG GTG ATG

V00555  ,IGHE*02,Cepsilon-1             --- --- --- --- --- --- --- --- --- --- --- --- --- --- --- --- --- --- --- ---

[9](1)  ,IGHE*03,epsilon-1              --- --- --- --- --- --- --- --- --- --- --- --- --- --- --- --- --- --- --- ---
                                       41  42  43  44  45  46  47  48  49  50  51  52  53  54  55  56  57  58  59  60
                                        V   T   C   D   T   G   S   L   N   G   T   T   M   T   L   P   A   T   T   L
J00222  ,IGHE*01,Cepsilon1 (CH1)       GTG ACC TGC GAC ACA GGC TCC CTC AAC GGG ACA ACT ATG ACC TTA CCA GCC ACC ACC CTC
                                                W
V00555  ,IGHE*02,Cepsilon-1             --- --- --G --- --- --- --- --- --- --- --- --- --- --- --- --- --- --- --- ---

[9](1)  ,IGHE*03,epsilon-1              --- --- --- --- --- --- --- --- --- --- --- --- --- --- --- --- --- --- --- ---
                                       61  62  63  64  65  66  67  68  69  70  71  72  73  74  75  76  77  78  79  80
                                        T   L   S   G   H   Y   A   T   I   S   L   L   T   V   S   G   A   W   A   K
J00222  ,IGHE*01,Cepsilon1 (CH1)       ACG CTC TCT GGT CAC TAT GCC ACC ATC AGC TTG CTG ACC GTC TCG GGT GCG TGG GCC AAG

V00555  ,IGHE*02,Cepsilon-1             --- --- --- --- --- --- --- --- --- --- --- --- --- --- --- --- --- --- --- ---

[9](1)  ,IGHE*03,epsilon-1              --- --- --- --- --- --- --- --- --- --- --- --- --- --- --- --- --- --- --- ---
                                       81  82  83  84  85  86  87  88  89  90  91  92  93  94  95  96  97  98  99 100
                                        Q   M   F   T   C   R   V   A   H   T   P   S   S   T   D   W   V   D   N   K
J00222  ,IGHE*01,Cepsilon1 (CH1)       CAG ATG TTC ACC TGC CGT GTG GCA CAC ACT CCA TCG TCC ACA GAC TGG GTC GAC AAC AAA

V00555  ,IGHE*02,Cepsilon-1             --- --- --- --- --- --- --- --- --- --- --- --- --- --- --- --- --- --- --- ---

[9](1)  ,IGHE*03,epsilon-1              --- --- --- --- --- --- --- --- --- --- --- --- --- --- --- --- --- --- --- ---
                                      101 102 103
                                        T   F   S
J00222  ,IGHE*01,Cepsilon1 (CH1)       ACC TTC AGC G

V00555  ,IGHE*02,Cepsilon-1             --- --- --- -

[9](1)  ,IGHE*03,epsilon-1              --- --  --- -
```

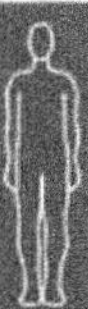

```
                                          1   2   3   4   5   6   7   8   9  10  11  12  13  14  15  16  17  18  19  20
                                          V   C   S   R   D   F   T   P   P   T   V   K   I   L   Q   S   S   C   D   G
J00222  ,IGHE*01,Cepsilon1 (CH2)         TC TGC TCC AGG GAC TTC ACC CCG CCC ACC GTG AAG ATC TTA CAG TCG TCC TGC GAC GGC
V00555  ,IGHE*02,Cepsilon-1              -- --- --- --- --- --- --- --- --- --- --- --- --- --- --- --- --- --- --- ---
[9](1)  ,IGHE*03,epsilon-1               -- --- --- --- --- --- --- --- --- --- --- --- --- --- --- --- --- --- --- ---
                                         21  22  23  24  25  26  27  28  29  30  31  32  33  34  35  36  37  38  39  40
                                          G   G   H   F   P   P   T   I   Q   L   L   C   L   V   S   G   Y   T   P   G
J00222  ,IGHE*01,Cepsilon1 (CH2)        GGC GGG CAC TTC CCC CCG ACC ATC CAG CTC CTG TGC CTC GTC TCT GGG TAC ACC CCA GGG
V00555  ,IGHE*02,Cepsilon-1             --- --- --- --- --- --- --- --- --- --- --- --- --- --- --- --- --- --- --- ---
[9](1)  ,IGHE*03,epsilon-1              --- --- --- --- --- --- --- --- --- --- --- --- --- --- --- --- --- --- --- ---
                                         41  42  43  44  45  46  47  48  49  50  51  52  53  54  55  56  57  58  59  60
                                          T   I   N   I   T   W   L   E   D   G   Q   V   M   D   V   D   L   S   T   A
J00222  ,IGHE*01,Cepsilon1 (CH2)        ACT ATC AAC ATC ACC TGG CTG GAG GAC GGG CAG GTC ATG GAC GTG GAC TTG TCC ACC GCC
V00555  ,IGHE*02,Cepsilon-1             --- --- --- --- --- --- --- --- --- --- --- --- --- --- --- --- --- --- --- ---
[9](1)  ,IGHE*03,epsilon-1              --- --- --- --- --- --- --- --- --- --- --- --- --- --- --- --- --- --- --- ---
                                         61  62  63  64  65  66  67  68  69  70  71  72  73  74  75  76  77  78  79  80
                                          S   T   T   Q   E   G   E   L   A   S   T   Q   S   E   L   T   L   S   Q   K
J00222  ,IGHE*01,Cepsilon1 (CH2)        TCT ACC ACG CAG GAG GGT GAG CTG GCC TCC ACA CAA AGC GAG CTC ACC CTC AGC CAG AAG
V00555  ,IGHE*02,Cepsilon-1             --- --- --- --- --- --- --- --- --- --- --- --- --- --- --- --- --- --- --- ---
[9](1)  ,IGHE*03,epsilon-1              --- --- --- --- --- --- --- --- --- --- --- --- --- --- --- --- --- --- --- ---
                                         81  82  83  84  85  86  87  88  89  90  91  92  93  94  95  96  97  98  99 100
                                          H   W   L   S   D   R   T   Y   T   C   Q   V   T   Y   Q   G   H   T   F   E
J00222  ,IGHE*01,Cepsilon1 (CH2)        CAC TGG CTG TCA GAC CGC ACC TAC ACC TGC CAG GTC ACC TAT CAA GGT CAC ACC TTT GAG
V00555  ,IGHE*02,Cepsilon-1             --- --- --- --- --- --- --- --- --- --- --- --- --- --- --- --- --- --- --- ---
[9](1)  ,IGHE*03,epsilon-1              --- --- --- --- --- --- --- --- --- --- --- --- --- --- --- --- --- --- --- ---
                                        101 102 103 104 105 106 107
                                          D   S   T   K   K   C   A
J00222  ,IGHE*01,Cepsilon1 (CH2)        GAC AGC ACC AAG AAG TGT GCA G
V00555  ,IGHE*02,Cepsilon-1             --- --- --- --- --- --- --- -
[9](1)  ,IGHE*03,epsilon-1              --- --- --- --- --- --- --- -

                                          1   2   3   4   5   6   7   8   9  10  11  12  13  14  15  16  17  18  19  20
                                          D   S   N   P   R   G   V   S   A   Y   L   S   R   P   S   P   F   D   L   F
J00222  ,IGHE*01,Cepsilon1 (CH3)         AT TCC AAC CCG AGA GGG GTG AGC GCC TAC CTA AGC CGG CCC AGC CCG TTC GAC CTG TTC
V00555  ,IGHE*02,Cepsilon-1              -- --- --- --- --- --- --- --- --- --- --- --- --- --- --- --- --- --- --- ---
[9](1)  ,IGHE*03,epsilon-1               -- --- --- --- --- --- --- --- --- --- --- --- --- --- --- --- --- --- --- ---
                                         21  22  23  24  25  26  27  28  29  30  31  32  33  34  35  36  37  38  39  40
                                          I   R   K   S   P   T   I   T   C   L   V   V   D   L   A   P   S   K   G   T
J00222  ,IGHE*01,Cepsilon1 (CH3)        ATC CGC AAG TCG CCC ACG ATC ACC TGT CTG GTG GTG GAC CTG GCA CCC AGC AAG GGG ACC
V00555  ,IGHE*02,Cepsilon-1             --- --- --- --- --- --- --- --- --- --- --- --- --- --- --- --- --- --- --- ---
[9](1)  ,IGHE*03,epsilon-1              --- --- --- --- --- --- --- --- --- --- --- --- --- --- --- --- --- --- --- ---
                                         41  42  43  44  45  46  47  48  49  50  51  52  53  54  55  56  57  58  59  60
                                          V   N   L   T   W   S   R   A   S   G   K   P   V   N   H   S   T   R   K   E
J00222  ,IGHE*01,Cepsilon1 (CH3)        GTG AAC CTG ACC TGG TCC CGG GCC AGT GGG AAG CCT GTG AAC CAC TCC ACC AGA AAG GAG
V00555  ,IGHE*02,Cepsilon-1             --- --- --- --- --- --- --- --- --- --- --- --- --- --- --- --- --- --- --- ---
[9](1)  ,IGHE*03,epsilon-1              --- --- --- --- --- --- --- --- --- --- --- --- --- --- --- --- --- --- --- ---
                                         61  62  63  64  65  66  67  68  69  70  71  72  73  74  75  76  77  78  79  80
                                          E   K   Q   R   N   G   T   L   T   V   T   S   T   L   P   V   G   T   R   D
J00222  ,IGHE*01,Cepsilon1 (CH3)        GAG AAG CAG CGC AAT GGC ACG TTA ACC GTC ACG TCC ACC CTG CCG GTG GGC ACC CGA GAC
V00555  ,IGHE*02,Cepsilon-1             --- --- --- --- --- --- --- --- --- --- --- --- --- --- --- --- --- --- --- ---
[9](1)  ,IGHE*03,epsilon-1              --- --- --- --- --- --- --- --- --- --- --- --- --- --- --- --- --- --- --- ---
                                         81  82  83  84  85  86  87  88  89  90  91  92  93  94  95  96  97  98  99 100
                                          W   I   E   G   E   T   Y   Q   C   R   V   T   H   P   H   L   P   R   A   L
J00222  ,IGHE*01,Cepsilon1 (CH3)        TGG ATC GAG GGG GAG ACC TAC CAG TGC AGG GTG ACC CAC CCC CAC CTG CCC AGG GCC CTC
V00555  ,IGHE*02,Cepsilon-1             --- --- --- --- --- --- --- --- --- --- --- --- --- --- --- --- --- --- --- ---
[9](1)  ,IGHE*03,epsilon-1              --- --- --- --- --- --- --- --- --- --- --- --- --- --- --- --- --- --- --- ---
                                        101 102 103 104 105 106 107 108
                                          M   R   S   T   T   K   T   S
J00222  ,IGHE*01,Cepsilon1 (CH3)        ATG CGG TCC ACG ACC AAG ACC AGC G
V00555  ,IGHE*02,Cepsilon-1             --- --- --- --- --- --- --- --- -
[9](1)  ,IGHE*03,epsilon-1              --- --- --- --- --- --- --- --- -
```

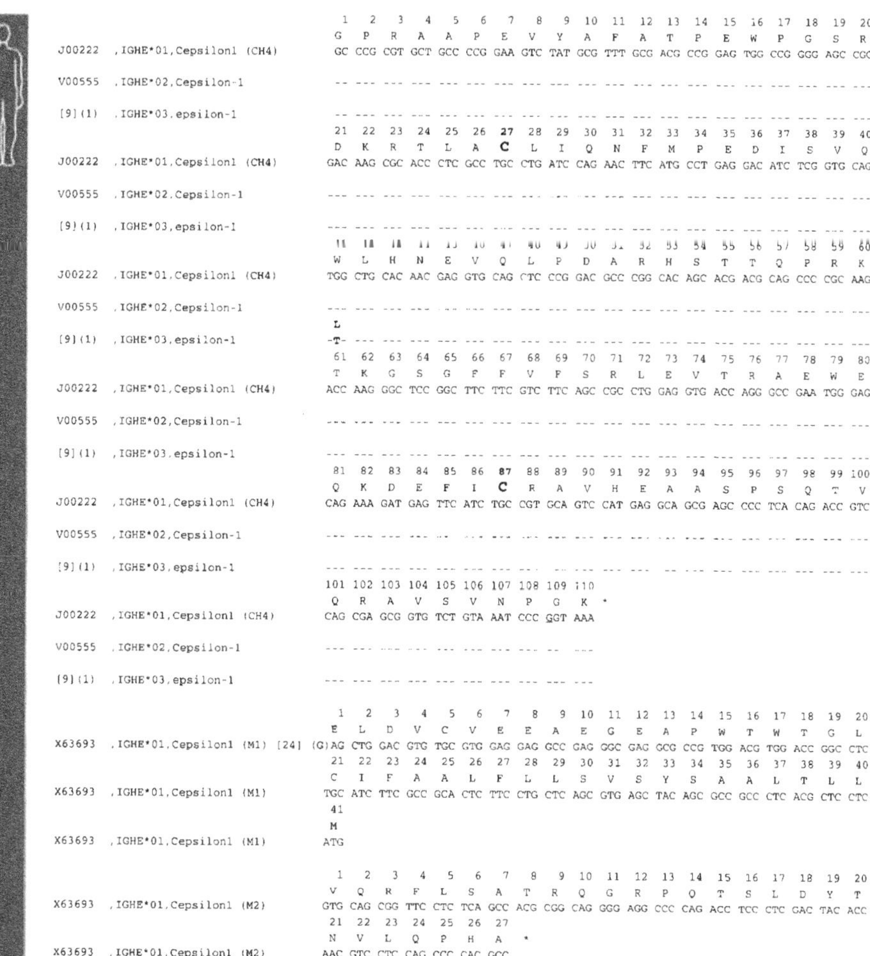

```
                                              1   2   3   4   5   6   7   8   9  10  11  12  13  14  15  16  17  18  19  20
                                              G   P   R   A   A   P   E   V   Y   A   F   A   T   P   E   W   P   G   S   R
J00222  ,IGHE*01,Cepsilon1 (CH4)             GC CCG CGT GCT GCC CCG GAA GTC TAT GCG TTT GCG ACG CCG GAG TGG CCG GGG AGC CGG

V00555  ,IGHE*02,Cepsilon-1                  -- --- --- --- --- --- --- --- --- --- --- --- --- --- --- --- --- --- --- ---

[9](1)  ,IGHE*03,epsilon-1                   -- --- --- --- --- --- --- --- --- --- --- --- --- --- --- --- --- --- --- ---
                                             21  22  23  24  25  26  27  28  29  30  31  32  33  34  35  36  37  38  39  40
                                              D   K   R   T   L   A   C   L   I   Q   N   F   M   P   E   D   I   S   V   Q
J00222  ,IGHE*01,Cepsilon1 (CH4)            GAC AAG CGC ACC CTC GCC TGC CTG ATC CAG AAC TTC ATG CCT GAG GAC ATC TCG GTG CAG

V00555  ,IGHE*02,Cepsilon-1                 --- --- --- --- --- --- --- --- --- --- --- --- --- --- --- --- --- --- --- ---

[9](1)  ,IGHE*03,epsilon-1                  --- --- --- --- --- --- --- --- --- --- --- --- --- --- --- --- --- --- --- ---
                                             41  42  43  44  45  46  47  48  49  50  51  52  53  54  55  56  57  58  59  60
                                              W   L   H   N   E   V   Q   L   P   D   A   R   H   S   T   T   Q   P   R   K
J00222  ,IGHE*01,Cepsilon1 (CH4)            TGG CTG CAC AAC GAG GTG CAG CTC CCG GAC GCC CGG CAC AGC ACG ACG CAG CCC CGC AAG

V00555  ,IGHE*02,Cepsilon-1                 --- --- --- --- --- --- --- --- --- --- --- --- --- --- --- --- --- --- --- ---
                                              L
[9](1)  ,IGHE*03,epsilon-1                  -T- --- --- --- --- --- --- --- --- --- --- --- --- --- --- --- --- --- --- ---
                                             61  62  63  64  65  66  67  68  69  70  71  72  73  74  75  76  77  78  79  80
                                              T   K   G   S   G   F   F   V   F   S   R   L   E   V   T   R   A   E   W   E
J00222  ,IGHE*01,Cepsilon1 (CH4)            ACC AAG GGC TCC GGC TTC TTC GTC TTC AGC CGC CTG GAG GTG ACC AGG GCC GAA TGG GAG

V00555  ,IGHE*02,Cepsilon-1                 --- --- --- --- --- --- --- --- --- --- --- --- --- --- --- --- --- --- --- ---

[9](1)  ,IGHE*03,epsilon-1                  --- --- --- --- --- --- --- --- --- --- --- --- --- --- --- --- --- --- --- ---
                                             81  82  83  84  85  86  87  88  89  90  91  92  93  94  95  96  97  98  99 100
                                              Q   K   D   E   F   I   C   R   A   V   H   E   A   A   S   P   S   Q   T   V
J00222  ,IGHE*01,Cepsilon1 (CH4)            CAG AAA GAT GAG TTC ATC TGC CGT GCA GTC CAT GAG GCA GCG AGC CCC TCA CAG ACC GTC

V00555  ,IGHE*02,Cepsilon-1                 --- --- --- --- --- --- --- --- --- --- --- --- --- --- --- --- --- --- --- ---

[9](1)  ,IGHE*03,epsilon-1                  --- --- --- --- --- --- --- --- --- --- --- --- --- --- --- --- --- --- --- ---
                                            101 102 103 104 105 106 107 108 109 110
                                              Q   R   A   V   S   V   N   P   G   K  *
J00222  ,IGHE*01,Cepsilon1 (CH4)            CAG CGA GCG GTG TCT GTA AAT CCC GGT AAA

V00555  ,IGHE*02,Cepsilon-1                 --- --- --- --- --- --- --- --- --  ---

[9](1)  ,IGHE*03,epsilon-1                  --- --- --- --- --- --- --- --- --- ---

                                              1   2   3   4   5   6   7   8   9  10  11  12  13  14  15  16  17  18  19  20
                                              E   L   D   V   C   V   E   E   A   E   G   E   A   P   W   T   W   T   G   L
X63693  ,IGHE*01,Cepsilon1 (M1) [24]     (G)AG CTG GAC GTG TGC GTG GAG GAG GCC GAG GGC GAG GCG CCG TGG ACG TGG ACC GGC CTC
                                             21  22  23  24  25  26  27  28  29  30  31  32  33  34  35  36  37  38  39  40
                                              C   I   F   A   A   L   F   L   L   S   V   S   Y   S   A   A   L   T   L   L
X63693  ,IGHE*01,Cepsilon1 (M1)             TGC ATC TTC GCC GCA CTC TTC CTG CTC AGC GTG AGC TAC AGC GCC GCC CTC ACG CTC CTC
                                             41
                                              M
X63693  ,IGHE*01,Cepsilon1 (M1)             ATG

                                              1   2   3   4   5   6   7   8   9  10  11  12  13  14  15  16  17  18  19  20
                                              V   Q   R   F   L   S   A   T   R   Q   G   R   P   Q   T   S   L   D   Y   T
X63693  ,IGHE*01,Cepsilon1 (M2)             GTG CAG CGG TTC CTC TCA GCC ACG CGG CAG GGG AGG CCC CAG ACC TCC CTC GAC TAC ACC
                                             21  22  23  24  25  26  27
                                              N   V   L   Q   P   H   A  *
X63693  ,IGHE*01,Cepsilon1 (M2)             AAC GTC CTC CAG CCC CAC GCC
```

Note:

(1) Not submitted but comment in the J00222 flat file.

Genome database accession numbers

GDB:119335 LocusLink: 3497 OMIM: 147180

Nomenclature

IGHG1: Immunoglobulin heavy constant gamma 1.

Definition and functionality

IGHG1 is one of the nine functional genes of the human IGHC group which comprises 11 mapped genes.

Gene location

IGHG1 is in the IGH locus on chromosome 14 at 14q32.33.

Gene structure

The IGHG1 gene comprises three main exons encoding the CH1, CH2 and CH3 domains, one separate hinge (H) exon and two transmembrane exons (M1 and M2).

Nucleotide and amino acid sequences for human IGHG1

The nucleotide between parentheses at the beginning of exons comes from a DONOR_SPLICE (n from ngt). The first nucleotide from an INT_DONOR_SPLICE is underlined (n from ngt).

The Cysteines involved in the CH1, CH2 and CH3 intrachain disulphide bridges are shown with their number and letter **C** in bold. The Cysteines involved in the H-L or H-H interchain disulphide bridges are shown with only the letter **C** in bold.

```
                                                1   2   3   4   5   6   7   8   9  10  11  12  13  14  15  16  17  18  19  20
                                                A   S   T   K   G   P   S   V   F   P   L   A   P   S   S   K   S   T   S   G
J00228 ,IGHG1*01,gamma-1 (CH1)   [5]  (G)CC TCC ACC AAG GGC CCA TCG GTC TTC CCC CTG GCA CCC TCC TCC AAG AGC ACC TCT GGG
Z17370 ,IGHG1*02,gamma1          [25] (·)-- --- --- --- --- --- --- --- --- --- --- --- --- --- --- --- --- --- --- ---
                                               21  22  23  24  25  26  27  28  29  30  31  32  33  34  35  36  37  38  39  40
                                                G   T   A   A   L   G   C   L   V   K   D   Y   F   P   E   P   V   T   V   S
J00228 ,IGHG1*01,gamma-1 (CH1)                GGC ACA GCG GCC CTG GGC TGC CTG GTC AAG GAC TAC TTC CCC GAA CCG GTG ACG GTG TCG
Z17370 ,IGHG1*02,gamma1                       --- --- --- --- --- --- --- --- --- --- --- --- --- --- --- --- --- --- --- ---
                                               41  42  43  44  45  46  47  48  49  50  51  52  53  54  55  56  57  58  59  60
                                                W   N   S   G   A   L   T   S   G   V   H   T   F   P   A   V   L   Q   S   S
J00228 ,IGHG1*01,gamma-1 (CH1)                TGG AAC TCA GGC GCC CTG ACC AGC GGC GTG CAC ACC TTC CCG GCT GTC CTA CAG TCC TCA
Z17370 ,IGHG1*02,gamma1                       --- --- --- --- --- --- --- --- --- --- --- --- --- --- --- --- --- --- --- ---
                                               61  62  63  64  65  66  67  68  69  70  71  72  73  74  75  76  77  78  79  80
                                                G   L   Y   S   L   S   S   V   V   T   V   P   S   S   S   L   G   T   Q   T
J00228 ,IGHG1*01,gamma-1 (CH1)                GGA CTC TAC TCC CTC AGC AGC GTG GTG ACC GTG CCC TCC AGC AGC TTG GGC ACC CAG ACC
Z17370 ,IGHG1*02,gamma1                       --- --- --- --- --- --- --- --- --- --- --- --- --- --- --- --- --- --- --- ---
                                               81  82  83  84  85  86  87  88  89  90  91  92  93  94  95  96  97  98
                                                Y   I   C   N   V   N   H   K   P   S   N   T   K   V   D   K   K   V
J00228 ,IGHG1*01,gamma-1 (CH1)                TAC ATC TGC AAC GTG AAT CAC AAG CCC AGC AAC ACC AAG GTG GAC AAG AAA GTT G
Z17370 ,IGHG1*02,gamma1                       --- --- --- --- --- --- --- --- --- --- --- --- --- --- --- --- --- --- -

                                                1   2   3   4   5   6   7   8   9  10  11  12  13  14  15
                                                E   P   K   S   C   D   K   T   H   T   C   P   P   C   P
J00228 ,IGHG1*01,gamma-1 (H)                  AG CCC AAA TCT TGT GAC AAA ACT CAC ACA TGC CCA CCG TGC CCA G
Z17370 ,IGHG1*02,gamma1                       -- --- --- --- --- --- --- --- --- --- --- --- --- --- --- -

                                                1   2   3   4   5   6   7   8   9  10  11  12  13  14  15  16  17  18  19  20
                                                A   P   E   L   L   G   G   P   S   V   F   L   F   P   P   K   P   K   D   T
J00228 ,IGHG1*01,gamma-1 (CH2)                CA CCT GAA CTC CTG GGG GGA CCG TCA GTC TTC CTC TTC CCC CCA AAA CCC AAG GAC ACC
Z17370 ,IGHG1*02,gamma1                       -- --- --- --- --- --- --- --- --- --- --- --- --- --- --- --- --- --- --- ---
                                               21  22  23  24  25  26  27  28  29  30  31  32  33  34  35  36  37  38  39  40
                                                L   M   I   S   R   T   P   E   V   T   C   V   V   V   D   V   S   H   E   D
J00228 ,IGHG1*01,gamma-1 (CH2)                CTC ATG ATC TCC CGG ACC CCT GAG GTC ACA TGC GTG GTG GTG GAC GTG AGC CAC GAA GAC
Z17370 ,IGHG1*02,gamma1                       --- --- --- --- --- --- --- --- --- --- --- --- --- --- --- --- --- --- --- ---
```

```
                                                41  42  43  44  45  46  47  48  49  50  51  52  53  54  55  56  57  58  59  60
                                                P   E   V   K   F   N   W   Y   V   D   G   V   E   V   H   N   A   K   T   K
J00228  ,IGHG1*01,gamma-1 (CH2)                CCT GAG GTC AAG TTC AAC TGG TAC GTG GAC GGC GTG GAG GTG CAT AAT GCC AAG ACA AAG
Z17370  ,IGHG1*02,gamma1                       --- --- --- --- --- --- --- --- --- --- --- --- --- --- --- --- --- --- --- ---
                                                61  62  63  64  65  66  67  68  69  70  71  72  73  74  75  76  77  78  79  80
                                                P   R   E   E   Q   Y   N   S   T   Y   R   V   V   S   V   L   T   V   L   H
J00228  ,IGHG1*01,gamma-1 (CH2)                CCG CGG GAG GAG CAG TAC AAC AGC ACG TAC CGG GTG GTC AGC GTC CTC ACC GTC CTG CAC
Z17370  ,IGHG1*02,gamma1                       --- --- --- --- --- --- --- --- --- --- --T --- --- --- --- --- --- --- --- ---
                                                81  82  83  84  85  86  87  88  89  90  91  92  93  94  95  96  97  98  99 100
                                                Q   D   W   L   N   G   K   E   Y   K   C   K   V   S   N   K   A   L   P   A
J00228  ,IGHG1*01,gamma-1 (CH2)                CAG GAC TGG CTG AAT GGC AAG GAG TAC AAG TGC AAG GTC TCC AAC AAA GCC CTC CCA GCC
Z17370  ,IGHG1*02,gamma1                       --- --- --- --- --- --- --- --- --- --- --- --- --- --- --- --- --- --- --- ---
                                               101 102 103 104 105 106 107 108 109 110
                                                P   I   E   K   T   I   S   K   A   K
J00228  ,IGHG1*01,gamma-1 (CH2)                CCC ATC GAG AAA ACC ATC TCC AAA GCC AAA G
Z17370  ,IGHG1*02,gamma1                       --- --- --- --- --- --- --- --- --- --- -

                                                 1   2   3   4   5   6   7   8   9  10  11  12  13  14  15  16  17  18  19  20
                                                G   Q   P   R   E   P   Q   V   Y   T   L   P   P   S   R   D   E   L   T   K
J00228  ,IGHG1*01,gamma-1 (CH3)                 GG CAG CCC CGA GAA CCA CAG GTG TAC ACC CTG CCC CCA TCC CGG GAT GAG CTG ACC AAG
Z17370  ,IGHG1*02,gamma1                        -- --- --- --- --- --- --- --- --- --- --- --- --- --- --- --- --- --- --- ---
                                                21  22  23  24  25  26  27  28  29  30  31  32  33  34  35  36  37  38  39  40
                                                N   Q   V   S   L   T   C   L   V   K   G   F   Y   P   S   D   I   A   V   E
J00228  ,IGHG1*01,gamma-1 (CH3)                AAC CAG GTC AGC CTG ACC TGC CTG GTC AAA GGC TTC TAT CCC AGC GAC ATC GCC GTG GAG
Z17370  ,IGHG1*02,gamma1                       --- --- --- --- --- --- --- --- --- --- --- --- --- --- --- --- --- --- --- ---
                                                41  42  43  44  45  46  47  48  49  50  51  52  53  54  55  56  57  58  59  60
                                                W   E   S   N   G   Q   P   E   N   N   Y   K   T   T   P   P   V   L   D   S
J00228  ,IGHG1*01,gamma-1 (CH3)                TGG GAG AGC AAT GGG CAG CCG GAG AAC AAC TAC AAG ACC ACG CCT CCC GTG CTG GAC TCC
Z17370  ,IGHG1*02,gamma1                       --- --- --- --- --- --- --- --- --- --- --- --- --- --- --- --- --- --- --- ---
                                                61  62  63  64  65  66  67  68  69  70  71  72  73  74  75  76  77  78  79  80
                                                D   G   S   F   F   L   Y   S   K   L   T   V   D   K   S   R   W   Q   Q   G
J00228  ,IGHG1*01,gamma-1 (CH3)                GAC GGC TCC TTC TTC CTC TAC AGC AAG CTC ACC GTG GAC AAG AGC AGG TGG CAG CAG GGG
Z17370  ,IGHG1*02,gamma1                       --- --- --- --- --- --- --- --- --- --- --- --- --- --- --- --- --- --- --- ---
                                                81  82  83  84  85  86  87  88  89  90  91  92  93  94  95  96  97  98  99 100
                                                N   V   F   S   C   S   V   M   H   E   A   L   H   N   H   Y   T   Q   K   S
J00228  ,IGHG1*01,gamma-1 (CH3)                AAC GTC TTC TCA TGC TCC GTG ATG CAT GAG GCT CTG CAC AAC CAC TAC ACG CAG AAG AGC
Z17370  ,IGHG1*02,gamma1                       --- --- --- --- --- --- --- --- --- --- --- --- --- --- --- --- --- --- --- ---
                                               101 102 103 104 105 106 107
                                                L   S   L   S   P   G   K   *
J00228  ,IGHG1*01,gamma-1 (CH3)                CTC TCC CTG TCT CCG GGT AAA
Z17370  ,IGHG1*02,gamma1                       --- --- --- --- --- --- ---

                                                 1   2   3   4   5   6   7   8   9  10  11  12  13  14  15  16  17  18  19  20
                                                E   L   Q   L   E   E   S   C   A   E   A   Q   D   G   E   L   D   G   L   W
X52847  ,IGHG1*01(cDNA)(m) (M1) [23]        (G)AG CTG CAA CTG GAG GAG AGC TGT GCG GAG GCG CAG GAC GGG GAG CTG GAC GGG CTG TGG
                                                21  22  23  24  25  26  27  28  29  30  31  32  33  34  35  36  37  38  39  40
                                                T   T   I   T   I   F   I   T   L   F   L   L   S   V   C   Y   S   A   T   V
X52847  ,IGHG1*01(cDNA)(m) (M1)                ACG ACC ATC ACC ATC TTC ATC ACA CTC TTC CTG TTA AGC GTG TGC TAC AGT GCC ACC GTC
                                                41  42  43  44
                                                T   F   F   K
X52847  ,IGHG1*01(cDNA)(m) (M1)                ACC TTC TTC AAG

                                                 1   2   3   4   5   6   7   8   9  10  11  12  13  14  15  16  17  18  19  20
                                                V   K   W   I   F   S   S   V   V   D   L   K   Q   T   I   I   P   D   Y   R
X52847  ,IGHG1*01(cDNA)(m) (M2)                GTG AAG TGG ATC TTC TCC TCG GTG GTG GAC CTG AAG CAG ACC ATC ATC CCC GAC TAC AGG
                                                21  22  23  24  25  26  27
                                                N   M   I   G   Q   G   A   *
X52847  ,IGHG1*01(cDNA)(m) (M2)                AAC ATG ATC GGA CAG GGG GCC
```

(m) Transcript of a membrane chain.

Genome database accession numbers

GDB:120085 LocusLink: 3500 OMIM: 147100

Nomenclature

IGHG2: Immunoglobulin heavy constant gamma 2.

Definition and functionality

IGHG2 is one of the nine functional genes of the human IGHC group which comprises 11 mapped genes.

Gene location

IGHG2 is in the IGH locus on chromosome 14 at 14q32.33.

Gene structure

The IGHG2 gene comprises three main exons encoding the CH1, CH2 and CH3 domains, one separate hinge (H) exon and two transmembrane exons (M1 and M2).

Nucleotide and amino acid sequences for human IGHG2

The nucleotide between parentheses at the beginning of exons comes from a DONOR_SPLICE (n from ngt). The first nucleotide from an INT_DONOR_SPLICE is underlined (n from ngt).

The Cysteines involved in the CH1, CH2 and CH3 intrachain disulphide bridges are shown with their number and letter **C** in bold. The Cysteines involved in the H-L or H-H interchain disulphide bridges are shown with only the letter **C** in bold.

```
                                                1   2   3   4   5   6   7   8   9   10  11  12  13  14  15  16  17  18  19  20
                                                A   S   T   K   G   P   S   V   F   P   L   A   P   C   S   R   S   T   S   E
J00230  ,IGHG2*01,Cgamma2 (CH1) [6]          (G)CC TCC ACC AAG GGC CCA TCG GTC TTC CCC CTG GCG CCC TGC TCC AGG AGC ACC TCC GAG
                                                21  22  23  24  25  26  27  28  29  30  31  32  33  34  35  36  37  38  39  40
                                                S   T   A   A   L   G   C   L   V   K   D   Y   F   P   E   P   V   T   V   S
J00230  ,IGHG2*01,Cgamma2 (CH1)                AGC ACA GCC GCC CTG GGC TGC CTG GTC AAG GAC TAC TTC CCC GAA CCG GTG ACG GTG TCG
                                                41  42  43  44  45  46  47  48  49  50  51  52  53  54  55  56  57  58  59  60
                                                W   N   S   G   A   L   T   S   G   V   H   T   F   P   A   V   L   Q   S   S
J00230  ,IGHG2*01,Cgamma2 (CH1)                TGG AAC TCA GGC GCT CTG ACC AGC GGC GTG CAC ACC TTC CCA GCT GTC CTA CAG TCC TCA
                                                61  62  63  64  65  66  67  68  69  70  71  72  73  74  75  76  77  78  79  80
                                                G   L   Y   S   L   S   S   V   V   T   V   P   S   S   N   F   G   T   Q   T
J00230  ,IGHG2*01,Cgamma2 (CH1)                GGA CTC TAC TCC CTC AGC AGC GTG GTG ACC GTG CCC TCC AGC AAC TTC GGC ACC CAG ACC
                                                81  82  83  84  85  86  87  88  89  90  91  92  93  94  95  96  97  98
                                                Y   T   C   N   V   D   H   K   P   S   N   T   K   V   D   K   T   V
J00230  ,IGHG2*01,Cgamma2 (CH1)                TAC ACC TGC AAC GTA GAT CAC AAG CCC AGC AAC ACC AAG GTG GAC AAG ACA GTT G

                                                1   2   3   4   5   6   7   8   9   10  11  12
                                                E   R   K   C   C   V   E   C   P   P   C   P
J00230  ,IGHG2*01,Cgamma2 (H)                  AG CGC AAA TGT TGT GTC GAG TGC CCA CCG TGC CCA G
Z49802  ,IGHG2*01                   [27]       -- --- --- --- --- --- --- --- --- --- --- --- -
Z49801  ,IGHG2*02                   [27]       -- --- --- --- --- --- --- --- --- --- --- --- -

                                                1   2   3   4   5   6   7   8   9   10  11  12  13  14  15  16  17  18  19  20
                                                A   P       P   V   A   G   P   S   V   F   L   F   P   P   K   P   K   D   T
J00230  ,IGHG2*01,Cgamma2 (CH2) (1)            CA CCA ... CCT GTG GCA GGA CCG TCA GTC TTC CTC TTC CCC CCA AAA CCC AAG GAC ACC
Z49802  ,IGHG2*01                              -- --- ... --- --- --- --- --- --- --- --- --- --- --- --- --- --- --- --- ---
Z49801  ,IGHG2*02                              -- --- ... --- --- --- --- --- --- --- --- --- --- --- --- --- --- --- --- ---
                                                21  22  23  24  25  26  27  28  29  30  31  32  33  34  35  36  37  38  39  40
                                                L   M   I   S   R   T   P   E   V   T   C   V   V   V   D   V   S   H   E   D
J00230  ,IGHG2*01,Cgamma2 (CH2)                CTC ATG ATC TCC CGG ACC CCT GAG GTC ACG TGC GTG GTG GTG GAC GTG AGC CAC GAA GAC
Z49802  ,IGHG2*01                              --- --- --- --- --- --- --- --- --- --- --- --- --- --- --- --- --- --- --- ---
Z49801  ,IGHG2*02                              --- --- --- --- --- --- --- --- --- --- --- --- --- --- --- --- --- --- --- ---
                                                41  42  43  44  45  46  47  48  49  50  51  52  53  54  55  56  57  58  59  60
                                                P   E   V   Q   F   N   W   Y   V   D   G   V   E   V   H   N   A   K   T   K
J00230  ,IGHG2*01,Cgamma2 (CH2)                CCC GAG GTC CAG TTC AAC TGG TAC GTG GAC GGC GTG GAG GTG CAT AAT GCC AAG ACA AAG
Z49802  ,IGHG2*01                              --- --- --- --- --- --- --- --- --- --- --- --- --- --- --- --- --- --- --- ---
                                                                                        M
Z49801  ,IGHG2*02                              --- --- --- --- --- --- --- --- --- --- A-- --- --- --- --- --- --- --- --- ---
```

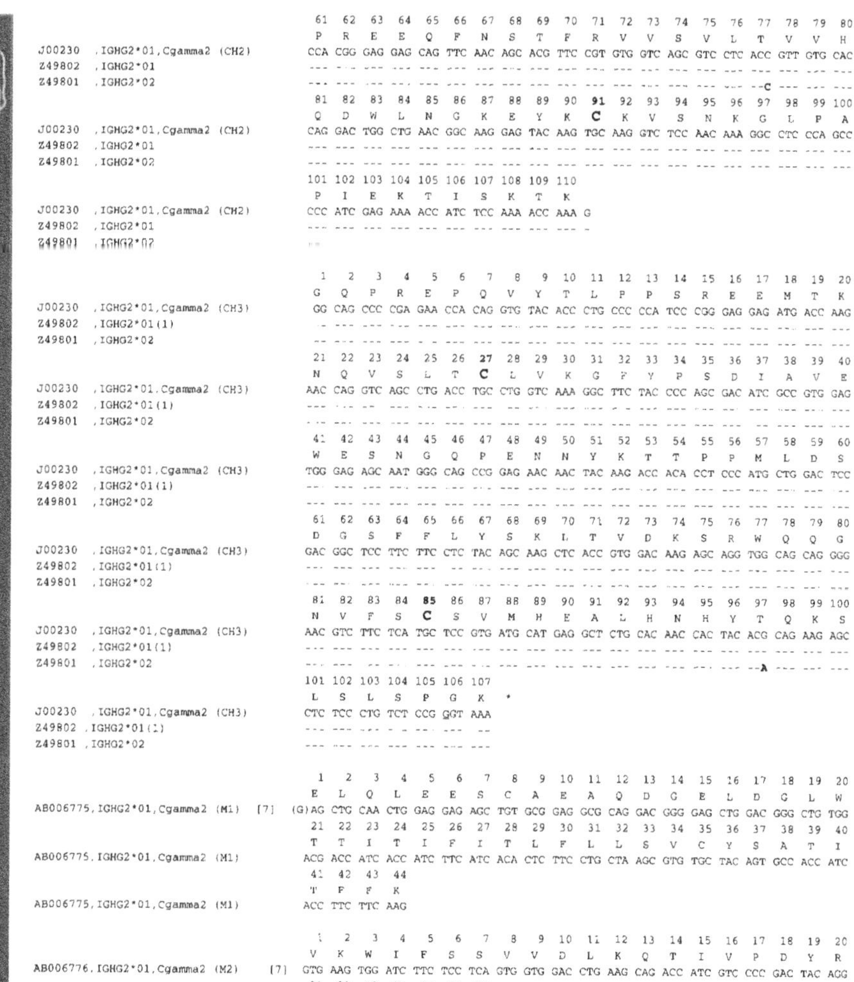

```
                                        61  62  63  64  65  66  67  68  69  70  71  72  73  74  75  76  77  78  79  80
                                         P   R   E   E   Q   F   N   S   T   F   R   V   V   S   V   L   T   V   V   H
J00230  ,IGHG2*01,Cgamma2 (CH2)         CCA CGG GAG GAG CAG TTC AAC AGC ACG TTC CGT GTG GTC AGC GTC CTC ACC GTT GTG CAC
Z49802  ,IGHG2*01                       --- --- --- --- --- --- --- --- --- --- --- --- --- --- --- --- --- --- --- ---
Z49801  ,IGHG2*02                       --- --- --- --- --- --- --- --- --- --- --- --- --- --- --- --- --C --- --- ---
                                        81  82  83  84  85  86  87  88  89  90  91  92  93  94  95  96  97  98  99 100
                                         Q   D   W   L   N   G   K   E   Y   K   C   K   V   S   N   K   G   L   P   A
J00230  ,IGHG2*01,Cgamma2 (CH2)         CAG GAC TGG CTG AAC GGC AAG GAG TAC AAG TGC AAG GTC TCC AAC AAA GGC CTC CCA GCC
Z49802  ,IGHG2*01                       --- --- --- --- --- --- --- --- --- --- --- --- --- --- --- --- --- --- --- ---
Z49801  ,IGHG2*02                       --- --- --- --- --- --- --- --- --- --- --- --- --- --- --- --- --- --- --- ---
                                        101 102 103 104 105 106 107 108 109 110
                                         P   I   E   K   T   I   S   K   T   K
J00230  ,IGHG2*01,Cgamma2 (CH2)         CCC ATC GAG AAA ACC ATC TCC AAA ACC AAA G
Z49802  ,IGHG2*01                       --- --- --- --- --- --- --- --- --- --- -
Z49801  ,IGHG2*02                       --

                                         1   2   3   4   5   6   7   8   9  10  11  12  13  14  15  16  17  18  19  20
                                         G   Q   P   R   E   P   Q   V   Y   T   L   P   P   S   R   E   E   M   T   K
J00230  ,IGHG2*01,Cgamma2 (CH3)         GG CAG CCC CGA GAA CCA CAG GTG TAC ACC CTG CCC CCA TCC CGG GAG GAG ATG ACC AAG
Z49802  ,IGHG2*01(1)                    -- --- --- --- --- --- --- --- --- --- --- --- --- --- --- --- --- --- --- ---
Z49801  ,IGHG2*02                       -- --- --- --- --- --- --- --- --- --- --- --- --- --- --- --- --- --- --- ---
                                        21  22  23  24  25  26  27  28  29  30  31  32  33  34  35  36  37  38  39  40
                                         N   Q   V   S   L   T   C   L   V   K   G   F   Y   P   S   D   I   A   V   E
J00230  ,IGHG2*01,Cgamma2 (CH3)         AAC CAG GTC AGC CTG ACC TGC CTG GTC AAA GGC TTC TAC CCC AGC GAC ATC GCC GTG GAG
Z49802  ,IGHG2*01(1)                    --- --- --- --- --- --- --- --- --- --- --- --- --- --- --- --- --- --- --- ---
Z49801  ,IGHG2*02                       --- --- --- --- --- --- --- --- --- --- --- --- --- --- --- --- --- --- --- ---
                                        41  42  43  44  45  46  47  48  49  50  51  52  53  54  55  56  57  58  59  60
                                         W   E   S   N   G   Q   P   E   N   N   Y   K   T   T   P   P   M   L   D   S
J00230  ,IGHG2*01,Cgamma2 (CH3)         TGG GAG AGC AAT GGG CAG CCG GAG AAC AAC TAC AAG ACC ACA CCT CCC ATG CTG GAC TCC
Z49802  ,IGHG2*01(1)                    --- --- --- --- --- --- --- --- --- --- --- --- --- --- --- --- --- --- --- ---
Z49801  ,IGHG2*02                       --- --- --- --- --- --- --- --- --- --- --- --- --- --- --- --- --- --- --- ---
                                        61  62  63  64  65  66  67  68  69  70  71  72  73  74  75  76  77  78  79  80
                                         D   G   S   F   F   L   Y   S   K   L   T   V   D   K   S   R   W   Q   Q   G
J00230  ,IGHG2*01,Cgamma2 (CH3)         GAC GGC TCC TTC TTC CTC TAC AGC AAG CTC ACC GTG GAC AAG AGC AGG TGG CAG CAG GGG
Z49802  ,IGHG2*01(1)                    --- --- --- --- --- --- --- --- --- --- --- --- --- --- --- --- --- --- --- ---
Z49801  ,IGHG2*02                       --- --- --- --- --- --- --- --- --- --- --- --- --- --- --- --- --- --- --- ---
                                        81  82  83  84  85  86  87  88  89  90  91  92  93  94  95  96  97  98  99 100
                                         N   V   F   S   C   S   V   M   H   E   A   L   H   N   H   Y   T   Q   K   S
J00230  ,IGHG2*01,Cgamma2 (CH3)         AAC GTC TTC TCA TGC TCC GTG ATG CAT GAG GCT CTG CAC AAC CAC TAC ACG CAG AAG AGC
Z49802  ,IGHG2*01(1)                    --- --- --- --- --- --- --- --- --- --- --- --- --- --- --- --- --- --- --- ---
Z49801  ,IGHG2*02                       --- --- --- --- --- --- --- --- --- --- --- --- --- --- --- --- --A --- --- ---
                                        101 102 103 104 105 106 107
                                         L   S   L   S   P   G   K   *
J00230  ,IGHG2*01,Cgamma2 (CH3)         CTC TCC CTG TCT CCG GGT AAA
Z49802  ,IGHG2*01(1)                    --- --- --- --- --- --- --
Z49801  ,IGHG2*02                       --- --- --- --- --- --- ---

                                              1   2   3   4   5   6   7   8   9  10  11  12  13  14  15  16  17  18  19  20
                                              E   L   Q   L   E   E   S   C   A   E   A   Q   D   G   E   L   D   G   L   W
AB006775,IGHG2*01,Cgamma2 (M1)   [7]     (G)AG CTG CAA CTG GAG GAG AGC TGT GCG GAG GCG CAG GAC GGG GAG CTG GAC GGG CTG TGG
                                        21  22  23  24  25  26  27  28  29  30  31  32  33  34  35  36  37  38  39  40
                                         T   T   I   T   I   F   I   T   L   F   L   L   S   V   C   Y   S   A   T   I
AB006775,IGHG2*01,Cgamma2 (M1)          ACG ACC ATC ACC ATC TTC ATC ACA CTC TTC CTG CTA AGC GTG TGC TAC AGT GCC ACC ATC
                                        41  42  43  44
                                         T   F   F   K
AB006775,IGHG2*01,Cgamma2 (M1)          ACC TTC TTC AAG

                                         1   2   3   4   5   6   7   8   9  10  11  12  13  14  15  16  17  18  19  20
                                         V   K   W   I   F   S   S   V   V   D   L   K   Q   T   I   V   P   D   Y   R
AB006776,IGHG2*01,Cgamma2 (M2)   [7]    GTG AAG TGG ATC TTC TCC TCA GTG GTG GAC CTG AAG CAG ACC ATC GTC CCC GAC TAC AGG
                                        21  22  23  24  25  26  27
                                         N   M   I   R   Q   G   A   *
AB006776,IGHG2*01,Cgamma2 (M2)          AAC ATG ATC AGG CAG GGG GCC
```

Notes:

(1) The IGHG2 gene has a codon missing at position 3 of CH2, compared to the other human IGHG genes.

(2) Z49802 previously described as a IGHG2*03 allele due to a289>g, g291>a, T97>A in CH3 is now considered as a IGHG2*01 allele, following the EMBL update of 20 September 1999 (a289, g291, T97).

Genome database accession numbers

GDB:119338 LocusLink: 3501 OMIM: 147110

IGHG3

Nomenclature

IGHG3: Immunoglobulin heavy constant gamma 3.

Definition and functionality

IGHG3 is one of the nine functional genes of the human IGHC group which comprises 11 mapped genes.

Gene location

IGHG3 is in the IGH locus on chromosome 14 at 14q32.33.

Gene structure

The IGHG3 gene comprises three main exons encoding the CH1, CH2 and CH3 domains, 2–5 hinge (H) exons and two transmembrane exons (M1 and M2). The IGHG3 hinge is made of two, three, four or five distinct hinge exons correspond to allelic polymorphisms.

Nucleotide and amino acid sequences for human IGHG3

The nucleotide between parentheses at the beginning of exons comes from a DONOR_SPLICE (n from ngt). The first nucleotide from an INT_DONOR_SPLICE is underlined (n from ngt).

The Cysteines involved in the CH1, CH2 and CH3 intrachain disulphide bridges are shown with their number and letter **C** in bold. The Cysteines involved in the H-L or H-H interchain disulphide bridges are shown with only the letter **C** in bold.

IGHG3*01 and *02 have 4 hinge exons, IGHG3*03 has 3 hinge exons, and IGHG3*04 has 2 hinge exons. The 5-exon hinge IGHG3 has been characterized (L. Osipova, unpublished data) but has not yet been sequenced.

			1	2	3	4	5	6	7	8	9	10	11	12	13	**14**	15	16	17	18	19	20
			A	S	T	K	G	P	S	V	F	P	L	A	P	**C**	S	R	S	T	S	G
X03604 (1)	,IGHG3*01,G3(Ezz) (CH1)	[17]	(G)CT	TCC	ACC	AAG	GGC	CCA	TCG	GTC	TTC	CCC	CTG	GCG	CCC	TGC	TCC	AGG	AGC	ACC	TCT	GGG
D78345	,IGHG3*01	[28]	(-)--	---	---	---	---	---	---	---	---	---	---	---	---	---	---	---	---	---	---	---
X16110	,IGHG3*03,IGHG3(LAT)	[18]	(-)--	---	---	---	---	---	---	---	---	---	---	---	---	---	---	---	---	---	---	---
X99549	,IGHG3*04,G3(Mand-541)	[29]	(-)-C	---	---	---	---	---	---	---	---	---	---	---	---	---	---	---	---	---	---	---

		21	22	23	24	25	26	**27**	28	29	30	31	32	33	34	35	36	37	38	39	40
		G	T	A	A	L	G	**C**	L	V	K	D	Y	F	P	E	P	V	T	V	S
X03604	,IGHG3*01,G3(Ezz) (CH1)	GGC	ACA	GCG	GCC	CTG	GGC	TGC	CTG	GTC	AAG	GAC	TAC	TTC	CCC	GAA	CCG	GTG	ACG	GTG	TCG
D78345	,IGHG3*01	---	---	---	---	---	---	---	---	---	---	---	---	---	---	---	---	---	---	---	---
X16110	,IGHG3*03,IGHG3(LAT)	---	---	---	---	---	---	---	---	---	---	---	---	---	---	---	---	---	---	---	---
X99549	,IGHG3*04,G3(Mand-541)	---	---	---	---	---	---	---	---	---	---	---	---	---	---	---	---	---	---	---	---

		41	42	43	44	45	46	47	48	49	50	51	52	53	54	55	56	57	58	59	60
		W	N	S	G	A	L	T	S	G	V	H	T	F	P	A	V	L	Q	S	S
X03604	,IGHG3*01,G3(Ezz) (CH1)	TGG	AAC	TCA	GGC	GCC	CTG	ACC	AGC	GGC	GTG	CAC	ACC	TTC	CCG	GCT	GTC	CTA	CAG	TCC	TCA
D78345	,IGHG3*01	---	---	---	---	---	---	---	---	---	---	---	---	---	---	---	---	---	---	---	---
X16110	,IGHG3*03,IGHG3(LAT)	---	---	---	---	---	---	---	---	---	---	---	---	---	---	---	---	---	---	---	---
X99549	,IGHG3*04,G3(Mand-541)	---	---	---	---	---	---	---	---	---	---	---	---	---	---	---	---	---	---	---	---

		61	62	63	64	65	66	67	68	69	70	71	72	73	74	75	76	77	78	79	80
		G	L	Y	S	L	S	S	V	V	T	V	P	S	S	S	L	G	T	Q	T
X03604	,IGHG3*01,G3(Ezz) (CH1)	GGA	CTC	TAC	TCC	CTC	AGC	AGC	GTG	GTG	ACC	GTG	CCC	TCC	AGC	AGC	TTG	GGC	ACC	CAG	ACC
D78345	,IGHG3*01	---	---	---	---	---	---	---	---	---	---	---	---	---	---	---	---	---	---	---	---
X16110	,IGHG3*03,IGHG3(LAT)	---	---	---	---	---	---	---	---	---	---	---	---	---	---	---	---	---	---	---	---
X99549	,IGHG3*04,G3(Mand-541)	---	---	---	---	---	---	---	---	---	---	---	---	---	---	---	---	---	---	---	---

```
                                                   81  82  83  84  85  86  87  88  89  90  91  92  93  94  95  96  97  98
                                                   Y   T   C   N   V   N   H   K   P   S   N   T   K   V   D   K   R   V
X03604  ,IGHG3*01,G3(Ezz) (CH1)                   TAC ACC TGC AAC GTG AAT CAC AAG CCC AGC AAC ACC AAG GTG GAC AAG AGA GTT G
X04646  ,IGHG3*01                    [16]                                   - --- --- --- --- --- --- --- --- --- --- -
D78345  ,IGHG3*01                                 --- --- --- --- --- --- --- --- --- --- --- --- --- --- --- --- --- --- -
K01313  ,IGHG3*02,gamma3             [10]                                   - --- --- --- --- --- --- --- --- --- --- -
X16110  ,IGHG3*03,IGHG3(LAT)                      --- --- --- --- --- --- --- --- --- --- --- --- --- --- --- --- --- --- -
X99549  ,IGHG3*04,G3(Mand-541)                    --- --- --- --- --- --- --- --- --- --- --- --- --- --- --- --- --- --- -

                                                   1   2   3   4   5   6   7   8   9   10  11  12  13  14  15  16  17
                                                   E   L   K   T   P   L   G   D   T   T   H   T   C   P   R   C   P
X03604  ,IGHG3*01,G3(Ezz) (H1)                    AG CTC AAA ACC CCA CTT GGT GAC ACA ACT CAC ACA TGC CCA CGG TGC CCA G
X04646  ,IGHG3*01                                 -- --- --- --- --- --- --- --- --- --- --- --- --- --- --- --- --- -
D78345  ,IGHG3*01                                                    --- --- --- --- --- --- --- --- --- --- --- --- -
K01313  ,IGHG3*02,gamma3                           -- --- --- --- --- --- --- --- --- --- --- --- --- --- --- --- --- -
X16110  ,IGHG3*03,IGHG3(LAT)                      -- --- --- --- --- --- --- --- --- --- --- --- --- --- --- --- --- -
X99549  ,IGHG3*04,G3(Mand-541)                    -- --- --- --- --- --- --- --- --- --- --- --- --- --- --- --- --- -

                                                   1   2   3   4   5   6   7   8   9   10  11  12  13  14  15
                                                   E   P   K   S   C   D   T   P   P   P   C   P   R   C   P
X03604  ,IGHG3*01,G3(Ezz) (H2)                    AG CCC AAA TCT TGT GAC ACA CCT CCC CCG TGC CCA CGG TGC CCA G
X04646  ,IGHG3*01                                 -- --- --- --- --- --- --- --- --- --- --- --- --- --- --- -
D78345  ,IGHG3*01                                 -- --- --- --- --- --- --- --- --- --- --- --- --- --- --- -
K01313  ,IGHG3*02,gamma3                          -- --T --- --- --- --- --- --- --- --- --- --- --- --- --- -

                                                   1   2   3   4   5   6   7   8   9   10  11  12  13  14  15
                                                   E   P   K   S   C   D   T   P   P   P   C   P   R   C   P
X03604  ,IGHG3*01,G3(Ezz) (H3)                    AG CCC AAA TCT TGT GAC ACA CCT CCC CCA TGC CCA CGG TGC CCA G
X04646  ,IGHG3*01                                 -- --- --- --- --- --- --- --- --- --- --- --- --- --- --- -
D78345  ,IGHG3*01                                 -- --- --- --- --- --- --- --- --- --- --- --- --- --- --- -
K01313  ,IGHG3*02,gamma3                          -- --T --- --- --- --- --- --- --- --G --- --- --- --- --- -
X16110  ,IGHG3*03,IGHG3(LAT)                      -- --- --- --- --- --- --- --- --- --- --- --- --- --- --- -

                                                   1   2   3   4   5   6   7   8   9   10  11  12  13  14  15
                                                   E   P   K   S   C   D   T   P   P   P   C   P   R   C   P
X03604  ,IGHG3*01,G3(Ezz) (H4)                    AG CCC AAA TCT TGT GAC ACA CCT CCC CCG TGC CCA AGG TGC CCA G
X04646  ,IGHG3*01                                 -- --- --- --- --- --- --- --- --- --- --- --- --- --- --- -
D78345  ,IGHG3*01                                 -- --- --- --- --- --- --- --- --- --- --- --- --- --- --- -
K01313  ,IGHG3*02,gamma3                          -- --- --- --- --- --- --- --- --- --- --- --- --- --- --- -
X16110  ,IGHG3*03,IGHG3(LAT)                      -- --- --- --- --- --- --- --- --- --- --- --- --- --- --- -
X99549  ,IGHG3*04,G3(Mand-541)                    -- --- --- --- --- --- --- --- --- --- --- --- --- --- --- -

                                                   1   2   3   4   5   6   7   8   9   10  11  12  13  14  15  16  17  18  19  20
                                                   A   P   E   L   L   G   G   P   S   V   F   L   F   P   P   K   P   K   D   T
X03604  ,IGHG3*01,G3(Ezz) (CH2)                   CA CCT GAA CTC CTG GGA GGA CCG TCA GTC TTC CTC TTC CCC CCA AAA CCC AAG GAT ACC
X04646  ,IGHG3*01                                 -- --- --- --- --- --- --- --- --- --- --- --- --- --- --- ---
D78345  ,IGHG3*01                                 -- --- --- --- --- --- --- --- --- --- --- --- --- --- --- --- --- --- --- ---
K01313  ,IGHG3*02,gamma3                          -- --- --- --- --- --- --- --- --- --- --- --- --- --- --T --- ---
X16110  ,IGHG3*03,IGHG3(LAT)                      -- --- --- --- --- --- --- --- --- --- --- --- --- --- --- --- --- --- --- ---
X99549  ,IGHG3*04,G3(Mand-541)                    -- --- --- --- --- --- --- --- --- --- --- --- --- --- --- --- --- --- --- ---
                                                   21  22  23  24  25  26  27  28  29  30  31  32  33  34  35  36  37  38  39  40
                                                   L   M   I   S   R   T   P   E   V   T   C   V   V   V   D   V   S   H   E   D
X03604  ,IGHG3*01,G3(Ezz) (CH2)                   CTT ATG ATT TCC CGG ACC CCT GAG GTC ACG TGC GTG GTG GTG GAC GTG AGC CAC GAA GAC
X04646  ,IGHG3*01
D78345  ,IGHG3*01                                 --- --- --- --- --- --- --- --- --- --- --- --- --- --- --- --- --- --- --- ---
K01313  ,IGHG3*02,gamma3
X16110  ,IGHG3*03,IGHG3(LAT)                      --- --- --- --- --- --- --- --- --- --- --- --- --- --- --- --- --- --- --- ---
X99549  ,IGHG3*04,G3(Mand-541)                    --- --- --- --- --- --- --- --- --- --- --- --- --- --- --- --- --- --- --- ---
                                                   41  42  43  44  45  46  47  48  49  50  51  52  53  54  55  56  57  58  59  60
                                                   P   E   V   Q   F   K   W   Y   V   D   G   V   E   V   H   N   A   K   T   K
X03604  ,IGHG3*01,G3(Ezz) (CH2)                   CCC GAG GTC CAG TTC AAG TGG TAC GTG GAC GGC GTG GAG GTG CAT AAT GCC AAG ACA AAG
X04646  ,IGHG3*01
D78345  ,IGHG3*01                                 --- --- --- --- --- --- --- --- --- --- --- --- --- --- --- --- --- --- --- ---
K01313  ,IGHG3*02,gamma3
X16110  ,IGHG3*03,IGHG3(LAT)                      --- --- --- --- --- --- --- --- --- --- --- --- --- --- --- --- --- --- --- ---
X99549  ,IGHG3*04,G3(Mand-541)                    --- --- --- --- --- --- --- --- --- --- --- --- --- --- --- --- --- --- --- ---
X74165  ,IGHG3*05                    [19]         --- --- --- --- --- --- --- --- --- --- --- --- --- --- --- --- --- --- --- ---
X74166  ,IGHG3*06                    [19]         --- --- --- --- --- --- --- --- --- --- --- --- --- --- --- --- --- --- --- ---
                                                   61  62  63  64  65  66  67  68  69  70  71  72  73  74  75  76  77  78  79  80
                                                   P   R   E   E   Q   Y   N   S   T   F   R   V   V   S   V   L   T   V   L   H
X03604  ,IGHG3*01,G3(Ezz) (CH2)                   CCG CGG GAG GAG CAG TAC AAC AGC ACG TTC CGT GTG GTC AGC GTC CTC ACC GTC CTG CAC
D78345  ,IGHG3*01                                 --- --- --- --- --- --- --- --- --- --- --- --- --- --- --- --- --- --- --- ---
X16110  ,IGHG3*03,IGHG3(LAT)                      --- --- --- --- --- --- --- --- --- --- --- --- --- --- --- --- --- --- --- ---
X99549  ,IGHG3*04,G3(Mand-541)                    --- --- --- --- --- --- --- --- --- --- --- --- --- --- --- --- --- --- --- ---
                                                                           F
X74165  ,IGHG3*05                                 --- --- --- --- --- -T- --- --- --- --- --- --- --- --- --- --- --- --- --- ---
                                                   L
X74166  ,IGHG3*06                                 -T- --- --- --- --- --- --- --- --- --- --- --- --- --- --- --- --- --- --- ---
                                                   81  82  83  84  85  86  87  88  89  90  91  92  93  94  95  96  97  98  99  100
                                                   Q   D   W   L   N   G   K   E   Y   K   C   K   V   S   N   K   A   L   P   A
X03604  ,IGHG3*01,G3(Ezz) (CH2)                   CAG GAC TGG CTG AAC GGC AAG GAG TAC AAG TGC AAG GTC TCC AAC AAA GCC CTC CCA GCC
D78345  ,IGHG3*01                                 --- --- --- --- --- --- --- --- --- --- --- --- --- --- --- --- --- --- --- ---
X16110  ,IGHG3*03,IGHG3(LAT)                      --- --- --- --- --- --- --- --- --- --- --- --- --- --- --- --- --- --- --- ---
X99549  ,IGHG3*04,G3(Mand-541)                    --- --- --- --- --- --- --- --- --- --- --- --- --- --- --- --- --- --- --- ---
X74165  ,IGHG3*05                                 --- --- --- --- --- --- --- --- --- --- --- --- --- --- --- --- --- --- --- ---
X74166  ,IGHG3*06                                 --- --- --- --- --- --- --- --- --- --- --- --- --- --- --- --- --- --- --- ---
                                                  101 102 103 104 105 106 107 108 109 110
                                                   P   I   E   K   T   I   S   K   T   K
X03604  ,IGHG3*01,G3(Ezz) (CH2)                   CCC ATC GAG AAA ACC ATC TCC AAA ACC AAA G
D78345  ,IGHG3*01                                 --- --- --- --- --- --- --- --- --- --- -
X16110  ,IGHG3*03,IGHG3(LAT)                      --- --- --- --- --- --- --- --- --- --- -
```

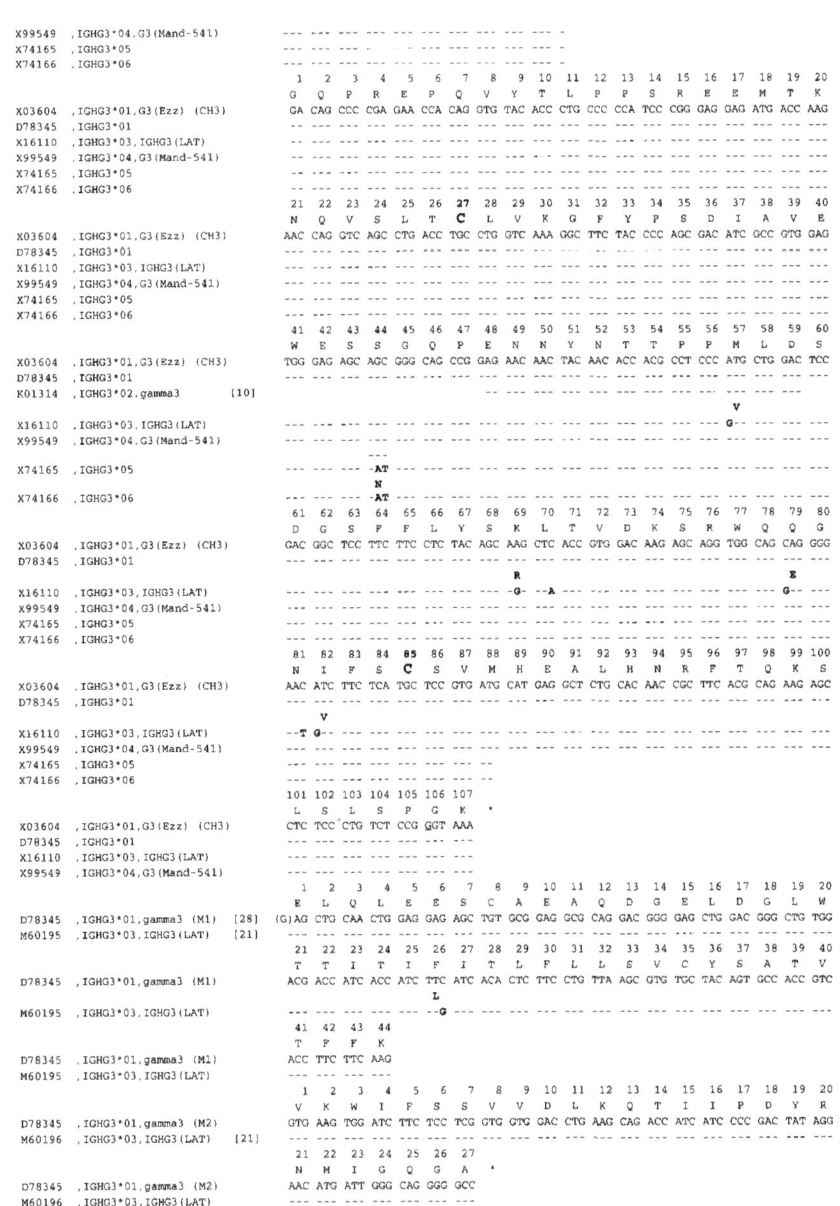

Note:

(1) This entry has M12958 as secondary accession number in the generalist databases. As this number M12958 was the original primary number, M12958 has also been kept as primary accession number in IMGT, in order not to lose information.

Genome database accession numbers

GDB:119339 LocusLink: 3502 OMIM: 147120

Nomenclature

IGHG4: Immunoglobulin heavy constant gamma 4.

Definition and functionality

IGHG4 is one of the nine functional genes of the human IGHC group which comprises 11 mapped genes.

Gene location

IGHG4 is in the IGH locus on chromosome 14 at 14q32.33.

Gene structure

The IGHG4 gene comprises three main exons encoding the CH1, CH2 and CH3 domains and one separate hinge (H) exon. The transmembrane M1 and M2 exons have not yet been sequenced.

Nucleotide and amino acid sequences for human IGHG4

The nucleotide between parentheses at the beginning of exons comes from a DONOR_SPLICE (n from ngt). The first nucleotide from an INT_DONOR_SPLICE is underlined (n from ngt).

The Cysteines involved in the CH1, CH2 and CH3 intrachain disulphide bridges are shown with their number and letter **C** in bold. The Cysteines involved in the H-L or H-H interchain disulphide bridges are shown with only the letter **C** in bold.

```
                                                1   2   3   4   5   6   7   8   9  10  11  12  13  14  15  16  17  18  19  20
                                                A   S   T   K   G   P   S   V   F   P   L   A   P   C   S   R   S   T   S   E
K01316  ,IGHG4*01,Cgamma4 (CH1) [2]         (G)CT TCC ACC AAG GGC CCA TCC GTC TTC CCC CTG GCG CCC TGC TCC AGG AGC ACC TCC GAG
                                               21  22  23  24  25  26  27  28  29  30  31  32  33  34  35  36  37  38  39  40
                                                S   T   A   A   L   G   C   L   V   K   D   Y   F   P   E   P   V   T   V   S
K01316  ,IGHG4*01,Cgamma4 (CH1)               AGC ACA GCC GCC CTG GGC TGC CTG GTC AAG GAC TAC TTC CCC GAA CCG GTG ACG GTG TCG
                                               41  42  43  44  45  46  47  48  49  50  51  52  53  54  55  56  57  58  59  60
                                                W   N   S   G   A   L   T   S   G   V   H   T   F   P   A   V   L   Q   S   S
K01316  ,IGHG4*01,Cgamma4 (CH1)               TGG AAC TCA GGC GCC CTG ACC AGC GGC GTG CAC ACC TTC CCG GCT GTC CTA CAG TCC TCA
                                               61  62  63  64  65  66  67  68  69  70  71  72  73  74  75  76  77  78  79  80
                                                G   L   Y   S   L   S   S   V   V   T   V   P   S   S   S   L   G   T   K   T
K01316  ,IGHG4*01,Cgamma4 (CH1)               GGA CTC TAC TCC CTC AGC AGC GTG GTG ACC GTG CCC TCC AGC AGC TTG GGC ACG AAG ACC
                                               81  82  83  84  85  86  87  88  89  90  91  92  93  94  95  96  97  98
                                                Y   T   C   N   V   D   H   K   P   S   N   T   K   V   D   K   R   V
K01316  ,IGHG4*01,Cgamma4 (CH1)               TAC ACC TGC AAC GTA GAT CAC AAG CCC AGC AAC ACC AAG GTG GAC AAG AGA GTT G

                                                1   2   3   4   5   6   7   8   9  10  11  12
                                                E   S   K   Y   G   P   P   C   P   S   C   P
K01316  ,IGHG4*01,Cgamma4 (H)                  AG TCC AAA TAT GGT CCC CCA TGC CCA TCA TGC CCA G
AJ001563,IGHG4*02                   [8]        -- --- --- --- --- --- --G --- --- --- --- --- -

                                                1   2   3   4   5   6   7   8   9  10  11  12  13  14  15  16  17  18  19  20
                                                A   P   E   F   L   G   G   P   S   V   F   L   F   P   P   K   P   K   D   T
K01316  ,IGHG4*01,Cgamma4 (CH2)                CA CCT GAG TTC CTG GGG GGA CCA TCA GTC TTC CTG TTC CCC CCA AAA CCC AAG GAC ACT
AJ001563,IGHG4*02                              -- --- --- --- --- --- --- --- --- --- --- --- --- --- --- --- --- --- --- ---
AJ001564,IGHG4*03                   [8]        -- --- --- --- --- --- --- --- --- --- --- --- --- --- --- --- --- --- --- ---
                                               21  22  23  24  25  26  27  28  29  30  31  32  33  34  35  36  37  38  39  40
                                                L   M   I   S   R   T   P   E   V   T   C   V   V   V   D   V   S   Q   E   D
K01316  ,IGHG4*01,Cgamma4 (CH2)               CTC ATG ATC TCC CGG ACC CCT GAG GTC ACG TGC GTG GTG GTG GAC GTG AGC CAG GAA GAC
AJ001563,IGHG4*02                             --- --- --- --- --- --- --- --- --- --- --- --- --- --- --- --- --- --- --- ---
AJ001564,IGHG4*03                             --- --- --- --- --- --- --- --- --- --- --- --- --- --- --- --- --- --- --- ---
```

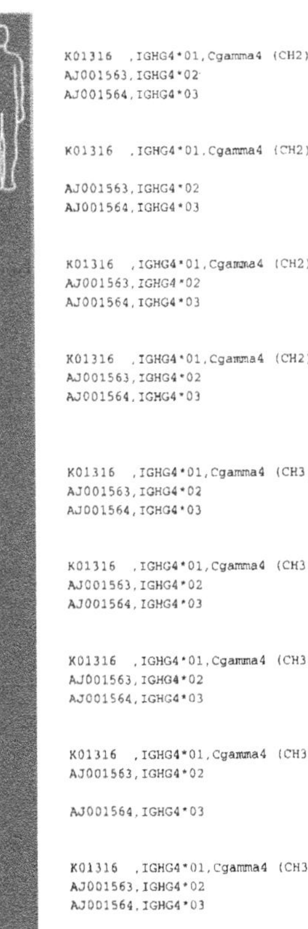

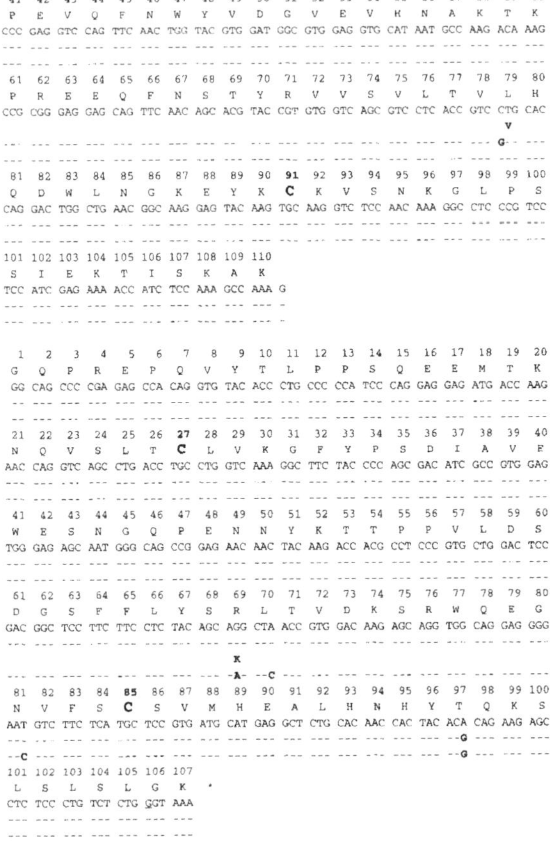

```
                                   41  42  43  44  45  46  47  48  49  50  51  52  53  54  55  56  57  58  59  60
                                   P   E   V   Q   F   N   W   Y   V   D   G   V   E   V   H   N   A   K   T   K
K01316   ,IGHG4*01,Cgamma4 (CH2)  CCC GAG GTC CAG TTC AAC TGG TAC GTG GAT GGC GTG GAG GTG CAT AAT GCC AAG ACA AAG
AJ001563,IGHG4*02                 --- --- --- --- --- --- --- --- --- --- --- --- --- --- --- --- --- --- --- ---
AJ001564,IGHG4*03                 --- --- --- --- --- --- --- --- --- --- --- --- --- --- --- --- --- --- --- ---
                                   61  62  63  64  65  66  67  68  69  70  71  72  73  74  75  76  77  78  79  80
                                   P   R   E   E   Q   F   N   S   T   Y   R   V   V   S   V   L   T   V   L   H
K01316   ,IGHG4*01,Cgamma4 (CH2)  CCG CGG GAG GAG CAG TTC AAC AGC ACG TAC CGT GTG GTC AGC GTC CTC ACC GTC CTG CAC
                                                                                                           V
AJ001563,IGHG4*02                 --- --- --- --- --- --- --- --- --- --- --- --- --- --- --- --- --- --- G-- ---
AJ001564,IGHG4*03                 --- --- --- --- --- --- --- --- --- --- --- --- --- --- --- --- --- --- --- ---
                                   81  82  83  84  85  86  87  88  89  90  91  92  93  94  95  96  97  98  99 100
                                   Q   D   W   L   N   G   K   E   Y   K   C   K   V   S   N   K   G   L   P   S
K01316   ,IGHG4*01,Cgamma4 (CH2)  CAG GAC TGG CTG AAC GGC AAG GAG TAC AAG TGC AAG GTC TCC AAC AAA GGC CTC CCG TCC
AJ001563,IGHG4*02                 --- --- --- --- --- --- --- --- --- --- --- --- --- --- --- --- --- --- --- ---
AJ001564,IGHG4*03                 --- --- --- --- --- --- --- --- --- --- --- --- --- --- --- --- --- --- --- ---
                                  101 102 103 104 105 106 107 108 109 110
                                   S   I   E   K   T   I   S   K   A   K
K01316   ,IGHG4*01,Cgamma4 (CH2)  TCC ATC GAG AAA ACC ATC TCC AAA GCC AAA G
AJ001563,IGHG4*02                 --- --- --- --- --- --- --- --- --- --- -
AJ001564,IGHG4*03                 --- --- --- --- --- --- --- --- --- --- -

                                    1   2   3   4   5   6   7   8   9  10  11  12  13  14  15  16  17  18  19  20
                                   G   Q   P   R   E   P   Q   V   Y   T   L   P   P   S   Q   E   E   M   T   K
K01316   ,IGHG4*01,Cgamma4 (CH3)   GG CAG CCC CGA GAG CCA CAG GTG TAC ACC CTG CCC CCA TCC CAG GAG GAG ATG ACC AAG
AJ001563,IGHG4*02                  -- --- --- --- --- --- --- --- --- --- --- --- --- --- --- --- --- --- --- ---
AJ001564,IGHG4*03                  -- --- --- --- --- --- --- --- --- --- --- --- --- --- --- --- --- --- --- ---
                                   21  22  23  24  25  26  27  28  29  30  31  32  33  34  35  36  37  38  39  40
                                   N   Q   V   S   L   T   C   L   V   K   G   F   Y   P   S   D   I   A   V   E
K01316   ,IGHG4*01,Cgamma4 (CH3)  AAC CAG GTC AGC CTG ACC TGC CTG GTC AAA GGC TTC TAC CCC AGC GAC ATC GCC GTG GAG
AJ001563,IGHG4*02                 --- --- --- --- --- --- --- --- --- --- --- --- --- --- --- --- --- --- --- ---
AJ001564,IGHG4*03                 --- --- --- --- --- --- --- --- --- --- --- --- --- --- --- --- --- --- --- ---
                                   41  42  43  44  45  46  47  48  49  50  51  52  53  54  55  56  57  58  59  60
                                   W   E   S   N   G   Q   P   E   N   N   Y   K   T   T   P   P   V   L   D   S
K01316   ,IGHG4*01,Cgamma4 (CH3)  TGG GAG AGC AAT GGG CAG CCG GAG AAC AAC TAC AAG ACC ACG CCT CCC GTG CTG GAC TCC
AJ001563,IGHG4*02                 --- --- --- --- --- --- --- --- --- --- --- --- --- --- --- --- --- --- --- ---
AJ001564,IGHG4*03                 --- --- --- --- --- --- --- --- --- --- --- --- --- --- --- --- --- --- --- ---
                                   61  62  63  64  65  66  67  68  69  70  71  72  73  74  75  76  77  78  79  80
                                   D   G   S   F   F   L   Y   S   R   L   T   V   D   K   S   R   W   Q   E   G
K01316   ,IGHG4*01,Cgamma4 (CH3)  GAC GGC TCC TTC TTC CTC TAC AGC AGG CTA ACC GTG GAC AAG AGC AGG TGG CAG GAG GGG
AJ001563,IGHG4*02                 --- --- --- --- --- --- --- --- --- --- --- --- --- --- --- --- --- --- --- ---
                                                                   K
AJ001564,IGHG4*03                 --- --- --- --- --- --- --- --- -A- --C --- --- --- --- --- --- --- --- --- ---
                                   81  82  83  84  85  86  87  88  89  90  91  92  93  94  95  96  97  98  99 100
                                   N   V   F   S   C   S   V   M   H   E   A   L   H   N   H   Y   T   Q   K   S
K01316   ,IGHG4*01,Cgamma4 (CH3)  AAT GTC TTC TCA TGC TCC GTG ATG CAT GAG GCT CTG CAC AAC CAC TAC ACA CAG AAG AGC
AJ001563,IGHG4*02                 --- --- --- --- --- --- --- --- --- --- --- --- --- --- --- --- --G --- --- ---
AJ001564,IGHG4*03                 --C --- --- --- --- --- --- --- --- --- --- --- --- --- --- --- --G --- --- ---
                                  101 102 103 104 105 106 107
                                   L   S   L   S   L   G   K   *
K01316   ,IGHG4*01,Cgamma4 (CH3)  CTC TCC CTG TCT CTG GGT AAA
AJ001563,IGHG4*02                 --- --- --- --- --- --- ---
AJ001564,IGHG4*03                 --- --- --- --- --- --- ---
```

Genome database accession numbers

GDB:119340 LocusLink: 3503 OMIM: 147130

Nomenclature

IGHM: Immunoglobulin heavy constant mu.

Definition and functionality

IGHM is one of the nine functional genes of the human IGHC group which comprises 11 mapped genes.

Gene location

IGHM is in the IGH locus on chromosome 14 at 14q32.33.

Gene structure

The IGHM gene comprises four exons encoding the CH1, CH2, CH3 and CH4 domains and two transmembrane exons (M1 and M2).

Nucleotide and amino acid sequences for human IGHM

The nucleotide between parentheses at the beginning of exons comes from a DONOR_SPLICE (n from ngt). The first nucleotide from an INT_DONOR_SPLICE is underlined (n from ngt).

The Cysteine involved in the intrachain disulphide bridges are shown with their number and their letter in bold. The Cysteine involved in the H-L or H-H interchain disulphide bridges are shown with only their letter in bold.

```
                                          1   2   3   4   5   6   7   8   9  10  11  12  13  14  15  16  17  18  19  20
                                          G   S   A   S   A   P   T   L   F   P   L   V   S   C   E   N   S   P   S   D
X14940  ,IGHM*01  (CH1)        [20]  (G)GG AGT GCA TCC GCC CCA ACC CTT TTC CCC CTC GTC TCC TGT GAG AAT TCC CCG TCG GAT
K01307  ,IGHM*02                [3]  (-)-- --- --- --- --- --- --- --- --- --- --- --- --- --- --- --- ---
X57331  ,IGHM*03               [22]  (-)-- --- --- --- --- --- --- --- --- --- --- --- --- --- --- --- --- --- --- ---
                                         21  22  23  24  25  26  27  28  29  30  31  32  33  34  35  36  37  38  39  40
                                          T   S   S   V   A   V   G   C   L   A   Q   D   F   L   P   D   S   I   T   L
X14940  ,IGHM*01  (CH1)              ACG AGC AGC GTG GCC GTT GGC TGC CTC GCA CAG GAC TTC CTT CCC GAC TCC ATC ACT TTG
                                                                                                                      F
X57331  ,IGHM*03                     --- --- --- --- --- --- --- --- --- --- --- --- --- --- --- --- --- --- --- --C
                                         41  42  43  44  45  46  47  48  49  50  51  52  53  54  55  56  57  58  59  60
                                          S   W   K   Y   K   N   N   S   D   I   S   S   T   R   G   F   P   S   V   L
X14940  ,IGHM*01  (CH1)              TCC TGG AAA TAC AAG AAC AAC TCT GAC ATC AGC AGT ACC CGG GGC TTC CCA TCA GTC CTG
X57331  ,IGHM*03                     --- --- --- --- --- --- --- --- --- --- --- --C --- --- --- --- --- --- --- ---
                                         61  62  63  64  65  66  67  68  69  70  71  72  73  74  75  76  77  78  79  80
                                          R   G   G   K   Y   A   A   T   S   Q   V   L   L   P   S   K   D   V   M   Q
X14940  ,IGHM*01  (CH1)              AGA GGG GGC AAG TAC GCA GCC ACC TCA CAG GTG CTG CTG CCT TCC AAG GAC GTC ATG CAG
X57331  ,IGHM*03                     --- --- --- --- --- --- --- --- --- --- --- --- --- --- --- --- --- --- --- ---
                                         81  82  83  84  85  86  87  88  89  90  91  92  93  94  95  96  97  98  99 100
                                          G   T   D   E   H   V   V   C   K   V   Q   H   P   N   G   N   K   E   K   N
X14940  ,IGHM*01  (CH1)              GGC ACA GAC GAA CAC GTG GTG TGC AAA GTC CAG CAC CCC AAC GGC AAC AAA GAA AAC AAC
X57331  ,IGHM*03                     --- --- --- --- --- --- --- --- --- --- --- --- --- --- --- --- --- --- --- ---
                                        101 102 103 104
                                          V   P   L   P
X14940  ,IGHM*01  (CH1)              GTG CCT CTT CCA G
X57331  ,IGHM*03                     --- --- --- --- -

                                          1   2   3   4   5   6   7   8   9  10  11  12  13  14  15  16  17  18  19  20
                                          V   I   A   E   L   P   P   K   V   S   V   F   V   P   P   R   D   G   F   F
X14940  ,IGHM*01, (CH2)               TG ATT GCC GAG CTG CCT CCC AAA GTG AGC GTC TTC GTC CCA CCC CGC GAC GGC TTC TTC
K01308  ,IGHM*02                [3]   -- --- --T --- --- --- --- --- --- --- --- --- --- --- --- --- --- --- --- ---
X57331  ,IGHM*03                      -- --- --T --- --- --- --- --- --- --- --- --- --- --- --- --- --- --- --- ---
                                         21  22  23  24  25  26  27  28  29  30  31  32  33  34  35  36  37  38  39  40
                                          G   N   P   R   K   S   K   L   I   C   Q   A   T   G   F   S   P   R   Q   I
X14940  ,IGHM*01, (CH2)              GGC AAC CCC CGC AAG TCC AAC CTC ATC TGC CAG GCC ACG GGT TTC AGT CCC CGG CAG ATT
K01308  ,IGHM*02                     --- --- --- --- --- --- --- --- --- --- --- --- --- --- --- --- --- --- --- ---
X57331  ,IGHM*03                     --- --- --- --- --- --- --- --- --- --- --- --- --- --- --- --- --- --- --- ---
                                         41  42  43  44  45  46  47  48  49  50  51  52  53  54  55  56  57  58  59  60
                                          Q   V   S   W   L   R   E   G   K   Q   V   G   S   G   V   T   T   D   Q   V
X14940  ,IGHM*01, (CH2)              CAG GTG TCC TGG CTG CGC GAG GGG AAG CAG GTG GGG TCT GGC GTC ACC ACG GAC CAG GTG
K01308  ,IGHM*02                     --- --- --- --- --- --- --- --- --- --- --- --- --- --- --- --- --- --- --- ---
X57331  ,IGHM*03                     --- --- --- --- --- --- --- --- --- --- --- --- --- --- --- --- --- --- --- ---
```

```
                                     61  62  63  64  65  66  67  68  69  70  71  72  73  74  75  76  77  78  79  80
                                     Q   A   E   A   K   E   S   G   P   T   T   Y   K   V   T   S   T   L   T   I
X14940  ,IGHM*01, (CH2)             CAG GCT GAG GCC AAA GAG TCT GGG CCC ACG ACC TAC AAG GTG ACC AGC ACA CTG ACC ATC
K01308  ,IGHM*02                    --- --- --- --- --- --- --- --- --- --- --- --- --- --- --- --- --- --- --- ---
X57331  ,IGHM*03                    --- --- --- --- --- --- --- --- --- --- --- --- --- --- --- --- --- --- --- ---
                                     81  82  83  84  85  86  87  88  89  90  91  92  93  94  95  96  97  98  99 100
                                     K   E   S   D   W   L   G   Q   S   M   F   T   C   R   V   D   H   R   G   L
X14940  ,IGHM*01, (CH2)             AAA GAG AGC GAC TGG CTC GGC CAG AGC ATG TTC ACC TGC CGC GTG GAT CAC AGG GGC CTG
K01308  ,IGHM*02                    --- ---
K01309  ,IGHM*02               [3]                                                              --- --- --- --- ---
                                                             S
X57331  ,IGHM*03                    --- --- --- --- --- --- A-- --- --- --- --- --- --- --- --- --- --- --- --- ---
                                    101 102 103 104 105 106 107 108 109 110 111 112
                                     T   F   Q   Q   N   A   S   S   M   C   V   P
X14940  ,IGHM*01, (CH2)             ACC TTC CAG CAG AAT GCG TCC TCC ATG TGT GTC CCC G
K01309  ,IGHM*02                    --- --- --- --- --- --- --- --- --- --- --- --- -
X57331  ,IGHM*03                    --- --- --- --- --- --- --- --- --- --- --- --- -

                                      1   2   3   4   5   6   7   8   9  10  11  12  13  14  15  16  17  18  19  20
                                     D   Q   D   T   A   I   R   V   F   A   I   P   P   S   F   A   S   I   F   L
X14940  ,IGHM*01, (CH3)              AT CAA GAC ACA GCC ATC CGG GTC TTC GCC ATC CCC CCA TCC TTT GCC AGC ATC TTC CTC
K01309  ,IGHM*02                     -- --- --- --- --- --- --- --- --- --- --- --- --- --- --- --- --- --- --- --G
X57331  ,IGHM*03                     -- --- --- --- --- --- --- --- --- --- --- --- --- --- --- --- --- --- --- ---
                                     21  22  23  24  25  26  27  28  29  30  31  32  33  34  35  36  37  38  39  40
                                     T   K   S   T   K   L   T   C   L   V   T   D   L   T   T   Y   D   S   V   T
X14940  ,IGHM*01, (CH3)             ACC AAG TCC ACC AAG TTG ACC TGC CTG GTC ACA GAC CTG ACC ACC TAT GAC AGC GTG ACC
K01309  ,IGHM*02                    --- --- --- --- --- --- --- --- --- --- --- --- --- --- --- --- --- --- --- ---
X57331  ,IGHM*03                    --- --- --- --- --- --- --- --- --- --- --- --- --- --- --- --- --- --- --- ---
                                     41  42  43  44  45  46  47  48  49  50  51  52  53  54  55  56  57  58  59  60
                                     I   S   W   T   R   Q   N   G   E   A   V   K   T   H   T   N   I   S   E   S
X14940  ,IGHM*01, (CH3)             ATC TCC TGG ACC CGC CAG AAT GGC GAA GCT GTG AAA ACC CAC ACC AAC ATC TCC GAG AGC
K01309  ,IGHM*02                    --- --- ---
X57331  ,IGHM*03                    --- --- --- --- --- --- --- --- --- --- --- --- --- --- --- --- --- --- --- ---
                                     61  62  63  64  65  66  67  68  69  70  71  72  73  74  75  76  77  78  79  80
                                     H   P   N   A   T   F   S   A   V   G   E   A   S   I   C   E   D   D   W   N
X14940  ,IGHM*01, (CH3)             CAC CCC AAT GCC ACT TTC AGC GCC GTG GGT GAG GCC AGC ATC TGC GAG GAT GAC TGG AAT
J00259  ,IGHM*02               [1]                                                                          - ---
X57331  ,IGHM*03                    --- --- --- --- --- --- --- --- --- --- --- --- --- --- --- --- --- --- --- ---
                                     81  82  83  84  85  86  87  88  89  90  91  92  93  94  95  96  97  98  99 100
                                     S   G   E   R   F   T   C   T   V   T   H   T   D   L   P   S   P   L   K   Q
X14940  ,IGHM*01, (CH3)             TCC GGG GAG AGG TTC ACG TGC ACC GTG ACC CAC ACA GAC CTG CCC TCG CCA CTG AAG CAG
V00561  ,IGHM*02               [1]      --- --- --- --- --- --- --- --- --- --- --- --- --- --- --- --- --- --- ---
J00259  ,IGHM*02                    --G --- --- --- --- --- --- --- --- --- --- --- --- --- --- --- --- --- --- ---
X57331  ,IGHM*03                    --- --- --- --- --- --- --- --- --- --- --- --- --- --- --- --- --- --- --- ---
                                    101 102 103 104 105 106
                                     T   I   S   R   P   K
X14940  ,IGHM*01, (CH3)             ACC ATC TCC CGG CCC AAG G
V00561  ,IGHM*02                    --- --- --- --- --- --- -
J00259  ,IGHM*02                    --- --- --- --- --- --- -
X57331  ,IGHM*03                    --- --- --- --- --- --- -

                                      1   2   3   4   5   6   7   8   9  10  11  12  13  14  15  16  17  18  19  20
                                     G   V   A   L   H   R   P   D   V   Y   L   L   P   P   A   R   E   Q   L   N
X14940  ,IGHM*01, (CH4)              GG GTG GCC CTG CAC AGG CCC GAT GTC TAC TTG CTG CCA CCA GCC CGG GAG CAG CTG AAC
J00260  ,IGHM*02               [3]                                                                  --- --- --- ---
X57331  ,IGHM*03                     -- --- --- --- --- --- --- --- --- --- --- --- --- --- --- --- --- --- --- ---
                                     21  22  23  24  25  26  27  28  29  30  31  32  33  34  35  36  37  38  39  40
                                     L   R   E   S   A   T   I   T   C   L   V   T   G   F   S   P   A   D   V   F
X14940  ,IGHM*01, (CH4)             CTG CGG GAG TCG GCC ACC ATC ACG TGC CTG GTG ACG GGC TTC TCT CCC GCG GAC GTC TTC
J00260  ,IGHM*02                    --- --- --- --- --- --- --- --- --- --- --- --- --- --- --- --- --- --- --- ---
V00562  ,IGHM*02               [1]                -- --- --- --- --- --- --- --- --- --- --- --- --- --- --- ---
X57331  ,IGHM*03                    --- --- --- --- --- --- --- --- --- --- --- --- --- --- --- --- --- --- --- ---
                                     41  42  43  44  45  46  47  48  49  50  51  52  53  54  55  56  57  58  59  60
                                     V   Q   W   M   Q   R   G   Q   P   L   S   P   E   K   Y   V   T   S   A   P
X14940  ,IGHM*01, (CH4)             GTG CAG TGG ATG CAG AGG GGG CAG CCC TTG TCC CCG GAG AAG TAT GTG ACC AGC GCC CCA
J00260  ,IGHM*02                    --- --- --- --- --- --- --- --- --- --- --- --- --- --- --- --- --- --- --- ---
V00562  ,IGHM*02                    --- --- --- --- --- --- --- ---
X57331  ,IGHM*03                    --- --- --- --- --- --- --- --- --- --- --- --- --- --- --- --- --- --- --- ---
                                     61  62  63  64  65  66  67  68  69  70  71  72  73  74  75  76  77  78  79  80
                                     M   P   E   P   Q   A   P   G   R   Y   F   A   H   S   I   L   T   V   S   E
X14940  ,IGHM*01, (CH4)             ATG CCT GAG CCC CAG GCC CCA GGC CGG TAC TTC GCC CAC AGC ATC CTG ACC GTG TCC GAA
J00260  ,IGHM*02                    --- --- --- --- --- --- --- --- --- --- --- --- --- --- --- --- --- --- --- --G
X57331  ,IGHM*03                    --- --- --- --- --- --- --- --- --- --- --- --- --- --- --- --- --- --- --- ---
                                     81  82  83  84  85  86  87  88  89  90  91  92  93  94  95  96  97  98  99 100
                                     E   E   W   N   T   G   E   T   Y   T   C   V       A   H   E   A   L   P   N
X14940  ,IGHM*01, (CH4)             GAG GAA TGG AAC ACG GGG GAG ACC TAC ACC TGC GTG ... GCC CAT GAG GCC CTG CCC AAC
                                                                                             D
J00260  ,IGHM*02                    --- --- --- --- --- --- --- --- --- --- --- --- ... --- --- --C --- --- --- ---
                                                                                 V
X57331  ,IGHM*03                    --- --- --- --- --- --- --- --- --- --- --- --- GTG --- --- --- --- --- --- ---
                                    101 102 103 104 105 106 107 108 109 110 111 112 113 114 115 116 117 118 119 120
                                     R   V   T   E   R   T   V   D   K   S   T   G   K   P   T   L   Y   N   V   S
X14940  ,IGHM*01, (CH4)             AGG GTC ACC GAG AGG ACC GTG GAC AAG TCC ACC GGT AAA CCC ACC CTG TAC AAC GTG TCC
J00260  ,IGHM*02                    --- --- --- --- --- --- --- --- --- --- --- --- --- --- --- --- --- --- --- ---
X57331  ,IGHM*03                    --- --- --- --- --- --- --- --- --- --- --- --- --- --- --- --- --- --- --- ---
```

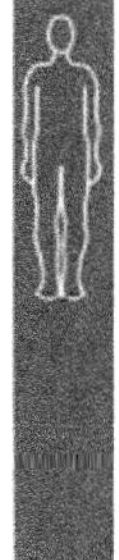

```
                            121 122 123 124 125 126 127 128 129 130 131
                             L   V   M   S   D   T   A   G   T   C   Y  *
X14940  ,IGHM*01, (CH4)     CTG GTC ATG TCC GAC ACA GCT GGC ACC TGC TAC
J00260  ,IGHM*02            --- --- --- --- --- --- --- --- --- --- ---
X57331  ,IGHM*03            --- --- --- --- --- --- --- --- --- --- ---

                              1   2   3   4   5   6   7   8   9  10  11  12  13  14  15  16  17  18  19  20
                              E   G   E   V   S   A   D   E   E   G   F   E   N   L   W   A   T   A   S   T
X14939  ,IGHM*01 (M1)      (G)AG GGG GAG GTG AGC GCC GAC GAG GAG GGC TTT GAG AAC CTG TGG GCC ACC GCC TCC ACC
X57331  ,IGHM*03           (-)-- --- --- --- --- --- --- --- --- --- --- --- --- --- --- --- --- --- --- ---
                             21  22  23  24  25  26  27  28  29  30  31  32  33  34  35  36  37  38  39
                              F   I   V   L   F   L   L   S   L   F   Y   S   T   T   V   T   L   F   K
X14939  ,IGHM*01 (M1)       TTC ATC GTC CTC TTC CTC CTG AGC CTC TTC TAC AGT ACC ACC GTC ACC TTG TTC AAG
X57331  ,IGHM*03            --- --- --- --- --- --- --- --- --- --- --- --- --- --- --- --- --- --- ---

                              1   2
                              V   K  *
X14939  ,IGHM*01 (M2)       GTG AAA
X57331  ,IGHM*03            --- ---
```

Genome database accession numbers

GDB:120086 LocusLink: 3507 OMIM: 147020

Protein display for the IGHC group

Figure

Amino acids resulting from the splicing are shown between parentheses.
N-glycosylation (NXS/T) and O-glycosylation sites (S or T) are boxed.
Amino acid sequences are deduced from the alleles *01.

Notes:

(1) The four IGHG3 hinge exons H1, H2, H3 and H4, observed in alleles*01 and *02, are shown on a same line. H1, H2 and H4 are present in allele*03, and H1 and H4 in allele*04.
(2) In the IGHA1 and IGHA2 genes, the hinge (H) and the CH2 are joined in a single exon, and there is a unique transmembrane M exon.
(3) The 9-amino acid CH-S of the IGHD gene is encoded on a separate exon.
(4) Extracellular, transmembrane and intracytoplasmic regions are according to [21].

```
IGHC
CH1
IGHA1   (A)SPTSPKVFPLSLCSTQP...DGNVVIACLVQGFFP.QEPLSVTWSESGQGV...TARNFPPSQDASGD...LYTTSSQLTLPATQC..LAGKSVTCHVKHYTN...PSQDVTVPCP
IGHA2   (A)SPTSPKVFPLSLDSTPQ...DGNVVVACLVQGFFP.QEPLSVTWSESGQNV...TARNFPPSQDASGD...LYTTSSQLTLPATQC..PDGKSVTCHVKHYTN...PSQDVTVPCP
IGHD    (A)PTKAPDVFPIISGCRHPKD.NSPVVLACLITGYHP..TSVTVTWYMGTQSQ...PQRTFPEIQRRDS....YYMTSSQLSTPLQQW...RQGEYKCVVQHTAS...KSKKEIFRWP
IGHE    (A)STQSPSVFPLTRCCKNIPSNATSVTLGCLATGYFP..EPVMVTCDTGSLNG...TTMTLPATTLTLSG...HYATISLLTVSGAWA....KQMFTCRVAHTPSSTDWVDNKTFS
IGHG1   (A)STKGPSVFPLAPSSKSTS..GGTAALGCLVKDYFP..EPVTVSWNSGALTS...GVHTFPAVLQSSG....LYSLSSVVTVPSSSL...GTQTYICNVNHKPSN..TKVDKKV
IGHG2   (A)STKGPSVFPLAPCSRSTS..ESTAALGCLVKDYFP..EPVTVSWNSGALTS...GVHTFPAVLQSSG....LYSLSSVVTVPSSNF...GTQTYTCNVDHKPSN..TKVDKTV
IGHG3   (A)STKGPSVFPLAPCSRSTS..GGTAALGCLVKDYFP..EPVTVSWNSGALTS...GVHTFPAVLQSSG....LYSLSSVVTVPSSSL...GTQTYTCNVNHKPSN..TKVDKRV
IGHG4   (A)STKGPSVFPLAPCSRSTS..ESTAALGCLVKDYFP..EPVTVSWNSGALTS...GVHTFPAVLQSSG....LYSLSSVVTVPSSSL...GTKTYTCNVDHKPSN..TKVDKRV
IGHM    (G)SASAPTLFPLVSCENSPSD.TSSVAVGCLAQDFLP..DSITFSWKYKNNSDIS.STRGFPSVLRGG.....KYAATSQVLLPSKDVMQGTDEHVVCKVQHPN....GNKEKNVPLP
Hinge
IGHA1   (V)PSTPPTPSPSTPPTPSPS
IGHA2   (V)PPPPP
IGHD    (E)SPKAQASSVPTAQPQAEGSLAKATTAPATTRNT   (G)RGGEEKKKEKEKEEQEERETKTP
IGHG1   (E)PKSCDKTHTCPPCP
IGHG2      (E)RKCCVECPPCP
IGHG3 (E)LKTPLGDTTHTCPRCP    (E)PKSCDTPPPCPRCP    (E)PKSCDTPPPCPRCP    (E)PKSCDTPPPCPRCP   (1)
IGHG4      (E)SKYGPPCPSCP
```

```
CH2
IGHA1  (2) CCHPRLSLHRPALEDLLL.GSEANLTCTLTGLRDASG.VTFTWTPSSGKS...AVQGPPERDLCG....CYSVSSVLPGCAEPW..NHGKTFTCTAAYPESK..TPLTATLSKS
IGHA2      CCHPRLSLHRPALEDLLL.GSEANLTCTLTGLRDASG.ATFTWTPSSGKS...AVQGPPERDLCG....CYSVSSVLPGCAQPW..NHGETFTCTAAHPELK..TPLTANITKS
IGHD  (R)CPSHTQPLGVYLLTPAVQDLW.LRDKATFTCFVVGSDLKD..AHLTWEVAGKVPTG.GVEEGLLERHSNG    SQSQHSRLTLPRSLW..NAGTSVTCTLNHPSL..PPQRLMALREP
IGHE  (V)CSRDFTPPTVKILQSSCDGGGHFPPTIQLLCLVSGYTPGT..INITWLEDGQVMD..VDLSTASTTQEGE....LASTQSELTLSQKHW..LSDRTYTCQVTYQG...HTFEDSTKKCA
IGHG1  (A)PELLGGPSVFLFPPKPKDTLMISRTPEVTCVVVDVSHEDPEVKFNWYVDGVEVH..NAKTKPREEQYNS    TYRVVSVLTVLHQDW..LNGKEYKCKVSNKALP..APIEKTISKAK
IGHG2  (A)P.PVAGPSVFLFPPKPKDTLMISRTPEVTCVVVDVSHEDPEVQFNWYVDGVEVH..NAKTKPREEQFNS    TFRVVSVLTVVHQDW..LNGKEYKCKVSNKGLP..APIEKTISKTK
IGHG3  (A)PELLGGPSVFLFPPKPKDTLMISRTPEVTCVVVDVSHEDPEVQFKWYVDGVEVH..NAKTKPREEQYNS    TFRVVSVLTVLHQDW..LNGKEYKCKVSNKALP..APIEKTISKTK
IGHG4  (A)PEFLGGPSVFLFPPKPKDTLMISRTPEVTCVVVDVSQEDPEVQFNWYVDGVEVH..NAKTKPREEQFNS    TYRVVSVLTVLHQDW..LNGKEYKCKVSNKGLP..SSIEKTISKAK
IGHM    (V)IAELPPKVSVFVPPRDGFFGNPRKSKLICQATGFSPRQ..IQVSWLREGKQVGS.GVTTDQVQAEAKESGPTTYKVTSTLTIKESDW..LSQSMFTCRVDHRG...LTFQQNASSMCVP
CH3
IGHA1     (G)NTFRPEVHLLPPPSEELAL.NELVTLTCLARGFSPKD..VLVRWLQGSQELPREKYLTWASRQEPSQGTT.TFAVTSILRVAAEDW..KKGDTFSCMVGHEALPL.AFTQKTIDRLAGKPTHVNVSVVMAEVDGTCY
IGHA2     (G)NTFRPEVHLLPPPSEELAL.NELVTLTCLARGFSPKD..VLVRWLQGSQELPREKYLTWASRQEPSQGTT.TFAVTSILRVAAEDW..KKGDTFSCMVGHEALPL.AFTQKTIDRLAGKPTHVNVSVVMAEVDGTCY
IGHD      (A)AQAPVKLSLNLLASSDPP..EAASWLLCEVSGFSPPN..ILLMWLEDQREVNTSGFAPARPPPQPRST...TFWAWSVLRVPAPPS..PQPATYTCVVSHEDSRTLLNASRSLEVS    (Y)VTDHGPMK  (3)
IGHE     (D)SNPRGVSAYLSRPSPFDLFI.RKSPTITCLVVDLAPSKGTVNLTWSRASGKPV..NHSTRKEEKQRNG      TLTVTSTLPVGTRDW..IEGETYQCRVTHPHLP..RALMRSTTKTS
IGHG1     (G)QPREPQVYTLPPSRDELT..KNQVSLTCLVKGFYPSD..IAVEWESNGQPEN..NYKTTPPVLDSDG....SFFLYSKLTVDKSRW..QQGNVFSCSVMHEALHN.HYTQKSLSLSPGK
IGHG2     (G)QPREPQVYTLPPSREEMT..KNQVSLTCLVKGFYPSD..IAVEWESNGQPEN..NYKTTPPMLDSDG....SFFLYSKLTVDKSRW..QQGNVFSCSVMHEALHN.HYTQKSLSLSPGK
IGHG3     (G)QPREPQVYTLPPSREEMT..KNQVSLTCLVKGFYPSD..IAVEWESSGQPEN..NYNTTPPMLDSDG....SFFLYSKLTVDKSRW..QQGNIFSCSVMHEALHN.RFTQKSLSLSPGK
IGHG4     (G)QPREPQVYTLPPSQEEMT..KNQVSLTCLVKGFYPSD..IAVEWESNGQPEN..NYKTTPPVLDSDG....SFFLYSRLTVDKSRW..QEGNVFSCSVMHEALHN.HYTQKSLSLSLGK
IGHM      (D)QDTAIRVFAIPPSFASIFL.TKSTKLTCLVTDLTTYDS.VTISWTRQNGEAV..KTHTNISESHPNA      TFSAVGEASICEDDW..NSGERFTCTVTHTDLP..SPLKQTISRPK
CH4
IGHE      (G)PRAAPEVYAFATPEWPGS..RDKRTLACLIQNFMPED..ISVQWLHNEVQLPDARHSTTQPRKTKGS....GFFVFSRLEVTRAEW..EQKDEFICRAVHEAASPSQTVQRAVSVNPGK
IGHM     (G)VALHRPDVYLLPPAREQLNL.RESATITCLVTGFSPAD..VFVQWMQRGQPLSPEKYVTSAPMPEPQAPG..RYFAHSILTVSEEEW..NTGETYTCV.AHEALPN.RVTERTVDKSTGKPTLYNVSLVMSDTAGTCY
M
IGHA1                  (G)SCSVADWQMPPPYVVLDLPQETLEEETPGANLWPTTITFLTLFLLSLFYSTALTVTSVRGPSGNREGPQY
IGHA2                  (G)SCCVADWQMPPPYVVLDLPQETLEEETPGANLWPTTITFLTLFLLSLFYSTALTVTSVRGPSGKREGPQY
                                                      M1                          M2
IGHD                        (Y)LAMTPLIPQSKDENSDDYTTFDDVGSLWTTLSTFVALFILTLLYSGIVTFIK       VK
IGHE                                    (E)LDVCVEEAEGEAPWTWTGLCIFAALFLLSVSYSAALTLLM       VQRFLSATRQGRPQTSLDYTNVLQPHA
IGHG1                                 (E)LQLEESCAEAQDGELDGLWTTITIFITLFLLSVCYSATVTFFK       VKWIFSSVVDLKQTIIPDYRNMIGQGA
IGHG2                                 (E)LQLEESCAEAQDGELDGLWTTITIFITLFLLSVCYSATITFFK       VKWIFSSVVDLKQTIVPDYRNMIRQGA
IGHG3                                 (E)LQLEESCAEAQDGELDGLWTTITIFITLFLLSVCYSATVTFFK       VKWIFSSVVDLKQTIIPDYRNMIGQGA
IGHM                                       (E)GEVSADEEGFENLWATASTFIVLFLLSLFYSTTVTLFK       VK
                                   Extracellular region | Transmembrane region   | intracytoplasmic region  (4)
```

IGLC

```
IGLC1   (G)QPKANPTVTLFPPSSEELQ..ANKATLVCLISDFYPGA..VTVAWKADGSPVKA.GVETTKPSKQSNN....KYAASSYLSLTPEQW..KSHRSYSCQVTHEG....STVEKTVAPTECS
IGLC2   (G)QPKAAPSVTLFPPSSEELQ..ANKATLVCLISDFYPGA..VTVAWKADSSPVKA.GVETTTPSKQSNN....KYAASSYLSLTPEQW..KSHRSYSCQVTHEG....STVEKTVAPTECS
IGLC3   (G)QPKAAPSVTLFPPSSEELQ..ANKATLVCLISDFYPGA..VTVAWKADSSPVKA.GVETTTPSKQSNN....KYAASSYLSLTPEQW..KSHKSYSCQVTHEG....STVEKTVAPTECS
IGLC6   (G)QPKAAPSVTLFPPSSEELQ..ANKATLVCLISDFYPGA..VKVAWKADGSPVNT.GVETTTPSKQSNN....KYAASSYLSLTPEQW..KSHRSYSCQVTHEG....STVEKTVAPAECS
IGLC7   (G)QPKAAPSVTLFPPSSEELQ..ANKATLVCLVSDFYPGA..VTVAWKADGSPVKV.GVETTKPSKQSNN....KYAASSYLSLTPEQW..KSHRSYSCRVTHEG....STVEKTVAPAECS
```

IGKC

```
IGKC    (R)TVAAPSVFIFPPSDEQLK..SGTASVVCLLNNFYPRE  AKVQWKVDNALQSG NSQESVTEQDSKDS...TYSLSSTLTLSKADY..EKHKVYACEVTHQGLS..SPVTKSFNRGEC
```

References

[1] Takahashi, N. et al. (1980) Nucleic Acids Res. 8, 5983–5991.
[2] Ellison, J.W. et al. (1981) DNA 1, 11–18.
[3] Rabbitts, T.H. et al. (1981) Nucleic Acids Res. 9, 4509–4524.
[4] Battey, J. et al. (1982) Proc. Natl Acad. Sci. USA 79, 5956–5960.
[5] Ellison, J.W. et al. (1982) Nucleic Acids Res. 10, 4071–4079.
[6] Ellison, J.W. and Hood, L.E. (1982) Proc. Natl Acad. Sci. USA 79, 1984–1988.
[7] Tashita, H. et al. (1998) J. Clin. Invest. 101, 677–681.
[8] Brusco, A. et al. (1998) Eur. J. Immunogenet. 25, 349–355.
[9] Max, E.E. et al. (1982) Cell 29, 691–699.
[10] Takahashi, N. et al. (1982) Cell 29, 671–679.
[11] Ueda, S. et al. (1982) EMBO J. 1, 1539–1544.
[12] Zheng, S. et al. unpublished.
[13] Flanagan, J.G. et al. (1984) Cell 36, 681–688.
[14] Lefranc, M.P. and Rabbitts, T.H. (1984) Nucleic Acids Res. 12, 1303–1311.
[15] White, M.B. et al. (1985) Science 228, 733–737.
[16] Huck, S. et al. (1986) FEBS Lett. 208, 221–230.
[17] Huck, S. et al. (1986) Nucleic Acids Res. 14, 1779–1789.
[18] Huck, S. et al. (1989) Immunogenetics 30, 250–257.
[19] Balbin, M. et al. (1994) Immunogenetics 39, 187–193.
[20] Dorai, H. and Gillies, S.D. (1989) Nucleic Acids Res. 17, 6412.
[21] Bensmana, M. and Lefranc, M.P. (1990) Immunogenetics 32, 321–330.
[22] Word, C.J. et al. (1989) Int. Immunol. 1, 296–309.
[23] Kinoshita, K. et al. (1991) Immunol. Lett. 27, 151–155.
[24] Zhang, K. et al. (1992) J. Exp. Med. 176, 233–243.
[25] Walls, M.A. et al. (1993) Nucleic Acids Res. 21, 2921–2929.
[26] Chintalacharuvu, K. R. et al. (1994) J. Immunol. 154, 5299–5304.
[27] Brusco, A. et al. (1995) Immunogenetics 42, 414–417.
[28] Akahori, Y. and Kurosawa, Y. (1997) Genomics 41, 100–104.
[29] Dard, P. et al. (1997) Hum. Genet. 99, 138–141.

Part 2

IGHD

Nomenclature

Immunoglobulin heavy diversity group.

Definition and functionality

The human IGHD group comprises 27 mapped genes, of which 23 are functional. Four IGHD which have non-conserved 5'D-HEPTAMER (and 3'D-HEPTAMER, for one of them) are considered as ORF. The IGHD genes have one, two or three ORF in both the direct and inverted orientations.

Gene location

The human IGHD genes are located in the IGH locus on chromosome 14 at 14q32.33, between the IGHV and IGHJ genes, and upstream from the IGHC genes. There are 54 kb between IGHD1-1 (the most 5' IGHD gene) and IGHD7-27 (the most 3' IGHD gene). IGHD7-27 is localized 63 bp downstream of the pseudogene IGHJ1P (the first most 5' IGHJ gene) and 93 bp upstream from IGHJ1 (the first most 5' functional IGHJ gene).

Nucleotide and amino acid sequences for the human IGHD genes with nomenclature

Some of the D-REGIONs having been found in inverted orientation [Yamada, M. et al., J. Exp. Med., 173, 395–407 (1991); Tuaillon, N. et al., J. Immunol., 154, 6453–6465 (1995)] the IGH D-REGIONs are shown in both direct and inverted orientations. References are shown in square brackets.

```
                                      Direct 5'- 3' orientation     Inverted orientation

IGHD1-1
Immunoglobulin heavy diversity 1-1
                                       G  T  T  G  T                 V  V  P  V  V
                                        V  Q  L  E  R                 S  F  Q  L  Y
                                         Y  N  W  N  D                 R  S  S  C  T
X97051  ,IGHD1-1*01, D1-1      [6]     ggtacaactggaacgac             gtcgttccagttgtacc

IGHD1-7
Immunoglobulin heavy diversity 1-7
                                       G  I  T  G  T                 V  V  P  V  I
                                        V  *  L  E  L                 *  F  Q  L  Y
                                         Y  N  W  N  Y                 S  S  S  Y  T
X13972  ,IGHD1-7*01, DM1       [5]     ggtataactggaactac             gtagttccagttatacc

IGHD1-14
Immunoglobulin heavy diversity 1-14
                                       G  I  T  G  T                 V  V  P  V  I
                                        V  *  P  E  P                 W  F  R  L  Y
                                         Y  N  R  N  H                 G  S  G  Y  T
X13972  ,IGHD1-14*01, DM2      [5]     ggtataaccggaaccac             gtggttccggttatacc

IGHD1-20
Immunoglobulin heavy diversity 1-20
                                       G  I  T  G  T                 V  V  P  V  I
                                        V  *  L  E  R                 S  F  Q  L  Y
                                         Y  N  W  N  D                 R  S  S  Y  T
X97051  ,IGHD1-20*01, D1-20    [6]     ggtataactggaacgac             gtcgttccagttatacc

IGHD1-26
Immunoglobulin heavy diversity 1-26
                                       G  I  V  G  A  T              V  V  A  P  T  I
                                        V  *  W  E  L  L              *  *  L  P  L  Y
                                         Y  S  G  S  Y  Y              S  S  S  H  Y  T
X97051  ,IGHD1-26*01, D1-26    [6]     ggtatagtgggagctactac          gtagtagctcccactatacc
```

IGHD2-2
Immunoglobulin heavy diversity 2-2

```
                                  R I L * * Y Q L L C          G I A A G T T T I S
                                   G Y C S S T S C Y A          A * Q L V L L Q Y P
                                    D I V V V P A A M            H S S W Y Y Y N I
J00232  ,IGHD2-2*01, D4     [1]   aggatattgtagtagtaccagctgctatgcc  ggcatagcagctggtactactacaatatcct
                                  R I L * * Y Q L L Y          G I A A G T T T I S
                                   G Y C S S T S C Y T          Y * Q L V L L Q Y P
                                    D I V V V P A A I            Y S S W Y Y Y N I
X97051  ,IGHD2-2*02, D2-2   [6]   aggatattgtagtagtaccagctgctatacc  ggtatagcagctggtactactacaatatcct
                                  W I L * * Y Q L L C          G I A A G T T T I S
                                   G Y C S S T S C Y A          A * Q L V L L Q Y P
                                    D I V V V P A A M            H S S W Y Y Y N I
M35648  ,IGHD2-2*03, D4     [3]   tggatattgtagtagtaccagctgctatgcc  ggcatagcagctggtactactacaatatcca
```

IGHD2-8
Immunoglobulin heavy diversity 2-8

```
                                  R I L Y * W C M L Y          G I A Y T I S T I S
                                   G Y C T N G V C Y T          V * H T P L V Q Y P
                                    D I V L M V Y A I            Y S I H H * Y N I
X13972  ,IGHD2-8*01, DLR1   [5]   aggatattgtactaatggtgtatgctatacc  ggtatagcatacaccattagtacaatatcct
                                  R I L Y W W C M L Y          G I A Y T T S T I S
                                   G Y C T G G V C Y T          V * H T P P V Q Y P
                                    D I V L V V Y A I            Y S I H H Q Y N I
J00233  ,IGHD2-8*02, D1     [1]   aggatattgtactggtggtgtatgctatacc  ggtatagcatacaccaccagtacaatatcct
```

IGHD2-15
Immunoglobulin heavy diversity 2-15

```
                                  R I L * W W * L L L          G V A A T T T T I S
                                   G Y C S G G S C Y S          E * Q L P P L Q Y P
                                    D I V V V V A A T            S S S Y H H Y N I
J00234  ,IGHD2-15*01, D2    [1]   aggatattgtagtggtggtagctgctactcc  ggagtagcagctaccaccactacaatatcct
```

IGHD2-21
Immunoglobulin heavy diversity 2-21

```
                                  S I L W W * L L F            G I A I T T T I C
                                   A Y C G G D C Y S            E * Q S P P Q Y A
                                    H I V V V I A I              N S N H H H N M
J00235  ,IGHD2-21*01, D3    [1]   agcatattgtggtggtgattgctattcc     ggaatagcaatcaccaccacaatatgct
                                  S I L W W * L L F            G I A V T T T I C
                                   A Y C G G D C Y S            E * Q S P P Q Y A
                                    H I V V V T A I              N S S H H H N M
X97051  ,IGHD2-21*02, D2-21 [6]   agcatattgtggtggtgactgctattcc     ggaatagcagtcaccaccacaatatgct
```

IGHD3-3
Immunoglobulin heavy diversity 3-3

```
                                  V L R F L E W L L Y          G I I T T P K I V I
                                   Y Y D F W S G Y Y T          V * * P L Q K S * Y
                                    I T I F G V V I I            Y N N H S K N R N
X13972  ,IGHD3-3*01, DXP4   [5]   gtattacgattttggagtggttattatacc   ggtataataaccactccaaaaatcgtaatac
                                  V L A F L E W L L Y          G I I T T P K M L I
                                   Y * H F W S G Y Y T          V * * P L Q K C * Y
                                    I S : F G V V I I            Y N N H S K N A N
X93618  ,IGHD3-3*02, D23/7  [4]   gtattagcattttggagtggttattatacc   ggtataataaccactccaaaaatgctaatac
```

IGHD3-9
Immunoglobulin heavy diversity 3-9

```
                                  V L R Y F D W L L *          V I I T S Q N I V I
                                   Y Y D I L T G Y Y N          L * * P V K I S * Y
                                    I T I F * L V I I            Y N N Q S K Y R N
X13972  ,IGHD3-9*01, DXP1   [5]   gtattacgatattttgactggttattataac  gttataataaccagtcaaaatatcgtaatac
```

IGHD3-10
Immunoglobulin heavy diversity 3-10

```
                                  V L L W F G E L L *          V I I T P R T I V I
                                   Y Y Y G S G S Y Y N          L * * L P E P * * Y
                                    I T M V R G V I I            Y N N S P N H S N
X13972  ,IGHD3-10*01,DXP'1  [5]   gtattactatggttcggggagttattataac  gttataataactccccgaaccatagtaatac
                                  V L L C S G S Y Y N          V I I T P R T * * Y
                                   Y Y Y V R G V I I            L * * L P E H S N
                                    I T M F G E L L *            Y N N S P N I Y I
X93615  ,IGHD3-10*02, D21/7 [4]   gtattactatg.ttcggggagttattataac  gttataataactccccgaac.atagtaatac
```

IGHD3-16
Immunoglobulin heavy diversity 3-16

```
                               V L * L R L G E L C L Y     G I S I T P P N V I I I
                                Y Y D Y V W G S Y A Y T     V * A * L P Q T * S * Y
                                 I M I T F G G V M L I       Y K H N S P K R N H N
X93614 ,IGHD3-16*01,D21/10[4] gtattatgattacgtttgggggagttatgcttatacc ggtataagcataactcccccaaacgtaatcataatac
```

IGHD3-22
Immunoglobulin heavy diversity 3-22

```
                                  V L L * * * W L L L          V V I T T T I I V I
                                   Y Y Y D S S G Y Y Y          * * * P L L S * * Y
                                    I T M I V V V I T            S N N H Y Y H S N
X93616  ,IGHD3-22*01, D21/9 [4]   gtattactatgatagtagtggttattactac  gtagtaataaccactactatcatagtaatac
```

IGHD4-4
Immunoglobulin heavy diversity 4-4

```
                                  * L Q * L                    V V T V V
                                   D Y S N Y                    * L L * S
                                    T T V T                      S Y C S
X13972  ,IGHD4-4*01, DA4    [5]   tgactacagtaactac                 gtagttactgtagtca
```

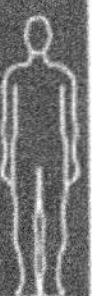

```
IGHD4-11
Immunoglobulin heavy diversity 4-11
                                             * L Q * L                      V V T V V
                                              D Y S N Y                      * L L * S
                                               T T V T                      S Y C S
X13972  ,IGHD4-11*01, DA1     [5]   tgactacagtaactac             gtagttactgtagtca

IGHD4-17
Immunoglobulin heavy diversity 4-17
                                             * L R * L                      V V T V V
                                              D Y G D Y                      * S P * S
                                               T T V T                      S H R S
X97051  ,IGHD4-17*01, D4-17   [6]   tgactacggtgactac             gtagtcaccgtagtca

IGHD4-23
Immunoglobulin heavy diversity 4-23
                                             * L R W * L                    G V T T V V
                                              D Y G G N S                    E L P P * S
                                               T T V V T                      S Y H R S
X97051  ,IGHD4-23*01, D4-23   [6]   tgactacggtggtaactcc          ggagttaccaccgtagtca

IGHD5-5
Immunoglobulin heavy diversity 5-5
                                             V D T A M V                    V T I A V S
                                              W I Q L W L                    * P * L Y P
                                               G Y S Y G Y                    N H S C I H
X13972  ,IGHD5-5*01, DK4      [5]   gtggatacagctatggttac         gtaaccatagctgtatccac

IGHD5-12
Immunoglobulin heavy diversity 5-12
                                             V D I V A T I                  V I V A T I S
                                              W I * W L R L                  * S * P L Y P
                                               G Y S G Y D Y                  N R S H Y I H
X13972  ,IGHD5-12*01, DK1     [5]   gtggatatagtggctacgattac      gtaatcgtagccactatatccac

IGHD5-18
Immunoglobulin heavy diversity 5-18
                                             V D T A M V                    V T I A V S
                                              W I Q L W L                    * P * L Y P
                                               G Y S Y G Y                    N H S C I H
X97051  ,IGHD5-18*01, D5-18   [6]   gtggatacagctatggttac         gtaaccatagctgtatccac

IGHD5-24
Immunoglobulin heavy diversity 5-24
                                             V E M A T I                    V I V A I S
                                              * R W L Q L                    * L * P S L
                                               R D G Y N Y                    N C S H L Y
X97051  ,IGHD5-24*01, D5-24   [6]   gtagagatggctacaattac         gtaattgtagccatctctac

IGHD6-6
Immunoglobulin heavy diversity 6-6
                                             E Y S S S S                    G R A A I L
                                              S I A A R                      D E L L Y
                                               V * Q L V                      T S C Y T
X13972  ,IGHD6-6*01, DN4      [5]   gagtatagcagctcgtcc           ggacgagctgctatactc

IGHD6-13
Immunoglobulin heavy diversity 6-13
                                             G Y S S S W Y                  V P A A A I P
                                              G I A A A G                    Y Q L L L Y
                                               V * Q Q L V                    T S C C Y T
X13972  ,IGHD6-13*01, DN1     [5]   gggtatagcagcagctggtac        gtaccagctgctgctataccc

IGHD6-19
Immunoglobulin heavy diversity 6-19
                                             G Y S S G W Y                  V P A T A I P
                                              G I A V A G                    Y Q P L L Y
                                               V * Q W L V                    T S H C Y T
X97051  ,IGHD6-19*01, D6-19   [6]   gggtatagcagtggctggtac        gtaccagccactgctataccc

IGHD6-25
Immunoglobulin heavy diversity 6-25
                                             G Y S S G Y                    V A A A I P
                                              G I A A A                      * P L L Y
                                               V * Q R L                      S R C Y T
X97051  ,IGHD6-25*01, D6-25   [6]   gggtatagcagcggctac           gtagccgctgctataccc

IGHD7-27
Immunoglobulin heavy diversity 7-27
                                             L T G                          S P V
                                              * L G                          P Q L
                                               N W G                          P S *
J00256  ,IGHD7-27*01, DHQ52   [2]   ctaactgggga                  tccccagttag
```

Recombination signals

Non-conserved nucleotides taken into account for the ORF functionality definition are shown in bold and italic.

5'D recombination signal			IGHD gene and allele name	3'D recombination signal		
5'D-NONAMER	(bp)	5'D-HEPTAMER		3'D-HEPTAMER	(bp)	3'D-NONAMER
AGATTCTGA	12	CACGGTG	IGHD1-1*01	CACCGTG	12	TCCAAAACT
GGATTCTGA	12	CACAGTG	IGHD1-7*01	CACTGTG	12	TCCAAAACG
GGATTCCGA	12	CACAG***C***G	IGHD1-14*01 (ORF)	CACTGT***C***	12	TCAAAAACT
GGATTCTGA	12	CACAGTG	IGHD1-20*01	CACCGTG	12	TCCAAAACT
GGATTCTGA	12	CACGGTG	IGHD1-26*01	CACTGTG	12	TCCAAAACT
GGATTTTGT	12	CACTGTG	IGHD2-2*01	CACAGTG	12	nd
GGATTTTGT	12	CACTGTG	IGHD2-2*02	CACAGTG	12	TCCCAAAGC
GGATTTTGT	12	CACTGTG	IGHD2-2*03	CACAGTG	12	TCCCAAAGC
GGATTTTGT	12	CACTGTG	IGHD2-8*01	CACAGTG	12	TCCCAAAGC
GGATTTTGT	12	CACTGTG	IGHD2-8*02	CACAGTG	12	nd
GGATTTTGT	12	CACTGTG	IGHD2-15*01	CACAGTG	12	TCCCAAAGC
GGATTTTGT	12	CACTGTG	IGHD2-21*01	CACAGTG	12	nd
GGATTTTGT	12	CACTGTG	IGHD2-21*02	CACAGTG	12	TCCTAAAGC
GGTTTGGGG	12	CACTGTG	IGHD3-3*01	CACAGTG	12	TCAAAAACC
GGTTTGGGG	12	CACTGTG	IGHD3-3*02	CACAGTG	12	TCAAAAACC
GGTTTAGAA	12	CACTGTG	IGHD3-9*01	CACAGTG	12	TCAAAAACC
GGTTTGGGG	12	CACTGTG	IGHD3-10*01	CACAGTG	12	TCAAAAACC
GGTTTGGGG	12	CACTGTG	IGHD3-10*02	CACAGTG	12	TCAAAAACC
GGTTTGAAG	12	CACTGTG	IGHD3-16*01	CACAGCA	12	TCAGAAACC
GGTTTGAAG	12	CACTGTG	IGHD3-22*01	CACAGTG	12	TCAAAAACT
GCTTTTTTGT	12	TACTGTG	IGHD4-4*01	CACAGTG	12	GCAAAAACT
GCTTTTTTGT	12	T***G***CTGTG	IGHD4-11*01 (ORF)	CATAGTG	12	GCAAAAACT
GCTTTTTTGT	12	TACTGTG	IGHD4-17*01	CACAGTG	12	GCAAAAACT
GCTTTTTTGT	12	T***G***CTGTG	IGHD4-23*01 (ORF)	CACAGTG	12	GCAAAAACT
GGTTATTGT	12	GACTGTG	IGHD5-5*01	CACAGTG	12	GCAGCAACC
GGTTATTGT	12	GACTGTG	IGHD5-12*01	CACAGTG	12	GCAGCAACC
GGTTATTGT	12	GACTGTG	IGHD5-18*01	CACAGTG	12	GCAGCAACC
GGTTATTGT	12	G***G***CCGTG	IGHD5-24*01 (ORF)	CACAGTG	12	GCAGCAACC
(C/A)GTTTCTGA	12	CACAGTG	IGHD6-6*01	CACAGTG	12	CCAGAAACC
GGTTTCTGA	12	CACAGTG	IGHD6-13*01	CACAGTG	12	CCAGAAACC
GGTTTCTGA	12	CACAGTG	IGHD6-19*01	CACAGTG	12	CCAGAAACC
GGTTTCTGA	12	CACAGTC	IGHD6-25*01	CACAATG	12	ACAGAAACC
GGTTTTGG(G/C)	12	CACTGTG	IGHD7-27*01	CACAGTG	12	ACAAAAACC

nd: not defined

IGHD group references

References

[1] Siebenlist, U. et al. (1981) Nature 294, 631–635.
[2] Ravetch, J.V. et al. (1981) Cell 27, 583–591.
[3] Zong, S.Q. et al. (1988) Immunol. Lett. 17, 329–334.
[4] Buluwela, L. et al. (1988) EMBO J. 7, 2003–2010.
[5] Ichihara, Y. et al. (1988) EMBO J. 7, 4141–4150.
[6] Corbett, S. et al. (1997) J. Mol. Biol. 270, 587–597.

Part 3

IGHJ

Nomenclature

Immunoglobulin heavy joining group.

Definition and functionality

The human IGHJ group comprises nine mapped genes of which six are functional (IGHJ1, IGHJ2, IGHJ3, IGHJ4, IGHJ5 and IGHJ6) and three are pseudogenes (IGHJ1P, IGHJ2P and IGHJ3P).

Gene location

The human IGHJ genes are located in the IGH locus on chromosome 14 at 14q32.33, upstream from the IGHC genes.

Nucleotide and amino acid sequences for the human functional IGHJ genes with nomenclature

The conserved **WGXG** motif, characteristic of the IGH J-REGION is underlined. References are shown in square brackets.

```
IGHJ1
Immunoglobulin heavy joining 1
                                           A   E   Y   F   Q   H   W   G   Q   G   T   L   V   T   V   S   S
J00256 ,IGHJ1*01 (1)[1]                   GCT GAA TAC TTC CAG CAC TGG GGC CAG GGC ACC CTG GTC ACC GTC TCC TCA G
IGHJ2
Immunoglobulin heavy joining 2
                                           Y   W   Y   F   D   L   W   G   R   G   T   L   V   T   V   S   S
J00256 ,IGHJ2*01 (1)[1]                 C TAC TGG TAC TTC GAT CTC TGG GGC CGT GGC ACC CTG GTC ACT GTC TCC TCA G
IGHJ3
Immunoglobulin heavy joining 3
                                               D   A   F   D   V   W   G   Q   G   T   M   V   T   V   S   S
J00256 ,IGHJ3*01 (1)[1]                     T GAT GCT TTT GAT GTC TGG GGC CAA GGG ACA ATG GTC ACC GTC TCT TCA G
                                                               I
X86355 ,IGHJ3*02 (2)[3]                     - --- --- --- --- A-- --- --- --- --- --- --- --- --- --- --- --- -
IGHJ4
Immunoglobulin heavy joining 4
                                                   Y   F   D   Y   W   G   Q   G   T   L   V   T   V   S   S
J00256 ,IGHJ4*01 (1)[1]                        AC TAC TTT GAC TAC TGG GGC CAA GGA ACC CTG GTC ACC GTC TCC TCA G
X86355 ,IGHJ4*02 (2)[3]                        -- --- --- --- --- --- --- --G --- --- --- --- --- --- --- --- -
M25625 ,IGHJ4*03 (5)[2]                        G- --- --- --- --- --- --- --- --G --- --- --- --- --- --- --- -
IGHJ5
Immunoglobulin heavy joining 5
                                               N   W   F   D   S   W   G   Q   G   T   L   V   T   V   S   S
J00256 ,IGHJ5*01 (1)[1]                    AC AAC TGG TTC GAC TCC TGG GGC CAA GGA ACC CTG GTC ACC GTC TCC TCA G
                                                               P
X86355 ,IGHJ5*02 (2)[3]                    -- --- --- --- --- C-- --- --- --G --- --- --- --- --- --- --- --- -
IGHJ6
Immunoglobulin heavy joining 6
                               Y   Y   Y   Y   Y   G   M   D   V   W   G   Q   G   T   T   V   T   V   S   S
J00256 ,IGHJ6*01 (1)[1]    AT TAC TAC TAC TAC TAC GGT ATG GAC GTC TGG GGG CAA GGG ACC ACG GTC ACC GTC TCC TCA G
X86355 ,IGHJ6*02 (3)[3]    -- --- --- --- --- --- --- --- --- --- --- --C --- --- --- --- --- --- --- --- --- .
                                                   Y                       K
X86356 ,IGHJ6*03 (4)[3]    -- --- --- --- --- --- TAC --- --- --- --- --C A-- --- --- --- --- --- --- --- --- .
```

Haplotypes

Haplotypes have been described in [3].

(1) sequenced in haplotype a.
(2) sequenced in haplotypes b1, b1c, b2, b3 and c, respectively.
(3) sequenced in haplotypes b1, b2 and b3, respectively.
(4) sequenced in haplotypes b1c and c, respectively.
(5) sequenced in haplotype d.

Recombination signals

J recombination signal			IGHJ gene and allele name
J-NONAMER	(bp)	J-HEPTAMER	
GTGTTTTGG	21	CACTGGC	IGHJ1P (1)
GGTTTCTGT	21 or 22	CACCGTG	IGHJ1*01
TGTTTTTGT	23	GACTGTG	IGHJ2*01
TGTTCATGT	20	TAGTGTG	IGHJ2P (1)
GGTTTTTGT	23	CCCTGTG	IGHJ3*01
GGTTTGTGT	23	CCCTGTG	IGHJ3*02
GGTTTTTGG	22	CAATGTG	IGHJ3P (1)
GGTTTTTGT	23	CAATGTG	IGHJ4*01
GGTTTTTGT	23	CAATGTG	IGHJ4*02
GGTTTTTGT	23	CAATGTG	IGHJ4*03
GTTCTTTGT	21	CAATGTG	IGHJ5*01
GTTCTTGCC	22	CAATGTG	IGHJ5*02
GGTTTTTGT	22	CATTGTG	IGHJ6*01
GGTTTTTGT	22	CATTGTG	IGHJ6*02
GGTTTTTGT	22	CATTGTG	IGHJ6*03

(1) Alleles are not described for the pseudogenes.

IGHJ group references

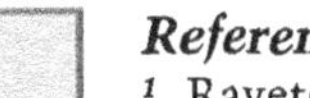

References

[1] Ravetch, J. V. et al. (1981) Cell 27, 583–591.

[2] Rabbitts, T. H. (1983) Biochem. Soc. Trans. 11, 119–126.

[3] Mattila, P. S. et al. (1995) Eur. J. Immunol. 25, 2578–2582.

Part 4

IGHV

IGHV1-2

Nomenclature

IGHV1-2: Immunoglobulin heavy variable 1-2.

Definition and functionality

IGHV1-2 is one of the nine mapped functional genes of the IGHV1 subgroup. This subgroup comprises 14 mapped genes in the IGH locus, two unmapped genes (of which one is an ORF and one is a functional gene) and 11 orphons.

Gene location

IGHV1-2 is in the IGH locus on chromosome 14 at 14q32.33.

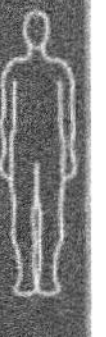

Nucleotide and amino acid sequences for human IGHV1-2

References are shown in square brackets.

```
                                        1   2   3   4   5   6   7   8   9   10  11  12  13  14  15  16  17  18  19  20
                                        Q   V   Q   L   V   Q   S   G   A       E   V   K   K   P   G   A   S   V   K
X07448,IGHV1-2*01,V35/VI-2b       [25]  CAG GTG CAG CTG GTG CAG TCT GGG GCT ... GAG GTG AAG AAG CCT GGG GCC TCA GTG AAG

X62106,IGHV1-2*02,VI-2            [39]  --- --- --- --- --- --- --- --- --- ... --- --- --- --- --- --- --- --- --- ---

Z14071,IGHV1-2*02,DP-75           [41]  --- --- --- --- --- --- --- --- --- ... --- --- --- --- --- --- --- --- --- ---

X59704,IGHV1-2*02,hv1L1           [29]  --- --- --- --- --- --- --- --- --- ... --- --- --- --- --- --- --- --- --- ---
                                                                                                L
X92208,IGHV1-2*03,1-1              [4]  --- --- --- --- --- --- --- --- --- ... --- --- --- --- T-- --- --- --- --- ---

Z12310,IGHV1-2*04,DP-8            [41]  --- --- --- --- --- --- --- --- --- ... --- --- --- --- --- --- --- --- --- ---

                                                                ___________________CDR1 IMGT____________________
                                        21  22  23  24  25  26  27  28  29  30  31  32  33  34  35  36  37  38  39  40
                                        V   S   C   K   A   S   G   Y   T   F   T   G   Y   Y                   M   H
X07448,IGHV1-2*01,V35/VI-2b             GTC TCC TGC AAG GCT TCT GGA TAC ACC TTC ACC GGC TAC TAT ... ... ... ... ATG CAC

X62106,IGHV1-2*02,VI-2                  --- --- --- --- --- --- --- --- --- --- --- --- --- --- ... ... ... ... --- ---

Z14071,IGHV1-2*02,DP-75                 --- --- --- --- --- --- --- --- --- --- --- --- --- --- ... ... ... ... --- ---

X59704,IGHV1-2*02,hv1L1                 --- --- --- --- --- --- --- --- --- --- --- --- --- --- ... ... ... ... --- ---

X92208,IGHV1-2*03,1-1                   --- --- --- --- --- --- --- --- --- --- --- --- --- --- ... ... ... ... --- ---

Z12310,IGHV1-2*04,DP-8                  --- --- --- --- --- --- --- --- --- --- --- --- --- --- ... ... ... ... --- ---

                                                                                                    _______________CDR2-
                                        41  42  43  44  45  46  47  48  49  50  51  52  53  54  55  56  57  58  59  60
                                        W   V   R   Q   A   P   G   Q   G   L   E   W   M   G   R   I   N   P   N   S
X07448,IGHV1-2*01,V35/VI-2b             TGG GTG CGA CAG GCC CCT GGA CAA GGG CTT GAG TGG ATG GGA CGG ATC AAC CCT AAC AGT
                                                                                                W
X62106,IGHV1-2*02,VI-2                  --- --- --- --- --- --- --- --- --- --- --- --- --- --- T-- --- --- --- --- ---
                                                                                                W
Z14071,IGHV1-2*02,DP-75                 --- --- --- --- --- --- --- --- --- --- --- --- --- --- T-- --- --- --- --- ---
                                                                                                W
X59704,IGHV1-2*02,hv1L1                 --- --- --- --- --- --- --- --- --- --- --- --- --- --- T-- --- --- --- --- ---
                                                                                                W
X92208,IGHV1-2*03,1-1                   --- --- -X- --- --- --- --- --- --- --- --- --- --- --- T-- --- --- --- --- ---
                                                                                                W
Z12310,IGHV1-2*04,DP-8                  --- --- --- --- --- --- --- --- --- --- --- --- --- --- T-- --- --- --- --- ---

                                        IMGT_______________
                                        61  62  63  64  65  66  67  68  69  70  71  72  73  74  75  76  77  78  79  80
                                        G   G   T           N   Y   A   Q   K   F   Q       G   R   V   T   S   T   R
X07448,IGHV1-2*01,V35/VI-2b             GGT GGC ACA ... ... AAC TAT GCA CAG AAG TTT CAG ... GGC AGG GTC ACC AGT ACC AGG
                                                                                                            M
X62106,IGHV1-2*02,VI-2                  --- --- --- ... ... --- --- --- --- --- --- --- ... --- --- --- --- -TG --- ---
                                                                                                            M
Z14071,IGHV1-2*02,DP-75                 --- --- --- ... ... --- --- --- --- --- --- --- ... --- --- --- --- -TG --- ---
                                                                                                            M
X59704,IGHV1-2*02,hv1L1                 --- --- --- ... ... --- --- --- --- --- --- --- ... --- --- --- --- -TG --- ---
                                                                                                            M
X92208,IGHV1-2*03,1-1                   --- --- --- ... ... --- --- --- --- --- --- --- ... --- --- --- --- -TG --- ---
                                                                                                W           M
Z12310,IGHV1-2*04,DP-8                  --- --- --- ... ... --- --- --- --- --- --- --- ... --- T-- --- --- -TG --- ---
```

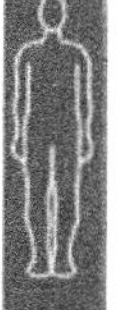

	81	82	83	84	85	86	87	88	89	90	91	92	93	94	95	96	97	98	99	100
	D	T	S	I	S	T	A	Y	M	E	L	S	R	L	R	S	D	D	T	V
X07448,IGHV1-2*01,V35/VI-2b	GAC	ACG	TCC	ATC	AGC	ACA	GCC	TAC	ATG	GAG	CTG	AGC	AGG	CTG	AGA	TCT	GAC	GAC	ACG	GTC
X62106,IGHV1-2*02,VI-2	---	---	---	---	---	---	---	---	---	---	---	---	---	---	---	---	---	---	---	A -C
Z14071,IGHV1-2*02,DP-75	---	---	---	---	---	---	---	---	---	---	---	---	---	---	---	---	---	---	---	A -C-
X59704,IGHV1-2*02,hv1L1	---	---	---	---	---	---	---	---	---	---	---	---	---	---	---	---	---	---	---	A -C
X92208,IGHV1-2*03,1-1	---	---	---	---	---	---	---	---	---	---	---	---	---	---	---	---	---	---	---	A -C-
Z12310,IGHV1-2*04,DP-8	---	---	---	---	---	---	---	---	---	---	---	---	---	---	---	---	---	---	---	A -C-

					CDR3-IMGT		
	101	102	103	104	105	106	
	V	Y	Y	C	A	R	
X07448,IGHV1-2*01,V35/VI-2b	GTG	TAT	TAC	TGT	GCG	AGA	GA
X62106,IGHV1-2*02,VI-2	---	---	---	---	---	---	--
Z14071,IGHV1-2*02,DP-75		---	---	---	---	---	
X59704,IGHV1-2*03,hv1L1	---	---	---	---	---	---	--
X92208,IGHV1-2*03,1-1	---	---	---	---	---	---	--
Z12310,IGHV1-2*04,DP-8		---	---	---	---	---	

Framework and complementarity determining regions

FR1-IMGT: 25 (-1 aa: 10)
FR2-IMGT: 17
FR3-IMGT: 38 (-1 aa: 73)

CDR1-IMGT: 8
CDR2-IMGT: 8
CDR3-IMGT: 2

Collier de Perles for human IGHV1-2*01

Accession number: IMGT X07448

EMBL/GenBank/DDBJ: X07448

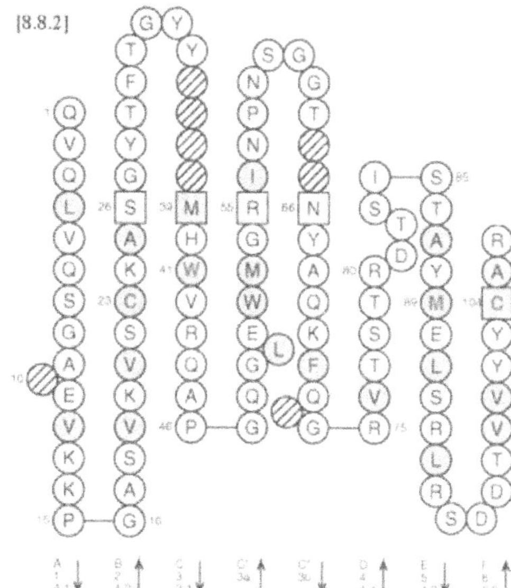

Genome database accession numbers

GDB:9931660

LocusLink: 28474

Nomenclature

IGHV1-3: Immunoglobulin heavy variable 1-3.

Definition and functionality

IGHV1-3 is one of the nine mapped functional genes of the IGHV1 subgroup. This subgroup comprises 14 mapped genes in the IGH locus, two unmapped genes (of which one is an ORF and one is a functional gene) and 11 orphons.

Gene location

IGHV1-3 is in the IGH locus on chromosome 14 at 14q32.33.

Nucleotide and amino acid sequences for human IGHV1-3

```
                                  1   2   3   4   5   6   7   8   9  10  11  12  13  14  15  16  17  18  19  20
                                  Q   V   Q   L   V   Q   S   G   A       E   V   K   K   P   G   A   S   V   K
X62109,IGHV1-3*01,VI-3b    [39]  CAG GTC CAG CTT GTG CAG TCT GGG GCT ... GAG GTG AAG AAG CCT GGG GCC TCA GTG AAG

Z12327,IGHV1-3*01,DP-25    [41]  --- --- --- --- --- --- --- --- --- ... --- --- --- --- --- --- --- --- --- ---

X62107,IGHV1-3*02,VI-3     [39]  --- --T --- --G --- --- --- --- --- ... --- --- --- --- --- --- --- --- --- ---

                                                          ___________CDR1-IMGT____________
                                 21  22  23  24  25  26  27  28  29  30  31  32  33  34  35  36  37  38  39  40
                                  V   S   C   K   A   S   G   Y   T   F   T   S   Y   A                   M   H
X62109,IGHV1-3*01,VI-3b          GTT TCC TGC AAG GCT TCT GGA TAC ACC TTC ACT AGC TAT GCT ... ... ... ... ATG CAT

Z12327,IGHV1-3*01,DP-25          --- --- --- --- --- --- --- --- --- --- --- --- --- --- ... ... ... ... --- ---

X62107,IGHV1-3*02,VI-3           --- --- --- --- --- --- --- --- --- --- --- --- --- --- ... ... ... ... --- ---

                                                                                          __________ CDR2-
                                 41  42  43  44  45  46  47  48  49  50  51  52  53  54  55  56  57  58  59  60
                                  W   V   R   Q   A   P   G   Q   R   L   E   W   M   G   W   I   N   A   G   N
X62109,IGHV1-3*01,VI-3b          TGG GTG CGC CAG GCC CCC GGA CAA AGG CTT GAG TGG ATG GGA TGG ATC AAC GCT GGC AAT

Z12327,IGHV1-3*01,DP-25          --- --- --- --- --- --- --- --- --- --- --- --- --- --- --- --- --- --- --- ---
                                                                                              S
X62107,IGHV1-3*02,VI-3           --- --- --- --- --- --- --- --- --- --- --- --- --- --- --- -G- --- --- --- ---

                                 IMGT_______________
                                 61  62  63  64  65  66  67  68  69  70  71  72  73  74  75  76  77  78  79  80
                                  G   N   T           K   Y   S   Q   K   F   Q       G   R   V   T   I   T   R
X62109,IGHV1-3*01,VI-3b          GGT AAC ACA ... ... AAA TAT TCA CAG AAG TTC CAG ... GGC AGA GTC ACC ATT ACC AGG

Z12327,IGHV1-3*01,DP-25          --- --- --- ... ... --- --- --- --- --- --- --- ... --- --- --- --- --- --- ---
                                                                      E
X62107,IGHV1-3*02,VI-3           --- --- --- ... ... --- --- --- --- G-- --- --- ... --- --- --- --- --- --- ---

                                 81  82  83  84  85  86  87  88  89  90  91  92  93  94  95  96  97  98  99 100
                                  D   T   S   A   S   T   A   Y   M   E   L   S   S   L   R   S   E   D   T   A
X62109,IGHV1-3*01,VI-3b          GAC ACA TCC GCG AGC ACA GCC TAC ATG GAG CTG AGC AGC CTG AGA TCT GAA GAC ACG GCT

Z12327,IGHV1-3*01,DP-25          --- --- --- --- --- --- --- --- --- --- --- --- --- --- --- --- --- --- --- ---
                                                                                                          M
X62107,IGHV1-3*02,VI-3           --- --- --- --- --- --- --- --- --- --- --- --- --- --- --- --- --G --- -T- ---

                                                    _CDR3-IMGT
                                 101 102 103 104 105 106
                                  V   Y   Y   C   A   R
X62109,IGHV1-3*01,VI-3b          GTG TAT TAC TGT GCG AGA GA

Z12327,IGHV1-3*01,DP-25          --- --- --- --- --- ---

X62107,IGHV1-3*02,VI-3           --- --- --- --- --- --- --
```

Framework and complementarity determining regions

FR1-IMGT: 25 (-1 aa: 10)
FR2-IMGT: 17
FR3-IMGT: 38 (-1 aa: 73)

CDR1-IMGT: 8
CDR2-IMGT: 8
CDR3-IMGT: 2

Collier de Perles for human IGHV1-3*01

Accession number: IMGT X62109 EMBL/GenBank/DDBJ: X62109

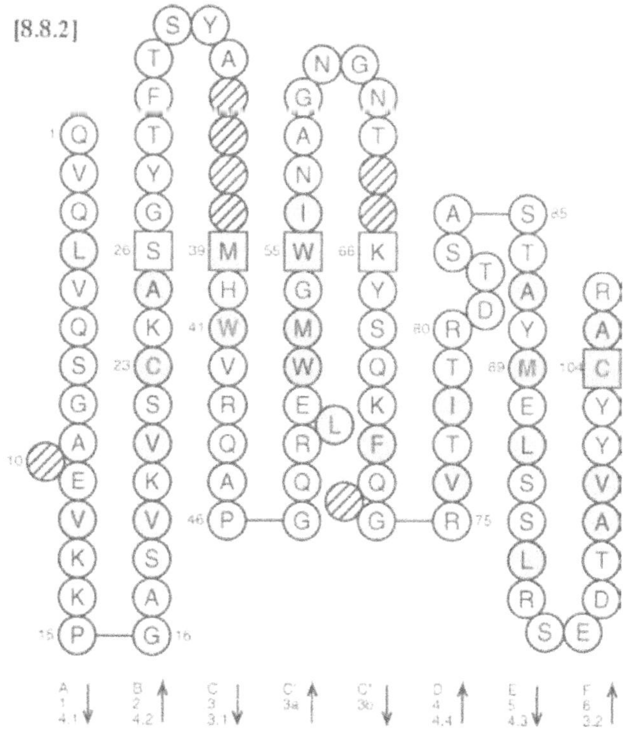

Genome database accession numbers

GDB:9931661 LocusLink: 28473

IGHV1-8

Nomenclature

IGHV1-8: Immunoglobulin heavy variable 1-8.

Definition and functionality

IGHV1-8 is one of the nine mapped functional genes of the IGHV1 subgroup. This subgroup comprises 14 mapped genes in the IGH locus, two unmapped genes (of which one is an ORF and one is a functional gene) and 11 orphons.

Gene location

IGHV1-8 is in the IGH locus on chromosome 14 at 14q32.33.

Nucleotide and amino acid sequences for human IGHV1-8

```
                                     1   2   3   4   5   6   7   8   9  10  11  12  13  14  15  16  17  18  19  20
                                     Q   V   Q   L   V   Q   S   G   A       E   V   K   K   P   G   A   S   V   K
M99637,IGHV1-8*01,V1-8        [26]  CAG GTG CAG CTG GTG CAG TCT GGG GCT ... GAG GTG AAG AAG CCT GGG GCC TCA GTG AAG

Z12317,IGHV1-8*01,DP-15       [41]  --- --- --- --- --- --- --- --- --- ... --- --- --- --- --- --- --- --- --- ---

                                                     ____________________CDR1-IMGT_______________________
                                    21  22  23  24  25  26  27  28  29  30  31  32  33  34  35  36  37  38  39  40
                                     V   S   C   K   A   S   G   Y   T   F   T   S   Y   D                   I   N
M99637,IGHV1-8*01,V1-8              GTC TCC TGC AAG GCT TCT GGA TAC ACC TTC ACC AGT TAT GAT ... ... ... ... ATC AAC

Z12317,IGHV1-8*01,DP-15             --- --- --- --- --- --- --- --- --- --- --- --- --- --- ... ... ... ... --- ---

                                                                                                ___________ CDR2
                                    41  42  43  44  45  46  47  48  49  50  51  52  53  54  55  56  57  58  59  60
                                     W   V   R   Q   A   T   G   Q   G   L   E   W   M   G   W   M   N   P   N   S
M99637,IGHV1-8*01,V1-8              TGG GTG CGA CAG GCC ACT GGA CAA GGG CTT GAG TGG ATG GGA TGG ATG AAC CCT AAC AGT

Z12317,IGHV1-8*01,DP-15             --- --- --- --- --- --- --- --- --- --- --- --- --- --- --- --- --- --- --- ---

                                    IMGT________________
                                    61  62  63  64  65  66  67  68  69  70  71  72  73  74  75  76  77  78  79  80
                                     G   N   T           G   Y   A   Q   K   F   Q       G   R   V   T   M   T   R
M99637,IGHV1-8*01,V1-8              GGT AAC ACA ... ... GGC TAT GCA CAG AAG TTC CAG ... GGC AGA GTC ACC ATG ACC AGG

Z12317,IGHV1-8*01,DP-15             --- --- --- ... ... --- --- --- --- --- --- --- ... --- --- --- --- --- --- ---

                                    81  82  83  84  85  86  87  88  89  90  91  92  93  94  95  96  97  98  99 100
                                     N   T   S   I   S   T   A   Y   M   E   L   S   S   L   R   S   E   D   T   A
M99637,IGHV1-8*01,V1-8              AAC ACC TCC ATA AGC ACA GCC TAC ATG GAG CTG AGC AGC CTG AGA TCT GAG GAC ACG GCC

Z12317,IGHV1-8*01,DP-15             --- --- --- --- --- --- --- --- --- --- --- --- --- --- --- --- --- --- --- ---

                                                        _CDR3-IMGT
                                    101 102 103 104 105 106
                                     V   Y   Y   C   A   R
M99637,IGHV1-8*01,V1-8              GTG TAT TAC TGT GCG AGA GG

Z12317,IGHV1-8*01,DP-15             --- --- --- --- --- --- --
```

Framework and complementarity determining regions

FR1-IMGT: 25 (-1 aa: 10)
FR2-IMGT: 17
FR3-IMGT: 38 (-1 aa: 73)

CDR1-IMGT: 8
CDR2-IMGT: 8
CDR3-IMGT: 2

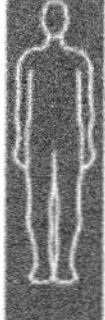

Collier de Perles for human IGHV1-8*01

Accession number: IMGT M99637 EMBL/GenBank/DDBJ: M99637

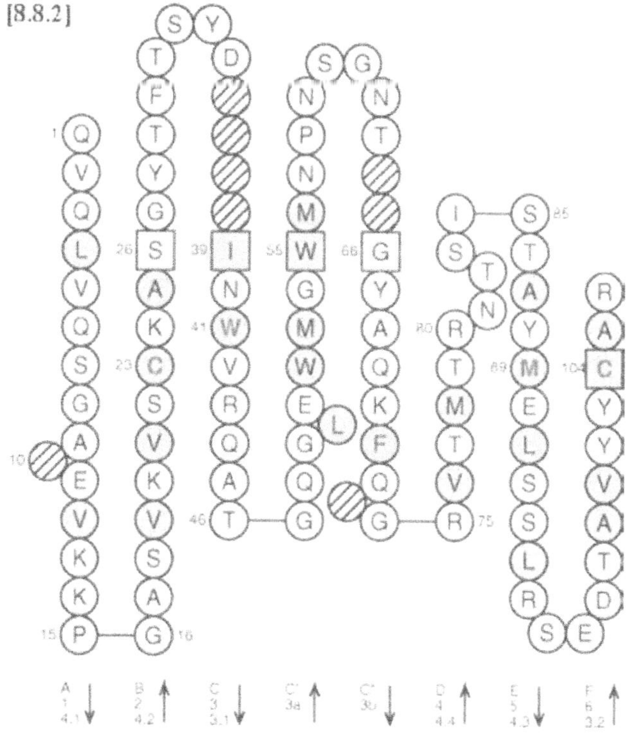

Genome database accession numbers

GDB:9931662 LocusLink: 28472

IGHV1-18

Nomenclature

IGHV1-18: Immunoglobulin heavy variable 1-18.

Definition and functionality

IGHV1-18 is one of the nine mapped functional genes of the IGHV1 subgroup. This subgroup comprises 14 mapped genes in the IGH locus, two unmapped genes (of which one is an ORF and one is a functional gene) and 11 orphons.

Gene location

IGHV1-18 is in the IGH locus on chromosome 14 at 14q32.33.

Nucleotide and amino acid sequences for human IGHV1-18

```
                                      1   2   3   4   5   6   7   8   9  10  11  12  13  14  15  16  17  18  19  20
                                      Q   V   Q   L   V   Q   S   G   A       E   V   K   K   P   G   A   S   V   K
M99641,IGHV1-18*01,V1-18       [26]  CAG GTT CAG CTG GTG CAG TCT GGA GCT ... GAG GTG AAG AAG CCT GGG GCC TCA GTG AAG

Z12316,IGHV1-18*01,DP-14       [41]  --- --- --- --- --- --- --- --- --- ... --- --- --- --- --- --- --- --- --- ---

X60503,IGHV1-18*02,VH1GRR      [16]  --- --- --- --- --- --- --- --- --- ... --- --- --- --- --- --- --- --- --- ---

                                                              ________________CDR1-IMGT_____________
                                     21  22  23  24  25  26  27  28  29  30  31  32  33  34  35  36  37  38  39  40
                                      V   S   C   K   A   S   G   Y   T   F   T   S   Y   G                   I   S
M99641,IGHV1-18*01,V1-18             GTC TCC TGC AAG GCT TCT GGT TAC ACC TTT ACC AGC TAT GGT ... ... ... ... ATC AGC

Z12316,IGHV1-18*01,DP-14             --- --- --- --- --- --- --- --- --- --- --- --- --- --- ... ... ... ... --- ---

X60503,IGHV1-18*02,VH1GRR            --- --- --- --- --- --- --- --- --- --- --- --- --- --- ... ... ... ... --- ---

                                                                                                   __________CDR2-
                                     41  42  43  44  45  46  47  48  49  50  51  52  53  54  55  56  57  58  59  60
                                      W   V   R   Q   A   P   G   Q   G   L   E   W   M   G   W   I   S   A   Y   N
M99641,IGHV1-18*01,V1-18             TGG GTG CGA CAG GCC CCT GGA CAA GGG CTT GAG TGG ATG GGA TGG ATC AGC GCT TAC AAT

Z12316,IGHV1-18*01,DP-14             --- --- --- --- --- --- --- --- --- --- --- --- --- --- --- --- --- --- --- ---

X60503,IGHV1-18*02,VH1GRR            --- --- --- --- --- --- --- --- --- --- --- --- --- --- --- --- --- --- --- ---

                                     IMGT_______________
                                     61  62  63  64  65  66  67  68  69  70  71  72  73  74  75  76  77  78  79  80
                                      G   N   T           N   Y   A   Q   K   L   Q       G   R   V   T   M   T   T
M99641,IGHV1-18*01,V1-18             GGT AAC ACA ... ... AAC TAT GCA CAG AAG CTC CAG ... GGC AGA GTC ACC ATG ACC ACA

Z12316,IGHV1-18*01,DP-14             --- --- --- ... ... --- --- --- --- --- --- --- ... --- --- --- --- --- --- ---

X60503,IGHV1-18*02,VH1GRR            --- --- --- ... ... --- --- --- --- --- --- --- ... --- --- --- --- --- --- ---

                                     81  82  83  84  85  86  87  88  89  90  91  92  93  94  95  96  97  98  99 100
                                      D   T   S   T   S   T   A   Y   M   E   L   R   S   L   R   S   D   D   T   A
M99641,IGHV1-18*01,V1-18             GAC ACA TCC ACG AGC ACA GCC TAC ATG GAG CTG AGG AGC CTG AGA TCT GAC GAC ACG GCC

Z12316,IGHV1-18*01,DP-14             --- --- --- --- --- --- --- --- --- --- --- --- --- --- --- --- --- --- --- ---

X60503,IGHV1-18*02,VH1GRR            --- --- --- --- --- --- --- --- --- --- --- --- --- --A --- --- --- --- --- ---

                                                         _CDR3-IMGT
                                     101 102 103 104 105 106
                                      V   Y   Y   C   A   R
M99641,IGHV1-18*01,V1-18             GTG TAT TAC TGT GCG AGA GA

Z12316,IGHV1-18*01,DP-14             --- --- --- --- --- ---

X60503,IGHV1-18*02,VH1GRR
```

Framework and complementarity determining regions

FR1-IMGT: 25 (-1 aa: 10)	CDR1-IMGT: 8
FR2-IMGT: 17	CDR2-IMGT: 8
FR3-IMGT: 38 (-1 aa: 73)	CDR3-IMGT: 2

Collier de Perles for human IGHV1-18*01

Accession number: IMGT M99641 EMBL/GenBank/DDBJ: M99641

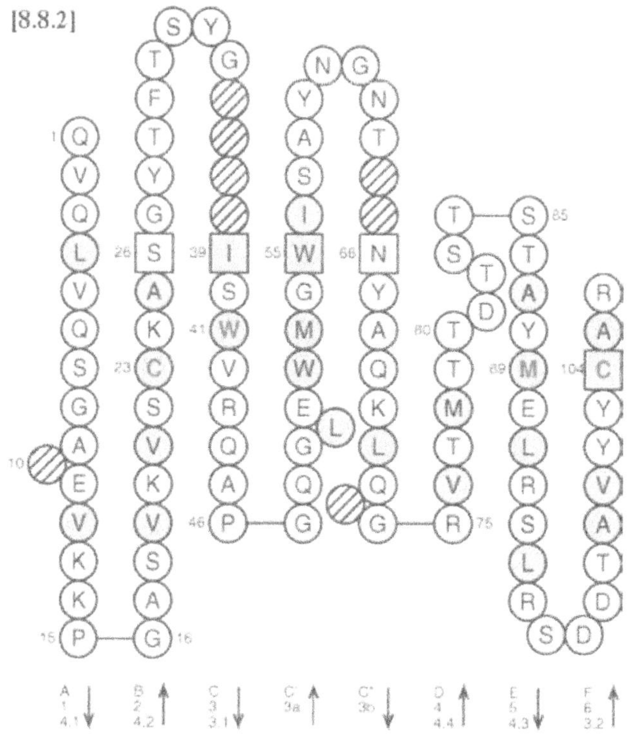

Genome database accession numbers

GDB:9931666 LocusLink: 28468

Nomenclature

IGHV1-24: Immunoglobulin heavy variable 1-24.

Definition and functionality

IGHV1-24 is one of the nine mapped functional genes of the IGHV1 subgroup. This subgroup comprises 14 mapped genes in the IGH locus, two unmapped genes (of which one is an ORF and one is a functional gene) and 11 orphons.

Gene location

IGHV1-24 is in the IGH locus on chromosome 14 at 14q32.33.

Nucleotide and amino acid sequences for human IGHV1-24

```
                                     1   2   3   4   5   6   7   8   9  10  11  12  13  14  15  16  17  18  19  20
                                     Q   V   Q   L   V   Q   S   G   A       E   V   K   K   P   G   A   S   V   K
M99642,IGHV1-24*01,V1-24P      (26) CAG GTC CAG CTG GTA CAG TCT GGG GCT ... GAG GTG AAG AAG CCT GGG GCC TCA GTG AAG

Z12307,IGHV1-24*01,DP-5        (41) --- --- --- --- --- --- --- --- --- ... --- --- --- --- --- --- --- --- ---

                                                             _____________CDR1-IMGT_____________
                                    21  22  23  24  25  26  27  28  29  30  31  32  33  34  35  36  37  38  39  40
                                     V   S   C   K   V   S   G   Y   T   L   T   E   L   S                   M   H
M99642,IGHV1-24*01,V1-24P           GTC TCC TGC AAG GTT TCC GGA TAC ACC CTC ACT GAA TTA TCC ... ... ... ... ATG CAC

Z12307,IGHV1-24*01,DP-5             --- --- --- --- --- --- --- --- --- --- --- --- --- --- ... ... ... ... --- ---

                                                                                                  ________ CDR2-
                                    41  42  43  44  45  46  47  48  49  50  51  52  53  54  55  56  57  58  59  60
                                     W   V   R   Q   A   P   G   K   G   L   E   W   M   G   G   F   D   P   E   D
M99642,IGHV1-24*01,V1-24P           TGG GTG CGA CAG GCT CCT GGA AAA GGG CTT GAG TGG ATG GGA GGT TTT GAT CCT GAA GAT

Z12307,IGHV1-24*01,DP-5             --- --- --- --- --- --- --- --- --- --- --- --- --- --- --- --- --- --- --- ---

                                    IMGT_______________
                                    61  62  63  64  65  66  67  68  69  70  71  72  73  74  75  76  77  78  79  80
                                     G   E   T           I   Y   A   Q   K   F   Q       G   R   V   T   M   T   E
M99642,IGHV1-24*01,V1-24P           GGT GAA ACA ... ... ATC TAC GCA CAG AAG TTC CAG ... GGC AGA GTC ACC ATG ACC GAG

Z12307,IGHV1-24*01,DP-5             --- --- --- ... ... --- --- --- --- --- --- --- ... --- --- --- --- --- --- ---

                                    81  82  83  84  85  86  87  88  89  90  91  92  93  94  95  96  97  98  99 100
                                     D   T   S   T   D   T   A   Y   M   E   L   S   S   L   R   S   E   D   T   A
M99642,IGHV1-24*01,V1-24P           GAC ACA TCT ACA GAC ACA GCC TAC ATG GAG CTG AGC AGC CTG AGA TCT GAG GAC ACG GCC

Z12307,IGHV1-24*01,DP-5             --- ---     --- --- --- --- --- --- --- --- --- --- --- --- --- --- --- --- --

                                                        CDR3-IMGT
                                   101 102 103 104 105 106
                                     V   Y   Y   C   A   T
M99642,IGHV1-24*01,V1-24P           GTG TAT TAC TGT GCA ACA GA

Z12307,IGHV1-24*01,DP-5             --- --- --- --- --- ---
```

Framework and complementarity determining regions

FR1-IMGT: 25 (-1 aa: 10)
FR2-IMGT: 17
FR3-IMGT: 38 (-1 aa: 73)

CDR1-IMGT: 8
CDR2-IMGT: 8
CDR3-IMGT: 2

Collier de Perles for human IGHV1-24*01

Accession number: IMGT M99642 EMBL/GenBank/DDBJ: M99642

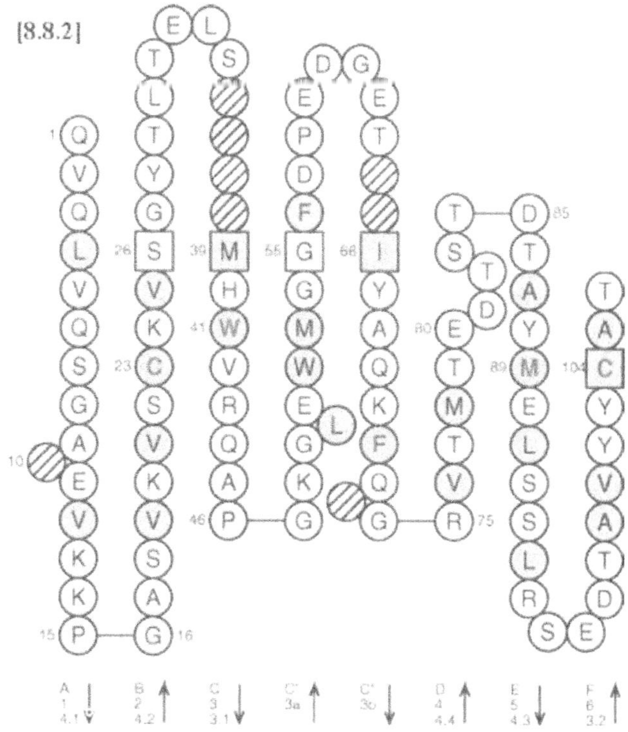

Genome database accession numbers

GDB:9931667 LocusLink: 28467

Nomenclature

IGHV1-45: Immunoglobulin heavy variable 1-45.

Definition and functionality

IGHV1-45 is one of the nine mapped functional genes of the IGHV1 subgroup. This subgroup comprises 14 mapped genes in the IGH locus, two unmapped genes (of which one is an ORF and one is a functional gene) and 11 orphons.

Gene location

IGHV1-45 is in the IGH locus on chromosome 14 at 14q32.33.

Nucleotide and amino acid sequences for human IGHV1-45

```
                                    1   2   3   4   5   6   7   8   9  10  11  12  13  14  15  16  17  18  19  20
                                    Q   M   Q   L   V   Q   S   G   A       E   V   K   K   T   G   S   S   V   K
X92209,IGHV1-45*01,7-2         [4]  CAG ATG CAG CTG GTG CAG TCT GGG GCT ... GAG GTG AAG AAG ACT GGG TCC TCA GTG AAG

M99645,IGHV1-45*01/*02,V1-45  [26]  --- --- --- --- --- --- --- --- --- ... --- --- --- --- --- --- --- --- --- ---

Z12306,IGHV1-45*02,DP-4       [41]  --- --- --- --- --- --- --- --- --- ... --- --- --- --- --- --- --- --- --- ---

Z17391,IGHV1-45*03,COS-5      [42]                                                   -- --- --- --- --- --- --- ---

                                                        _______________CDR1-IMGT________________
                                   21  22  23  24  25  26  27  28  29  30  31  32  33  34  35  36  37  38  39  40
                                    V   S   C   K   A   S   G   Y   T   F   T   Y   R   Y                   L   H
X92209,IGHV1-45*01,7-2              GTT TCC TGC AAG GCT TCC GGA TAC ACC TTC ACC TAC CGC TAC ... ... ... ... CTG CAC

M99645,IGHV1-45*01/*02,V1-45        --- --- --- --- --- --- --- --- --- --- --- --- --- --- ... ... ... ... --- ---

Z12306,IGHV1-45*02,DP-4             --- --- --- --- --- --- --- --- --- --- --- --- --- --- ... ... ... ... --- ---

Z17391,IGHV1-45*03,COS-5            --- --- --- --- --- --- --- --- --- --- --- --- --- --- ... ... ... ... --- ---

                                                                                            ___________ CDR2-
                                   41  42  43  44  45  46  47  48  49  50  51  52  53  54  55  56  57  58  59  60
                                    W   V   R   Q   A   P   G   Q   A   L   E   W   M   G   W   I   T   P   F   N
X92209,IGHV1-45*01,7-2              TGG GTG CGA CAG GCC CCC GGA CAA GCG CTT GAG TGG ATG GGA TGG ATC ACA CCT TTC AAT

M99645,IGHV1-45*01/*02,V1-45        --- --- --- --- --- --- --- --- --- --- --- --- --- --- --- --- --- --- --- ---

Z12306,IGHV1-45*02,DP-4             --- --- --- --- --- --- --- --- --- --- --- --- --- --- --- --- --- --- --- ---
                                                            R
Z17391,IGHV1-45*03,COS-5            --- --- --- --- --- --- A-- --- --- --- --- --- --- --- --- --- --- --- --- ---

                                    IMGT_______________
                                   61  62  63  64  65  66  67  68  69  70  71  72  73  74  75  76  77  78  79  80
                                    G   N   T           N   Y   A   Q   K   F   Q       D   R   V   T   I   T   R
X92209,IGHV1-45*01,7-2              GGT AAC ACC ... ... AAC TAC GCA CAG AAA TTC CAG ... GAC AGA GTC ACC ATT ACT AGG

M99645,IGHV1-45*01/*02,V1-45 (1)    --- --- --- ... ... --- --- --- ---(... ... ... ... ... ... ... ... ... ... ...

Z12306,IGHV1-45*02,DP-4             --- --- --- ... ... --- --- --- --- --- --- --- ... --- --- --- --- --- --C ---

Z17391,IGHV1-45*03,COS-5            --- --- --- ... ... --- --- --- --- --- --- --- ... --- --- --- --- --- --C ---

                                   81  82  83  84  85  86  87  88  89  90  91  92  93  94  95  96  97  98  99 100
                                    D   R   S   M   S   T   A   Y   M   E   L   S   S   L   R   S   E   D   T   A
X92209,IGHV1-45*01,7-2              GAC AGG TCT ATG AGC ACA GCC TAC ATG GAG CTG AGC AGC CTG AGA TCT GAG GAC ACA GCC

M99645,IGHV1-45*01/*02,V1-45        ... ... .)- --- --- --- --- --- --- --- --- --- --- --- --- --- --- --- --- ---

Z12306,IGHV1-45*02,DP-4             --- --- --- --- --- --- --- --- --- --- --- --- --- --- --- --- --- --- ---  --

Z17391,IGHV1-45*03,COS-5            --- --- --- --- --- --- --- --- --- --- --- --- --- --- --- --- --- --- --- ---

                                                    _CDR3-IMGT
                                   101 102 103 104 105 106
                                    M   Y   Y   C   A   R
X92209,IGHV1-45*01,7-2              ATG TAT TAC TGT GCA AGA XA

M99645,IGHV1-45*01/*02,V1-45        --- --- --- --- --- --- T-

Z12306,IGHV1-45*02,DP-4             --- --- --- --- --- ---

Z17391,IGHV1-45*03,COS-5            --- --- --- --- --- ---
```

Note:

(1) The internal DELETION of nucleotides 208 to 248 (IMGT numbering) in M99645 EMBL/Genbank flat file is typing error. (No internal DELETION in the amino acid sequence in [26].) The assignment of the sequence to allele *01 or *02 is not possible.

Framework and complementarity determining regions

FR1-IMGT: 25 (-1 aa: 10)
FR2-IMGT: 17
FR3-IMGT: 38 (-1 aa: 73)

CDR1-IMGT: 8
CDR2-IMGT: 8
CDR3-IMGT: 2

Collier de Perles for human IGHV1-45*01

Accession number: IMGT X92209

EMBL/GenBank/DDBJ: X92209

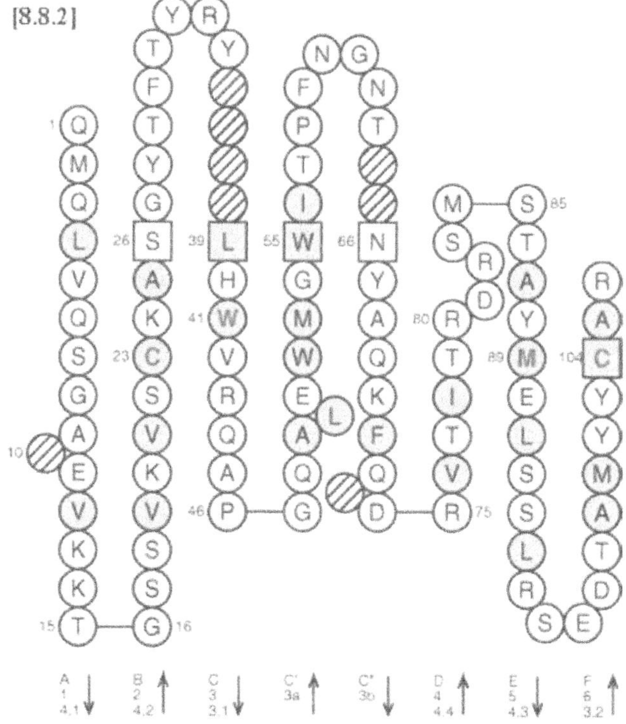

Genome database accession numbers

GDB:9931668

LocusLink: 28466

IGHV1-46

Nomenclature

IGHV1-46: Immunoglobulin heavy variable 1-46.

Definition and functionality

IGHV1-46 is one of the nine mapped functional genes of the IGHV1 subgroup. This subgroup comprises 14 mapped genes in the IGH locus, two unmapped genes (of which one is an ORF and one is a functional gene) and 11 orphons.

Gene location

IGHV1-46 is in the IGH locus on chromosome 14 at 14q32.33.

Nucleotide and amino acid sequences for human IGHV1-46

		1	2	3	4	5	6	7	8	9	10	11	12	13	14	15	16	17	18	19	20
		Q	V	Q	L	V	Q	S	G	A		E	V	K	K	P	G	A	S	V	K
X92343,IGHV1-46*01,21-2	[4]	CAG	GTG	CAG	CTG	GTG	CAG	TCT	GGG	GCT	...	GAG	GTG	AAG	AAG	CCT	GGG	GCC	TCA	GTG	AAG
Z12309,IGHV1-46*01,DP-7	[41]	---	---	---	---	---	---	---	---	---	...	---	---	---	---	---	---	---	---	---	---
X92207,IGHV1-46*01,3-1	[4]	---	---	---	---	---	---	---	---	---	...	---	---	---	---	---	---	---	---	---	---
L06611,IGHV1-46*01,hv1f10	[40]	---	---	---	---	---	---	---	---	---	...	---	---	---	---	---	---	---	---	---	---
M99646,IGHV1-46*01,V1-46	[26]	---	---	---	---	---	---	---	---	---	...	---	---	---	---	---	---	---	---	---	---
J00240,IGHV1-46*02,HG3	[33]	---	---	---	---	---	---	---	--	---	...	---	---	---	---	---	---	---	---	---	---
L06612,IGHV1-46*03,hv1f10t	[40]	---	---	---	---	---	---	---	---	---	...	---	---	---	---	---	---	---	---	---	---

							CDR1-IMGT													
	21	22	23	24	25	26	27	28	29	30	31	32	33	34	35	36	37	38	39	40
	V	S	C	K	A	S	G	Y	T	F	T	S	Y	Y					M	H
X92343,IGHV1-46*01,21-2	GTT	TCC	TGC	AAG	GCA	TCT	GGA	TAC	ACC	TTC	ACC	AGC	TAC	TAT	...	...	...	...	ATG	CAC
Z12309,IGHV1-46*01,DP-7	---	---	---	---	---	---	---	---	---	---	---	---	---	---	...	...	...	...		
X92207,IGHV1-46*01,3-1	---	---	---	---	---	---	---	---	---	---	---	---	---	---	...	...	...	...	---	---
L06611,IGHV1-46*01,hv1f10	---	---	---	---	---	---	---	---	---	---	---	---	---	---	...	...	...	...	---	---
M99646,IGHV1-46*01,V1-46	---	---	---	---	---	---	---	---	---	---	---	---	---	---	...	...	...	...	---	---
											N									
J00240,IGHV1-46*02,HG3	---	---	---	---	---			---	---	---	-A-	---	---		...	...	...	...	---	---
L06612,IGHV1-46*03,hv1f10t	---	---	---	---	---	---	---	---	---	---	---	-	---	---	...	...	...	...	---	---

																CDR2-				
	41	42	43	44	45	46	47	48	49	50	51	52	53	54	55	56	57	58	59	60
	W	V	R	Q	A	P	G	Q	G	L	E	W	M	G	I	I	N	P	S	G
X92343,IGHV1-46*01,21-2	TGG	GTG	CGA	CAG	GCC	CCT	GGA	CAA	GGG	CTT	GAG	TGG	ATG	GGA	ATA	ATC	AAC	CCT	AGT	GGT
Z12309,IGHV1-46*01,DP-7	---	---	---	---	---	---	---	---	---	---	---	---	---	---	---	---	---	---	---	---
X92207,IGHV1-46*01,3-1	---	---	--	---	---	---	---	---	---	---	---	---	---	---	---	---	---	---	---	---
L06611,IGHV1-46*01,hv1f10	---	---	---	---	---	---	---	---	---	---	---	---	---	---	---	---	---	---	---	---
M99646,IGHV1-46*01,V1-46	---	---	---	---	---	---	---	---	---	---	---	---	---	---	---	---	---	---	---	---
J00240,IGHV1-46*02,HG3	---	---	---	---	---	---	---	---	---	---	---	---	---	---	---	---	---	---	---	---
L06612,IGHV1-46*03,hv1f10t	---	---	---	---	---	---	---	---	---	---	---	---	---	---	---	---	---	---	---	---

```
                                  IMGT________________
                                  61  62  63  64  65  66  67  68  69  70  71  72  73  74  75  76  77  78  79  80
                                  G   S   T           S   Y   A   Q   K   F   Q       G   R   V   T   M   T   R
X92343,IGHV1-46*01,21-2           GGT AGC ACA ... ... AGC TAC GCA CAG AAG TTC CAG ... GGC AGA GTC ACC ATG ACC AGG

Z12309,IGHV1-46*01,DP-7           --- --- --- ... ... --- --- --- --- --- --- --- ... --- --- --- --- --- --- ---

X92207,IGHV1-46*01,3-1            --- --- --- ... ... --- --- --- --- --- --- --- ... --- --- --- --- --- --- ---

L06611,IGHV1-46*01,hv1f10         --- --- --- ... ... --- --- --- --- --- --- --- ... --- --- --- --- --- --- ---

M99646,IGHV1-46*01,V1-46      (1) --- --- --- ... ... --- --- --- --- --- --- --- ...         --- --- --- --- ---

J00240,IGHV1-46*02,HG3            --- --- --- ... ... --- --- --- --- --- --- --- ... --- --- --- --- --- --- ---

L06612,IGHV1-46*03,hv1f10t        --- --- --- ... ... --- --- --- --- --- --- --- ... --- --- --- --- --- --- --

                                  81  82  83  84  85  86  87  88  89  90  91  92  93  94  95  96  97  98  99 100
                                  D   T   S   T   S   T   V   Y   M   E   L   S   S   L   R   S   E   D   T   A
X92343,IGHV1-46*01,21-2           GAC ACG TCC ACG AGC ACA GTC TAC ATG GAG CTG AGC AGC CTG AGA TCT GAG GAC ACG GCC

Z12309,IGHV1-46*01,DP-7           --- --- --- --- --- --- --- --- --- --- --- --- --- --- --- --- --- --- --- ---

X92207,IGHV1-46*01,3-1            --- --- --- --- --- --- --- --- --- --- --- --- --- --- --- --- --- --- --- ---

L06611,IGHV1-46*01,hv1f10         --- --- --- --- --- --- --- --- --- --- --- --- --- --- --- --- --- --- --- ---

M99646,IGHV1-46*01,V1-46          --- --- --- --- --- --- --- --- --- --- --- --- --- --- --- --- --- --- --- ---

J00240,IGHV1-46*02,HG3            --- --- --- --- --- --- --- --- --- --- --- --- --- --- --- --- --- --- --- ---

L06612,IGHV1-46*03,hv1f10t        --- --- --- --- --- --- --- --- --- --- --- --- --- --- --- --- --- --- --- ---

                                                  _CDR3-IMGT
                                  101 102 103 104 105 106
                                  V   Y   Y   C   A   R
X92343,IGHV1-46*01,21-2           GTG TAT TAC TGT GCG AGA GA

Z12309,IGHV1-46*01,DP-7           --- --- --- --- --- ---

X92207,IGHV1-46*01,3-1            --- --- --- --- --- --- --

L06611,IGHV1-46*01,hv1f10         --- --- --- --- --- --- --

M99646,IGHV1-46*01,V1-46          --- --- --- --- --- --- --

J00240,IGHV1-46*02,HG3            --- --- --- --- --- --- --

L06612,IGHV1-46*03,hv1f10t        --- --- --- --- --T --- --
```

Note:

(1) The INSERTION of one "a" nucleotide between position 195 and 196 (IMGT numbering), in M99646 EMBL/Genbank flat file is a typing error. (No INSERTION in [26].)

Framework and complementarity determining regions

FR1-IMGT: 25 (-1 aa: 10)
FR2-IMGT: 17
FR3-IMGT: 38 (-1 aa: 73)

CDR1-IMGT: 8
CDR2-IMGT: 8
CDR3-IMGT: 2

Collier de Perles for human IGHV1-46*01

Accession number: IMGT X92343 EMBL/GenBank/DDBJ: X92343

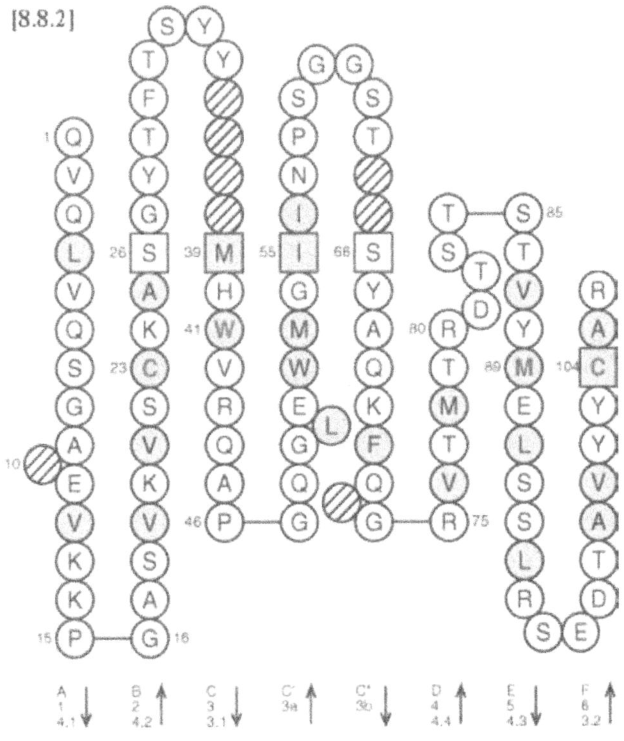

Genome database accession numbers

GDB:9931669 LocusLink: 28465

Nomenclature

IGHV1-58: Immunoglobulin heavy variable 1-58.

Definition and functionality

IGHV1-58 is one of the nine mapped functional genes of the IGHV1 subgroup. This subgroup comprises 14 mapped genes in the IGH locus, two unmapped genes (of which one is an ORF and one is a functional gene) and 11 orphons.

Gene location

IGHV1-58 is in the IGH locus on chromosome 14 at 14q32.33.

Nucleotide and amino acid sequences for human IGHV1-58

```
                                        1   2   3   4   5   6   7   8   9  10  11  12  13  14  15  16  17  18  19  20
                                        Q   M   Q   L   V   Q   S   G   P       E   V   K   K   P   G   T   S   V   K
M29809,IGHV1-58*01,V71-5         (22)  CAA ATG CAG CTG GTG CAG TCT GGG CCT ... GAG GTG AAG AAG CCT GGG ACC TCA GTG AAG

Z12304,IGHV1-58*01,DP-2          (41)  --- --- --- --- --- --- --- --- --- ... --- --- --- --- --- --- --- --- --- ---

                                                                   _________CDR1-IMGT_________________
                                       21  22  23  24  25  26  27  28  29  30  31  32  33  34  35  36  37  38  39  40
                                        V   S   C   K   A   S   G   F   T   F   T   S   S   A                   V   Q
M29809,IGHV1-58*01,V71-5               GTC TCC TGC AAG GCT TCT GGA TTC ACC TTT ACT AGC TCT GCT ... ... ... ... GTG CAG

Z12304,IGHV1-58*01,DP-2                --- --- --- --- --- --- --- --- --- --- --- --- --- --- ... ... ... ... --- ---

                                                                                                  ___________ CDR2-
                                       41  42  43  44  45  46  47  48  49  50  51  52  53  54  55  56  57  58  59  60
                                        W   V   R   Q   A   R   G   Q   R   L   E   W   I   G   W   I   V   V   G   S
M29809,IGHV1-58*01,V71-5               TGG GTG CGA CAG GCT CGT GGA CAA CGC CTT GAG TGG ATA GGA TGG ATC GTC GTT GGC AGT

Z12304,IGHV1-58*01,DP-2                --- --- --- --- --- --- --- --- --- --- --- --- --- --- --- --- --- --- --- ---

                                       IMGT_______________
                                       61  62  63  64  65  66  67  68  69  70  71  72  73  74  75  76  77  78  79  80
                                        G   N   T           N   Y   A   Q   K   F   Q       E   R   V   T   I   T   R
M29809,IGHV1-58*01,V71-5               GGT AAC ACA ... ... AAC TAC GCA CAG AAG TTC CAG ... GAA AGA GTC ACC ATT ACC AGG

Z12304,IGHV1-58*01,DP-2                --- --- --- ... ... --- --- --- --- --- --- --- ... --- --- --- --- --- --- ---

                                       81  82  83  84  85  86  87  88  89  90  91  92  93  94  95  96  97  98  99 100
                                        D   M   S   T   S   T   A   Y   M   E   L   S   S   L   R   S   E   D   T   A
M29809,IGHV1-58*01,V71-5               GAC ATG TCC ACA AGC ACA GCC TAC ATG GAG CTG AGC AGC CTG AGA TCC GAG GAC ACG GCC

Z12304,IGHV1-58*01,DP-2                --- --- --- --- --- --- --- --- --- --- --- --- --- --- --- --- --- --- --- ---

                                                       _CDR3-IMGT
                                       101 102 103 104 105 106
                                        V   Y   Y   C   A   A
M29809,IGHV1-58*01,V71-5               GTG TAT TAC TGT GCG GCA GA

Z12304,IGHV1-58*01,DP-2                --- --- --- --- --- --- --
```

Framework and complementarity determining regions

FR1-IMGT: 25 (-1 aa: 10)	CDR1-IMGT: 8
FR2-IMGT: 17	CDR2-IMGT: 8
FR3-IMGT: 38 (-1 aa: 73)	CDR3-IMGT: 2

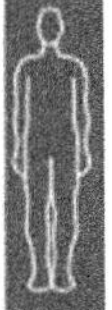

Collier de Perles for human IGHV1-58*01

Accession number: IMGT M29809 EMBL/GenBank/DDBJ: M29809

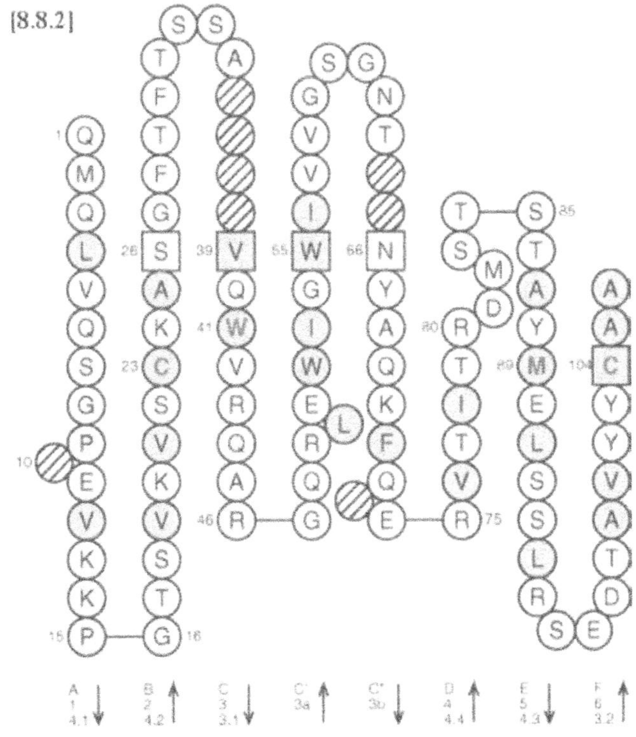

Genome database accession numbers

GDB:9931670 LocusLink: 28464

Nomenclature

IGHV1-69: Immunoglobulin heavy variable 1-69.

Definition and functionality

IGHV1-69 is one of the nine mapped functional genes of the IGHV1 subgroup. This subgroup comprises 14 mapped genes in the IGH locus, two unmapped genes (of which one is an ORF and one is a functional gene) and 11 orphons. Alleles IGHV1-69*06 and IGHV1-69*07 correspond to sequences previously designated as IGHV1-e.

Gene location

IGHV1-69 are in the IGH locus on chromosome 14 at 14q32.33.

Nucleotide and amino acid sequences for human IGHV1-69

```
                                   1   2   3   4   5   6   7   8   9  10  11  12  13  14  15  16  17  18  19  20
                                   Q   V   Q   L   V   Q   S   G   A       E   V   K   K   P   G   S   S   V   K
L22582,IGHV1-69*01,YAC-7     [12] CAG GTG CAG CTG GTG CAG TCT GGG GCT ... GAG GTG AAG AAG CCT GGG TCC TCG GTG AAG

Z12312,IGHV1-69*01,DP-10     [41] --- --- --- --- --- --- --- --- --- ... --- --- --- --- --- --- --- --- --- ---

X92298,IGHV1-69*01,HULGLVH1  [13] --- --- --- --- --- --- --- --- --- ... --- --- --- --- --- --- --- --- --- ---

Z29982,IGHV1-69*01,DA-6      [12]                                                 -- --- --- --- --- --- --- ---

Z27506,IGHV1-69*02,hv1051    [54] --- --C --- --- --- --A --- --- --- ... --- --- --- --- --- --- --- --- --- ---

X67902,IGHV1-69*02,RR.VH1.1  [44] --- --C --- --- --- --A --- --- --- ... --- --- --- --- --C --- --- --- --- ---

X92340,IGHV1-69*03,57GTA8    [19] --- --- --- --- --- --- --- --- --- ... --- --- --- --- --- --- --- --- --- ---

M83132,IGHV1-69*04,hv1263    [11] --- --C --- --- --- --- --- --- --- ... --- --- --- --- --- --- --- --- --- ---

X67905,IGHV1-69*05,RR.VH1.2  [44] --- --C --- --- --- --- --- --- --- ... --- --- --- --- --- --- --- --- --- ---

L22583,IGVH1-69*06,hv1051K   [54] --- --- --- --- --- --- --- --- --- ... --- --- --- --- --- --- --- --- --- ---

Z49804,IGHV1-69*06,DP-88     [53] --- --- --- --- --- --- --- --- --- ... --- --- --- --- --- --- --- --- --- ---

Z29978,IGHV1-69*07,DA-2      [12]                                                 -- --- --- --- --- --- --- ---


                                                          _____________________CDR1-IMGT_________________
                                  21  22  23  24  25  26  27  28  29  30  31  32  33  34  35  36  37  38  39  40
                                   V   S   C   K   A   S   G   G   T   F   S   S   Y   A                   I   S
L22582,IGHV1-69*01,YAC-7          GTC TCC TGC AAG GCT TCT GGA GGC ACC TTC AGC AGC TAT GCT ... ... ... ... ATC AGC

Z12312,IGHV1-69*01,DP-10          --- --- --- --- --- --- --- --- --- --- --- --- --- --- ... ... ... ... --- ---

X92298,IGHV1-69*01,HULGLVH1       --- --- --- --- --- --- --- --- --- --- --- --- --- --- ... ... ... ... --- ---

Z29982,IGHV1-69*01,DA-6           --- --- --- --- --- --- --- --- --- --- --- --- --- --- ... ... ... ... --- ---
                                                                                       T
Z27506,IGHV1-69*02,hv1051         --- --- --- --- --- --- --- --- --- --- --- --- --- A-- ... ... ... ... --- ---
                                                                                       T
X67902,IGHV1-69*02,RR.VH1.1       --- --- --- --- --- --- --- --- --- --- --- --- --- A-- ... ... ... ... --- ---

X92340,IGHV1-69*03,57GTA8         --- --- --- --- --- --- --- --- --- --- --- --- --- --- ... ... ... ... --- ---

M83132,IGHV1-69*04,hv1263         --- --- --- --- --- --- --- --- --- --- --- --- --- --- ... ... ... ... --- ---

X67905,IGHV1-69*05,RR.VH1.2       --- --- --- --- --- --- --- --- --- --- --- --- --- --- ... ... ... ... --- ---

L22583,IGVH1-69*06,hv1051K        --- --- --- --- --- --- --- --- --- --- --- --- --- --- ... ... ... ... --- ---

Z49804,IGHV1-69*06,DP-88          --- --- --- --- --- --- --- --- --- --- --- --- --- --- ... ... ... ... --- ---

Z29978,IGHV1-69*07,DA-2           --- --- --- --- --- --- --- --- --- --- --- --- --- --- ... ... ... ... --- ---


                                                                                              ________________CDR2-
                                  41  42  43  44  45  46  47  48  49  50  51  52  53  54  55  56  57  58  59  60
                                   W   V   R   Q   A   P   G   Q   G   L   E   W   M   G   G   I   I   P   I   F
L22582,IGHV1-69*01,YAC-7          TGG GTG CGA CAG GCC CCT GGA CAA GGG CTT GAG TGG ATG GGA GGG ATC ATC CCT ATC TTT

Z12312,IGHV1-69*01,DP-10          --- --- --- --- --- --- --- --- --- --- --- --- --- --- --- --- --- --- --- ---

X92298,IGHV1-69*01,HULGLVH1       --- --- --- --- --- --- --- --- --- --- --- --- --- --- --- --- --- --- --- ---

Z29982,IGHV1-69*01,DA-6           --- --- --- --- --- --- --- --- --- --- --- --- --- --- --- --- --- --- --- ---
                                                                                           R                   L
Z27506,IGHV1-69*02,hv1051         --- --- --- --- --- --- --- --- --- --- --- --- --- --- A-- --- --- --- --- C--
                                                                                           R                   L
X67902,IGHV1-69*02,RR.VH1.1       --- --- --- --- --- --- --- --- --- --- --- --- --- --- A-- --- --- --- --- C--
```

```
X92340,IGHV1-69*03,57GTA8           --- --- --- --- --- --- --- --- --- --- --- --- --- --- --- --- --- --- --- ---
                                                                                             R                   L
M83132,IGHV1-69*04,hv1263           --- --- --- --- --- --- --- --- --- --- --- --- --- --- A-- --- --- --- --- C--
X67905,IGHV1-69*05,RR.VH1.2         --- --- --- --- --- --- --- --- --- --- --- --- --- --- --- --- --- --- --- ---
L22583,IGVH1-69*06,hv1051K          --- --- --- --- --- --- --- --- --- --- --- --- --- --- --- --- --- --- --- ---
Z49804,IGHV1-69*06,DP-88            --- --- --- --- --- --- --- --- --- --- --- --- --- --- --- --- --- --- --- ---
                                                                                             R
Z29978,IGHV1-69*07,DA-2             --- --- --- --- --- --- --- --- --- --- --- --- --- --- A-- --- --- --- --- ---

                                    IMGT_______________
                                    61  62  63  64  65  66  67  68  69  70  71  72  73  74  75  76  77  78  79  80
                                     G   T   A           N   Y   A   Q   K   F   Q       G   R   V   T   I   T   A
L22582,IGHV1-69*01,YAC-7            GGT ACA GCA ... ... AAC TAC GCA CAG AAG TTC CAG ... GGC AGA GTC ACG ATT ACC GCG
Z12312,IGHV1-69*01,DP-10            --- --- --- ... ... --- --- --- --- --- --- --- ... --- --- --- --- --- --- ---
X92298,IGHV1-69*01,HULGLVH1         --- --- --- ... ... --- --- --- --- --- --- --- ... --- --- --- --- --- --- ---
Z29982,IGHV1-69*01,DA-6             --- --- --- ... ... --- --- --- --- --- --- --- ... --- --- --- --- --- --- ---
                                         I
Z27506,IGHV1-69*02,hv1051           --- -T- --- ... ... --- --- --- --- --- --- --- ... --- --- --- --- --- --- ---
                                         I
X67902,IGHV1-69*02,RR.VH1.1         --- -T- --- ... ... --- --- --- --- --- --- --- ... --- --- --- --- --- --- ---
X92340,IGHV1-69*03,57GTA8           --- --- --- ... ... --- --- --- --- --- --- --- ... --- --- --- --- --- --- ---
                                         I
M83132,IGHV1-69*04,hv1263           --- -T- --- ... ... --- --- --- --- --- --- --- ... --- --- --- --- --- --- ---
                                                                                                                 T
X67905,IGHV1-69*05,RR.VH1.2         --- --- --- ... ... --- --- --- --- --- --- --- ... --- --- --- --- --- --- A--
L22583,IGVH1-69*06,hv1051K          --- --- --- ... ... --- --- --- --- --- --- --- ... --- --- --- --- --- --- ---
Z49804,IGHV1-69*06,DP-88            --- --- --- ... ... --- --- --- --- --- --- --- ... --- --- --- --- --- --- ---
Z29978,IGHV1-69*07,DA-2             --- --- --- ... ... --- --- --- --- --- --- --- ... --- --- --- --- --- --- ---

                                    81  82  83  84  85  86  87  88  89  90  91  92  93  94  95  96  97  98  99  100
                                     D   E   S   T   S   T   A   Y   M   E   L   S   S   L   R   S   E   D   T   A
L22582,IGHV1-69*01,YAC-7            GAC GAA TCC ACG AGC ACA GCC TAC ATG GAG CTG AGC AGC CTG AGA TCT GAG GAC ACG GCC
Z12312,IGHV1-69*01,DP-10            --- --- --- --- --- --- --- --- --- --- --- --- --- --- --- --- --- --- --- ---
X92298,IGHV1-69*01,HULGLVH1         --- --- --- --- --- --- --- --- --- --- --- --- --- --- --- --- --- --- --- ---
Z29982,IGHV1-69*01,DA-6             --- --- --- --- --- --- --- --- --- --- --- --- --- --- --- --- ---
                                         K
Z27506,IGHV1-69*02,hv1051           --- A-- --- --- --- --- --- --- --- --- --- --- --- --- --- --- --- --- --- ---
                                         K
X67902,IGHV1-69*02,RR.VH1.1         --- A-- --- --- --- --- --- --- --- --- --- --- --- --- --- --- --- --- --- ---
                                                                                                     D
X92340,IGHV1-69*03,57GTA8           --- --- --- --- --- --- --- --- --- --- --- --- --- --- --- --- --T --- --- --
                                         K
M83132,IGHV1-69*04,hv1263           --- A-- --- --- --- --- --- --- --- --- --- --- --- --- --- --- --- --- --- ---
X67905,IGHV1-69*05,RR.VH1.2         --- --- --- --- --- --- --- --- --- --- --- --- --- --- --- --- --- --- --- ---
                                         K
L22583,IGVH1-69*06,hv1051K          --- A-- --- --- --- --- --- --- --- --- --- --- --- --- --- --- --- --- --- ---
                                         K
Z49804,IGHV1-69*06,DP-88            --- A-- --- --- --- --- --- --- --- --- --- --- --- --- --- --- --- --- --- ---
Z29978,IGHV1-69*07,DA-2             --- --- --- --- --- --- --- --- --- --- --- --- --- --- --- --- ---

                                                   _CDR3-IMGT
                                    101 102 103 104 105 106
                                     V   Y   Y   C   A   R
L22582,IGHV1-69*01,YAC-7            GTG TAT TAC TGT GCG AGA GA
Z12312,IGHV1-69*01,DP-10            --- --- --- --- --- ---
X92298,IGHV1-69*01,HULGLVH1         --- --- --- --- --- ---
Z29982,IGHV1-69*01,DA-6
Z27506,IGHV1-69*02,hv1051           --- --- --- --- --- --- --
X67902,IGHV1-69*02,RR.VH1.1         --- --- --- --- --- ---
X92340,IGHV1-69*03,57GTA8
M83132,IGHV1-69*04,hv1263           --- --- --- --- --- --- --
X67905,IGHV1-69*05,RR.VH1.2         --- --- --- --- --- ---
L22583,IGVH1-69*06,hv1051K          --- --- --- --- --- --- --
Z49804,IGHV1-69*06,DP-88            --- --- --- --- --- ---
Z29978,IGHV1-69*07,DA-2
```

Framework and complementarity determining regions

FR1-IMGT: 25 (-1 aa: 10)

FR2-IMGT: 17

FR3-IMGT: 38 (-1 aa: 73)

CDR1-IMGT: 8

CDR2-IMGT: 8

CDR3-IMGT: 2

Collier de Perles for human IGHV1-69*01

Accession number: IMGT L22582 EMBL/GenBank/DDBJ:L22582

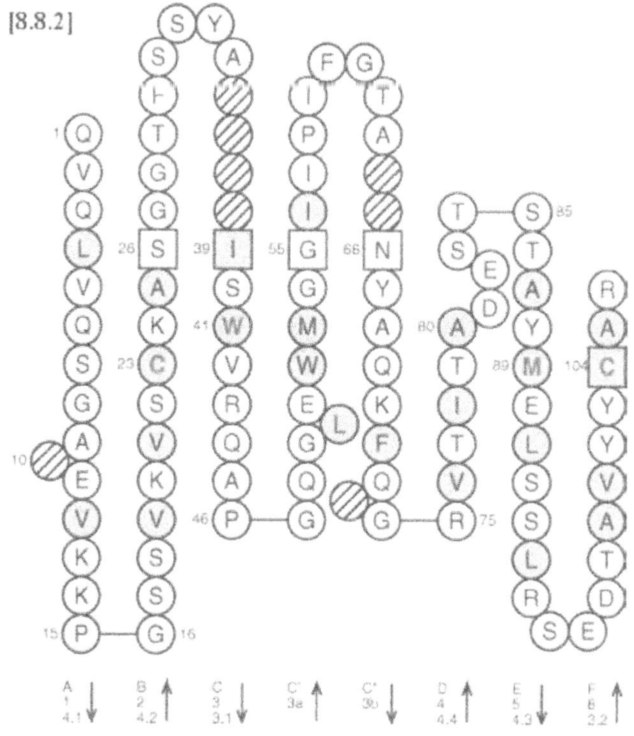

Genome database accession numbers

GDB:9931673 LocusLink: 28461

Nomenclature

IGHV1-c: Immunoglobulin heavy variable 1-c (provisional).

Definition and functionality

IGHV1-c is an unmapped ORF which belongs to the IGHV1 subgroup. This subgroup comprises 14 mapped genes in the IGH locus (of which nine are functional), two unmapped genes (the ORF IGHV1-c and a functional gene) and 11 orphons. IGHV1-c is an ORF due to a Tyrosine (tat) instead of the 2nd_CYS at position 104.

Gene location

IGHV1-c is in the IGH locus on chromosome 14 at 14q32.33.

Nucleotide and amino acid sequences for human IGHV1-c

```
                                   1   2   3   4   5   6   7   8   9  10  11  12  13  14  15  16  17  18  19  20
                                                                                   K   S   G   A   S   V   K
Z18904,IGHV1-c*01,COS-19   (42)                                                GG AAG TCT GGG GCC TCA GTG AAA

                                                          ________________CDR1-IMGT_______________
                                  21  22  23  24  25  26  27  28  29  30  31  32  33  34  35  36  37  38  39  40
                                   V   S   C   S   F   S   G   F   T   I   T   S   Y   G                       H
Z18904,IGHV1-c*01,COS-19          GTC TCC TGT AGT TTT TCT GGG TTT ACC ATC ACC AGC TAC GGT ... ... ... ... ATA CAT

                                                                                              ____________CDR2-
                                  41  42  43  44  45  46  47  48  49  50  51  52  53  54  55  56  57  58  59  60
                                   W   V   Q   Q   S   P   G   Q   G   L   E   W   M   G   W   I   N   P   G   N
Z18904,IGHV1-c*01,COS-19          TGG GTG CAA CAG TCC CCT GGA CAA GGG CTT GAG TGG ATG GGA TGG ATC AAC CCT GGC AAT

                                  IMGT_____________
                                  61  62  63  64  65  66  67  68  69  70  71  72  73  74  75  76  77  78  79  80
                                   G   S   P           S   Y   A   K   K   F   Q       G   R   F   T   M   T   R
Z18904,IGHV1-c*01,COS-19          GGT AGC CCA ... ... AGC TAT GCC AAG AAG TTT CAG ... GGC AGA TTC ACC ATG ACC AGG

                                  81  82  83  84  85  86  87  88  89  90  91  92  93  94  95  96  97  98  99 100
                                   D   M   S   T   T   T   A   Y   T   D   L   S   S   L   T   S   E   D   M   A
Z18904,IGHV1-c*01,COS-19          GAC ATG TCC ACA ACC ACA GCC TAC ACA GAC CTG AGC AGC CTG ACA TCT GAG GAC ATG GCT

                                                      _CDR3-IMGT
                                 101 102 103 104 105 106
                                   V   Y   Y   Y   A   R
Z18904,IGHV1-c*01,COS-19          GTG TAT TAC TAT GCA AGA
```

Framework and complementarity determining regions

FR1-IMGT: 25 (-1aa: 10)	CDR1-IMGT: 8
FR2-IMGT: 17	CDR2-IMGT: 8
FR3-IMGT: 38 (-1 aa: 73)	CDR3-IMGT: 2

Collier de Perles for human IGHV1-c*01

Accession number: IMGT Z18904 EMBL/GenBank/DDBJ: Z18904

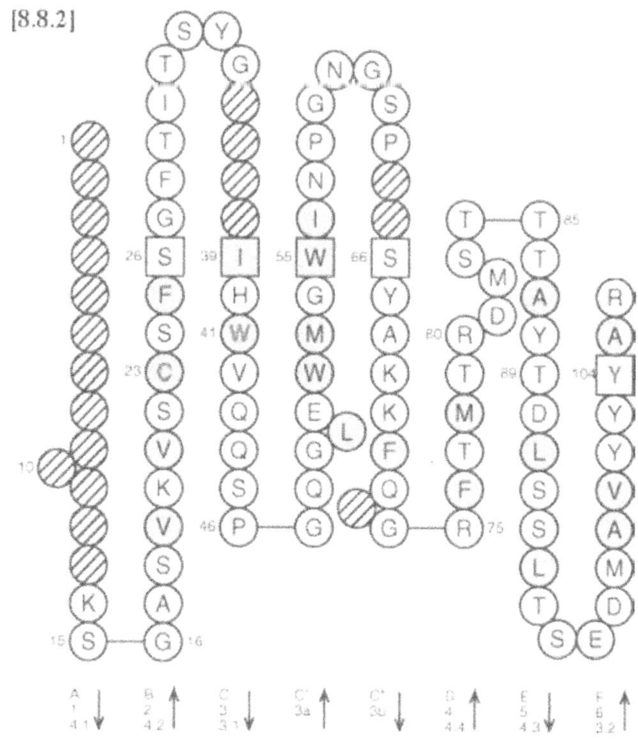

Genome database accession numbers

GDB:9931674 LocusLink: 28460

IGHV1-f

Nomenclature

IGHV1-f: Immunoglobulin heavy variable 1-f (provisional).

Definition and functionality

IGHV1-f is an unmapped functional gene which belongs to the IGHV1 subgroup. This subgroup comprises 14 mapped genes in the IGH locus (of which nine are functional), two unmapped genes (the functional IGHV1-f and an ORF) and 11 orphons.

Gene location

IGHV1-f is in the IGH locus on chromosome 14 at 14q32.33.

Nucleotide and amino acid sequences for human IGHV1-f

```
                                    1   2   3   4   5   6   7   8   9  10  11  12  13  14  15  16  17  18  19  20
                                    E   V   Q   L   V   Q   S   G   A       E   V   K   K   P   G   A   T   V   K
Z12305,IGHV1-f*01,DP-3      [40]  GAG GTC CAG CTG GTA CAG TCT GGG GCT ... GAG GTG AAG AAG CCT GGG GCT ACA GTG AAA

Z29977,IGHV1-f*02,DA-1      [12]                                                  --- --- --- --- --- --- --- ---

                                                        _______________CDR1-IMGT________________
                                   21  22  23  24  25  26  27  28  29  30  31  32  33  34  35  36  37  38  39  40
                                    I   S   C   K   V   S   G   Y   T   F   T   D   Y   Y                   M   H
Z12305,IGHV1-f*01,DP-3            ATC TCC TGC AAG GTT TCT GGA TAC ACC TTC ACC GAC TAC TAC ... ... ... ... ATG CAC

Z29977,IGHV1-f*02,DA-1            --- --- --- --- --- --- --- --- --- --- --- --- --- --- ... ... ... ... --- ---

                                                                                           ____________ CDR2-
                                   41  42  43  44  45  46  47  48  49  50  51  52  53  54  55  56  57  58  59  60
                                    W   V   Q   Q   A   P   G   K   G   L   E   W   M   G   L   V   D   P   E   D
Z12305,IGHV1-f*01,DP-3            TGG GTG CAA CAG GCC CCT GGA AAA GGG CTT GAG TGG ATG GGA CTT GTT GAT CCT GAA GAT

Z29977,IGHV1-f*02,DA-1            --- --- --- --- --- --- --- --- --- --- --- --- --- --- --- --- --- --- --- ---

                                  IMGT____________
                                   61  62  63  64  65  66  67  68  69  70  71  72  73  74  75  76  77  78  79  80
                                    G   E   T           I   Y   A   E   K   F   Q       G   R   V   T   I   T   A
Z12305,IGHV1-f*01,DP-3            GGT GAA ACA ... ... ATA TAC GCA GAG AAG TTC CAG ... GGC AGA GTC ACC ATA ACC GCG

Z29977,IGHV1-f*02,DA-1            --- --- --- ... ... --- --T --- --- --- --- --- ... --- --- --- --- --- --- ---

                                   81  82  83  84  85  86  87  88  89  90  91  92  93  94  95  96  97  98  99 100
                                    D   T   S   T   D   T   A   Y   M   E   L   S   S   L   R   S   E   D   T   A
Z12305,IGHV1-f*01,DP-3            GAC ACG TCT ACA GAC ACA GCC TAC ATG GAG CTG AGC AGC CTG AGA TCT GAG GAC ACG GCC

Z29977,IGHV1-f*02,DA-1            --- --- --- --- --- --- --- --- --- --- --- --- --- --- --- --- ---

                                                    _CDR3-IMGT
                                  101 102 103 104 105 106
                                    V   Y   Y   C   A   T
Z12305,IGHV1-f*01,DP-3            GTG TAT TAC TGT GCA ACA

Z29977,IGHV1-f*02,DA-1
```

Framework and complementarity determining regions

FR1-IMGT: 25 (-1 aa: 10)
FR2-IMGT: 17
FR3-IMGT: 38 (-1 aa: 73)

CDR1-IMGT: 8
CDR2-IMGT: 8
CDR3-IMGT: 2

Collier de Perles for human IGHV1-f*01

Accession number: IMGT Z12305 EMBL/GenBank/DDBJ: Z12305

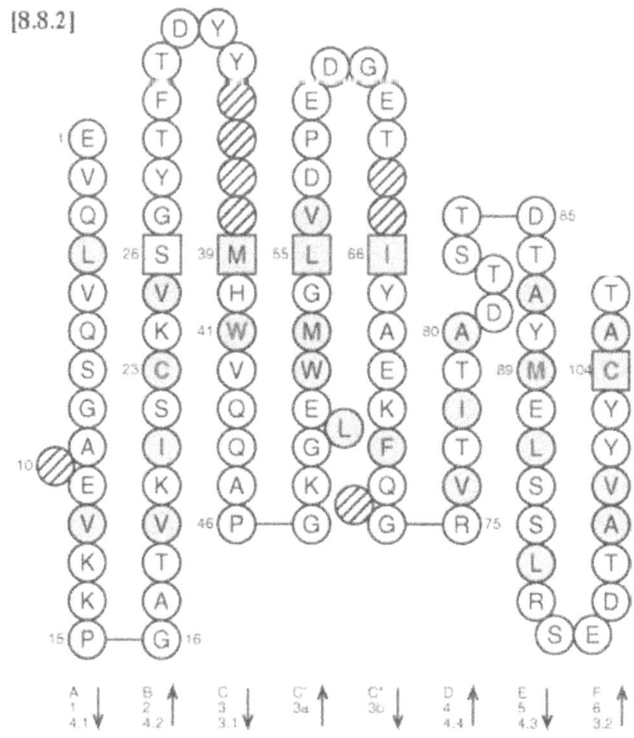

Genome database accession numbers

GDB:9931676 LocusLink: 28458

IGHV2-5

Nomenclature

IGHV2-5: Immunoglobulin heavy variable 2-5.

Definition and functionality

IGHV2-5 is one of the three mapped functional genes of the IGHV2 subgroup which comprises four mapped genes in the IGH locus and one orphon.

Gene location

IGHV2-5 is in the IGH locus on chromosome 14 at 14q32.33.

Nucleotide and amino acid sequences for human IGHV2-5

```
                                        1   2   3   4   5   6   7   8   9   10  11  12  13  14  15  16  17  18  19  20
                                        Q   I   T   L   K   E   S   G   P       T   L   V   K   P   T   Q   T   L   T
X62111,IGHV2-5*01,VII-5          [39]  CAG ATC ACC TTG AAG GAG TCT GGT CCT ... ACG CTG GTG AAA CCC ACA CAG ACC CTC ACG

Z14072,IGHV2-5*02,DP76           [41]

X62108,IGHV2-5*02,VII-5b         [39]  --- --- --- --- --- --- --- --- --- ... --- --- --- --- --- --- --- --- --- ---

X93619,IGHV2-5*03,WAD4GL         [48]                                              --- --- --- --- --- --- --- --- ---

L21963,IGHV2-5*04,S12.2           [2]  --- --- --- --- --- --- --- --- --- ... --- --- --- --- --- --- --- --- --- ---

L21964,IGHV2-5*05,S12.4           [2]  --- --- --- --- --- --- --- --- --- ... --- --- --- --- --- --- --- --- --- ---

L21966,IGHV2-5*06,S12.6           [2]  --- --- --- --- --- --- --- --- --- ... --- --- --A --- --- --- --- --- --- ---

L21968,IGHV2-5*07,S12.8           [2]  --- --- --- --- --- --- --- --- --- ... --- --- --- --- --- --- --- --- --- ---
                                            V                                   A
L21971,IGHV2-5*08,S12.12          [2]  --- G-- --- --- --- --- --- --- --- ... G-- --- --- --- --- --- --- --- --- --A
                                            V
L21972,IGHV2-5*09,S12.14          [2]  --- G-- --- --- --- --- --- --- --- ... --- --- --- --- --- --- --- --- --- ---

                                                                ______________CDR1-IMGT______________
                                        21  22  23  24  25  26  27  28  29  30  31  32  33  34  35  36  37  38  39  40
                                        L   T   C   T   F   S   G   F   S   L   S   T   S   G   V   G           V   G
X62111,IGHV2-5*01,VII-5                CTG ACC TGC ACC TTC TCT GGG TTC TCA CTC AGC ACT AGT GGA GTG GGT ... ... CTG GGC

Z14072,IGHV2-5*02,DP76                                                         --- --- --- --- --- ... ... --- ---

X62108,IGHV2-5*02,VII-5b          (1)  --- --- --- --- --- --- --- --- --- --- --- --- --- --- --- --- ... ... --- ---

X93619,IGHV2-5*03,WAD4GL               --- --- --- --- --- --- --- --- --- --- --- --- --- --- --- --- ... ... --- ---

L21963,IGHV2-5*04,S12.2                --- --- --- --- --- --- --- --- --- --- --- --- --- --- --- --- ... ... --- ---

L21964,IGHV2-5*05,S12.4                --- --- --- --- --- --- --- --- --- --- --- --- --- --- --- --- ... ... --- ---

L21966,IGHV2-5*06,S12.6                --- --- --- --- --- --- --- --- --- --- --- --- --- --- --- --- ... ... --- ---

L21968,IGHV2-5*07,S12.8                --- --- --- --- --- --- --- --- --- --- --- --- --- --- --- --- ... ... --- ---
                                                                                                M   R               S
L21971,IGHV2-5*08,S12.12               --- --- --- --- --- --- --- --- --- --- --- --- --- --- A-- C-- ... ... --- A--

L21972,IGHV2-5*09,S12.14               --- --- --- --- --- --- --- --- --- --- --- --- --- --- --- --- ... ... --- ---

                                                                                                    __________CDR2-
                                        41  42  43  44  45  46  47  48  49  50  51  52  53  54  55  56  57  58  59  60
                                        W   I   R   Q   P   P   G   K   A   L   E   W   L   A   L   I   Y   W   N   D
X62111,IGHV2-5*01,VII-5                TGG ATC CGT CAG CCC CCA GGA AAG GCC CTG GAG TGG CTT GCA CTC ATT TAT TGG AAT GAT
                                                                                                                D
Z14072,IGHV2-5*02,DP76                 --- --- --- --- --- --- --- --- --- --- --- --- --- --- --- --- --- --- G-- ---
                                                                                                                D
X62108,IGHV2-5*02,VII-5b               --- --- --- --- --- --- --- --- --- --- --- --- --- --- --- --- --- --- G-- ---
                                                                                                                D
X93619,IGHV2-5*03,WAD4GL               --- --- --- --- --- --- --- --- --- --- --- --- --- --- --- --- --- --- G-- ---

L21963,IGHV2-5*04,S12.2                --- --- --- --- --- --- --- --- --- --- --- --- --- --- --- --- --- --- --- ---
                                                                                                                D
L21964,IGHV2-5*05,S12.4                --- --- --- --- --- --- --- --- --- --- --- --- --- --- --- --- --- --- G-- ---
                                                                                                                D
L21966,IGHV2-5*06,S12.6                --- --- --- --- --- --- --- --- --- --- --- --- --- --- --- --- --- --- G-- ---

L21968,IGHV2-5*07,S12.8                --- --- --- --- --- --- --- --- --- --- --- --- --- --- --- --- --- --- --- ---
                                                                                                                D
L21971,IGHV2-5*08,S12.12               --- --- --- --- --- --- --- --- --- --- --- --- --- --- --- --- --- --- G-- ---
                                                                                                                D
L21972,IGHV2-5*09,S12.14               --- --- --- --- --- --- --- --- --- --- --- --- --- --- --- --- --- --- G-- ---
```

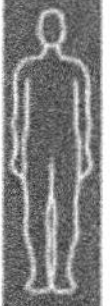

```
                          IMGT_______________
                           61  62  63  64  65  66  67  68  69  70  71  72  73  74  75  76  77  78  79  80
                           D   K               R   Y   S   P   S   L   K       S   R   L   T   I   T   K
X62111,IGHV2-5*01,VII-5   GAT AAG ... ... ... CGC TAC AGC CCA TCT CTG AAG ... AGC AGG CTC ACC ATC ACC AAG

Z14072,IGHV2-5*02,DP76    --- --- ... ... ... --- --- --- --- --- --- --- ... --- --- --- --- --- --

X62108,IGHV2-5*02,VII-5b  --- --- ... ... ... --- --- --- --- --- --- --- ... --- --- --- --- --- --- ---

X93619,IGHV2-5*03,WAD4GL  --- --- ... ... ... --- --- --- --- --- --- --- ... --- --- --- --- --T --- ---

L21963,IGHV2-5*04,S12.2   --- --- ... ... ... --- --- --- --- --- --- --- ... --- --- --- --- --- --- ---
                                                      G
L21964,IGHV2-5*05,S12.4   --- --- ... ... ... --- --- G-- --- --- --- --- ... --- --- --- --- --- --- ---
                                                      G
L21966,IGHV2-5*06,S12.6   --- --- ... ... ... --- --- G-- --- --- --- --- ... --- --- --- --- --- --- ---

L21968,IGHV2-5*07,S12.8   --- --- ... ... ... --- --- --- --- --- --- --- ... --- --- --- --- --- --- ---

L21971,IGHV2-5*08,S12.12  --- --- ... ... ... --- --- --- --- --- --- --- ... --- --- --- --- --- --- ---
                                                      G
L21972,IGHV2-5*09,S12.14  --- --- ... ... ... --- --- G-- --- --- --- --- ... --- --- --- --- --- --- ---

                           81  82  83  84  85  86  87  88  89  90  91  92  93  94  95  96  97  98  99 100
                           D   T   S   K   N   Q   V   V   L   T   M   T   N   M   D   P   V   D   T   A
X62111,IGHV2-5*01,VII-5   GAC ACC TCC AAA AAC CAG GTG GTC CTT ACA ATG ACC AAC ATG GAC CCT GTG GAC ACA GCC

Z14072,IGHV2-5*02,DP76

X62108,IGHV2-5*02,VII-5b  --- --- --- --- --- --- --- --- --- --- --- --- --- --- --- --- --- --- --- ---

X93619,IGHV2-5*03,WAD4GL  --- --- --- --- --- --- --
                                                                                                       G
L21963,IGHV2-5*04,S12.2   --- --- --- --- --- --- --- --- --- --- --- --- --- --- --- --- --- --- --- -G-

L21964,IGHV2-5*05,S12.4   --- --- --- --- --- --- --- --- --- --- --- --- --- --- --- --- --- --- --- ---

L21966,IGHV2-5*06,S12.6   --- --- --- --- --- --- --- --- --- --- --- --- --- --- --- --- --- --- --- ---
                                                                                                       G
L21968,IGHV2-5*07,S12.8   --- --- --- --- --- --- --- --- --- --- --- --- --- --- --- --- --- --- --- -G-

L21971,IGHV2-5*08,S12.12  --- --- --- --- --- --- --- --- --- --- --- --- --- --- --- --- --- --- --- ---

L21972,IGHV2-5*09,S12.14  --- --- --- --- --- --- --- --- --- --- --- --- --- --- --- --- --- --- --- ---

                                          __CDR3-IMGT__
                          101 102 103 104 105 106 107
                           T   Y   Y   C   A   H   R
X62111,IGHV2-5*01,VII-5   ACA TAT TAC TGT GCA CAC AGA CC

Z14072,IGHV2-5*02,DP76

X62108,IGHV2-5*02,VII-5b  --- --- --- --- --- --- --- --

X93619,IGHV2-5*03,WAD4GL
                                           V   R
L21963,IGHV2-5*04,S12.2   --- --- --- --- -T- -GG

L21964,IGHV2-5*05,S12.4   --- --- --- --- --- --- --- -

L21966,IGHV2-5*06,S12.6   --- --- --- --- --- --- ---
                                           V
L21968,IGHV2-5*07,S12.8   --- --- --- ---  T

L21971,IGHV2-5*08,S12.12  --- --- --- --- --- --- ---

L21972,IGHV2-5*09,S12.14  --- --- --- --- --- --- --- -
```

Note:

Three nucleotides 'g' DELETION at position 100, 107 and 117 may result from sequencing errors in X62108, instead of having one amino acid missing as suggested in [39]. The sequence is probably identical to allele *02 with a CDR1-IMGT of 10 amino acids, like the other IGHV2 subgroup genes.

Framework and complementarity determining regions

FR1-IMGT: 25 (-1 aa: 10)
FR2-IMGT: 17
FR3-IMGT: 38 (-1 aa: 73)

CDR1-IMGT: 10
CDR2-IMGT: 7
CDR3-IMGT: 3

Collier de Perles for human IGHV2-5*01

Accession number: IMGT X62111 EMBL/GenBank/DDBJ: X62111

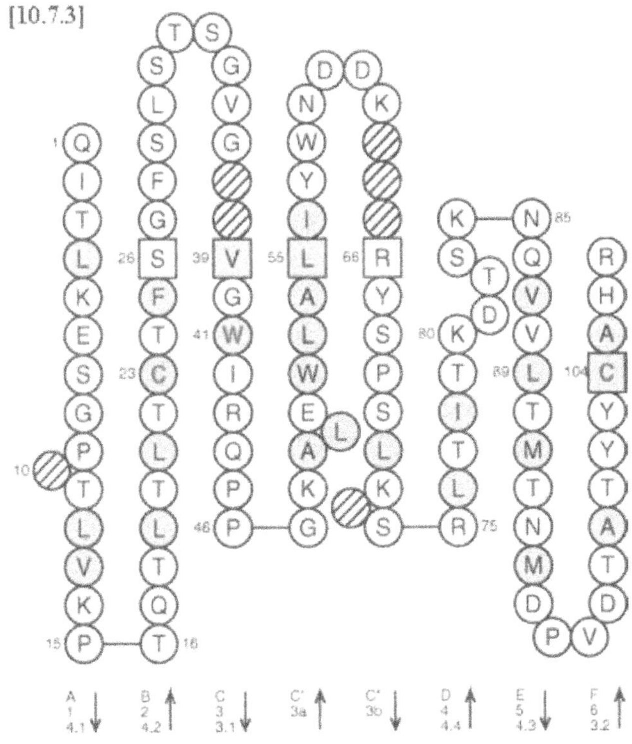

Genome database accession numbers

GDB:9931677 LocusLink: 28457

Nomenclature

IGHV2-26: Immunoglobulin heavy variable 2-26.

Definition and functionality

IGHV2-26 is one of the three mapped functional genes of the IGHV2 subgroup which comprises four mapped genes in the IGH locus and one orphon.

Gene location

IGHV2-26 is in the IGH locus on chromosome 14 at 14q32.33.

Nucleotide and amino acid sequences for human IGHV2-26

```
                                     1   2   3   4   5   6   7   8   9  10  11  12  13  14  15  16  17  18  19  20
                                     Q   V   T   L   K   E   S   G   P       V   L   V   K   P   T   E   T   L   T
M99648,IGHV2-26*01,V2-26     [26]   CAG GTC ACC TTG AAG GAG TCT GGT CCT ... GTG CTG GTG AAA CCC ACA GAG ACC CTC ACG

Z12328,IGHV2-26*01,DP-26     [41]   --- --- --- --- --- --- --- --- --- ... --- --- --- --- --- --- --- --- --- ---

                                                             ________________CDR1-IMGT_______________
                                    21  22  23  24  25  26  27  28  29  30  31  32  33  34  35  36  37  38  39  40
                                     L   T   C   T   V   S   G   F   S   L   S   N   A   R   M   G           V   S
M99648,IGHV2-26*01,V2-26            CTG ACC TGC ACC GTC TCT GGG TTC TCA CTC AGC AAT GCT AGA ATG GGT ... ... GTG AGC

Z12328,IGHV2-26*01,DP-26            --- --- --- --- --- --- --- --- --- --- --- --- --- --- --- --- ... ... --- ---

                                                                                              ___________ CDR2-
                                    41  42  43  44  45  46  47  48  49  50  51  52  53  54  55  56  57  58  59  60
                                     W   I   R   Q   P   P   G   K   A   L   E   W   L   A   H   I   F   S   N   D
M99648,IGHV2-26*01,V2-26            TGG ATC CGT CAG CCC CCA GGG AAG GCC CTG GAG TGG CTT GCA CAC ATT TTT TCG AAT GAC

Z12328,IGHV2-26*01,DP-26            --- --- --- --- --- --- --- --- --- --- --- --- --- --- --- --- --- --- --- ---

                                    IMGT______________
                                    61  62  63  64  65  66  67  68  69  70  71  72  73  74  75  76  77  78  79  80
                                     E   K               S   Y   S   T   S   L   K       S   R   L   T   I   S   K
M99648,IGHV2-26*01,V2-26            GAA AAA ... ... ... TCC TAC AGC ACA TCT CTG AAG ... AGC AGG CTC ACC ATC TCC AAG

Z12328,IGHV2-26*01,DP-26            --- --- ... ... ... --- --- --- --- --- --- --- ... --- --- --- --- --- --- ---

                                    81  82  83  84  85  86  87  88  89  90  91  92  93  94  95  96  97  98  99 100
                                     D   T   S   K   S   Q   V   V   L   T   M   T   N   M   D   P   V   D   T   A
M99648,IGHV2-26*01,V2-26            GAC ACC TCC AAA AGC CAG GTG GTC CTT ACC ATG ACC AAC ATG GAC CCT GTG GAC ACA GCC

Z12328,IGHV2-26*01,DP-26            --- --- --- --- --- --- --- --- --- --- --- --- --- --- --- --- --- --- --- ---

                                                    __CDR3-IMGT__
                                    101 102 103 104 105 106 107
                                     T   Y   Y   C   A   R   I
M99648,IGHV2-26*01,V2-26            ACA TAT TAC TGT GCA CGG ATA C

Z12328,IGHV2-26*01,DP-26            --- --- ---
```

Framework and complementarity determining regions

FR1-IMGT: 25 (-1 aa: 10)	CDR1-IMGT: 10
FR2-IMGT: 17	CDR2-IMGT: 7
FR3-IMGT: 38 (-1 aa: 73)	CDR3-IMGT: 3

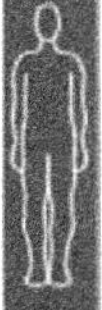

Collier de Perles for human IGHV2-26*01

Accession number: IMGT M99648 EMBL/GenBank/DDBJ: M99648

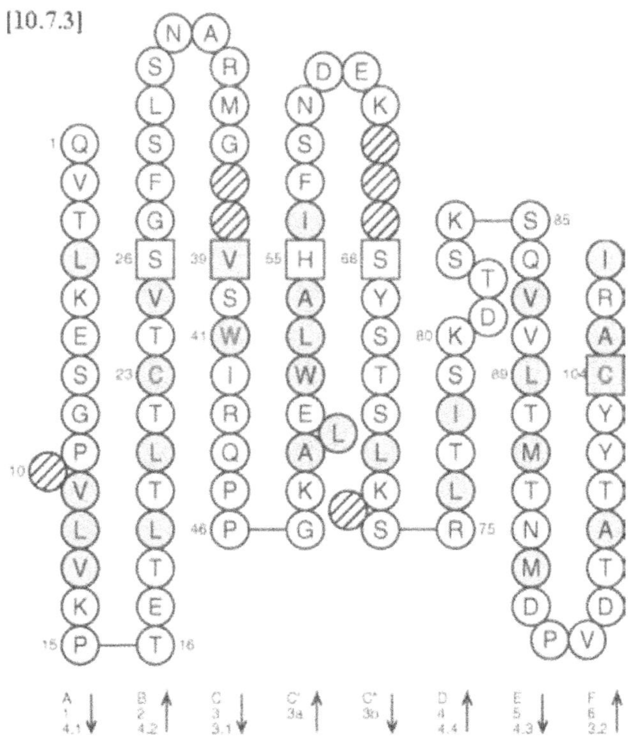

Genome database accession numbers

GDB:9931679 LocusLink: 28455

IGHV2-70

Nomenclature

IGHV2-70: Immunoglobulin heavy variable 2-70.

Definition and functionality

IGHV2-70 is one of the three mapped functional genes of the IGHV2 subgroup which comprises four mapped genes in the IGH locus and one orphon.

Gene location

IGHV2-70 is in the IGH locus on chromosome 14 at 14q32.33.

Nucleotide and amino acid sequences for human IGHV2-70

```
                                          1   2   3   4   5   6   7   8   9   10  11  12  13  14  15  16  17  18  19  20
                                          Q   V   T   L   R   E   S   G   P       A   L   V   K   P   T   Q   T   L   T
L21969,IGHV2-70*01,S12-9            [2]   CAG GTC ACC TTG AGG GAG TCT GGT CCT ... GCG CTG GTG AAA CCC ACA CAG ACC CTC ACA
Z12329,IGHV2-70*01,DP-27            [41]  --- --- --- --- --- --- --- --- --- ... --- --- --- --- --- ---     ---
X92241,IGHV2-70*02,VH2-MC2          [7]   --- --- --- --- --- --- --- --- --- ... --- --- --- --- --- --- --- --- --- ---
X92242,IGHV2-70*02,VH2-MC2a         [7]   --- --- --- --- --- --- --- --- --- ... --- --- --- --- --- --- --- ---     ---
X92244,IGHV2-70*02,VH2-MC2c         [7]   --- --- --- --- --- --- --- --- --- ... --- --- --- --- --- --- --- --- --- ---
X92246,IGHV2-70*02,VH2-MC2e         [7]   --- --- --- --- --- --- --- --- --- ... --- --- --- ---     --- --- --- --- ---
                                                          K
X92238,IGHV2-70*03,VH2-MC1          [7]   --- --- --- --- A-- --- --- --- --- ... --- --- --- --- --- --- --- --- --- ---
                                                          K
X92240,IGHV2-70*03,VH2-MC1b         [7]       --- --- --- -A- --- --- --- --- ... --- --- --- --- --- --- --- --- --- ---
                                                          K
Z12330,IGHV2-70*04,DP-28            [41]  --- --- --- --- -A- --- --- --- --- ... --- --- --- --- --- --- --- --- --- ---
Z29983,IGHV2-70*03 or 04,DA-7       [12]                                                                          --- ---
Z27502,IGHV2-70*05,YAC-3            [12]                                    - ... --- --- --- --- --- --- --- --- --- ---
                                                          K
X92239,IGHV2-70*6,VH2-MC1a          [7]               --- -A- --- --- --- --- ... --- --- --- --- --- --- --- --- --- ---
X92243,IGHV2-70*7,VH2-MC2b          [7]   --- --- --- --- --- --- --- --- --- ... --- --- --- --- --- --- --- --- --- ---
X92245,IGHV2-70*8,VH2-MC2d          [8]   --- --- --- --- --- --- --- --- --- ... --- --- --- --- --- --- --- --- --- ---
                                              I           K                       T
L21962,IGHV2-70*09,S12-1            [2]       A-- --- --- -A- --- --- --- --- ... A-- --- --- --- --- --- --- --- --- --G
                                                          K
L21965,IGHV2-70*10,S12-5            [2]   --- --- --- --- -A- --- --- --- --- ... --- --- --- --- --- --- --- --- --- ---
                                          R
L21967,IGHV2-70*11,S12-7            [2]   -G- --- --- --- --- --- --- --- --- ... --- --- --- --- --- --- --- --- --- ---
                                              I           K                       T
L21970,IGHV2-70*12,S12-10           [2]       A--     --- -A- --- --- --- --- ... A-- --- --- --- --- --- --- --- --- --G

                                                                  ______________CDR1 IMGT______________
                                          21  22  23  24  25  26  27  28  29  30  31  32  33  34  35  36  37  38  39  40
                                          L   T   C   T   F   S   G   F   S   L   S   T   S   G   M   C           V   S
L21969,IGHV2-70*01,S12-9                  CTG ACC TGC ACC TTC TCT GGG TTC TCA CTC AGC ACT AGT GGA ATG TGT ... ... GTG ACC
Z12329,IGHV2-70*01,DP-27                  --- --- --- --- --- --- --- --- --- --- --- --- --- --- --- --- ... ... --- ---
X92241,IGHV2-70*02,VH2-MC2                --- --- --- --- --- --- --- --- --- --- --- --- --- --- --- --- ... ...
X92242,IGHV2-70*02,VH2-MC2a               --- --- --- --- --- --- --- --- --- --- --- --- --- --- --- --- ... ... --- ---
X92244,IGHV2-70*02,VH2-MC2c               --- --- --- --- --- --- --- --- --- --- --- --- --- --- --- --- ... ... --- ---
X92246,IGHV2-70*02,VH2-MC2e               --- --- --- --- --- --- --- --- --- --- --- --- --- --- --- --- ... ... --- ---
                                                                                              R
X92238,IGHV2-70*03,VH2-MC1                --- --- --- --- --- --- --- --- --- --- --- --- --- --- --- C-- ... ... --- ---
                                                                                              R
X92240,IGHV2-70*03,VH2-MC1b               --- --- --- --- --- --- --- --- --- --- --- --- --- --- --- C-- ... ... --- ---
                                                                                              R
Z12330,IGHV2-70*04,DP-28                  --- --- --- --- --- --- --- --- --- --- --- --- --- --- --- C-- ... ... --- ---
                                                                                              R
Z29983,IGHV2-70*03 or 04,DA-7             --- --- --- --- --- --- --- --- --- --- --- --- --- --- --- C-- ... ... --- ---
                                                                                              R           A
Z27502,IGHV2-70*05,YAC-3                  --- --- --- --- --- --- --- --- --- --- --- --- --- --- --- C-- ... ... -C- ---
                                                                                              R
X92239,IGHV2-70*06,VH2-MC1a               --- --- --- --- --- --- --- --- --- --- --- --- --- --- --- C-- ... ... --- ---
X92243,IGHV2-70*07,VH2-MC2b               --- --- --- --- --- --- --- --- --- --- --- --- --- --- --- --- ... ... --- ---
                                                          A
X92245,IGHV2-70*08,VH2-MC2d               --- --- --- G-- --- --- --- --- --- --- --- --- --- --- --- --- ... ... --- ---
                                                  R
L21962,IGHV2-70*09,S12-1                  --- --- C-- --- --- --- --- --- --- --- --- --- --- --- --- --- ... ... --- ---
                                                                                              R
L21965,IGHV2-70*10,S12-5                  --- --- --- --- --- --- --- --- --- --- --- --- --- --- --- C-- ... ...
L21967,IGHV2-70*11,S12-7                  --- --- --- --- --- --- --- --- --- --- --- --- --- --- --- --- ... ... --- ---
L21970,IGHV2-70*12,S12-10                 --- --- --- --- --- --- --- --- --- --- --- --- --- --- --- --- ... ... --- ---

                                                                                           ______________CDR2-
                                          41  42  43  44  45  46  47  48  49  50  51  52  53  54  55  56  57  58  59  60
                                          W   I   R   Q   P   P   G   K   A   L   E   W   L   A   L   I   D   W   D   D
L21969,IGHV2-70*01,S12-9                  TGG ATC CGT CAG CCC CCA GGG AAG GCC CTG GAG TGG CTT GCA CTC ATT GAT TGG GAT GAT
Z12329,IGHV2-70*01,DP-27                  --- --- --- --- --- --- --- --- --- --- --- --- --- --- --- --- --- --- --- ---
X92241,IGHV2-70*02,VH2-MC2                --- --- --- --- --- --- --- --- --- --- --- --- --- --- --- --- --- --- --- ---
X92242,IGHV2-70*02,VH2-MC2a               --- --- --- --- --- --- --- --- --- --- --- --- --- --- --- --- --- --- --- ---
X92244,IGHV2-70*02,VH2-MC2c               --- --- --- --- --- --- --- --- --- --- --- --- --- --- --- --- --- --- --- ---
```

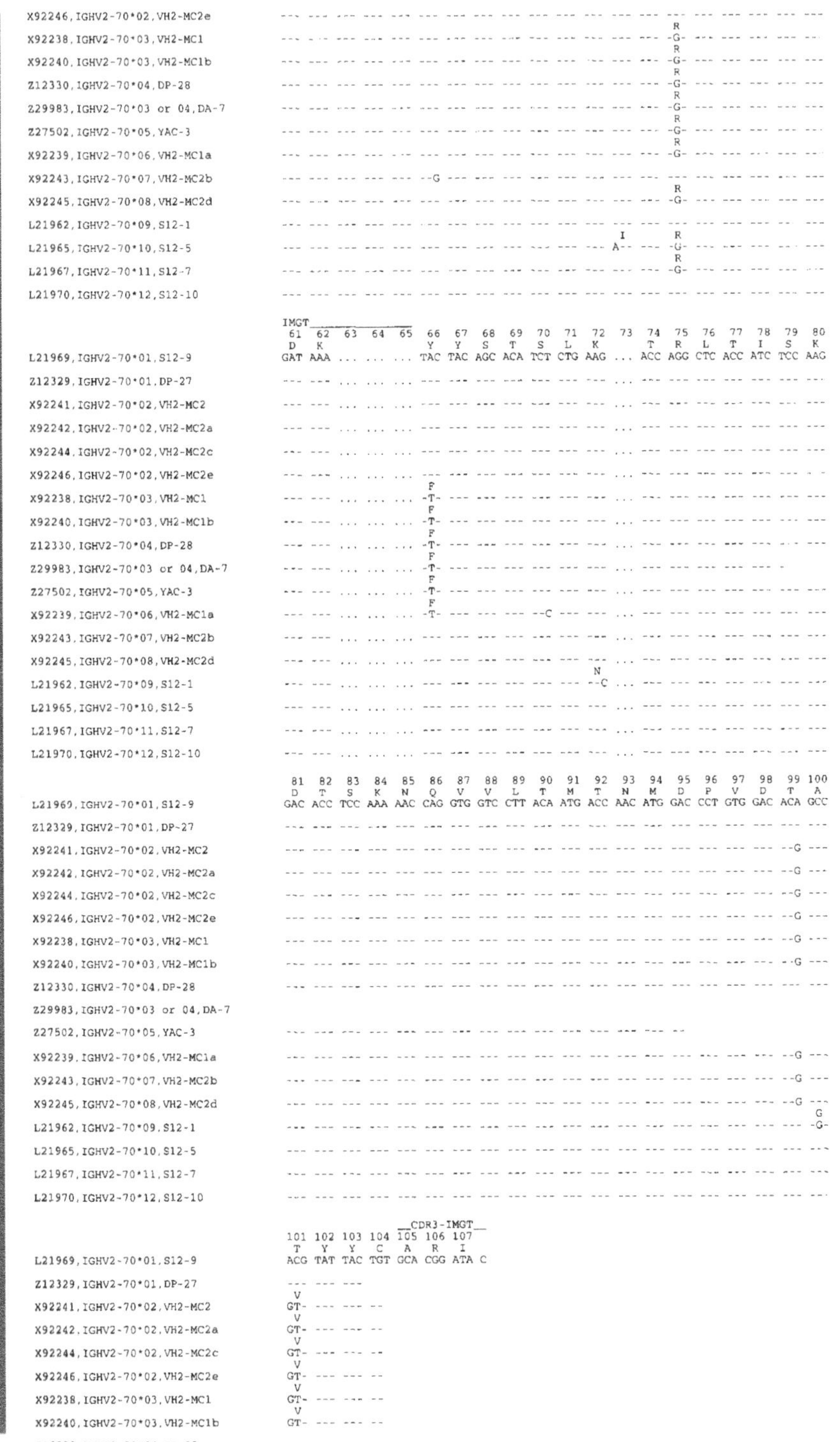

```
X92246,IGHV2-70*02,VH2-MC2e          --- --- --- --- --- --- --- --- --- --- --- --- --- --- --- --- --- --- --- ---
                                                                                       R
X92238,IGHV2-70*03,VH2-MC1           --- --- --- --- --- --- --- --- --- --- --- --- --- --- -G- --- --- --- --- ---
                                                                                       R
X92240,IGHV2-70*03,VH2-MC1b          --- --- --- --- --- --- --- --- --- --- --- --- --- --- -G- --- --- --- --- ---
                                                                                       R
Z12330,IGHV2-70*04,DP-28             --- --- --- --- --- --- --- --- --- --- --- --- --- --- -G- --- --- --- --- ---
                                                                                       R
Z29983,IGHV2-70*03 or 04,DA-7        --- --- --- --- --- --- --- --- --- --- --- --- --- --- -G- --- --- --- --- ---
                                                                                       R
Z27502,IGHV2-70*05,YAC-3             --- --- --- --- --- --- --- --- --- --- --- --- --- --- -G- --- --- --- --- ---
                                                                                       R
X92239,IGHV2-70*06,VH2-MC1a          --- --- --- --- --- --- --- --- --- --- --- --- --- --- -G- --- --- --- --- ---
X92243,IGHV2-70*07,VH2-MC2b          --- --- --- --- --- --G --- --- --- --- --- --- --- --- --- --- --- --- --- ---
                                                                                       R
X92245,IGHV2-70*08,VH2-MC2d          --- --- --- --- --- --- --- --- --- --- --- --- --- --- -G- --- --- --- --- ---
L21962,IGHV2-70*09,S12-1             --- --- --- --- --- --- --- --- --- --- --- --- --- --- --- --- --- --- --- ---
                                                                                   I       R
L21965,IGHV2-70*10,S12-5             --- --- --- --- --- --- --- --- --- --- --- --- A-- --- -G- --- --- --- --- ---
                                                                                       R
L21967,IGHV2-70*11,S12-7             --- --- --- --- --- --- --- --- --- --- --- --- --- --- -G- --- --- --- --- ---
L21970,IGHV2-70*12,S12-10            --- --- --- --- --- --- --- --- --- --- --- --- --- --- --- --- --- --- --- ---

                                     IMGT_______________
                                     61  62  63  64  65  66  67  68  69  70  71  72  73  74  75  76  77  78  79  80
                                      D   K               Y   Y   S   T   S   L   K       T   R   L   T   I   S   K
L21969,IGHV2-70*01,S12-9             GAT AAA ... ... ... TAC TAC AGC ACA TCT CTG AAG ... ACC AGG CTC ACC ATC TCC AAG
Z12329,IGHV2-70*01,DP-27             --- --- ... ... ... --- --- --- --- --- --- --- ... --- --- --- --- --- --- ---
X92241,IGHV2-70*02,VH2-MC2           --- --- ... ... ... --- --- --- --- --- --- --- ... --- --- --- --- --- --- ---
X92242,IGHV2-70*02,VH2-MC2a          --- --- ... ... ... --- --- --- --- --- --- --- ... --- --- --- --- --- --- ---
X92244,IGHV2-70*02,VH2-MC2c          --- --- ... ... ... --- --- --- --- --- --- --- ... --- --- --- --- --- --- ---
X92246,IGHV2-70*02,VH2-MC2e          --- --- ... ... ... --- --- --- --- --- --- --- ... --- --- --- --- --- --- ---
                                                          F
X92238,IGHV2-70*03,VH2-MC1           --- --- ... ... ... -T- --- --- --- --- --- --- ... --- --- --- --- --- --- ---
                                                          F
X92240,IGHV2-70*03,VH2-MC1b          --- --- ... ... ... -T- --- --- --- --- --- --- ... --- --- --- --- --- --- ---
                                                          F
Z12330,IGHV2-70*04,DP-28             --- --- ... ... ... -T- --- --- --- --- --- --- ... --- --- --- --- --- --- ---
                                                          F
Z29983,IGHV2-70*03 or 04,DA-7        --- --- ... ... ... -T- --- --- --- --- --- --- ... --- --- --- --- --- -
                                                          F
Z27502,IGHV2-70*05,YAC-3             --- --- ... ... ... -T- --- --- --- --- --- --- ... --- --- --- --- --- --- ---
                                                          F
X92239,IGHV2-70*06,VH2-MC1a          --- --- ... ... ... -T- --- --- --- --C --- --- ... --- --- --- --- --- --- ---
X92243,IGHV2-70*07,VH2-MC2b          --- --- ... ... ... --- --- --- --- --- --- --- ... --- --- --- --- --- --- ---
X92245,IGHV2-70*08,VH2-MC2d          --- --- ... ... ... --- --- --- --- --- --- --- ... --- --- --- --- --- --- ---
                                                                                  N
L21962,IGHV2-70*09,S12-1             --- --- ... ... ... --- --- --- --- --- --- --C ... --- --- --- --- --- --- ---
L21965,IGHV2-70*10,S12-5             --- --- ... ... ... --- --- --- --- --- --- --- ... --- --- --- --- --- --- ---
L21967,IGHV2-70*11,S12-7             --- --- ... ... ... --- --- --- --- --- --- --- ... --- --- --- --- --- --- ---
L21970,IGHV2-70*12,S12-10            --- --- ... ... ... --- --- --- --- --- --- --- ... --- --- --- --- --- --- ---

                                     81  82  83  84  85  86  87  88  89  90  91  92  93  94  95  96  97  98  99  100
                                      D   T   S   K   N   Q   V   V   L   T   M   T   N   M   D   P   V   D   T   A
L21969,IGHV2-70*01,S12-9             GAC ACC TCC AAA AAC CAG GTG GTC CTT ACA ATG ACC AAC ATG GAC CCT GTG GAC ACA GCC
Z12329,IGHV2-70*01,DP-27             --- --- --- --- --- --- --- --- --- --- --- --- --- --- --- --- --- --- --- ---
X92241,IGHV2-70*02,VH2-MC2           --- --- --- --- --- --- --- --- --- --- --- --- --- --- --- --- --- --- --G ---
X92242,IGHV2-70*02,VH2-MC2a          --- --- --- --- --- --- --- --- --- --- --- --- --- --- --- --- --- --- --G ---
X92244,IGHV2-70*02,VH2-MC2c          --- --- --- --- --- --- --- --- --- --- --- --- --- --- --- --- --- --- --G ---
X92246,IGHV2-70*02,VH2-MC2e          --- --- --- --- --- --- --- --- --- --- --- --- --- --- --- --- --- --- --G ---
X92238,IGHV2-70*03,VH2-MC1           --- --- --- --- --- --- --- --- --- --- --- --- --- --- --- --- --- --- --G ---
X92240,IGHV2-70*03,VH2-MC1b          --- --- --- --- --- --- --- --- --- --- --- --- --- --- --- --- --- --- --G ---
Z12330,IGHV2-70*04,DP-28             --- --- --- --- --- --- --- --- --- --- --- --- --- --- --- --- --- --- --- ---
Z29983,IGHV2-70*03 or 04,DA-7
Z27502,IGHV2-70*05,YAC-3             --- --- --- --- --- --- --- --- --- --- --- --- --- --- --
X92239,IGHV2-70*06,VH2-MC1a          --- --- --- --- --- --- --- --- --- --- --- --- --- --- --- --- --- --- --G ---
X92243,IGHV2-70*07,VH2-MC2b          --- --- --- --- --- --- --- --- --- --- --- --- --- --- --- --- --- --- --G ---
X92245,IGHV2-70*08,VH2-MC2d          --- --- --- --- --- --- --- --- --- --- --- --- --- --- --- --- --- --- --G ---
                                                                                                              G
L21962,IGHV2-70*09,S12-1             --- --- --- --- --- --- --- --- --- --- --- --- --- --- --- --- --- --- --- -G-
L21965,IGHV2-70*10,S12-5             --- --- --- --- --- --- --- --- --- --- --- --- --- --- --- --- --- --- --- ---
L21967,IGHV2-70*11,S12-7             --- --- --- --- --- --- --- --- --- --- --- --- --- --- --- --- --- --- --- ---
L21970,IGHV2-70*12,S12-10            --- --- --- --- --- --- --- --- --- --- --- --- --- --- --- --- --- --- --- ---

                                                         __CDR3-IMGT__
                                     101 102 103 104 105 106 107
                                      T   Y   Y   C   A   R   I
L21969,IGHV2-70*01,S12-9             ACG TAT TAC TGT GCA CGG ATA C
Z12329,IGHV2-70*01,DP-27             --- --- ---
                                      V
X92241,IGHV2-70*02,VH2-MC2           GT- --- --- --
                                      V
X92242,IGHV2-70*02,VH2-MC2a          GT- --- --- --
                                      V
X92244,IGHV2-70*02,VH2-MC2c          GT- --- --- --
                                      V
X92246,IGHV2-70*02,VH2-MC2e          GT- --- --- --
                                      V
X92238,IGHV2-70*03,VH2-MC1           GT- --- --- --
                                      V
X92240,IGHV2-70*03,VH2-MC1b          GT- --- --- --
Z12330,IGHV2-70*04,DP-28             --- --- ---
```

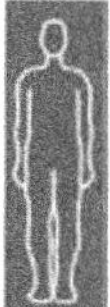

Z29983,IGHV2-70*03 or 04,DA-7
Z27502,IGHV2-70*05,YAC-3
X92239,IGHV2-70*06,VH2-MC1a
X92243,IGHV2-70*07,VH2-MC2b
X92245,IGHV2-70*08,VH2-MC2d
L21962,IGHV2-70*09,S12-1
L21965,IGHV2-70*10,S12-5
L21967,IGHV2-70*11,S12-7
L21970,IGHV2-70*12,S12-10

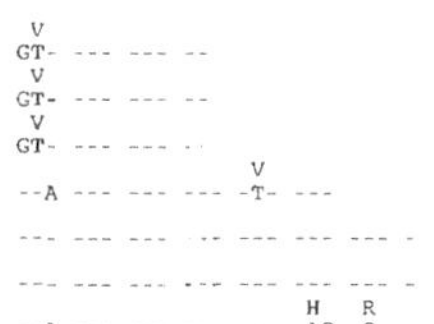

Framework and complementarity determining regions

FR1-IMGT: 25 (-1 aa: 10) CDR1-IMGT: 10
FR2-IMGT: 17 CDR2-IMGT: 7
FR3-IMGT: 38 (-1 aa: 73) CDR3-IMGT: 3

Collier de Perles for human IGHV2-70*01

Accession number: IMGT L21969 EMBL/GenBank/DDBJ: L21969

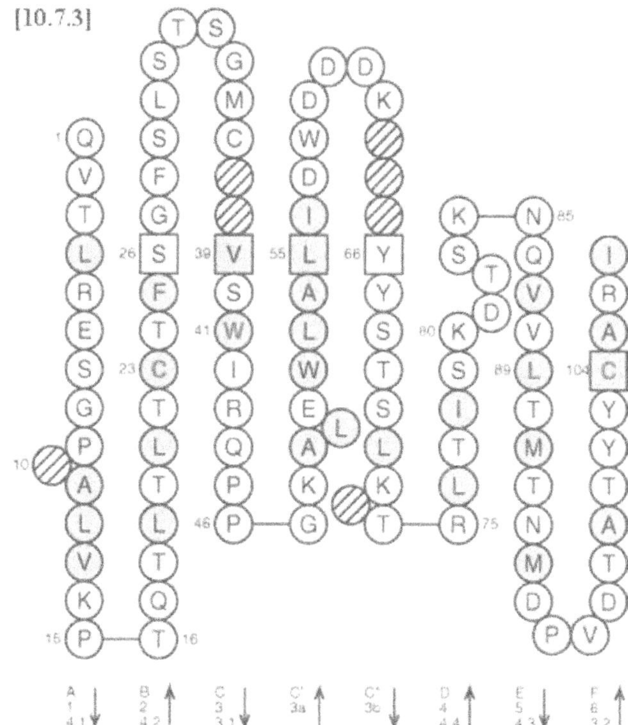

Genome database accession numbers

GDB:9931680 LocusLink: 28454

Nomenclature

IGHV3-7: Immunoglobulin heavy variable 3-7.

Definition and functionality

IGHV3-7 is one of the 18–21 mapped functional genes of the IGHV3 subgroup. This subgroup comprises 47–49 mapped genes in the IGH locus, depending on the haplotypes, three unmapped genes (one functional, one pseudogene and one not defined) and 12 orphons.

Gene location

IGHV3-7 is in the IGH locus on chromosome 14 at 14q32.33.

Nucleotide and amino acid sequences for human IGHV3-7

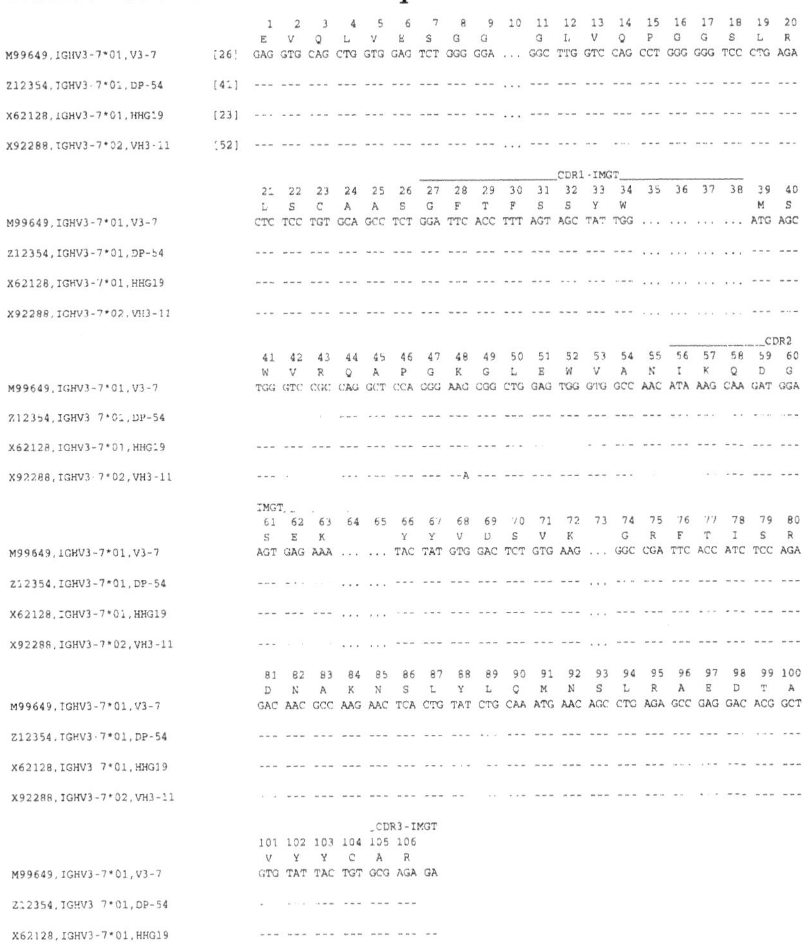

Framework and complementarity determining regions

FR1-IMGT: 25 (-1 aa: 10)	CDR1-IMGT: 8
FR2-IMGT: 17	CDR2-IMGT: 8
FR3-IMGT: 38 (-1 aa: 73)	CDR3-IMGT: 2

Collier de Perles for human IGHV3-7*01

Accession number: IMGT M99649 EMBL/GenBank/DDBJ: M99649

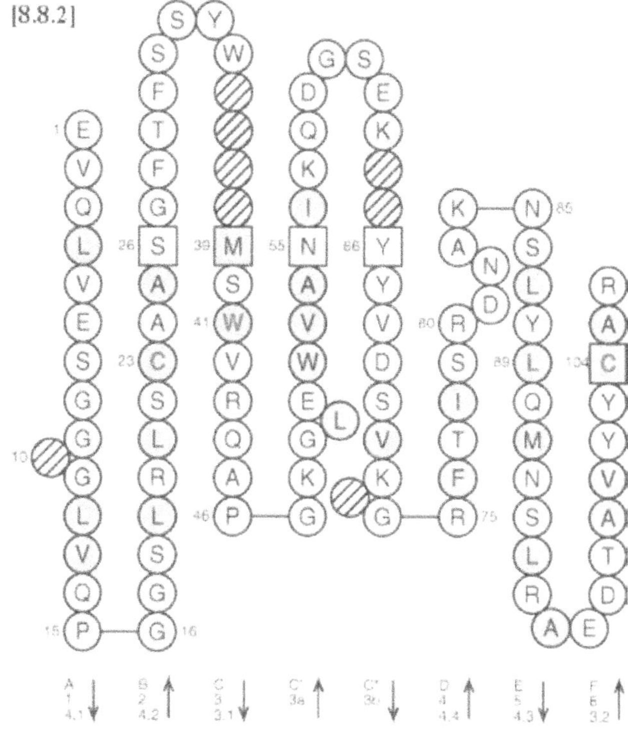

Genome database accession numbers

GDB:9931682 LocusLink: 28452

Nomenclature

IGHV3-9: Immunoglobulin heavy variable 3-9.

Definition and functionality

IGHV3-9 is one of the 18–21 mapped functional genes of the IGHV3 subgroup. This subgroup comprises 47–49 mapped genes in the IGH locus, depending on the haplotypes, three unmapped genes (one functional, one pseudogene and one not defined) and 12 orphons.

Gene location

IGHV3-9 is in the IGH locus on chromosome 14 at 14q32.33.

Nucleotide and amino acid sequences for human IGHV3-9

```
                                    1   2   3   4   5   6   7   8   9  10  11  12  13  14  15  16  17  18  19  20
                                    E   V   Q   L   V   E   S   G   G       G   L   V   Q   P   G   R   S   L   R
M99651,IGHV3-9*01,V3-9       [26]  GAA GTG CAG CTG GTG GAG TCT GGG GGA ... GGC TTG GTA CAG CCT GGC AGG TCC CTG AGA

Z12333,IGHV3-9*01,DP-31      [41]  --- --- --- --- --- --- --- --- --- ... --- --- --- --- --- --- --- --- --- ---

                                                        ______________________CDR1-IMGT____________________
                                   21  22  23  24  25  26  27  28  29  30  31  32  33  34  35  36  37  38  39  40
                                    L   S   C   A   A   S   G   F   T   F   D   D   Y   A                   M   H
M99651,IGHV3-9*01,V3-9             CTC TCC TGT GCA GCC TCT GGA TTC ACC TTT GAT GAT TAT GCC ... ... ... ... ATG CAC

Z12333,IGHV3-9*01,DP-31            --- --- --- --- --- --- --- --- --- --- --- --- --- --- ... ... ... ... --- ---

                                                                                                   ____________CDR2-
                                   41  42  43  44  45  46  47  48  49  50  51  52  53  54  55  56  57  58  59  60
                                    W   V   R   Q   A   P   G   K   G   L   E   W   V   S   G   I   S   W   N   S
M99651,IGHV3-9*01,V3-9             TGG GTC CGG CAA GCT CCA GGG AAG GGC CTG GAG TGG GTC TCA GGT ATT AGT TGG AAT AGT

Z12333,IGHV3-9*01,DP-31            --- --- --- --- --- --- --- --- --- --- --- --- --- --- --- --- --- --- --- ---

                                   IMGT______________
                                   61  62  63  64  65  66  67  68  69  70  71  72  73  74  75  76  77  78  79  80
                                    G   S   I           G   Y   A   D   S   V   K       G   R   F   T   I   S   R
M99651,IGHV3-9*01,V3-9             GGT AGC ATA ... ... GGC TAT GCG GAC TCT GTG AAG ... GGC CGA TTC ACC ATC TCC AGA

Z12333,IGHV3-9*01,DP-31            --- --- --- ... ... --- --- --- --- --- --- --- ... --- --- --- --- --- --- ---

                                   81  82  83  84  85  86  87  88  89  90  91  92  93  94  95  96  97  98  99 100
                                    D   N   A   K   N   S   L   Y   L   Q   M   N   S   L   R   A   E   D   T   A
M99651,IGHV3-9*01,V3-9             GAC AAC GCC AAG AAC TCC CTG TAT CTG CAA ATG AAC AGT CTG AGA GCT GAG GAC ACG GCC

Z12333,IGHV3-9*01,DP-31            --- --- --- --- --- --- --- --- --- --- --- --- --- --- --- --- --- --- --- ---

                                                    ____CDR3-IMGT__
                                  101 102 103 104 105 106 107
                                    L   Y   Y   C   A   K   D
M99651,IGHV3-9*01,V3-9             TTG TAT TAC TGT GCA AAA GAT A

Z12333,IGHV3-9*01,DP-31            --- --- --- ... --- ---
```

Framework and complementarity determining regions

FR1-IMGT: 25 (-1 aa: 10)
FR2-IMGT: 17
FR3-IMGT: 38 (-1 aa: 73)

CDR1-IMGT: 8
CDR2-IMGT: 8
CDR3-IMGT: 3

Collier de Perles for human IGHV3-9*01

Accession number: IMGT M99651 EMBL/GenBank/DDBJ: M99651

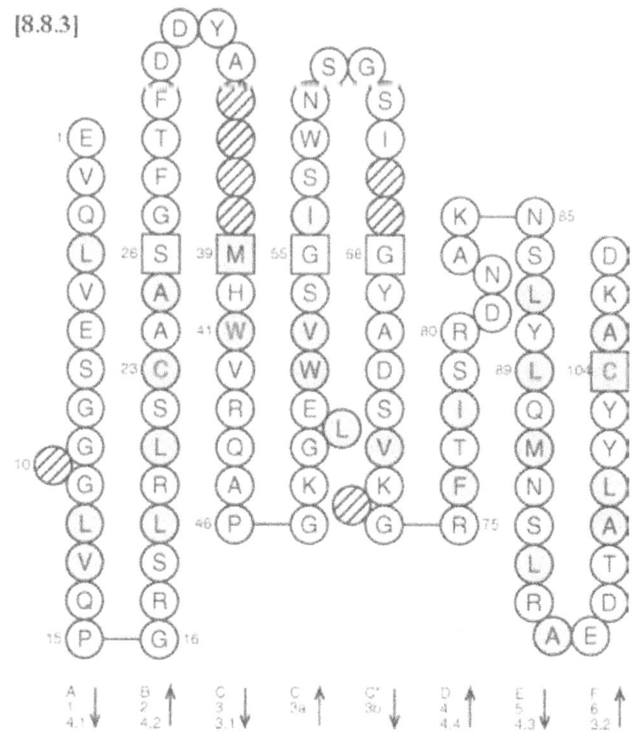

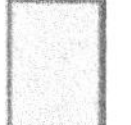

Genome database accession numbers

GDB:9931683 LocusLink: 28451

Nomenclature

IGHV3-11: Immunoglobulin heavy variable 3-11.

Definition and functionality

IGHV3-11 is a mapped functional gene (alleles *01 and *03) or a pseudogene (allele *02). IGHV3-11 belongs to the IGHV3 subgroup which comprises 47–49 mapped genes (of which 18–21 are functional) in the IGH locus, depending on the haplotypes, three unmapped genes (one functional, one pseudogene and one not defined) and 12 orphons. IGHV3-11*02 is a pseudogene due to 1 nt INSERTION and 2 nt DELETIONs in FR1-IMGT and 2 nt DELETIONs in FR3-IMGT, leading to frameshifts.

Gene location

IGHV3-11 is in the IGH locus on chromosome 14 at 14q32.33.

Nucleotide and amino acid sequences for human IGHV3-11

```
                                    1   2   3   4   5   6   7   8   9  10  11  12  13  14  15  16  17  18  19  20
                                    Q   V   Q   L   V   E   S   G   G       G   L   V   K   P   G   G   S   L   R
M99652,IGHV3-11*01,V3-11     [26]  CAG GTG CAG CTG GTG GAG TCT GGG GGA ... GGC TTG GTC AAG CCT GGA GGG TCC CTG AGA
Z12337,IGHV3-11*01,DP-35     [41]  --- --- --- --- --- --- --- --- --- ... --- --- --- --- --- --- --- --- --- ---
X92220,IGHV3-11*01,22-2B      [4]  --- --- --- --- --- --- --- --- --- ... --- --- --- --- --- --- --- --- --- ---
                                          #                             #                   #
M15496,IGHV3-11*02,hv3.3     [43]  --- --- ---T-- --- --- --- --- --- ... -.- --- --- --- --- -.- --- --- --- ---
                                                L
X92287,IGHV3-11*03,VH3-8     [52]  --- --- --- --- T-- --- --- --- --- ... --- --- --- --- --- --- --- --- --- ---

                                                            ____________CDR1-IMGT____________________
                                   21  22  23  24  25  26  27  28  29  30  31  32  33  34  35  36  37  38  39  40
                                    L   S   C   A   A   S   G   F   T   F   S   D   Y   Y                   M   S
M99652,IGHV3-11*01,V3-11           CTC TCC TGT GCA GCC TCT GGA TTC ACC TTC AGT GAC TAC TAC ... ... ... ... ATG AGC
Z12337,IGHV3-11*01,DP-35           --- --- --- --- --- --- --- --- --- --- --- --- --- --- ... ... ... ... --- ---
X92220,IGHV3-11*01,22-2B           --- --- --- --- --- --- --- --- --- --- --- --- --- --- ... ... ... ... --- ---
M15496,IGHV3-11*02,hv3.3           --- --- --- --- --- --- --- --- --- --- --- --- --- --- ... ... ... ... --- ---
X92287,IGHV3-11*03,VH3-8           --- --- --- --- --- --- --- --- --- --- --- --- --- --- ... ... ... ... --- ---

                                                                                       ________________CDR2-
                                   41  42  43  44  45  46  47  48  49  50  51  52  53  54  55  56  57  58  59  60
                                    W   I   R   Q   A   P   G   K   G   L   E   W   V   S   Y   I   S   S   S   G
M99652,IGHV3-11*01,V3-11           TGG ATC CGC CAG GCT CCA GGG AAG GGG CTG GAG TGG GTT TCA TAC ATT AGT AGT AGT GGT
Z12337,IGHV3-11*01,DP-35           --- --- --- --- --- --- --- --- --- --- --- --- --- --- --- --- --- --- --- ---
X92220,IGHV3-11*01,22-2B           --- --- --- --- --- --- --- --- --- --- --- --- --- --- --- --- --- --- --- ---
M15496,IGHV3-11*02,hv3.3           --- --- --- --- --- --- --- --- --- --- --- --- --- --- --- --- --- --- --- ---
                                                                                                            S
X92287,IGHV3-11*03,VH3-8           --- --- --- --- --- --- --- --- --- --- --- --- --- --- --- --- --- --- --- A--

                                   IMGT_______________
                                   61  62  63  64  65  66  67  68  69  70  71  72  73  74  75  76  77  78  79  80
                                    S   T   I           Y   Y   A   D   S   V   K       G   R   F   T   I   S   R
M99652,IGHV3-11*01,V3-11           AGT ACC ATA ... ... TAC TAC GCA GAC TCT GTG AAG ... GGC CGA TTC ACC ATC TCC AGG
Z12337,IGHV3-11*01,DP-35           --- --- --- ... ... --- --- --- --- --- --- --- ... --- --- --- --- --- --- ---
X92220,IGHV3-11*01,22-2B           --- --- --- ... ... --- --- --- --- --- --- --- ... --- --- --- --- --- --- ---
                                                                    #
M15496,IGHV3-11*02,hv3.3           --- --- --- ... ... --- --- --- .-- --- --- --- ... --- --- --- --- --- --- ---
                                        Y   T           N
X92287,IGHV3-11*03,VH3-8           --- TA- C-- ... ... A-- --- --- --- --- --- --- ... --- --- --- --- --- --- --A

                                   81  82  83  84  85  86  87  88  89  90  91  92  93  94  95  96  97  98  99 100
                                    D   N   A   K   N   S   L   Y   L   Q   M   N   S   L   R   A   E   D   T   A
M99652,IGHV3-11*01,V3-11           GAC AAC GCC AAG AAC TCA CTG TAT CTG CAA ATG AAC AGC CTG AGA GCC GAG GAC ACG GCC
Z12337,IGHV3-11*01,DP-35           --- --- --- --- --- --- --- --- --- --- --- --- --- --- --- --- --- --- --- ---
X92220,IGHV3-11*01,22-2B           --- --- --- --- --- --- --- --- --- --- --- --- --- --- --- --- --- --- --- ---
                                    E                                                                       #
M15496,IGHV3-11*02,hv3.3           -G- --- --- --- --- --- --- --- --- --- --- --- --- --- --- --- --- --- -.. ---
X92287,IGHV3-11*03,VH3-8           --- --- --- --- --- --- --- --- --- --- --- --- --- --- --- --- --- --- --- ---
```

```
                                              _CDR3 IMGT
                              101 102 103 104 105 106
                               V   Y   Y   C   A   R
M99652,IGHV3-11*01,V3-11      GTG TAT TAC TGT GCG AGA GA

Z12337,IGHV3-11*01,DP-35      --- --- -       --- ---

X92220,IGHV3-11*01,22-2B       -- --- --- --- --- --- --
                               L   L   T       E
M15496,IGHV3-11*02,hv3.3      T-A CT- ACT -C  -A- --  --

X92287,IGHV3-11*03,VH3-8        -  -- --- --- --- ---
```

Framework and complementarity determining regions

FR1-IMGT: 25 (-1 aa: 10)
FR2-IMGT: 17
FR3-IMGT: 38 (-1 aa: 73)

CDR1-IMGT: 8
CDR2-IMGT: 8
CDR3-IMGT: 2

Collier de Perles for human IGHV3-11*01

Accession number: IMGT M99652

EMBL/GenBank/DDBJ: M99652

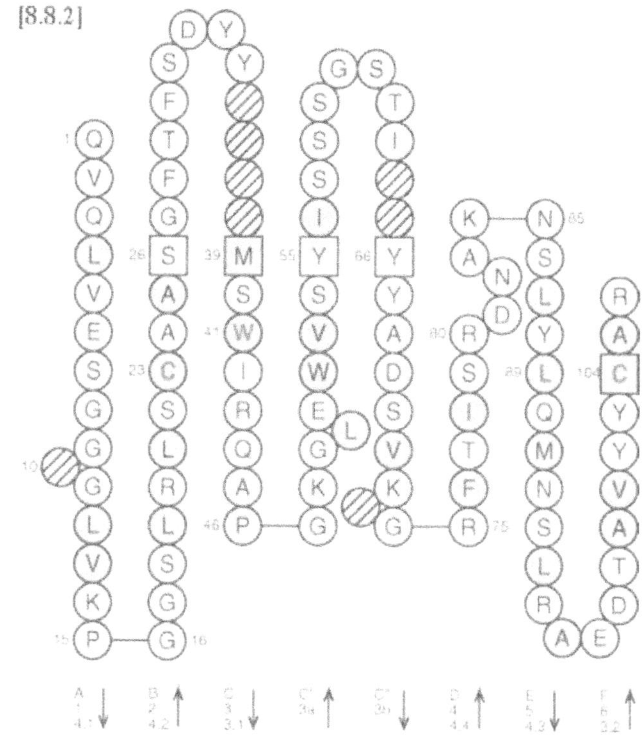

Genome database accession numbers

GDB:9931684

LocusLink: 28450

IGHV3-13

Nomenclature

IGHV3-13: Immunoglobulin heavy variable 3-13.

Definition and functionality

IGHV3-13 is one of the 18–21 mapped functional genes of the IGHV3 subgroup. This subgroup comprises 47–49 mapped genes in the IGH locus, depending on the haplotypes, three unmapped genes (one functional, one pseudogene and one not defined) and 12 orphons.

Gene location

IGHV3-13 is in the IGH locus on chromosome 14 at 14q32.33.

Nucleotide and amino acid sequences for human IGHV3-13

```
                                 1   2   3   4   5   6   7   8   9  10  11  12  13  14  15  16  17  18  19  20
                                 E   V   Q   L   V   E   S   G   G       G   L   V   Q   P   G   G   S   L   R
X92217,IGHV3-13*01,13-2      [4] GAG GTG CAG CTG GTG GAG TCT GGG GGA ... GGC TTG GTA CAG CCT GGG GGG TCC CTG AGA

Z12348,IGHV3-13*01,DP-48    [41] --- --- --- --- --- --- --- --- --- ... --- --- --- --- --- --- --- --- --- ---
                                         H                                                           A
M99653,IGHV3-13*02,V3-13    [26] --- --- --T --- --- --- --- --- --- ... --- --- --- --- --- --- --- G-- --- ---

                                                        ________CDR1-IMGT________________________
                                21  22  23  24  25  26  27  28  29  30  31  32  33  34  35  36  37  38  39  40
                                 L   S   C   A   A   S   G   F   T   F   S   S   Y   D                   M   H
X92217,IGHV3-13*01,13-2          CTC TCC TGT GCA GCC TCT GGA TTC ACC TTC AGT AGC TAC GAC ... ... ... ... ATG CAC

Z12348,IGHV3-13*01,DP-48         --- --- --- --- --- --- --- --- --- --- --- --- --- --- ... ... ... ... --- ---
                                                                             N
M99653,IGHV3-13*02,V3-13         --- --- --- --- --- --- --- --- --- --- --- -A- --- --- ... ... ... ... --- ---

                                                                                             ________________ CDR2-
                                41  42  43  44  45  46  47  48  49  50  51  52  53  54  55  56  57  58  59  60
                                 W   V   R   Q   A   T   G   K   G   L   E   W   V   S   A   I   G   T   A   G
X92217,IGHV3-13*01,13-2          TGG GTC CGC CAA GCT ACA GGA AAA GGT CTG GAG TGG GTC TCA GCT ATT GGT ACT GCT GGT

Z12348,IGHV3-13*01,DP-48         --- --- --- --- --- --- --- --- --- --- --- --- --- --- --- --- --- --- --- ---
                                                                                             N
M99653,IGHV3-13*02,V3-13         --- --- --- --- --- --- --- --- --- --- --- --- --- --- --C -A- --- --- --- ---

                                 IMGT_______________
                                61  62  63  64  65  66  67  68  69  70  71  72  73  74  75  76  77  78  79  80
                                 D   T               Y   Y   P   G   S   V   K       G   R   F   T   I   S   R
X92217,IGHV3-13*01,13-2          GAC ACA ... ... ... TAC TAT CCA GGC TCC GTG AAG ... GGC CGA TTC ACC ATC TCC AGA

Z12348,IGHV3-13*01,DP-48         --- --- ... ... ... --- --- --- --- --- --- --- ... --- --- --- --- --- --- ---

M99653,IGHV3-13*02,V3-13         --- --- ... ... ... --- --- --- --- --- --- --- ... --G --- --- --- --- --- ---

                                81  82  83  84  85  86  87  88  89  90  91  92  93  94  95  96  97  98  99 100
                                 E   N   A   K   N   S   L   Y   L   Q   M   N   S   L   R   A   G   D   T   A
X92217,IGHV3-13*01,13-2          GAA AAT GCC AAG AAC TCC TTG TAT CTT CAA ATG AAC AGC CTG AGA GCC GGG GAC ACG GCT

Z12348,IGHV3-13*01,DP-48         --- --- --- --- --- --- --- --- --- --- --- --- --- --- --- --- --- --- --- ---

M99653,IGHV3-13*02,V3-13         --- --- --- --- --- --- --- --- --- --- --- --- --- --- --- --- --- --- --- ---

                                                 _CDR3-IMGT
                                 101 102 103 104 105 106
                                  V   Y   Y   C   A   R
X92217,IGHV3-13*01,13-2          GTG TAT TAC TGT GCA AGA GA

Z12348,IGHV3-13*01,DP-48         --- --- --- --- --- ---

M99653,IGHV3-13*02,V3-13         --- --- --- --- --- --- --
```

Framework and complementarity determining regions

FR1-IMGT: 25 (-1 aa: 10)
FR2-IMGT: 17
FR3-IMGT: 38 (-1 aa: 73)

CDR1-IMGT: 8
CDR2-IMGT: 7
CDR3-IMGT: 2

Collier de Perles for human IGHV3-13*01

Accession number: IMGT X92217 EMBL/GenBank/DDBJ: X92217

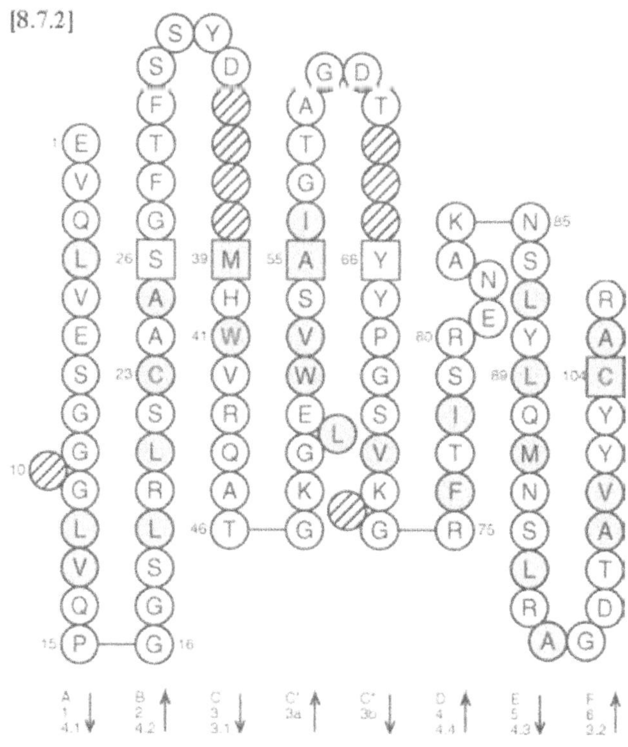

Genome database accession numbers

GDB:9931685 LocusLink: 28449

Nomenclature

IGHV3-15: Immunoglobulin heavy variable 3-15.

Definition and functionality

IGHV3-15 is one of the 18–21 mapped functional genes of the IGHV3 subgroup. This subgroup comprises 47–49 mapped genes in the IGH locus, depending on the haplotypes, three unmapped genes (one functional, one pseudogene and one not defined) and 12 orphons.

Gene location

IGHV3-15 is in the IGH locus on chromosome 14 at 14q32.33.

Nucleotide and amino acid sequences for human IGHV3-15

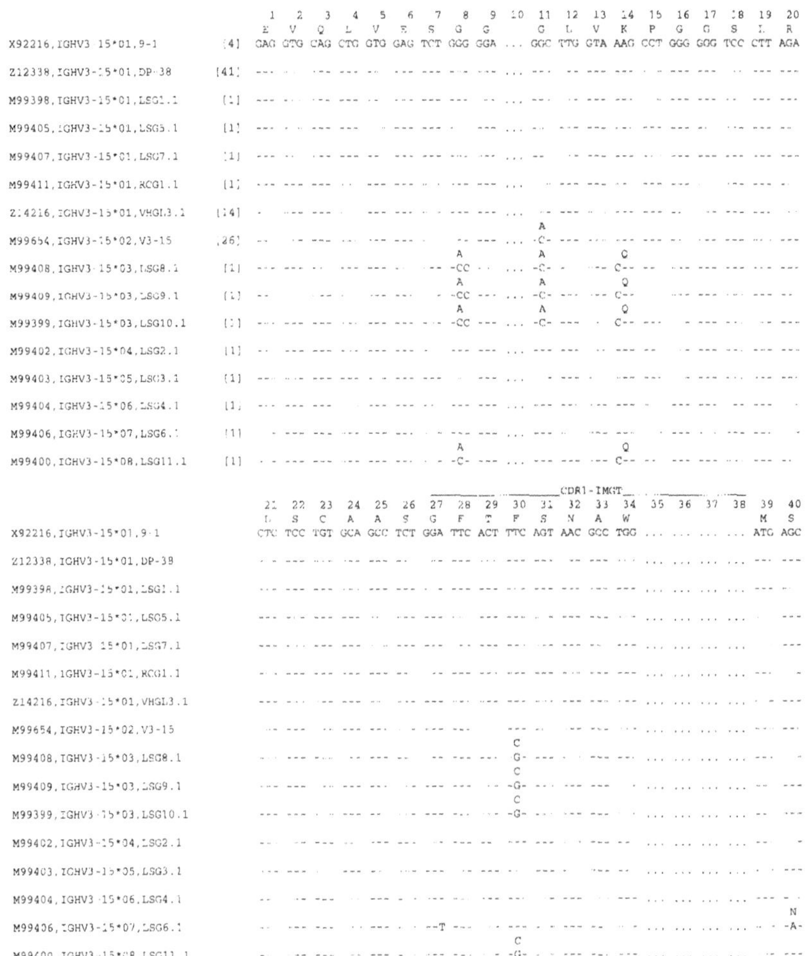

```
                                                                                      _____________CDR2-
                              41  42  43  44  45  46  47  48  49  50  51  52  53  54  55  56  57  58  59  60
                              W   V   R   Q   A   P   G   K   G   L   E   W   V   G   R   I   K   S   K   T
X92216,IGHV3-15*01,9-1        TGG GTC CGC CAG GCT CCA GGG AAG GGG CTG GAG TGG GTT GGC CGT ATT AAA AGC AAA ACT
Z12338,IGHV3-15*01,DP-38      --- --- --- --- --- --- --- --- --- --- --- --- --- --- --- --- --- --- --- ---
M99398,IGHV3-15*01,LSG1.1     --- --- --- --- --- --- --- --- --- --- --- --- --- --- --- --- --- --- --- ---
M99405,IGHV3-15*01,LSG5.1     --- --- --- --- --- --- --- --- --- --- --- --- --- --- --- --- --- --- --- ---
M99407,IGHV3-15*01,LSG7.1     --- --- --- --- --- --- --- --- --- --- --- --- --- --- --- --- --- --- --- ---
M99411,IGHV3-15*01,RCG1.1     --- --- --- --- --- --- --- --- --- --- --- --- --- --- --- --- --- --- --- ---
Z14216,IGHV3-15*01,VHGL3.1    --- --- --- --- --- --- --- --- --- --- --- --- --- --- --- --- --- --- --- ---
M99654,IGHV3-15*02,V3-15      --- --- --- --- --- --- --- --- --- --- --- --- --- --- --- --- --- --- --- ---
                                                                                                         A
M99408,IGHV3-15*03,LSG8.1     --- --- --- --- --- --- --- --- --- --- --- --- --- --- --- --- --- --- --- G--
                                                                                                         A
M99409,IGHV3-15*03,LSG9.1     --- --- --- --- --- --- --- --- --- --- --- --- --- --- --- --- --- --- --- G--
                                                                                                         A
M99399,IGHV3-15*03,LSG10.1    --- --- --- --- --- --- --- --- --- --- --- --- --- --- --- --- --- --- --- G--
                                                                                             E
M99402,IGHV3-15*04,LSG2.1     --- --- --- --- --- --- --- --- --- --- --- --- --- --- --- --- G-- --- --- ---
M99403,IGHV3-15*05,LSG3.1     --- --- --- --- --- --- --- --- --- --- --- --- --- --- --- --- --- --- --- ---
M99404,IGHV3-15*06,LSG4.1     --- --- --- --- --- --- --- --- --- --- --- --- --C --- --- --- --- --- --- ---
M99406,IGHV3-15*07,LSG6.1     --- --- --- --- --- --- --- --- --- --- --- --- --C --- --- --- --- --- --- ---
                                                                                 C                       A
M99400,IGHV3-15*08,LSG11.1    --- --- --- --- --- --- --- --- --- --- --- --- --- --- T-- --- --- --- --- G--


                              IMGT_______________
                              61  62  63  64  65  66  67  68  69  70  71  72  73  74  75  76  77  78  79  80
                              D   G   G   T   T   D   Y   A   A   P   V   K       G   R   F   T   I   S   R
X92216,IGHV3-15*01,9-1        GAT GGT GGG ACA ACA GAC TAC GCT GCA CCC GTG AAA ... GGC AGA TTC ACC ATC TCA AGA
Z12338,IGHV3-15*01,DP-38      --- --- --- --- --- --- --- --- --- --- --- --- ... --- --- --- --- --- --- ---
M99398,IGHV3-15*01,LSG1.1     --- --- --- --- --- --- --- --- --- --- --- --- ... --- --- --- --- --- --- ---
M99405,IGHV3-15*01,LSG5.1     --- --- --- --- --- --- --- --- --- --- --- --- ... --- --- --- --- --- --- ---
M99407,IGHV3-15*01,LSG7.1     --- --- --- --- --- --- --- --- --- --- --- --- ... --- --- --- --- --- --- ---
M99411,IGHV3-15*01,RCG1.1     --- --- --- --- --- --- --- --- --- --- --- --- ... --- --- --- --- --- --- ---
Z14216,IGHV3-15*01,VHGL3.1    --- --- --- --- --- --- --- --- --- --- --- --- ... --- --- --- --- --- --- ---
M99654,IGHV3-15*02,V3-15      --- --- --- --- --- --- --- --- --- --- --- --- ... --- --- --- --- --- --- ---
                              N
M99408,IGHV3-15*03,LSG8.1     A-- --- --- --- --- --- --- --- --- --T --- --- ... --- --- --- --- --- --- ---
                              N
M99409,IGHV3-15*03,LSG9.1     A-- --- --- --- --- --- --- --- --- --T --- --- ... --- --- --- --- --- --- ---
                              N
M99399,IGHV3-15*03,LSG10.1    A-- --- --- --- --- --- --- --- --- --T --- --- ... --- --- --- --- --- --- ---
M99402,IGHV3-15*04,LSG2.1     --- --- --- --- --- --- --- --- --- --- --- --- ... --- --- --- --- --- --- ---
M99403,IGHV3-15*05,LSG3.1     --- --- --- --- --- --- --- --- --- --- --- --- ... --- --- --- --- --- --- ---
                                                  N
M99404,IGHV3-15*06,LSG4.1     --- --- --- --- --- A-- --- --- --- --- --- --- ... --- --- --- --- --- ---
M99406,IGHV3-15*07,LSG6.1     --- --- --- --- --- --- --- --- --- --- --- --- ... --- --- --- --- --- --- ---
                              N
M99400,IGHV3-15*08,LSG11.1    A-- --- --- --- --- --- --- --- --- --T --- --- ... --- --- --- --- --- --- ---


                              81  82  83  84  85  86  87  88  89  90  91  92  93  94  95  96  97  98  99  100
                              D   D   S   K   N   T   L   Y   L   Q   M   N   S   L   K   T   E   D   T   A
X92216,IGHV3-15*01,9-1        GAT GAT TCA AAA AAC ACG CTG TAT CTG CAA ATG AAC AGC CTG AAA ACC GAG GAC ACA GCC
Z12338,IGHV3-15*01,DP-38      --- --- --- --- --- --- --- --- --- --- --- --- --- --- --- --- --- --- --- ---
M99398,IGHV3-15*01,LSG1.1     --- --- --- --- --- --- --- --- --- --- --- --- --- --- --- --- --- --- --- ---
M99405,IGHV3-15*01,LSG5.1     --- --- --- --- --- --- --- --- --- --- --- --- --- --- --- --- --- --- --- ---
M99407,IGHV3-15*01,LSG7.1     --- --- --- --- --- --- --- --- --- --- --- --- --- --- --- --- --- --- --- ---
M99411,IGHV3-15*01,RCG1.1     --- --- --- --- --- --- --- --- --- --- --- --- --- --- --- --- --- --- --- ---
Z14216,IGHV3-15*01,VHGL3.1    --- --- --- --- --- --- --- --- --- --- --- --- --- --- --- --- --- --- --- ---
M99654,IGHV3-15*02,V3-15      --- --- --- --- --- --- --- --- --- --- --- --- --- --- --- --- --- --- --- ---
                              V
M99408,IGHV3-15*03,LSG8.1     -T- --- --- --- --- --- --- --- --- --- --- --- --- --- --- --- --- --- --- ---
                              V
M99409,IGHV3-15*03,LSG9.1     -T- --- --- --- --- --- --- --- --- --- --- --- --- --- --- --- --- --- --- ---
                              V
M99399,IGHV3-15*03,LSG10.1    -T- --- --- --- --- --- --- --- --- --- --- --- --- --- --- --- --- --- --- ---
M99402,IGHV3-15*04,LSG2.1     --- --- --- --- --- --- --- --- --- --- --- --- --- --- --- --- --- --- --- ---
M99403,IGHV3-15*05,LSG3.1     --- --- --- --- --- --- --- --- --- --- --- --- --T --- --- --- --- --- --- ---
M99404,IGHV3-15*06,LSG4.1     --- --- --- --- --- --- --- --- --- --- --- --- --- --- --- --- --- --- --- ---
M99406,IGHV3-15*07,LSG6.1     --- --- --- --- --- --- --- --- --- --- --- --- --- --- --- --- --- --- --- ---
                                                                              I
M99400,IGHV3-15*08,LSG11.1    --- --- --- --- --- --- --- --- --- --- --- -T- --- --- --- --- --- --- --G ---
```

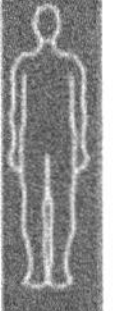

```
                                            _CDR3-IMGT
                              101 102 103 104 105 106
                               V   Y   Y   C   T   T
X92216,IGHV3-15*01,9-1        GTG TAT TAC TGT ACC ACA GA
Z12338,IGHV3-15*01,DP-38      --- --- --- --- --- ---
M99398,IGHV3-15*01,LSG1.1     --- --- --- --- --- --- --
M99405,IGHV3-15*01,LSG5.1     --- --- --- --- --- --- --
M99407,IGHV3-15*01,LSG7.1     --- --- --- --- --- --- --
M99411,IGHV3-15*01,RCG1.1     --- --- --- --- --- --- --
Z14216,IGHV3-15*01,VHGL3.1    --- --- --- --- --- --- --
M99654,IGHV3-15*02,V3-15      --- --- --- --- --- --- --
M99408,IGHV3-15*03,LSG8.1     --- --- --- --- --- --- --
M99409,IGHV3-15*03,LSG9.1     --- --- --- --- --- --- --
M99399,IGHV3-15*03,LSG10.1    --- --- --- --- --- --- --
M99402,IGHV3-15*04,LSG2.1     --- --- --- --- --- --- --
M99403,IGHV3-15*05,LSG3.1     --- --- --- --- --- --- --
M99404,IGHV3-15*06,LSG4.1     --- --- --- --- --- --- --
M99406,IGHV3-15*07,LSG6.1     --- --- --- --- --- --- --
M99400,IGHV3-15*08,LSG11.1    --- --- --- --- --- --- -G
```

Framework and complementarity determining regions

FR1-IMGT: 25 (-1 aa: 10)
FR2-IMGT: 17
FR3-IMGT: 38 (-1 aa: 73)

CDR1-IMGT: 8
CDR2-IMGT: 10
CDR3-IMGT: 2

Collier de Perles for human IGHV3-15*01

Accession number: IMGT X92216   EMBL/GenBank/DDBJ: X92216

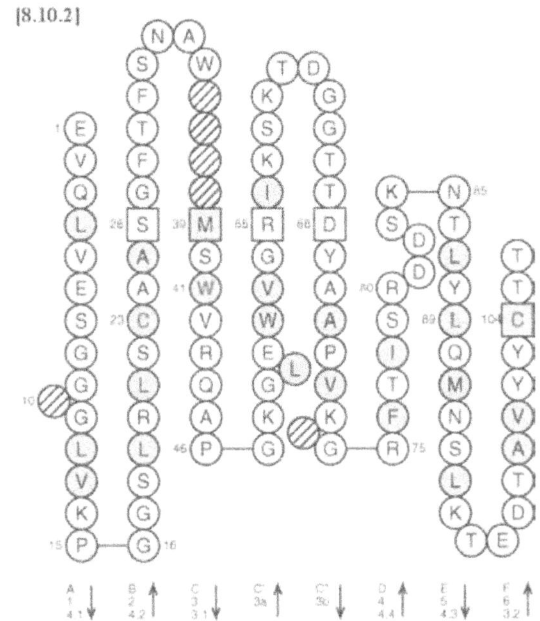

Genome database accession numbers

GDB:9931686   LocusLink: 28448

Nomenclature

IGHV3-16: Immunoglobulin heavy variable 3-16.

Definition and functionality

IGHV3-16 is one of the three–four mapped ORF of the IGHV3 subgroup. This subgroup comprises 47–49 mapped genes (of which 18–21 are functional) in the IGH locus, depending on the haplotypes, three unmapped genes (one functional, one pseudogene and one not defined) and 12 orphons. IGHV3-16 is an ORF due to an unusual V-HEPTAMER sequence: tcctgtg instead of cacagtg.

Gene location

IGHV3-16 is in the IGH locus on chromosome 14 at 14q32.33.

Nucleotide and amino acid sequences for human IGHV3-16

```
                                  1   2   3   4   5   6   7   8   9  10  11  12  13  14  15  16  17  18  19  20
                                  E   V   Q   L   V   E   S   G   G       G   L   V   Q   P   G   G   S   L   R
M99655,IGHV3-16*01,V3-16P   [26] GAG GTA CAA CTG GTG GAG TCT GGG GGA ... GGC TTG GTA CAG CCT GGG GGG TCC CTG AGA

                                                      ________  ____________CDR1-IMGT_____________________
                                 21  22  23  24  25  26  27  28  29  30  31  32  33  34  35  36  37  38  39  40
                                  L   S   C   A   A   S   G   F   T   F   S   N   S   D                   M   N
M99655,IGHV3-16*01,V3-16P        CTC TCC TGT GCA GCC TCT GGA TTC ACC TTC AGT AAC AGT GAC ... ... ... ... ATG AAC

                                                                                              ____________CDR2
                                 41  42  43  44  45  46  47  48  49  50  51  52  53  54  55  56  57  58  59  60
                                  W   A   R   K   A   P   G   K   G   L   E   W   V   S   G   V   S   W   N   G
M99655,IGHV3-16*01,V3-16P        TGG GCC CGC AAG GCT CCA GGA AAG GGG CTC GAG TGG GTA TCG GGT GTT AGT TGG AAT GGC

                                 IMGT_______  ______
                                 61  62  63  64  65  66  67  68  69  70  71  72  73  74  75  76  77  78  79  80
                                  S   R   T           H   Y   V   D   S   V   K       R   R   F   I   I   S   R
M99655,IGHV3-16*01,V3-16P        AGT AGG ACG ... ... CAC TAT GTG GAC TCC GTG AAG ... CGC CGA TTC ATC ATC TCC AGA

                                 81  82  83  84  85  86  87  88  89  90  91  92  93  94  95  96  97  98  99 100
                                  D   N   S   R   N   S   L   Y   L   Q   K   N   R   R   R   A   E   D   M   A
M99655,IGHV3-16*01,V3-16P        GAC AAT TCC AGG AAC TCC CTG TAT CTG CAA AAG AAC AGA CGC AGA GCC GAG GAC ATG GCT

                                                 CDR3-IMGT
                                101 102 103 104 105 106
                                  V   Y   Y   C   V   R
M99655,IGHV3-16*01,V3-16P        GTG TAT TAC TGT GTG AGA AA
```

Framework and complementarity determining regions

FR1-IMGT: 25 (-1 aa: 10)
FR2-IMGT: 17
FR3-IMGT: 38 (-1 aa: 73)

CDR1-IMGT: 8
CDR2-IMGT: 8
CDR3-IMGT: 2

Collier de Perles for human IGHV3-16*01

Accession number: IMGT M99655 EMBL/GenBank/DDBJ: M99655

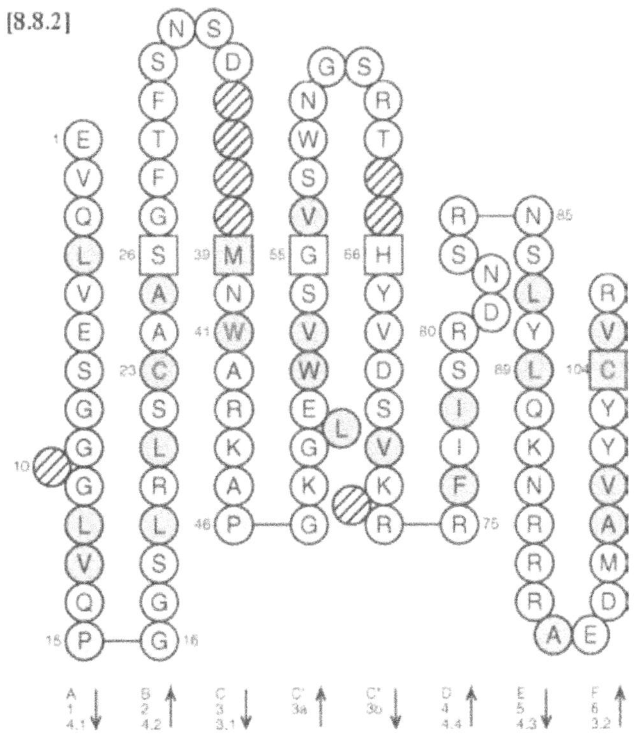

Genome database accession numbers

GDB:9931687 LocusLink: 28447

Nomenclature

IGHV3-20: Immunoglobulin heavy variable 3-20.

Definition and functionality

IGHV3-20 is one of the 18–21 mapped functional genes of the IGHV3 subgroup. This subgroup comprises 47–49 mapped genes in the IGH locus, depending on the haplotypes, three unmapped genes (one functional, one pseudogene and one not defined) and 12 orphons.

Gene location

IGHV3-20 is in the IGH locus on chromosome 14 at 14q32.33.

Nucleotide and amino acid sequences for human IGHV3-20

```
                                     1   2   3   4   5   6   7   8   9   10  11  12  13  14  15  16  17  18  19  20
                                     E   V   Q   L   V   E   S   G   G       G   V   V   R   P   G   G   S   L   R
M99657,IGHV3-20*01,V3-20      (26)  GAG GTG CAG CTG GTG GAG TCT GGG GGA ... GGT GTG GTA CGG CCT GGG GGG TCC CTG AGA
Z12334,IGHV3-20*01,DP-32      (41)  --- --- --- --- --- --- --- --- --- ... --- --- --- --- --- --- --- --- --- ---

                                                             ______________CDR1-IMGT_____________________
                                     21  22  23  24  25  26  27  28  29  30  31  32  33  34  35  36  37  38  39  40
                                     L   S   C   A   A   S   G   F   T   F   D   D   Y   G                   M   S
M99657,IGHV3-20*01,V3-20            CTC TCC TGT GCA GCC TCT GGA TTC ACC TTT GAT GAT TAT GGC ... ... ... ... ATG AGC
Z12334,IGHV3-20*01,DP-32            --- --- --- --- --- --- --- --- --- --- --- --- --- --- ... ... ... ... --- ---

                                                                                                   ___________CDR2
                                     41  42  43  44  45  46  47  48  49  50  51  52  53  54  55  56  57  58  59  60
                                     W   V   R   Q   A   P   G   K   G   L   E   W   V   S   G   I   N   W   N   G
M99657,IGHV3-20*01,V3-20            TGG GTC CGC CAA GCT CCA GGG AAG GGG CTG GAG TGG GTC TCT GGT ATT AAT TGG AAT GGT
Z12334,IGHV3-20*01,DP-32            --- --- --- --- --- --- --- --- --- --- --- --- --- --- --- --- --- --- --- ---

                                    IMGT_______________
                                     61  62  63  64  65  66  67  68  69  70  71  72  73  74  75  76  77  78  79  80
                                     G   S   T           G   Y   A   D   S   V   K       G   R   F   T   I   S   R
M99657,IGHV3-20*01,V3-20            GGT AGC ACA ... ... GGT TAT GCA GAC TCT GTG AAG ... GGC CGA TTC ACC ATC TCC AGA
Z12334,IGHV3-20*01,DP-32            --- --- --- ... ... --- --- --- --- --- --- --- ... --- --- --- --- --- --- ---

                                     81  82  83  84  85  86  87  88  89  90  91  92  93  94  95  96  97  98  99 100
                                     D   N   A   K   N   S   L   Y   L   Q   M   N   S   L   R   A   E   D   T   A
M99657,IGHV3-20*01,V3-20            GAC AAC GCC AAG AAC TCC CTG TAT CTG CAA ATG AAC AGT CTG AGA GCC GAG GAC ACG GCC
Z12334,IGHV3-20*01,DP-32            --- --- --- --- --- --- --- --- --- --- --- --- --- --- --- --- --- --- --- ---

                                                        _CDR3-IMGT
                                    101 102 103 104 105 106
                                     L   Y   H   C   A   R
M99657,IGHV3-20*01,V3-20            TTG TAT CAC TGT GCG AGA GA
Z12334,IGHV3-20*01,DP-32            --- --- --- --- --- ---
```

Framework and complementarity determining regions

FR1-IMGT: 25 (-1 aa: 10)	CDR1-IMGT: 8
FR2-IMGT: 17	CDR2-IMGT: 8
FR3-IMGT: 38 (-1 aa: 73)	CDR3-IMGT: 2

Collier de Perles for human IGHV3-20*01

Accession number: IMGT M99657 EMBL/GenBank/DDBJ: M99657

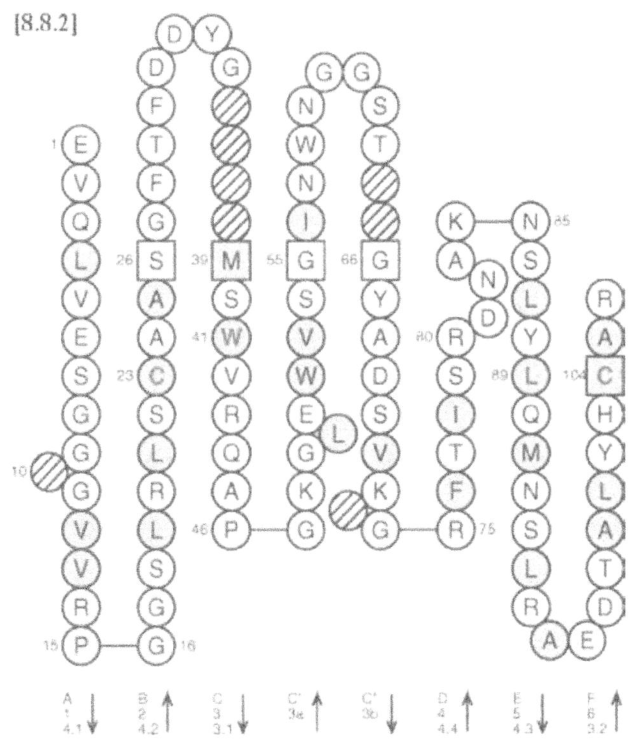

Genome database accession numbers

GDB:9931689 LocusLink: 28445

Nomenclature

IGHV3-21: Immunoglobulin heavy variable 3-21.

Definition and functionality

IGHV3-21 is one of the 18–21 mapped functional genes of the IGHV3 subgroup. This subgroup comprises 47–49 mapped genes in the IGH locus, depending on the haplotypes, three unmapped genes (one functional, one pseudogene, and one not defined) and 12 orphons.

Gene location

IGHV3-21 is in the IGH locus on chromosome 14 at 14q32.33.

Nucleotide and amino acid sequences for human IGHV3-21

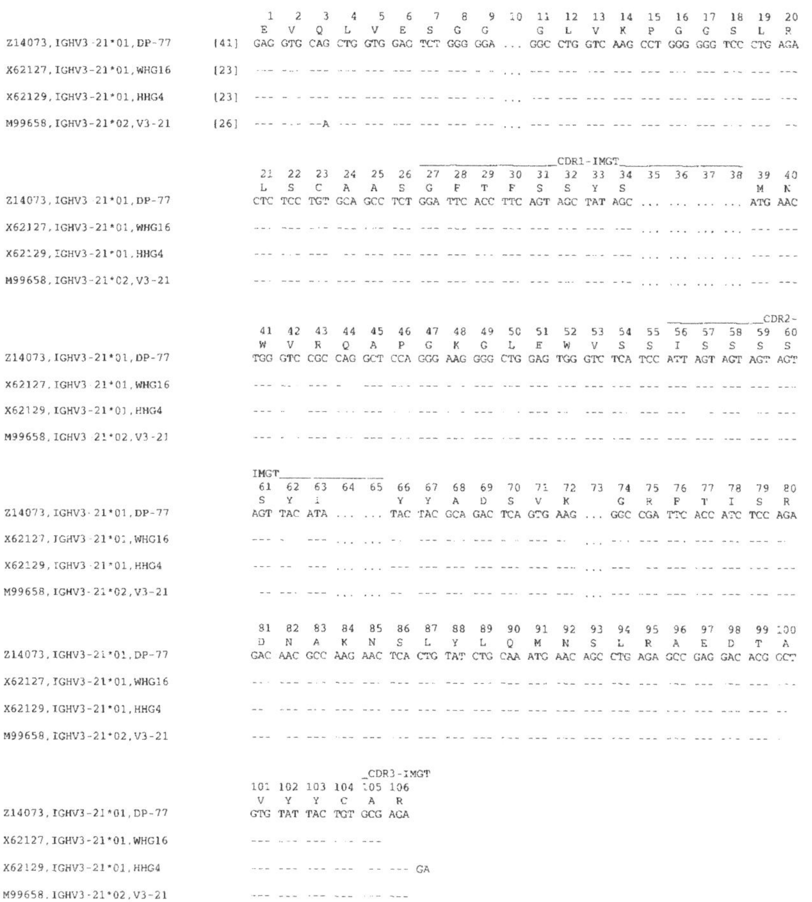

```
                                        1   2   3   4   5   6   7   8   9   10  11  12  13  14  15  16  17  18  19  20
                                        E   V   Q   L   V   E   S   G   G       G   L   V   K   P   G   G   S   L   R
Z14073,IGHV3-21*01,DP-77          [41]  GAG GTG CAG CTG GTG GAG TCT GGG GGA ... GGC CTG GTC AAG CCT GGG GGG TCC CTG AGA
X62127,IGHV3-21*01,WHG16          [23]  --- --- --- --- --- --- --- --- --- ... --- --- --- --- --- --- --- --- --- ---
X62129,IGHV3-21*01,HHG4           [23]  --- --- --- --- --- --- --- --- --- ... --- --- --- --- --- --- --- --- --- ---
M99658,IGHV3-21*02,V3-21          [26]  --- --- --A --- --- --- --- --- --- ... --- --- --- --- --- --- --- --- --- ---

                                                                    _____________CDR1-IMGT_______________
                                        21  22  23  24  25  26  27  28  29  30  31  32  33  34  35  36  37  38  39  40
                                        L   S   C   A   A   S   G   F   T   F   S   S   Y   S                   M   N
Z14073,IGHV3-21*01,DP-77                CTC TCC TGT GCA GCC TCT GGA TTC ACC TTC AGT AGC TAT AGC ... ... ... ... ATG AAC
X62127,IGHV3-21*01,WHG16                --- --- --- --- --- --- --- --- --- --- --- --- --- --- ... ... ... ... --- ---
X62129,IGHV3-21*01,HHG4                 --- --- --- --- --- --- --- --- --- --- --- --- --- --- ... ... ... ... --- ---
M99658,IGHV3-21*02,V3-21                --- --- --- --- --- --- --- --- --- --- --- --- --- --- ... ... ... ... --- ---

                                                                                                ______________CDR2-
                                        41  42  43  44  45  46  47  48  49  50  51  52  53  54  55  56  57  58  59  60
                                        W   V   R   Q   A   P   G   K   G   L   E   W   V   S   S   I   S   S   S   S
Z14073,IGHV3-21*01,DP-77                TGG GTC CGC CAG GCT CCA GGG AAG GGG CTG GAG TGG GTC TCA TCC ATT AGT AGT AGT AGT
X62127,IGHV3-21*01,WHG16                --- --- --- --- --- --- --- --- --- --- --- --- --- --- --- --- --- --- --- ---
X62129,IGHV3-21*01,HHG4                 --- --- --- --- --- --- --- --- --- --- --- --- --- --- --- --- --- --- --- ---
M99658,IGHV3-21*02,V3-21                --- --- --- --- --- --- --- --- --- --- --- --- --- --- --- --- --- --- --- ---

                                        IMGT___________
                                        61  62  63  64  65  66  67  68  69  70  71  72  73  74  75  76  77  78  79  80
                                        S   Y   I           Y   Y   A   D   S   V   K       G   R   F   T   I   S   R
Z14073,IGHV3-21*01,DP-77                AGT TAC ATA ... ... TAC TAC GCA GAC TCA GTG AAG ... GGC CGA TTC ACC ATC TCC AGA
X62127,IGHV3-21*01,WHG16                --- --- --- ... ... --- --- --- --- --- --- --- ... --- --- --- --- --- --- ---
X62129,IGHV3-21*01,HHG4                 --- --- --- ... ... --- --- --- --- --- --- --- ... --- --- --- --- --- --- ---
M99658,IGHV3-21*02,V3-21                --- --- --- ... ... --- --- --- --- --- --- --- ... --- --- --- --- --- --- ---

                                        81  82  83  84  85  86  87  88  89  90  91  92  93  94  95  96  97  98  99  100
                                        D   N   A   K   N   S   L   Y   L   Q   M   N   S   L   R   A   E   D   T   A
Z14073,IGHV3-21*01,DP-77                GAC AAC GCC AAG AAC TCA CTG TAT CTG CAA ATG AAC AGC CTG AGA GCC GAG GAC ACG GCT
X62127,IGHV3-21*01,WHG16                --- --- --- --- --- --- --- --- --- --- --- --- --- --- --- --- --- --- --- ---
X62129,IGHV3-21*01,HHG4                 --- --- --- --- --- --- --- --- --- --- --- --- --- --- --- --- --- --- --- ---
M99658,IGHV3-21*02,V3-21                --- --- --- --- --- --- --- --- --- --- --- --- --- --- --- --- --- --- --- ---

                                                        _CDR3-IMGT
                                        101 102 103 104 105 106
                                        V   Y   Y   C   A   R
Z14073,IGHV3-21*01,DP-77                GTG TAT TAC TGT GCG AGA
X62127,IGHV3-21*01,WHG16                --- --- --- --- ---
X62129,IGHV3-21*01,HHG4                 --- --- --- --- --- --- GA
M99658,IGHV3-21*02,V3-21                --- --- --- --- --- ---
```

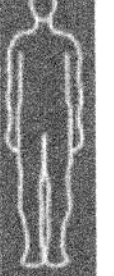

Framework and complementarity determining regions

FR1-IMGT: 25 (-1 aa: 10)
FR2-IMGT: 17
FR3-IMGT: 38 (-1 aa: 73)

CDR1-IMGT: 8
CDR2-IMGT: 8
CDR3-IMGT: 2

Collier de Perles for human IGHV3-21*01

Accession number: IMGT Z14073 EMBL/GenBank/DDBJ: Z14073

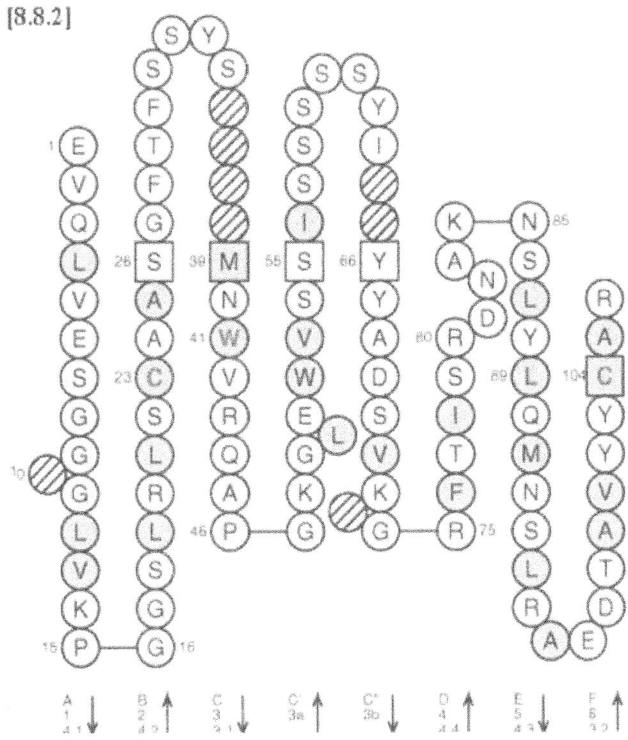

Genome database accession numbers

GDB:9931690 LocusLink: 28444

Nomenclature

IGHV3-23: Immunoglobulin heavy variable 3-23.

Definition and functionality

IGHV3-23 is one of the 18–21 mapped functional genes of the IGHV3 subgroup. This subgroup comprises 47–49 mapped genes in the IGH locus, depending on the haplotypes, three unmapped genes (one functional, one pseudogene and one not defined) and 12 orphons.

Gene location

IGHV3-23 is in the IGH locus on chromosome 14 at 14q32.33.

Nucleotide and amino acid sequences for human IGHV3-23

```
                                     1   2   3   4   5   6   7   8   9  10  11  12  13  14  15  16  17  18  19  20
                                     E   V   Q   L   L   E   S   G   G       G   L   V   Q   P   G   G   S   L   R
M99660,IGHV3-23*01,V3-23       [26] GAG GTG CAG CTG TTG GAG TCT GGG GGA ... GGC TTG GTA CAG CCT GGG GGG TCC CTG AGA
Z12347,IGHV3-23*01,DP-47       [41] --- --- --- --- --- --- --- --- --- ... --- --- --- --- --- --- --- --- --- ---
M83136,IGHV3-23*01,VH26         [9] --- --- --- --- --- --- --- --- --- ... --- --- --- --- --- --- --- --- --- ---
U29482,IGHV3-23*01,VH26-5.0    [36] --- --- --- --- --- --- --- --- --- ... --- --- --- --- --- --- --- --- --- ---
J00236,IGHV3-23*02,VH26        [28] --- --- --- --- --- --- --- --- --- ... --- --- --- --- --- --- --- --- --- ---
U29481,IGHV3-23*03,VH26-3.7    [36] --- --- --- --- --- --- --- --- --- ... --- --- --- --- --- --- --- --- --- ---

                                                             ____________CDR1-IMGT____________
                                    21  22  23  24  25  26  27  28  29  30  31  32  33  34  35  36  37  38  39  40
                                     L   S   C   A   A   S   G   F   T   F   S   S   Y   A                   M   S
M99660,IGHV3-23*01,V3-23            CTC TCC TGT GCA GCC TCT GGA TTC ACC TTT AGC AGC TAT GCC ... ... ... ... ATG AGC
Z12347,IGHV3-23*01,DP-47            --- --- --- --- --- --- --- --- --- --- --- --- --- --- ... ... ... ... --- ---
M83136,IGHV3-23*01,VH26             --- --- --- --- --- --- --- --- --- --- --- --- --- --- ... ... ... ... --- ---
U29482,IGHV3-23*01,VH26-5.0         --- --- --- --- --- --- --- --- --- --- --- --- --- --- ... ... ... ... --- ---
J00236,IGHV3-23*02,VH26             --- --- --- --- --- --- --- --- --- --- --- --- --- --- ... ... ... ... --- ---
U29481,IGHV3-23*03,VH26-3.7         --- --- --- --- --- --- --- --- --- --- --- --- --- --- ... ... ... ... --- ---

                                                                                             __________CDR2-
                                    41  42  43  44  45  46  47  48  49  50  51  52  53  54  55  56  57  58  59  60
                                     W   V   R   Q   A   P   G   K   G   L   E   W   V   S   A   I   S   G   S   G
M99660,IGHV3-23*01,V3-23            TGG GTC CGC CAG GCT CCA GGG AAG GGG CTG GAG TGG GTC TCA GCT ATT AGT GGT AGT GGT
Z12347,IGHV3-23*01,DP-47            --- --- --- --- --- --- --- --- --- --- --- --- --- --- --- --- --- --- --- ---
M83136,IGHV3-23*01,VH26             --- --- --- --- --- --- --- --- --- --- --- --- --- --- --- --- --- --- --- ---
U29482,IGHV3-23*01,VH26-5.0         --- --- --- --- --- --- --- --- --- --- --- --- --- --- --- --- --- --- --- ---
J00236,IGHV3-23*02,VH26             --- --- --- --- --- --- --- --- --- --- --- --- --- --- --- --- --- --- --- ---
                                                                                     V       Y   S   G
U29481,IGHV3-23*03,VH26-3.7         --- --- --- --- --- --- --- --- --- --- --- --- --- --- -T- --- TA- A-C G-- ---

                                    IMGT_______________
                                    61  62  63  64  65  66  67  68  69  70  71  72  73  74  75  76  77  78  79  80
                                     G   S   T           Y   Y   A   D   S   V   K       G   R   F   T   I   S   R
M99660,IGHV3-23*01,V3-23            GGT AGC ACA ... ... TAC TAC GCA GAC TCC GTG AAG ... GGC CGG TTC ACC ATC TCC AGA
Z12347,IGHV3-23*01,DP-47            --- --- --- ... ... --- --- --- --- --- --- --- ... --- --- --- --- --- --- ---
M83136,IGHV3-23*01,VH26             --- --- --- ... ... --- --- --- --- --- --- --- ... --- --- --- --- --- --- ---
U29482,IGHV3-23*01,VH26-5.0         --- --- --- ... ... --- --- --- --- --- --- --- ... --- --- --- --- --- --- ---
                                                             G
J00236,IGHV3-23*02,VH26             --- --- --- ... ... --- --- -G- --- --- --- --- ... --- --- --- --- --- --A ---
                                     S
U29481,IGHV3-23*03,VH26-3.7         A-- --- --- ... ... --- --T --- --- --- --- --- ... --- --- --- --- --- --- ---

                                    81  82  83  84  85  86  87  88  89  90  91  92  93  94  95  96  97  98  99 100
                                     D   N   S   K   N   T   L   Y   L   Q   M   N   S   L   R   A   E   D   T   A
M99660,IGHV3-23*01,V3-23            GAC AAT TCC AAG AAC ACG CTG TAT CTG CAA ATG AAC AGC CTG AGA GCC GAG GAC ACG GCC
Z12347,IGHV3-23*01,DP-47            --- --- --- --- --- --- --- --- --- --- --- --- --- --- --- --- --- --- --- ---
M83136,IGHV3-23*01,VH26             --- --- --- --- --- --- --- --- --- --- --- --- --- --- --- --- --- --- --- ---
U29482,IGHV3-23*01,VH26-5.0         --- --- --- --- --- --- --- --- --- --- --- --- --- --- --- --- --- --- --- ---
J00236,IGHV3-23*02,VH26             --- --- --- --- --- --- --- --- --- --- --- --- --- --- --- --- --- --- --- ---
U29481,IGHV3-23*03,VH26-3.7         --T --- --- --- --- --- --- --- --- --- --- --- --- --- --- --- --- --- --- ---
```

```
                                     _CDR3-IMGT
                          101 102 103 104 105 106
                           V   Y   Y   C   A   K
M99660,IGHV3-23*01,V3-23  GTA TAT TAC TGT GCG AAA GA

Z12347,IGHV3-23*01,DP-47  --- --- --- --- --- ---

M83136,IGHV3-23*01,VH26   --- --- --- --- --- --- --

U29482,IGHV3-23*01,VH26-5.0  --- --- --- --- --- ---

J00236,IGHV3-23*02,VH26   --- --- --- --- --- --- --

U29481,IGHV3-23*03,VH26-3.7  --- --- --- --- --- ---
```

Framework and complementarity determining regions

FR1-IMGT: 25 (-1 aa: 10)
FR2-IMGT: 17
FR3-IMGT: 38 (-1 aa: 73)

CDR1-IMGT: 8
CDR2-IMGT: 8
CDR3-IMGT: 2

Collier de Perles for human IGHV3-23*01

Accession number: IMGT M99660 EMBL/GenBank/DDBJ: M99660

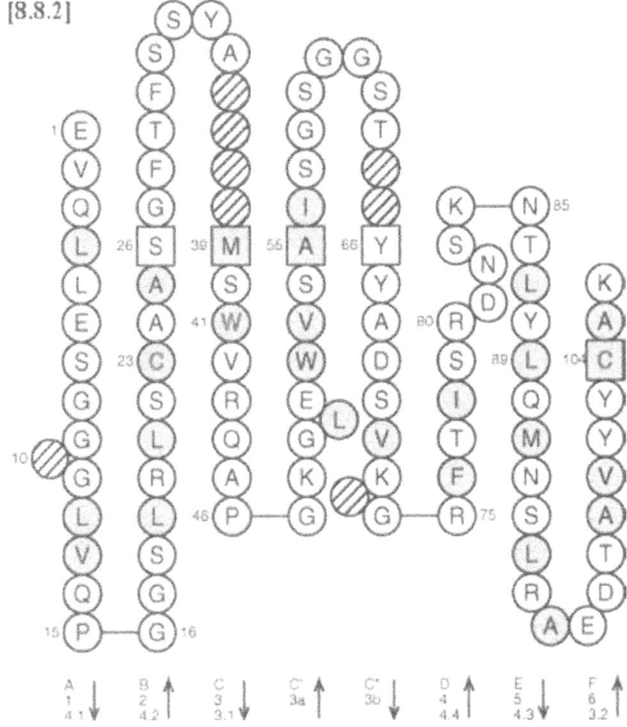

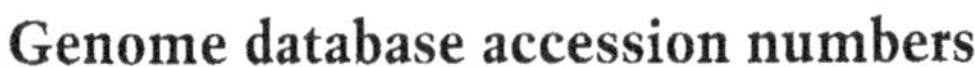

Genome database accession numbers

GDB:9931692 LocusLink: 28442

Nomenclature

IGHV3-30: Immunoglobulin heavy variable 3-30.
IGHV3-30-5: Immunoglobulin heavy variable 3-30-5.
Sequences of the polymorphic IGHV3-30-5 gene cannot be differentiated from those of the IGHV3-30 gene. All sequences are described therefore as "IGHV3-30 alleles" and compared to the allele *01 of IGHV3-30 (M83134). However, it is not excluded that some of these "alleles" belong exclusively to IGH3-30-5.

Definition and functionality

IGHV3-30 is one of the 18–21 mapped functional genes of the IGHV3 subgroup. This subgroup comprises 47–49 mapped genes in the IGH locus, depending on the haplotypes, three unmapped genes (one functional, one pseudogene and one not defined) and 12 orphons.

Gene location

IGHV3-30 is in the IGH locus on chromosome 14 at 14q32.33.

Nucleotide and amino acid sequences for human IGHV3-30

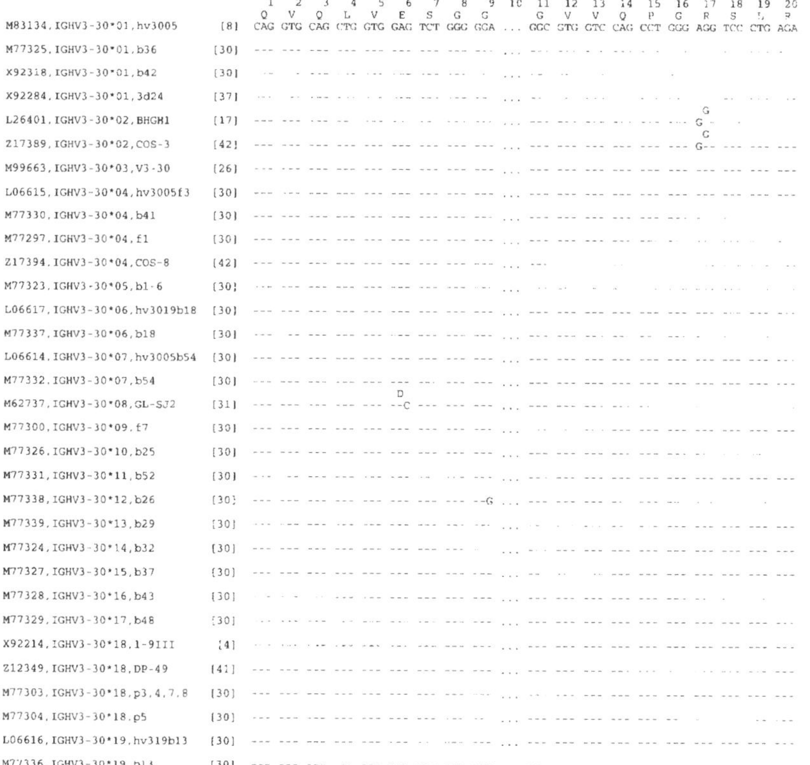

```
                                         1   2   3   4   5   6   7   8   9  10  11  12  13  14  15  16  17  18  19  20
                                         Q   V   Q   L   V   E   S   G   G       G   V   V   Q   P   G   R   S   L   R
M83134,IGHV3-30*01,hv3005         [8]   CAG GTG CAG CTG GTG GAG TCT GGG GGA ... GGC GTG GTC CAG CCT GGG AGG TCC CTG AGA
M77325,IGHV3-30*01,b36           [30]   --- --- --- --- --- --- --- --- --- ... --- --- --- --- --- --- --- --- --- ---
X92318,IGHV3-30*01,b42           [30]   --- --- --- --- --- --- --- --- --- ... --- --- --- --- --- --- --- --- --- ---
X92284,IGHV3-30*01,3d24          [37]   --- --- --- --- --- --- --- --- --- ... --- --- --- --- --- --- --- --- --- ---
                                                                                                         G
L26401,IGHV3-30*02,BHGH1         [17]   --- --- --- --- --- --- --- --- --- ... --- --- --- --- --- --- G-- --- --- ---
                                                                                                         G
Z17389,IGHV3-30*02,COS-3         [42]   --- --- --- --- --- --- --- --- --- ... --- --- --- --- --- --- G-- --- --- ---
M99663,IGHV3-30*03,V3-30         [26]   --- --- --- --- --- --- --- --- --- ... --- --- --- --- --- --- --- --- --- ---
L06615,IGHV3-30*04,hv3005f3      [30]   --- --- --- --- --- --- --- --- --- ... --- --- --- --- --- --- --- --- --- ---
M77330,IGHV3-30*04,b41           [30]   --- --- --- --- --- --- --- --- --- ... --- --- --- --- --- --- --- --- --- ---
M77297,IGHV3-30*04,f1            [30]   --- --- --- --- --- --- --- --- --- ... --- --- --- --- --- --- --- --- --- ---
Z17394,IGHV3-30*04,COS-8         [42]   --- --- --- --- --- --- --- --- --- ... --- --- --- --- --- --- --- --- --- ---
M77323,IGHV3-30*05,b1-6          [30]   --- --- --- --- --- --- --- --- --- ... --- --- --- --- --- --- --- --- --- ---
L06617,IGHV3-30*06,hv3019b18     [30]   --- --- --- --- --- --- --- --- --- ... --- --- --- --- --- --- --- --- --- ---
M77337,IGHV3-30*06,b18           [30]   --- --- --- --- --- --- --- --- --- ... --- --- --- --- --- --- --- --- --- ---
L06614,IGHV3-30*07,hv3005b54     [30]   --- --- --- --- --- --- --- --- --- ... --- --- --- --- --- --- --- --- --- ---
M77332,IGHV3-30*07,b54           [30]   --- --- --- --- --- --- --- --- --- ... --- --- --- --- --- --- --- --- --- ---
                                                             D
M62737,IGHV3-30*08,GL-SJ2        [31]   --- --- --- --- --- --C --- --- --- ... --- --- --- --- --- --- --- --- --- ---
M77300,IGHV3-30*09,f7            [30]   --- --- --- --- --- --- --- --- --- ... --- --- --- --- --- --- --- --- --- ---
M77326,IGHV3-30*10,b25           [30]   --- --- --- --- --- --- --- --- --- ... --- --- --- --- --- --- --- --- --- ---
M77331,IGHV3-30*11,b52           [30]   --- --- --- --- --- --- --- --- --- ... --- --- --- --- --- --- --- --- --- ---
M77338,IGHV3-30*12,b26           [30]   --- --- --- --- --- --- --- --- --G ... --- --- --- --- --- --- --- --- --- ---
M77339,IGHV3-30*13,b29           [30]   --- --- --- --- --- --- --- --- --- ... --- --- --- --- --- --- --- --- --- ---
M77324,IGHV3-30*14,b32           [30]   --- --- --- --- --- --- --- --- --- ... --- --- --- --- --- --- --- --- --- ---
M77327,IGHV3-30*15,b37           [30]   --- --- --- --- --- --- --- --- --- ... --- --- --- --- --- --- --- --- --- ---
M77328,IGHV3-30*16,b43           [30]   --- --- --- --- --- --- --- --- --- ... --- --- --- --- --- --- --- --- --- ---
M77329,IGHV3-30*17,b48           [30]   --- --- --- --- --- --- --- --- --- ... --- --- --- --- --- --- --- --- --- ---
X92214,IGHV3-30*18,1-9III         [4]   --- --- --- --- --- --- --- --- --- ... --- --- --- --- --- --- --- --- --- ---
Z12349,IGHV3-30*18,DP-49         [41]   --- --- --- --- --- --- --- --- --- ... --- --- --- --- --- --- --- --- --- ---
M77303,IGHV3-30*18,p3,4,7,8      [30]   --- --- --- --- --- --- --- --- --- ... --- --- --- --- --- --- --- --- --- ---
M77304,IGHV3-30*18,p5            [30]   --- --- --- --- --- --- --- --- --- ... --- --- --- --- --- --- --- --- --- ---
L06616,IGHV3-30*19,hv319b13      [30]   --- --- --- --- --- --- --- --- --- ... --- --- --- --- --- --- --- --- --- ---
M77336,IGHV3-30*19,b13           [30]   --- --- --- --- --- --- --- --- --- ... --- --- --- --- --- --- --- --- --- ---
```

												CDR1-IMGT								
	21	22	23	24	25	26	27	28	29	30	31	32	33	34	35	36	37	38	39	40
	L	S	C	A	A	S	G	F	T	F	S	S	Y	A					M	H
M83134,IGHV3-30*01,hv3005	CTC	TCC	TGT	GCA	GCC	TCT	GGA	TTC	ACC	TTC	AGT	AGC	TAT	GCT	...	...	...	...	ATG	CAC
M77325,IGHV3-30*01,b36	---	---	---	---	---	---	---	---	---	---	---	---	---	---	...	...	...	...	---	---
X92318,IGHV3-30*01,b42	---	---	---	---	---	---	---	---	---	---	---	---	---	---	...	...	...	...	---	---
X92284,IGHV3-30*01,3d24	---	---	---	---	---	---	---	---	---	---	---	---	---	---	...	...	...	...	---	---
L26401,IGHV3-30*02,BHGH1	---	---	---	---	--G	---	---	---	---	---	---	---	---	-GC (G)	...	...	...	...	---	---
Z17389,IGHV3-30*02,COS-3	---	---	---	---	--G	---	---	---	---	---	---	---	---	-GC (G)	...	...	...	...	---	---
M99663,IGHV3-30*03,V3-30	---	---	---	---	---	---	---	---	---	---	---	---	---	-GC (G)	...	...	...	...	---	---
L06615,IGHV3-30*04,hv3005f3	---	---	---	---	---	---	---	---	---	---	---	---	---	---	...	...	...	...	---	---
M77330,IGHV3-30*04,b41	---	---	---	---	---	---	---	---	---	---	---	---	---	---	...	...	...	...	---	---
M77297,IGHV3-30*04,f1	---	---	---	---	---	---	---	---	---	---	---	---	---	---	...	...	...	...	---	
Z17394,IGHV3-30*04,COS-8	---	---	---	---	---	---	---	---	---	---	---	---	---	---	...	...	...	...	---	---
M77323,IGHV3-30*05,b1-6	---	---	---	---	---	---	---	---	---	---	---	---	---	-GC (G)	...	...	...	...	---	---
L06617,IGHV3-30*06,hv3019b18	---	---	---	---	--G	---	---	---	---	---	---	---	---	-GC (G)	...	...	...	...	---	---
M77337,IGHV3-30*06,b18	---	---	---	---	--G	---	---	---	---	---	---	---	---	-GC (G)	...	...	...	...	---	---
L06614,IGHV3-30*07,hv3005b54	---	---	---	---	---	---	---	---	---	---	---	---	---	---	...	...	...	...	---	---
M77332,IGHV3-30*07,b54	---	---	---	---	---	---	---	---	---	---	---	---	---	---	...	...	...	...	---	---
M62737,IGHV3-30*08,GL-SJ2	---	---	---	---	---	---	-C- (A)	---	---	---	---	---	---	---	...	...	...	...	---	---
M77300,IGHV3-30*09,f7	---	---	---	---	---	---	---	---	---	---	---	---	---	---	...	...	...	...	---	---
M77326,IGHV3-30*10,b25	---	---	---	---	---	---	---	---	---	---	---	---	---	---	...	...	...	...	---	---
M77331,IGHV3-30*11,b52	---	---	---	---	--G	---	---	---	---	---	---	---	---	---	...	...	...	...	---	---
M77338,IGHV3-30*12,b26	---	---	---	---	--G	---	---	---	---	---	---	---	---	-GC (G)	...	...	...	...	---	---
M77339,IGHV3-30*13,b29	---	---	---	---	---	---	---	---	---	---	---	---	---	-GC (G)	...	...	...	...	---	---
M77324,IGHV3-30*14,b32	---	---	---	---	---	---	---	---	---	---	---	---	---	---	...	...	...	...	---	---
M77327,IGHV3-30*15,b37	---	---	---	---	---	---	---	---	---	---	---	---	---	---	...	...	...	...	---	---
M77328,IGHV3-30*16,b43	---	---	---	---	---	---	---	---	---	---	---	---	---	---	...	...	...	...	---	---
M77329,IGHV3-30*17,b48	---	---	---	---	---	---	---	---	---	---	---	---	---	---	...	...	...	...	---	---
X92214,IGHV3-30*18,1-9III	---	---	---	---	---	---	---	---	---	---	---	---	---	-GC (G)	...	...	...	...	---	---
Z12349,IGHV3-30*18,DP-49	---	---	---	---	---	---	---	---	---	...	---	---	---	-GC (G)	...	...	...	...	---	---
M77303,IGHV3-30*18,p3,4,7,8	---	---	---	---	---	---	---	---	---	...	---	---	---	-GC (G)	...	...	...	...	---	---
M77304,IGHV3-30*18,p5	---	---	---	---	---	---	---	---	---	...	---	---	---	-GC (G)	...	...	...	...	---	---
L06616,IGHV3-30*19,hv319b13	---	---	---	---	--G	---	---	---	---	...	---	---	---	-GC (G)	...	...	...	...	---	---
M77336,IGHV3-30*19,b13	---	---	---	---	--G	---	---	---	---	...	---	---	---	-GC (G)	...	...	...	...	---	---

																		CDR2-		
	41	42	43	44	45	46	47	48	49	50	51	52	53	54	55	56	57	58	59	60
	W	V	R	Q	A	P	G	K	G	L	E	W	V	A	V	I	S	Y	D	G
M83134,IGHV3-30*01,hv3005	TGG	GTC	CGC	CAG	GCT	CCA	GGC	AAG	GGG	CTA	GAG	TGG	GTG	GCA	GTT	ATA	TCA	TAT	GAT	GGA
M77325,IGHV3-30*01,b36	---	---	---	---	---	---	---	---	---	---	---	---	---	---	---	---	---	---	---	---
X92318,IGHV3-30*01,b42	---	---	---	---	---	---	---	---	---	---	---	---	---	---	---	---	---	---	---	---
X92284,IGHV3-30*01,3d24	---	---	---	---	---	---	---	---	---	---	---	---	---	---	---	---	---	---	---	---
L26401,IGHV3-30*02,BHGH1	---	---	---	---	---	---	---	---	---	--G	---	---	---	---	T-- (F)	---	CGG (R)	---	---	---
Z17389,IGHV3-30*02,COS-3	---	---	---	---	---	---	---	---	---	--G	---	---	---	---	T-- (F)	---	CGG (R)	---	---	---
M99663,IGHV3-30*03,V3-30	---	---	---	---	---	---	---	---	---	--G	---	---	---	---	---	---	---	---	---	---
L06615,IGHV3-30*04,hv3005f3	---	---	---	---	---	---	---	---	---	--G	---	---	---	---	---	---	---	---	---	---
M77330,IGHV3-30*04,b41	---	---	---	---	---	---	---	---	---	--G	---	---	---	---	---	---	---	---	---	---
M77297,IGHV3-30*04,f1	---	---	---	---	---	---	---	---	---	--G	---	---	---	---	---	---	---	---	---	---
Z17394,IGHV3-30*04,COS-8	---	---	---	---	---	---	---	---	---	--G	---	---	---	---	---	---	---	---	---	---
M77323,IGHV3-30*05,b1-6	---	---	---	---	---	---	---	---	---	---	---	---	---	---	---	---	---	---	---	---
L06617,IGHV3-30*06,hv3019b18	---	---	---	---	---	---	---	---	---	---	---	---	---	---	---	---	---	---	---	---
M77337,IGHV3-30*06,b18	---	---	---	---	---	---	---	---	---	---	---	---	---	---	---	---	---	---	---	---
L06614,IGHV3-30*07,hv3005b54	---	---	---	---	---	---	---	---	---	---	---	---	---	---	---	---	---	---	---	---
M77332,IGHV3-30*07,b54	---	---	---	---	---	---	---	---	---	---	---	---	---	---	---	---	---	---	---	---
M62737,IGHV3-30*08,GL-SJ2	---	---	---	---	---	---	---	---	---	---	---	---	---	---	---	---	---	---	---	---
M77300,IGHV3-30*09,f7	---	---	---	---	---	---	---	---	---	--G	---	---	---	---	---	---	---	---	---	---
M77326,IGHV3-30*10,b25	---	---	---	---	---	---	---	---	---	---	---	---	---	---	---	---	---	---	---	---
M77331,IGHV3-30*11,b52	---	---	---	---	---	---	---	---	---	---	---	---	---	---	---	---	---	---	---	---
M77338,IGHV3-30*12,b26	---	---	---	---	---	---	---	---	---	---	---	---	---	---	---	---	---	---	---	---
M77339,IGHV3-30*13,b29	---	---	---	---	---	---	---	---	---	---	---	---	---	---	---	---	---	---	---	---
M77324,IGHV3-30*14,b32	---	---	---	---	---	---	---	---	---	--G	---	---	---	---	---	---	---	---	---	---
M77327,IGHV3-30*15,b37	---	---	---	---	---	---	---	---	---	---	---	---	---	---	---	---	---	---	---	---
M77328,IGHV3-30*16,b43	---	---	---	---	--C	---	---	---	---	---	---	---	---	---	---	---	---	---	---	---
M77329,IGHV3-30*17,b48	---	---	---	---	---	--G	---	---	---	---	---	---	---	---	---	---	---	---	---	---
X92214,IGHV3-30*18,1-9III	---	---	---	---	---	---	---	---	---	--G	---	---	---	---	---	---	---	---	---	---
Z12349,IGHV3-30*18,DP-49	---	---	---	---	---	---	---	---	---	--G	---	---	---	---	---	---	---	---	---	---
M77303,IGHV3-30*18,p3,4,7,8	---	---	---	---	---	---	---	---	---	--G	---	---	---	---	---	---	---	---	---	---
M77304,IGHV3-30*18,p5	---	---	---	---	---	---	---	---	---	--G	---	---	---	---	---	---	---	---	---	--
L06616,IGHV3-30*19,hv319b13	---	---	---	---	---	---	---	---	---	--G	---	---	---	---	---	---	---	---	---	---
M77336,IGHV3-30*19,b13	---	---	---	---	---	---	---	---	---	--G	---	---	---	---	---	---	---	---	---	---

IMGT	61	62	63	64	65	66	67	68	69	70	71	72	73	74	75	76	77	78	79	80
	S	N	K			Y	Y	A	D	S	V	K		G	R	F	T	I	S	R
M83134,IGHV3-30*01,hv3005	AGT	AAT	AAA	...	...	TAC	TAC	GCA	GAC	TCC	GTG	AAG	...	GGC	CGA	TTC	ACC	ATC	TCC	AGA
M77325,IGHV3-30*01,b36	---	---	---	...	...	---	---	---	---	---	---	---	...	---	---	---	---	---	---	---
X92318,IGHV3-30*01,b42	---	---	---	...	...	---	---	---	---	---	---	---	...	---	---	---	---	---	---	---
X92284,IGHV3-30*01,3d24	---	---	---	...	...	---	---	---	---	---	---	---	...	---	---	---	---	---	---	---
L26401,IGHV3-30*02,BHGH1	---	---	---	...	...	---	--T	---	---	---	---	---	...	---	---	---	---	---	---	---
Z17389,IGHV3-30*02,COS-3	---	---	---	...	...	---	--T	---	---	---	---	---	...	---	---	---	---	---	---	---
M99663,IGHV3-30*03,V3-30	---	---	---	...	...	---	--T	---	---	---	---	---	...	---	---	---	---	---	---	---
L06615,IGHV3-30*04,hv3005f3	---	---	---	...	...	---	---	---	---	---	---	---	...	---	---	---	---	---	---	---
M77330,IGHV3-30*04,b41	---	---	---	...	...	---	---	---	---	---	---	---	...	---	---	---	---	---	---	---
M77297,IGHV3-30*04,f1	---	---	---	...	...	---	---	---	---	---	---	---	...	---	---	---	---	---	---	---
Z17394,IGHV3-30*04,COS-8	---	---	---	...	...	---	---	---	---	---	---	---	...	---	---	---	---	---	---	---
M77323,IGHV3-30*05,b1-6	---	---	---	...	...	---	---	---	---	---	---	---	...	---	---	---	---	---	---	---
L06617,IGHV3-30*06,hv3019b18	---	---	---	...	...	---	---	---	---	---	---	---	...	---	---	---	---	---	---	---
M77337,IGHV3-30*06,b18			---	...	...	---	---	---	---	---	---	---	...	---	---	---	---	---	---	
L06614,IGHV3-30*07,hv3005b54		---	---	...	...	---	---	---	---	---	---	---	...	---	---	---	---	---	---	
M77332,IGHV3-30*07,b54	---	---	---	...	...	---	---	---	---	---	---	---	...	---	---	---	---	---	---	---
M62737,IGHV3-30*08,GL SJ2	---	---	---	...	...	---	---	---	---	---	---	---	...	---	---	---	---	---	---	---
M77300,IGHV3-30*09,f7		---	---	...	...	---	---	---	---	---	---	---	...	---	---	---	A G--	---	---	---
M77326,IGHV3-30*10,b25		---	---	...	...		T A--	---	---	---	---	---	...	---	---	---	---	---	---	---
M77331,IGHV3-30*11,b52	---	---	---	...	...	---	---	---	---	---	---	---	...	---	---	---	---	---	---	
M77338,IGHV3-30*12,b26	---	---	---	...	...	---	---	---	---	---	---	---	...	---	---	---	---	---	---	---
M77339,IGHV3-30*13,b29	---	---	---	...	...	---	---	---	---	---	---	---	...	---	---	---	---	---	---	---
M77324,IGHV3-30*14,b32		---	---	...	...	---	---	---	---	---	---	---	...	---	---	---	---	---	---	---
M77327,IGHV3-30*15,b37	---	---	---	...	...	---	---	---	---	---	---	---	...	---	---	---	---	---	---	
M77328,IGHV3-30*16,b43	---	---	---	...	...	---	---	---	---	---	---	---	...	---	---	---	---	---	---	---
M77329,IGHV3-30*17,b48		---	---	...	...	---	---	---	---	---	---	---	...	---	---	---	---	---		
X92214,IGHV3-30*18,1-9III	---	---	---	...	...	---	--T	---	---	---	---	---	...	---	---	---	---	---	---	---
Z12349,IGHV3-30*18,DP-49	---	---	---	...	...	---	--T	---	---	---	---	---	...	---	---	---	---	---	---	
M77303,IGHV3-30*18,p3,4,7,8	---	---	---	...	...	---	--T	---	---	---	---	---	...	---	---	---	---	---	---	---
M77304,IGHV3-30*18,p5	---	---	---	...	...	---	--T	---	---	---	---	---	...	---	---	---	---	---	---	---
L06616,IGHV3-30*19,hv319b13	---	---	---	...	...	---	---	---	---	---	---	---	...	---	---	---	---	---	---	---
M77336,IGHV3-30*19,b13	---	---	---	...	...	---	---	---	---	---	---	---	...	---	---	---	---	---	---	---

	81	82	83	84	85	86	87	88	89	90	91	92	93	94	95	96	97	98	99	100
	D	N	S	K	N	T	L	Y	L	Q	M	N	S	L	R	A	E	D	T	A
M83134,IGHV3-30*01,hv3005	GAC	AAT	TCC	AAG	AAC	ACG	CTG	TAT	CTG	CAA	ATG	AAC	AGC	CTG	AGA	GCT	GAG	GAC	ACG	GCT
M77325,IGHV3-30*01,b36	---	---	---	---	---	---	---	---	---	---	---	---	---	---	---	---	---	---	---	---
X92318,IGHV3-30*01,b42	---	---	---	---	---	---	---	---	---	---	---	---	---	---	---	---	---	---	---	---
X92284,IGHV3-30*01,3d24	---	---	---	---	---	---	---	---	---	---	---	---	---	---	---	---	---	---	---	---
L26401,IGHV3-30*02,BHGH1	---	---	---	---	---	---	---	---	---	---	---	---	---	---	---	---	---	---	---	---
Z17389,IGHV3-30*02,COS-3	---	---	---	---	---	---	---	---	---	---	---	---	---	---	---	---	---	---	---	---
M99663,IGHV3-30*03,V3-30	---	---	---	---	---	---	---	---	---	---	---	---	---	---	---	---	---	---	---	---
L06615,IGHV3-30*04,hv3005f3			---	---	---	---	---	---	---	---	---	---	---	---	---	---	---	---	---	---
M77330,IGHV3-30*04,b41	---	---	---	---	---	---	---	---	---	---	---	---	---	---	---	---	---	---	---	---
M77297,IGHV3-30*04,f1	---	---	---	---	---	---	---	---	---	---	---	---	---	---	---	---	---	---	---	---
Z17394,IGHV3-30*04,COS-8	---	---	---	---	---	---	---	---	---	---	---	---	---	---	---	---	---	---	---	---
M77323,IGHV3-30*05,b1-6	---	---	---	---	---	---	---	---	---	---	---	---	---	---	---	---	---	G -G-	---	---
L06617,IGHV3-30*06,hv3019b18	---	---	---	---	---	---	---	---	---	---	---	---	---	---	---	---	---	---	---	---
M77337,IGHV3-30*06,b18	---	---	---	---	---	---	---	---	---	---	---	---	---	---	---	---	---	---	---	---
L06614,IGHV3-30*07,hv3005b54	---	---	---	---	---	---	---	---	---	---	---	---	---	---	---	--C	---	---	---	---
M77332,IGHV3-30*07,b54	---	---	---	---	---	---	---	---	---	---	---	---	---	---	---	--C	---	---	---	---
M62737,IGHV3-30*08,GL SJ2	---	---	---	---	---	---	---	---	---	---	---	---	---	---	---	---	---	---	---	---
M77300,IGHV3-30*09,f7	---	---	---	---	---	---	---	---	---	---	---	---	---	---	---	---	---	---	---	---
M77326,IGHV3-30*10,b25	---	---	---	---	---	---	---	---	---	---	---	---	---	---	---	---	---	---	---	---

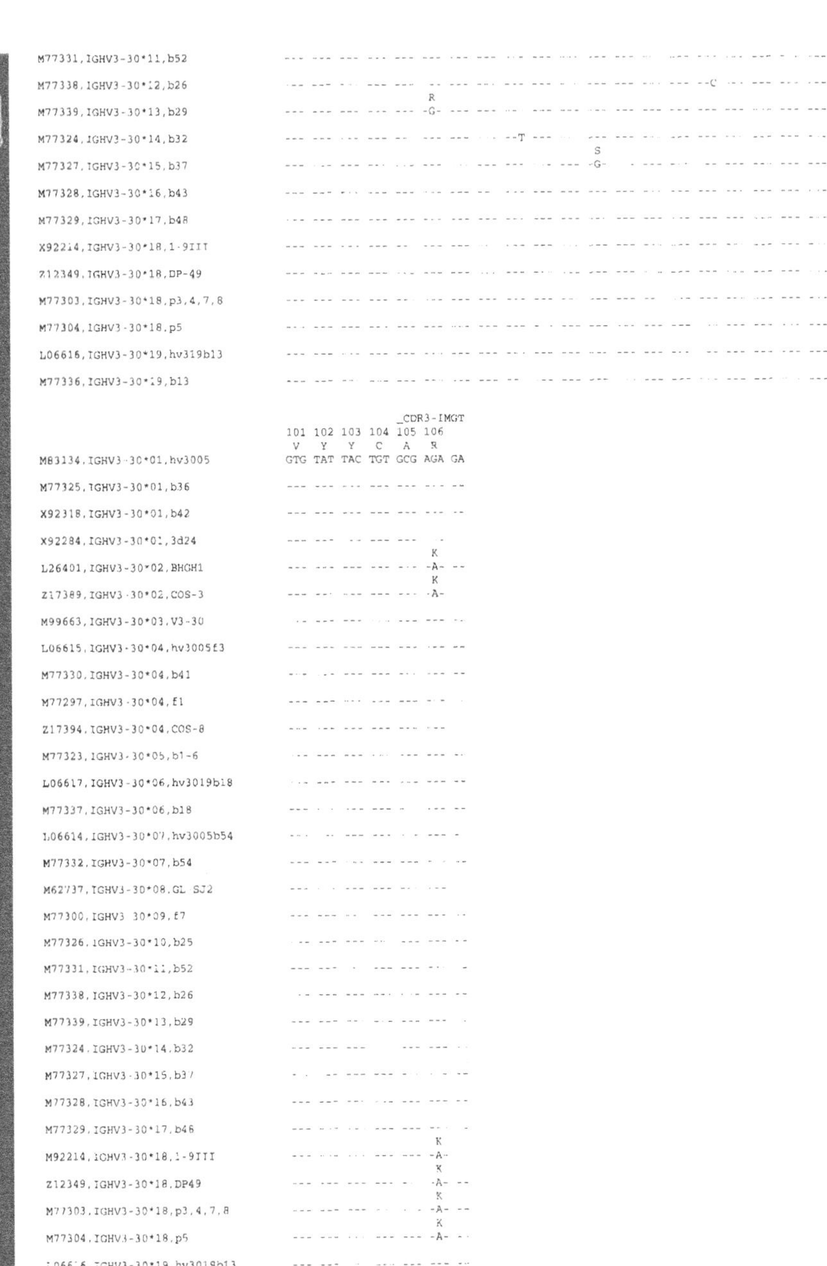

```
M77331,IGHV3-30*11,b52         --- --- --- --- --- --- --- --- --- --- --- --- --- --- --- --- --- --- --- ---
M77338,IGHV3-30*12,b26         --- --- --- --- --- --- --- --- --- --- --- --- --- --- --- --- --C --- --- ---
                                                    R
M77339,IGHV3-30*13,b29         --- --- --- --- --- -G- --- --- --- --- --- --- --- --- --- --- --- --- --- ---
M77324,IGHV3-30*14,b32         --- --- --- --- --- --- --- --- --T --- --- --- --- --- --- --- --- --- --- ---
                                                                            S
M77327,IGHV3-30*15,b37         --- --- --- --- --- --- --- --- --- --- --- -G- --- --- --- --- --- --- --- ---
M77328,IGHV3-30*16,b43         --- --- --- --- --- --- --- --- --- --- --- --- --- --- --- --- --- --- --- ---
M77329,IGHV3-30*17,b48         --- --- --- --- --- --- --- --- --- --- --- --- --- --- --- --- --- --- --- ---
X92214,IGHV3-30*18,1-9III      --- --- --- --- --- --- --- --- --- --- --- --- --- --- --- --- --- --- --- ---
Z12349,IGHV3-30*18,DP-49       --- --- --- --- --- --- --- --- --- --- --- --- --- --- --- --- --- --- --- ---
M77303,IGHV3-30*18,p3,4,7,8    --- --- --- --- --- --- --- --- --- --- --- --- --- --- --- --- --- --- --- ---
M77304,IGHV3-30*18,p5          --- --- --- --- --- --- --- --- --- --- --- --- --- --- --- --- --- --- --- ---
L06616,IGHV3-30*19,hv319b13    --- --- --- --- --- --- --- --- --- --- --- --- --- --- --- --- --- --- --- ---
M77336,IGHV3-30*19,b13         --- --- --- --- --- --- --- --- --- --- --- --- --- --- --- --- --- --- --- ---

                                                   _CDR3-IMGT
                               101 102 103 104 105 106
                                V   Y   Y   C   A   R
M83134,IGHV3-30*01,hv3005      GTG TAT TAC TGT GCG AGA GA
M77325,IGHV3-30*01,b36         --- --- --- --- --- --- --
X92318,IGHV3-30*01,b42         --- --- --- --- --- --- --
X92284,IGHV3-30*01,3d24        --- --- --- --- --- ---
                                                    K
L26401,IGHV3-30*02,BHGH1       --- --- --- --- --- -A- --
                                                    K
Z17389,IGHV3-30*02,COS-3       --- --- --- --- --- -A-
M99663,IGHV3-30*03,V3-30       --- --- --- --- --- --- --
L06615,IGHV3-30*04,hv3005f3    --- --- --- --- --- --- --
M77330,IGHV3-30*04,b41         --- --- --- --- --- --- --
M77297,IGHV3-30*04,f1          --- --- --- --- --- --- --
Z17394,IGHV3-30*04,COS-8       --- --- --- --- --- ---
M77323,IGHV3-30*05,b1-6        --- --- --- --- --- --- --
L06617,IGHV3-30*06,hv3019b18   --- --- --- --- --- --- --
M77337,IGHV3-30*06,b18         --- --- --- --- --- --- --
L06614,IGHV3-30*07,hv3005b54   --- --- --- --- --- --- -
M77332,IGHV3-30*07,b54         --- --- --- --- --- --- --
M62737,IGHV3-30*08,GL SJ2      --- --- --- --- --- ---
M77300,IGHV3-30*09,f7          --- --- --- --- --- --- --
M77326,IGHV3-30*10,b25         --- --- --- --- --- --- --
M77331,IGHV3-30*11,b52         --- --- --- --- --- --- -
M77338,IGHV3-30*12,b26         --- --- --- --- --- --- --
M77339,IGHV3-30*13,b29         --- --- --- --- --- --- -
M77324,IGHV3-30*14,b32         --- --- --- --- --- --- --
M77327,IGHV3-30*15,b37         --- --- --- --- --- --- --
M77328,IGHV3-30*16,b43         --- --- --- --- --- --- --
M77329,IGHV3-30*17,b48         --- --- --- --- --- --- --
                                                    K
M92214,IGHV3-30*18,1-9III      --- --- --- --- --- -A-
                                                    K
Z12349,IGHV3-30*18,DP49        --- --- --- --- --- -A- --
                                                    K
M77303,IGHV3-30*18,p3,4,7,8    --- --- --- --- --- -A- --
                                                    K
M77304,IGHV3-30*18,p5          --- --- --- --- --- -A- --
L06616,IGHV3-30*19,hv3019b13   --- --- --- --- --- --- --
M77336,IGHV3-30*19,hv3019b13   --- --- --- --- --- --- --
```

Framework and complementarity determining regions

FR1-IMGT: 25 (-1 aa: 10)
FR2-IMGT: 17
FR3-IMGT: 38 (-1 aa: 73)

CDR1-IMGT: 8
CDR2-IMGT: 8
CDR3-IMGT: 2

Collier de Perles for human IGHV3-30*01

Accession number: IMGT M83134 EMBL/GenBank/DDBJ: M83134

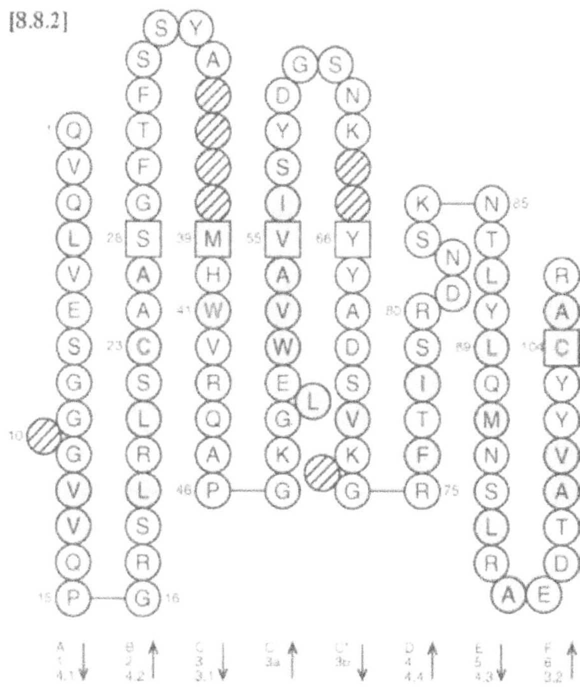

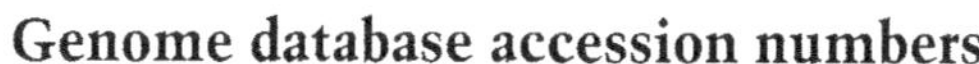

Genome database accession numbers

GDB:9931735 LocusLink: 28439

Nomenclature

IGHV3-30-3: Immunoglobulin heavy variable 3-30-3.

Definition and functionality

IGHV3-30-3 is a mapped functional gene, which may, or may not, be present due to a polymorphism by insertion/deletion. IGHV3-30-3 belongs to the IGHV3 subgroup which comprises 47–49 mapped genes (of which 18–21 are functional) in the IGH locus, depending on the haplotypes, three unmapped genes (one functional, one pseudogene and one not defined) and 12 orphons.

Gene location

IGHV3-30-3, in the haplotypes where it is present, is in the IGH locus on chromosome 14 at 14q32.33.

Nucleotide and amino acid sequences for human IGHV3-30-3

```
                                         1   2   3   4   5   6   7   8   9   10  11  12  13  14  15  16  17  18  19  20
                                         Q   V   Q   L   V   E   S   G   G       G   V   V   Q   P   G   R   S   L   R
X92283,IGHV3-30-3*01,3d216        [37]  CAG GTG CAG CTG GTG GAG TCT GGG GGA ... GGC GTG GTC CAG CCT GGG AGG TCC CTG AGA
Z12346,IGHV3-30-3*01,DP-46        [41]  --- --- --- --- --- --- --- --- --- ... --- --- --- --- --- --- --- --- ---
M77302,IGHV3-30-3*02,p2           [30]  --- --- --- --- --- --- --- --- --- ... --- --- --- --- --- --- --- --- --- ---

                                                                          ____________CDR1-IMGT____________
                                         21  22  23  24  25  26  27  28  29  30  31  32  33  34  35  36  37  38  39  40
                                         L   S   C   A   A   S   G   F   T   F   S   S   Y   A                   M   H
X92283,IGHV3-30-3*01,3d216              CTC TCC TGT GCA GCC TCT GGA TTC ACC TTC AGT AGC TAT GCT ... ... ... ... ATG CAC
Z12346,IGHV3-30-3*01,DP-46                  --- --- --- --- --- --- --- --- --- --- --- --- --- ... ... ... ... --- -
M77302,IGHV3-30-3*02,p2                 --- --- --- --- G-- --- --- --- --- --- --- --- --- --- ... ... ... ... ---

                                                                                                 ___________CDR2
                                         41  42  43  44  45  46  47  48  49  50  51  52  53  54  55  56  57  58  59  60
                                         W   V   R   Q   A   P   G   K   G   L   E   W   V   A   V   I   S   Y   D   G
X92283,IGHV3-30-3*01,3d216              TGG GTC CGC CAG GCT CCA GGC AAG GGG CTG GAG TGG GTG GCA GTT ATA TCA TAT GAT GGA
Z12346,IGHV3-30-3*01,DP-46              --- --- --- --- --- --- --- --- --- --- --- --- --- --- --- --- --- ---
M77302,IGHV3-30-3*02,p2                 ---     --- --- --- --- --- --- --- --- --- --- --- --- --- --- --- ---     ---

                                        IMGT____________
                                         61  62  63  64  65  66  67  68  69  70  71  72  73  74  75  76  77  78  79  80
                                         S   N   K           Y   Y   A   D   S   V   K       G   R   F   T   I   S   R
X92283,IGHV3-30-3*01,3d216              AGC AAT AAA ... ... TAC TAC GCA GAC TCC GTG AAG ... GGC CGA TTC ACC ATC TCC AGA
Z12346,IGHV3-30-3*01,DP-46              ---     --- ... ... ---     --- --- ---     --- ... --- ---     --- --- ---
M77302,IGHV3-30-3*02,p2                 --- --- --- ... ... --- --- --- --- --- --- --- ...     --- --- --- ---     ---

                                         81  82  83  84  85  86  87  88  89  90  91  92  93  94  95  96  97  98  99 100
                                         D   N   S   K   N   T   L   Y   L   Q   M   N   S   L   R   A   E   D   T   A
X92283,IGHV3-30-3*01,3d216              GAC AAT TCC AAG AAC ACG CTG TAT CTG CAA ATG AAC AGC CTG AGA GCT GAG GAC ACG GCT
Z12346,IGHV3-30-3*01,DP-46              --- --- --- --- --- ---     --- --- --- ---     --- --- --- --- --- --- --- ---
M77302,IGHV3-30-3*02,p2                 --- --- --- --- --- --- ---     --- --- --- --- --- ---     --- --- --- ---  ---

                                                            _CDR3-IMGT
                                        101 102 103 104 105 106
                                         V   Y   Y   C   A   R
X92283,IGHV3-30-3*01,3d216              GTG TAT TAC TGT GCG AGA
Z12346,IGHV3-30-3*01,DP-46              --- --- --- --- --- --
                                                             K
M77302,IGHV3-30-3*02,p2                 --- --- --- ---     A-- GA
```

Framework and complementarity determining regions

FR1-IMGT: 25 (-1 aa: 10)
FR2-IMGT: 17
FR3-IMGT: 38 (-1 aa: 73)

CDR1-IMGT: 8
CDR2-IMGT: 8
CDR3-IMGT: 2

Collier de Perles for human IGHV3-30-3*01

Accession number: IMGT X92283 EMBL/GenBank/DDBJ: X92283

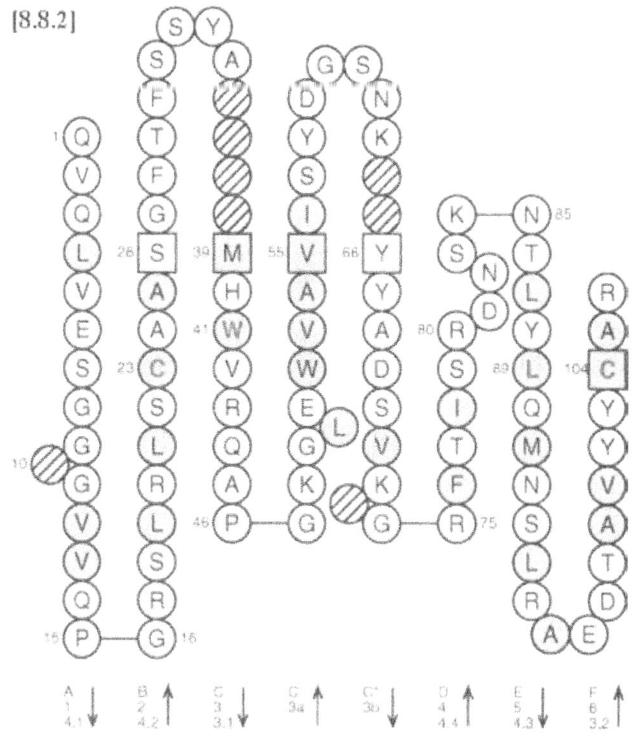

Genome database accession numbers

GDB:9931696 LocusLink: 28437

Nomenclature

IGHV3-33: Immunoglobulin heavy variable 3-33.

Definition and functionality

IGHV3-33 is one of the 18–21 mapped functional genes of the IGHV3 subgroup. This subgroup comprises 47–49 mapped genes in the IGH locus, depending on the haplotypes, three unmapped genes (one functional, one pseudogene and one not defined) and 12 orphons.

Gene location

IGHV3-33 is in the IGH locus on chromosome 14 at 14q32.33.

Nucleotide and amino acid sequences for human IGHV3-33

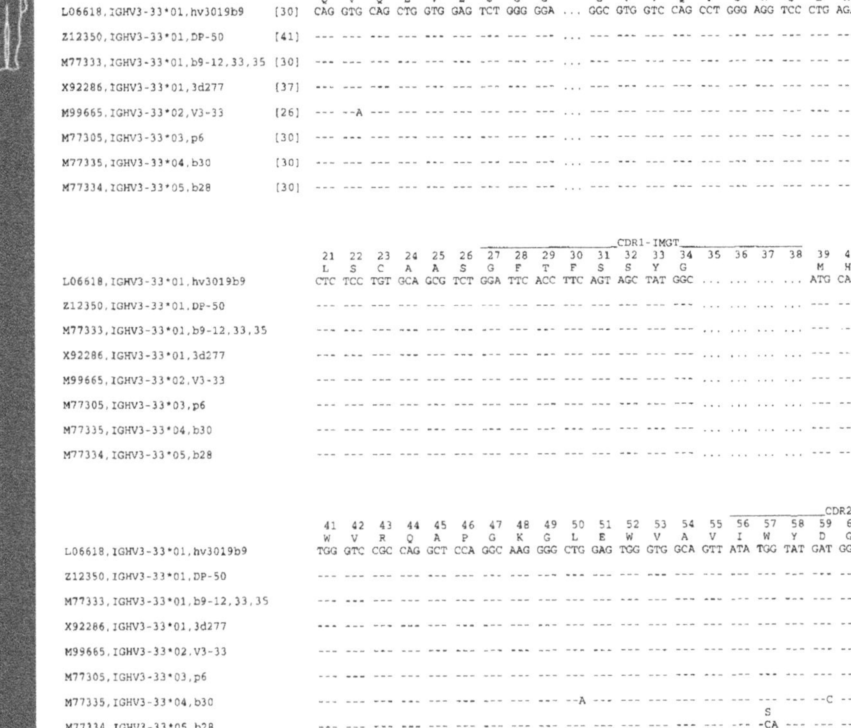

```
                                        1   2   3   4   5   6   7   8   9  10  11  12  13  14  15  16  17  18  19  20
                                        Q   V   Q   L   V   E   S   G   G       G   V   V   Q   P   G   R   S   L   R
L06618,IGHV3-33*01,hv3019b9       [30] CAG GTG CAG CTG GTG GAG TCT GGG GGA ... GGC GTG GTC CAG CCT GGG AGG TCC CTG AGA

Z12350,IGHV3-33*01,DP-50          [41] --- --- --- --- --- --- --- --- --- ... --- --- --- --- --- --- --- --- --- ---

M77333,IGHV3-33*01,b9-12,33,35    [30] --- --- --- --- --- --- --- --- --- ... --- --- --- --- --- --- --- --- --- ---

X92286,IGHV3-33*01,3d277          [37] --- --- --- --- --- --- --- --- --- ... --- --- --- --- --- --- --- --- --- ---

M99665,IGHV3-33*02,V3-33          [26] --- --A --- --- --- --- --- --- --- ... --- --- --- --- --- --- --- --- --- ---

M77305,IGHV3-33*03,p6             [30] --- --- --- --- --- --- --- --- --- ... --- --- --- --- --- --- --- --- --- ---

M77335,IGHV3-33*04,b30            [30] --- --- --- --- --- --- --- --- --- ... --- --- --- --- --- --- --- --- --- ---

M77334,IGHV3-33*05,b28            [30] --- --- --- --- --- --- --- --- --- ... --- --- --- --- --- --- --- --- --- ---


                                                                 _______________CDR1-IMGT_______________
                                       21  22  23  24  25  26  27  28  29  30  31  32  33  34  35  36  37  38  39  40
                                        L   S   C   A   A   S   G   F   T   F   S   S   Y   G                   M   H
L06618,IGHV3-33*01,hv3019b9            CTC TCC TGT GCA GCG TCT GGA TTC ACC TTC AGT AGC TAT GGC ... ... ... ... ATG CAC

Z12350,IGHV3-33*01,DP-50               --- --- --- --- --- --- --- --- --- --- --- --- --- --- ... ... ... ... --- ---

M77333,IGHV3-33*01,b9-12,33,35         --- --- --- --- --- --- --- --- --- --- --- --- --- --- ... ... ... ... --- ---

X92286,IGHV3-33*01,3d277               --- --- --- --- --- --- --- --- --- --- --- --- --- --- ... ... ... ... --- ---

M99665,IGHV3-33*02,V3-33               --- --- --- --- --- --- --- --- --- --- --- --- --- --- ... ... ... ... --- ---

M77305,IGHV3-33*03,p6                  --- --- --- --- --- --- --- --- --- --- --- --- --- --- ... ... ... ... --- ---

M77335,IGHV3-33*04,b30                 --- --- --- --- --- --- --- --- --- --- --- --- --- --- ... ... ... ... --- ---

M77334,IGHV3-33*05,b28                 --- --- --- --- --- --- --- --- --- --- --- --- --- --- ... ... ... ... --- ---


                                                                                                ___________CDR2-
                                       41  42  43  44  45  46  47  48  49  50  51  52  53  54  55  56  57  58  59  60
                                        W   V   R   Q   A   P   G   K   G   L   E   W   V   A   V   I   W   Y   D   G
L06618,IGHV3-33*01,hv3019b9            TGG GTC CGC CAG GCT CCA GGC AAG GGG CTG GAG TGG GTG GCA GTT ATA TGG TAT GAT GGA

Z12350,IGHV3-33*01,DP-50               --- --- --- --- --- --- --- --- --- --- --- --- --- --- --- --- --- --- --- ---

M77333,IGHV3-33*01,b9-12,33,35         --- --- --- --- --- --- --- --- --- --- --- --- --- --- --- --- --- --- --- ---

X92286,IGHV3-33*01,3d277               --- --- --- --- --- --- --- --- --- --- --- --- --- --- --- --- --- --- --- ---

M99665,IGHV3-33*02,V3-33               --- --- --- --- --- --- --- --- --- --- --- --- --- --- --- --- --- --- --- ---

M77305,IGHV3-33*03,p6                  --- --- --- --- --- --- --- --- --- --- --- --- --- --- --- --- --- --- --- ---

M77335,IGHV3-33*04,b30                 --- --- --- --- --- --- --- --- --- --A --- --- --- --- --- --- --- --- --C ---
                                                                                                    S
M77334,IGHV3-33*05,b28                 --- --- --- --- --- --- --- --- --- --- --- --- --- --- --- --- -CA --- --- ---


                                       IMGT________________
                                       61  62  63  64  65  66  67  68  69  70  71  72  73  74  75  76  77  78  79  80
                                        S   N   K           Y   Y   A   D   S   V   K       G   R   F   T   I   S   R
L06618,IGHV3-33*01,hv3019b9            AGT AAT AAA ... ... TAC TAT GCA GAC TCC GTG AAG ... GGC CGA TTC ACC ATC TCC AGA

Z12350,IGHV3-33*01,DP-50               --- --- --- ... ... --- --- --- --- --- --- --- ... --- --- --- --- --- --- ---
```

```
M77333,IGHV3-33*01,b9-12,33,35   --- --- --- ... ... --- --- --- --- --- --- --- ... --- --- --- --- --- --- ---
X92286,IGHV3-33*01,3d277         --- --- --- ... ... --- --- --- --- --- --- --- ... --- --- --- --- --- --- ---
                                                                         A
M99665,IGHV3-33*02,V3-33         --- --- --- ... ... --- --- --- --- --- -C- --- ... --- --- --- --- --- --- ---
M77305,IGHV3-33*03,p6            --- --- --- ... ... --- --- --- --- --- --- --- ... --- --- --- --- --- --- ---
M77335,IGHV3-33*04,b30           --- --- --- ... ... --- --- --- --- --- --- --- ... --- --- --- --- --- --- ---
M77334,IGHV3-33*05,b28           --- --- --- ... ... --- --- --- --- --- --- --- ... --- --- --- --- --- --- ---

                                  81  82  83  84  85  86  87  88  89  90  91  92  93  94  95  96  97  98  99 100
                                  D   N   S   K   N   T   L   Y   L   Q   M   N   S   L   R   A   E   D   T   A
L06618,IGHV3-33*01,hv3019b9      GAC AAT TCC AAG AAC ACG CTG TAT CTG CAA ATG AAC AGC CTG AGA GCC GAG GAC ACG GCT
Z12350,IGHV3-33*01,DP-50                                                                     --- --- --- --- ---
M77333,IGHV3-33*01,b9-12,33,35   --- --- --- --- --- --- --- --- --- --- --- --- --- --- --- --- --- --- --- ---
X92286,IGHV3-33*01,3d277         --- --- --- --- --- --- --- --- --- --- --- --- --- --- --- --- --- --- --- ---
                                              T               F
M99665,IGHV3-33*02,V3-33         --- --- --- -C- --- --- --- -T- --- --- --- --- --- --- --- --- --- --- --- ---
M77305,IGHV3-33*03,p6            --- --C --- --- --- --- --- --- --- --- --- --- --- --- --- --- --- --- --- ---
M77335,IGHV3-33*04,b30           --- --- --- --- --- --- --- --- --- --- --- --- --- --- --- --- --- --- --- ---
M77334,IGHV3-33*05,b28           --- --- --- --- --- --- --- --- --- --- --- --- --- --- --- --- --- --- --- ---

                                                 _CDR3-IMGT
                                 101 102 103 104 105 106
                                  V   Y   Y   C   A   R
L06618,IGHV3-33*01,hv3019b9      GTG TAT TAC TGT GCG AGA GA
Z12350,IGHV3-33*01,DP-50         --- --- --- --- --- ---
M77333,IGHV3-33*01,b9-12,33,35   --- --- --- --- --- --- --
X92286,IGHV3-33*01,3d277         --- --- --- --- --- ---
M99665,IGHV3-33*02,V3-33         --- --- --- --- --- --- --
                                                      K
M77305,IGHV3-33*03,p6            --- --- --- --- --- -A- --
M77335,IGHV3-33*04,b30           --- --- --- --- --- --- --
M77334,IGHV3-33*05,b28           --- --- --- --- --- --- --
```

Framework and complementarity determining regions

FR1-IMGT: 25 (-1 aa: 10)
FR2-IMGT: 17
FR3-IMGT: 38 (-1 aa: 73)

CDR1-IMGT: 8
CDR2-IMGT: 8
CDR3-IMGT: 2

Collier de Perles for human IGHV3-33*01

Accession number: IMGT L06618 EMBL/GenBank/DDBJ: L06618

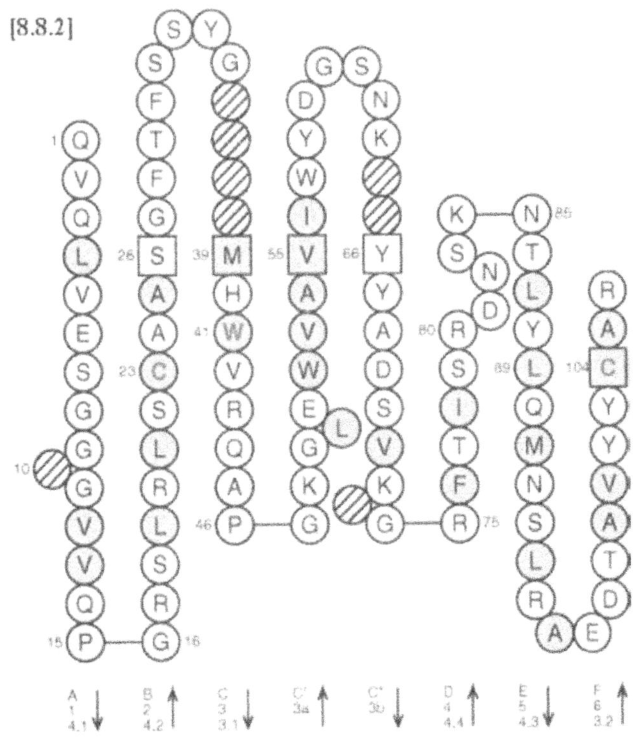

Genome database accession numbers

GDB:9931699 LocusLink: 28434

Nomenclature

IGHV3-35: Immunoglobulin heavy variable 3-35.

Definition and functionality

IGHV3-35 is one of the three–four mapped ORF of the IGHV3 subgroup. This subgroup comprises 47–49 mapped genes (of which 18–21 are functional) in the IGH locus, depending on the haplotypes, three unmapped genes (one functional, one pseudogene and one not defined) and 12 orphons. IGHV3-35 is an ORF due to an unusual V-HEPTAMER sequence: cactgtg instead of cacagtg.

Gene location

IGHV3-35 is in the IGH locus on chromosome 14 at 14q32.33.

Nucleotide and amino acid sequences for human IGHV3-35

```
                                    1   2   3   4   5   6   7   8   9  10  11  12  13  14  15  16  17  18  19  20
                                    E   V   Q   L   V   E   S   G   G       G   L   V   Q   P   G   G   S   L   R
M99666,IGHV3-35*01,V3-35     [26]  GAG GTG CAG CTG GTG GAG TCT GGG GGA ... GGC TTG GTA CAG CCT GGG GGA TCC CTG AGA

Z12359,IGHV3-35*01,DP-59     [41]  --- --- --- --- --- --- --- --- --- ... --- --- --- --- --- --- --- --- --- ---

X92276,IGHV3-35*01,VH19       [3]  --- --- --- --- --- --- --- --- --- ... --- --- --- --- --- --- --- --- --- ---


                                                            _______________CDR1-IMGT_______________
                                   21  22  23  24  25  26  27  28  29  30  31  32  33  34  35  36  37  38  39  40
                                    L   S   C   A   A   S   G   F   T   F   S   N   S   D                   M   N
M99666,IGHV3-35*01,V3-35           CTC TCC TGT GCA GCC TCT GGA TTC ACC TTC AGT AAC AGT GAC ... ... ... ... ATG AAC

Z12359,IGHV3-35*01,DP-59           --- --- --- --- --- --- --- --- --- --- --- --- --- --- ... ... ... ... --- ---

X92276,IGHV3-35*01,VH19            --- --- --- --- --- --- --- --- --- --- --- --- --- --- ... ... ... ... --- ---


                                                                                                ________ CDR2-
                                   41  42  43  44  45  46  47  48  49  50  51  52  53  54  55  56  57  58  59  60
                                    W   V   H   Q   A   P   G   K   G   L   E   W   V   S   G   V   S   W   N   G
M99666,IGHV3-35*01,V3-35           TGG GTC CAT CAG GCT CCA GGA AAG GGG CTG GAG TGG GTA TCG GGT GTT AGT TGG AAT GGC

Z12359,IGHV3-35*01,DP-59           --- --- --- --- --- --- --- --- --- --- --- --- --- --- --- --- --- --- --- ---

X92276,IGHV3-35*01,VH19            --- --- --- --- --- --- --- --- --- --- --- --- --- --- --- --- --- --- --- ---


                                   IMGT_______________
                                   61  62  63  64  65  66  67  68  69  70  71  72  73  74  75  76  77  78  79  80
                                    S   R   T           H   Y   A   D   S   V   K       G   R   F   T   I   S   R
M99666,IGHV3-35*01,V3-35           AGT AGG ACG ... ... CAC TAT GCA GAC TCT GTG AAG ... GGC CGA TTC ATC ATC TCC AGA

Z12359,IGHV3-35*01,DP-59           --- --- --- ... ... --- --- --- --- --- --- --- ... --- --- --- --- --- --- ---

X92276,IGHV3-35*01,VH19            --- --- --- ... ... --- --- --- --- --- --- --- ... --- --- --- --- --- --- ---


                                   81  82  83  84  85  86  87  88  89  90  91  92  93  94  95  96  97  98  99 100
                                    D   N   S   R   N   T   L   Y   L   Q   T   N   S   L   R   A   E   D   T   A
M99666,IGHV3-35*01,V3-35           GAC AAT TCC AGG AAC ACC CTG TAT CTG CAA ACG AAT AGC CTG AGG GCC GAG GAC ACG GCT

Z12359,IGHV3-35*01,DP-59           --- --- --- --- --- --- --- --- --- --- --- --- --- --- --- --- --- --- --- ---

X92276,IGHV3-35*01,VH19            --- --- --- --- --- --- --- --- --- --- --- --- --- --- --- --- --- --- --- ---


                                                     _CDR3 IMGT
                                   101 102 103 104 105 106
                                    V   Y   Y   C   V   R
M99666,IGHV3-35*01,V3-35           GTG TAT TAC TGT GTG AGA AA

Z12359,IGHV3-35*01,DP-59

X92276,IGHV3-35*01,VH19            --- --- --- --- --- --- --
```

Framework and complementarity determining regions

FR1-IMGT: 25 (-1 aa: 10)
FR2-IMGT: 17
FR3-IMGT: 38 (-1 aa: 73)

CDR1-IMGT: 8
CDR2-IMGT: 8
CDR3-IMGT: 2

Collier de Perles for human IGHV3-35*01

Accession number: IMGT M99666

EMBL/GenBank/DDBJ: M99666

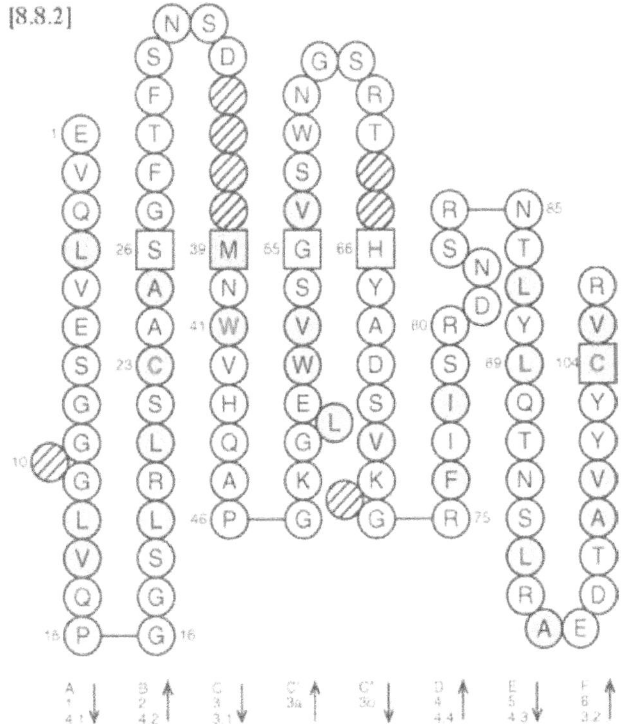

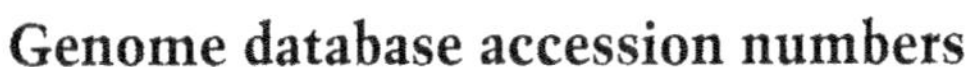

Genome database accession numbers

GDB:9931701

LocusLink: 28432

Nomenclature

IGHV3-38: Immunoglobulin heavy variable 3-38.

Definition and functionality

IGHV3-38 is one of the three–four mapped ORF of the IGHV3 subgroup. This subgroup comprises 47–49 mapped genes (of which 18–21 are functional) in the IGH locus, depending on the haplotypes, three unmapped genes (one functional, one pseudogene and one not defined) and 12 orphons. IGHV3-38 is an ORF due to an unusual V-HEPTAMER sequence: cacagag instead of cacagtg.

Gene location

IGHV3-38 is in the IGH locus on chromosome 14 at 14q32.33.

Nucleotide and amino acid sequences for human IGHV3-38

```
                                          1   2   3   4   5   6   7   8   9   10  11  12  13  14  15  16  17  18  19  20
                                          E   V   Q   L   V   E   S   G   G       G   L   V   Q   P   R   G   S   L   R
M99669,IGHV3-38*01,V3-38P          [26]  GAG GTG CAG CTG GTG GAG TCT GGG GGA ... GGC TTG GTA CAG CCT AGG GGG TCC CTG AGA

Z27447,IGHV3-38*02,COS-23          [42]  -   --- --- --- --- --- --- --- --- ... --- --- --- --- --- --- --- --- --- ---

                                                                  _______________CDR1-IMGT_______________
                                          21  22  23  24  25  26  27  28  29  30  31  32  33  34  35  36  37  38  39  40
                                          L   S   C   A   A   S   G   F   T   V   S   S   N   E                   M   S
M99669,IGHV3-38*01,V3-38P                CTC TCC TGT GCA GCC TCT GGA TTC ACC GTC AGT AGC AAT GAG ... ... ... ... ATG AGC

Z27447,IGHV3-38*02,COS-23                --- --- --- --- ---  -- --- --- --- --- --- --- --- --- ... ... ... ... --- ---

                                                                                                  _______________CDR2-
                                          41  42  43  44  45  46  47  48  49  50  51  52  53  54  55  56  57  58  59  60
                                          W   I   R   Q   A   P   G   K   G   L   E   W   V   S   S   I   S   G   G   S
M99669,IGHV3-38*01,V3-38P                TGG ATC CGC CAG GCT CCA GGG AAG GGG CTG GAG TGG GTC TCA TCC ATT AGT GGT GGT AGC

Z27447,IGHV3-38*02,COS-23                --- --- --- --- --- --- --- --- --- --- --- --- --- --- --- --- --- --- --- ---

                                         IMGT_______________
                                          61  62  63  64  65  66  67  68  69  70  71  72  73  74  75  76  77  78  79  80
                                          T                   Y   Y   A   D   S   R   K       G   R   F   T   I   S   R
M99669,IGHV3-38*01,V3-38P                ACA ... ... ... ... TAC TAC GCA GAC TCC AGG AAG ... GGC AGA TTC ACC ATC TCC AGA

Z27447,IGHV3-38*02,COS-23                --- ... ... ... ... --- --- --- --- --- --- --- ... --- --- --- --- --- - . .

                                          81  82  83  84  85  86  87  88  89  90  91  92  93  94  95  96  97  98  99 100
                                          D   N   S   K   N   T   L   Y   L   Q   M   N   N   L   R   A   E   G   T   A
M99669,IGHV3-38*01,V3-38P                GAC AAT TCC AAG AAC ACG CTG TAT CTT CAA ATG AAC AAC CTG AGA GCT GAG GGC ACG GCC

Z27447,IGHV3-38*02,COS-23                --- --- --- --- --- --- --- --- --  --- --- --- --- ---   . --- --- --- --- ---

                                                             _CDR3-IMGT
                                         101 102 103 104 105 106 107
                                          A   Y   Y   C   A   R   Y
M99669,IGHV3-38*01,V3-38P                GCG TAT TAC TGT GCC AGA TAT A
                                          V
Z27447,IGHV3-38*02,COS-23                -T- --- --- --- --- ---
```

Framework and complementarity determining regions

FR1-IMGT: 25 (-1 aa: 10)
FR2-IMGT: 17
FR3-IMGT: 38 (-1 aa: 73)

CDR1-IMGT: 8
CDR2-IMGT: 6
CDR3-IMGT: 3

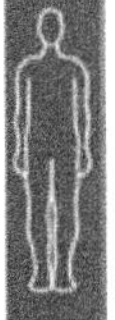

Collier de Perles for human IGHV3-38*01

Accession number: IMGT M99669 EMBL/GenBank/DDBJ: M99669

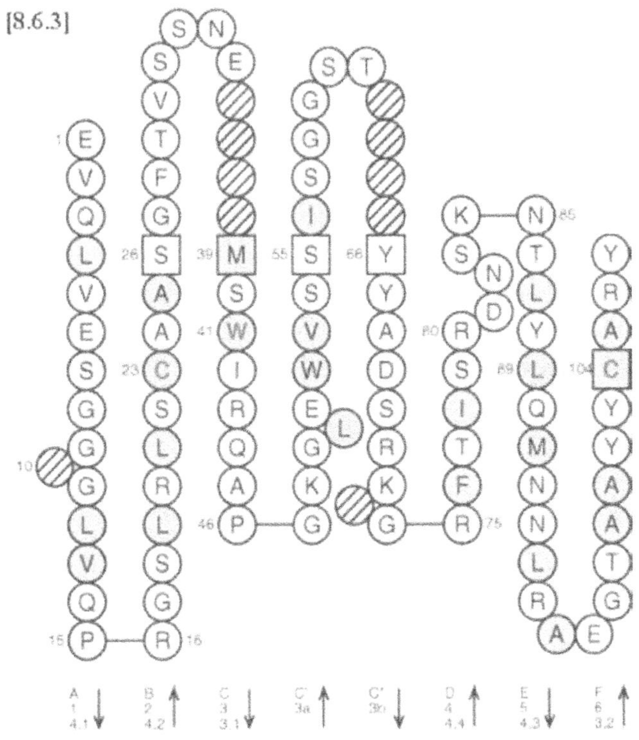

Genome database accession numbers

GDB:9931704 LocusLink: 28429

Nomenclature

IGHV3-43: Immunoglobulin heavy variable 3-43.

Definition and functionality

IGHV3-43 is one of the 18–21 mapped functional genes of the IGHV3 subgroup. This subgroup comprises 47–49 mapped genes in the IGH locus, depending on the haplotypes, three unmapped genes (one functional, one pseudogene and one not defined) and 12 orphons.

Gene location

IGHV3-43 is in the IGH locus on chromosome 14 at 14q32.33.

Nucleotide and amino acid sequences for human IGHV3-43

```
                                     1   2   3   4   5   6   7   8   9  10  11  12  13  14  15  16  17  18  19  20
                                     E   V   Q   L   V   E   S   G   G       V   V   V   Q   P   G   G   S   L   R
M99672,IGHV3-43*01,V3-43      [26]  GAA GTG CAG CTG GTG GAG TCT GGG GGA ... GTC GTG GTA CAG CCT GGG GGG TCC CTG AGA

Z12335,IGHV3-43*01,DP-33      [41]  --- --- --- --- --- --- --- --- --- ... --- --- --- --- --- --- --- --- --- ---
                                                                             G
Z18901,IGHV3-43*02,COS-16     [42]  --- --- --- --- --- --- --- --- --- ... -G- --- --- --- --- --- --- --- --- ---

                                                         ____________ _______CDR1-IMGT____________________
                                    21  22  23  24  25  26  27  28  29  30  31  32  33  34  35  36  37  38  39  40
                                     L   S   C   A   A   S   G   F   T   F   D   D   Y   T                   M   H
M99672,IGHV3-43*01,V3-43            CTC TCC TGT GCA GCC TCT GGA TTC ACC TTT GAT GAT TAT ACC ... ... ... ... ATG CAC

Z12335,IGHV3-43*01,DP-33            --- --- --- --- --- --- --- --- --- --- --- --- --- --- ... ... ... ... --- ---
                                                                                         A
Z18901,IGHV3-43*02,COS-16           --- --- --- --- --- --- --- --- --- --- --- --- --- G-- ... ... ... ... --- ---

                                                                                             ____________ CDR2-
                                    41  42  43  44  45  46  47  48  49  50  51  52  53  54  55  56  57  58  59  60
                                     W   V   R   Q   A   P   G   K   G   L   E   W   V   S   L   I   S   W   D   G
M99672,IGHV3-43*01,V3-43            TGG GTC CGT CAA GCT CCG GGG AAG GGT CTG GAG TGG GTC TCT CTT ATT AGT TGG GAT GGT

Z12335,IGHV3-43*01,DP-33            --- --- --- --- --- --- --- --- --- --- --- --- --- --- --- --- --- --- --- ---
                                                                                                     G
Z18901,IGHV3-43*02,COS-16           --- --- --- --- --- --A --- --- --- --- --- --- --- --- --- --- --- G-- --- ---

                                    IMGT_______ _______
                                    61  62  63  64  65  66  67  68  69  70  71  72  73  74  75  76  77  78  79  80
                                     G   S   T           Y   Y   A   D   S   V   K       G   R   F   T   I   S   R
M99672,IGHV3-43*01,V3-43            GGT AGC ACA ... ... TAC TAT GCA GAC TCT GTG AAG ... GGC CGA TTC ACC ATC TCC AGA

Z12335,IGHV3-43*01,DP-33            --- --- --- ... ... --- --- --- --- --- --- --- ... --- --- --- --- --- --- ---

Z18901,IGHV3-43*02,COS-16           --- --- --- ... ... --- --- --- --- --- --- --- ... --- --- --- --- --- --- ---

                                    81  82  83  84  85  86  87  88  89  90  91  92  93  94  95  96  97  98  99 100
                                     D   N   S   K   N   S   L   Y   L   Q   M   N   S   L   R   T   E   D   T   A
M99672,IGHV3-43*01,V3-43            GAC AAC AGC AAA AAC TCC CTG TAT CTG CAA ATG AAC AGT CTG AGA ACT GAG GAC ACC GCC

Z12335,IGHV3-43*01,DP-33            --- --- --- --- --- --- --- --- --- --- --- --- --- --- --- --- --- --- --- ---

Z18901,IGHV3-43*02,COS-16           --- --- --- --- --- --- --- --- --- --- --- --- --- --- --- --- --- --- --- ---

                                                    __CDR3-IMGT__
                                    101 102 103 104 105 106 107
                                     L   Y   Y   C   A   K   D
M99672,IGHV3-43*01,V3-43            TTG TAT TAC TGT GCA AAA GAT A

Z12335,IGHV3-43*01,DP-33            --- --- --- --- --- ...

Z18901,IGHV3-43*02,COS-16           --- --- --- --- --- ---
```

Framework and complementarity determining regions

FR1-IMGT: 25 (-1 aa: 10)

FR2-IMGT: 17

FR3-IMGT: 38 (-1 aa: 73)

CDR1-IMGT: 8

CDR2-IMGT: 8

CDR3-IMGT: 3

Collier de Perles for human IGHV3-43*01

Accession number: IMGT M99672 EMBL/GenBank/DDBJ: M99672

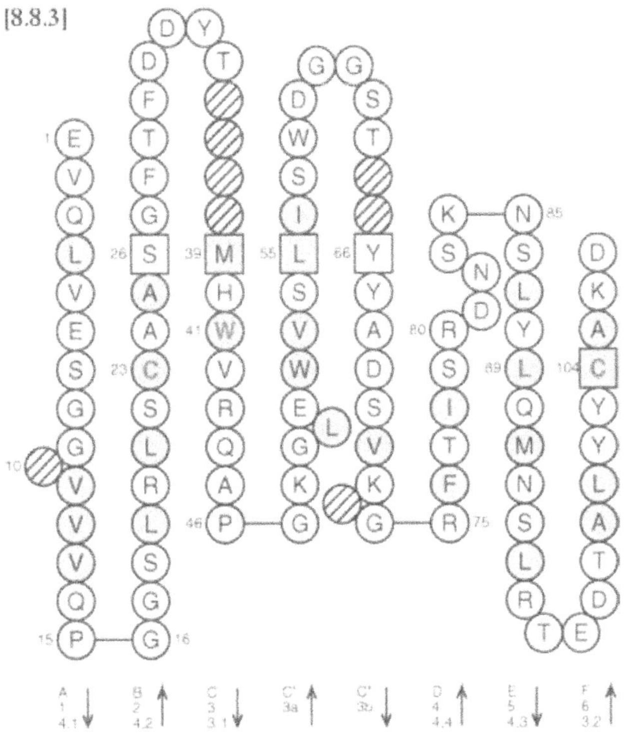

Genome database accession numbers

GDB:9931707 LocusLink: 28426

IGHV3-47

Nomenclature

IGHV3-47: Immunoglobulin heavy variable 3-47.

Definition and functionality

IGHV3-47 is a mapped ORF (alleles *01 and *02) or a pseudogene (allele *03). IGHV3-47 belongs to the IGHV3 subgroup which comprises 47–49 mapped genes (of which 18–21 are functional) in the IGH locus, depending on the haplotypes, three unmapped genes (one functional, one pseudogene, and one not defined) and 12 orphons.

IGHV3-47*01 and IGHV3-47*02 are considered as ORF because L-PART1, L-PART2, V-HEPTAMER and V-NONAMER have not yet been sequenced and there is so far no known rearranged transcript, suggesting some defect outside the V-REGION. IGHV3-47*03 is a pseudogene due to a 1 nt DELETION in codon 89 leading to a frameshift in FR3-IMGT.

Gene location

IGHV3-47 is in the IGH locus on chromosome 14 at 14q32.33.

Nucleotide and amino acid sequences for human IGHV3-47

```
                                      1   2   3   4   5   6   7   8   9  10  11  12  13  14  15  16  17  18  19  20
                                      E   D   Q   L   V   E   S   G   G       G   L   V   Q   P   G   G   S   L   R
Z18900,IGHV3-47*01,COS15        [42] GAG GAT CAG CTG CTG GAG TCT GGG GGA ... GGC TTG GTA CAG CCT GGG GGG TCC CTG CGA
Z12352,IGHV3-47*02,DP-52        [41] --- --- --- --- --- --- --- --- --- ... --- --- --- --- --- --- --- --- --- A--
M99674,IGHV3-47*03,V3-47P       [26] --- --- --- --- --- --- --- --- --- ... --- --- --- --- --- --- --- --- --- A--

                                                              ______________ CDR1-IMGT ___________________
                                     21  22  23  24  25  26  27  28  29  30  31  32  33  34  35  36  37  38  39  40
                                      P   S   C   A   A   S   G   F   A   F   S   S   Y   A                   L   H
Z18900,IGHV3-47*01,COS15             CCC TCC TGT GCA GCC TCT GGA TTC GCC TTC AGT AGC TAT GCT ... ... ... ... CTG CAC
                                                                                      V
Z12352,IGHV3-47*02,DP-52             --- --- --- --- --- --- --- --- --- --- --- --- --- -T- ... ... ... ... --- ---
                                                                                      V
M99674,IGHV3-47*03,V3-47P            --- --- --- --- --- --- --- --- --- --- --- --- --- -T- ... ... ... ... --- ---

                                                                                             ______________ CDR2-
                                     41  42  43  44  45  46  47  48  49  50  51  52  53  54  55  56  57  58  59  60
                                      W   V   R   R   A   P   G   K   G   L   E   W   V   S   A   I   G   T   G   G
Z18900,IGHV3-47*01,COS15             TGG GTT CGC CGG GCT CCA GGG AAG GGT CTG GAG TGG GTA TCA GCT ATT GGT ACT GGT GGT
                                                                          P
Z12352,IGHV3-47*02,DP-52             --- --- --- --- --- --- --- --- --- -C- --- --- --- --- --- --- --- --- --- ---
                                                                          P
M99674,IGHV3-47*03,V3-47P            --- --- --- --- --- --- --- --- --- -C- --- --- --- --- --- --- --- --- --- ---

                                     IMGT________________
                                     61  62  63  64  65  66  67  68  69  70  71  72  73  74  75  76  77  78  79  80
                                      D   T               Y   Y   A   D   S   V   M       G   R   F   T   I   S   R
Z18900,IGHV3-47*01,COS15             GAT ACA ... ... ... TAC TAT GCA GAC TCC GTG ATG ... GGC CGA TTC ACC ATC TCC AGA
Z12352,IGHV3-47*02,DP-52             --- --- ... ... ... --- --- --- --- --- --- --- ... --- --- --- --- --- --- ---
M99674,IGHV3-47*03,V3-47P            --- --- ... ... ... --- --- --- --- --- --- --- ... --- --- --- --- --- --- ---

                                     81  82  83  84  85  86  87  88  89  90  91  92  93  94  95  96  97  98  99 100
                                      D   N   A   K   K   S   L   Y   L   H   M   N   S   L   I   A   E   D   M   A
Z18900,IGHV3-47*01,COS15             GAC AAC GCC AAG AAG TCC TTG TAT CTT CAT ATG AAC AGC CTG ATA GCT GAG GAC ATG GCT
                                                                          Q
Z12352,IGHV3-47*02,DP-52             --- --- --- --- --- --- --- --- --- --A --- --- --- --- --- --- --- --- --- ---
                                                                      #   Q
M99674,IGHV3-47*03,V3-47P            --- --- --- --- --- --- --- --- --. --A --- --- --- --- --- --- --- --- --- ---

                                                         _CDR3-IMGT
                                     101 102 103 104 105 106
                                      V   Y   Y   C   A   R
Z18900,IGHV3-47*01,COS15             GTG TAT TAT TGT GCA AGA
Z12352,IGHV3-47*02,DP-52             --- --- --- --- --- ---
M99674,IGHV3-47*03,V3-47P            --- --- --- G
```

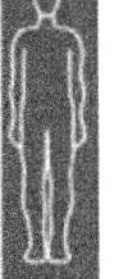

Framework and complementarity determining regions

FR1-IMGT: 25 (-1 aa: 10)
FR2-IMGT: 17
FR3-IMGT: 38 (-1 aa: 73)

CDR1-IMGT: 8
CDR2-IMGT: 7
CDR3-IMGT: 2

Collier de Perles for human IGHV3-47*01

Accession number: IMGT Z18900

EMBL/GenBank/DDBJ: Z18900

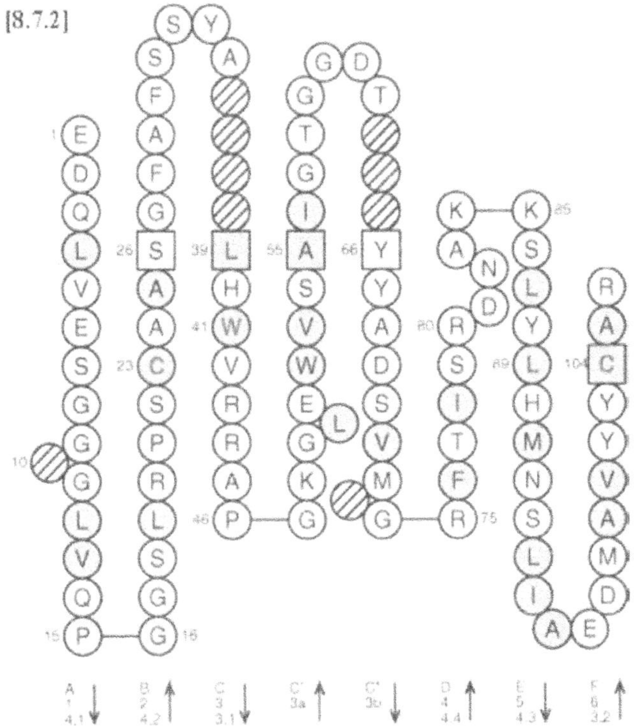

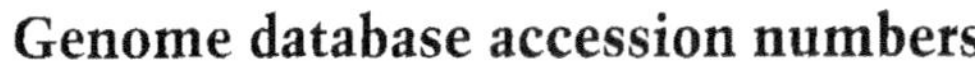

Genome database accession numbers

GDB:9931708

LocusLink: 28425

Nomenclature

IGHV3-48: Immunoglobulin heavy variable 3-48.

Definition and functionality

IGHV3-48 is one of the 18–21 mapped functional genes of the IGHV3 subgroup. This subgroup comprises 47–49 mapped genes in the IGH locus, depending on the haplotypes, three unmapped genes (one functional, one pseudogene and one not defined) and 12 orphons.

Gene location

IGHV3-48 is in the IGH locus on chromosome 14 at 14q32.33.

Nucleotide and amino acid sequences for human IGHV3-48

```
                                          1   2   3   4   5   6   7   8   9  10  11  12  13  14  15  16  17  18  19  20
                                          E   V   Q   L   V   E   S   G   G       G   L   V   Q   P   G   G   S   L   R
M99675,IGHV3-48*01,V3-48           [26]  GAG GTG CAG CTG GTG GAG TCT GGG GGA ... GGC TTG GTA CAG CCT GGG GGG TCC CTG AGA
X92299,IGHV3-48*01,hv3d1           [15]  --- --- --- --- --- --- --- --- --- ... --- --- --- --- --- --- --- --- --- ---
X62130,IGHV3-48*01,WHG26           [23]  --- --- --- --- --- --- --- --- --- ... --- --- --- --- --- --- --- --- --- ---
Z12351,IGHV3-48*02,DP-51           [41]  --- --- --- --- --- --- --- --- --- ... --- --- --- --- --- --- --- --- --- ---
Z12358,IGHV3-48*03,DP-58           [41]  --- --- --- --- --- --- --- --- --- ... --- --- --- --- --- --A --- --- --- ---
U03893,IGHV3-48*03,hv3dEG          [15]  --- --- --- --- --- --- --- --- --- ... --- --- --- --- --- --A --- --- --- ---

                                                                  _______________CDR1-IMGT_______________
                                         21  22  23  24  25  26  27  28  29  30  31  32  33  34  35  36  37  38  39  40
                                          L   S   C   A   A   S   G   F   T   F   S   S   Y   S                   M   N
M99675,IGHV3-48*01,V3-48                 CTC TCC TGT GCA GCC TCT GGA TTC ACC TTC AGT AGC TAT AGC ... ... ... ... ATG AAC
X92299,IGHV3-48*01,hv3d1                 --- --- --- --- --- --- --- --- --- --- --- --- --- --- ... ... ... ... --- ---
X62130,IGHV3-48*01,WHG26                 --- --- --- --- --- --- --- --- --- --- --- --- --- --- ... ... ... ... --- ---
Z12351,IGHV3-48*02,DP-51                 --- --- --- --- --- --- --- --- --- --- --- --- --- --- ... ... ... ... --- ---
                                                                                          E
Z12358,IGHV3-48*03,DP-58                 --- --- --- --- --- --- --- --- --- --- --- --T --- GAA ... ... ... ... --- ---
                                                                                          E
U03893,IGHV3-48*03,hv3dEG                --- --- --- --- --- --- --- --- --- --- --- --T --- GAA ... ... ... ... --- ---

                                                                                             ____________CDR2-
                                         41  42  43  44  45  46  47  48  49  50  51  52  53  54  55  56  57  58  59  60
                                          W   V   R   Q   A   P   G   K   G   L   E   W   V   S   Y   I   S   S   S   S
M99675,IGHV3-48*01,V3-48                 TGG GTC CGC CAG GCT CCA GGG AAG GGG CTG GAG TGG GTT TCA TAC ATT AGT AGT AGT AGT
X92299,IGHV3-48*01,hv3d1                 --- --- --- --- --- --- --- --- --- --- --- --- --- --- --- --- --- --- --- ---
X62130,IGHV3-48*01,WHG26                 --- --- --- --- --- --- --- --- --- --- --- --- --- --- --- --- --- --- --- ---
Z12351,IGHV3-48*02,DP-51                 --- --- --- --- --- --- --- --- --- --- --- --- --- --- --- --- --- --- --- ---
                                                                                                              G
Z12358,IGHV3-48*03,DP-58                 --- --- --- --- --- --- --- --- --- --- --- --- --- --- --- --- --- --- --- G--
                                                                                                              G
U03893,IGHV3-48*03,hv3dEG                --- --- --- --- --- --- --- --- --- --- --- --- --- --- --- --- --- --- --- G--

                                         IMGT_______________
                                         61  62  63  64  65  66  67  68  69  70  71  72  73  74  75  76  77  78  79  80
                                          S   T   I           Y   Y   A   D   S   V   K       G   R   F   T   I   S   R
M99675,IGHV3-48*01,V3-48                 AGT ACC ATA ... ... TAC TAC GCA GAC TCT GTG AAG ... GGC CGA TTC ACC ATC TCC AGA
X92299,IGHV3-48*01,hv3d1                 --- --- --- ... ... --- --- --- --- --- --- --- ... --- --- --- --- --- --- ---
X62130,IGHV3-48*01,WHG26                 --- --- --- ... ... --- --- --- --- --- --- --- ... --- --- --- --- --- --- ---
Z12351,IGHV3-48*02,DP-51                 --- --- --- ... ... --- --- --- --- --- --- --- ... --- --- --- --- --- --- ---
Z12358,IGHV3-48*03,DP-58                 --- --- --- ... ... --- --- --- --- --- --- --- ... --- --- --- --- --- --- ---
U03893,IGHV3-48*03,hv3dEG                --- --- --- ... ... --- --- --- --- --- --- --- ... --- --- --- --- --- --- ---

                                         81  82  83  84  85  86  87  88  89  90  91  92  93  94  95  96  97  98  99 100
                                          D   N   A   K   N   S   L   Y   L   Q   M   N   S   L   R   A   E   D   T   A
M99675,IGHV3-48*01,V3-48                 GAC AAT GCC AAG AAC TCA CTG TAT CTG CAA ATG AAC AGC CTG AGA GCC GAG GAC ACG GCT
X92299,IGHV3-48*01,hv3d1                 --- --- --- --- --- --- --- --- --- --- --- --- --- --- --- --- --- --- --- ---
X62130,IGHV3-48*01,WHG26                 --- --- --- --- --- --- --- --- --- --- --- --- --- --- --- --- --- --- --- ---
                                                                                                  D
Z12351,IGHV3-48*02,DP-51                 --- --- --- --- --- --- --- --- --- --- --- --- --- --- --- -A- --- --- --- ---
```

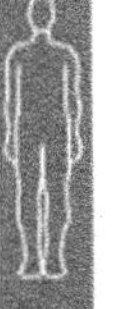

```
Z12358,IGHV3-48*03,DP-58      --- --C --- --- --- --- --- --- --- --- --- --- --- --- --- --- --- ---
U03893,IGHV3-48*03,hv3dEG     --- --C --- --- --- --- --- --- --- --- --- --- --- --- --- --- --- ---

                                                  _CDR3-IMGT
                              101 102 103 104 105 106
                               V   Y   Y   C   A   R
M99675,IGHV3-48*01,V3-48      GTG TAT TAC TGT GCG AGA GA

X92299,IGHV3-48*01,hv3d1      --- --- --- --- --- --- --

X62130,IGHV3-48*01,WHG26      --- --- --- --- --- ---

Z12351,IGHV3-48*02,DP-51      --- --- --- --- --- ---

Z12358,IGHV3-48*03,DP-58      --T --- --- --- --- ---

U03893,IGHV3-48*03,hv3dEG     --T --- --- --- --- --- --
```

Framework and complementarity determining regions

FR1-IMGT: 25 (-1 aa: 10)
FR2-IMGT: 17
FR3-IMGT: 38 (-1 aa: 73)

CDR1-IMGT: 8
CDR2-IMGT: 8
CDR3-IMGT: 2

Collier de Perles for human IGHV3-48*01

Accession number: IMGT M99675 EMBL/GenBank/DDBJ: M99675

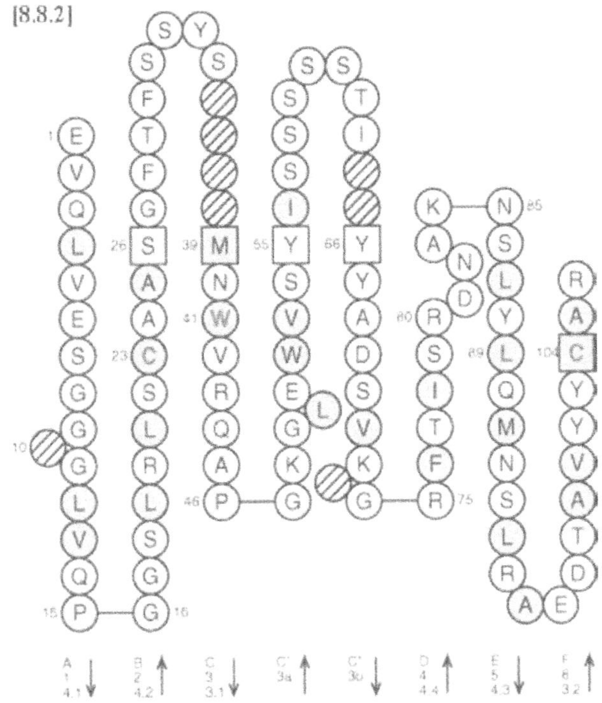

Genome database accession numbers

GDB:9931709 LocusLink: 28424

Nomenclature

IGHV3-49: Immunoglobulin heavy variable 3-49.

Definition and functionality

IGHV3-49 is one of the 18–21 mapped functional genes of the IGHV3 subgroup. This subgroup comprises 47–49 mapped genes in the IGH locus, depending on the haplotypes, three unmapped genes (one functional, one pseudogene and one not defined) and 12 orphons.

Gene location

IGHV3-49 is in the IGH locus on chromosome 14 at 14q32.33.

Nucleotide and amino acid sequences for human IGHV3-49

```
                                       1   2   3   4   5   6   7   8   9  10  11  12  13  14  15  16  17  18  19  20
                                       E   V   Q   L   V   E   S   G   G       G   L   V   Q   P   G   R   S   L   R
M99676,IGHV3-49*01,V3-49         [26] GAG GTG CAG CTG GTG GAG TCT GGG GGA ... GGC TTG GTA CAG CCA GGG CGG TCC CTG AGA
                                                                                                     P
M99401,IGHV3-49*02,LSG12.1        [1] --- --- --- --- --- --- --- --- --- ... --- --- --- --- --- --- -C- --- --- ---

X87090,IGHV3-49*03,3-49RB         [5]

                                                           ______________________CDR1-IMGT____________________
                                      21  22  23  24  25  26  27  28  29  30  31  32  33  34  35  36  37  38  39  40
                                       L   S   C   T   A   S   G   F   T   F   G   D   Y   A                   M   S
M99676,IGHV3-49*01,V3-49              CTC TCC TGT ACA GCT TCT GGA TTC ACC TTT GGT GAT TAT GCT ... ... ... ... ATG AGC
                                                                                   Y       P
M99401,IGHV3-49*02,LSG12.1            --- --- --- --- --- --- --- --- --- --- --G T-- --- C-- ... ... ... ... --- ---

X87090,IGHV3-49*03,3-49RB             --- --- --- --- --- --- --- --- --- --- --- --- --- --- ... ... ... ... --- ---

                                                                                                 _____________CDR2-
                                      41  42  43  44  45  46  47  48  49  50  51  52  53  54  55  56  57  58  59  60
                                       W   F   R   Q   A   P   G   K   G   L   E   W   V   G   F   I   R   S   K   A
M99676,IGHV3-49*01,V3-49              TGG TTC CGC CAG GCT CCA GGG AAG GGG CTG GAG TGG GTA GGT TTC ATT AGA AGC AAA GCT
                                           V
M99401,IGHV3-49*02,LSG12.1            --- G-- --- --- --- --- --- --- --- --- --- --- --- --- --- --- --- --- --- ---

X87090,IGHV3-49*03,3-49RB             --- --- --- --- --- --- --- --- --- --- --- --- --- --- --- --- --- --- --- ---

                                      IMGT_______________
                                      61  62  63  64  65  66  67  68  69  70  71  72  73  74  75  76  77  78  79  80
                                       Y   G   G   T   T   E   Y   T   A   S   V   K       G   R   F   T   I   S   R
M99676,IGHV3-49*01,V3-49              TAT GGT GGG ACA ACA GAA TAC ACC GCG TCT GTG AAA ... GGC AGA TTC ACC ATC TCA AGA
                                                                   A
M99401,IGHV3-49*02,LSG12.1            --- --- --- --- --- --- --- G-- --- --- --- --- ... --- --- --- --- --- --- ---
                                                                   A
X87090,IGHV3-49*03,3-49RB             --- --- --- --- --- --- --- G-- --- --- --- --- ... --- --- --- --- --- --- ---

                                      81  82  83  84  85  86  87  88  89  90  91  92  93  94  95  96  97  98  99 100
                                       D   G   S   K   S   I   A   Y   L   Q   M   N   S   L   K   T   E   D   T   A
M99676,IGHV3-49*01,V3-49              GAT GGT TCC AAA AGC ATC GCC TAT CTG CAA ATG AAC AGC CTG AAA ACC GAG GAC ACA GCC
                                           D
M99401,IGHV3-49*02,LSG12.1            --- -A- --- --- --- --- --- --- --- --- --- --- --- --- --- --- --- --- --- ---
                                           D
X87090,IGHV3-49*03,3-49RB             --- -A- --- --- --- --- -X- --- --- --- --- --- --- --- --- --- --- --- --- ---

                                                          _CDR3-IMGT
                                      101 102 103 104 105 106
                                       V   Y   Y   C   T   R
M99676,IGHV3-49*01,V3-49              GTG TAT TAC TGT ACT AGA GA

M99401,IGHV3-49*02,LSG12.1            --- --- --- --- --- --- --

X87090,IGHV3-49*03,3-49RB             --- --- --- --- X-- --
```

Framework and complementarity determining regions

FR1-IMGT: 25 (-1 aa: 10)
FR2-IMGT: 17
FR3-IMGT: 38 (-1 aa: 73)

CDR1-IMGT: 8
CDR2-IMGT: 10
CDR3-IMGT: 2

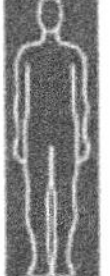

Collier de Perles for human IGHV3-49*01

Accession number: IMGT M99676 EMBL/GenBank/DDBJ: M99676

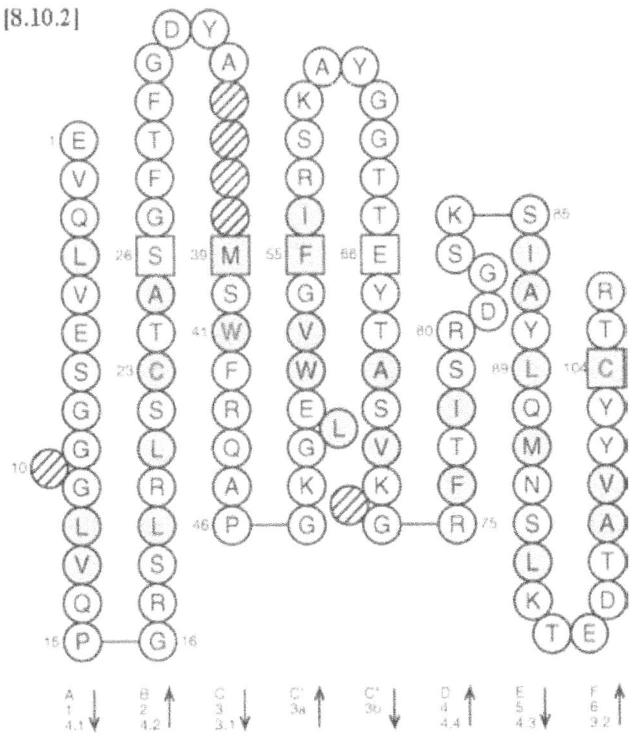

Genome database accession numbers

GDB:9931710 LocusLink: 28423

Nomenclature

IGHV3-53: Immunoglobulin heavy variable 3-53.

Definition and functionality

IGHV3-53 is one of the 18–21 mapped functional genes of the IGHV3 subgroup. This subgroup comprises 47–49 mapped genes in the IGH locus, depending on the haplotypes, three unmapped genes (one functional, one pseudogene and one not defined) and 12 orphons.

Gene location

IGHV3-53 is in the IGH locus on chromosome 14 at 14q32.33.

Nucleotide and amino acid sequences for human IGHV3-53

```
                                     1   2   3   4   5   6   7   8   9  10  11  12  13  14  15  16  17  18  19  20
                                     E   V   Q   L   V   E   S   G   G       G   L   I   Q   P   G   G   S   L   R
M99679,IGHV3-53*01,V3-53      (26)  GAG GTG CAG CTG GTG GAG TCT GGA GGA ... GGC TTG ATC CAG CCT GGG GGG TCC CTG AGA
                                                             T
Z12342,IGHV3-53*02,DP-42      (41)  --- --- --- --- --- --- A-- --- --- ... --- --- --- --- --- --- --- --- --- ---

                                                            ____________________CDR1-IMGT______________________
                                    21  22  23  24  25  26  27  28  29  30  31  32  33  34  35  36  37  38  39  40
                                     L   S   C   A   A   S   G   F   T   V   S   S   N   Y                   M   S
M99679,IGHV3-53*01,V3-53            CTC TCC TGT GCA GCC TCT GGG TTC ACC GTC AGT AGC AAC TAC ... ... ... ... ATG AGC

Z12342,IGHV3-53*02,DP-42            --- --- --- --- --- --- --- --- --- --- --- --- --- --- ... ... ... ... --- ---

                                                                                                _____________ CDR2-
                                    41  42  43  44  45  46  47  48  49  50  51  52  53  54  55  56  57  58  59  60
                                     W   V   R   Q   A   P   G   K   G   L   E   W   V   S   V   I   Y   S   G   G
M99679,IGHV3-53*01,V3-53            TGG GTC CGC CAG GCT CCA GGG AAG GGG CTG GAG TGG GTC TCA GTT ATT TAT AGC GGT GGT

Z12342,IGHV3-53*02,DP-42            --- --- --- --- --- --- --- --- --- --- --- --- --- --- --- --- --- --- --- ---

                                    IMGT_______________
                                    61  62  63  64  65  66  67  68  69  70  71  72  73  74  75  76  77  78  79  80
                                     S   T               Y   Y   A   D   S   V   K       G   R   F   T   I   S   R
M99679,IGHV3-53*01,V3-53            AGC ACA ... ... ... TAC TAC GCA GAC TCC GTG AAG ... GGC CGA TTC ACC ATC TCC AGA

Z12342,IGHV3-53*02,DP-42            --- --- ... ... ... --- --- --- --- --- --- --- ... --- --- --- --- --- --- ---

                                    81  82  83  84  85  86  87  88  89  90  91  92  93  94  95  96  97  98  99 100
                                     D   N   S   K   N   T   L   Y   L   Q   M   N   S   L   R   A   E   D   T   A
M99679,IGHV3-53*01,V3-53            GAC AAT TCC AAG AAC ACG CTG TAT CTT CAA ATG AAC AGC CTG AGA GCC GAG GAC ACG GCC

Z12342,IGHV3-53*02,DP-42            --- --- --- --- --- --- --- --- --- --- --- --- --- --- --- --- --- --- --- ---

                                                       _CDR3-IMGT
                                    101 102 103 104 105 106
                                     V   Y   Y   C   A   R
M99679,IGHV3-53*01,V3-53            GTG TAT TAC TGT GCG AGA GA

Z12342,IGHV3-53*02,DP-42            --- --- --- --- --- ---
```

Framework and complementarity determining regions

FR1-IMGT: 25 (-1 aa: 10)	CDR1-IMGT: 8
FR2-IMGT: 17	CDR2-IMGT: 7
FR3-IMGT: 38 (-1 aa: 73)	CDR3-IMGT: 2

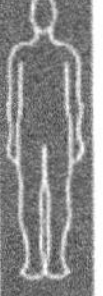

Collier de Perles for human IGHV3-53*01

Accession number: IMGT M99679 EMBL/GenBank/DDBJ: M99679

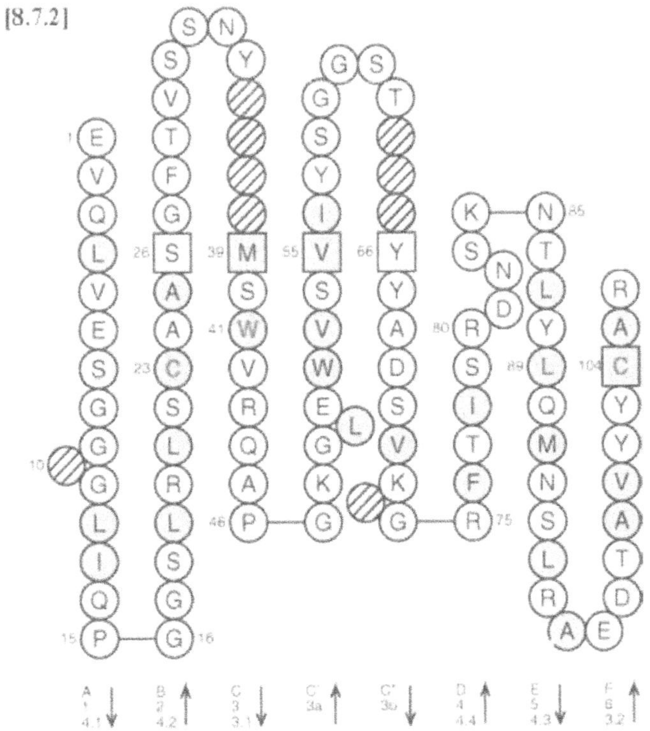

Genome database accession numbers

GDB:9931713 LocusLink: 28420

Nomenclature

IGHV3-64: Immunoglobulin heavy variable 3-64.

Definition and functionality

IGHV3-64 is one of the 18–21 mapped functional genes of the IGHV3 subgroup. This subgroup comprises 47–49 mapped genes in the IGH locus, depending on the haplotypes, three unmapped genes (one functional, one pseudogene and one not defined) and 12 orphons.

Gene location

IGHV3-64 is in the IGH locus on chromosome 14 at 14q32.33.

Nucleotide and amino acid sequences for human IGHV3-64

```
                                        1   2   3   4   5   6   7   8   9  10  11  12  13  14  15  16  17  18  19  20
                                        E   V   Q   L   V   E   S   G   G       G   L   V   Q   P   G   G   S   L   R
M99682,IGHV3-64*01,V3-64          [26] GAG GTG CAG CTG GTG GAG TCT GGG GGA ... GGC TTG GTC CAG CCT GGG GGG TCC CTG AGA
Z27505,IGHV3-64*01,YAC-6          [12]
Z12361,IGHV3-64*02,DP-61          [41]
M77298,IGHV3-64*03,f2             [30] --- --- --- --- --- --- --- --- --- ... --- --- --- --- --- --- --- --- --- ---
                                        Q
M77299,IGHV3-64*04,f3             [30] C-- --- --- --- --- --- --- --- --- ... --- --- --- --- --- --- --- --- --- ---
M77301,IGHV3-64*05,p1             [30] --- --- --- --- --- --- --- --- --- ... --- --- --- --- --- --- --- --- --- ---

                                                               _______________CDR1-IMGT________________________
                                       21  22  23  24  25  26  27  28  29  30  31  32  33  34  35  36  37  38  39  40
                                        L   S   C   A   A   S   G   F   T   F   S   S   Y   A                   M   H
M99682,IGHV3-64*01,V3-64               CTC TCC TGT GCA GCC TCT GGA TTC ACC TTC AGT AGC TAT GCT ... ... ... ... ATG CAC
Z27505,IGHV3-64*01,YAC-6                                               --- --- --- --- --- --- ... ... ... ... --- ---
Z12361,IGHV3-64*02,DP-61                                               --- --- --- --- --- --- ... ... ... ... --- ---
                                                    S
M77298,IGHV3-64*03,f2                  --- --- --- T-- --- --- --- --- --- --- --- --- --- --- ... ... ... ... --- ---
                                                    S
M77299,IGHV3-64*04,f3                  --- --- --- T-- --- --- --- --- --- --- --- --- --- --- ... ... ... ... --- ---
                                                    S
M77301,IGHV3-64*05,p1                  --- --- --- T-- --- --- --- --- --- --- --- --- --- --- ... ... ... ... --- ---

                                                                                                   _______________CDR2-
                                       41  42  43  44  45  46  47  48  49  50  51  52  53  54  55  56  57  58  59  60
                                        W   V   R   Q   A   P   G   K   G   L   E   Y   V   S   A   I   S   S   N   G
M99682,IGHV3-64*01,V3-64               TGG GTC CGC CAG GCT CCA GGG AAG GGA CTG GAA TAT GTT TCA GCT ATT AGT AGT AAT GGG
Z27505,IGHV3-64*01,YAC-6               --- --- --- --- --- --- --- --- --- --- --- --- --- --- --- --- --- --- --- ---
Z12361,IGHV3-64*02,DP-61               --- --- --- --- --- --- --- --- --- --- --- --- --- --- --- --- --- --- --- ---
M77298,IGHV3-64*03,f2                  --- --- --- --- --- --- --- --- --- --- --- --- --- --- --- --- --- --- --- ---
M77299,IGHV3-64*04,f3                  --- --- --- --- --- --- --- --- --- --- --- --- --- --- --- --- --- --- --- ---
M77301,IGHV3-64*05,p1                  --- --- --- --- --- --- --- --- --- --- --- --- --- --- --- --- --- --- --- ---

                                       IMGT________________
                                       61  62  63  64  65  66  67  68  69  70  71  72  73  74  75  76  77  78  79  80
                                        G   S   T           Y   Y   A   N   S   V   K       G   R   F   T   I   S   R
M99682,IGHV3-64*01,V3-64               GGT AGC ACA ... ... TAT TAT GCA AAC TCT GTG AAG ... GGC AGA TTC ACC ATC TCC AGA
Z27505,IGHV3-64*01,YAC-6               --- --- --- ... ... --- --- --- --- -
                                                                        D
Z12361,IGHV3-64*02,DP-61               --- --- --- ... ... --- --- --- G--
                                                                        D
M77298,IGHV3-64*03,f2                  --- --- --- ... ... --C --C --- G-- --A --- --- ... --- --- --- --- --- --- ---
                                                                        D
M77299,IGHV3-64*04,f3                  --- --- --- ... ... --C --C --- G-- --A --- --- ... --- --- --- --- --- --- ---
                                                                        D
M77301,IGHV3-64*05,p1                  --- --- --- ... ... --C --C --- G-- --A --- --- ... --- --- --- --- --- --- ---

                                       81  82  83  84  85  86  87  88  89  90  91  92  93  94  95  96  97  98  99 100
                                        D   N   S   K   N   T   L   Y   L   Q   M   G   S   L   R   A   E   D   M   A
M99682,IGHV3-64*01,V3-64               GAC AAT TCC AAG AAC ACG CTG TAT CTT CAA ATG GGC AGC CTG AGA GCT GAG GAC ATG GCT
Z27505,IGHV3-64*01,YAC-6
Z12361,IGHV3-64*02,DP-61
                                                                        V           S                           T
M77298,IGHV3-64*03,f2                  --- --- --- --- --- --- --- --- G-C --- --- A-- -T- --- --- --- --- --- -C- ---
                                                                                    N                           T
M77299,IGHV3-64*04,f3                  --- --- --- --- --- --- --- --- --G --- --- AA- --- --- --- --- --- --- -C- ---
                                                                        V           S                           T
M77301,IGHV3-64*05,p1                  --- --- --- --- --- --- --- --- G-- --- --- A-- --T --- --- --- --- --- -C- ---
```

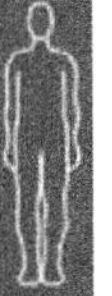

```
                                                 _CDR3-IMGT_
                             101 102 103 104 105 106
                              V   Y   Y   C   A   R
M99682,IGHV3-64*01,V3-64     GTG TAT TAC TGT GCG AGA GA

Z27505,IGHV3-64*01,YAC-6

Z12361,IGHV3-64*02,DP-61
                                              V   K
M77298,IGHV3-64*03,f2        --- --- --- --- -T- -A- --

M77299,IGHV3-64*04,f3        --- --- --- --- --- ---
                                              V   K
M77301,IGHV3-64*05,p1        --- --- --- --- -T- -A- --
```

Framework and complementarity determining regions

FR1-IMGT: 25 (-1 aa: 10)
FR2-IMGT: 17
FR3-IMGT: 38 (-1 aa: 73)

CDR1-IMGT: 8
CDR2-IMGT: 8
CDR3-IMGT: 2

Collier de Perles for human IGHV3-64*01

Accession number: IMGT M99682 EMBL/GenBank/DDBJ: M99682

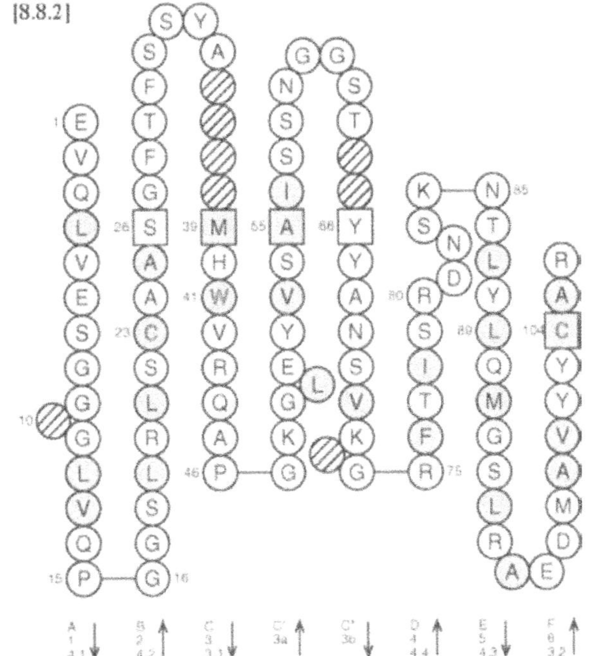

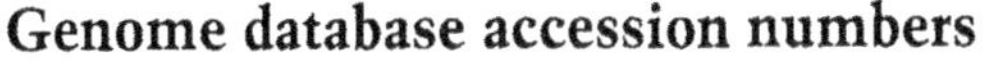

Genome database accession numbers

GDB:9931719 LocusLink: 28414

Nomenclature

IGHV3-66: Immunoglobulin heavy variable 3-66.

Definition and functionality

IGHV3-66 is one of the 18–21 mapped functional genes of the IGHV3 subgroup. This subgroup comprises 47–49 mapped genes in the IGH locus, depending on the haplotypes, three unmapped genes (one functional, one pseudogene and one not defined) and 12 orphons.

Gene location

IGHV3-66 is in the IGH locus on chromosome 14 at 14q32.33.

Nucleotide and amino acid sequences for human IGHV3-66

```
                                   1   2   3   4   5   6   7   8   9   10  11  12  13  14  15  16  17  18  19  20
                                   E   V   Q   L   V   E   S   G   G       G   L   V   Q   P   G   G   S   L   R
X92218,IGHV3-66*01,8-1B      (4)   GAG GTG CAG CTG GTG GAG TCT GGG GGA ... GGC TTG GTC CAG CCT GGG GGC TCC CTG AGA
Z27504,IGHV3-66*02,YAC-5     (12)  --- --- --- --- --- --- --- --- --- ... --- --- --- --- --- --- --- --- --- ---
Z29984,IGHV3-66*03,DA-9      (12)
Z27455,IGHV3-66*03,DP-86     (42)

                                                           _______________CDR1-IMGT________________
                                   21  22  23  24  25  26  27  28  29  30  31  32  33  34  35  36  37  38  39  40
                                   L   S   C   A   A   S   G   F   T   V   S   S   N   Y                   M   S
X92218,IGHV3-66*01,8-1B            CTC TCC TGT GCA GCC TCT GGA TTC ACC GTC AGT AGC AAC TAC ... ... ... ... ATG AGC
Z27504,IGHV3-66*02,YAC-5           --- --- --- --- --- --- --- --- --- --- --- --- --- --- ... ... ... ... --- ---
Z29984,IGHV3-66*03,DA-9                                            --- --- --- --- --- --- ... ... ... ... --- ---
Z27455,IGHV3-66*03,DP-86                                           --- --- --- --- --- --- ... ... ... ... --- ---

                                                                                               ____________CDR2-
                                   41  42  43  44  45  46  47  48  49  50  51  52  53  54  55  56  57  58  59  60
                                   W   V   R   Q   A   P   G   K   G   L   E   W   V   S   V   I   Y   S   G   G
X92218,IGHV3-66*01,8-1B            TGG GTC CGC CAG GCT CCA GGG AAG GGG CTG GAG TGG GTC TCA GTT ATT TAT AGC GGT GGT
Z27504,IGHV3-66*02,YAC-5           --- --- --- --- --- --- --- --- --- --- --- --- --- --- --- --- --- --- --- ---
                                                                                                        C
Z29984,IGHV3-66*03,DA-9            --- --- --- --- --- --- --- --- --- --- --- --- --- --- --- --- --- --- T--
                                                                                                        C
Z27455,IGHV3-66*03,DP-86           --- --- --- --- --- --- --- --- --- --- --- --- --- --- --- --- --- --- T   ---

                                   IMGT________________
                                   61  62  63  64  65  66  67  68  69  70  71  72  73  74  75  76  77  78  79  80
                                   S   T               Y   Y   A   D   S   V   K       G   R   F   T   I   S   R
X92218,IGHV3-66*01,8-1B            AGC ACA ... ... ... TAC TAC GCA GAC TCC GTG AAG ... GGC AGA TTC ACC ATC TCC AGA
Z27504,IGHV3-66*02,YAC-5           --- --- ... ... ... --- --- --- --- --- --- --- ... --- C-- --- --- --- --- ---
Z29984,IGHV3-66*03,DA-9            --- --- ... ... ... --- --- --- --
Z27455,IGHV3-66*03,DP-86           --- --- ... ... ... --- --- --- --

                                   81  82  83  84  85  86  87  88  89  90  91  92  93  94  95  96  97  98  99  100
                                   D   N   S   K   N   T   L   Y   L   Q   M   N   S   L   R   A   E   D   T   A
X92218,IGHV3-66*01,8-1B            GAC AAT TCC AAG AAC ACG CTG TAT CTT CAA ATG AAC AGC CTG AGA GCC GAG GAC ACG GCT
Z27504,IGHV3-66*02,YAC-5           --- --- --- --- --- --- --- --- --- --- --- --- --- --- --- --T --- --- --- ---
Z29984,IGHV3-66*03,DA-9
Z27455,IGHV3-66*03,DP-86

                                                   _CDR3 IMGT
                                   101 102 103 104 105 106
                                   V   Y   Y   C   A   R
X92218,IGHV3-66*01,8-1B            GTG TAT TAC TGT GCG AGA GA
Z27504,IGHV3-66*02,YAC-5           --- --- --- --- --- ---
Z29984,IGHV3-66*03,DA-9
Z27455,IGHV3-66*03,DP-86
```

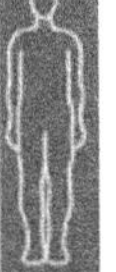

Framework and complementarity determining regions

FR1-IMGT: 25 (-1 aa: 10)
FR2-IMGT: 17
FR3-IMGT: 38 (-1 aa: 73)

CDR1-IMGT: 8
CDR2-IMGT: 8
CDR3-IMGT: 2

Collier de Perles for human IGHV3-66*01

Accession number: IMGT X92218 EMBL/GenBank/DDBJ: X92218

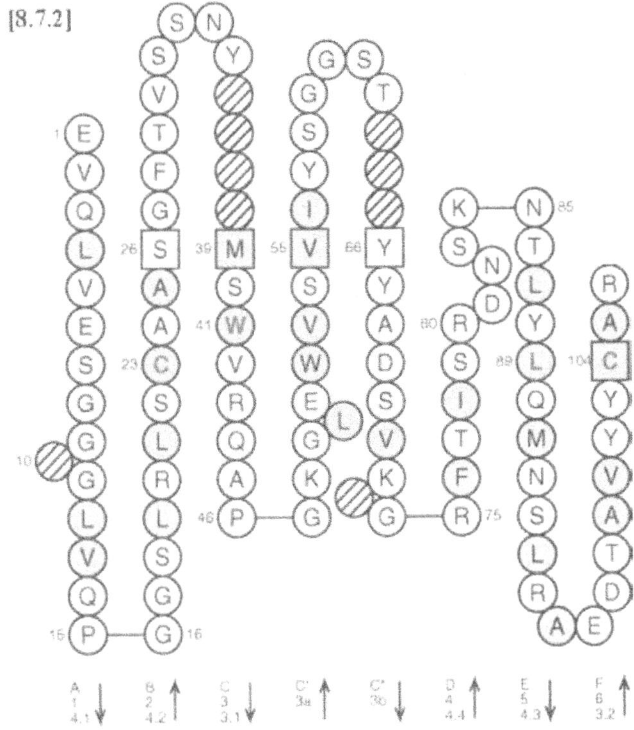

Genome database accession numbers

GDB:9931736 LocusLink: 28412

IGHV3-72

Nomenclature

IGHV3-72: Immunoglobulin heavy variable 3-72.

Definition and functionality

IGHV3-72 is one of the 18–21 mapped functional genes of the IGHV3 subgroup. This subgroup comprises 47–49 mapped genes in the IGH locus, depending on the haplotypes, three unmapped genes (one functional, one pseudogene and one not defined) and 12 orphons.

Gene location

IGHV3-72 is in the IGH locus on chromosome 14 at 14q32.33.

Nucleotide and amino acid sequences for human IGHV3-72

```
                                     1   2   3   4   5   6   7   8   9  10  11  12  13  14  15  16  17  18  19  20
                                     E   V   Q   L   V   E   S   G   G       G   L   V   Q   P   G   G   S   L   R
X92206,IGHV3-72*01,12-2       [4]   GAG GTG CAG CTG GTG GAG TCT GGG GGA ... GGC TTG GTC CAG CCT GGA GGG TCC CTG AGA
Z12331,IGHV3-72*01,DP-29     [41]   --- --- --- --- --- --- --- --- --- ... --- --- --- --- --- --- --- --- --- ---
Z29979,IGHV3-72*02,DA-3      [12]

                                                             ____________________CDR1-IMGT____________________
                                    21  22  23  24  25  26  27  28  29  30  31  32  33  34  35  36  37  38  39  40
                                     L   S   C   A   A   S   G   F   T   F   S   D   H   Y                   M   D
X92206,IGHV3-72*01,12-2             CTC TCC TGT GCA GCC TCT GGA TTC ACC TTC AGT GAC CAC TAC ... ... ... ... ATG GAC
Z12331,IGHV3-72*01,DP-29            --- --- --- --- --- --- --- --- --- --- --- --- --- --- ... ... ... ... --- ---
Z29979,IGHV3-72*02,DA-3                                             --- --- --- --- --- --- ... ... ... ... --- ---

                                                                                                 ________________ CDR2-
                                    41  42  43  44  45  46  47  48  49  50  51  52  53  54  55  56  57  58  59  60
                                     W   V   R   Q   A   P   G   K   G   L   E   W   V   G   R   T   R   N   K   A
X92206,IGHV3-72*01,12-2             TGG GTC CGC CAG GCT CCA GGG AAG GGG CTG GAG TGG GTT GGC CGT ACT AGA AAC AAA GCT
Z12331,IGHV3-72*01,DP-29            --- --- --- --- --- --- --- --- --- --- --- --- --- --- --- --- --- --- --- ---
Z29979,IGHV3-72*02,DA-3             --- --- --- --- --- --- --- --- --- --- --- --- --- --- --- --- --- --- --- ---

                                    IMGT_______________
                                    61  62  63  64  65  66  67  68  69  70  71  72  73  74  75  76  77  78  79  80
                                     N   S   Y   T   T   E   Y   A   A   S   V   K       G   R   F   T   I   S   R
X92206,IGHV3-72*01,12-2             AAC AGT TAC ACC ACA GAA TAC GCC GCG TCT GTG AAA ... GGC AGA TTC ACC ATC TCA AGA
Z12331,IGHV3-72*01,DP-29            --- --- --- --- --- --- --- --- --- --- --- --- ... --- --- --- --- --- --- ---
Z29979,IGHV3-72*02,DA-3             --- --C --- --- --- --- --- --- --- --- --- --- ... --- --- --- --- --- --- ---

                                    81  82  83  84  85  86  87  88  89  90  91  92  93  94  95  96  97  98  99 100
                                     D   D   S   K   N   S   L   Y   L   Q   M   N   S   L   K   T   E   D   T   A
X92206,IGHV3-72*01,12-2             GAT GAT TCA AAG AAC TCA CTG TAT CTG CAA ATG AAC AGC CTG AAA ACC GAG GAC ACG GCC
Z12331,IGHV3-72*01,DP-29            --- --- --- --- --- --- --- --- --- --- --- --- --- --- --- --- --- --- --- ---
Z29979,IGHV3-72*02,DA-3             --- --- --- --- --- --- --- ---

                                                    _CDR3-IMGT
                                    101 102 103 104 105 106
                                     V   Y   Y   C   A   R
X92206,IGHV3-72*01,12-2             GTG TAT TAC TGT GCT AGA GA
Z12331,IGHV3-72*01,DP-29            --- --- --- --- --- ---
Z29979,IGHV3-72*02,DA-3
```

Framework and complementarity determining regions

FR1-IMGT: 25 (-1 aa: 10)
FR2-IMGT: 17
FR3-IMGT: 38 (-1 aa: 73)

CDR1-IMGT: 8
CDR2-IMGT: 10
CDR3-IMGT: 2

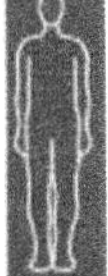

Collier de Perles for human IGHV3-72*01

Accession number: IMGT X92206 EMBL/GenBank/DDBJ: X92206

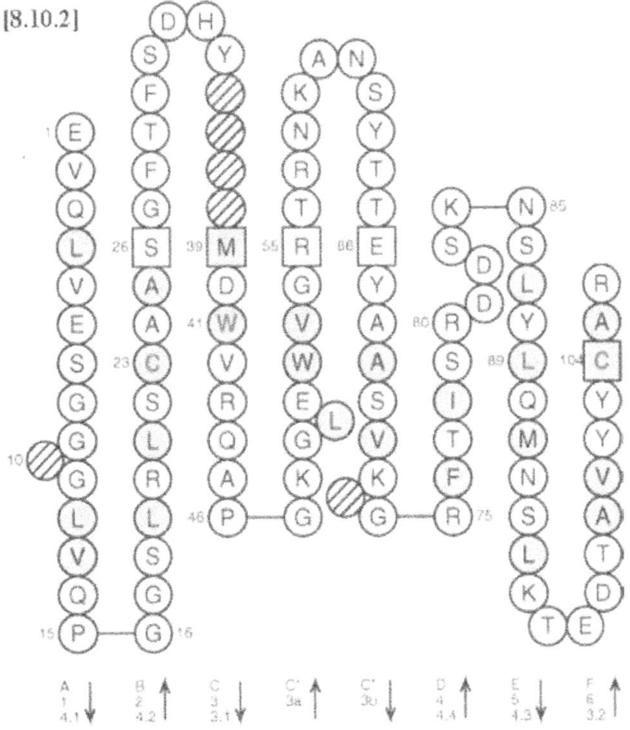

Genome database accession numbers

GDB:9931722 LocusLink: 28410

IGHV3-73

Nomenclature

IGHV3-73: Immunoglobulin heavy variable 3-73.

Definition and functionality

IGHV3-73 is one of the 18–21 mapped functional genes of the IGHV3 subgroup. This subgroup comprises 47–49 mapped genes in the IGH locus, depending on the haplotypes, three unmapped genes (one functional, one pseudogene and one not defined) and 12 orphons.

Gene location

IGHV3-73 is in the IGH locus on chromosome 14 at 14q32.33.

Nucleotide and amino acid sequences for human IGHV3-73

```
                                      1   2   3   4   5   6   7   8   9  10  11  12  13  14  15  16  17  18  19  20
                                     E   V   Q   L   V   E   S   G   G       G   L   V   Q   P   G   G   S   L   K
Z27508,IGHV3-73*01,YAC-9       [12] GAG GTG CAG CTG GTG GAG TCT GGG GGA ... GGC TTG GTC CAG CCT GGG GGG TCC CTG AAA
L15467,IGHV3-73*01,MTGL        [55] --- --- --- --- --- --- --- --- --- ... --- --- --- --- --- --- --- --- --- ---
Z29986,IGHV3-73*01,DA-11       [12]
Z27451,IGHV3-73*01,COS-27      [42]

                                                            _______________CDR1-IMGT____________________
                                     21  22  23  24  25  26  27  28  29  30  31  32  33  34  35  36  37  38  39  40
                                     L   S   C   A   A   S   G   F   T   F   S   G   S   A                   M   H
Z27508,IGHV3-73*01,YAC-9            CTC TCC TGT GCA GCC TCT GGG TTC ACC TTC AGT GGC TCT GCT ... ... ... ... ATG CAC
L15467,IGHV3-73*01,MTGL             --- --- --- --- --- --- --- --- --- --- --- --- --- --- ... ... ... ... --- ---
Z29986,IGHV3-73*01,DA-11                                            --- --- --- --- --- --- ... ... ... ... --- ---
Z27451,IGHV3-73*01,COS-27                                           --- --- --- --- --- --- ... ... ... ... --- ---

                                                                                                ____________CDR2-
                                     41  42  43  44  45  46  47  48  49  50  51  52  53  54  55  56  57  58  59  60
                                     W   V   R   Q   A   S   G   K   G   L   E   W   V   G   R   I   R   S   K   A
Z27508,IGHV3-73*01,YAC-9            TGG GTC CGC CAG GCT TCC GGG AAA GGG CTG GAG TGG GTT GGC CGT ATT AGA AGC AAA GCT
L15467,IGHV3-73*01,MTGL             --- --- --- --- --- --- --- --- --- --- --- --- --- --- --- --- --- --- --- ---
Z29986,IGHV3-73*01,DA-11            --- --- --- --- --- --- --- --- --- --- --- --- --- --- --- --- --- --- --- ---
Z27451,IGHV3-73*01,COS-27           --- --- --- --- --- --- --- --- --- --- --- --- --- --- --- --- --- --- --- ---

                                    IMGT________________
                                     61  62  63  64  65  66  67  68  69  70  71  72  73  74  75  76  77  78  79  80
                                     N   S   Y   A   T   A   Y   A   A   S   V   K       G   R   F   T   I   S   R
Z27508,IGHV3-73*01,YAC-9            AAC AGT TAC GCG ACA GCA TAT GCT GCG TCG GTG AAA ... GGC AGG TTC ACC ATC TCC AGA
L15467,IGHV3-73*01,MTGL             --- --- --- --- --- --- --- --- --- --- --- --- ... --- --- --- --- --- --- ---
Z29986,IGHV3-73*01,DA-11            --- --- --- --- --- --- --- --- --- --- --- --- ... --- --- --- --- --- --- ---
Z27451,IGHV3-73*01,COS-27           --- --- --- --- --- --- --- --- --- --- --- --- ... --- --- --- --- --- --- ---

                                     81  82  83  84  85  86  87  88  89  90  91  92  93  94  95  96  97  98  99 100
                                     D   D   S   K   N   T   A   Y   L   Q   M   N   S   L   K   T   E   D   T   A
Z27508,IGHV3-73*01,YAC-9            GAT GAT TCA AAG AAC ACG GCG TAT CTG CAA ATG AAC AGC CTG AAA ACC GAG GAC ACG GCC
L15467,IGHV3-73*01,MTGL             --- --- --- --- --- --- --- --- --- --- --- --- --- --- --- --- --- --- --- ---
Z29986,IGHV3-73*01,DA-11            --- --- --- --- --- --- --- ---
Z27451,IGHV3-73*01,COS-27           --- --- --- --- --- --- --- ---

                                                    _CDR3-IMGT
                                    101 102 103 104 105 106
                                     V   Y   Y   C   T   R
Z27508,IGHV3-73*01,YAC-9            GTG TAT TAC TGT ACT AGA
L15467,IGHV3-73*01,MTGL             --- --- --- --- --- --- CA
Z29986,IGHV3-73*01,DA-11
Z27451,IGHV3-73*01,COS-27
```

Framework and complementarity determining regions

FR1-IMGT: 25 (-1 aa: 10)
FR2-IMGT: 17
FR3-IMGT: 38 (-1 aa: 73)

CDR1-IMGT: 8
CDR2-IMGT: 10
CDR3-IMGT: 2

Collier de Perles for human IGHV3-73*01

Accession number: IMGT Z27508 EMBL/GenBank/DDBJ: Z27508

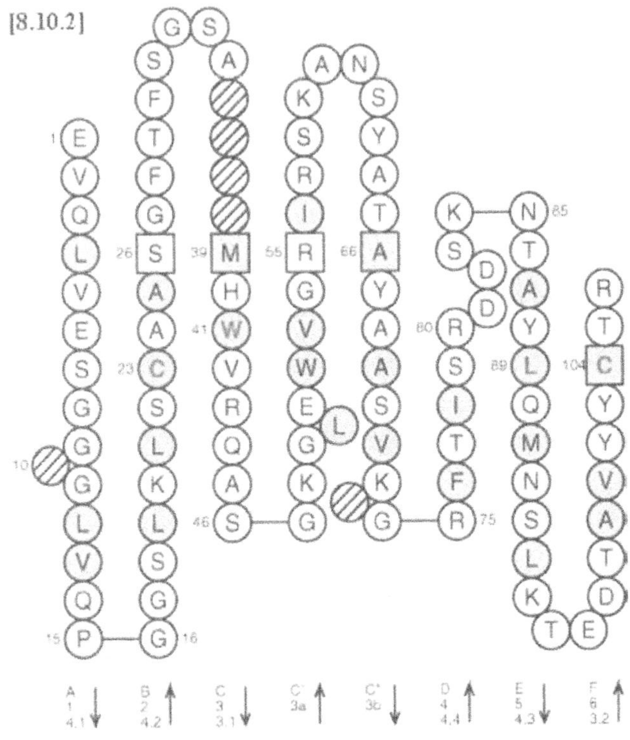

Genome database accession numbers

GDB:9931723 LocusLink: 28409

Nomenclature

IGHV3-74: Immunoglobulin heavy variable 3-74.

Definition and functionality

IGHV3-74 is one of the 18–21 mapped functional genes of the IGHV3 subgroup. This subgroup comprises 47–49 mapped genes in the IGH locus, depending on the haplotypes, three unmapped genes (one functional, one pseudogene and one not defined) and 12 orphons.

Gene location

IGHV3-74 is in the IGH locus on chromosome 14 at 14q32.33.

Nucleotide and amino acid sequences for human IGHV3-74

```
                                     1   2   3   4   5   6   7   8   9  10  11  12  13  14  15  16  17  18  19  20
                                     E   V   Q   L   V   E   S   G   G       G   L   V   Q   P   G   G   S   L   R
Z12353,IGHV3-74*01,DP-53       [41] GAG GTG CAG CTG GTG GAG TCC GGG GGA ... GGC TTA GTT CAG CCT GGG GGG TCC CTG AGA
D16832,IGHV3-74*01,13G12       [20] --- --- --- --- --- --- --- --- --- ... --- --- --- --- --- --- --- --- --- ---
Z30082,IGHV3-74*01/02,DA-8     [12]
Z17392,IGHV3-74*02,COS-6       [42] --- --- --- --- --- --- --T --- --- ... --- --- --- --- --- --- --- --- --- ---
J00239,IGHV3-74*03,H11         [32] --- --- --- --- --- --- --- --- --- ... --- --- --- --- --- --- --- --- --- ---

                                                             ______________CDR1-IMGT_______________
                                    21  22  23  24  25  26  27  28  29  30  31  32  33  34  35  36  37  38  39  40
                                     L   S   C   A   A   S   G   F   T   F   S   S   Y   W                   M   H
Z12353,IGHV3-74*01,DP-53            CTC TCC TGT GCA GCC TCT GGA TTC ACC TTC AGT AGC TAC TGG ... ... ... ... ATG CAC
D16832,IGHV3-74*01,13G12            --- --- --- --- --- --- --- --- --- --- --- --- --- --- ... ... ... ... --- ---
Z30082,IGHV3-74*01/02,DA-8                                          --- --- --- --- --- --- ... ... ... ... --- ---
Z17392,IGHV3-74*02,COS-6            --- --- --- --- --- --- --- --- --- --- --- --- --- --- ... ... ... ... --- ---
J00239,IGHV3-74*03,H11              --- --- --- --- --- --- --- --- --- --- --- --- --- --- ... ... ... ... --- ---

                                                                                              ______________CDR2
                                    41  42  43  44  45  46  47  48  49  50  51  52  53  54  55  56  57  58  59  60
                                     W   V   R   Q   A   P   G   K   G   L   V   W   V   S   R   I   N   S   D   G
Z12353,IGHV3-74*01,DP-53            TGG GTC CGC CAA GCT CCA GGG AAG GGG CTG GTG TGG GTC TCA CGT ATT AAT AGT GAT GGC
D16832,IGHV3-74*01,13G12            --- --- --- --- --- --- --- --- --- --- --- --- --- --- --- --- --- --- --- ---
Z30082,IGHV3-74*01/02,DA-8          --- --- --- --- --- --- --      -- --- --- --- --- --- --- --- --- --- --- ---
Z17392,IGHV3-74*02,COS-6            --- --- --- --- --- --- --- --- --- --- --- --- --- --- --- --- --- --- - - -
J00239,IGHV3-74*03,H11              --- --- --- --- --- --- --- --- --- --- --- --- --- --- --- --- --- --- --- ---

                                    IMGT______________
                                    61  62  63  64  65  66  67  68  69  70  71  72  73  74  75  76  77  78  79  80
                                     S   S   T           S   Y   A   D   S   V   K       G   R   F   T   I   S   R
Z12353,IGHV3-74*01,DP-53            AGT AGC ACA ... ... AGC TAC GCG GAC TCC GTG AAG ... GGC CGA TTC ACC ATC TCC AGA
D16832,IGHV3-74*01,13G12            --- --- --- ... ... --- --- --- --- --- --- --- ... --- --- --- --- --- --- ---
Z30082,IGHV3-74*01/02,DA-8          --- --- --- ... ... --- --- --- --- -
Z17392,IGHV3-74*02,COS-6            --- --- --- ... ... --- --- --- --- --- --- --- ... --- --- --- --- --- --- ---
                                                         T
J00239,IGHV3-74*03,H11              --- --- --- ... ... -CG --- --- --- --- --- --- ... --- --- --- --- --- --- ---

                                    81  82  83  84  85  86  87  88  89  90  91  92  93  94  95  96  97  98  99 100
                                     D   N   A   K   N   T   L   Y   L   Q   M   N   S   L   R   A   E   D   T   A
Z12353,IGHV3-74*01,DP-53            GAC AAC GCC AAG AAC ACG CTG TAT CTG CAA ATG AAC AGT CTG AGA GCC GAG GAC ACG GCT
D16832,IGHV3-74*01,13G12            --- --- --- --- --- --- --- --- --- --- --- --- --- --- -
Z30082,IGHV3-74*01/02,DA-8
Z17392,IGHV3-74*02,COS-6            --- --- --- --- --- --- --- --- --- --- --- --- --- --- --- --- --- --- --- ---
J00239,IGHV3-74*03,H11              --- --- --- --- --- --- --- --- --- --- --- --- --- --- --- --- --- --- --- ---

                                                      _CDR3-IMGT
                                    101 102 103 104 105 106
                                     V   Y   Y   C   A   R
Z12353,IGHV3-74*01,DP-53            GTG TAT TAC TGT GCA AGA
D16832,IGHV3-74*01,13G12
Z30082,IGHV3-74*01/02,DA-8
Z17392,IGHV3-74*02,COS-6            --- --- --- --- --- ---
J00239,IGHV3-74*03,H11              --- --- --- --- --- --- GA
```

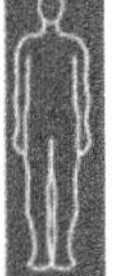

Framework and complementarity determining regions

FR1-IMGT: 25 (-1 aa: 10)
FR2-IMGT: 17
FR3-IMGT: 38 (-1 aa: 73)

CDR1-IMGT: 8
CDR2-IMGT: 8
CDR3-IMGT: 2

Collier de Perles for human IGHV3-74*01

Accession number: IMGT Z12353   EMBL/GenBank/DDBJ: Z12353

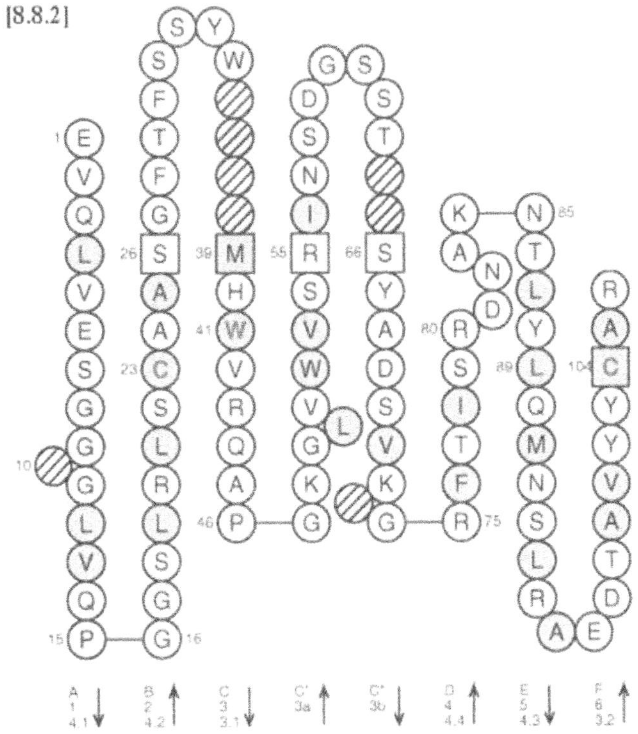

Genome database accession numbers

GDB:9931724   LocusLink: 28408

Nomenclature

IGHV3-d: Immunoglobulin heavy variable 3-d (provisional).

Definition and functionality

IGHV3-d is an unmapped functional gene of the IGHV3 subgroup. This subgroup comprises 47–49 mapped genes (of which 18–21 are functional) in the IGH locus, depending on the haplotypes, three unmapped genes (one functional, one pseudogene and one not defined) and 12 orphons.

Gene location

IGHV3-d is in the IGH locus on chromosome 14 at 14q32.33.

Nucleotide and amino acid sequences for human IGHV3-d

```
                                     1   2   3   4   5   6   7   8   9  10  11  12  13  14  15  16  17  18  19  20
                                     E   V   Q   L   V   E   S   R   G       V   L   V   Q   P   G   G   S   L   R
Z18898,IGHV3-d*01,COS-12      [42]  GAG GTG CAG CTG GTG GAG TCT CGG GGA ... GTC TTG GTA CAG CCT GGG GGG TCC CTG AGA

                                                             ______________CDR1-IMGT______________
                                    21  22  23  24  25  26  27  28  29  30  31  32  33  34  35  36  37  38  39  40
                                     L   S   C   A   A   S   G   F   T   V   S   S   N   E                   M   S
Z18898,IGHV3-d*01,COS-12            CTC TCC TGT GCA GCC TCT GGA TTC ACC GTC AGT AGC AAT GAG ... ... ... ... ATG ACC

                                                                                             ____________CDR2
                                    41  42  43  44  45  46  47  48  49  50  51  52  53  54  55  56  57  58  59  60
                                     W   V   R   Q   A   P   G   K   G   L   E   W   V   S   S   I   S   G   G   S
Z18898,IGHV3-d*01,COS-12            TGG GTC CGC CAG GCT CCA GGG AAG GGT CTG GAG TGG GTC TCA TCC ATT AGT GGT GGT AGC

                                    IMGT_______________
                                    61  62  63  64  65  66  67  68  69  70  71  72  73  74  75  76  77  78  79  80
                                     T                   Y   Y   A   D   S   R   K       G   R   F   T   I   S   R
Z18898,IGHV3-d*01,COS-12            ACA ... ... ... ... TAC TAC GCA GAC TCC AGG AAG ... GGC AGA TTC ACC ATC TCC AGA

                                    81  82  83  84  85  86  87  88  89  90  91  92  93  94  95  96  97  98  99 100
                                     D   N   S   K   N   T   L   H   L   Q   M   N   S   L   R   A   E   D   T   A
Z18898,IGHV3-d*01,COS-12            GAC AAT TCC AAG AAC ACG CTG CAT CTT CAA ATG AAC AGC CTG AGA GCT GAG GAC ACG GCT

                                                        CDR3-IMGT
                                    101 102 103 104 105 106
                                     V   Y   Y   C   K   K
Z18898,IGHV3-d*01,COS-12            GTG TAT TAC TGT AAG AAA
```

Framework and complementarity determining regions

FR1-IMGT: 25 (-1 aa: 10)
FR2-IMGT: 17
FR3-IMGT: 38 (-1 aa: 73)

CDR1-IMGT: 8
CDR2-IMGT: 6
CDR3-IMGT: 2

Collier de Perles for human IGHV3-d*01

Accession number: IMGT Z18898 EMBL/GenBank/DDBJ: Z18898

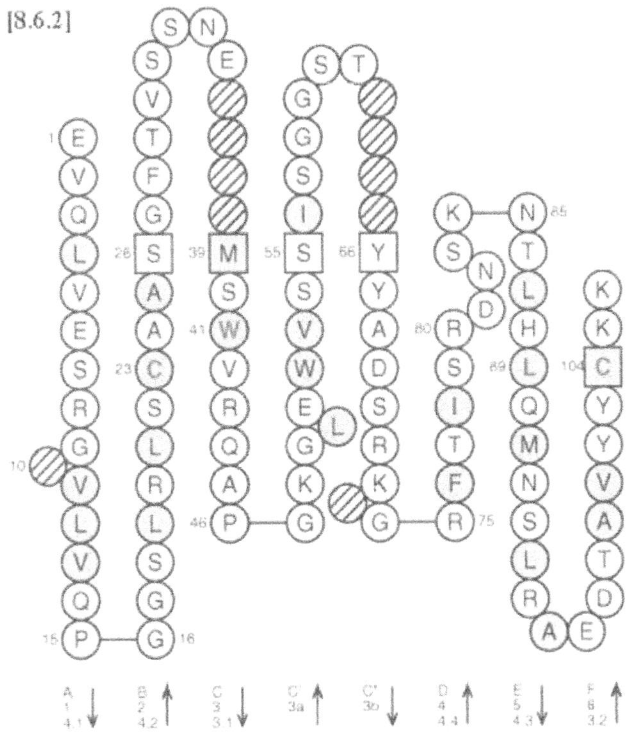

Genome database accession numbers

GDB:9931729 LocusLink: 28404

IGHV4-4

Nomenclature

IGHV4-4: Immunoglobulin heavy variable 4-4.

Definition and functionality

IGHV4-4 is one of the 6–10 mapped functional genes of the IGHV4 subgroup. This subgroup comprises 9–12 mapped genes in the IGH locus, depending on the haplotypes, one unmapped functional gene and one orphon.

Gene location

IGHV4-4 is in the IGH locus on chromosome 14 at 14q32.33.

Nucleotide and amino acid sequences for human IGHV4-4

```
                                     1   2   3   4   5   6   7   8   9   10  11  12  13  14  15  16  17  18  19  20
                                     Q   V   Q   L   Q   E   S   G   P       G   L   V   K   P   P   G   T   L   S
X05713,IGHV4-4*01,V79/VIV-4b  [24]  CAG CTG CAG CTG CAG GAG TCG GGC CCA ... GGA CTG GTG AAG CCT CCG GGG ACC CTG TCC
X56363,IGHV4-4*01,VH4.19      [35]  --- --- --- --- --- --- --- --- --- ... --- --- --- --- --- --- --- --- --- ---
                                                                                 S
X92232,IGHV4-4*02,4.41        [50]  --- --- --- --- --- --- --- --- --- ... --- --- --- --- --- T-- --- --- --- ---
                                                                                 S
Z12370,IGHV4-4*02,DP-70       [41]  --- --- --- --- --- --- --- --- --- ... --- --- --- --- --- T-- --- --- --- ---
                                                                                 S
L10091,IGHV4-4*02,4d68        [47]  --- --- --- --- --- --- --- --- --- ... --- --- --- --- --- T-- --- --- --- ---
Z75350,IGHV4-4*02,VH4-GL15    [49]
X92252,IGHV4-4*03,VH4-MC4     [7]   --- --- --- --- --- --- --- --- --- ... --- --- --- --- --- --- --- --- --- ---
X92253,IGHV4-4*04,VH4-MC4a    [7]   --- --- --- --- --- --- --- --- --- ... --- --- --- --- --- --- --- --- --- ---
                                                             L
X92254,IGHV4-4*05,VH4-MC4b    [7]   --- --- --- --- --- --- T-- --- --- ... --- --- --- --- --- --- --- --- --- ---
Z75355,IGHV4-4*06,VH4-GL3     [49]
                                                                                 S   E
X62112,IGHV4-4*07,VIV-4       [39]  --- --- --- --- --- --- --- --- --- ... --- --- --- --- --- T-- -A- --- --- ---
                                                                                 S   E
Z14240,IGHV4-4*07,4.35        [50]  --- --- --- --- --- --- --- --- --- ... --- --- --- --- --- T-- -A- --- --- ---
                                                         Q   W       A               L           S   E
Z14243,IGHV4-4*08,4.38        [50]  --- --- --- --A --- C-- -G- --- G-- ... --- --- T-- --- --- T-- -A- --- --- ---

                                                            ___________________CDR1-IMGT____________________
                                     21  22  23  24  25  26  27  28  29  30  31  32  33  34  35  36  37  38  39  40
                                     L   T   C   A   V   S   G   G   S   I   S   S   S   N   W               W   S
X05713,IGHV4-4*01,V79/VIV-4b        CTC ACC TGC GCT GTC TCT GGT GGC TCC ATC AGC AGT AGT AAC TGG ... ... ... TGG AGT
X56363,IGHV4-4*01,VH4.19            --- --- --- --- --- --- --- --- --- --- --- --- --- ---     ... ... ...
X92232,IGHV4-4*02,4.41              --- --- --- --- --- --- --- --- --- --- --- --- --- --- --- ... ... ... --- ---
Z12370,IGHV4-4*02,DP-70             --- --- --- --- --- --- --- --- --- --- --- --- --- --- --- ... ... ... --- ---
L10091,IGHV4-4*02,4d68              --- --- --- --- --- --- --- --- --- --- --- --- --- --- --- ... ... ... --- ---
Z75350,IGHV4-4*02,VH4-GL15                              --- --- --- --- --- --- --- --- --- --- ... ... ... --- ---
X92252,IGHV4-4*03,VH4-MC4           --- --- --- --- --- --- --- --- --- --- --- --- --- --- --- ... ... ... --- ---
                                                     I
X92253,IGHV4-4*04,VH4-MC4a          --- --- --- --- A-- --- --- --- --- --- --- --- --- --- --- ... ... ... --- ---
X92254,IGHV4-4*05,VH4-MC4b          --- --- --- --- --- --- --- --- --- --- --- --- --- --- --- ... ... ... --- ---
Z75355,IGHV4-4*06,VH4-GL3                               --- --- --- --- --- --- --- --- --- --- ... ... ... --- ---
                                                 T                                   Y   Y
X62112,IGHV4-4*07,VIV-4             --- --- --- A-- --- --- --- --- --- --- --T --- TAC T-- ... ... ... ... --- --C
                                                 T                                   Y   Y
S14240,IGHV4-4*07,4.35              --- --- --- A-- --- --- --- --- --- --- --T --- TAC T-- ... ... ... ... --- --C
                                                         Y                           Y   Y
S14243,IGHV4-4*08,4.38              --- --- --- --- --- -A- --- --- --- --- --T --- TAC T-- ... ... ... ... --- --C

                                                                                                _______________CDR2-
                                     41  42  43  44  45  46  47  48  49  50  51  52  53  54  55  56  57  58  59  60
                                     W   V   R   Q   P   P   G   K   G   L   E   W   I   G   E   I   Y   H   S   G
X05713,IGHV4-4*01,V79/VIV-4b        TGG GTC CGC CAG CCC CCA GGG AAG GGG CTG GAG TGG ATT GGG GAA ATC TAT CAT AGT GGG
X56363,IGHV4-4*01,VH4.19            --- --- --- --- --- --- --- --- --- --- --- --- --- --- --- --- --- --- --- ---
X92232,IGHV4-4*02,4.41              --- --- --- --- --- --- --- --- --- --- --- --- --- --- --- --- --- --- --- ---
Z12370,IGHV4-4*02,DP-70             --- --- --- --- --- --- --- --- --- --- --- --- --- --- --- --- --- --- --- ---
L10091,IGHV4-4*02,4d68              --- --- --- --- --- --- --- --- --- --- --- --- --- --- --- --- --- --- --- ---
Z75350,IGHV4-4*02,VH4-GL15          X--                 --- -XX -XX XX- --- --- --- --- --- --- --- --- --- --- ---
X92252,IGHV4-4*03,VH4-MC4           --- --- --- --- --- --- --- --- --- --- --- --- --- --- --- --- --- --- --- ---
X92253,IGHV4-4*04,VH4-MC4a          --- --- --- --- --- --- --- --- --- --- --- --- --- --- --- --- --- --- --- ---
X92254,IGHV4-4*05,VH4-MC4b          --- --- --- --- --- --- --- --- --- --- --- --- --- --- --- --- --- --- --- ---
Z75355,IGHV4-4*06,VH4-GL3           --- --- --- --- --- --- --- -XX X-- --- --- --- --- --- --- --- --- --- --- ---
```

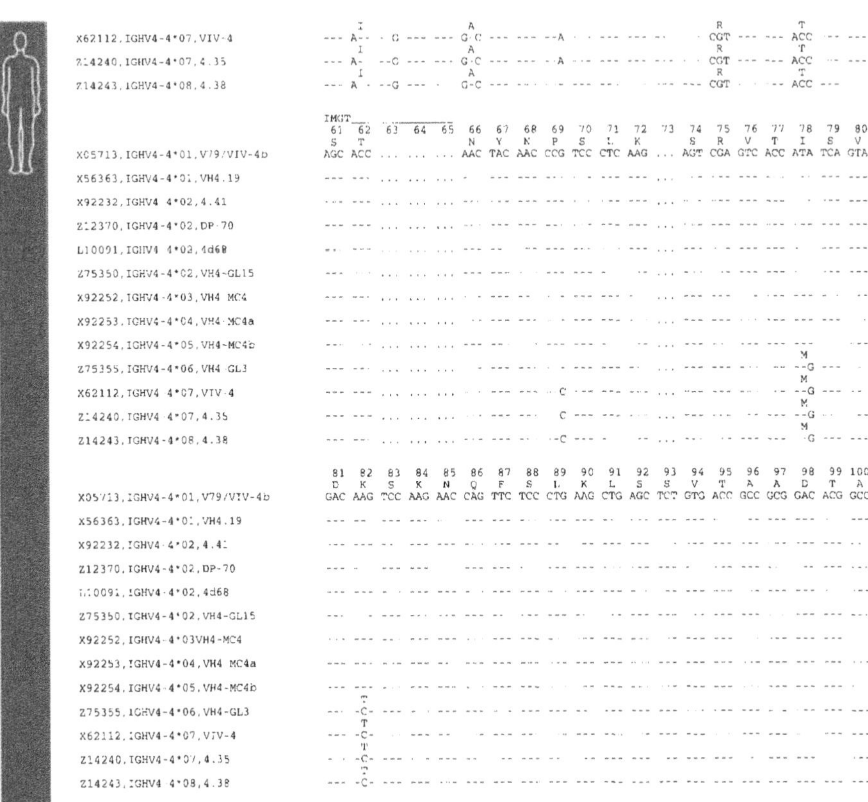

Framework and complementarity determining regions

FR1-IMGT: 25 (-1 aa: 10)
FR2-IMGT: 17
FR3-IMGT: 38 (-1 aa: 73)

CDR1-IMGT: 9
CDR2-IMGT: 7
CDR3-IMGT: 2

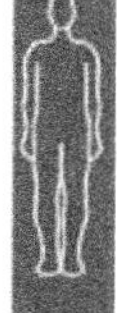

Collier de Perles for human IGHV4-4*01

Accession number: IMGT X05713 EMBL/GenBank/DDBJ: X05713

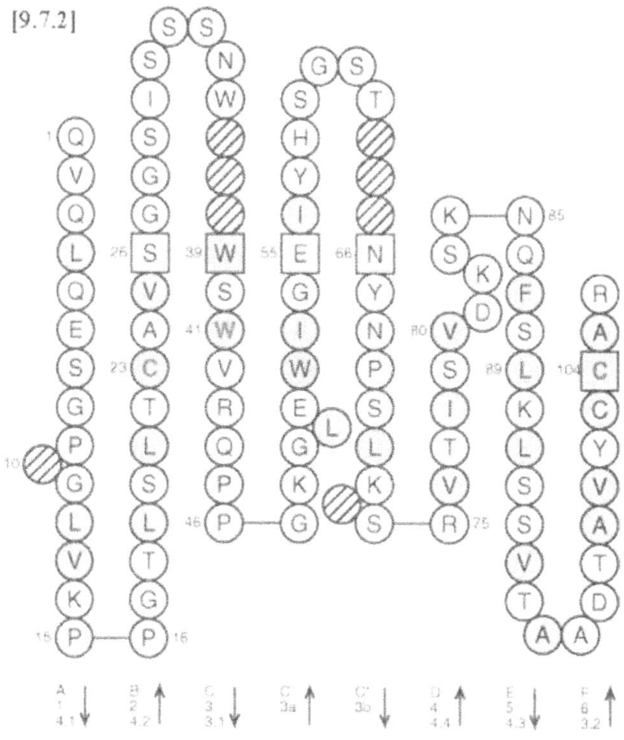

Genome database accession numbers

GDB:9931731 LocusLink: 28401

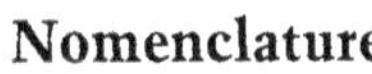

Nomenclature

IGHV4-28: Immunoglobulin heavy variable 4-28.

Definition and functionality

IGHV4-28 is one of the 6–10 mapped functional genes of the IGHV4 subgroup. This subgroup comprises 9–12 mapped genes in the IGH locus, depending on the haplotypes, one unmapped functional gene and one orphon.

Gene location

IGHV4-28 is in the IGH locus on chromosome 14 at 14q32.33.

Nucleotide and amino acid sequences for human IGHV4-28

```
                                       1   2   3   4   5   6   7   8   9  10  11  12  13  14  15  16  17  18  19  20
                                       Q   V   Q   L   Q   E   S   G   P       G   L   V   K   P   S   D   T   L   S
X05714,IGHV4-28*01,V12G-1        [24] CAG GTG CAG CTG CAG GAG TCG GGC CCA ... GGA CTG GTG AAG CCT TCG GAC ACC CTG TCC
X92222,IGHV4-28*01,1.9II          [4] --- --- --- --- --- --- --- --- --- ... --- --- --- --- --- --- --- --- --- ---
Z12368,IGHV4-28*01,DP-68         [41] --- --- --- --- --- --- --- --- --- ... --- --- --- --- --- --- --- --- --- ---
L10096,IGHV4-28*01,3d28d         [37] --- --- --- --- --- --- --- --- --- ... --- --- --- --- --- --- --- --- --- ---
X56357,IGHV4-28*01,VH4.13        [35] --- --- --- --- --- --- --- --- --- ... --- --- --- --- --- --- --- --- --- ---
Z75357,IGHV4-28*01,VH4-GL5       [49]
M95112,IGHV4-28*01 or *03,H2 (1) [45] --- --- --- --- --- --- --- --- --- ... --- --- --- --- --- --- --- --- --- ---
                                                                                                       Q
M83133,IGHV4-28*02,hv4005        [10] --- --- --- --- --- --- --- --- --- ... --- --- --- --- --- --A C-G --- --- ---
                                                                                                       Q
L10099,IGHV4-28*02,3d24d         [35] --- --- --- --- --- --- --- --- --- ... --- --- --- --- --- --A C-G --- --- ---
X92233,IGHV4-28*03,4.42          [50] --- --- --- --- --- --- --- --- --- ... --- --- --- --- --- --- --- --- --- ---
X56358,IGHV4-28*04,VH4.14        [50] --- --- --- --- --- --- --- --- --- ... --- --- --- --- --- --- --- --- --- ---
X92260,IGHV4-28*05,VH4-MC7        [7] --- --- --- --- --- --- --- --- --- ... --- --- --- --- --- --- --- --- --- ---

                                                               _______________CDR1-IMGT_______________
                                      21  22  23  24  25  26  27  28  29  30  31  32  33  34  35  36  37  38  39  40
                                       L   T   C   A   V   S   G   Y   S   I   S   S   S   N   W               W   G
X05714,IGHV4-28*01,V12G-1             CTC ACC TGC GCT GTC TCT GGT TAC TCC ATC AGC AGT AGT AAC TGG ... ... ... TGG GGC
X92222,IGHV4-28*01,1.9II              --- --- --- --- --- --- --- --- --- --- --- --- --- --- --- ... ... ... --- ---
Z12368,IGHV4-28*01,DP-68              --- --- --- --- --- --- --- --- --- --- --- --- --- --- --- ... ... ... --- ---
L10096,IGHV4-28*01,3d28d              --- --- --- --- --- --- --- --- --- --- --- --- --- --- --- ... ... ... --- ---
X56357,IGHV4-28*01,VH4.13             --- --- --- --- --- --- --- --- --- --- --- --- --- --- --- ... ... ... --- ---
Z75357,IGHV4-28*01,VH4-GL5                                  - --- --- --- --- --- --- --- --- --- ... ... ... --- ---
M95112,IGHV4-28*01 or *03,H2          --- --- --- --- --- --- --- --- --- --- --- --- --- --- --- ... ... ... --- ---
M83133,IGHV4-28*02,hv4005             --- --- --- --- --- --- --- --- --- --- --- --- --- --- --- ... ... ... --- ---
L10099,IGHV4-28*02,3d24d              --- --- --- --- --- --- --- --- --- --- --- --- --- --- --- ... ... ... --- ---
X92233,IGHV4-28*03,4.42               --- --- --- --- --- --- --- --- --- --- --- --- --- --- --- ... ... ... --- ---
X56358,IGHV4-28*04,VH4.14             --- --- --- --- --- --- --- --- --- --- --- --- --- --- --- ... ... ... --- ---
X92260,IGHV4-28*05,VH4-MC7            --- --- --- --- --- --- --- --- --- --- --- --- --- --- --- ... ... ... --- ---

                                                                                          _________________CDR2-
                                      41  42  43  44  45  46  47  48  49  50  51  52  53  54  55  56  57  58  59  60
                                       W   I   R   Q   P   P   G   K   G   L   E   W   I   G   Y   I   Y   Y   S   G
X05714,IGHV4-28*01,V12G-1             TGG ATC CGG CAG CCC CCA GGG AAG GGA CTG GAG TGG ATT GGG TAC ATC TAT TAT AGT GGG
X92222,IGHV4-28*01,1.9II              --- --- --- --- --- --- --- --- --- --- --- --- --- --- --- --- --- --- --- ---
Z12368,IGHV4-28*01,DP-68              --- --- --- --- --- --- --- --- --- --- --- --- --- --- --- --- --- --- --- ---
L10096,IGHV4-28*01,3d28d              --- --- --- --- --- --- --- --- --- --- --- --- --- --- --- --- --- --- --- ---
X56357,IGHV4-28*01,VH4.13             --- --- --- --- --- --- --- --- --- --- --- --- --- --- --- --- --- --- --- ---
Z75357,IGHV4-28*01,VH4-GL5            --- --- --- --- --- --- --- --- --- --- --- --- --- --- --- --- --- --- --- ---
M95112,IGHV4-28*01 or *03,H2          --- --- --- --- --- --- --- --- --- --- --- --- --- --- --- --- --- --- --- ---
M83133,IGHV4-28*02,hv4005             --- --- --- --- --- --- --- --- --- --- --- --- --- --- --- --- --- --- --- ---
L10099,IGHV4-28*02,3d24d              --- --- --- --- --- --- --- --- --- --- --- --- --- --- --- --- --- --- --- ---
X92233,IGHV4-28*03,4.42               --- --- --- --- --- --- --- --- --- --- --- --- --- --- --- --- --- --- --- ---
X56358,IGHV4-28*04,VH4.14             --- --- --- --- --- --- --- --- --- --- --- --- --- --- --- --- --- --- --- ---
X92260,IGHV4-28*05,VH4-MC7            --- --- --- --- --- --- --- --- --- --- --- --- --- --- --- --- --- --- --- ---
```

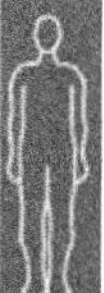

```
                                     IMGT_____________
                                     61  62  63  64  65  66  67  68  69  70  71  72  73  74  75  76  77  78  79  80
                                     S   T               Y   Y   N   P   S   L   K       S   R   V   T   M   S   V
X05714,IGHV4-28*01,V12G-1            AGC ACC ... ... ... TAC TAC AAC CCG TCC CTC AAG ... AGT CGA GTC ACC ATG TCA GTA
X92222,IGHV4-28*01,1.9II             --- --- ... ... ... --- --- --- --- --- --- --- ... --- --- --- --- --- --- ---
Z12368,IGHV4-28*01,DP-68             --- --- ... ... ... --- --- --- --- --- --- --- ... --- --- --- --- --- --- ---
L10096,IGHV4-28*01,3d28d             --- --- ... ... ... --- --- --- --- --- --- --- ... --- --- --- --- --- --- ---
X56357,IGHV4-28*01,VH4.13            --- --- ... ... ... --- --- --- --- --- --- --- ... --- --- --- --- --- --- ---
Z75357,IGHV4-28*01,VH4-GL5           --- --- ... ... ... --- --- --- --- --- --- --- ... --- --- --- --- --- --- ---
M95112,IGHV4-28*01 or *03,H2         --- --- ... ... ... --- --- --- --- --- --- --- ... --- --- --- --- --- --- ---
                                         I
M83133,IGHV4-28*02,hv4005            --- -T- ... ... ... --- --- --- --- --- --- --- ... --- --- --- --- --- --- ---
                                         I
[illegible]                              [illegible] ... ... [illegible] --- --- ... --- --- --- --- --- --- ---
X92233,IGHV4-28*03,4.42              --- --- ... ... ... --- --- --- --- --- --- --- ... --- --- --- --- --- --- ---
X56358,IGHV4-28*04,VH4.14            --- --- ... ... ... --- --- --- --- --- --- --- ... --- --- --- --- --- --- ---
                                         I
X92260,IGHV4-28*05,VH4-MC7           --- -T- ... ... ... --- --- --- --- --- --- --- ... --- --- --- --- --- --- ---

                                     81  82  83  84  85  86  87  88  89  90  91  92  93  94  95  96  97  98  99 100
                                     D   T   S   K   N   Q   F   S   L   K   L   S   S   V   T   A   V   D   T   A
X05714,IGHV4-28*01,V12G-1            GAC ACG TCC AAG AAC CAG TTC TCC CTG AAG CTG AGC TCT GTG ACC GCC GTG GAC ACG GCC
X92222,IGHV4-28*01,1.9II             --- --- --- --- --- --- --- --- --- --- --- --- --- --- --- --- --- --- --- ---
Z12368,IGHV4-28*01,DP-68             --- --- --- --- --- --- --- --- --- --- --- --- --- --- --- --- --- --- --- ---
L10096,IGHV4-28*01,3d28d             --- --- --- --- --- --- --- --- --- --- --- --- --- --- --- --- --- --- --- ---
X56357,IGHV4-28*01,VH4.13            --- --- --- --- --- --- --- --- --- --- --- --- --- --- --- --- --- --- --- ---
Z75357,IGHV4-28*01,VH4-GL5           --- --- --- --- --- --- --- --- --- --- --- --- --- --- --- --- --- --- --- ---
M95112,IGHV4-28*01 or *03,H2         --- --- --- --- --- --- --- --- --- --- --- --- --- --- --- --- --- --- --- ---
M83133,IGHV4-28*02,hv4005            --- --- --- --- --- --- --- --- --- --- --- --- --- --- --- --- --- --- --- ---
L10099,IGHV4-28*02,3d24d             --- --- --- --- --- --- --- --- --- --- --- --- --- --- --- --- --- --- --- ---
X92233,IGHV4-28*03,4.42              --- --- --- --- --- --- --- --- --- --- --- --- --- --- --- --- --- --- --- ---
                                                                                                                 G
X56358,IGHV4-28*04,VH4.14            --- --- --- --- --- --- --- --- --- --- --- --- --- --- --- --- --- --- --C -G-
X92260,IGHV4-28*05,VH4-MC7           --- --- --- --- --- --- --- --- --- --- --- --- --- --- --- --- --- --- --- ---

                                                    _CDR3-IMGT
                                     101 102 103 104 105 106 107
                                     V   Y   Y   C   A   R
X05714,IGHV4-28*01,V12G-1            GTG TAT TAC TGT GCG AGA AA
X92222,IGHV4-28*01,1.9II             --- --- --- --- --- --- --
Z12368,IGHV4-28*01,DP-68             --- --- --- --- --- ---
L10096,IGHV4-28*01,3d28d             --- --- --- --- --- --- --
X56357,IGHV4-28*01,VH4.13            --- --- --- --- --- ---
Z75357,IGHV4-28*01,VH4-GL5           --- --- --- --- --- --- --
M95112,IGHV4-28*01 or *03,H2         --- --- --- --- ---
M83133,IGHV4-28*02,hv4005            --- --- --- --- --- --- --
L10099,IGHV4-28*02,3d24d             --- --- --- --- --- --- --
X92233,IGHV4-28*03,4.42              --- --- --- --- --- --- G-
X56358,IGHV4-28*04,VH4.14            --- --- --- --- --- ---
X92260,IGHV4-28*05,VH4-MC7           --- --- --- --
```

Note:

(1) It can not be determined whether the M95112 sequence belongs to IGHV4-28*01 or IGHV4-28*03 alleles that only differ by a single nucleotide at position 319.

Framework and complementarity determining regions

FR1-IMGT: 25 (-1 aa: 10)
FR2-IMGT: 17
FR3-IMGT: 38 (-1 aa: 73)

CDR1-IMGT: 9
CDR2-IMGT: 7
CDR3-IMGT: 2

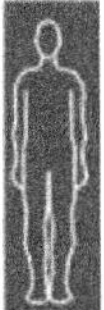

Collier de Perles for human IGHV4-28*01

Accession number: IMGT X05714 EMBL/GenBank/DDBJ: X05714

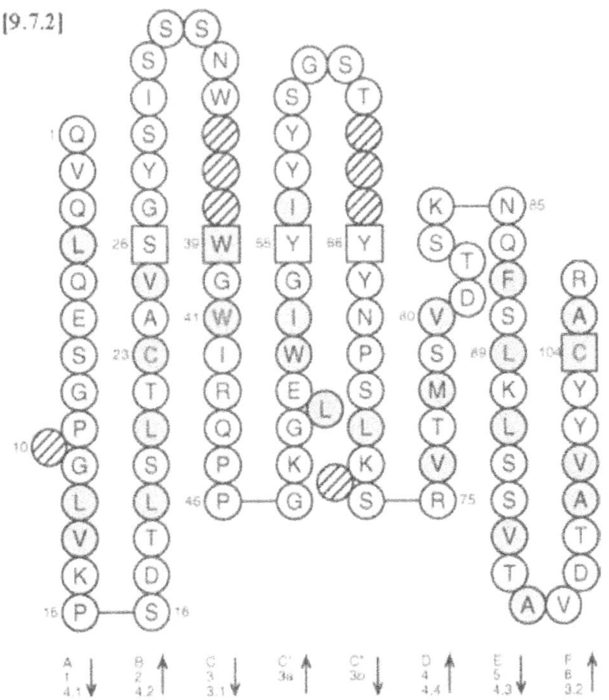

Genome database accession numbers

GDB:9931732 LocusLink: 28400

Nomenclature

IGHV4-30-2: Immunoglobulin heavy variable 4-30-2.

Definition and functionality

IGHV4-30-2 is a mapped functional gene, which may, or may not, be present due to a polymorphism by insertion/deletion. IGHV4-30-2 belongs to the IGHV4 subgroup which comprises 9–12 mapped genes (of which 6–10 are functional) in the IGH locus, depending on the haplotypes, one unmapped functional gene and one orphon.

Gene location

IGHV4-30-2, in the haplotypes where it is present, is in the IGH locus on chromosome 14 at 14q32.33.

Nucleotide and amino acid sequences for human IGHV4-30-2

```
                                            1   2   3   4   5   6   7   8   9   10  11  12  13  14  15  16  17  18  19  20
                                            Q   L   Q   L   Q   E   S   G   S       G   L   V   K   P   S   Q   T   L   S
L10089,IGHV4-30-2*01,3d216d          [37]   CAG CTG CAG CTG CAG GAG TCC GGC TCA ... GGA CTG GTG AAG CCT TCA CAG ACC CTG TCC
Z12364,IGHV4-30-2*01,DP-64           [41]   --- --- --- --- --- --- --- --- --- ... --- --- --- --- --- --- --- --- --- ---
Z75349,IGHV4-30-2*01,VH4-GL12        [49]
M95122,IGHV4-30-2*02,H12             [45]   --- --- --- --- --- --- --- --- --- ... --- --- --- --- --- --- --- --- --- ---
X92229,IGHV4-30-2*03,4.31            [50]   --- --- --- --- --- --- --- --- --- ... --- --- --- --- --- --- --- --- --- ---
Z75351,IGHV4-30-2*04,VH4-GL17        [49]

                                                                    __________CDR1-IMGT_____________
                                            21  22  23  24  25  26  27  28  29  30  31  32  33  34  35  36  37  38  39  40
                                            L   T   C   A   V   S   G   G   S   I   S   S   G   G   Y   S           W   S
L10089,IGHV4-30-2*01,3d216d                 CTC ACC TGC GCT GTC TCT GGT GGC TCC ATC AGC AGT GGT GGT TAC TCC ... ... TGG AGC
Z12364,IGHV4-30-2*01,DP-64                  --- --- --- --- --- --- --- --- --- --- --- --- --- --- --- --- ... ... --- ---
Z75349,IGHV4-30-2*01,VH4-GL12                                       --- --- --- --- --- --- --- --- --- --- ... ... --- ---
M95122,IGHV4-30-2*02,H12                    --- --- --- --- --- --- --- --- --- --- --- --- --- --- --- --- ... ... ---
X92229,IGHV4-30-2*03,4.31                       --- --- --- --- --- --- --- --- --- --- --- --- --- --- --- ... ... --- ---
Z75351,IGHV4-30-2*04,VH4-GL17                                   --- --- --- --- --- --- --- --- --- --- --- ... ... --- ---

                                                                                                    ______________CDR2-
                                            41  42  43  44  45  46  47  48  49  50  51  52  53  54  55  56  57  58  59  60
                                            W   I   R   Q   P   P   G   K   G   L   E   W   I   G   Y   I   Y   H   S   G
L10089,IGHV4-30-2*01,3d216d                 TGG ATC CGG CAG CCA CCA GGG AAG GGC CTG GAG TGG ATT GGG TAC ATC TAT CAT AGT GGG
Z12364,IGHV4-30-2*01,DP-64                  --- --- --- --- --- --- --- --- --- --- --- --- --- --- --- --- --- --- --- ---
Z75349,IGHV4-30-2*01,VH4-GL12               --- --- --- --- --- --- --- --- --- --- --- --- --- --- --- --- --- --- --- ---
M95122,IGHV4-30-2*02,H12                    --- --- --- --- --- --- --- --- --- --- --- --- --- --- --- --- --- --- --- ---
                                                                                            S               Y
X92229,IGHV4-30-2*03,4.31                   --- --- --- --- --- --- --- --- --- --- --- --- --- --- AGT --- --- T-- --- ---
Z75351,IGHV4-30-2*04,VH4-GL17               --- --- --- --- --- --- --- --- --- --- --- --- --- --- --- --- --- --- --- ---

                                            IMGT_________
                                            61  62  63  64  65  66  67  68  69  70  71  72  73  74  75  76  77  78  79  80
                                            S   T               Y   Y   N   P   S   L   K       S   R   V   T   I   S   V
L10089,IGHV4-30-2*01,3d216d                 AGC ACC ... ... ... TAC TAC AAC CCG TCC CTC AAG ... AGT CGA GTC ACC ATA TCA GTA
Z12364,IGHV4-30-2*01,DP-64                  --- --- ... ... ... --- --- --- --- --- --- --- ... --- --- --- --- --- --- ---
Z75349,IGHV4-30-2*01,VH4-GL12               --- --- ... ... ... --- --- --- --- --- --- --- ... --- --- --- --- --- --- ---
M95122,IGHV4-30-2*02,H12                    --- --- ... ... ... --- --- --- --- --- --- --- ... --- --- --- --- --- --- ---
X92229,IGHV4-30-2*03,4.31                   --- --- ... ... ... --- --- --- --- --- --- --- ... --- --- --- --- --- -C- ---
Z75351,IGHV4-30-2*04,VH4-GL17               --- --- ... ... ... --- --- --- --- --- --- --- ... --- --- --- --- --- --- ---

                                            81  82  83  84  85  86  87  88  89  90  91  92  93  94  95  96  97  98  99  100
                                            D   R   S   K   N   Q   F   S   L   K   L   S   S   V   T   A   A   D   T   A
L10089,IGHV4-30-2*01,3d216d                 GAC AGG TCC AAG AAC CAG TTC TCC CTG AAG CTG AGC TCT GTG ACC GCC GCG GAC ACG GCC
Z12364,IGHV4-30-2*01,DP-64                  --- --- --- --- --- --- --- --- --- --- --- --- --- --- --- --- --- --- --- ---
Z75349,IGHV4-30-2*01,VH4-GL12               --- --- --- --- --- --- --- --- --- --- --- --- --- --- --- --- --- --- --- ---
M95122,IGHV4-30-2*02,H12                    --- --- --- --- --- --- --- --- --- --- --- --- --- --- --- --T --- --- --- ---
                                                T
X92229,IGHV4-30-2*03,4.31                   --- -C- --- --- --- --- --- --- --- --- --- --- --- --- --- --T -A- --- --- --T
                                                T
Z75351,IGHV4-30-2*04,VH4-GL17               --- -C- --- --- --- --- --- --- --- --- --- --- --- --- --- --- --A --- --- ---
```

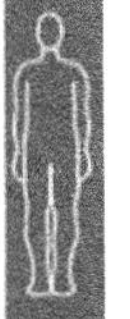

```
                                                  _CDR3-IMGT
                               101 102 103 104 105 106
                                V   Y   Y   C   A   R
L10089,IGHV4-30-2*01,3d216d    GTG TAT TAC TGT GCC AGA GA
Z12364,IGHV4-30-2*01,DP-64     --- --- --- --- --- ---
Z75349,IGHV4-30-2*01,VH4-GL12  --- --- --- --- --- --- --
M95122,IGHV4-30-2*02,H12       --- --- --- --- --G
X92229,IGHV4-30-2*03,4.31      --- - - --- --- --G --- CA
Z75351,IGHV4-30-2*04,VH4-GL17  --- --- --- --- --G --- --
```

Framework and complementarity determining regions

FR1-IMGT: 25 (-1 aa: 10)
FR2-IMGT: 17
FR3-IMGT: 38 (-1 aa: 73)

CDR1-IMGT: 10
CDR2-IMGT: 7
CDR3-IMGT: 2

Collier de Perles for human IGHV4-30-2*01

Accession number: IMGT L10089

EMBL/GenBank/DDBJ: L10089

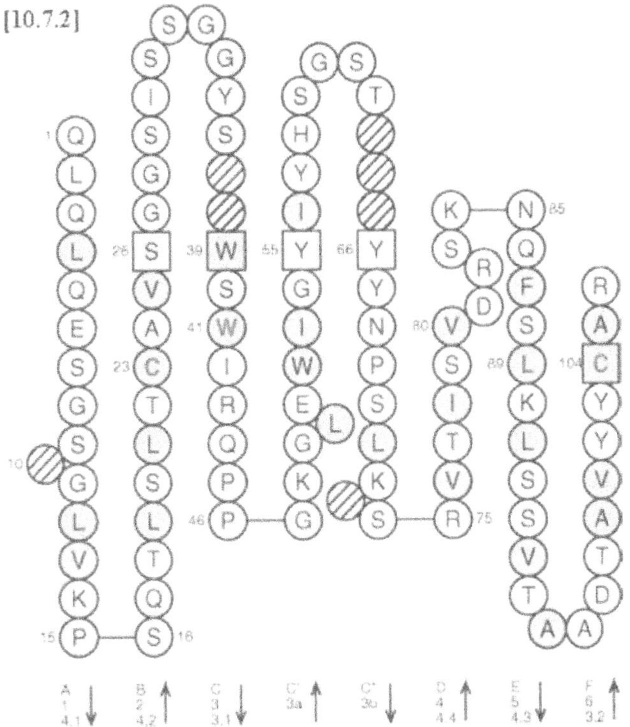

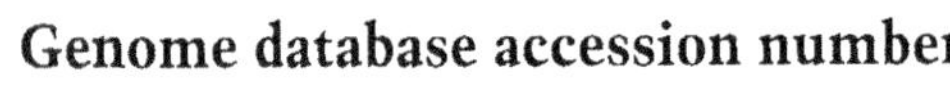

Genome database accession numbers

GDB:9953352

LocusLink: 28398

IGHV4-30-4

Nomenclature

IGHV4-30-4: Immunoglobulin heavy variable 4-30-4.

Definition and functionality

IGHV4-30-4 is a mapped functional gene, which may, or may not, be present due to a polymorphism by insertion/deletion. IGHV4-30-4 belongs to the IGHV4 subgroup which comprises 9–12 mapped genes (of which 6–10 are functional) in the IGH locus, depending on the haplotypes, one unmapped functional gene and one orphon.

Gene location

IGHV4-30-4, in the haplotypes where it is present, is in the IGH locus on chromosome 14 at 14q32.33.

Nucleotide and amino acid sequences for human IGHV4-30-4

```
                                      1   2   3   4   5   6   7   8   9  10  11  12  13  14  15  16  17  18  19  20
                                      Q   V   Q   L   Q   E   S   G   P       G   L   V   K   P   S   Q   T   L   S
Z14238,IGHV4-30-4*01,4.34        [50] CAG GTG CAG CTG CAG GAG TCG GGC CCA ... GGA CTG GTG AAG CCT TCA CAG ACC CTG TCC
Z14074,IGHV4-30-4*01,DP-78       [41]
L10100,IGHV4-30-4*01,3d230d      [47] --- --- --- --- --- --- --- --- --- ... --- --- --- --- --- --- --- --- --- ---
                                                                                                      D
Z14239,IGHV4-30-4*02,4.34.2      [50] --- --- --- --- --- --- --- --- --- ... --- --- --- --- --- --G G-C --- --- ---
X92274,IGHV4-30-4*03,VH4-MC9e     [7] --- --- --- --- --- --- --- --- --- ... --- --- --- --- --- --- --- --- --- ---
                                                          D
X92275,IGHV4-30-4*04,VH4-MC9f     [7] --- --- --- --- --- --C --- --- --- ... --- --- --- --- --- --- --- --- --- ---
Z75353,IGHV4-30-4*05,VH4-GL2     [49]
Z75360,IGHV4-30-4*06,VH4-GL8     [49]

                                                          ______________CDR1-IMGT______________
                                     21  22  23  24  25  26  27  28  29  30  31  32  33  34  35  36  37  38  39  40
                                      L   T   C   T   V   S   G   G   S   I   S   S   G   D   Y   Y           W   S
Z14238,IGHV4-30-4*01,4.34             CTC ACC TGC ACT GTC TCT GGT GGC TCC ATC AGC AGT GGT GAT TAC TAC ... ... TGG AGT
Z14074,IGHV4-30-4*01,DP-78                        --- --- --- --- --- --- --- --- --- --- --- --- --- ... ... --- ---
L10100,IGHV4-30-4*01,3d230d           --- --- --- --- --- --- --- --- --- --- --- --- --- --- --- --- ... ... --- ---
Z14239,IGHV4-30-4*02,4.34.2           --- --- --- --- --- --- --- --- --- --- --- --- --- --- --- --- ... ... --- ---
X92274,IGHV4-30-4*03,VH4-MC9e         --- --- --- --- --- --- --- --- --- --- --- --- --- --- --- --- ... ... --- ---
X92275,IGHV4-30-4*04,VH4-MC9f         --- --- --- --- --- --- --- --- --- --- --- --- --- --- --- --- ... ... --- ---
Z75353,IGHV4-30-4*05,VH4-GL2                              --- --- --- --- --- --- --- --- --- --- --- ... ... --- ---
Z75360,IGHV4-30-4*06,VH4-GL8                          --- --- --- --- --- --- --- --- --- --- --- --- ... ... --- ---

                                                                                                ________CDR2-
                                     41  42  43  44  45  46  47  48  49  50  51  52  53  54  55  56  57  58  59  60
                                      W   I   R   Q   P   P   G   K   G   L   E   W   I   G   Y   I   Y   Y   S   G
Z14238,IGHV4-30-4*01,4.34             TGG ATC CGC CAG CCC CCA GGG AAG GGC CTG GAG TGG ATT GGG TAC ATC TAT TAC AGT GGG
Z14074,IGHV4-30-4*01,DP-78            --- --- --- --- --- --- --- --- --- --- --- --- --- --- --- --- --- --- --- ---
L10100,IGHV4-30-4*01,3d230d           --- --- --- --- --- --- --- --- --- --- --- --- --- --- --- --- --- --- --- ---
Z14239,IGHV4-30-4*02,4.34.2           --- --- --- --- --- --- --- --- --- --- --- --- --- --- --- --- --- --- --- ---
X92274,IGHV4-30-4*03,VH4-MC9e         --- --- --- --- --- --- --- --- --- --- --- --- --- --- --- --- --- --- --- ---
                                                                                              F
X92275,IGHV4-30-4*04,VH4-MC9f         --- --- --- --- --- --- --- --- --- --- --- --- --- --- --- T-- --- --- --- ---
Z75353,IGHV4-30-4*05,VH4-GL2          --- --- --- --- -X- --- --- --- --- --- --- --- --- --- --- --- --- --- --- ---
                                                       H
Z75360,IGHV4-30-4*06,VH4-GL8          --- --- --- --- -A- --- --- --- --- --- --- --- --- --- --- --- --- --- --- ---

                                      IMGT_______________
                                     61  62  63  64  65  66  67  68  69  70  71  72  73  74  75  76  77  78  79  80
                                      S   T               Y   Y   N   P   S   L   K       S   R   V   T   I   S   V
Z14238,IGHV4-30-4*01,4.34             AGC ACC ... ... ... TAC TAC AAC CCG TCC CTC AAG ... AGT CGA GTT ACC ATA TCA GTA
Z14074,IGHV4-30-4*01,DP-78            --- --- ... ... ... --- --- ---
L10100,IGHV4-30-4*01,3d230d           --- --- ... ... ... --- --- --- --- --- --- --- ... --- --- --- --- --- --- ---
Z14239,IGHV4-30-4*02,4.34.2           --- --- ... ... ... --- --- --- --- --- --- --- ... --- --- --- --- --- --- ---
X92274,IGHV4-30-4*03,VH4-MC9e         --- --- ... ... ... --- --- --- --- --- --- --- ... --- --- --- --- --- --- ---
X92275,IGHV4-30-4*04,VH4-MC9f         --- --- ... ... ... --- --- --- --- --- --- --- ... --- --- --- --- --- --- ---
Z75353,IGHV4-30-4*05,VH4-GL2          --- --- ... ... ... --- --- --- --- --- --- --- ... --- --- --C --- --- --- ---
Z75360,IGHV4-30-4*06,VH4-GL8          --- --- ... ... ... --- --- --- --- --- --- --- ... --- --- --- --- --- --- ---
```

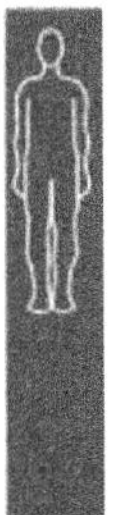

```
                                    81  82  83  84  85  86  87  88  89  90  91  92  93  94  95  96  97  98  99 100
                                     D   T   S   K   N   Q   F   S   L   K   L   S   S   V   T   A   A   D   T   A
Z14238,IGHV4-30-4*01,4.34           GAC ACG TCC AAG AAC CAG TTC TCC CTG AAG CTG AGC TCT GTG ACT GCC GCA GAC ACG GCC
Z14074,IGHV4-30-4*01,DP-78
L10100,IGHV4-30-4*01,3d230d         --- --- --- --- --- --- --- --- --- --- --- --- --- --- --- --- --- --- --- ---
Z14239,IGHV4-30-4*02,4.34.2         --- --- --- --- --- --- --- --- --- --- --- --- --- --- --- --A --- --- --- ---
X92274,IGHV4-30-4*03,VH4-MC9e       --- --- --- --- --- --- --- --- --- --- --- --- --- --- --- --- --G --- --- ---
X92275,IGHV4-30-4*04,VH4-MC9f       --- --- --- --- --- --- --- --- --- --- --- --  --- --- --- --- --- --- --- ---
Z75353,IGHV4-30-4*05,VH4-GL2        --- --- --  --- --- --- --- --- --- --- --- --- --- --- --- --- --- --- --- ---
Z75360,IGHV4-30-4*06,VH4-GL8        --- --- --- --- --- --- --- --- --- ---  -- --- --- --- --- --- --- --- --- ---
```

```
                                                        _CDR3-IMGT
                                    101 102 103 104 105 106
                                     V   Y   Y   C   A   R
Z14238,IGHV4-30-4*01,4.34           GTG TAT TAC TGT GCC AGA GA
Z14074,IGHV4-30-4*01,DP-78
L10100,IGHV4-30-4*01,3d230d         --- --- --- --- --- --- --
Z14239,IGHV4-30-4*02,4.34.2         --- --- --- --- --- ---
X92274,IGHV4-30-4*03,VH4-MC9e       --- --- --- --
X92275,IGHV4-30-4*04,VH4-MC9f       --- --- --- --
Z75353,IGHV4-30-4*05,VH4-GL2        --- --- --- --- --- --- --
Z75360,IGHV4-30-4*06,VH4-GL8        --- --- --- --- --- --- --
```

Framework and complementarity determining regions

FR1-IMGT: 25 (-1 aa: 10)
FR2-IMGT: 17
FR3-IMGT: 38 (-1 aa: 73)

CDR1-IMGT: 10
CDR2-IMGT: 7
CDR3-IMGT: 2

Collier de Perles for human IGHV4-30-4*01

Accession number: IMGT Z14238 EMBL/GenBank/DDBJ: Z14238

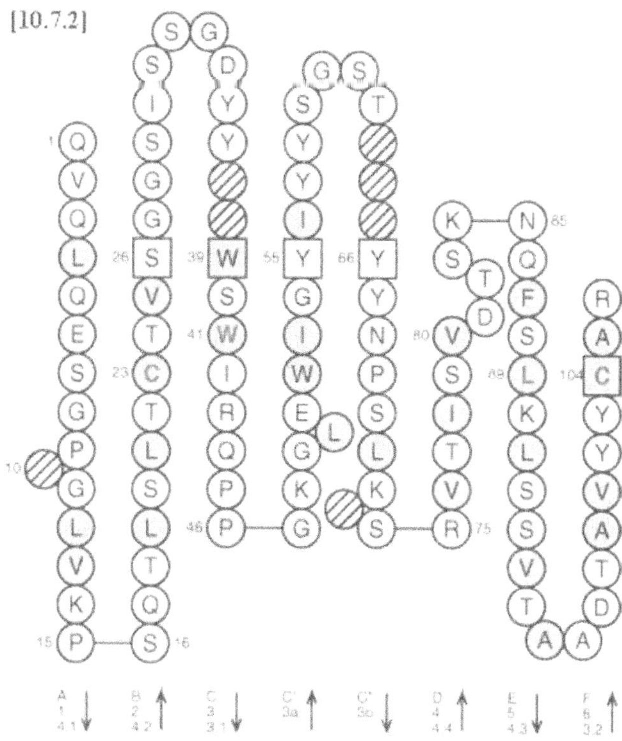

Genome database accession numbers

GDB:9953354 LocusLink: 28397

Nomenclature

IGHV4-31: Immunoglobulin heavy variable 4-31.
IGHV4-30-1: Immunoglobulin heavy variable 4-30-1.
Sequences of the polymorphic IGHV4-30-1 gene cannot be differentiated from those of the IGHV4-31 gene. All sequences are described therefore as "IGHV4-31 alleles" and compare to the allele *01 of IGHV4-31 (L10098). However, it is not excluded that some of these "alleles" belong exclusively to IGH4-30-1.

Definition and functionality

IGHV4-31 is one of the 6–10 mapped functional genes of the IGHV4 subgroup. This subgroup comprises in addition to 9–12 mapped genes in the IGH locus, depending on the haplotypes, one unmapped functional gene and one orphon.

Gene location

IGHV4-31 is in the IGH locus on chromosome 14 at 14q32.33.

Nucleotide and amino acid sequences for human IGHV4-31

```
                                   1   2   3   4   5   6   7   8   9  10  11  12  13  14  15  16  17  18  19  20
                                   Q   V   Q   L   Q   E   S   G   P       G   L   V   K   P   S   Q   T   L   S
L10098,IGHV4-31*01,3d277d   [37]  CAG GTG CAG CTG CAG GAG TCG GGC CCA ... GGA CTG GTG AAG CCT TCA CAG ACC CTG TCC
M99683,IGHV4-31*02,V4-31    [26]  --- --- --- --- --- --- --- --- --- ... --- --- --- --- --- --- --- --- --- ---
Z75358,IGHV4-31*02,VH4-GL6  [49]
Z14237,IGVH4-31*03,4.33     [50]  --- --- --- --- --- --- --- --- --- ... --- --- --- --- --- --- --- --- --- ---
Z12365,IGVH4-31*03,DP65     [41]  --- --- --- --- --- --- --- --- --- ... --- --- --- --- --- --- --- --- --- ---
L10095,IGVH4-31*03,3d75d    [47]  --- --- --- --- --- --- --- --- --- ... --- --- --- --- --- --- --- --- --- ---
X92269,IGVH4-31*03,VH4-MC9   [7]  --- --- --- --- --- --- --- --- --- ... --- --- --- --- --- --- --- --- --- ---
                                           R
M95120,IGVH4-31*04,H10      [45]  --- --- -G- --- --- --- --- --- --- ... --- --- --- --- --- --- --- --- --- ---
M95121,IGVH4-31*05,H11      [45]  --- --- --- --- --- --- --- --- --- ... --- --- --- --- --- --- --- --- --- ---
X92270,IGVH4-31*06,VH4-MC9a  [7]  --- --- --- --- --- --- --- --- --- ... --- --- --- --- --- --- --- --- --- ---
X92271,IGVH4-31*07,VH4-MC9b  [7]  --- --- --- --- --- --- --- --- --- ... --- --- --- --- --- --- --- --- --- ---
X92272,IGVH4-31*08,VH4-MC9c  [7]  --- --- --- --- --- --- --- --- --- ... --- --- --- --- --- --- --- --- --- ---
X92273,IGVH4-31*09,VH4-MC9d  [7]  --- --- --- --- --- --- --- --- --- ... --- --- --- --- --- --- --- --- --- ---
Z14235,IGVH4-31*10,4.32     [50]  --- --- --- --- --- --- --- --- --- ... --- --- --- --- --- --- --- --- --- ---

                                                          ____________________CDR1-IMGT____________________
                                  21  22  23  24  25  26  27  28  29  30  31  32  33  34  35  36  37  38  39  40
                                   L   T   C   T   V   S   G   G   S   I   S   S   G   G   Y   Y           W   S
L10098,IGHV4-31*01,3d277d         CTC ACC TGC ACT GTC TCT GGT GGC TCC ATC AGC AGT GGT GGT TAC TAC ... ... TGG AGC
M99683,IGHV4-31*02,V4-31          --- --- --T --- --- --- --- --- --- --- --- --- --- --- --- --- ... ... --- ---
Z75358,IGHV4-31*02,VH4-GL6                        - --- --- --- --- --- --- --- --- --- --- --- ... ... --- ---
Z14237,IGVH4-31*03,4.33           --- --- --- --- --- --- --- --- --- --- --- --- --- --- --- --- ... ... --- ---
Z12365,IGVH4-31*03,DP65           --- --- --- --- --- --- --- --- --- --- --- --- --- --- --- --- ... ... --- ---
L10095,IGVH4-31*03,3d75d          --- --- --- --- --- --- --- --- --- --- --- --- --- --- --- --- ... ... --- ---
X92269,IGVH4-31*03,VH4-MC9        --- --- --- --- --- --- --- --- --- --- --- --- --- --- --- --- ... ... --- ---
M95120,IGVH4-31*04,H10            --- --- --- --- --- --- --- --- --- --- --- --- --- --- --- --- ... ... --- ---
M95121,IGVH4-31*05,H11            --- --- --- --- --- --- --- --- --- --- --- --- --- --- --- --- ... ... --- ---
                                                                                   S
X92270,IGVH4-31*06,VH4-MC9a       --- --- --- --- --- --- --- --- --- --- --- --- --- A-- --- --- ... ... --- ---
X92271,IGVH4-31*07,VH4-MC9b       --- --- --- --- --- --- --- --A --- --- --- --- --- --- --- --- ... ... --- ---
X92272,IGVH4-31*08,VH4-MC9c       --- --- --- --- --- --- --- --- --- --- --- --- --- --- --- --- ... ... --- ---
X92273,IGVH4-31*09,VH4-MC9d       --- --- --- --- --- --- --- --- --- --- --- --- --- --- --- --- ... ... --- ---
Z14235,IGVH4-31*10,4.32           --- --- --- --- --- --- --- --- --- --- --- --- --- --- --- --- ... ... --- ---
```

```
                                                                                               ___________________CDR2-
                                   41  42  43  44  45  46  47  48  49  50  51  52  53  54  55  56  57  58  59  60
                                    W   I   R   Q   H   P   G   K   G   L   E   W   I   G   Y   I   Y   Y   S   G
L10098,IGHV4-31*01,3d277d          TGC ATC CGC CAG CAC CCA GGG AAG GGC CTG GAG TGG ATT GGG TAC ATC TAT TAC AGT GGG
M99683,IGHV4-31*02,V4-31           --- --- --- --- --- --- --- --- --- --- --- --- --- --- --- --- --- --- --- ---
Z75358,IGHV4-31*02,VH4-GL6         --- --- --- --- --- --- --- --- --- --- --- --- --- --- --- --- --- --- --- ---
Z14237,IGVH4-31*03,4.33            --- --- --- --- --- --- --- --- --- --- --- --- --- --- --- --- --- --- --- ---
Z12365,IGVH4-31*03,DP65            --- --- --- --- --- --- --- --- --- --- --- --- --- --- --- --- --- --- --- ---
L10095,IGVH4-31*03,3d75d           --- --- --- --- --- --- --- --- --- --- --- --- --- --- --- --- --- --- --- ---
X92269,IGVH4-31*03,VH4-MC9         --- --- --- --- --- --- --- --- --- --- --- --- --- --- --- --- --- --- --- ---
M95120,IGVH4-31*04,H10             --- --- --- --- --- --- --- --- --- --- --- --- --- --- --- --- --- --- --- ---
M95121,IGVH4-31*05,H11             --- --- --- --- --- ### ### ### ###
X92270,IGVH4-31*06,VH4 MC9a        --- --- --- --- --- --- --- --- --- --- --- --- --- --- --- --- --- --- --- ---
X92271,IGVH4-31*07,VH4-MC9b        --- --- --- --- --- --- --- --- --- --- --- --- --- --- --- --- --- --- --- ---
X92272,IGVH4-31*08,VH4-MC9c        --- --- --- --- --- --- --- --- --- --- --- --- --- --- --- --- --- --- --- ---
X92273,IGVH4-31*09,VH4-MC9d        --- --- --- --- --- --- --- --- --- --- --- --- --- --- --- --- --- --- --- ---
                                                                                            C
Z14235,IGVH4-31*10,4.32            --- --- --- --- --- --- --- --- --- --- --- --- --- --- -G- --- --- --- --- ---

                                   IMGT________________
                                   61  62  63  64  65  66  67  68  69  70  71  72  73  74  75  76  77  78  79  80
                                    S   T               Y   Y   N   P   S   L   K       S   L   V   T   I   S   V
L10098,IGHV4-31*01,3d277d          AGC ACC ... ... ... TAC TAC AAC CCG TCC CTC AAG ... AGT CTA GTT ACC ATA TCA GTA
                                                                                            R
M99683,IGHV4-31*02,V4-31           --- --- ... ... ... --- --- --- --- --- --- --- ... --- -G- --- --- --- --- ---
                                                                                            R
Z75358,IGHV4-31*02,VH4-GL6         --- --- ... ... ... --- --- --- --- --- --- --- ... --- -G- --- --- --- --- ---
                                                                                            R
Z14237,IGVH4-31*03,4.33            --- --- ... ... ... --- --- --- --- --- --- --- ... --- -G- --- --- --- --- ---
                                                                                            R
Z12365,IGVH4-31*03,DP65            --- --- ... ... ... --- --- --- --- --- --- --- ... --- -G- --- --- --- --- ---
                                                                                            R
L10095,IGVH4-31*03,3d75d           --- --- ... ... ... --- --- --- --- --- --- --- ... --- -G- --- --- --- --- ---
                                                                                            R
X92269,IGVH4-31*03,VH4-MC9         --- --- ... ... ... --- --- --- --- --- --- --- ... --- -G- --- --- --- --- ---
                                                                                            R
M95120,IGVH4-31*04,H10             --- --- ... ... ... --- --- --- --- --- --- --- ... --- -G- --- --- --- --- ---
                                                                                            R
M95121,IGVH4-31*05,H11             --- --- ... ... ... --- --- --- --- --- --- --- ... --- -G- --- --- --- --- ---
                                                                                            R
X92270,IGVH4-31*06,VH4-MC9a        --- --- ... ... ... --- --- --- --- --- --- --- ... --- -G- --- --- --- --- ---
                                                                                            R
X92271,IGVH4-31*07,VH4-MC9b        --- --- ... ... ... --- --- --- --- --- --- --- ... --- -G- --- --- --- --- ---
                                                                                            R
X92272,IGVH4-31*08,VH4-MC9c        --- --- ... ... ... --- --- --- --- --- --- --- ... --- -G- --- --- --- --C ---
                                                                                            R
X92273,IGVH4-31*09,VH4-MC9d        --- --- ... ... ... --- --- --- --- --- --- --- ... --- -G- --- --- --- --- ---
                                                                                            R
Z14235,IGVH4-31*10,4.32            --- --- ... ... ... - - --- --- --- --- --- --- ... --- -G- --- --- --- --- ---

                                   81  82  83  84  85  86  87  88  89  90  91  92  93  94  95  96  97  98  99  100
                                    D   T   S   K   N   Q   F   S   L   K   L   S   S   V   T   A   A   D   T   A
L10098,IGHV4-31*01,3d277d          GAC ACG TCT AAG AAC CAG TTC TCC CTG AAG CTG AGC TCT GTG ACT GCC GCG GAC ACG GCC
M99683,IGHV4-31*02,V4-31           --- --- --- --- --- --- --- --- --- --- --- --- --- --- --- --- --- --- --- ---
Z75358,IGHV4-31*02,VH4-GL6         --- --- --- --- --- --- --- --- --- --- --- --- --- --- --- --- --- --- --- ---
Z14237,IGVH4-31*03,4.33            --- --- --- --- --- --- --- --- --- --- --- --- --- --- --- --- --- --- --- ---
Z12365,IGVH4-31*03,DP65            --- --- --- --- --- --- --- --- --- --- --- --- --- --- --- --- --- --- --- ---
L10095,IGVH4-31*03,3d75d           --- --- --- --- --- --- --- --- --- --- --- --- --- --- --- --- --- --- --- ---
X92269,IGVH4-31*03,VH4-MC9         --- --- --- --- --- --- --- --- --- --- --- --- --- --- --- --- --- --- --- ---
M95120,IGVH4-31*04,H10             --- --- --- --- --- --- --- --- --- --- --- --- --- --- --- --- --- --- --- ---
M95121,IGVH4-31*05,H11         (1) --- --- --- --- --- --- --- --- --- --- --- --- --- --- --- --- --- --- G-- ---
X92270,IGVH4-31*06,VH4-MC9a        --- --- --- --- --- --- --- --- --- --- --- --- --- --- --- --- --- --- --- ---
X92271,IGVH4-31*07,VH4-MC9b        --- --- --- --- --- --- --- --- --- --- --- --- --- --- --- --- --- --- --- ---
X92272,IGVH4-31*08,VH4-MC9c        --- --- --C --- --- --- --- --- --- --- --- --- --- --- --- --- --- --- --- ---
                                        K
X92273,IGVH4-31*09,VH4-MC9d        --- -A- --C --- --- --- --- --- --- --- --- --- --- --- --C --- --- --- --- ---
                                        P                           P
Z14235,IGVH4-31*10,4.32            --- C-- --C --- --- --- --- --- -C- --- --- --- --- --- --- --- --- --- --- ---

                                                   _CDR3-IMGT
                                   101 102 103 104 105 106
                                    V   Y   Y   C   A   R
L10098,IGHV4-31*01,3d277d          GTG TAT TAC TGT GCG AGA GA
M99683,IGHV4-31*02,V4-31           --  --- --- --- --- --- --
Z75358,IGHV4-31*02,VH4-GL6         --- --- --- --- --- --- --
Z14237,IGVH4-31*03,4.33            --- --- --- --- --- --- --
Z12365,IGVH4-31*03,DP65            --- --- --- --- --- --- --
L10095,IGVH4-31*03,3d75d           --- --- --- --- --- ---
X92269,IGVH4-31*03,VH4-MC9         --- --- --- --
M95120,IGVH4-31*04,H10             --- --- --- --- ---
M95121,IGVH4-31*05,H11             --- --- ---
X92270,IGVH4-31*06,VH4-MC9a        --- --- --- --
X92271,IGVH4-31*07,VH4-MC9b        --- --- --- --
X92272,IGVH4-31*08,VH4-MC9c        --- --- --- --
X92273,IGVH4-31*09,VH4-MC9d        --- --- --- --
                                       D
Z14235,IGVH4-31*10,4.32            --- G-- --- --- --- ---
```

Note:

(1) It is not known if the 3 nt DELETION at position 285 to 287 (IMGT numbering) in M95121 is a typing error.

Framework and complementarity determining regions

FR1-IMGT: 25 (-1 aa: 10)
FR2-IMGT: 17
FR3-IMGT: 38 (-1 aa: 73)

CDR1-IMGT: 10
CDR2-IMGT: 7
CDR3-IMGT: 2

Collier de Perles for human IGHV4-31*01

Accession number: IMGT L10098 EMBL/GenBank/DDBJ: L10098

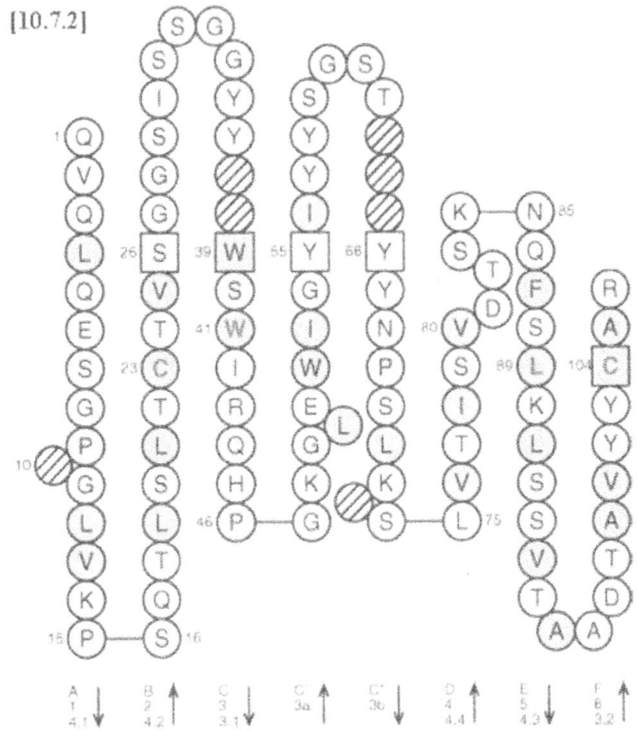

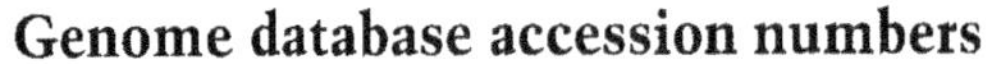

Genome database accession numbers

GDB:9931737 LocusLink: 28396

Nomenclature

IGHV4-34: Immunoglobulin heavy variable 4-34.

Definition and functionality

IGHV4-34 is one of the 6–10 mapped functional genes of the IGHV4 subgroup. This subgroup comprises 9–12 mapped genes in the IGH locus, depending on the haplotypes, one unmapped functional gene and one orphon.

Gene location

IGHV4-34 is in the IGH locus on chromosome 14 at 14q32.33.

Nucleotide and amino acid sequences for human IGHV4-34

```
                                        1   2   3   4   5   6   7   8   9  10  11  12  13  14  15  16  17  18  19  20
                                        Q   V   Q   L   Q   Q   W   G   A       G   L   L   K   P   S   E   T   L   S
X92278,IGHV4-34*01,VH5            [3]  CAG GTG CAG CTA CAG CAG TGG GGC GCA ... GGA CTG TTG AAG CCT TCG GAG ACC CTG TCC
Z12363,IGHV4-34*01,DP-63          [47] --- --- --- --- --- --- --- --- --- ... --- --- --- --- --- --- --- --- --- ---
X56364,IGHV4-34*01,VH4.21         [35] --- --- --- --- --- --- --- --- --- ... --- --- --- --- --- --- --- --- --- ---
L10090,IGHV4-34*01,4d76           [41] --- --- --- --- --- --- --- --- --- ... --- --- --- --- --- --- --- --- --- ---
Z75354,IGHV4-34*01,VH4-GL20       [49]
M99684,IGHV4-34*02,V4-34          [26] --- --- --- --- --A --- --- --- --- ... --- --- --- --- --- --- --- --- --- ---
X92255,IGHV4-34*03,VH4-MC5        [7]  --- --- --- --- --- --- --- --- --- ... --- --- --- --- --- --- --- --- --- ---
X92257,IGHV4-34*03,VH4-MC5b       [7]  --- --- --- --- --- --- --- --- --- ... --- --- --- --- --- --- --- --- --- ---
X92236,IGHV4-34*04,4.44           [50] --- --- --- --- --- --- --- --- --- ... --- --- --- --- --- --- --- --- --- ---
X92237,IGHV4-34*05,4.44.3         [50] --- --- --- --- --- --- --- --- --- ... --- --- --- --- --- --- --- --- --- ---
X92256,IGHV4-34*06,VH4-MC5a       [7]  --- --- --- --- --- --- --- --- --- ... --- --- --- --- --- --- --- --- --- ---
X92258,IGHV4-34*07,VH4-MC5c       [7]  --- --- --- --- --- --- --- --- --- ... --- --- --- --- --- --- --- --- --- ---
M95113,IGHV4-34*08,H3             [45] --- --- --- --- --- --- --- --- --- ... --- --- --- --- --- --- --- --- --- ---
                                                            E   S       P               V               Q
Z14241,IGHV4-34*09,4.36           [50] --- --- --- --G --- G-- -C- --- C-- ... --- --- G-- --- --- --A C-- --- --- ---
                                                            E   S       P               V
Z14242,IGHV4-34*10,4.37           [50] --- --- --- --G --- G-- -C- --- C-- ... --- --- G-- --- --- --- --- --- --- ---
X05716,IGHV4-34*11,V58            [24] --- --- --- --- --- --- --- --- --- ... --- --- --- --- --- --- --- --- --- ---
X56591,IGHV4-34*12,TouVH4.21      [46] --- --- --- --- --- --- --- --- --- ... --- --- --- --- --- --- --- --- --- ---
Z75356,IGHV4-34*13,VH4-GL4        [49]

                                                                         ______CDR1-IMGT______________________
                                       21  22  23  24  25  26  27  28  29  30  31  32  33  34  35  36  37  38  39  40
                                        L   T   C   A   V   Y   G   G   S   F   S   G   Y   Y                   W   S
X92278,IGHV4-34*01,VH5                 CTC ACC TGC GCT GTC TAT GGT GGG TCC TTC AGT GGT TAC TAC ... ... ... ... TGG AGC
Z12363,IGHV4-34*01,DP-63               --- --- --- --- --- --- --- --- --- --- --- --- --- --- ... ... ... ... --- ---
X56364,IGHV4-34*01,VH4.21              --- --- --- --- --- --- --- --- --- --- --- --- --- --- ... ... ... ... --- ---
L10090,IGHV4-34*01,4d76                --- --- --- --- --- --- --- --- --- --- --- --- --- --- ... ... ... ... --- ---
Z75354,IGHV4-34*01,VH4-GL20                                --- --- --- --- --- --- --- --- --- ... ... ... ... --- ---
M99684,IGHV4-34*02,V4-34               --- --- --- --- --- --- --- --- --- --- --- --- --- --- ... ... ... ... --- ---
X92255,IGHV4-34*03,VH4-MC5             --- --- --- --- --- --- --- --- --- --- --- --- --- --- ... ... ... ... --- ---
X92257,IGHV4-34*03,VH4-MC5b            --- --- --- --- --- --- --- --- --- --- --- --- --- --- ... ... ... ... --- ---
X92236,IGHV4-34*04,4.44                --- --- --- --- --- --- --- --- --- --- --- --- --- --- ... ... ... ... --- ---
                                                                                                            C
X92237,IGHV4-34*05,4.44.3              --- --- --- --- --- --- --- --- --- --- --- --- --- --- ... ... ... ... --- T--
X92256,IGHV4-34*06,VH4-MC5a            --- --- --- --- --- --- --- --- --- --- --- --- --- --- ... ... ... ... --- ---
X92258,IGHV4-34*07,VH4-MC5c            --- --- --- --- --- --- --- --- --- --- --- --- --- --- ... ... ... ... --- ---
                                                                        T
M95113,IGHV4-34*08,H3                  --- --- --- --- --- --- --- --- A-- --- --- --- --- --- ... ... ... ... --- ---
Z14241,IGHV4-34*09,4.36                --- --- --- --- --- --- --- --- --- --- --- --- --- --- ... ... ... ... --- ---
Z14242,IGHV4-34*10,4.37                --- --- --- --- --- --- --- --- --- --- --- --- --- --- ... ... ... ... --- ---
                                                                            V
X05716,IGHV4-34*11,V58                 --- --- --- --- --- --- --- --- --- G-- --- --- --- --- ... ... ... ... --- ---
X56591,IGHV4-34*12,TouVH4.21           --- --- --- --- --- --- --- --- --- --- --- --- --- --- ... ... ... ... --- ---
Z75356,IGHV4-34*13,VH4-GL4                                 --- --- --- --- --- --- --- --- --- ... ... ... ... --- ---

                                                                                                   _________CDR2-
                                       41  42  43  44  45  46  47  48  49  50  51  52  53  54  55  56  57  58  59  60
                                        W   I   R   Q   P   P   G   K   G   L   E   W   I   G   E   I   N   H   S   G
X92278,IGHV4-34*01,VH5                 TGG ATC CGC CAG CCC CCA GGG AAG GGG CTG GAG TGG ATT GGG GAA ATC AAT CAT AGT GGA
Z12363,IGHV4-34*01,DP-63               --- --- --- --- --- --- --- --- --- --- --- --- --- --- --- --- --- --- --- ---
X56364,IGHV4-34*01,VH4.21              --- --- --- --- --- --- --- --- --- --- --- --- --- --- --- --- --- --- --- ---
L10090,IGHV4-34*01,4d76                --- --- --- --- --- --- --- --- --- --- --- --- --- --- --- --- --- --- --- ---
```

```
Z75354,IGHV4-34*01,VH4-GL20         --- --- --- --- --- --- --- --- --- --- --- --- --- --- --- --- --- --- --- ---
M99684,IGHV4-34*02,V4-34            --- --- --- --- --- --- --- --- --- --- --- --- --- --- --- --- --- --- --- ---
X92255,IGHV4-34*03,VH4-MC5          --- --- --- --- --- --- --- --- --- --- --- --- --- --- --- --- --- --- --- ---
X92257,IGHV4-34*03,VH4-MC5b         --- --- --- --- --- --- --- --- --- --- --- --- --- --- --- --- --- --- --- ---
X92236,IGHV4-34*04,4.44             --- --- --- --- --- --- --- --- --- --- --- --- --- --- --- --- --- --- --- ---
                                                         L
X92237,IGHV4-34*05,4.44.3           --- --- --- --- --- -T- --- --- --- --- --- --- --- --- --- --- --- --- --- ---
X92256,IGHV4-34*06,VH4-MC5a         --- --- --- --- --- --- --- --- --- --- --- --- --- --- --- --- --- --- --- ---
X92258,IGHV4-34*07,VH4-MC5c         --- --- --- --- --- --- --- --- --- --- --- --- --- --- --- --- --C --- --- ---
M95113,IGHV4-34*08,H3               --- --- --- --- --- --- --- --- --- --- --- --- --- --- --- --- --- --- --- ---
Z14241,IGHV4-34*09,4.36             --- --- --- --- --- --- --- --- --A --- --- --- --- --- --- --- --- --- --- ---
Z14242,IGHV4-34*10,4.37             --- --- --- --- --- --- --- --- --A --- --- --- --- --- --- --- --- --- --- ---
                                                                                             Y       Y   Y
X05716,IGHV4-34*11,V58              --- --- --G --- --- --- --- --- --- --- --- --- --- --- T-T --- T-- T-- --- --G
                                                                                                     I
X56591,IGHV4-34*12,TouVH4.21        --- --- --- --- --- --- --- --- --- --- --- --- --- --- --- --- -T- --- --- ---
Z75356,IGHV4-34*13,VH4-GL4          --- --- --- --- --- --- --- --- --- --- --- --- --- --- --- --- --- --- --- ---

                                    IMGT________________
                                     61  62  63  64  65  66  67  68  69  70  71  72  73  74  75  76  77  78  79  80
                                     S   T               N   Y   N   P   S   L   K       S   R   V   T   I   S   V
X92278,IGHV4-34*01,VH5              AGC ACC ... ... ... AAC TAC AAC CCG TCC CTC AAG ... AGT CGA GTC ACC ATA TCA GTA
Z12363,IGHV4-34*01,DP-63            --- --- ... ... ... --- --- --- --- --- --- --- ... --- --- --- --- --- --- ---
X56364,IGHV4-34*01,VH4.21           --- --- ... ... ... --- --- --- --- --- --- --- ... --- --- --- --- --- --- ---
L10090,IGHV4-34*01,4d76             --- --- ... ... ... --- --- --- --- --- --- --- ... --- --- --- --- --- --- ---
M99684,IGHV4-34*02,V4-34            --- --- ... ... ... --- --- --- --- --- --- --- ... --- --- --- --- --- --- ---
X92255,IGHV4-34*03,VH4-MC5          --- --- ... ... ... --- --- --- --- --- --- --- ... --- --- --- --- --- --- ---
X92257,IGHV4-34*03,VH4-MC5b         --- --- ... ... ... --- --- --- --- --- --- --- ... --- --- --- --- --- --- ---
                                                             N                                   A
X92236,IGHV4-34*04,4.44             --- --- ... ... ... --- A-- --- --- --- --- --- ... --- --- -C- --- --- --- ---
                                                             N                                   A
X92237,IGHV4-34*05,4.44.3           --- --- ... ... ... --- A-- --- --- --- --- --- ... --- --- -C- --- --- --- ---
X92256,IGHV4-34*06,VH4-MC5a         --- --- ... ... ... --- --- --- --- --- --- --- ... --- --- --- --- --- --- ---
X92258,IGHV4-34*07,VH4-MC5c         --- --- ... ... ... --- --- --- --- --- --- --- ... --- --- --- --- --- --- ---
M95113,IGHV4-34*08,H3               --- --- ... ... ... --- --- --- --- --- --- --- ... --- --- --- --- --- --- ---
Z14241,IGHV4-34*09,4.36             --- --- ... ... ... --- --- --- --- --- --- --- ... --- --- --T --- --- --- ---
                                                                                                 I       M
Z14242,IGHV4-34*10,4.37             --- --- ... ... ... --- --- --- --- --- --- --- ... --- --- A-- --- --G --- ---
                                                             N                                   A
X05716,IGHV4-34*11,V58              --- --- ... ... ... --- A-- --- --C --- --- --- ... --- --- -C- --- --- --- ---
X56591,IGHV4-34*12,TouVH4.21        --- --- ... ... ... --- --- --- --- --- --- --- ... --- --- --- --- --- --- ---
Z75356,IGHV4-34*13,VH4-GL4          --- --- ... ... ... --- --- --- --- --- --- --- ... --- --- --- --- --- --- ---

                                     81  82  83  84  85  86  87  88  89  90  91  92  93  94  95  96  97  98  99 100
                                     D   T   S   K   N   Q   F   S   L   K   L   S   S   V   T   A   A   D   T   A
X92278,IGHV4-34*01,VH5              GAC ACG TCC AAG AAC CAG TTC TCC CTG AAG CTG AGC TCT GTG ACC GCC GCG GAC ACG GCT
Z12363,IGHV4-34*01,DP-63            --- --- --- --- --- --- --- --- --- --- --- --- --- --- --- --- --- --- --- ---
X56364,IGHV4-34*01,VH4.21           --- --- --- --- --- --- --- --- --- --- --- --- --- --- --- --- --- --- --- ---
L10090,IGHV4-34*01,4d76             --- --- --- --- --- --- --- --- --- --- --- --- --- --- --- --- --- --- --- ---
Z75354,IGHV4-34*01,VH4-GL20         --- --- --- --- --- --- --- --- --- --- --- --- --- --- --- --- --- --- --- ---
M99684,IGHV4-34*02,V4-34            --- --- --- --- --- --- --- --- --- --- --- --- --- --- --- --- --- --- --- ---
X92255,IGHV4-34*03,VH4-MC5          --- --- --- --- --- --- --- --- --- --- --- --- --- --- --- --- --- --- --- --C
X92257,IGHV4-34*03,VH4-MC5b         --- --- --- --- --- --- --- --- --- --- --- --- --- --- --- --- --- --- --- --C
X92236,IGHV4-34*04,4.44             --- --- --- --- --- --- --- --- --- --- --- --- --- --- --- --- --- --- --- ---
X92237,IGHV4-34*05,4.44.3           --- --- --- --- --- --- --- --- --- --- --- --- --- --- --- --- --- --- --- ---
                                                                                 G
X92256,IGHV4-34*06,VH4-MC5a         --- --- --- --- --- --- --- --- --- --- --- G-- --- --- --- --- --- --- --- --C
X92258,IGHV4-34*07,VH4-MC5c         --- --- --- --- --- --- --- --- --- --- --- --- --- --- --- --- --- --- --- --C
M95113,IGHV4-34*08,H3               --- --- --- --- --- --- --- --- --- --- --- --- --- --- --- --- --- --- --- ---
Z14241,IGHV4-34*09,4.36             --- --- --T --- --- --- --- --- --- --- --- --- --- --- --T --- --- --- --- --C
                                                                 Y
Z14242,IGHV4-34*10,4.37             --- --- --- --- --- --- --- -A- --- --- --- --- --- --- --- --- --- --- --- --C
                                                                         N
X05716,IGHV4-34*11,V58              --- --- --- --- --- --- --- --- --- --C --- --- --- --- --- --- --- --- --- --C
X56591,IGHV4-34*12,TouVH4.21        --- --- --- --- --- --- --- --- --- --- --- --- --- --- --- --- --- --- --- ---
Z75356,IGHV4-34*13,VH4-GL4          --- --- --- --- --- --- --- --- --C --- --- --- --- --- --- --- --- --- --- ---

                                                    _CDR3-IMGT
                                    101 102 103 104 105 106
                                     V   Y   Y   C   A   R
X92278,IGHV4-34*01,VH5              GTG TAT TAC TGT GCG AGA GG
Z12363,IGHV4-34*01,DP-63            --- --- --- --- --- ---
X56364,IGHV4-34*01,VH4.21           --- --- --- --- --- ---
L10090,IGHV4-34*01,4d76             --- --- --- --- --- --- --
Z75354,IGHV4-34*01,VH4-GL20         --- --- --- -
M99684,IGHV4-34*02,V4-34            --- --- --- --- --- --- --
X92255,IGHV4-34*03,VH4-MC5          --- --- --- --
X92257,IGHV4-34*03,VH4-MC5b         --- --- --- --
```

```
X92236,IGHV4-34*04,4.44             --- --- --- --- --- --- --
X92237,IGHV4-34*05,4.44.3           --- --- --- --- --- --- --
X92256,IGHV4-34*06,VH4-MC5a         --- --- --- --
X92258,IGHV4-34*07,VH4-MC5c         --- --- --- --
M95113,IGHV4-34*08,H3               --- --- --- --- ---
Z14241,IGHV4-34*09,4.36             --- --- --- --- --- ---
Z14242,IGHV4-34*10,4.37             --- --- --- --- --- ---
                                            C
X05716,IGHV4-34*11,V58              --- --- -G- --- --- ---
X56591,IGHV4-34*12,TouVH4.21        --- --- --- --- --- ---
Z75356,IGHV4-34*13,VH4-GL4          --- --- --- --- --- --- --
```

Framework and complementarity determining regions

FR1-IMGT: 25 (-1 aa: 10)
FR2-IMGT: 17
FR3-IMGT: 38 (-1 aa: 73)

CDR1-IMGT: 8
CDR2-IMGT: 7
CDR3-IMGT: 2

Collier de Perles for human IGHV4-34*01

Accession number: IMGT X92278

EMBL/GenBank/DDBJ: X92278

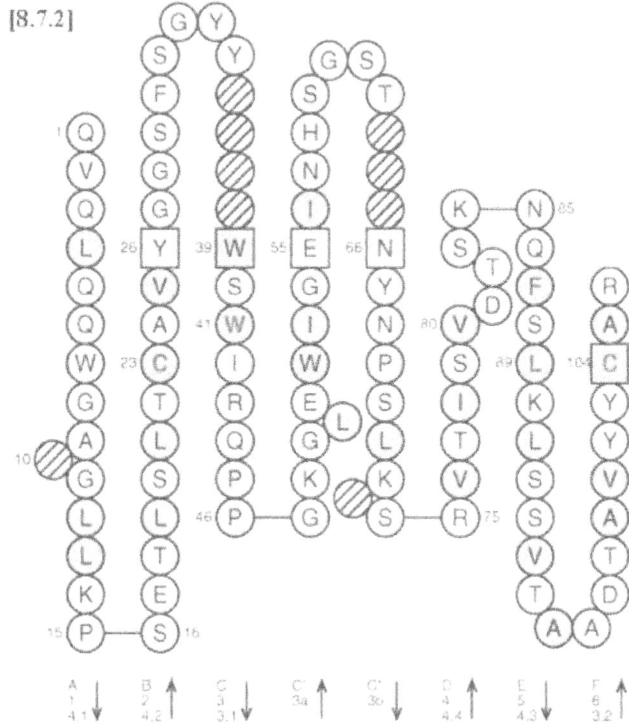

Genome database accession numbers

GDB:9931738

LocusLink: 28395

Nomenclature

IGHV4-39: Immunoglobulin heavy variable 4-39.

Definition and functionality

IGHV4-39 is one of the 6–10 mapped functional genes of the IGHV4 subgroup. This subgroup comprises 9–12 mapped genes in the IGH locus, depending on the haplotypes, one unmapped functional gene and one orphon.

Gene location

IGHV4-39 is in the IGH locus on chromosome 14 at 14q32.33.

Nucleotide and amino acid sequences for human IGHV4-39

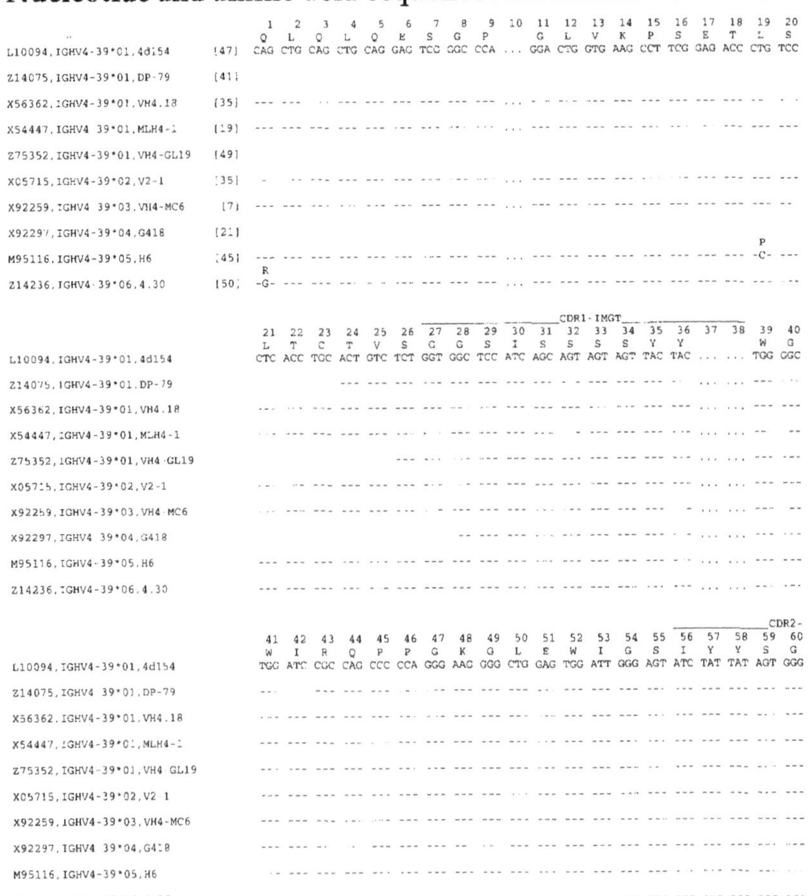

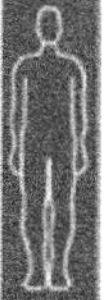

```
                                IMGT________________
                                61  62  63  64  65  66  67  68  69  70  71  72  73  74  75  76  77  78  79  80
                                S   T               Y   Y   N   P   S   L   K       S   R   V   T   I   S   V
L10094,IGHV4-39*01,4d154        AGC ACC ... ... ... TAC TAC AAC CCG TCC CTC AAG ... AGT CGA GTC ACC ATA TCC GTA

Z14075,IGHV4-39*01,DP-79        --- --- ... ... ... --- --- ---

X56362,IGHV4-39*01,VH4.18       --- --- ... ... ... --- --- --- --- --- --- --- ... --- --- --- --- --- --- ---

X54447,IGHV4-39*01,MLH4-1       --- --- ... ... ... --- --- --- --- --- --- --- ... --- --- --- --- --- --- ---

Z75352,IGHV4-39*01,VH4-GL19     --- --- ... ... ... --- --- --- --- --- --- --- ... --- --- --- --- --- --- ---

X05715,IGHV4-39*02,V2-1         --- --- ... ... ... --- --- --- --- --- --- --- ... --- --- --- --- --- --- ---

X92259,IGHV4-39*03,VH4-MC6      --- --- ... ... ... --- --- --- --- --- --- --- ... --- --- --- --- --- --- ---

X92297,IGHV4-39*04,G418         --- ---

M95116,IGHV4-39*05,H6           --- --- ... ... ... --- --- --- --- --- --- --- ... --- --- --- --- --- --- ---

Z14236,IGHV4-39*06,4.30         --- --- ... ... ... --- --- --- --- --- --- --- ... --- --- --- --- --- --A ---
```

```
                                81  82  83  84  85  86  87  88  89  90  91  92  93  94  95  96  97  98  99  100
                                D   T   S   K   N   Q   F   S   L   K   L   S   S   V   T   A   A   D   T   A
L10094,IGHV4-39*01,4d154        GAC ACG TCC AAG AAC CAG TTC TCC CTG AAG CTG AGC TCT GTG ACC GCC GCA GAC ACG GCT

Z14075,IGHV4-39*01,DP-79

X56362,IGHV4-39*01,VH4.18       --- --- --- --- --- --- --- --- --- --- --- --- --- --- --- --- --- --- --- ---

X54447,IGHV4-39*01,MLH4-1       --- --- --- --- --- --- --- --- --- --- --- --- --- --- --- --- --- --- --- ---

Z75352,IGHV4-39*01,VH4-GL19     --- --- --- --- --- --- --- --- --- --- --- --- --- --- --- --- --- --- --- ---
                                                     H
X05715,IGHV4-39*02,V2-1         --- --- --- --- --- --C --- --- --- --- --- --- --- --- --- --- --- --- --- ---

X92259,IGHV4-39*03,VH4-MC6      --- --- --- --- --- --- --- --- --- --- --- --- --- --- --- --- --- --- --- --C

X92297,IGHV4-39*04,G418         --- --- --- --- --- --- --- --- --- --- --- --- --- --- --- --- --G --- --

M95116,IGHV4-39*05,H6           --- --- --- --- --- --- --- --- --- --- --- --- --- --- --- --- --- --- --- ---
                                                             P
Z14236,IGHV4-39*06,4.30         --- --- --- --- --- --- --- C-- --- --- --- --- --- --- --- --- --G --- --- --C
```

```
                                                _CDR3-IMGT
                                101 102 103 104 105 106
                                V   Y   Y   C   A   R
L10094,IGHV4-39*01,4d154        GTG TAT TAC TGT GCG AGA CA

Z14075,IGHV4-39*01,DP-79

X56362,IGHV4-39*01,VH4.18       --- --- --- --- --- ---

X54447,IGHV4-39*01,MLH4-1       --- --- --- --

Z75352,IGHV4-39*01,VH4-GL19     --- --- --- --- --- --- --

X05715,IGHV4-39*02,V2-1         --- --- --- --- --- --- G-

X92259,IGHV4-39*03,VH4-MC6      --- --- --- --

X92297,IGHV4-39*04,G418

M95116,IGHV4-39*05,H6           --- --- --- --- ---

Z14236,IGHV4-39*06,4.30         --- --- --- --- --- ---
```

Framework and complementarity determining regions

FR1-IMGT: 25 (-1 aa: 10)
FR2-IMGT: 17
FR3-IMGT: 38 (-1 aa: 73)

CDR1-IMGT: 10
CDR2-IMGT: 7
CDR3-IMGT: 2

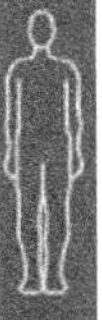

Collier de Perles for human IGHV4-39*01

Accession number: IMGT L10094 EMBL/GenBank/DDBJ: L10094

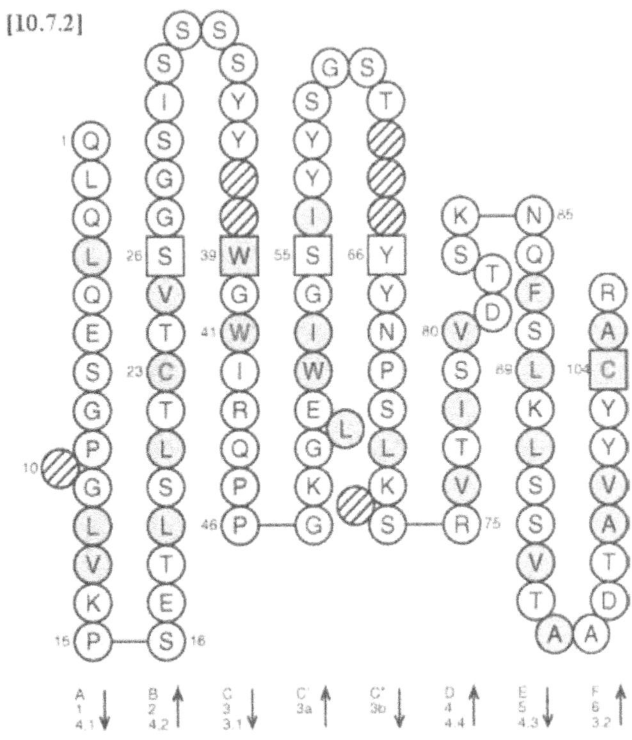

Genome database accession numbers

GDB:9931739 LocusLink: 28394

Nomenclature

IGHV4-59: Immunoglobulin heavy variable 4-59.

Definition and functionality

IGHV4-59 is one of the 6–10 mapped functional genes of the IGHV4 subgroup. This subgroup comprises 9–12 mapped genes in the IGH locus, depending on the haplotypes, one unmapped functional gene and one orphon.

Gene location

IGHV4-59 is in the IGH locus on chromosome 14 at 14q32.33.

Nucleotide and amino acid sequences for human IGHV4-59

```
                                    1   2   3   4   5   6   7   8   9  10  11  12  13  14  15  16  17  18  19  20
                                  Q   V   Q   L   Q   E   S   G   P       G   L   V   K   P   S   E   T   L   S
L10088,IGHV4-59*01,3d197d   [47]  CAG GTG CAG CTG CAG GAG TCG GGC CCA ... GGA CTG GTG AAG CCT TCG GAG ACC CTG TCC
X92248,IGHV4-59*01,VH4-MC2   [7]  --- --- --- --- --- --- --- --- --- ... --- --- --- --- --- --- --- --- --- ---
Z12371,IGHV4-59*01,DP-71    [41]  --- --- --- --- --- --- --- --- --- ... --- --- --- --- --- --- --- --- --- ---
X56355,IGHV4-59*01,VH4.11   [35]  --- --- --- --- --- --- --- --- --- ... --- --- --- --- --- --- --- --- --- ---
U03896,IGHV4-59*01,hv4c2    [15]  --- --- --- --- --- --- --- --- --- ... --- --- --- --- --- --- --- --- --- ---
X56359,IGHV4-59*01,VH4.15   [35]  --- --- --- --- --- --- --- --- --- ... --- --- --- --- --- --- --- --- --- ---
X92296,IGHV4-59*01,G411     [21]
M29812,IGHV4-59*02,V71.4    [22]  --- --- --- --- --- --- --- --- --- ... --- --- --- --- --- --- --- --- --- ---
M95114,IGHV4-59*03,H4       [45]  --- --- --- --- --- --- --- --- --- ... --- --- --- --- --- --- --- --- --- ---
M95117,IGHV4-59*04,H7       [45]  --- --- --- --- --- --- --- --- --- ... --- --- --- --- --- --- --- --- --- ---
M95118,IGHV4-59*05,H8       [45]  --- --- --- --- --- --- --- --- --- ... --- --- --- --- --- --- --- --- --- ---
M95119,IGHV4-59*06,H9       [45]  --- --- --- --- --- --- --- --- --- ... --- --- --- --- --- --- --- --- --- ---
                                                                                   D
X56360,IGHV4-59*07,VH4.16   [50]  --- --- --- --- --- --- --- --- --- ... --- --- --- --- --- --- --C --- --- ---
X87091,IGHV4-59*08,DP-71RB   [5]                                                                              ---
Z75359,IGHV4-59*09,VH4-GL7  [49]

                                                          ___________________CDR1-IMGT____________________
                                   21  22  23  24  25  26  27  28  29  30  31  32  33  34  35  36  37  38  39  40
                                  L   T   C   T   V   S   G   G   S   I   S   S   Y   Y                   W   S
L10088,IGHV4-59*01,3d197d         CTC ACC TGC ACT GTC TCT GGT GGC TCC ATC AGT AGT TAC TAC ... ... ... ... TGG AGC
X92248,IGHV4-59*01,VH4-MC2        --- --- --- --- --- --- --- --- --- --- --- --- --- --- ... ... ... ... --- ---
Z12371,IGHV4-59*01,DP-71          --- --- --- --- --- --- --- --- --- --- --- --- --- --- ... ... ... ... --- ---
X56355,IGHV4-59*01,VH4.11         --- --- --- --- --- --- --- --- --- --- --- --- --- --- ... ... ... ... --- ---
U03896,IGHV4-59*01,hv4c2          --- --- --- --- --- --- --- --- --- --- --- --- --- --- ... ... ... ... --- ---
X56359,IGHV4-59*01,VH4.15         --- --- --- --- --- --- --- --- --- --- --- --- --- --- ... ... ... ... --- ---
X92296,IGHV4-59*01,G411                    -- --- --- --- --- --- --- --- --- --- --- --- ... ... ... ... --- ---
                                                                       V
M29812,IGHV4-59*02,V71.4          --- --- --- --- --- --- --- --- --- G-- --- --- --- --- ... ... ... ... --- ---
M95114,IGHV4-59*03,H4             --- --- --- --- --- --- --- --- --- --- --- --- --- --- ... ... ... ... --- ---
M95117,IGHV4-59*04,H7             --- --- --- --- --- --- --- --- --- --- --- --- --- --- ... ... ... ... --- ---
M95118,IGHV4-59*05,H8             --- --- --- --- --- --- --- --- --- --- --- --- --- --- ... ... ... ... --- ---
                                                       T
M95119,IGHV4-59*06,H9             --- --- --- --- --- A-- --- --- --- --- --- --- --- --- ... ... ... ... --- ---
X56360,IGHV4-59*07,VH4.16         --- --- --- --- --- --- --- --- --- --- --- --- --- --- ... ... ... ... --- ---
X87091,IGHV4-59*08,DP-71RB        --- --- --- --- --- --- --- --- --- --- --X --- --- --- ... ... ... ... --- ---
Z75359,IGHV4-59*09,VH4-GL7                            --- --- --- --- --- --- --- --- --- ... ... ... ... --- ---

                                                                                              ____________CDR2-
                                   41  42  43  44  45  46  47  48  49  50  51  52  53  54  55  56  57  58  59  60
                                  W   I   R   Q   P   P   G   K   G   L   E   W   I   G   Y   I   Y   Y   S   G
L10088,IGHV4-59*01,3d197d         TGG ATC CGG CAG CCC CCA GGG AAG GGA CTG GAG TGG ATT GGG TAT ATC TAT TAC AGT GGG
X92248,IGHV4-59*01,VH4-MC2        --- --- --- --- --- --- --- --- --- --- --- --- --- --- --- --- --- --- --- ---
Z12371,IGHV4-59*01,DP-71          --- --- --- --- --- --- --- --- --- --- --- --- --- --- --- --- --- --- --- ---
X56355,IGHV4-59*01,VH4.11         --- --- --- --- --- --- --- --- --- --- --- --- --- --- --- --- --- --- --- ---
U03896,IGHV4-59*01,hv4c2          --- --- --- --- --- --- --- --- --- --- --- --- --- --- --- --- --- --- --- ---
```

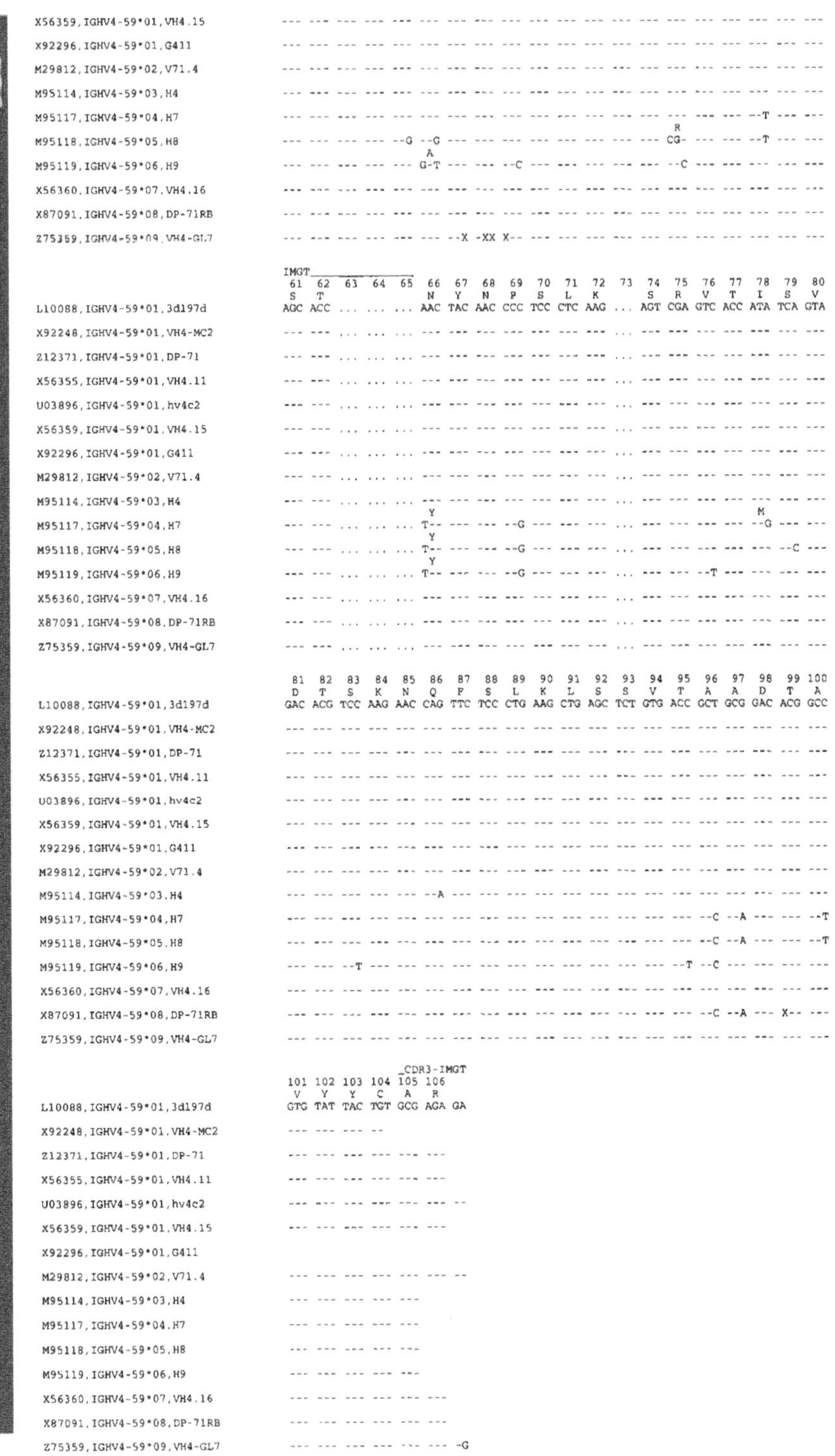

```
X56359,IGHV4-59*01,VH4.15          --- --- --- --- --- --- --- --- --- --- --- --- --- --- --- --- --- --- --- ---
X92296,IGHV4-59*01,G411            --- --- --- --- --- --- --- --- --- --- --- --- --- --- --- --- --- --- --- ---
M29812,IGHV4-59*02,V71.4           --- --- --- --- --- --- --- --- --- --- --- --- --- --- --- --- --- --- --- ---
M95114,IGHV4-59*03,H4              --- --- --- --- --- --- --- --- --- --- --- --- --- --- --- --- --- --- --- ---
M95117,IGHV4-59*04,H7              --- --- --- --- --- --- --- --- --- --- --- --- --- --- --- --- --- --T --- ---
                                                                                    R
M95118,IGHV4-59*05,H8              --- --- --- --- --G --G --- --- --- --- --- --- --- --- CG- --- --- --T --- ---
                                                        A
M95119,IGHV4-59*06,H9              --- --- --- --- --- G-T --- --- --C --- --- --- --- --- --C --- --- --- --- ---
X56360,IGHV4-59*07,VH4.16          --- --- --- --- --- --- --- --- --- --- --- --- --- --- --- --- --- --- --- ---
X87091,IGHV4-59*08,DP-71RB         --- --- --- --- --- --- --- --- --- --- --- --- --- --- --- --- --- --- --- ---
Z75359,IGHV4-59*09,VH4-GL7         --- --- --- --- --- --- --X -XX X-- --- --- --- --- --- --- --- --- --- --- ---

                                   IMGT_______________
                                    61  62  63  64  65  66  67  68  69  70  71  72  73  74  75  76  77  78  79  80
                                    S   T               N   Y   N   P   S   L   K       S   R   V   T   I   S   V
L10088,IGHV4-59*01,3d197d          AGC ACC ... ... ... AAC TAC AAC CCC TCC CTC AAG ... AGT CGA GTC ACC ATA TCA GTA
X92248,IGHV4-59*01,VH4-MC2         --- --- ... ... ... --- --- --- --- --- --- --- ... --- --- --- --- --- --- ---
Z12371,IGHV4-59*01,DP-71           --- --- ... ... ... --- --- --- --- --- --- --- ... --- --- --- --- --- --- ---
X56355,IGHV4-59*01,VH4.11          --- --- ... ... ... --- --- --- --- --- --- --- ... --- --- --- --- --- --- ---
U03896,IGHV4-59*01,hv4c2           --- --- ... ... ... --- --- --- --- --- --- --- ... --- --- --- --- --- --- ---
X56359,IGHV4-59*01,VH4.15          --- --- ... ... ... --- --- --- --- --- --- --- ... --- --- --- --- --- --- ---
X92296,IGHV4-59*01,G411            --- --- ... ... ... --- --- --- --- --- --- --- ... --- --- --- --- --- --- ---
M29812,IGHV4-59*02,V71.4           --- --- ... ... ... --- --- --- --- --- --- --- ... --- --- --- --- --- --- ---
M95114,IGHV4-59*03,H4              --- --- ... ... ... --- --- --- --- --- --- --- ... --- --- --- --- --- --- ---
                                                        Y                                               M
M95117,IGHV4-59*04,H7              --- --- ... ... ... T-- --- --- --G --- --- --- ... --- --- --- --- --G --- ---
                                                        Y
M95118,IGHV4-59*05,H8              --- --- ... ... ... T-- --- --- --G --- --- --- ... --- --- --- --- --- --C ---
                                                        Y
M95119,IGHV4-59*06,H9              --- --- ... ... ... T-- --- --- --G --- --- --- ... --- --- --T --- --- --- ---
X56360,IGHV4-59*07,VH4.16          --- --- ... ... ... --- --- --- --- --- --- --- ... --- --- --- --- --- --- ---
X87091,IGHV4-59*08,DP-71RB         --- --- ... ... ... --- --- --- --- --- --- --- ... --- --- --- --- --- --- ---
Z75359,IGHV4-59*09,VH4-GL7         --- --- ... ... ... --- --- --- --- --- --- --- ... --- --- --- --- --- --- ---

                                    81  82  83  84  85  86  87  88  89  90  91  92  93  94  95  96  97  98  99 100
                                    D   T   S   K   N   Q   F   S   L   K   L   S   S   V   T   A   A   D   T   A
L10088,IGHV4-59*01,3d197d          GAC ACG TCC AAG AAC CAG TTC TCC CTG AAG CTG AGC TCT GTG ACC GCT GCG GAC ACG GCC
X92248,IGHV4-59*01,VH4-MC2         --- --- --- --- --- --- --- --- --- --- --- --- --- --- --- --- --- --- --- ---
Z12371,IGHV4-59*01,DP-71           --- --- --- --- --- --- --- --- --- --- --- --- --- --- --- --- --- --- --- ---
X56355,IGHV4-59*01,VH4.11          --- --- --- --- --- --- --- --- --- --- --- --- --- --- --- --- --- --- --- ---
U03896,IGHV4-59*01,hv4c2           --- --- --- --- --- --- --- --- --- --- --- --- --- --- --- --- --- --- --- ---
X56359,IGHV4-59*01,VH4.15          --- --- --- --- --- --- --- --- --- --- --- --- --- --- --- --- --- --- --- ---
X92296,IGHV4-59*01,G411            --- --- --- --- --- --- --- --- --- --- --- --- --- --- --- --- --- --- --- ---
M29812,IGHV4-59*02,V71.4           --- --- --- --- --- --- --- --- --- --- --- --- --- --- --- --- --- --- --- ---
M95114,IGHV4-59*03,H4              --- --- --- --- --- --A --- --- --- --- --- --- --- --- --- --- --- --- --- ---
M95117,IGHV4-59*04,H7              --- --- --- --- --- --- --- --- --- --- --- --- --- --- --- --C --A --- --- --T
M95118,IGHV4-59*05,H8              --- --- --- --- --- --- --- --- --- --- --- --- --- --- --- --C --A --- --- --T
M95119,IGHV4-59*06,H9              --- --- --T --- --- --- --- --- --- --- --- --- --- --- --T --C --- --- --- ---
X56360,IGHV4-59*07,VH4.16          --- --- --- --- --- --- --- --- --- --- --- --- --- --- --- --- --- --- --- ---
X87091,IGHV4-59*08,DP-71RB         --- --- --- --- --- --- --- --- --- --- --- --- --- --- --- --C --A --- X-- ---
Z75359,IGHV4-59*09,VH4-GL7         --- --- --- --- --- --- --- --- --- --- --- --- --- --- --- --- --- --- --- ---

                                                       _CDR3-IMGT
                                   101 102 103 104 105 106
                                    V   Y   Y   C   A   R
L10088,IGHV4-59*01,3d197d          GTG TAT TAC TGT GCG AGA GA
X92248,IGHV4-59*01,VH4-MC2         --- --- --- --
Z12371,IGHV4-59*01,DP-71           --- --- --- --- --- ---
X56355,IGHV4-59*01,VH4.11          --- --- --- --- --- ---
U03896,IGHV4-59*01,hv4c2           --- --- --- --- --- --- --
X56359,IGHV4-59*01,VH4.15          --- --- --- --- --- ---
X92296,IGHV4-59*01,G411
M29812,IGHV4-59*02,V71.4           --- --- --- --- --- --- --
M95114,IGHV4-59*03,H4              --- --- --- --- ---
M95117,IGHV4-59*04,H7              --- --- --- --- ---
M95118,IGHV4-59*05,H8              --- --- --- --- ---
M95119,IGHV4-59*06,H9              --- --- --- --- ---
X56360,IGHV4-59*07,VH4.16          --- --- --- --- --- ---
X87091,IGHV4-59*08,DP-71RB         --- --- --- --- --- ---
Z75359,IGHV4-59*09,VH4-GL7         --- --- --- --- --- --- -G
```

Framework and complementarity determining regions

FR1-IMGT: 25 (-1 aa: 10)
FR2-IMGT: 17
FR3-IMGT: 38 (-1 aa: 73)

CDR1-IMGT: 8
CDR2-IMGT: 7
CDR3-IMGT: 2

Collier de Perles for human IGHV4-59*01

Accession number: IMGT L10088

EMBL/GenBank/DDBJ: L10088

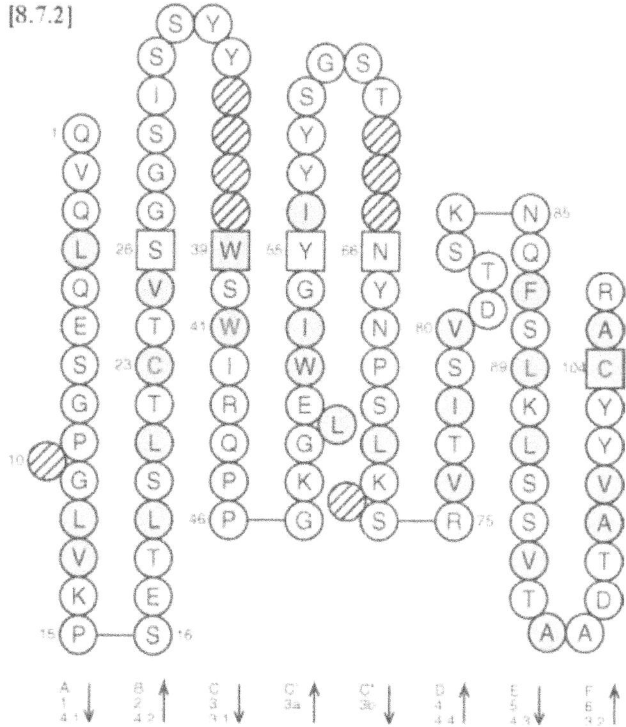

Genome database accession numbers

GDB:9931741

LocusLink: 28392

Nomenclature

IGHV4-61: Immunoglobulin heavy variable 4-61.

Definition and functionality

IGHV4-61 is a mapped functional gene (alleles *01, *02, *03, *04, *05 and *07) or an ORF (allele *06). IGHV4-61 belongs to the IGHV4 subgroup, which comprises 9–12 mapped genes (of which 6–10 are functional), in the IGH locus, depending on the haplotypes, one unmapped functional gene and one orphon. IGHV4-61*06 is an ORF due to an unusual V-HEPTAMER sequence: cacaatg instead of cacagtg.

Gene location

IGHV4-61 is in the IGH locus on chromosome 14 at 14q32.33.

Nucleotide and amino acid sequences for human IGHV4-61

```
                                  1   2   3   4   5   6   7   8   9  10  11  12  13  14  15  16  17  18  19  20
                                  Q   V   Q   L   Q   E   S   G   P       G   L   V   K   P   S   E   T   L   S
M29811,IGHV4-61*01,V71-2    [22] CAG GTG CAG CTG CAG GAG TCG GGC CCA ... GGA CTG GTG AAG CCT TCG GAG ACC CTG TCC
Z12366,IGHV4-61*01,DP-66    [41] --- --- --- --- --- --- --- --- --- ... --- --- --- --- --- --- --- --- --- ---
M95111,IGHV4-61*01,H1       [45] --- --- --- --- --- --- --- --- --- ... --- --- --- --- --- --- --- --- --- ---
X92249,IGHV4-61*01,VH4-MC3   [7] --- --- --- --- --- --- --- --- --- ... --- --- --- --- --- --- --- --- --- ---
X92251,IGHV4-61*01,VH4-MC3b  [7] --- --- --- --- --- --- --- --- --- ... --- --- --- --- --- --- --- --- --- ---
Z75346,IGHV4-61*01,VH4-GL1  [49]
                                                                                                  Q
L10097,IGHV4-61*02,3d279d   [47] --- --- --- --- --- --- --- --- --- ... --- --- --- --- --- --A C-- --- --- ---
X92230,IGHV4-61*03,4.39     [50] --- --- --- --- --- --- --- --- --- ... --- --- --- --- --- --- --- --- --- ---
X92250,IGHV4-61*04,VH4-MC3a  [7] --- --- --- --- --- --- --- --- --- ... --- --- --- --- --- --- --- --- --- ---
                                      L
X56356,IGHV4-61*05,VH4.12   [50] --- C-- --- --- --- --- --- --- --- ... --- --- --- --- --- --- --- --- --- ---
Z75347,IGHV4-61*06,VH4-GL10 [49]
Z75361,IGHV4-61*06,VH4-GL9  [49]
Z75348,IGHV4-61*07,VH4-GL11 [49]

                                                         ___________________CDR1-IMGT____________________
                                 21  22  23  24  25  26  27  28  29  30  31  32  33  34  35  36  37  38  39  40
                                  L   T   C   T   V   S   G   G   S   V   S   S   G   S   Y   Y           W   S
M29811,IGHV4-61*01,V71-2         CTC ACC TGC ACT GTC TCT GGT GGC TCC GTC AGC AGT GGT AGT TAC TAC ... ... TGG AGC
Z12366,IGHV4-61*01,DP-66         --- --- --- --- --- --- --- --- --- --- --- --- --- --- --- --- ... ... --- ---
M95111,IGHV4-61*01,H1            --- --- --- --- --- --- --- --- --- --- --- --- --- --- --- --- ... ... --- ---
X92249,IGHV4-61*01,VH4-MC3       --- --- --- --- --- --- --- --- --- --- --- --- --- --- --- --- ... ... --- ---
X92251,IGHV4-61*01,VH4-MC3b      --- --- --- --- --- --- --- --- --- --- --- --- --- --- --- --- ... ... --- ---
Z75346,IGHV4-61*01,VH4-GL1                         - --- --- --- --- --- --- --- --- --- --- --- ... ... --- ---
                                                                      I
L10097,IGHV4-61*02,3d279d        --- --- --- --- --- --- --- --- --- A-- --- --- --- --- --- --- ... ... --- ---
X92230,IGHV4-61*03,4.39          --- --- --- --- --- --- --- --- --- --- --- --- --- --- --- --- ... ... --- ---
X92250,IGHV4-61*04,VH4-MC3a      --- --- --- --- --- --- --- --- --- --- --- --- --- --- --- --- ... ... --- ---
                                                                      I           S                           G
X56356,IGHV4-61*05,VH4.12        --- --- --- --- --- --- --- --- --- A-- --- --- A-- --- --- --- ... ... --- G--
Z75347,IGHV4-61*06,VH4-GL10                          --- --- --- --- --- --- --- --- --- --- --- ... ... --- ---
Z75361,IGHV4-61*06,VH4-GL9                           --- --- --- --- --- --- --- --- --- --- --- ... ... --- ---
Z75348,IGHV4-61*07,VH4-GL11                          --- --- --- --- --- --- --- --- --- --- --- ... ... --- ---

                                                                                             _______________CDR2-
                                 41  42  43  44  45  46  47  48  49  50  51  52  53  54  55  56  57  58  59  60
                                  W   I   R   Q   P   P   G   K   G   L   E   W   I   G   Y   I   Y   Y   S   G
M29811,IGHV4-61*01,V71-2         TGG ATC CGG CAG CCC CCA GGG AAG GGA CTC GAG TGG ATT GGG TAT ATC TAT TAC AGT GGG
Z12366,IGHV4-61*01,DP-66         --- --- --- --- --- --- --- --- --- --- --- --- --- --- --- --- --- --- --- ---
M95111,IGHV4-61*01,H1            --- --- --- --- --- --- --- --- --- --- --- --- --- --- --- --- --- --- --- ---
X92249,IGHV4-61*01,VH4-MC3       --- --- --- --- --- --- --- --- --- --- --- --- --- --- --- --- --- --- --- ---
X92251,IGHV4-61*01,VH4-MC3b      --- --- --- --- --- --- --- --- --- --- --- --- --- --- --- --- --- --- --- ---
Z75346,IGHV4-61*01,VH4-GL1       --- --- --- --- --- --- --- --- --- --- --- --- --- --- --- --- --- --- --- ---
                                                      A                                   R           T
L10097,IGHV4-61*02,3d279d        --- --- --- --- --- G-C --- --- --- --- --- --- --- --- CG- --- --- AC- --- ---
```

```
X92230,IGHV4-61*03,4.39          --- --- --- --- --- --- --- --- --- --- --- --- --- --- --- --- --- --- --- ---
X92250,IGHV4-61*04,VH4-MC3a      --- --- --- --- --- --- --- --- --- --- --- --- --- --A --- --- --- --- --- ---
X56356,IGHV4-61*05,VH4.12        --- --- --- --- --- --- --- --- --- --- --- --- --- --- --- --- --- --- --- ---
Z75347,IGHV4-61*06,VH4-GL10      --- --- --- --- --- --- --- --- --- --- --- --- --- --- --- --- --- --- --- ---
Z75361,IGHV4-61*06,VH4-GL9       --- --- --- --- --- --- --- --- --- --- --- --- --- --- --- --- --- --- --- ---
Z75348,IGHV4-61*07,VH4-GL11      --- --- --- --- --- --- --- --- --- --- --- --- --- --- --- --- --- --- --- ---

                                 IMGT_______________
                                 61  62  63  64  65  66  67  68  69  70  71  72  73  74  75  76  77  78  79  80
                                 S   T               N   Y   N   P   S   L   K       S   R   V   T   I   S   V
M29811,IGHV4-61*01,V71-2         AGC ACC ... ... ... AAC TAC AAC CCC TCC CTC AAG ... AGT CGA GTC ACC ATA TCA GTA
Z12366,IGHV4-61*01,DP-66         --- --- ... ... ... --- --- --- --- --- --- --- ... --- --- --- --- --- --- ---
M95111,IGHV4-61*01,H1            --- --- ... ... ... --- --- --- --- --- --- --- ... --- --- --- --- --- --- ---
X92249,IGHV4-61*01,VH4-MC3       --- --- ... ... ... --- --- --- --- --- --- --- ... --- --- --- --- --- --- ---
X92251,IGHV4-61*01,VH4-MC3b      --- --- ... ... ... --- --- --- --- --- --- --- ... --- --- --- --- --- --- ---
Z75346,IGHV4-61*01,VH4-GL1       --- --- ... ... ... --- --- --- --- --- --- --- ... --- --- --- --- --- --- ---
L10097,IGHV4-61*02,3d279d        --- --- ... ... ... --- --- --- --- --- --- --- ... --- --- --- --- --- --- ---
X92230,IGHV4-61*03,4.39          --- --- ... ... ... --- --- --- --- --- --- --- ... --- --- --- --- --- --- ---
X92250,IGHV4-61*04,VH4-MC3a      --- --- ... ... ... --- --- --- --- --- --- --- ... --- --- --- --- --- --- ---
X56356,IGHV4-61*05,VH4.12        --- --- ... ... ... --- --- --- --- --- --- --- ... --- --- --- --- --- --- ---
Z75347,IGHV4-61*06,VH4-GL10      --- --- ... ... ... --- --- --- --- --- --- --- ... --- --- --- --- --- --- ---
Z75361,IGHV4-61*06,VH4-GL9       --- --- ... ... ... --- --- --- --- --- --- --- ... --- --- --- --- --- --- ---
Z75348,IGHV4-61*07,VH4-GL11      --- --- ... ... ... --- --- --- --- --- --- --- ... --- --- --- --- --- --- ---

                                 81  82  83  84  85  86  87  88  89  90  91  92  93  94  95  96  97  98  99 100
                                 D   T   S   K   N   Q   F   S   L   K   L   S   S   V   T   A   A   D   T   A
M29811,IGHV4-61*01,V71-2         GAC ACG TCC AAG AAC CAG TTC TCC CTG AAG CTG AGC TCT GTG ACC GCT GCG GAC ACG GCC
Z12366,IGHV4-61*01,DP-66         --- --- --- --- --- --- --- --- --- --- --- --- --- --- --- --- --- --- --- ---
M95111,IGHV4-61*01,H1            --- --- --- --- --- --- --- --- --- --- --- --- --- --- --- --- --- --- --- ---
X92249,IGHV4-61*01,VH4-MC3       --- --- --- --- --- --- --- --- --- --- --- --- --- --- --- --- --- --- --- ---
X92251,IGHV4-61*01,VH4-MC3b      --- --- --- --- --- --- --- --- --- --- --- --- --- --- --- --- --- --- --- ---
Z75346,IGHV4-61*01,VH4-GL1       --- --- --- --- --- --- --- --- --- --- --- --- --- --- --- --- --- --- --- ---
L10097,IGHV4-61*02,3d279d        --- --- --- --- --- --- --- --- --- --- --- --- --- --- --- --C --A --- --- ---
                                                         H
X92230,IGHV4-61*03,4.39          --- --- --- --- --- --C --- --- --- --- --- --- --- --- --- --- --- --- --- ---
X92250,IGHV4-61*04,VH4-MC3a  (1) --- --- --- --- --- --- --- --- --- --- --- --- --- --- --- --- ... --- --- ---
                                     K
X56356,IGHV4-61*05,VH4.12        --- -A- --- --- --- --- --- --- --- --- --- --- --- --- --- --C --- --- --- ---
Z75347,IGHV4-61*06,VH4-GL10      --- --- --- --- --- --- --- --- --- --- --- --- --- --- --- --C --- --- --- ---
Z75361,IGHV4-61*06,VH4-GL9       --- --- --- --- --- --- --- --- --- --- --- --- --- --- --- --C --- --- --- ---
Z75348,IGHV4-61*07,VH4-GL11      --- --- --- --- --- --- --- --- --- --- --- --- --- --- --- --- --- --- --- ---

                                                     _CDR3-IMGT
                                 101 102 103 104 105 106
                                  V   Y   Y   C   A   R
M29811,IGHV4-61*01,V71-2         GTG TAT TAC TGT GCG AGA GA
Z12366,IGHV4-61*01,DP-66         --- --- --- --- --- ---
M95111,IGHV4-61*01,H1            --- --- --- --- ---
X92249,IGHV4-61*01,VH4-MC3       --- --- --- --
X92251,IGHV4-61*01,VH4-MC3b      --- --- --- --
Z75346,IGHV4-61*01,VH4-GL1       --- --- --- --- --- --- --
L10097,IGHV4-61*02,3d279d        --- --- --- --- --- --- --
X92230,IGHV4-61*03,4.39          --- --- --- --- --- --- --
X92250,IGHV4-61*04,VH4-MC3a      --- --- -G
X56356,IGHV4-61*05,VH4.12        --- --- --- --- --- ---
Z75347,IGHV4-61*06,VH4-GL10      --- --- --- --- --C --- --
Z75361,IGHV4-61*06,VH4-GL9       --- --- --- --- --C --- --
Z75348,IGHV4-61*07,VH4-GL11      --- --- --- --- --- --- C-
```

Note:

(1) It is not known if the 3 nt DELETION at codon position 97 in X92250 is typing error.

Framework and complementarity determining regions

FR1-IMGT: 25 (-1 aa: 10) CDR1-IMGT: 10
FR2-IMGT: 17 CDR2-IMGT: 7
FR3-IMGT: 38 (-1 aa: 73) CDR3-IMGT: 2

Collier de Perles for human IGHV4-61*01

Accession number: IMGT M29811 EMBL/GenBank/DDBJ: M29811

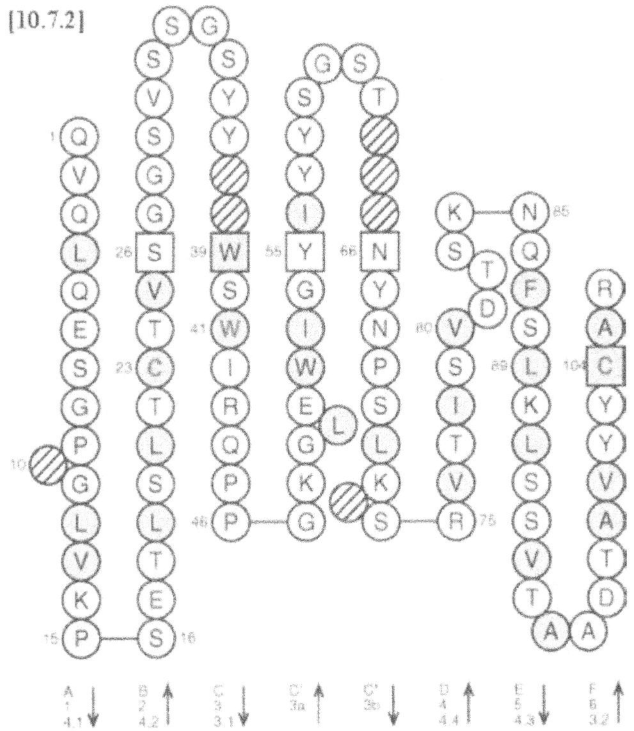

Genome database accession numbers

GDB:9931742 LocusLink: 28391

Nomenclature

IGHV4-b: Immunoglobulin heavy variable 4-b (provisional).

Definition and functionality

IGHV4-b is an unmapped functional gene of the IGHV4 subgroup. This subgroup comprises 9–12 mapped genes (of which 6–12 are functional) in the IGH locus, depending to the haplotypes, one unmapped functional gene and one orphon.

Gene location

IGHV4-b is in the IGH locus on chromosome 14 at 14q32.33.

Nucleotide and amino acid sequences for human IGHV4-b

```
                                   1   2   3   4   5   6   7   8   9  10  11  12  13  14  15  16  17  18  19  20
                                   Q   V   Q   L   Q   E   S   G   P       G   L   V   K   P   S   E   T   L   S
Z12367,IGHV4-b*01,DP-67     [41]  CAG GTG CAG CTG CAG GAG TCG GGC CCA ... GGA CTG GTG AAG CCT TCG GAG ACC CTG TCC
X92289,IGHV4-b*01,VH4-4B    [52]  --- --- --- --- --- --- --- --- --- ... --- --- --- --- --- --- --- --- --- ---
X56365,IGHV4-b*02,VH4.22    [35]  --- --- --- --- --- --- --- --- --- ... --- --- --- --- --- --- --- --- --- ---

                                                          _____________CDR1-IMGT____________________
                                  21  22  23  24  25  26  27  28  29  30  31  32  33  34  35  36  37  38  39  40
                                   L   T   C   A   V   S   G   Y   S   I   S   S   G   Y   Y               W   G
Z12367,IGHV4-b*01,DP-67           CTC ACC TGC GCT GTC TCT GGT TAC TCC ATC AGC AGT GGT TAC TAC ... ... ... TGG GGC
X92289,IGHV4-b*01,VH4-4B          --- --- --- --- --- --- --- --- --- --- --- --- --- --- --- ... ... ... --- ---
                                               T
X56365,IGHV4-b*02,VH4.22          --- --- --- A-- --- --- --- --- --- --- --- --- --- --- --- ... ... ... --- ---

                                                                                          ______________ CDR2-
                                  41  42  43  44  45  46  47  48  49  50  51  52  53  54  55  56  57  58  59  60
                                   W   I   R   Q   P   P   G   K   G   L   E   W   I   G   S   I   Y   H   S   G
Z12367,IGHV4-b*01,DP-67           TGG ATC CGG CAG CCC CCA GGG AAG GGG CTG GAG TGG ATT GGG AGT ATC TAT CAT AGT GGG
X92289,IGHV4-b*01,VH4-4B          --- --- --- --- --- --- --- --- --- --- --- --- --- --- --- --- --- --- --- ---
X56365,IGHV4-b*02,VH4.22          --- --- --- --- --- --- --- --- --- --- --- --- --- --- --- --- --- --- --- ---

                                  IMGT_______________
                                  61  62  63  64  65  66  67  68  69  70  71  72  73  74  75  76  77  78  79  80
                                   S   T               Y   Y   N   P   S   L   K       S   R   V   T   I   S   V
Z12367,IGHV4-b*01,DP-67           AGC ACC ... ... ... TAC TAC AAC CCG TCC CTC AAG ... AGT CGA GTC ACC ATA TCA GTA
X92289,IGHV4-b*01,VH4-4B          --- --- ... ... ... --- --- --- --- --- --- --- ... --- --- --- --- --- --- ---
X56365,IGHV4-b*02,VH4.22          --- --- ... ... ... --- --- --- --- --- --- --- ... --- --- --- --- --- --- ---

                                  81  82  83  84  85  86  87  88  89  90  91  92  93  94  95  96  97  98  99 100
                                   D   T   S   K   N   Q   F   S   L   K   L   S   S   V   T   A   A   D   T   A
Z12367,IGHV4-b*01,DP-67           GAC ACG TCC AAG AAC CAG TTC TCC CTG AAG CTG AGC TCT GTG ACC GCC GCA GAC ACG GCC
X92289,IGHV4-b*01,VH4-4B          --- --- --- --- --- --- --- --- --- --- --- --- --- --- --- --- --- --- --- ---
X56365,IGHV4-b*02,VH4.22          --- --- --- --- --- --- --- --- --- --- --- --- --- --- --- --- --- --- --- ---

                                                     _CDR3-IMGT
                                  101 102 103 104 105 106
                                   V   Y   Y   C   A   R
Z12367,IGHV4-b*01,DP-67           GTG TAT TAC TGT GCG AGA
X92289,IGHV4-b*01,VH4-4B          --- --- --- --- --- ---
X56365,IGHV4-b*02,VH4.22          --- --- --- --- --- ---
```

Framework and complementarity determining regions

FR1-IMGT: 25 (-1 aa: 10)	CDR1-IMGT: 9
FR2-IMGT: 17	CDR2-IMGT: 7
FR3-IMGT: 38 (-1 aa: 73)	CDR3-IMGT: 2

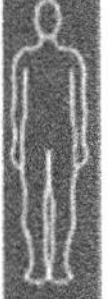

Collier de Perles for human IGHV4-b*01

Accession number: IMGT Z12367 EMBL/GenBank/DDBJ: Z12367

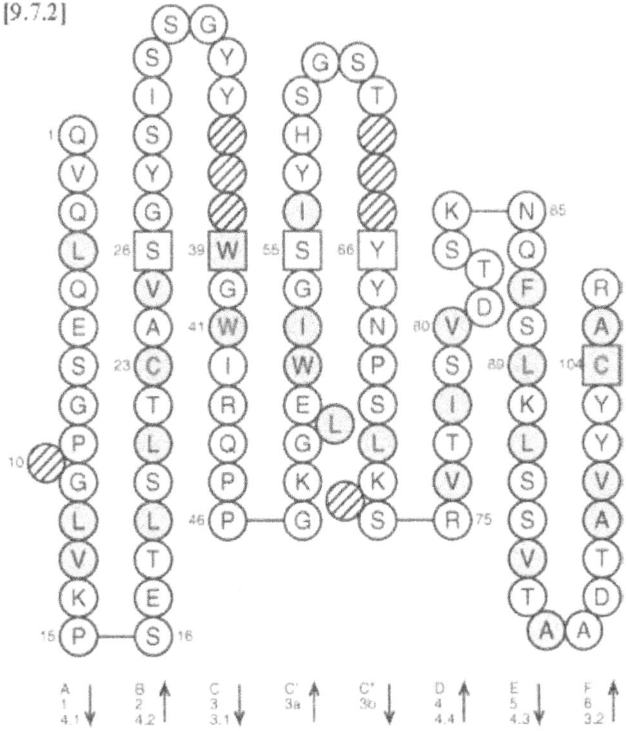

Genome database accession numbers

GDB:9931744 LocusLink: 28389

IGHV5-51

Nomenclature

IGHV5-51: Immunoglobulin heavy variable 5-51.

Definition and functionality

IGHV5-51 is the only mapped functional gene of the IGHV5 subgroup which comprises three genes in the IGH locus: the IGHV5-51 gene, one pseudogene and one unmapped gene (functional or pseudogene).

Gene location

IGHV5-51 is in the IGH locus on chromosome 14 at 14q32.33.

Nucleotide and amino acid sequences for human IGHV5-51

```
                                    1   2   3   4   5   6   7   8   9  10  11  12  13  14  15  16  17  18  19  20
                                    E   V   Q   L   V   Q   S   G   A       E   V   K   K   P   G   E   S   L   K
M99686,IGHV5-51*01,V5-51      [26] GAG GTG CAG CTG GTG CAG TCT GGA GCA ... GAG GTG AAA AAG CCC GGG GAG TCT CTG AAG
Z12373,IGHV5-51*01,DP-73      [41] --- --- --- --- --- --- --- --- --- ... --- --- --- --- --- --- --- --- --- ---
X92226,IGHV5-51*01,VH251      [35] --- --- --- --- --- --- --- --- --- ... --- --- --- --- --- --- --- --- --- ---
X56373,IGHV5-51*01,VHVBLK     [35] --- --- --- --- --- --- --- --- --- ... --- --- --- --- --- --- --- --- --- ---
X56372,IGHV5-51*01,VHVAU      [35] --- --- --- --- --- --- --- --- --- ... --- --- --- --- --- --- --- --- --- ---
X56370,IGHV5-51*01,VHVCH      [35] --- --- --- --- --- --- --- --- --- ... --- --- --- --- --- --- --- --- --- ---
X56369,IGHV5-51*01,VHVLB      [35] --- --- --- --- --- --- --- --- --- ... --- --- --- --- --- --- --- --- --- ---
X56371,IGHV5-51*01,VHVTT      [35] --- --- --- --- --- --- --- --- --- ... --- --- --- --- --- --- --- --- --- ---
M18806,IGHV5-51*02,VH251      [38]  -- --- --- --- --- --- --- --- --- ... --- --- --- --- --- --- --- --- --- ---
X56368,IGHV5-51*03,VHVCW      [35]  -- --- --- --- --- --- --- --- --- ... --- --- --- --- --G --- --- --- --- ---
Z27448,IGHV5-51*03,COS-24     [42]                                                  -- --- --G --- --- --- --- ---
X56367,IGHV5-51*04,VHVJB      [35] --- --- --- --- --- --- --- --- --- ... --- --- --- --- --G --- --- --- --- ---
Z27449,IGHV5-51*05,COS-25     [42]                                                  -- --- --- --- --- --- --- ---

                                                           _______________CDR1-IMGT_______________________
                                   21  22  23  24  25  26  27  28  29  30  31  32  33  34  35  36  37  38  39  40
                                    I   S   C   K   G   S   G   Y   S   F   T   S   Y   W                   I   G
M99686,IGHV5-51*01,V5-51           ATC TCC TGT AAG GGT TCT GGA TAC AGC TTT ACC AGC TAC TGG ... ... ... ... ATC GGC
Z12373,IGHV5-51*01,DP-73           --- --- --- --- --- --- --- --- --- --- --- --- --- --- ... ... ... ... --- ---
X92226,IGHV5-51*01,VH251           --- --- --- --- --- --- --- --- --- --- --- --- --- --- ... ... ... ... --- ---
X56373,IGHV5-51*01,VHVBLK          --- --- --- --- --- --- --- --- --- --- --- --- --- --- ... ... ... ... --- ---
X56372,IGHV5-51*01,VHVAU           --- --- --- --- --- --- --- --- --- --- --- --- --- --- ... ... ... ... --- ---
X56370,IGHV5-51*01,VHVCH           --- --- --- --- --- --- --- --- --- --- --- --- --- --- ... ... ... ... --- ---
X56369,IGHV5-51*01,VHVLB           --- --- --- --- --- --- --- --- --- --- --- --- --- --- ... ... ... ... --- ---
X56371,IGHV5-51*01,VHVTT           --- --- --- --- --- --- --- --- --- --- --- --- --- --- ... ... ... ... --- ---
                                                                                                            T
M18806,IGHV5-51*02,VH251           --- --- --- --- --- --- --- --- --- --- --- --- --- --- ... ... ... ... -C- ---
X56368,IGHV5-51*03,VHVCW           --- --- --- --- --- --- --- --- --- --- --- --- --- --- ... ... ... ... --- ---
Z27448,IGHV5-51*03,COS-24          --- --- --- --- --- --- --- --- --- --- --- --- --- --- ... ... ... ... --- ---
X56367,IGHV5-51*04,VHVJB           --- --- --- --- --- --- --- --- --- --- --- --- --- --- ... ... ... ... --- ---
Z27449,IGHV5-51*05,COS-25          --- --- --- --- --- --- --- --- --- --- --- --- --- --- ... ... ... ... --- ---

                                                                                               _________________CDR2-
                                   41  42  43  44  45  46  47  48  49  50  51  52  53  54  55  56  57  58  59  60
                                    W   V   R   Q   M   P   G   K   G   L   E   W   M   G   I   I   Y   P   G   D
M99686,IGHV5-51*01,V5-51           TGG GTG CGC CAG ATG CCC GGG AAA GGC CTG GAG TGG ATG GGG ATC ATC TAT CCT GGT GAC
Z12373,IGHV5-51*01,DP-73           --- --- --- --- --- --- --- --- --- --- --- --- --- --- --- --- --- --- --- ---
X92226,IGHV5-51*01,VH251           --- --- --- --- --- --- --- --- --- --- --- --- --- --- --- --- --- --- --- ---
X56373,IGHV5-51*01,VHVBLK          --- --- --- --- --- --- --- --- --- --- --- --- --- --- --- --- --- --- --- ---
X56372,IGHV5-51*01,VHVAU           --- --- --- --- --- --- --- --- --- --- --- --- --- --- --- --- --- --- --- ---
X56370,IGHV5-51*01,VHVCH           --- --- --- --- --- --- --- --- --- --- --- --- --- --- --- --- --- --- --- ---
X56369,IGHV5-51*01,VHVLB           --- --- --- --- --- --- --- --- --- --- --- --- --- --- --- --- --- --- --- ---
X56371,IGHV5-51*01,VHVTT           --- --- --- --- --- --- --- --- --- --- --- --- --- --- --- --- --- --- --- ---
M18806,IGHV5-51*02,VH251           --- --- --- --- --- --- --- --- --- T-- --- --- --- --- --- --- --- --- --- ---
X56368,IGHV5-51*03,VHVCW           --- --- --- --- --- --- --- --- --- --- --- --- --- --- --- --- --- --- --- ---
Z27448,IGHV5-51*03,COS-24          --- --- --- --- --- --- --- --- --- --- --- --- --- --- --- --- --- --- --- ---
X56367,IGHV5-51*04,VHVJB           --- --- --- --- --- --- --- --- --- --- --- --- --- --- --- --- --- --- --- ---
                                                            R
Z27449,IGHV5-51*05,COS-25          --- --- --- --- --- --- A-- --- --- --- --- --- --- --- --- --- --- --- --- ---
```

```
                                  IMGT_______________
                                  61  62  63  64  65  66  67  68  69  70  71  72  73  74  75  76  77  78  79  80
                                  S   D   T           R   Y   S   P   S   F   Q       G   Q   V   T   I   S   A
M99686,IGHV5-51*01,V5-51          TCT GAT ACC ... ... AGA TAC AGC CCG TCC TTC CAA ... GGC CAG GTC ACC ATC TCA GCC
Z12373,IGHV5-51*01,DP-73          --- --- --- ... ... --- --- --- --- --- --- --- ... --- --- --- --- --- --- ---
X92226,IGHV5-51*01,VH251          --- --- --- ... ... --- --- --- --- --- --- --- ... --- --- --- --- --- --- ---
X56373,IGHV5-51*01,VHVBLK         --- --- --- ... ... --- --- --- --- --- --- --- ... --- --- --- --- --- --- ---
X56372,IGHV5-51*01,VHVAU          --- --- --- ... ... --- --- --- --- --- --- --- ... --- --- --- --- --- --- ---
X56370,IGHV5-51*01,VHVCH          --- --- --- ... ... --- --- --- --- --- --- --- ... --- --- --- --- --- --- ---
X56369,IGHV5-51*01,VHVLB          --- --- --- ... ... --- --- --- --- --- --- --- ... --- --- --- --- --- --- ---
X56371,IGHV5-51*01,VHVTT          --- --- --- ... ... --- --- --- --- --- --- --- ... --- --- --- --- --- --- ---
M18806,IGHV5-51*02,VH251          --- --- --- ... ... --- --- --- --- --- --- --- ... --- --- --- --- --- --- ---
X56368,IGHV5-51*03,VHVCW          --- --- --- ... ... --- --- --- --- --- --- --- ... --- --- --- --- --- --- ---
Z27448,IGHV5-51*03,COS-24         --- --- --- ... ... --- --- --- --- --- --- --- ... --- --- --- --- --- --- ---
X56367,IGHV5-51*04,VHVJB          --- --- --- ... ... --- --- --- --- --- --- --- ... --- --- --- --- --- --- ---
Z27449,IGHV5-51*05,COS-25         --- --- --- ... ... --- --- --- --- --- --- --- ... --- --- --- --- --- --- ---

                                  81  82  83  84  85  86  87  88  89  90  91  92  93  94  95  96  97  98  99  100
                                  D   K   S   I   S   T   A   Y   L   Q   W   S   S   L   K   A   S   D   T   A
M99686,IGHV5-51*01,V5-51          GAC AAG TCC ATC AGC ACC GCC TAC CTG CAG TGG AGC AGC CTG AAG GCC TCG GAC ACC GCC
Z12373,IGHV5-51*01,DP-73          --- --- --- --- --- --- --- --- --- --- --- --- --- --- --- --- --- --- --- ---
X92226,IGHV5-51*01,VH251          --- --- --- --- --- --- --- --- --- --- --- --- --- --- --- --- --- --- --- ---
X56373,IGHV5-51*01,VHVBLK         --- --- --- --- --- --- --- --- --- --- --- --- --- --- --- --- --- --- --- ---
X56372,IGHV5-51*01,VHVAU          --- --- --- --- --- --- --- --- --- --- --- --- --- --- --- --- --- --- --- ---
X56370,IGHV5-51*01,VHVCH          --- --- --- --- --- --- --- --- --- --- --- --- --- --- --- --- --- --- --- ---
X56369,IGHV5-51*01,VHVLB          --- --- --- --- --- --- --- --- --- --- --- --- --- --- --- --- --- --- --- ---
X56371,IGHV5-51*01,VHVTT          --- --- --- --- --- --- --- --- --- --- --- --- --- --- --- --- --- --- --- ---
M18806,IGHV5-51*02,VH251          --- --- --- --- --- --- --- --- --- --- --- --- --- --- --- --- --- --- --- ---
X56368,IGHV5-51*03,VHVCW          --- --- --- --- --- --- --- --- --- --- --- --- --- --- --- --- --- --- --- ---
Z27448,IGHV5-51*03,COS-24         --- --- --- --- --- --- --- --- --- --- --- --- --- --- --- --- --- --- --- ---
                                          P
X56367,IGHV5-51*04,VHVJB          --- --- C-- --- --- --- --- --- --- --- --- --- --- --- --- --- --- --- --- ---
Z27449,IGHV5-51*05,COS-25         --- --- --- --- --- --- --- --- --- --- --- --- --- --- --- --- --- --- --- ---

                                                      _CDR3-IMGT
                                  101 102 103 104 105 106
                                  M   Y   Y   C   A   R
M99686,IGHV5-51*01,V5-51          ATG TAT TAC TGT GCG AGA CA
Z12373,IGHV5-51*01,DP-73          --- --- --- --- --- ---
X92226,IGHV5-51*01,VH251          --- --- --- --- --- ---
X56373,IGHV5-51*01,VHVBLK         --- --- --- --- --- ---
X56372,IGHV5-51*01,VHVAU          --- --- --- --- --- ---
X56370,IGHV5-51*01,VHVCH          --- --- --- --- --- ---
X56369,IGHV5-51*01,VHVLB          --- --- --- --- --- ---
X56371,IGHV5-51*01,VHVTT          --- --- --- --- --- ---
M18806,IGHV5-51*02,VH251          --- --- --- --- --- --- --
X56368,IGHV5-51*03,VHVCW          --- --- --- --- --- ---
Z27448,IGHV5-51*03,COS-24         ---
X56367,IGHV5-51*04,VHVJB          --- --- --- --- --- ---
Z27449,IGHV5-51*05,COS-25         ---
```

Framework and complementarity determining regions

FR1-IMGT: 25 (-1 aa: 10)
FR2-IMGT: 17
FR3-IMGT: 38 (-1 aa: 73)

CDR1-IMGT: 8
CDR2-IMGT: 8
CDR3-IMGT: 2

Collier de Perles for human IGHV5-51*01

Accession number: IMGT M99686 EMBL/GenBank/DDBJ: M99686

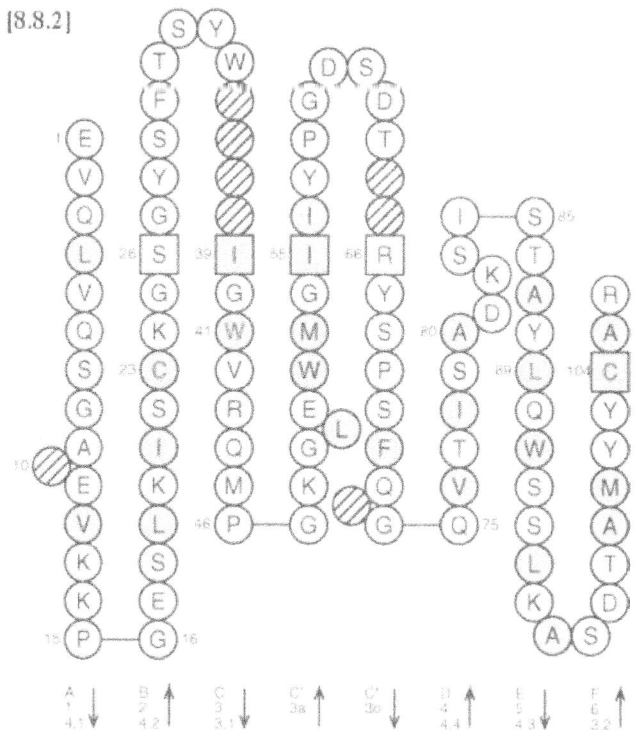

Genome database accession numbers

GDB:9931745 LocusLink: 28388

IGHV5-a

Nomenclature

IGHV5-a: Immunoglobulin heavy variable 5-a (provisional).

Definition and functionality

IGHV5-a is an unmapped functional gene (alleles *01, *03 and *04) or a pseudogene (allele *02). IGHV5-a belongs to the IGHV5 subgroup which comprises three genes in the IGH locus: the unmapped gene IGHV5-a, one mapped functional gene IGHV5-51 and a pseudogene. IGHV5-a is a pseudogene due to a 1 nt DELETION in codon 96 leading to a frameshift in FR3-IMGT.

Gene location

IGHV5-a is in the IGH locus on chromosome 14 at 14q32.33.

Nucleotide and amino acid sequences for human IGHV5-a

```
                                   1   2   3   4   5   6   7   8   9  10  11  12  13  14  15  16  17  18  19  20
                                   E   V   Q   L   V   Q   S   G   A       E   V   K   K   P   G   E   S   L   R
X92227,IGHV5-a*01,VH32      [35]  GAA GTG CAG CTG GTG CAG TCT GGA GCA ... GAG GTG AAA AAG CCC GGG GAG TCT CTG AGG
X92279,IGHV5-a*02,VH32      [18]  --- --- --- --- --- --- --- --- --- ... --- --- --- --- --- --- --- --- --- ---
X56375,IGHV5-a*03,VHVMW     [35]  --- --- --- --- --- --- --C --- --- ... --- --- --- --- --- --- --- --- --- ---
X56374,IGHV5-a*03,VHVRG     [35]  --- --- --- --- --- --- --C --- --- ... --- --- --- --- --- --- --- --- --- ---
X56376,IGHV5-a*04,VHVBLK32  [35]  --- --- --- --- --- --- --- --- --- ... --- --- --- --- --- --- --- --- --- ---

                                                          ___________________CDR1-IMGT___________________
                                  21  22  23  24  25  26  27  28  29  30  31  32  33  34  35  36  37  38  39  40
                                   I   S   C   K   G   S   G   Y   S   F   T   S   Y   W                   I   S
X92227,IGHV5-a*01,VH32            ATC TCC TGT AAG GGT TCT GGA TAC AGC TTT ACC AGC TAC TGG ... ... ... ... ATC AGC
X92279,IGHV5-a*02,VH32            --- --- --- --- --- --- --- --- --- --- --- --- --- --- ... ... ... ... --- ---
X56375,IGHV5-a*03,VHVMW           --- --- --- --- --- --- --- --- --- --- --- --- --- --- ... ... ... ... --- ---
X56374,IGHV5-a*03,VHVRG           --- --- --- --- --- --- --- --- --- --- --- --- --- --- ... ... ... ... --- ---
X56376,IGHV5-a*04,VHVBLK32        --- --- --- --- --- --- --- --- --- --- --- --- --- --- ... ... ... ... --- ---

                                                                                              _______________CDR2-
                                  41  42  43  44  45  46  47  48  49  50  51  52  53  54  55  56  57  58  59  60
                                   W   V   R   Q   M   P   G   K   G   L   E   W   M   G   R   I   D   P   S   D
X92227,IGHV5-a*01,VH32            TGG GTG CGC CAG ATG CCC GGG AAA GGC CTG GAG TGG ATG GGG AGG ATT GAT CCT AGT GAC
X92279,IGHV5-a*02,VH32            --- --- --- --- --- --- --- --- --- T-- --- --- --- --- --- --- --- --- --- ---
X56375,IGHV5-a*03,VHVMW           --- --- --- --- --- --- --- --- --- --- --- --- --- --- --- --- --- --- --- ---
X56374,IGHV5-a*03,VHVRG           --- --- --- --- --- --- --- --- --- --- --- --- --- --- --- --- --- --- --- ---
X56376,IGHV5-a*04,VHVBLK32        --- --- --- --- --- --- --- --- --- --- --- --- --- --- --- --- --- --- --- ---

                                  IMGT_______________
                                  61  62  63  64  65  66  67  68  69  70  71  72  73  74  75  76  77  78  79  80
                                   S   Y   T           N   Y   S   P   S   F   Q       G   H   V   T   I   S   A
X92227,IGHV5-a*01,VH32            TCT TAT ACC ... ... AAC TAC AGC CCG TCC TTC CAA ... GGC CAC GTC ACC ATC TCA GCT
X92279,IGHV5-a*02,VH32            --- --- --- ... ... --- --- --- --- --- --- --- ... --- --- --- --- --- --- ---
X56375,IGHV5-a*03,VHVMW           --- --- --- ... ... --- --- --- --- --- --- --- ... --- --- --- --- --- --- ---
X56374,IGHV5-a*03,VHVRG           --- --- --- ... ... --- --- --- --- --- --- --- ... --- --- --- --- --- --- ---
                                                                                           Q
X56376,IGHV5-a*04,VHVBLK32        --- --- --- ... ... --- --- --- --- --- --- --- ... --- --G --- --- --- --- ---

                                  81  82  83  84  85  86  87  88  89  90  91  92  93  94  95  96  97  98  99 100
                                   D   K   S   I   S   T   A   Y   L   Q   W   S   S   L   K   A   S   D   T   A
X92227,IGHV5-a*01,VH32            GAC AAG TCC ATC AGC ACT GCC TAC CTG CAG TGG AGC AGC CTG AAG GCC TCG GAC ACC GCC
                                                                                               #
X92279,IGHV5-a*02,VH32            --- --- --- --- --- --- --- --- --- --- --- --- --- --- --- --. --- --- --- ---
X56375,IGHV5-a*03,VHVMW           --- --- --- --- --- --- --- --- --- --- --- --- --- --- --- --- --- --- --- ---
X56374,IGHV5-a*03,VHVRG           --- --- --- --- --- --- --- --- --- --- --- --- --- --- --- --- --- --- --- ---
X56376,IGHV5-a*04,VHVBLK32        --- --- --- --- --- --- --- --- --- --- --- --- --- --- --- --- --- --- --- ---

                                                 _CDR3-IMGT
                                  101 102 103 104 105 106
                                   M   Y   Y   C   A   R
X92227,IGHV5-a*01,VH32            ATG TAT TAC TGT GCG AGA
X92279,IGHV5-a*02,VH32            --- --- --- --- --- --- CA
X56375,IGHV5-a*03,VHVMW           --- --- --- --- --- ---
X56374,IGHV5-a*03,VHVRG           --- --- --- --- --- ---
X56376,IGHV5-a*04,VHVBLK32        --- --- --- --- --- ---
```

Framework and complementarity determining regions

FR1-IMGT: 25 (-1 aa: 10)
FR2-IMGT: 17
FR3-IMGT: 38 (-1 aa: 73)

CDR1-IMGT: 8
CDR2-IMGT: 8
CDR3-IMGT: 2

Collier de Perles for human IGHV5-a*01

Accession number: IMGT X92227
EMBL/GenBank/DDBJ: X92227

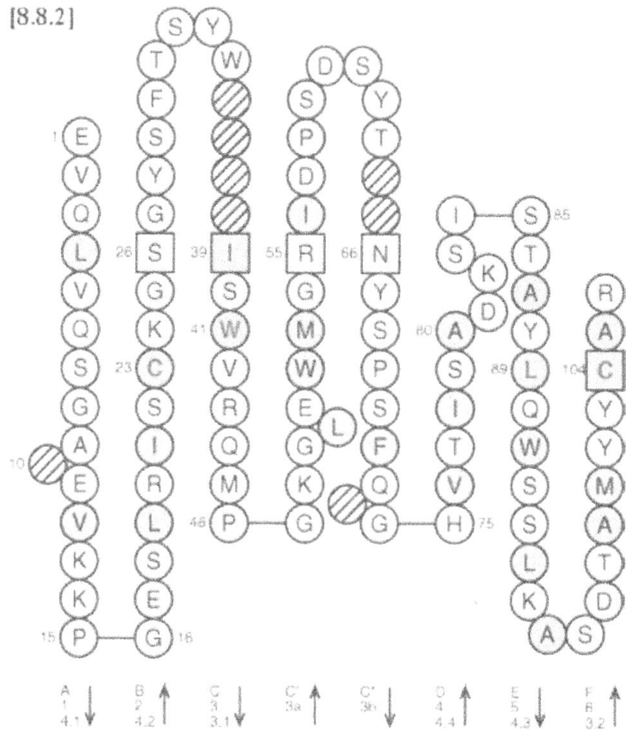

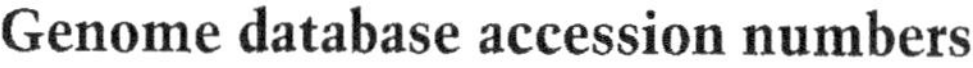

Genome database accession numbers

GDB:9931747
LocusLink: 28386

IGHV6-1

Nomenclature

IGHV6-1: Immunoglobulin heavy variable 6-1.

Definition and functionality

IGHV6-1 is the unique mapped functional gene of the IGHV6 subgroup which only comprises that gene.

Gene location

IGHV6-1 is in the IGH locus on chromosome 14 at 14q32.33.

Nucleotide and amino acid sequences for human IGHV6-1

		1	2	3	4	5	6	7	8	9	10	11	12	13	14	15	16	17	18	19	20
		Q	V	Q	L	Q	Q	S	G	P		G	L	V	K	P	S	Q	T	L	S
X92224/J04097,IGHV6-1*01,6-1G1/VH6	[4]	CAG	GTA	CAG	CTG	CAG	CAG	TCA	GGT	CCA	...	GGA	CTG	GTG	AAG	CCC	TCG	CAG	ACC	CTC	TCA
Z12374,IGHV6-1*01,DP-74	[41]	---	---	---	---	---	---	---	---	---	...	---	---	---	---	---	---	---	---	---	---
X92228,IGHV6-1*01,VHVI	[35]																				
X56382,IGHV6-1*01,VHVIBLK	[35]	---	---	---	---	---	---	---	---	---	...	---	---	---	---	---	---	---	---	---	---
X56380,IGHV6-1*01,VHVICH	[35]	---	---	---	---	---	---	---	---	---	...	---	---	---	---	---	---	---	---	---	---
X56381,IGHV6-1*01,VHVICW	[35]	---	---	---	---	---	---	---	---	---	...	---	---	---	---	---	---	---	---	---	---
X56377,IGHV6-1*01,VHVIIS	[35]	---	---	---	---	---	---	---	---	---	...	---	---	---	---	---	---	---	---	---	---
X56379,IGHV6-1*01,VHVIJB	[35]	---	---	---	---	---	---	---	---	---	...	---	---	---	---	---	---	---	---	---	---
X56383,IGHV6-1*01,VHIMW	[35]	---	---	---	---	---	---	---	---	---	...	---	---	---	---	---	---	---	---	---	---
X56378,IGHV6-1*01,VHITE	[35]	---	---	---	---	---	---	---	---	---	...	---	---	---	---	---	---	---	---	---	---
X14089,IGHV6-1*01,VH-VI	[6]	---	---	---	---	---	---	---	---	---	...	---	---	---	---	---	---	---	---	---	---
Z14223,IGHV6-1*02,VHGL6.3	[7]	---	---	---	---	---	---	---	---	--G	...	---	---	---	---	---	---	---	---	---	---

	21	22	23	24	25	26	27	28	29	30	31	32	33	34	35	36	37	38	39	40
						CDR1-IMGT														
	L	T	C	A	I	S	G	D	S	V	S	S	N	S	A	A			W	N
X92224/J04097,IGHV6-1*01,6-1G1/VH6	CTC	ACC	TGT	GCC	ATC	TCC	GGG	GAC	AGT	GTC	TCT	AGC	AAC	AGT	GCT	GCT	...	...	TGG	AAC
Z12374,IGHV6-1*01,DP-74	---	---	---	---	---	---	---	---	---	---	---	---	---	---	---	---	...	...	---	---
X92228,IGHV6-1*01,VHVI					--	---	---	---	---	---	---	---	---	---	---	---	...	...	---	---
X56382,IGHV6-1*01,VHVIBLK	---	---	---	---	---	---	---	---	---	---	---	---	---	---	---	---	...	...	---	---
X56380,IGHV6-1*01,VHVICH	---	---	---	---	---	---	---	---	---	---	---	---	---	---	---	---	...	...	---	---
X56381,IGHV6-1*01,VHVICW	---	---	---	---	---	---	---	---	---	---	---	---	---	---	---	---	...	...	---	---
X56377,IGHV6-1*01,VHVIIS	---	---	---	---	---	---	---	---	---	---	---	---	---	---	---	---	...	...	---	---
X56379,IGHV6-1*01,VHVIJB	---	---	---	---	---	---	---	---	---	---	---	---	---	---	---	---	...	...	---	---
X56383,IGHV6-1*01,VHIMW	---	---	---	---	---	---	---	---	---	---	---	---	---	---	---	---	...	...	---	---
X56378,IGHV6-1*01,VHITE	---	---	---	---	---	---	---	---	---	---	---	---	---	---	---	---	...	...	---	---
X14089,IGHV6-1*01,VH-VI	---	---	---	---	---	---	---	---	---	---	---	---	---	---	---	---	...	...	---	---
Z14223,IGHV6-1*02,VHGL6.3	---	---	---	---	---	---	---	---	---	---	---	---	---	---	---	---	...	...	---	---

	41	42	43	44	45	46	47	48	49	50	51	52	53	54	55	56	57	58	59	60
																CDR2-				
	W	I	R	Q	S	P	S	R	G	L	E	W	L	G	R	T	Y	Y	R	S
X92224/J04097,IGHV6-1*01,6-1G1/VH6	TGG	ATC	AGG	CAG	TCC	CCA	TCG	AGA	GGC	CTT	GAG	TGG	CTG	GGA	AGG	ACA	TAC	TAC	AGG	TCC
Z12374,IGHV6-1*01,DP-74	---	---	---	---	---	---	---	---	---	---	---	---	---	---	---	---	---	---	---	---
X92228,IGHV6-1*01,VHVI	---	---	---	---	---	---	---	---	---	---	---	---	---	---	---	---	---	---	---	---
X56382,IGHV6-1*01,VHVIBLK	---	---	---	---	---	---	---	---	---	---	---	---	---	---	---	---	---	---	---	---
X56380,IGHV6-1*01,VHVICH	---	---	---	---	---	---	---	---	---	---	---	---	---	---	---	---	---	---	---	---
X56381,IGHV6-1*01,VHVICW	---	---	---	---	---	---	---	---	---	---	---	---	---	---	---	---	---	---	---	---
X56377,IGHV6-1*01,VHVIIS	---	---	---	---	---	---	---	---	---	---	---	---	---	---	---	---	---	---	---	---
X56379,IGHV6-1*01,VHVIJB	---	---	---	---	---	---	---	---	---	---	---	---	---	---	---	---	---	---	---	---
X56383,IGHV6-1*01,VHIMW	---	---	---	---	---	---	---	---	---	---	---	---	---	---	---	---	---	---	---	---
X56378,IGHV6-1*01,VHITE	---	---	---	---	---	---	---	---	---	---	---	---	---	---	---	---	---	---	---	---
X14089,IGHV6-1*01,VH-VI	---	---	---	---	---	---	---	---	---	---	---	---	---	---	---	---	---	---	---	---
Z14223,IGHV6-1*02,VHGL6.3	---	---	---	---	---	---	---	---	---	---	---	---	---	---	---	---	---	---	---	---

```
                                     IMGT_______________
                                      61  62  63  64  65  66  67  68  69  70  71  72  73  74  75  76  77  78  79  80
                                      K   W   Y   N       D   Y   A   V   S   V   K       S   R   I   T   I   N   P
X92224/J04097,IGHV6-1*01,6-1G1/VH6   AAG TGG TAT AAT ... GAT TAT GCA GTA TCT GTG AAA ... AGT CGA ATA ACC ATC AAC CCA
Z12374,IGHV6-1*01,DP-74              --- --- --- --- ... --- --- --- --- --- --- --- ... --- --- --- --- --- --- ---
X92228,IGHV6-1*01,VHVI               --- --- --- --- ... --- --- --- --- --- --- --- ... --- --- --- --- --- --- ---
X56382,IGHV6-1*01,VHVIBLK            --- --- --- --- ... --- --- --- --- --- --- --- ... --- --- --- --- --- --- ---
X56380,IGHV6-1*01,VHVICH             --- --- --- --- ... --- --- --- --- --- --- --- ... --- --- --- --- --- --- ---
X56381,IGHV6-1*01,VHVICW             --- --- --- --- ... --- --- --- --- --- --- --- ... --- --- --- --- --- --- ---
X56377,IGHV6-1*01,VHVIIS             --- --- --- --- ... --- --- --- --- --- --- --- ... --- --- --- --- --- --- ---
X56379,IGHV6-1*01,VHVIJB             --- --- --- --- ... --- --- --- --- --- --- --- ... --- --- --- --- --- --- ---
X56383,IGHV6-1*01,VHIMW              --- --- --- --- ... --- --- --- --- --- --- --- ... --- --- --- --- --- --- ---
X56378,IGHV6-1*01,VHITE              --- --- --- --- ... --- --- --- --- --- --- --- ... --- --- --- --- --- --- ---
X14089,IGHV6-1*01,VH-VI              --- --- --- --- ... --- --- --- --- --- --- --- ... --- --- --- --- --- --- ---
Z14223,IGHV6-1*02,VHGL6.3            --- --- --- --- ... --- --- --- --- --- --- --- ... --- --- --- --- --- --- ---

                                      81  82  83  84  85  86  87  88  89  90  91  92  93  94  95  96  97  98  99 100
                                      D   T   S   K   N   Q   F   S   L   Q   L   N   S   V   T   P   E   D   T   A
X92224/J04097,IGHV6-1*01,6-1G1/VH6   GAC ACA TCC AAG AAC CAG TTC TCC CTG CAG CTG AAC TCT GTG ACT CCC GAG GAC ACG GCT
Z12374,IGHV6-1*01,DP-74              --- --- --- --- --- --- --- --- --- --- --- --- --- --- --- --- --- --- --- ---
X92228,IGHV6-1*01,VHVI               --- --- --- --- --- --- --- --- --- --- --- --- --- --- --- --- --- --- --- ---
X56382,IGHV6-1*01,VHVIBLK            --- --- --- --- --- --- --- --- --- --- --- --- --- --- --- --- --- --- --- ---
X56380,IGHV6-1*01,VHVICH             --- --- --- --- --- --- --- --- --- --- --- --- --- --- --- --- --- --- --- ---
X56381,IGHV6-1*01,VHVICW             --- --- --- --- -
X56377,IGHV6-1*01,VHVIIS             --- --- --- --- --- --- --- --- --- --- --- --- --- --- --- --- --- --- --- ---
X56379,IGHV6-1*01,VHVIJB             --- --- --- --- --- --- --- --- --- --- --- --- --- --- --- --- --- --- --- ---
X56383,IGHV6-1*01,VHIMW              --- --- --- --- --- --- --- --- --- --- --- --- --- --- --- --- --- --- --- ---
X56378,IGHV6-1*01,VHITE              --- --- --- --- --- --- --- --- --- --- --- --- --- --- --- --- --- --- --- ---
X14089,IGHV6-1*01,VH-VI              --- --- --- --- --- --- --- --- --- --- --- --- --- --- --- --- --- --- --- ---
Z14223,IGHV6-1*02,VHGL6.3            --- --- --- --- --- --- --- --- --- --- --- --- --- --- --- --- --- --- --- ---

                                                     _CDR3-IMGT
                                     101 102 103 104 105 106
                                      V   Y   Y   C   A   R
X92224/J04097,IGHV6-1*01,6-1G1/VH6   GTG TAT TAC TGT GCA AGA GA
Z12374,IGHV6-1*01,DP-74              --- --- --- --- --- ---
X92228,IGHV6-1*01,VHVI               --- --- --- --- ...
X56382,IGHV6-1*01,VHVIBLK            --- --- --- --- --- ---
X56380,IGHV6-1*01,VHVICH             --- --- --- --- --- ---
X56381,IGHV6-1*01,VHVICW
X56377,IGHV6-1*01,VHVIIS             --- --- --- --- --- ---
X56379,IGHV6-1*01,VHVIJB             --- --- --- --- --- ---
X56383,IGHV6-1*01,VHIMW              ---  .  --- --- --- -
X56378,IGHV6-1*01,VHITE              --- --- --- --- --- ---
X14089,IGHV6-1*01,VH-VI              --- --- --- --- --- ---
Z14223,IGHV6-1*02,VHGL6.3            --- --- --- --- --- --- --
```

Framework and complementarity determining regions

FR1-IMGT: 25 (-1 aa: 10)
FR2-IMGT: 17
FR3-IMGT: 38 (-1 aa: 73)

CDR1-IMGT: 10
CDR2-IMGT: 9
CDR3-IMGT: 2

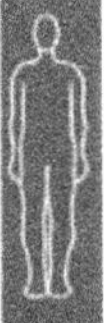

Collier de Perles for human IGHV6-1*01

Accession number: IMGT X92224/J04097 EMBL/GenBank/DDBJ: X92224/J04097

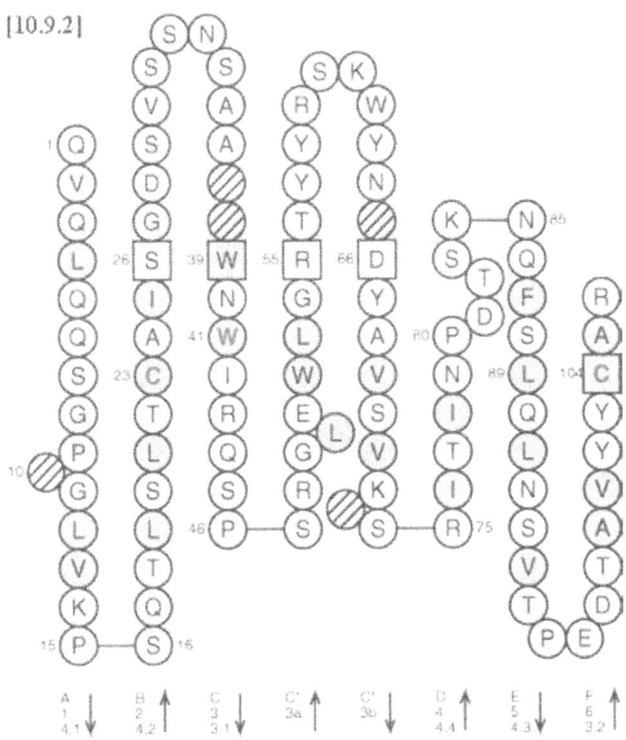

Genome database accession numbers

GDB:9931748 LocusLink: 28385

Nomenclature

IGHV7-4-1: Immunoglobulin heavy variable 7-4-1.

Definition and functionality

IGHV7-4-1 is the only mapped functional gene of the IGHV7 subgroup, IGHV7-4-1 may, or may not, be present due to a polymorphism by insertion/deletion. The IGHV7 subgroup comprises, in addition to the functional and polymorphic IGHV7-4-1 gene, four pseudogenes, and one ORF.

Gene location

IGHV7-4-1, in the haplotypes where it is present, is in the IGH locus on chromosome 14 at 14q32.33.

Nucleotide and amino acid sequences for human IGHV7-4-1

```
                                           1   2   3   4   5   6   7   8   9  10  11  12  13  14  15  16  17  18  19  20
                                           Q   V   Q   L   V   Q   S   G   S       E   L   K   K   P   G   A   S   V   K
L10057,IGHV7-4-1*01,4d275a         [51]   CAG GTG CAG CTG GTG CAA TCT GGG TCT ... GAG TTG AAG AAG CCT GGG GCC TCA GTG AAG
Z12323,IGHV7-4-1*01,DP-21          [41]   --- --- --- --- --- --- --- --- --- ... --- --- --- --- --- --- --- --- --- ---
X62110,IGHV7-4-1*02,VI-4.1b        [39]   --- --- --- --- --- --- --- --- --- ... --- --- --- --- --- --- --- --- --- ---
X92290,IGHV7-4-1*03,7A.4           [34]   --- --- --- --- --- --- --- --- --- ... --- --- --- --- --- --- --- --- --- ---

                                                                   ______________CDR1-IMGT______________
                                          21  22  23  24  25  26  27  28  29  30  31  32  33  34  35  36  37  38  39  40
                                           V   S   C   K   A   S   G   Y   T   F   T   S   Y   A                   M   N
L10057,IGHV7-4-1*01,4d275a                GTT TCC TGC AAG GCT TCT GGA TAC ACC TTC ACT AGC TAT GCT ... ... ... ... ATG AAT
Z12323,IGHV7-4-1*01,DP-21                 --- --- --- --- --- --- --- --- --- --- --- --- --- --- ... ... ... ... --- ---
X62110,IGHV7-4-1*02,VI-4.1b               --- --- --- --- --- --- --- --- --- --- --- --- --- --- ... ... ... ... --- ---
X92290,IGHV7-4-1*03,7A.4                  --- --- --- --- --- --- --- --- --- --- --- --- --- --- ... ... ... ... --- ---

                                                                                                   ______________CDR2-
                                          41  42  43  44  45  46  47  48  49  50  51  52  53  54  55  56  57  58  59  60
                                           W   V   R   Q   A   P   G   Q   G   L   E   W   M   G   W   I   N   T   N   T
L10057,IGHV7-4-1*01,4d275a                TGG GTG CGA CAG GCC CCT GGA CAA GGG CTT GAG TGG ATG GGA TGG ATC AAC ACC AAC ACT
Z12323,IGHV7-4-1*01,DP-21                 --- --- --- --- --- --- --- --- --- --- --- --- --- --- --- --- --- --- --- ---
X62110,IGHV7-4-1*02,VI-4.1b               --- --- --- --- --- --- --- --- --- --- --- --- --- --- --- --- --- --- --- ---
X92290,IGHV7-4-1*03,7A.4                  --- --- --- --- --- --- --- --- --- --- --- --- --- --- --- --- --- --- --- ---

                                          IMGT______________
                                          61  62  63  64  65  66  67  68  69  70  71  72  73  74  75  76  77  78  79  80
                                           G   N   P           T   Y   A   Q   G   F   T       G   R   F   V   F   S   L
L10057,IGHV7-4-1*01,4d275a                GGG AAC CCA ... ... ACG TAT GCC CAG GGC TTC ACA ... GGA CGG TTT GTC TTC TCC TTG
Z12323,IGHV7-4-1*01,DP-21                 --- --- --- ... ... --- --- --- --- --- --- --- ... --- --- --- --- --- --- ---
X62110,IGHV7-4-1*02,VI-4.1b               --- --- --- ... ... --- --- --- --- --- --- --- ... --- --- --- --- --- --- ---
X92290,IGHV7-4-1*03,7A.4                  --- --- --- ... ... --- --- --- --- --- --- --- ... --- --- --- --- --- --- ---

                                          81  82  83  84  85  86  87  88  89  90  91  92  93  94  95  96  97  98  99 100
                                           D   T   S   V   S   T   A   Y   L   Q   I   C   S   L   K   A   E   D   T   A
L10057,IGHV7-4-1*01,4d275a                GAC ACC TCT GTC AGC ACG GCA TAT CTG CAG ATC TGC AGC CTA AAG GCT GAG GAC ACT GCC
Z12323,IGHV7-4-1*01,DP-21                 --- --- --- --- --- --- --- --- --- --- --- --- --- --- --- --- --- --- --- ---
                                                                                       S
X62110,IGHV7-4-1*02,VI-4.1b               --- --- --- --- --- --- --- --- --- --- --- A-- --- --- --- --- --- --- --- ---
                                                                                       S   T
X92290,IGHV7-4-1*03,7A.4                  --- --- --- --- --- --- --- --- --- --- --- A-- -CG --- --- --- --- --- --- -

                                                          _CDR3-IMGT
                                          101 102 103 104 105 106
                                           V   Y   Y   C   A   R
L10057,IGHV7-4-1*01,4d275a                GTG TAT TAC TGT GCG AGA
Z12323,IGHV7-4-1*01,DP-21                 --- --- --- --- --- ---
X62110,IGHV7-4-1*02,VI-4.1b               --- --- --- --- --- --- GA
X92290,IGHV7-4-1*03,7A.4
```

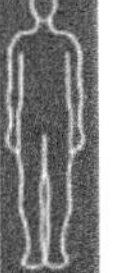

Framework and complementarity determining regions

FR1-IMGT: 25 (-1 aa: 10)
FR2-IMGT: 17
FR3-IMGT: 38 (-1 aa: 73)

CDR1-IMGT: 8
CDR2-IMGT: 8
CDR3-IMGT: 2

Collier de Perles for human IGHV7-4-1*01

Accession number: IMGT L10057

EMBL/GenBank/DDBJ: L10057

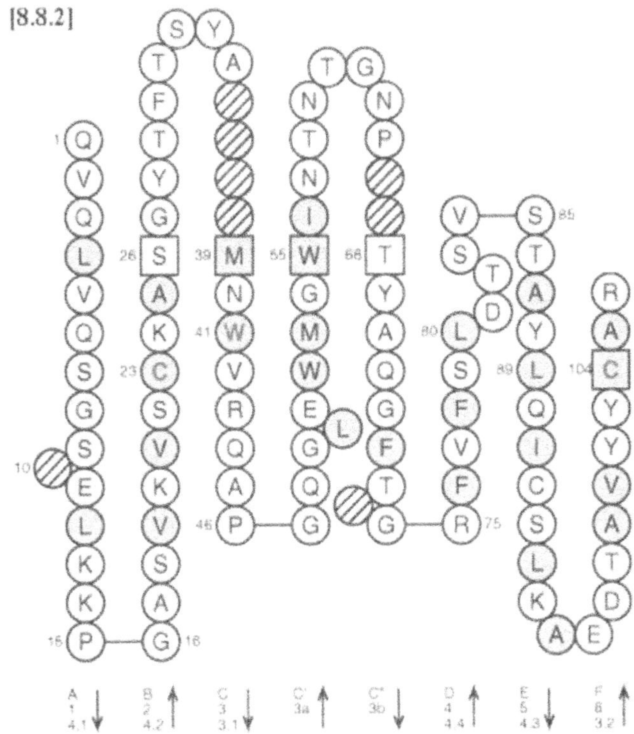

Genome database accession numbers

GDB:9931749

LocusLink: 28384

Nomenclature

IGHV7-81: Immunoglobulin heavy variable 7-81.

Definition and functionality

IGHV7-81 is a mapped ORF of the IGHV7 subgroup. The IGHV7 subgroup comprises, in addition to the ORF IGHV7-81, the functional and polymorphic IGHV7-4-1 gene, and four pseudogenes. IGHV7-81 is considered as ORF because L-PART1, V-HEPTAMER and V-NONAMER have not yet been sequenced and there is so far no known rearranged transcript, suggesting some defect outside the V-REGION.

Gene location

IGHV7-81 is in the IGH locus on chromosome 14 at 14q32.33.

Nucleotide and amino acid sequences for human IGHV7-81

```
                                         1   2   3   4   5   6   7   8   9   10  11  12  13  14  15  16  17  18  19  20
                                         Q   V   Q   L   V   Q   S   G   H       E   V   K   Q   P   G   A   S   V   K
Z27509,IGHV7-81*01,YAC-10          [12] CAG GTG CAG CTG GTG CAG TCT GGC CAT ... GAG GTG AAG CAG CCT GGG GCC TCA GTG AAG
L10058,IGHV7-81*01,1d37            [51] --- --- --- --- --- --- --- --- --- ... --- --- --- --- --- --- --- --- --- ---

                                                                ____________CDR1-IMGT____________
                                         21  22  23  24  25  26  27  28  29  30  31  32  33  34  35  36  37  38  39  40
                                         V   S   C   K   A   S   G   Y   S   F   T   T   Y   G                   M   N
Z27509,IGHV7-81*01,YAC-10               GTC TCC TGC AAG GCT TCT GGT TAC AGT TTC ACC ACC TAT GGT ... ... ... ... ATG AAT
L10058,IGHV7-81*01,1d37                 --- --- --- --- --- --- --- --- --- --- --- --- --- --- ... ... ... ... --- ---

                                                                                                    ______________CDR2-
                                         41  42  43  44  45  46  47  48  49  50  51  52  53  54  55  56  57  58  59  60
                                         W   V   P   Q   A   P   G   Q   G   L   E   W   M   G   W   F   N   T   Y   T
Z27509,IGHV7-81*01,YAC-10               TGG GTG CCA CAG GCC CCT GGA CAA GGG CTT GAG TGG ATG GGA TGG TTC AAC ACC TAC ACT
L10058,IGHV7-81*01,1d37                 --- --- --- --- --- --- --- --- --- --- --- --- --- --- --- --- --- --- --- ---

                                        IMGT_______________
                                         61  62  63  64  65  66  67  68  69  70  71  72  73  74  75  76  77  78  79  80
                                         G   N   P           T   Y   A   Q   G   F   T       G   R   F   V   F   S   M
Z27509,IGHV7-81*01,YAC-10               GGG AAC CCA ... ... ACA TAT GCC CAG GGC TTC ACA ... GGA CGG TTT GTC TTC TCC ATG
L10058,IGHV7-81*01,1d37                 --- --- --- ... ... --- --- --- --- --- --- --- ... --- --- --- --- --- --- ---

                                         81  82  83  84  85  86  87  88  89  90  91  92  93  94  95  96  97  98  99 100
                                         D   T   S   A   S   T   A   Y   L   Q   I   S   S   L   K   A   E   D   M   A
Z27509,IGHV7-81*01,YAC-10               GAC ACC TCT GCC AGC ACA GCA TAC CTG CAG ATC AGC AGC CTA AAG GCT GAG GAC ATG GCC
L10058,IGHV7-81*01,1d37                 --- --- --- --- --- --- --- --- --- --- --- --- --- --- --- --- --- --- --- ---

                                                        _CDR3-IMGT
                                        101 102 103 104 105 106
                                         M   Y   Y   C   A   R
Z27509,IGHV7-81*01,YAC-10               ATG TAT TAC TGT GCG AGA
L10058,IGHV7-81*01,1d37
```

Framework and complementarity determining regions

FR1-IMGT: 25 (-1 aa: 10)	CDR1-IMGT: 8
FR2-IMGT: 17	CDR2-IMGT: 8
FR3-IMGT: 38 (-1 aa: 73)	CDR3-IMGT: 2

Collier de Perles for human IGHV7-81*01

Accession number: IMGT Z27509 EMBL/GenBank/DDBJ: Z27509

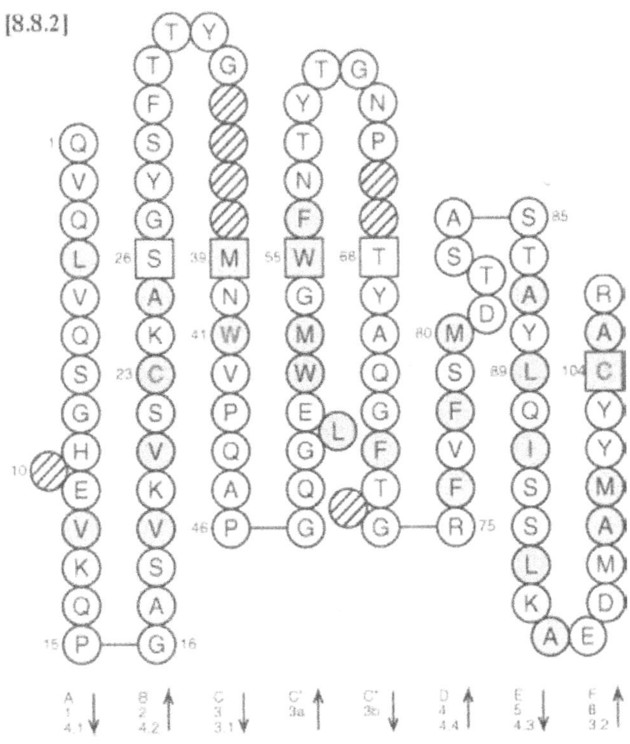

Genome database accession numbers

GDB:9931754 LocusLink: 28378

Protein display of the human IGH V-REGIONs

Only the *01 allele of each functional or ORF V-REGION is shown. IGHV genes are listed for each subgroup, according to their position from 3′ to 5′ in the locus.

IGHV gene	FR1-IMGT (1-26)	CDR1-IMGT (27-38)	FR2-IMGT (39-55)	CDR2-IMGT (56-65)	FR3-IMGT (66-104)	CDR3-IMGT (105-115)
	1 10 20	30	40 50	60	70 80 90 100	110
	.........|.........|......	...|........	.|.........|.....	....|.....	....|.........|.........|.........|....	....|.....
X07448,IGHV1-2	QVQLVQSGA.EVKKPGASVKVSCKAS	GYTFTGYY....	MHWVRQAPGQGLEWMGR	INPNSGGT..	NYAQKFQ.GRVTSTRDTSISTAYMELSRLRSDDTVVYYC	AR.........
X62109,IGHV1-3	QVQLVQSGA.EVKKPGASVKVSCKAS	GYTFTSYA....	MHWVRQAPGQRLEWMGW	INAGNGNT..	KYSQKFQ.GRVTITRDTSASTAYMELSSLRSEDTAVYYC	AR.........
M99637,IGHV1-8	QVQLVQSGA.EVKKPGASVKVSCKAS	GYTFTSYD....	INWVRQATGQGLEWMGW	MNPNSGNT..	GYAQKFQ.GRVTMTRNTSISTAYMELSSLRSEDTAVYYC	AR.........
M99641,IGHV1-18	QVQLVQSGA.EVKKPGASVKVSCKAS	GYTFTSYG....	ISWVRQAPGQGLEWMGW	ISAYNGNT..	NYAQKLQ.GRVTMTTDTSTSTAYMELRSLRSDDTAVYYC	AR.........
M99642,IGHV1-24	QVQLVQSGA.EVKKPGASVKVSCKVS	GYTLTELS....	MHWVRQAPGKGLEWMGG	FDPEDGET..	IYAQKFQ.GRVTMTEDTSTDTAYMELSSLRSEDTAVYYC	AT.........
X92209,IGHV1-45	QMQLVQSGA.EVKKTGSSVKVSCKAS	GYTFTYRY....	LHWVRQAPGQALEWMGW	ITPFNGNT..	NYAQKFQ.DRVTITRDRSMSTAYMELSSLRSEDTAMYYC	AR.........
X92343,IGHV1-46	QVQLVQSGA.EVKKPGASVKVSCKAS	GYTFTSYY....	MHWVRQAPGQGLEWMGI	INPSGGST..	SYAQKFQ.GRVTMTRDTSTSTVYMELSSLRSEDTAVYYC	AR.........
M29809,IGHV1-58	QMQLVQSGP.EVKKPGTSVKVSCKAS	GFTFTSSA....	VQWVRQARGQRLEWIGW	IVVGSGNT..	NYAQKFQ.ERVTITRDMSTSTAYMELSSLRSEDTAVYYC	AA.........
L22582,IGHV1-69	QVQLVQSGA.EVKKPGSSVKVSCKAS	GGTFSSYA....	ISWVRQAPGQGLEWMGG	IIPIFGTA..	NYAQKFQ.GRVTITADESTSTAYMELSSLRSEDTAVYYC	AR.........
Z18904,IGHV1-c	.............KSGASVKVSCSFS	GFTITSYG....	IHWVQQSPGQGLEWMGW	INPGNGSP..	SYAKKFQ.GRFTMTRDMSTTTAYTDLSSLTSEDMAVYYY	AR.........
Z12305,IGHV1-f	EVQLVQSGA.EVKKPGATVKISCKVS	GYTFTDYY....	MHWVQQAPGKGLEWMGL	VDPEDGET..	IYAEKFQ.GRVTITADTSTDTAYMELSSLRSEDTAVYYC	AT.........
X62111,IGHV2-5	QITLKESGP.TLVKPTQTLTLTCTFS	GFSLSTSGVG..	VGWIRQPPGKALEWLAL	IYWNDDK...	RYSPSLK.SRLTITKDTSKNQVVLTMTNMDPVDTATYYC	AHR........
M99648,IGHV2-26	QVTLKESGP.VLVKPTETLTLTCTVS	GFSLSNARMG..	VSWIRQPPGKALEWLAH	IFSNDEK...	SYSTSLK.SRLTISKDTSKSQVVLTMTNMDPVDTATYYC	ARI........
L21969,IGHV2-70	QVTLRESGP.ALVKPTQTLTLTCTFS	GFSLSTSGMC..	VSWIRQPPGKALEWLAL	IDWDDDK...	YYSTSLK.TRLTISKDTSKNQVVLTMTNMDPVDTATYYC	ARI........
M99649,IGHV3-7	EVQLVESGG.GLVQPGGSLRLSCAAS	GFTFSSYW....	MSWVRQAPGKGLEWVAN	IKQDGSEK..	YYVDSVK.GRFTISRDNAKNSLYLQMNSLRAEDTAVYYC	AR.........
M99651,IGHV3-9	EVQLVESGG.GLVQPGRSLRLSCAAS	GFTFDDYA....	MHWVRQAPGKGLEWVSG	ISWNSGSI..	GYADSVK.GRFTISRDNAKNSLYLQMNSLRAEDTALYYC	AKD........
M99652,IGHV3-11	QVQLVESGG.GLVKPGGSLRLSCAAS	GFTFSDYY....	MSWIRQAPGKGLEWVSY	ISSSGSTI..	YYADSVK.GRFTISRDNAKNSLYLQMNSLRAEDTAVYYC	AR.........
X92217,IGHV3-13	EVQLVESGG.GLVQPGGSLRLSCAAS	GFTFSSYD....	MHWVRQATGKGLEWVSA	IGTAGDT...	YYPGSVK.GRFTISRENAKNSLYLQMNSLRAGDTAVYYC	AR.........
X92216,IGHV3-15	EVQLVESGG.GLVKPGGSLRLSCAAS	GFTFSNAW....	MSWVRQAPGKGLEWVGR	IKSKTDGGTT	DYAAPVK.GRFTISRDDSKNTLYLQMNSLKTEDTAVYYC	TT.........
M99655,IGHV3-16	EVQLVESGG.GLVQPGGSLRLSCAAS	GFTFSNSD....	MNWARKAPGKGLEWVSG	VSWNGSRT..	HYVDSVK.RRFIISRDNSRNSLYLQKNRRRAEDMAVYYC	VR.........
M99657,IGHV3-20	EVQLVESGG.GVVRPGGSLRLSCAAS	GFTFDDYG....	MSWVRQAPGKGLEWVSG	INWNGGST..	GYADSVK.GRFTISRDNAKNSLYLQMNSLRAEDTALYHC	AR.........
Z14073,IGHV3-21	EVQLVESGG.GLVKPGGSLRLSCAAS	GFTFSSYS....	MNWVRQAPGKGLEWVSS	ISSSSSYI..	YYADSVK.GRFTISRDNAKNSLYLQMNSLRAEDTAVYYC	AR.........
M99660,IGHV3-23	EVQLLESGG.GLVQPGGSLRLSCAAS	GFTFSSYA....	MSWVRQAPGKGLEWVSA	ISGSGGST..	YYADSVK.GRFTISRDNSKNTLYLQMNSLRAEDTAVYYC	AK.........
M83134,IGHV3-30	QVQLVESGG.GVVQPGRSLRLSCAAS	GFTFSSYA....	MHWVRQAPGKGLEWVAV	ISYDGSNK..	YYADSVK.GRFTISRDNSKNTLYLQMNSLRAEDTAVYYC	AR.........
X92283,IGHV3-30-3	QVQLVESGG.GVVQPGRSLRLSCAAS	GFTFSSYA....	MHWVRQAPGKGLEWVAV	ISYDGSNK..	YYADSVK.GRFTISRDNSKNTLYLQMNSLRAEDTAVYYC	AR.........
L06618,IGHV3-33	QVQLVESGG.GVVQPGRSLRLSCAAS	GFTFSSYG....	MHWVRQAPGKGLEWVAV	IWYDGSNK..	YYADSVK.GRFTISRDNSKNTLYLQMNSLRAEDTAVYYC	AR.........
M99666,IGHV3-35	EVQLVESGG.GLVQPGGSLRLSCAAS	GFTFSNSD....	MNWVHQAPGKGLEWVSG	VSWNGSRT..	HYADSVK.GRFIISRDNSRNTLYLQTNSLRAEDTAVYYC	VR.........
M99669,IGHV3-38	EVQLVESGG.GLVQPRGSLRLSCAAS	GFTVSSNE....	MSWIRQAPGKGLEWVSS	ISGGST....	YYADSRK.GRFTISRDNSKNTLYLQMNNLRAEGTAAYYC	ARY........
M99672,IGHV3-43	EVQLVESGG.VVVQPGGSLRLSCAAS	GFTFDDYT....	MHWVRQAPGKGLEWVSL	ISWDGGST..	YYADSVK.GRFTISRDNSKNSLYLQMNSLRTEDTALYYC	AKD........

```
Z18900,IGHV3-47     EDQLVESGG.GLVQPGGSLRPSCAAS GFAFSSYA.... LHWVRRAPGKGLEWVSA IGTGGDT... YYADSVM.GRFTISRDNAKKSLYLHMNSLIAEDMAVYYC AR.........
M99675,IGHV3-48     EVQLVESGG.GLVQPGGSLRLSCAAS GFTFSSYS.... MNWVRQAPGKGLEWVSY ISSSSSTI.. YYADSVK.GRFTISRDNAKNSLYLQMNSLRAEDTAVYYC AR.........
M99676,IGHV3-49     EVQLVESGG.GLVQPGRSLRLSCTAS GFTFGDYA.... MSWFRQAPGKGLEWVGF IRSKAYGGTT EYTASVK.GRFTISRDGSKSIAYLQMNSLKTEDTAVYYC TR.........
M99679,IGHV3-53     EVQLVESGG.GLIQPGGSLRLSCAAS GFTVSSNY.... MSWVRQAPGKGLEWVSV IYSGGST... YYADSVK.GRFTISRDNSKNTLYLQMNSLRAEDTAVYYC AR.........
M99682,IGHV3-64     EVQLVESGG.GLVQPGGSLRLSCAAS GFTFSSYA.... MHWVRQAPGKGLEYVSA ISSNGGST.. YYANSVK.GRFTISRDNSKNTLYLQMGSLRAEDMAVYYC AR.........
X92218,IGHV3-66     EVQLVESGG.GLVQPGGSLRLSCAAS GFTVSSNY.... MSWVRQAPGKGLEWVSV IYSGGST... YYADSVK.GRFTISRDNSKNTLYLQMNSLRAEDTAVYYC AR.........
X92206,IGHV3-72     EVQLVESGG.GLVQPGGSLRLSCAAS GFTFSDHY.... MDWVRQAPGKGLEWVGR TRNKANSYTT EYAASVK.GRFTISRDDSKNSLYLQMNSLKTEDTAVYYC AR.........
Z27508,IGHV3-73     EVQLVESGG.GLVQPGGSLKLSCAAS GFTFSGSA.... MHWVRQASGKGLEWVGR IRSKANSYAT AYAASVK.GRFTISRDDSKNTAYLQMNSLKTEDTAVYYC TR.........
Z12353,IGHV3-74     EVQLVESGG.GLVQPGGSLRLSCAAS GFTFSSYW.... MHWVRQAPGKGLVWVSR INSDGSST.. SYADSVK.GRFTISRDNAKNTLYLQMNSLRAEDTAVYYC AR.........
Z18898,IGHV3-d      EVQLVESRG.VLVQPGGSLRLSCAAS GFTVSSNE.... MSWVRQAPGKGLEWVSS ISGGST.... YYADSRK.GRFTISRDNSKNTLHLQMNSLRAEDTAVYYC KK.........

X05713,IGHV4-4      QVQLQESGP.GLVKPPGTLSLTCAVS GGSISSSNW... WSWVRQPPGKGLEWIGE IYHSGST... NYNPSLK.SRVTISVDKSKNQFSLKLSSVTAADTAVYCC AR.........
X05714,IGHV4-28     QVQLQESGP.GLVKPSDTLSLTCAVS GYSISSSNW... WGWIRQPPGKGLEWIGY IYYSGST... YYNPSLK.SRVTMSVDTSKNQFSLKLSSVTAVDTAVYYC AR.........
L10089,IGHV4-30-2   QLQLQESGS.GLVKPSQTLSLTCAVS GGSISSGGYS.. WSWIRQPPGKGLEWIGY IYHSGST... YYNPSLK.SRVTISVDRSKNQFSLKLSSVTAADTAVYYC AR.........
Z14238,IGHV4-30-4   QVQLQESGP.GLVKPSQTLSLTCTVS GGSISSGDYY.. WSWIRQPPGKGLEWIGY IYYSGST... YYNPSLK.SRVTISVDTSKNQFSLKLSSVTAADTAVYYC AR.........
L10098,IGHV4-31     QVQLQESGP.GLVKPSQTLSLTCTVS GGSISSGGYY.. WSWIRQHPGKGLEWIGY IYYSGST... YYNPSLK.SLVTISVDTSKNQFSLKLSSVTAADTAVYYC AR.........
X92278,IGHV4-34     QVQLQQWGA.GLLKPSETLSLTCAVY GGSFSGYY.... WSWIRQPPGKGLEWIGE INHSGST... NYNPSLK.SRVTISVDTSKNQFSLKLSSVTAADTAVYYC AR.........
L10094,IGHV4-39     QLQLQESGP.GLVKPSETLSLTCTVS GGSISSSSYY.. WGWIRQPPGKGLEWIGS IYYSGST... YYNPSLK.SRVTISVDTSKNQFSLKLSSVTAADTAVYYC AR.........
L10088,IGHV4-59     QVQLQESGP.GLVKPSETLSLTCTVS GGSISSYY.... WSWIRQPPGKGLEWIGY IYYSGST... NYNPSLK.SRVTISVDTSKNQFSLKLSSVTAADTAVYYC AR.........
M29811,IGHV4-61     QVQLQESGP.GLVKPSETLSLTCTVS GGSVSSGSYY.. WSWIRQPPGKGLEWIGY IYYSGST... NYNPSLK.SRVTISVDTSKNQFSLKLSSVTAADTAVYYC AR.........
Z12367,IGHV4-b      QVQLQESGP.GLVKPSETLSLTCAVS GYSISSGYY... WGWIRQPPGKGLEWIGS IYHSGST... YYNPSLK.SRVTISVDTSKNQFSLKLSSVTAADTAVYYC AR.........

M99686,IGHV5-51     EVQLVQSGA.EVKKPGESLKISCKGS GYSFTSYW.... IGWVRQMPGKGLEWMGI IYPGDSDT.. RYSPSFQ.GQVTISADKSISTAYLQWSSLKASDTAMYYC AR.........
X92227,IGHV5-a      EVQLVQSGA.EVKKPGESLRISCKGS GYSFTSYW.... ISWVRQMPGKGLEWMGR IDPSDSYT.. NYSPSFQ.GHVTISADKSISTAYLQWSSLKASDTAMYYC AR.........

X92224,IGHV6-1      QVQLQQSGP.GLVKPSQTLSLTCAIS GDSVSSNSAA.. WNWIRQSPSRGLEWLGR TYYRSKWYN. DYAVSVK.SRITINPDTSKNQFSLQLNSVTPEDTAVYYC AR.........

L10057,IGHV7-4-1    QVQLVQSGS.ELKKPGASVKVSCKAS GYTFTSYA.... MNWVRQAPGQGLEWMGW INTNTGNP.. TYAQGFT.GRFVFSLDTSVSTAYLQICSLKAEDTAVYYC AR.........
Z27509,IGHV7-81     QVQLVQSGH.EVKQPGASVKVSCKAS GYSFTTYG.... MNWVPQAPGQGLEWMGW FNTYTGNP.. TYAQGFT.GRFVFSMDTSASTAYLQISSLKAEDMAMYYC AR.........
```

Recombination signals

Only the recombination signals of the allele *01 of each functional or ORF V-REGION are shown. Non-conserved nucleotides taken into account for the ORF functionality definition are shown in bold and italics.

IGHV gene name	V recombination signal: V-HEPTAMER	(bp)	V-NONAMER
IGHV1-2*01	CACAGTG	23	TCAGAAACC
IGHV1-3*01	CACAGTG	23	TCAGAAACC
IGHV1-8*01	CACAGTG	23	TCAGAAACC
IGHV1-18*01	CACAGTG	23	TCAGAAACC
IGHV1-24*01	CACAGTG	23	TCAGAAACC
IGHV1-45*01	CACAGTG	23	TCAGAAACC
IGHV1-46*01	CACAGTG	23	TCAGAAACC
IGHV1-58*01	CACAGTG	23	TCAGAAACG
IGHV1-69*01	CACAGTG	23	TCAGAAACC
IGHV1-c*01 (ORF)	nd	nd	nd
IGHV1-f*01	nd	nd	nd
IGHV2-5*01	CACAAAG	23	ACAAAAACC
IGHV2-26*01	CACAGAG	23	ACAAGAACC
IGHV2-70*01	CACAGAG	nd	nd
IGHV3-7*01	CACAGTG	23	ACACAAACC
IGHV3-9*01	CACAGTG	23	ACAAAAACC
IGHV3-11*01	CACAGTG	23	ACACAAACC
IGHV3-13*01	CACAGTG	23	ACACAAACC
IGHV3-15*01	CACAGTG	23	ACACAAACC
IGHV3-16*01 (ORF)	***TCCT***GTG	23	ACACAAACC
IGHV3-20*01	CACAGTG	23	ACACAAACG
IGHV3-21*01	nd	nd	nd
IGHV3-23*01	CACAGTG	23	ACACAAACC
IGHV3-30*01	CACAGTG	23	ACACAAACC
IGHV3-30-3*01	nd	nd	nd
IGHV3-33*01	CACAG (p)	nd	nd
IGHV3-35*01 (ORF)	CAC***T***GTG	23	ACACAAACC
IGHV3-38*01 (ORF)	CACAG***A***G	21	ACACAAACC
IGHV3-43*01	CACAGTG	23	ACAAAAACC
IGHV3-47*01 (ORF)	nd	nd	nd
IGHV3-48*01	nd	nd	nd
IGHV3-49*01	CACAGTG	23	ACACAGACC
IGHV3-53*01	CACAGTG	23	ACACAAACC
IGHV3-64*01	CACAGTG	23	GCAGAAACC
IGHV3-66*01	CACAGTG	23	ACACAAACC
IGHV3-72*01	CACAGCG	23	ACACAAACC
IGHV3-73*01	nd	nd	nd
IGHV3-74*01	nd	nd	nd
IGHV3-d*01	nd	nd	nd
IGHV4-4*01	CACAGTG	23	ACACAAACC
IGHV4-28*01	CACAGTG	23	ACACAAACC
IGHV4-30-2*01	nd	nd	nd
IGHV4-30-4*01	CACAATG	23	nd
IGHV4-31*01	CACA (p)	nd	nd
IGHV4-34*01	CACAGTG	23	ACCAAAACC
IGHV4-39*01	CACAGTG	nd	nd
IGHV4-59*01	CACAGTG	nd	nd
IGHV4-61*01	CACAGTG	24	ACACAAACC
IGHV4-61*06 (ORF)	CACA***A***TG	nd	nd
IGHV4b*01	nd	nd	nd
IGHV5-51*01	CACAGTG	22	TCTAAAACC
IGHV5-a*01	nd	nd	nd
IGHV6-1*01	CACAGTG	23	ACACAAACC
IGHV7-4-1*01	nd	nd	nd
IGHV7-81*01 (ORF)	nd	nd	nd

nd: not defined
(p): partial

IGHV references

1. Adderson, E.E. et al. (1993) J. Immunol. 151, 800–809.
2. Andris, J.S. et al. (1993) Mol. Immunol. 30, 1601–1616.
3. Baer, R. et al. (1988) J. Exp. Med. 167, 2011–2016.
4. Berman, J.E. et al. (1988) EMBO J. 7, 727–738.
5. Brezinschek, H.P. et al. (1995) J. Immunol. 155, 190–202.
6. Buluwela, L. et al. (1988) Eur. J. Immunol. 18, 1843–1845.
7. Campbell, M.J. et al. (1992) Mol. Immunol. 29, 193–203.
8. Chen, P.P. et al. (1990) Scand. J. Immunol. 31, 257–267.
9. Chen, P.P. et al. (1988) Arthritis Rheum. 31, 1429–1431.
10. Chen, P.P. et al. (1990) Scand. J. Immunol. 31, 593–599.
11. Chen, P.P. et al. (1989) Arthritis Rheum. 32, 72–76.
12. Cook, G.P. et al. (1994) Nature Genetics 7, 162–168.
13. Crouzier, R. et al. (1995) J. Immunol. 154, 412–421.
14. Cuisinier, A.M. et al. (1993) Eur. J. Immunol. 23, 110–118.
15. Deftos, M. et al. (1994) J. Clin. Invest. 93, 2545–2553.
16. Friedman, D.F. et al. (1991) J. Exp. Med. 174, 525–537.
17. Harmer, I.J. et al. (1995) Arthritis Rheum. 38, 1068–1076.
18. Humphries, G. et al. (1988) Nature 331, 446–449.
19. Ikematsu, H. et al. (1992) Ann. NY Acad. Sci. 651, 319–327.
20. Ikematsu, H. et al. (1994) J. Immunol. 152, 1430–1441.
21. Ikematsu, H. et al. (1993) J. Immunol. 151, 3604–3616.
22. Kodaira, M. et al. (1986) J. Mol. Biol. 190, 529–541.
23. Küppers, R. et al. (1992) Immunology Letters 34, 57–62.
24. Lee, R.H. et al. (1987) J. Mol. Biol. 195, 761–768.
25. Matsuda, F. et al. (1988) EMBO J. 7, 1047–1051.
26. Matsuda, F. et al. (1993) Nature Genetics 3, 88–94.
27. Matsuda, F. et al. (1998) J. Exp. Med. 188, 1–15.
28. Matthyssen, G. et al. (1980) Proc. Natl Acad. Sci. USA 77, 6561–6565.
29. Olee, T. et al. (1992) J. Exp. Med. 175, 831–842.
30. Olee, T. et al. (1991) J. Clin. Invest. 88, 193–203.
31. Pascual, V. et al. (1990) J. Clin. Invest. 86, 1320–1328.
32. Rechavi, G. et al. (1990) Proc. Natl Acad. Sci. USA 79, 4405–4409.
33. Rechavi, G. et al. (1983) Proc. Natl Acad. Sci. USA 80, 855–859.
34. Rubinstein, D.B. et al. (1994) Mol. Immunol. 31, 713–721.
35. Sanz, I. et al. (1989) EMBO J. 8, 3741–3748.
36. Sasso, E.H et al. (1995) J. Clin. Invest. 96, 1591–1600.
37. Sasso, E.H. et al. (1992) J. Immunol. 149, 1230–1236.
38. Shen, A. et al. (1987) Proc. Natl Acad. Sci. USA 84, 8563–8567.
39. Shin, E.K. et al. (1991) EMBO J. 10, 3641–3645.
40. Soto-Gil, R.W. et al. (1992) Arthritis Rheum. 35, 356–363.
41. Tomlinson, I.M. et al. (1992) J. Mol. Biol. 227, 776–798.
42. Tomlinson, I.M. Unpublished.
43. Turnbull, I.F. et al. (1987) Immunogenetics 25, 184–192.
44. vans Es, J.H. et al. (1992) J. Immunol. 149, 2234–2240.
45. vans Es, J.H. et al. (1992) J. Immunol. 149, 492–497.
46. vans Es, J.H. et al. (1991) J. Exp. Med. 173, 461–470.
47. van der Maarel, S. et al. (1993) J. Immunol. 150, 2858–2868.
48. Vescio, R.A. et al. (1995) J. Immunol. 155, 2487–2497.
49. Voswinkel, J. et al. (1997) Ann. NY Acad. Sci. 815, 312–315.

[50] Weng, N. et al. (1992) Eur. J. Immunol. 22, 1075–1082.
[51] Willems van Dijk, K. et al. (1993) Eur. J. Immunol. 23, 832–839.
[52] Winkler, T.H. et al. (1992) Eur. J. Mol. 22, 1719–1728.
[53] Winter, G. et al. (1992) J. Mol. Biol. 227, 799–817.
[54] Yang, P.M. et al. (1993) Scand. J. Immunol. 37, 504–508.
[55] Zelenetz, A.D. et al. (1992) J. Exp. Med. 176, 1137–1748.

Section III

THE HUMAN IMMUNOGLOBULIN IGK GENES

Part 1

IGKC

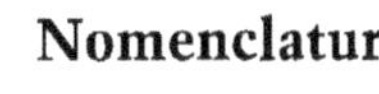

Nomenclature

IGKC: Immunoglobulin kappa constant.

Definition and functionality

IGKC is the functional and unique constant gene in the IGK locus.

Gene location

IGKC is in the IGK locus on chromosome 2 at 2p11.2.

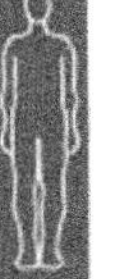

Nucleotide and amino acid sequences for human IGKC

The nucleotide between parentheses at the beginning of exons comes from a DONOR_SPLICE (n from ngt).

The Cysteines involved in the intrachain disulphide bridges are shown with their number and letter **C** in bold. The Cysteines involved in the H-L interchain disulphide bridges are shown with only the letter **C** in bold.

```
                                   1   2   3   4   5   6   7   8   9  10  11  12  13  14  15  16  17  18  19  20
                                   R   T   V   A   A   P   S   V   F   I   F   P   P   S   D   E   Q   L   K   S
J00241   ,IGKC*01           [1]  (C)GA ACT GTG GCT GCA CCA TCT GTC TTC ATC TTC CCG CCA TCT GAT GAG CAG TTG AAA TCT
X67858   ,IGKC*01           [3]  (-)-- --- --- --- --- --- --- --- --- --- --- --- --- --- --- --- --- --- --- ---
U72063   ,IGKC*01           [4]  (-)-- --- --- --- --- --- --- --- --- --- --- --- --- --- --- --- --- --- --- ---
M11736   ,IGKC*02           [2]  (-)-- --- --- --- --- --- --- --- --- --- --- --- --- --- --- --- --- --- --- ---
M11737   ,IGKC*03           [2]  (-)-- --- --- --- --- --- --- --- --- --- --- --- --- --- --- --- --- --- --- ---
AF017732,IGKC*04            [6]  (-)-- --- --- --- --- --- --- --- --- --- --- --- --- --- --- --- --- --- --- ---
AF113887,IGKC*05 (cDNA)     [5]  (-)-- --- --- --- --- --- --- --- --- --- --- --- --- --- --- --- --- --- --- ---

                                  21  22  23  24  25  26  27  28  29  30  31  32  33  34  35  36  37  38  39  40
                                   G   T   A   S   V   V   C   L   L   N   N   F   Y   P   R   E   A   K   V   Q
J00241   ,IGKC*01                 GGA ACT GCC TCT GTT GTG TGC CTG CTG AAT AAC TTC TAT CCC AGA GAG GCC AAA GTA CAG
X67858   ,IGKC*01                 --- --- --- --- --- --- --- --- --- --- --- --- --- --- --- --- --- --- --- ---
U72063   ,IGKC*01                 --- --- --- --- --- --- --- --- --- --- --- --- --- --- --- --- --- --- --- ---
M11736   ,IGKC*02                 --- --- --- --- --- --- --- --- --- --- --- --- --- --- --- --- --- --- --- ---
M11737   ,IGKC*03                 --- --- --- --- --- --- --- --- --- --- --- --- --- --- --- --- --- --- --- ---
AF017732,IGKC*04                  --- --- --- --- --- --- --- --- --- --- --- --- --- --- --- --- --- --- --- ---
AF113887,IGKC*05 (cDNA)           --- --- --- --- --- --- --- --- --- --- --- --- --- --- --- --- --- --- --- ---

                                  41  42  43  44  45  46  47  48  49  50  51  52  53  54  55  56  57  58  59  60
                                   W   K   V   D   N   A   L   Q   S   G   N   S   Q   E   S   V   T   E   Q   D
J00241   ,IGKC*01                 TGG AAG GTG GAT AAC GCC CTC CAA TCG GGT AAC TCC CAG GAG AGT GTC ACA GAG CAG GAC
X67858   ,IGKC*01                 --- --- --- --- --- --- --- --- --- --- --- --- --- --- --- --- --- --- --- ---
U72063   ,IGKC*01                 --- --- --- --- --- --- --- --- --- --- --- --- --- --- --- --- --- --- --- ---
                                                                                                            E
M11736   ,IGKC*02                 --- --- --- --- --- --- --- --- --- --- --- --- --- --- --- --- --- --- --- --G
                                  R                                                                         E
M11737   ,IGKC*03                 C-- --- --- --- --- --- --- --- --- --- --- --- --- --- --- --- --- --- --- --G
AF017732,IGKC*04                  --- --- --- --- --- --- --- --- --- --- --- --- --- --- --- --- --- --- --- ---
AF113887,IGKC*05 (cDNA)           --- --- --- --- --- --- --- --- --- --- --- --- --- --- --- --- --- --- --- ---

                                  61  62  63  64  65  66  67  68  69  70  71  72  73  74  75  76  77  78  79  80
                                   S   K   D   S   T   Y   S   L   S   S   T   L   T   L   S   K   A   D   Y   E
J00241   ,IGKC*01                 AGC AAG GAC AGC ACC TAC AGC CTC AGC AGC ACC CTG ACG CTG AGC AAA GCA GAC TAC GAG
X67858   ,IGKC*01                 --- --- --- --- --- --- --- --- --- --- --- --- --- --- --- --- --- --- --- ---
U72063   ,IGKC*01                 --- --- --- --- --- --- --- --- --- --- --- --- --- --- --- --- --- --- --- ---
M11736   ,IGKC*02                 --- --- --- --- --- --- --- --- --- --- --- --- --- --- --- --- --- --- --- ---
M11737   ,IGKC*03                 --- --- --- --- --- --- --- --- --- --- --- --- --- --- --- --- --- --- --- ---
AF017732,IGKC*04                  --- --- --- --- --- --- --- --- --- --- --- --- --- --- --- --- --- --- --- ---
                                                                       N
AF113887,IGKC*05 (cDNA)           --- --- --- --- --- --- --- --- --- -A- --- --- --- --- --- --- --- --- --- ---
```

```
                              81  82  83  84  85  86  87  88  89  90  91  92  93  94  95  96  97  98  99 100
                              K   H   K   V   Y   A   C   E   V   T   H   Q   G   L   S   S   P   V   T   K
J00241  ,IGKC*01              AAA CAC AAA GTC TAC GCC TGC GAA GTC ACC CAT CAG GGC CTG AGC TCG CCC GTC ACA AAG

X67858  ,IGKC*01              --- --- --- --- --- --- --- --- --- --- --- --- --- --- --- --- --- --- --- ---

U72063  ,IGKC*01              --- --- --- --- --- --- --- --- --- --- --- --- --- --- --- --- --- --- --- ---
                                                      G
M11736  ,IGKC*02              --- --- --- --- --- --- G-- --- --- --- --- --- --- --- --- --- --- --- --- ---

M11737  ,IGKC*03              --- --- --- --- --- --- --- --- --- --- --- --- --- --- --- --- --- --- --- ---
                                          L
AF017732,IGKC*04              --- --- --- C-- --- --- --- --- --- --- --- --- --- --- --- --- --- --- --- ---

AF113887,IGKC*05 (cDNA)       --- --- --- --- --- --- --- --- --- --- --- --- --- --- --- --- --- --- --- ---


                              101 102 103 104 105 106 107
                              S   F   N   R   G   E   C   *
J00241  ,IGKC*01              AGC TTC AAC AGG GGA GAG TGT

X67858  ,IGKC*01              --- --- --- --- --- --- ---

U72063  ,IGKC*01              --- --- --- --- --- --- ---

M11736  ,IGKC*02              --- --- --- --- --- --- ---

M11737  ,IGKC*03              --- --- --- --- --- --- ---

AF017732,IGKC*04              --- --- --- --- --- --- ---

AF113887,IGKC*05 (cDNA)       --- --- --- --- --- --- --C
```

Genome database accession numbers

GDB:120088 LocusLink: 3514

References

[1] Hieter, P.A. et al. (1980) Cell 22, 197–207.
[2] Stavnezer-Nordgren, J. et al. (1985) Science 230, 458–461.
[3] Whitehurst, C. et al. (1992) Nucleic Acids Res. 20, 4929–4930.
[4] Hu, W.X. et al. (1995) Sheng Wu Hua Hsueh Yu Sheng Wu Wu Li Hsueh Pao 27, 215–221.
[5] Wally, J. et al. (1999) Biochim. Biophys. Acta 1454, 49–56.
[6] Brandt, P. et al., unpublished.

Protein display

The Protein display of IGKC gene is shown pages 94 and 95.

Part 2

IGKJ

Nomenclature

Immunoglobulin kappa joining group.

Definition and functionality

The human IGKJ group comprises five functional mapped genes.

Gene location

The IGKJ genes are located in IGK locus on chromosome 2 at 2p11.2, upstream from the IGKC gene.

Nucleotide and amino acid sequences for the human functional IGKJ genes

The conserved FGXG motif, characteristic of the IGK (and IGL) J-REGION is underlined.

Rearranged genomic or cDNA have been included in the alignments of alleles when several sequences from independent sources have been found with the same mutations. These alleles need to be confirmed by complete germline genomic sequences.

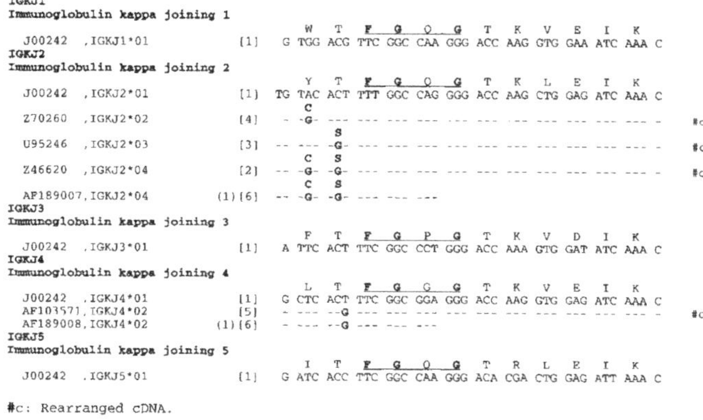

```
IGKJ1
Immunoglobulin kappa joining 1
                                       W   T   F   G   Q   G   T   K   V   E   I   K
  J00242   ,IGKJ1*01           [1]   G TGG ACG TTC GGC CAA GGG ACC AAG GTG GAA ATC AAA C
IGKJ2
Immunoglobulin kappa joining 2
                                       Y   T   F   G   Q   G   T   K   L   E   I   K
  J00242   ,IGKJ2*01           [1]  TG TAC ACT TTT GGC CAG GGG ACC AAG CTG GAG ATC AAA C
                                       C
  Z70260   ,IGKJ2*02           [4]   - -G- --- --- --- --- --- --- --- --- --- --- --- -   #c
                                           S
  U95246   ,IGKJ2*03           [3]  -- --- -G- --- --- --- --- --- --- --- --- --- --- -   #c
                                       C   S
  Z46620   ,IGKJ2*04           [2]  -- -G- -G- --- --- --- --- --- --- --- --- --- --- -   #c
                                       C   S
  AF189007,IGKJ2*04         (1)[6]  -- -G- -G- --- --- ---
IGKJ3
Immunoglobulin kappa joining 3
                                       F   T   F   G   P   G   T   K   V   D   I   K
  J00242   ,IGKJ3*01           [1]   A TTC ACT TTC GGC CCT GGG ACC AAA GTG GAT ATC AAA C
IGKJ4
Immunoglobulin kappa joining 4
                                       L   T   F   G   G   G   T   K   V   E   I   K
  J00242   ,IGKJ4*01           [1]   G CTC ACT TTC GGC GGA GGG ACC AAG GTG GAG ATC AAA C
  AF103571,IGKJ4*02            [5]   - --- --G --- --- --- --- --- --- --- --- --- --- -   #c
  AF189008,IGKJ4*02         (1)[6]   - --- --G --- --- ---
IGKJ5
Immunoglobulin kappa joining 5
                                       I   T   F   G   Q   G   T   R   L   E   I   K
  J00242   ,IGKJ5*01           [1]   G ATC ACC TTC GGC CAA GGG ACA CGA CTG GAG ATT AAA C

#c: Rearranged cDNA.
```

Note:

(1) partial sequence in 3′.

Recombination signals

J recombination signal (J-RS)			IGKJ gene and allele name
J-NONAMER	(bp)	J-HEPTAMER	
GGTTTCTGT	23	CACTGTG	IGKJ1*01
AGTTTTTGT	23	CATTGTG	IGKJ2*01 (1)
AGTTTTTGT	23	CATTGTG	IGKJ2*04 (1)
GGTTTTTGT	22	CACTGTG	IGKJ3*01
GGTTTTTGT	23	CACTGTG	IGKJ4*01
GGTTTTTGT	23	CACTGTG	IGKJ4*02
GATTTTTGT	23	CACTGTG	IGKJ5*01

Note:
(1) not determined for the IGKJ2*02 and IGKJ2*03.

References

1 Hieter, P.A. et al. (1981) J. Biol. Chem. 257, 1516–1522.
2 Giachino, C. et al. (1995) J. Exp. Med. 181,1245–1250.
3 Manheimer-Lory, A.J. et al. (1997) J. Clin. Invest. 100, 2538–46.
4 Sahota, S.S. et al. (1997) Blood 89, 219–226.
5 Wildt, R.M.T. et al. (1999) J. Mol. Biol. 285, 895–901.
6 Feeney, A.J. et al. (2000) Immunogenetics, 51, 487–488.

Part 3

IGKV

Nomenclature

IGKV1-5: Immunoglobulin kappa variable 1-5.

Definition and functionality

IGKV1-5 is one of the 15–16 functional genes of the IGKV1 subgroup which comprises 29 mapped genes in the IGK locus and 14–15 orphons.

Gene location

IGKV1-5 is in the proximal cluster of the IGK locus on chromosome 2 at 2p11.2.

Nucleotide and amino acid sequences for human IGKV1-5

References are shown in square brackets.

```
                                     1   2   3   4   5   6   7   8   9  10  11  12  13  14  15  16  17  18  19  20
                                     D   I   Q   M   T   Q   S   P   S   T   L   S   A   S   V   G   D   R   V   T
Z00001,IGKV1-5*01,L12       [2]     GAC ATC CAG ATG ACC CAG TCT CCT TCC ACC CTG TCT GCA TCT GTA GGA GAC AGA GTC ACC
J00245,IGKV1-5*01,HK102     [2]     --- --- --- --- --- --- --- --- --- --- --- --- --- --- --- --- --- --- --- ---
M23851,IGKV1-5*02,V1        [12]    --- --- --- --- --- --- --- --- --- --- --- --- --- --- --- --- --- --- --- ---
X72813,IGKV1-5*03,L12a      [10]    --- --- --- --- --- --- --- --- --- --- --- --- --- --- --- --- --- --- --- ---

                                                            _________________CDR1-IMGT_______________________
                                    21  22  23  24  25  26  27  28  29  30  31  32  33  34  35  36  37  38  39  40
                                     I   T   C   R   A   S   Q   S   I   S   S   W                           L   A
Z00001,IGKV1-5*01,L12               ATC ACT TGC CGG GCC AGT CAG AGT ATT AGT AGC TGG ... ... ... ... ... ... TTG GCC
J00245,IGKV1-5*01,HK102             --- --- --- --- --- --- --- --- --- --- --- --- ... ... ... ... ... ... --- ---
                                         I
M23851,IGKV1-5*02,V1                --- -T- --- --- --- --- --- --- --- --- --- --- ... ... ... ... ... ... --- ---
X72813,IGKV1-5*03,L12a              --- --- --- --- --- --- --- --- --- --- --- --- ... ... ... ... ... ... --- ---

                                                                                              ___________CDR2-
                                    41  42  43  44  45  46  47  48  49  50  51  52  53  54  55  56  57  58  59  60
                                     W   Y   Q   Q   K   P   G   K   A   P   K   L   L   I   Y   D   A   S
Z00001,IGKV1-5*01,L12               TGG TAT CAG CAG AAA CCA GGG AAA GCC CCT AAG CTC CTG ATC TAT GAT GCC TCC ... ...
J00245,IGKV1-5*01,HK102             --- --- --- --- --- --- --- --- --- --- --- --- --- --- --- --- --- --- ... ...
M23851,IGKV1-5*02,V1                --- --- --- --- --- --- --- --- --- --- --- --- --- --- --- --- --- --- ... ...
                                                                                             K
X72813,IGKV1-5*03,L12a              --- --- --- --- --- --- --- --- --- --- --- --- --- --- --- A-G --G --T ... ...

                                    IMGT________________
                                    61  62  63  64  65  66  67  68  69  70  71  72  73  74  75  76  77  78  79  80
                                                         S   L   E   S   G   V   P       S   R   F   S   G   S   G
Z00001,IGKV1-5*01,L12               ... ... ... ... ... AGT TTG GAA AGT GGG GTC CCA ... TCA AGG TTC AGC GGC AGT GGA
J00245,IGKV1-5*01,HK102             ... ... ... ... ... --- --- --- --- --- --- --- ... --- --- --- --- --- --- ---
M23851,IGKV1-5*02,V1                ... ... ... ... ... --- --- --- --- --- --- --- ... --- --- --- --- --- --- ---
X72813,IGKV1-5*03,L12a              ... ... ... ... ... --- --A --- --- --- --- --- ... --- --- --- --- --- --- ---

                                    81  82  83  84  85  86  87  88  89  90  91  92  93  94  95  96  97  98  99 100
                                             S   G   T   E   F   T   L   T   I   S   S   L   Q   P   D   D   F   A
Z00001,IGKV1-5*01,L12               ... ... TCT GGG ACA GAA TTC ACT CTC ACC ATC AGC AGC CTG CAG CCT GAT GAT TTT GCA
J00245,IGKV1-5*01,HK102             ... ... --- --- --- --- --- --- --- --- --- --- --- --- --- --- --- --- --- ---
M23851,IGKV1-5*02,V1                ... ... --- --- --- --- --- --- --- --- --- --- --- --- --- --- --- --- --- ---
X72813,IGKV1-5*03,L12a              ... ... --- --- --- --- --- --- --- --- --- --- --- --- --- --- --- --- --- ---

                                                    __________CDR3-IMGT_______________
                                    101 102 103 104 105 106 107 108 109 110 111
                                     T   Y   Y   C   Q   Q   Y   N   S   Y   S
Z00001,IGKV1-5*01,L12               ACT TAT TAC TGC CAA CAG TAT AAT AGT TAT TCT CC
J00245,IGKV1-5*01,HK102             --- --- --- --- --- --- --- --- --- --- --- --
M23851,IGKV1-5*02,V1                --- --- --- --- --- --- --- --- --- --- --- --
X72813,IGKV1-5*03,L12a              --- --- --- --- --- --- --- --- --- --- --- --
```

Framework and complementarity determining regions

FR1-IMGT: 26
FR2-IMGT: 17
FR3-IMGT: 36 (-3 aa: 73, 81, 82)

CDR1-IMGT: 6
CDR2-IMGT: 3
CDR3-IMGT: 7

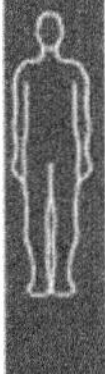

Collier de Perles for human IGKV1-5*01

Accession number: IMGT Z00001 EMBL/GenBank/DDBJ: Z00001

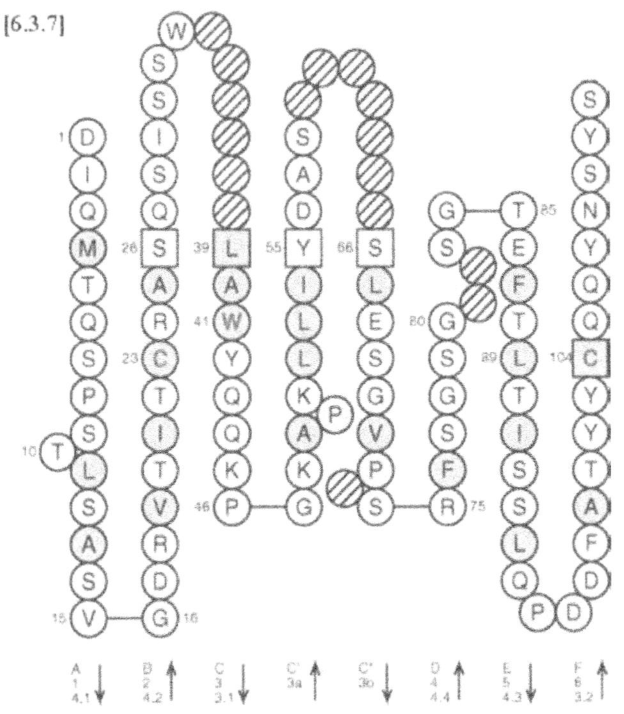

Genome database accession numbers

GDB:9953432 LocusLink: 28944

Nomenclature

IGKV1-6: Immunoglobulin kappa variable 1-6.

Definition and functionality

IGKV1-6 is one of the 15–16 functional genes of the IGKV1 subgroup which comprises 29 mapped genes in the IGK locus and 14–15 orphons.

Gene location

IGKV1-6 is in the proximal cluster of the IGK locus on chromosome 2 at 2p11.2.

Nucleotide and amino acid sequences for human IGKV1-6

```
                                    1   2   3   4   5   6   7   8   9  10  11  12  13  14  15  16  17  18  19  20
                                    A   I   Q   M   T   Q   S   P   S   S   L   S   A   S   V   G   D   R   V   T
M64858,IGKV1-6*01,L11         [26] GCC ATC CAG ATG ACC CAG TCT CCA TCC TCC CTG TCT GCA TCT GTA GGA GAC AGA GTC ACC
X93621,IGKV1-6*01,DPK3        [6]  --- --- --- --- --- --- --- --- --- --- --- --- --- --- --- --- --- --- --- ---

                                                           __________________CDR1-IMGT_____________________
                                   21  22  23  24  25  26  27  28  29  30  31  32  33  34  35  36  37  38  39  40
                                    I   T   C   R   A   S   Q   G   I   R   N   D                           L   G
M64858,IGKV1-6*01,L11              ATC ACT TGC CGG GCA AGT CAG GGC ATT AGA AAT GAT ... ... ... ... ... ... TTA GGC
X93621,IGKV1-6*01,DPK3             --- --- --- --- --- --- --- --- --- --- --- --- ... ... ... ... ... ... --- ---

                                                                                               _______________CDR2-
                                   41  42  43  44  45  46  47  48  49  50  51  52  53  54  55  56  57  58  59  60
                                    W   Y   Q   Q   K   P   G   K   A   P   K   L   L   I   Y   A   A   S
M64858,IGKV1-6*01,L11              TGG TAT CAG CAG AAA CCA GGG AAA GCC CCT AAG CTC CTG ATC TAT GCT GCA TCC ... ...
X93621,IGKV1-6*01,DPK3             --- --- --- --- --- --- --- --- --- --- --- --- --- --- --- --- --- --- ... ...

                                   IMGT________________
                                   61  62  63  64  65  66  67  68  69  70  71  72  73  74  75  76  77  78  79  80
                                                        S   L   Q   S   G   V   P       S   R   F   S   G   S   G
M64858,IGKV1-6*01,L11              ... ... ... ... ... AGT TTA CAA AGT GGG GTC CCA ... TCA AGG TTC AGC GGC AGT GGA
X93621,IGKV1-6*01,DPK3             ... ... ... ... ... --- --- --- --- --- --- --- ... --- --- --- --- --- --- ---

                                   81  82  83  84  85  86  87  88  89  90  91  92  93  94  95  96  97  98  99 100
                                            S   G   T   D   F   T   L   T   I   S   S   L   Q   P   E   D   F   A
M64858,IGKV1-6*01,L11              ... ... TCT GGC ACA GAT TTC ACT CTC ACC ATC AGC AGC CTG CAG CCT GAA GAT TTT GCA
X93621,IGKV1-6*01,DPK3             ... ... --- --- --- --- --- --- --- --- --- --- --- --- --- --- --- --- --- ---

                                                   __________CDR3-IMGT___________
                                   101 102 103 104 105 106 107 108 109 110 111
                                    T   Y   Y   C   L   Q   D   Y   N   Y   P
M64858,IGKV1-6*01,L11              ACT TAT TAC TGT CTA CAA GAT TAC AAT TAC CCT CC
X93621,IGKV1-6*01,DPK3             --- --- --- --- --- --- --- --- --- --- --- --
```

Framework and complementarity determining regions

FR1-IMGT: 26	CDR1-IMGT: 6
FR2-IMGT: 17	CDR2-IMGT: 3
FR3-IMGT: 36 (-3 aa: 73, 81, 82)	CDR3-IMGT: 7

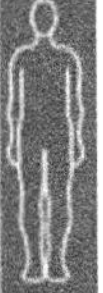

Collier de Perles for human IGKV1-6*01

Accession number: IMGT M64858 EMBL/GenBank/DDBJ: M64858

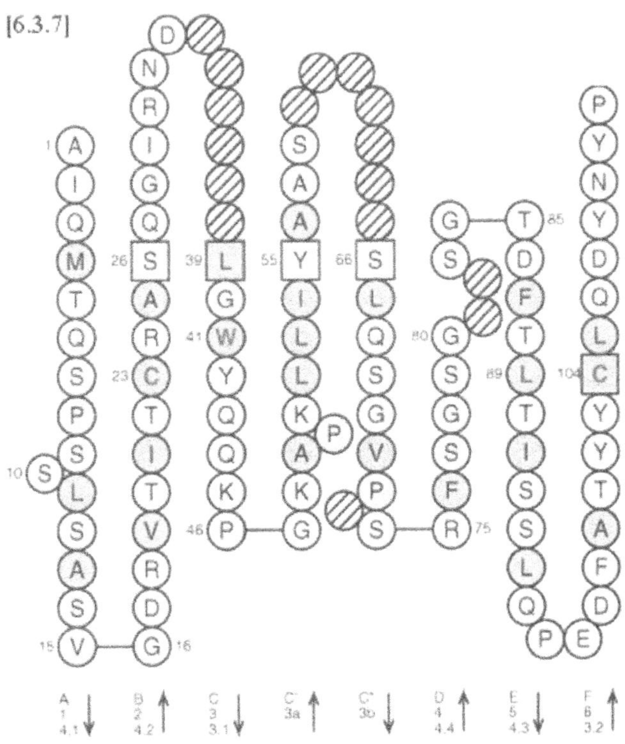

Genome database accession numbers

GDB:9953434 LocusLink: 28943

IGKV1-8

Nomenclature

IGKV1-8: Immunoglobulin kappa variable 1-8.

Definition and functionality

IGKV1-8 is one of the five ORF genes of the IGKV1 subgroup, in the IGK locus. The IGKV1 subgroup comprises 29 mapped genes (of which 15–16 are functional) in the IGK locus and 14–15 orphons.
IGKV1-8 is an ORF due to a 21 nt DELETION starting at nucleotide 4 of the DECAMER in the promoter[24].

Gene location

IGKV1-8 is in the proximal cluster of the IGK locus on chromosome 2 at 2p11.2.

Nucleotide and amino acid sequences for human IGKV1-8

```
                                 1   2   3   4   5   6   7   8   9  10  11  12  13  14  15  16  17  18  19  20
                                 A   I   R   M   T   Q   S   P   S   S   F   S   A   S   T   G   D   R   V   T
Z00014,IGKV1-8*01,L9       [20] GCC ATC CGG ATG ACC CAG TCT CCA TCC TCA TTC TCT GCA TCT ACA GGA GAC AGA GTC ACC

                                                        __________________CDR1-IMGT________________
                                21  22  23  24  25  26  27  28  29  30  31  32  33  34  35  36  37  38  39  40
                                 I   T   C   R   A   S   Q   G   I   S   S   Y                           L   A
Z00014,IGKV1-8*01,L9            ATC ACT TGT CGG GCG AGT CAG GGT ATT AGC AGT TAT ... ... ... ... ... ... TTA GCC

                                                                                            ___________CDR2-
                                41  42  43  44  45  46  47  48  49  50  51  52  53  54  55  56  57  58  59  60
                                 W   Y   Q   Q   K   P   G   K   A   P   K   L   L   I   Y   A   A   S
Z00014,IGKV1-8*01,L9            TGG TAT CAG CAA AAA CCA GGG AAA GCC CCT AAG CTC CTG ATC TAT GCT GCA TCC ... ...

                                IMGT_______________
                                61  62  63  64  65  66  67  68  69  70  71  72  73  74  75  76  77  78  79  80
                                                     T   L   Q   S   G   V   P       S   R   F   S   G   S   G
Z00014,IGKV1-8*01,L9            ... ... ... ... ... ACT TTG CAA AGT GGG GTC CCA ... TCA AGG TTC AGC GGC AGT GGA

                                81  82  83  84  85  86  87  88  89  90  91  92  93  94  95  96  97  98  99 100
                                         S   G   T   D   F   T   L   T   I   S   C   L   Q   S   E   D   F   A
Z00014,IGKV1-8*01,L9            ... ... TCT GGG ACA GAT TTC ACT CTC ACC ATC AGC TGC CTG CAG TCT GAA GAT TTT GCA

                                                ________CDR3-IMGT__________
                                101 102 103 104 105 106 107 108 109 110 111
                                 T   Y   Y   C   Q   Q   Y   Y   S   Y   P
Z00014,IGKV1-8*01,L9            ACT TAT TAC TGT CAA CAG TAT TAT AGT TAC CCT CC
```

Framework and complementarity determining regions

FR1-IMGT: 26 CDR1-IMGT: 6
FR2-IMGT: 17 CDR2-IMGT: 3
FR3-IMGT: 36 (-3 aa: 73, 81, 82) CDR3-IMGT: 7

Collier de Perles for human IGKV1-8*01

Accession number: IMGT Z00014 EMBL/GenBank/DDBJ: Z00014

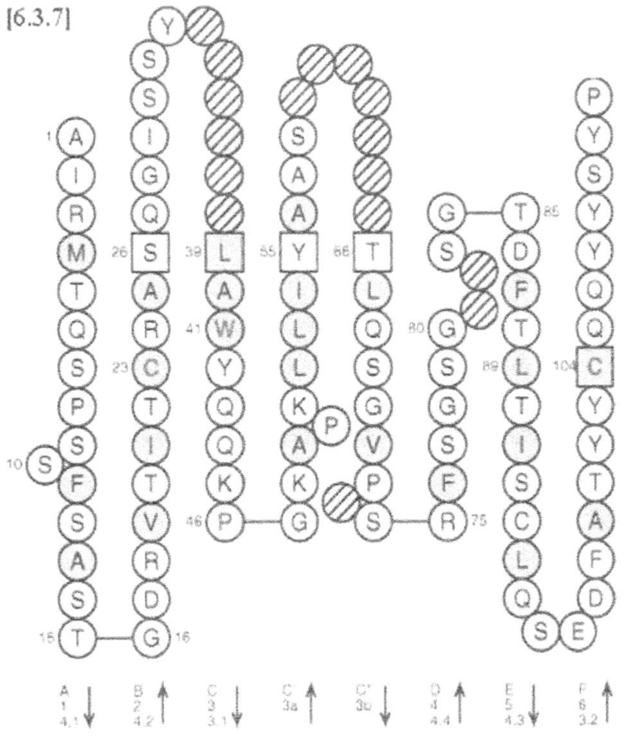

Genome database accession numbers

GDB:9953436 LocusLink: 28942

Nomenclature

IGKV1-9: Immunoglobulin kappa variable 1-9.

Definition and functionality

IGKV1-9 is one of the 15–16 functional genes of the IGKV1 subgroup which comprises 29 mapped genes in the IGK locus and 14–15 orphons.

Gene location

IGKV1-9 is in the proximal cluster of the IGK locus on chromosome 2 at 2p11.2.

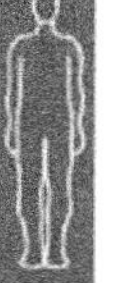

Nucleotide and amino acid sequences for human IGKV1-9

```
                                 1   2   3   4   5   6   7   8   9  10  11  12  13  14  15  16  17  18  19  20
                                 D   I   Q   L   T   Q   S   P   S   F   L   S   A   S   V   G   D   R   V   T
Z00013,IGKV1-9*01,L8      [20]  GAC ATC CAG TTG ACC CAG TCT CCA TCC TTC CTG TCT GCA TCT GTA GGA GAC AGA GTC ACC
X93626,IGKV1-9*01,DPK8    [6]   --- --- --- --- --- --- --- --- --- --- --- --- --- --- --- --- --- --- --- ---

                                                         _____________CDR1-IMGT______________
                                21  22  23  24  25  26  27  28  29  30  31  32  33  34  35  36  37  38  39  40
                                 I   T   C   R   A   S   Q   G   I   S   S   Y                           L   A
Z00013,IGKV1-9*01,L8            ATC ACT TGC CGG GCC AGT CAG GGC ATT AGC AGT TAT ... ... ... ... ... ... TTA GCC
X93626,IGKV1-9*01,DPK8          --- --- --- --- --- --- --- --- --- --- --- --- ... ... ... ... ... ... --- ---

                                                                                            _________CDR2-
                                41  42  43  44  45  46  47  48  49  50  51  52  53  54  55  56  57  58  59  60
                                 W   Y   Q   Q   K   P   G   K   A   P   K   L   L   I   Y   A   A   S
Z00013,IGKV1-9*01,L8            TGG TAT CAG CAA AAA CCA GGG AAA GCC CCT AAG CTC CTG ATC TAT GCT GCA TCC ... ...
X93626,IGKV1-9*01,DPK8          --- --- --- --- --- --- --- --- --- --- --- --- --- --- --- --- --- --- ... ...

                                IMGT_______________
                                61  62  63  64  65  66  67  68  69  70  71  72  73  74  75  76  77  78  79  80
                                                     T   L   Q   S   G   V   P       S   R   F   S   G   S   G
Z00013,IGKV1-9*01,L8            ... ... ... ... ... ACT TTG CAA AGT GGG GTC CCA ... TCA AGG TTC AGC GGC AGT GGA
X93626,IGKV1-9*01,DPK8          ... ... ... ... ... --- --- --- --- --- --- --- ... --- --- --- --- --- --- ---

                                81  82  83  84  85  86  87  88  89  90  91  92  93  94  95  96  97  98  99 100
                                         S   G   T   E   F   T   L   T   I   S   S   L   Q   P   E   D   F   A
Z00013,IGKV1-9*01,L8            ... ... TCT GGG ACA GAA TTC ACT CTC ACA ATC AGC AGC CTG CAG CCT GAA GAT TTT GCA
X93626,IGKV1-9*01,DPK8          ... ... --- --- --- --- --- --- --- --- --- --- --- --- --- --- --- --- --- ---

                                                ________CDR3-IMGT__________
                                101 102 103 104 105 106 107 108 109 110 111
                                 T   Y   Y   C   Q   Q   L   N   S   Y   P
Z00013,IGKV1-9*01,L8            ACT TAT TAC TGT CAA CAG CTT AAT AGT TAC CCT CC
X93626,IGKV1-9*01,DPK8          --- --- --- --- --- --- --- --- --- --- --- --
```

Framework and complementarity determining regions

FR1-IMGT: 26	CDR1-IMGT: 6
FR2-IMGT: 17	CDR2-IMGT: 3
FR3-IMGT: 36 (-3 aa: 73, 81, 82)	CDR3-IMGT: 7

Collier de Perles for human IGKV1-9*01

Accession number: IMGT Z00013 EMBL/GenBank/DDBJ: Z00013

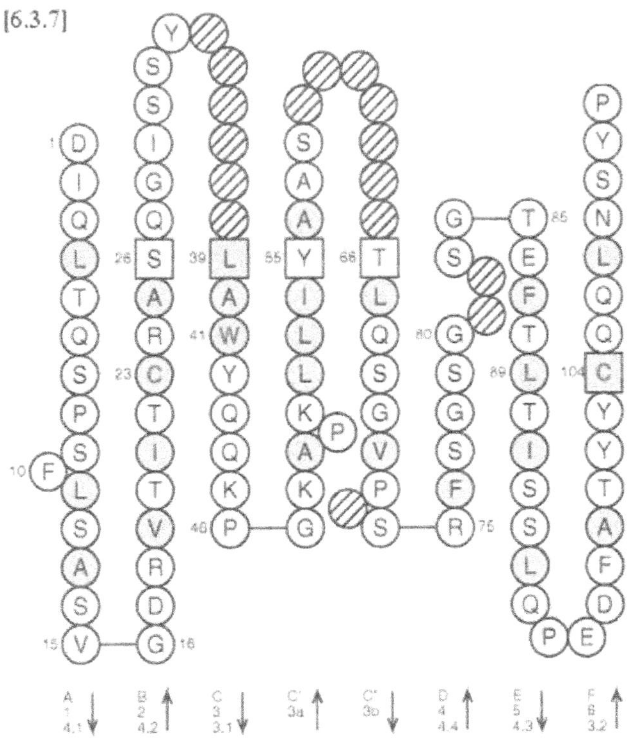

Genome database accession numbers

GDB:9953438 LocusLink: 28941

IGKV1-12 and IGKV1D-12

Nomenclature

IGKV1-12: Immunoglobulin kappa variable 1-12.
IGKV1D-12: Immunoglobulin kappa variable 1D-12.

Definition and functionality

IGKV1-12 and IGKV1D-12 are two of the 15–16 functional genes of the IGKV1 subgroup which comprises 29 mapped genes in the IGK locus and 14–15 orphons.

Gene location

IGKV1-12 is in the proximal cluster, and IGKV1D-12 is in the distal cluster of the IGK locus on chromosome 2 at 2p11.2.

Nucleotide and amino acid sequences for human IGKV1-12 and IGKV1D-12

```
                                         1   2   3   4   5   6   7   8   9  10  11  12  13  14  15  16  17  18  19  20
                                         D   I   Q   M   T   Q   S   P   S   S   V   S   A   S   V   G   D   R   V   T
V01577,IGKV1-12*01,L5               [20] GAC ATC CAG ATG ACC CAG TCT CCA TCT TCC GTG TCT GCA TCT GTA GGA GAC AGA GTC ACC
K02094,IGKV1-12*01,Vb               [20] --- --- --- --- --- --- --- --- --- --- --- --- --- --- --- --- --- --- --- ---
X93623,IGKV1-12*01,DPK5              [6] --- --- --- --- --- --- --- --- --- --- --- --- --- --- --- --- --- --- --- ---
X17263,IGKV1D-12*01,L19             [21] --- --- --- --- --- --- --- --- --- --T --- --- --- --- --- --- --- --- --- ---
X93624,IGKV1D-12*01,DPK6             [6] --- --- --- --- --- --- --- --- --- --T --- --- --- --- --- --- --- --- --- ---
V01576,IGKV1-12*02/1D-12*02,L5/19a  [20] --- --- --- --- --- --- --- --- --- --- --- --- --- --- --- --- --- --- --- ---

                                                                 _____________CDR1-IMGT_______________
                                        21  22  23  24  25  26  27  28  29  30  31  32  33  34  35  36  37  38  39  40
                                         I   T   C   R   A   S   Q   G   I   S   S   W                           L   A
V01577,IGKV1-12*01,L5                    ATC ACT TGT CGG GCG AGT CAG GGT ATT AGC AGC TGG ... ... ... ... ... ... TTA GCC
K02094,IGKV1-12*01,Vb                    --- --- --- --- --- --- --- --- --- --- --- --- ... ... ... ... ... ... --- ---
X93623,IGKV1-12*01,DPK5                  --- --- --- --- --- --- --- --- --- --- --- --- ... ... ... ... ... ... --- ---
X17263,IGKV1D-12*01,L19                  --- --- --- --- --- --- --- --- --- --- --- --- ... ... ... ... ... ... --- ---
X93624,IGKV1D-12*01,DPK6                 --- --- --- --- --- --- --- --- --- --- --- --- ... ... ... ... ... ... --- ---
V01576,IGKV1-12*02/1D-12*02,L5/19a       --- --- --- --- --- --- --- --- --- --- --- --- ... ... ... ... ... ... --- ---

                                                                                                 _____________CDR2-
                                        41  42  43  44  45  46  47  48  49  50  51  52  53  54  55  56  57  58  59  60
                                         W   Y   Q   Q   K   P   G   K   A   P   K   L   L   I   Y   A   A   S
V01577,IGKV1-12*01,L5                    TGG TAT CAG CAG AAA CCA GGG AAA GCC CCT AAG CTC CTG ATC TAT GCT GCA TCC ... ...
K02094,IGKV1-12*01,Vb                    --- --- --- --- --- --- --- --- --- --- --- --- --- --- --- --- --- --- ... ...
X93623,IGKV1-12*01,DPK5                  --- --- --- --- --- --- --- --- --- --- --- --- --- --- --- --- --- --- ... ...
X17263,IGKV1D-12*01,L19                  --- --- --- --- --- --- --- --- --- --- --- --- --- --- --- --- --- --- ... ...
X93624,IGKV1D-12*01,DPK6                 --- --- --- --- --- --- --- --- --- --- --- --- --- --- --- --- --- --- ... ...
V01576,IGKV1-12*02/1D-12*02,L5/19a       --- --- --- --- --- --- --- --- --- --- --- --- --- --- --- --- --- --- ... ...

                                         IMGT_______________
                                        61  62  63  64  65  66  67  68  69  70  71  72  73  74  75  76  77  78  79  80
                                                             S   L   Q   S   G   V   P       S   R   F   S   G   S   G
V01577,IGKV1-12*01,L5                    ... ... ... ... ... AGT TTG CAA AGT GGG GTC CCA ... TCA AGG TTC AGC GGC AGT GGA
K02094,IGKV1-12*01,Vb                    ... ... ... ... ... --- --- --- --- --- --- --- ... --- --- --- --- --- --- ---
X93623,IGKV1-12*01,DPK5                  ... ... ... ... ... --- --- --- --- --- --- --- ... --- --- --- --- --- --- ---
X17263,IGKV1D-12*01,L19                  ... ... ... ... ... --- --- --- --- --- --- --- ... --- --- --- --- --- --- ---
X93624,IGKV1D-12*01,DPK6                 ... ... ... ... ... --- --- --- --- --- --- --- ... --- --- --- --- --- --- ---
V01576,IGKV1-12*02/1D-12*02,L5/19a       ... ... ... ... ... --- --- --- --- --- --- --- ... --- --- --- --- --- --- ---

                                        81  82  83  84  85  86  87  88  89  90  91  92  93  94  95  96  97  98  99 100
                                                 S   G   T   D   F   T   L   T   I   S   S   L   Q   P   E   D   F   A
V01577,IGKV1-12*01,L5                    ... ... TCT GGG ACA GAT TTC ACT CTC ACC ATC AGC AGC CTG CAG CCT GAA GAT TTT GCA
K02094,IGKV1-12*01,Vb                    ... ... --- --- --- --- --- --- --- --- --- --- --- --- --- --- --- --- --- ---
X93623,IGKV1-12*01,DPK5                  ... ... --- --- --- --- --- --- --- --- --- --- --- --- --- --- --- --- --- ---
X17263,IGKV1D-12*01,L19                  ... ... --- --- --- --- --- --- --- --T --- --- --- --- --- --- --- --- --- ---
X93624,IGKV1D-12*01,DPK6                 ... ... --- --- --- --- --- --- --- --T --- --- --- --- --- --- --- --- --- ---
V01576,IGKV1-12*02/1D-12*02,L5/19a       ... ... --- --- --- --- --- --- --- --- --- --- --- --- --- --- --- --- --- ---
```

```
                                                      ______CDR3-IMGT__________
                                   101 102 103 104 105 106 107 108 109 110 111
                                    T   Y   Y   C   Q   Q   A   N   S   F   P
V01577,IGKV1-12*01,L5              ACT TAC TAT TGT CAA CAG GCT AAC AGT TTC CCT CC
K02094,IGKV1-12*01,Vb              --- --- --- --- --- --- --- --- --- --- --- --
X93623,IGKV1-12*01,DPK5            --- --- --- --- --- --- --- --- --- --- --- --
X17263,IGKV1D-12*01,L19            --- --- --- --- --- --- --- --- --- --- --- --
X93624,IGKV1D-12*01,DPK6           --- --- --- --- --- --- --- --- --- --- --- --
V01576,IGKV1-12*02/1D-12*02,L5/19a --- --- --- --- --- --- --- --- --- --- --- T-
```

Framework and complementarity determining regions

FR1-IMGT: 26
FR2-IMGT: 17
FR3-IMGT: 36 (-3 aa: 73, 81, 82)

CDR1-IMGT: 6
CDR2-IMGT: 3
CDR3-IMGT: 7

Collier de Perles for human IGKV1-12*01 and IGKV1D-12*01

IGKV1-12*01
Accession number: IMGT V01577 EMBL/GenBank/DDBJ: V01577
IGKV1D-12*01
Accession number: IMGT X17263 EMBL/GenBank/DDBJ: X17263

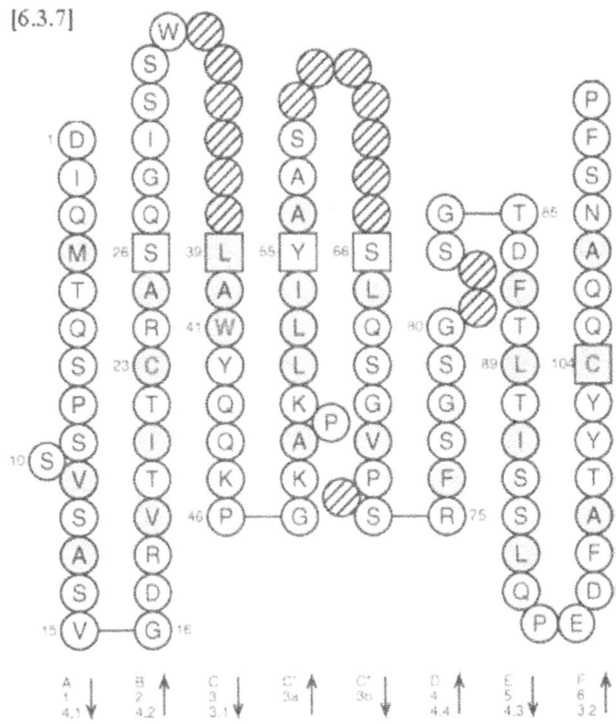

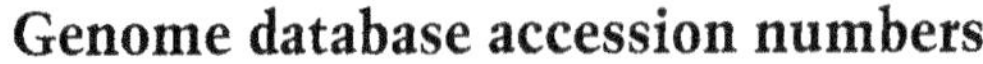

Genome database accession numbers

IGKV1-12:	GDB:9953440	LocusLink: 28940
IGKV1D-12:	GDB:9953514	LocusLink: 28903

IGKV1-13 and IGKV1D-13

Nomenclature

IGKV1-13: Immunoglobulin kappa variable 1-13.
IGKV1D-13: Immunoglobulin kappa variable 1D-13.

Definition and functionality

IGKV1-13 is a pseudogene, and IGKV1D-13 is one of the five ORF genes of the IGKV1 subgroup in the IGK locus. The IGKV1 subgroup comprises 29 mapped genes (of which 15–16 are functional) in the IGK locus and 14–15 orphons.
The IGKV1-13 gene is a pseudogene due to the CONSERVED_TRP (position 41 in FR2-IMGT) being replaced by a STOP-CODON[24].
The IGKV1D-13 gene is an ORF due to a non-canonical V-HEPTAMER (catagtg instead of cacagtg)[24].
Sequences which are non-mapped cannot be assigned with certainty to one or the other gene. It is therefore not known if the non-mapped Z00006 sequence in the alignment below represents an ORF allele of IGKV1-13 or of IGKV1D-13.

Gene location

IGKV1-13 is in the proximal cluster, and IGKV1D-13 is in the distal cluster of the IGK locus on chromosome 2 at 2p11.2.

Nucleotide and amino acid sequences for human IGKV1-13 and IGKV1D-13

```
                                         1   2   3   4   5   6   7   8   9   10  11  12  13  14  15  16  17  18  19  20
                                         A   I   Q   L   T   Q   S   P   S   S   L   S   A   S   V   G   D   R   V   T
Z00010,IGKV1-13*01,L4              [20] GCC ATC CAG TTG ACC CAG TCT CCA TCC TCC CTG TCT GCA TCT GTA GGA GAC AGA GTC ACC
X93647,IGKV1-13*01,DPK31            [6] --- --- --- --- --- --- --- --- --- --- --- --- --- --- --- --- --- --- --- ---
X17262,IGKV1D-13*01,L18            [21] --- --- --- --- --- --- --- --- --- --- --- --- --- --- --- --- --- --- --- ---
Z00006,IGKV1-13*02/1D-13*02,L4/18a [20] --- --- --- --- --- --- --- --- --- --- --- --- --- --- --- --- --- --- --- ---

                                                                 ____________________CDR1-IMGT___________________
                                         21  22  23  24  25  26  27  28  29  30  31  32  33  34  35  36  37  38  39  40
                                         I   T   C   R   A   S   Q   G   I   S   S   A                           L   A
Z00010,IGKV1-13*01,L4                   ATC ACT TGC CGG GCA AGT CAG GGC ATT AGC AGT GCT ... ... ... ... ... ... TTA GCC
X93647,IGKV1-13*01,DPK31                --- --- --- --- --- --- --- --- --- --- --- --- ... ... ... ... ... ... --- ---
X17262,IGKV1D-13*01,L18                 --- --- --- --- --- --- --- --- --- --- --- --- ... ... ... ... ... ... --- ---
Z00006,IGKV1-13*02/1D-13*02,L4/18a      --- --- --- --- --- --- --- --- --- --- --- --- ... ... ... ... ... ... --- ---

                                                                                                     ______________CDR2-
                                         41  42  43  44  45  46  47  48  49  50  51  52  53  54  55  56  57  58  59  60
                                         *   Y   Q   Q   K   P   G   K   A   P   K   L   L   I   Y   D   A   S
Z00010,IGKV1-13*01,L4                   TGA TAT CAG CAG AAA CCA GGG AAA GCT CCT AAG CTC CTG ATC TAT GAT GCC TCC ... ...
X93647,IGKV1-13*01,DPK31                --- --- --- --- --- --- --- --- --- --- --- --- --- --- --- --- --- --- ... ...
                                         W
X17262,IGKV1D-13*01,L18                 --G --- --- --- --- --- --- --- --- --- --- --- --- --- --- --- --- --- ... ...
                                         W
Z00006,IGKV1-13*02/1D-13*02,L4/18a      --G --- --- --- --- --- --- --- --- --- --- --- --- --- --- --- --- --- ... ...

                                        IMGT________________
                                         61  62  63  64  65  66  67  68  69  70  71  72  73  74  75  76  77  78  79  80
                                                             S   L   E   S   G   V   P       S   R   F   S   G   S   G
Z00010,IGKV1-13*01,L4                   ... ... ... ... ... AGT TTG GAA AGT GGG GTC CCA ... TCA AGG TTC AGC GGC AGT GGA
X93647,IGKV1-13*01,DPK31                ... ... ... ... ... --- --- --- --- --- --- --- ... --- --- --- --- --- --- ---
X17262,IGKV1D-13*01,L18                 ... ... ... ... ... --- --- --- --- --- --- --- ... --- --- --- --- --- --- ---
Z00006,IGKV1-13*02/1D-13*02,L4/18a      ... ... ... ... ... --- --- --- --- --- --- --- ... --- --- --- --- --- --- ---

                                         81  82  83  84  85  86  87  88  89  90  91  92  93  94  95  96  97  98  99 100
                                                 S   G   T   D   F   T   L   T   I   S   S   L   Q   P   E   D   F   A
Z00010,IGKV1-13*01,L4                   ... ... TCT GGG ACA GAT TTC ACT CTC ACC ATC AGC AGC CTG CAG CCT GAA GAT TTT GCA
X93647,IGKV1-13*01,DPK31                ... ... --- --- --- --- --- --- --- --- --- --- --- --- --- --- --- --- --- ---
X17262,IGKV1D-13*01,L18                 ... ... --- --- --- --- --- --- --- --- --- --- --- --- --- --- --- --- --- ---
Z00006,IGKV1-13*02/1D-13*02,L4/18a      ... ... --- --- --- --- --- --- --- --- --- --- --- --- --- --- --- --- --- ---
```

```
                                           ________CDR3-IMGT____________
                                101 102 103 104 105 106 107 108 109 110 111
                                 T   Y   Y   C   Q   Q   F   N   N   Y   P
Z00010,IGKV1-13*01,L4           ACT TAT TAC TGT CAA CAG TTT AAT AAT TAC CCT CA

X93647,IGKV1-13*01,DPK31        --- --- --- --- --- --- --- --- --- --- --- --

X17262,IGKV1D-13*01,L18         --- --- --- --- --- --- --- --- --- --- --- --
                                                                 S
Z00006,IGKV1-13*02/1D-13*02,L4/18a --- --- --- --- --- --- --- --- -G- --- --- --
```

Framework and complementarity determining regions

FR1-IMGT: 26
FR2-IMGT: 17
FR3-IMGT: 36 (-3 aa: 73, 81, 82)

CDR1-IMGT: 6
CDR2-IMGT: 3
CDR3-IMGT: 7

Collier de Perles for human IGKV1-13*01 and IGKV1D-13*01

IGKV1-13*01
Accession number: IMGT Z00010 EMBL/GenBank/DDBJ: Z00010
IGKV1D-13*01
Accession number: IMGT X17262 EMBL/GenBank/DDBJ: X17262

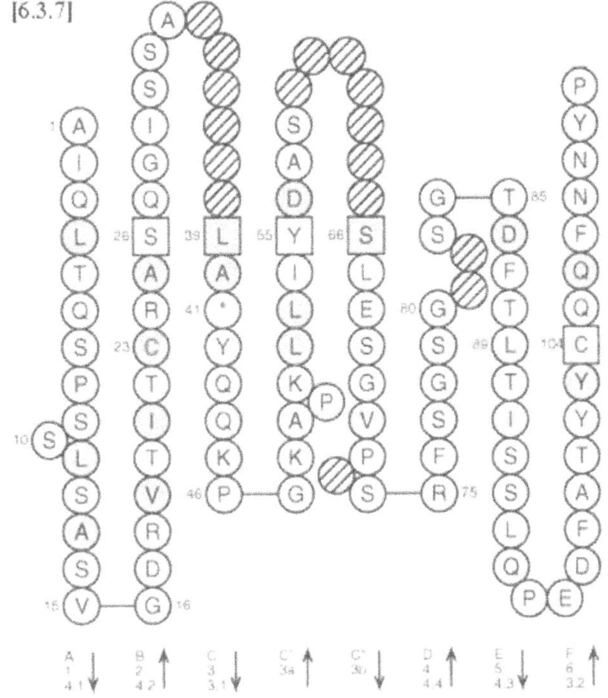

Genome database accession numbers

IGKV1-13: GDB:9953442 LocusLink: 28939
IGKV1D-13: GDB:9953516 LocusLink: 28902

IGKV1-16

Nomenclature

IGKV1-16: Immunoglobulin kappa variable 1-16.

Definition and functionality

IGKV1-16 is one of the 15–16 functional genes of the IGKV1 subgroup which comprises 29 mapped genes in the IGK locus and 14–15 orphons.

Gene location

IGKV1-16 is in the proximal cluster of the IGK locus on chromosome 2 at 2p11.2.

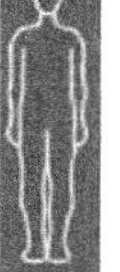

Nucleotide and amino acid sequences for human IGKV1-16

```
                              1   2   3   4   5   6   7   8   9  10  11  12  13  14  15  16  17  18  19  20
                              D   I   Q   M   T   Q   S   P   S   S   L   S   A   S   V   G   D   R   V   T
J00248,IGKV1-16*01,L1    [3] GAC ATC CAG ATG ACC CAG TCT CCA TCC TCA CTG TCT GCA TCT GTA GGA GAC AGA GTC ACC

                                                 ________________CDR1-IMGT_________________
                             21  22  23  24  25  26  27  28  29  30  31  32  33  34  35  36  37  38  39  40
                              I   T   C   R   A   S   Q   G   I   S   N   Y                           L   A
J00248,IGKV1-16*01,L1        ATC ACT TGT CGG GCG AGT CAG GGC ATT AGC AAT TAT ... ... ... ... ... ... TTA GCC

                                                                                     ____________CDR2-
                             41  42  43  44  45  46  47  48  49  50  51  52  53  54  55  56  57  58  59  60
                              W   F   Q   Q   K   P   G   K   A   P   K   S   L   I   Y   A   A   S
J00248,IGKV1-16*01,L1        TGG TTT CAG CAG AAA CCA GGG AAA GCC CCT AAG TCC CTG ATC TAT GCT GCA TCC ... ...

                             IMGT________________
                             61  62  63  64  65  66  67  68  69  70  71  72  73  74  75  76  77  78  79  80
                                                  S   L   Q   S   G   V   P       S   R   F   S   G   S   G
J00248,IGKV1-16*01,L1        ... ... ... ... ... AGT TTG CAA AGT GGG GTC CCA ... TCA AGG TTC AGC GGC AGT GGA

                             81  82  83  84  85  86  87  88  89  90  91  92  93  94  95  96  97  98  99 100
                                      S   G   T   D   F   T   L   T   I   S   S   L   Q   P   E   D   F   A
J00248,IGKV1-16*01,L1        ... ... TCT GGG ACA GAT TTC ACT CTC ACC ATC AGC AGC CTG CAG CCT GAA GAT TTT GCA

                                              _________CDR3-IMGT___________
                            101 102 103 104 105 106 107 108 109 110 111
                              T   Y   Y   C   Q   Q   Y   N   S   Y   P
J00248,IGKV1-16*01,L1        ACT TAT TAC TGC CAA CAG TAT AAT AGT TAC CCT CC
```

Framework and complementarity determining regions

FR1-IMGT: 26
FR2-IMGT: 17
FR3-IMGT: 36 (-3 aa: 73, 81, 82)

CDR1-IMGT: 6
CDR2-IMGT: 3
CDR3-IMGT: 7

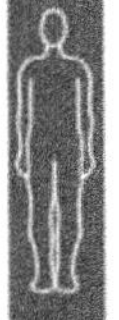

Collier de Perles for human IGKV1-16*01

Accession number: IMGT J00248 EMBL/GenBank/DDBJ: J00248

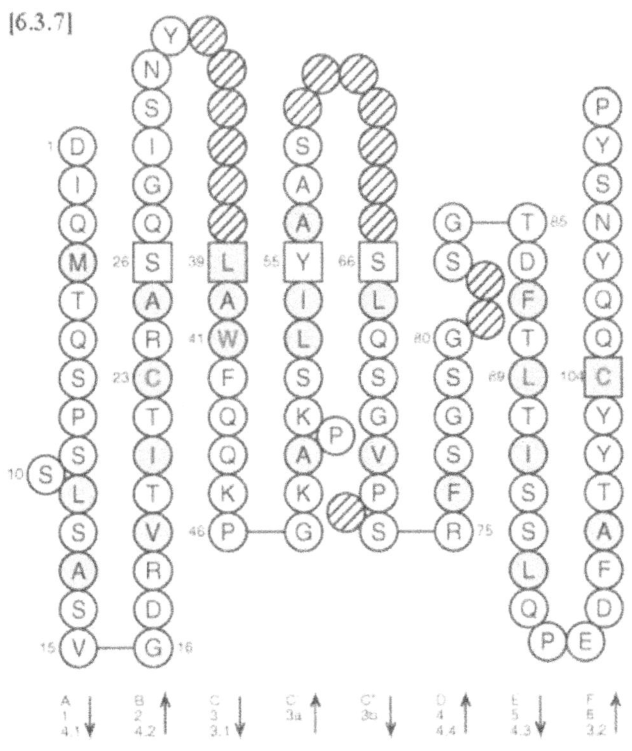

Genome database accession numbers

GDB:9953444 LocusLink: 28938

IGKV1-17

Nomenclature

IGKV1-17: Immunoglobulin kappa variable 1-17.

Definition and functionality

IGKV1-17 is one of the 15–16 functional genes of the IGKV1 subgroup which comprises 29 mapped genes in the IGK locus and 14–15 orphons.

Gene location

IGKV1-17 is in the proximal cluster of the IGK locus on chromosome 2 at 2p11.2.

Nucleotide and amino acid sequences for human IGKV1-17

```
                                  1   2   3   4   5   6   7   8   9  10  11  12  13  14  15  16  17  18  19  20
                                  D   I   Q   M   T   Q   S   P   S   S   L   S   A   S   V   G   D   R   V   T
X72808,IGKV1-17*01,A30     [10]  GAC ATC CAG ATG ACC CAG TCT CCA TCC TCC CTG TCT GCA TCT GTA GGA GAC AGA GTC ACC
X92334,IGKV1-17*01,SG3     [9]                           --- --- --- --- --- --- --- --- --- --- --- --- --- ---

                                                         ________________CDR1-IMGT________________
                                 21  22  23  24  25  26  27  28  29  30  31  32  33  34  35  36  37  38  39  40
                                  I   T   C   R   A   S   Q   G   I   R   N   D                           L   G
X72808,IGKV1-17*01,A30           ATC ACT TGC CGG GCA AGT CAG GGC ATT AGA AAT GAT ... ... ... ... ... ... TTA GGC
X92334,IGKV1-17*01,SG3           --- --- --- --- --- --- --- --- --- --- --- --- ... ... ... ... ... ... --- ---

                                                                                             __________CDR2-
                                 41  42  43  44  45  46  47  48  49  50  51  52  53  54  55  56  57  58  59  60
                                  W   Y   Q   Q   K   P   G   K   A   P   K   R   L   I   Y   A   A   S
X72808,IGKV1-17*01,A30           TGG TAT CAG CAG AAA CCA GGG AAA GCC CCT AAG CGC CTG ATC TAT GCT GCA TCC ... ...
X92334,IGKV1-17*01,SG3           --- --- --- --- --- --- --- --- --- --- --- --- --- --- --- --- --- --- ... ...

                                 IMGT________________
                                 61  62  63  64  65  66  67  68  69  70  71  72  73  74  75  76  77  78  79  80
                                                      S   L   Q   S   G   V   P       S   R   F   S   G   S   G
X72808,IGKV1-17*01,A30           ... ... ... ... ... AGT TTG CAA AGT GGG GTC CCA ... TCA AGG TTC AGC GGC AGT GGA
X92334,IGKV1-17*01,SG3           ... ... ... ... ... --- --- --- --- --- --- --- ... --- --- --- --- --- --- ---

                                 81  82  83  84  85  86  87  88  89  90  91  92  93  94  95  96  97  98  99 100
                                          S   G   T   E   F   T   L   T   I   S   S   L   Q   P   E   D   F   A
X72808,IGKV1-17*01,A30           ... ... TCT GGG ACA GAA TTC ACT CTC ACA ATC AGC AGC CTG CAG CCT GAA GAT TTT GCA
X92334,IGKV1-17*01,SG3           ... ... --- --- --- --- --- --- --- --- --- --- --- --- --- --- --- --- --- ---

                                                     __________CDR3-IMGT__________
                                 101 102 103 104 105 106 107 108 109 110 111
                                  T   Y   Y   C   L   Q   H   N   S   Y   P
X72808,IGKV1-17*01,A30           ACT TAT TAC TGT CTA CAG CAT AAT AGT TAC CCT CC
X92334,IGKV1-17*01,SG3           --- --- --- --- --- --- --- --- --- --- --- --
```

Framework and complementarity determining regions

FR1-IMGT: 26
FR2-IMGT: 17
FR3-IMGT: 36 (-3 aa: 73, 81, 82)

CDR1-IMGT: 6
CDR2-IMGT: 3
CDR3-IMGT: 7

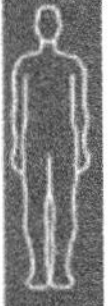

Collier de Perles for human IGKV1-17*01

Accession number: IMGT X72808 EMBL/GenBank/DDBJ: X72808

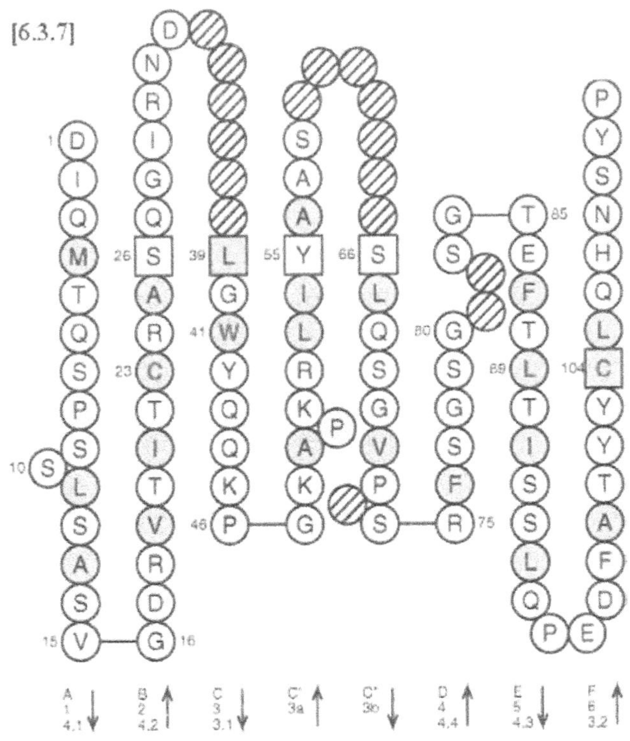

Genome database accession numbers

GDB:9953446 LocusLink: 28937

Nomenclature

IGKV1-27: Immunoglobulin kappa variable 1-27.

Definition and functionality

IGKV1-27 is one of the 15–16 functional genes of the IGKV1 subgroup which comprises 29 mapped genes in the IGK locus and 14–15 orphons.

Gene location

IGKV1-27 is in the proximal cluster of the IGK locus on chromosome 2 at 2p11.2.

Nucleotide and amino acid sequences for human IGKV1-27

```
                                 1   2   3   4   5   6   7   8   9  10  11  12  13  14  15  16  17  18  19  20
                                D   I   Q   M   T   Q   S   P   S   S   L   S   A   S   V   G   D   R   V   T
X63398,IGKV1-27*01,A20    [15] GAC ATC CAG ATG ACC CAG TCT CCA TCC TCC CTG TCT GCA TCT GTA GGA GAC AGA GTC ACC
X93622,IGKV1-27*01,DPK4   [6]  --- --- --- --- --- --- --- --- --- --- --- --- --- --- --- --- --- --- --- ---

                                                       _____________CDR1-IMGT________________________
                                21  22  23  24  25  26  27  28  29  30  31  32  33  34  35  36  37  38  39  40
                                I   T   C   R   A   S   Q   G   I   S   N   Y                           L   A
X63398,IGKV1-27*01,A20         ATC ACT TGC CGG GCG AGT CAG GGC ATT AGC AAT TAT ... ... ... ... ... ... TTA GCC
X93622,IGKV1-27*01,DPK4        --- --- --- --- --- --- --- --- --- --- --- --- ... ... ... ... ... ... --- ---

                                                                                           _____________CDR2-
                                41  42  43  44  45  46  47  48  49  50  51  52  53  54  55  56  57  58  59  60
                                W   Y   Q   Q   K   P   G   K   V   P   K   L   L   I   Y   A   A   S
X63398,IGKV1-27*01,A20         TGG TAT CAG CAG AAA CCA GGG AAA GTT CCT AAG CTC CTG ATC TAT GCT GCA TCC ... ...
X93622,IGKV1-27*01,DPK4        --- --- --- --- --- --- --- --- --- --- --- --- --- --- --- --- --- --- ... ...

                               IMGT_______________
                                61  62  63  64  65  66  67  68  69  70  71  72  73  74  75  76  77  78  79  80
                                                    T   L   Q   S   G   V   P       S   R   F   S   G   S   G
X63398,IGKV1-27*01,A20         ... ... ... ... ... ACT TTC CAA TCA GGG GTC CCA ... TCT CGG TTC AGT GGC AGT GGA
X93622,IGKV1-27*01,DPK4        ... ... ... ... ... --- --- --- --- --- --- --- ... --- --- --- --- --- --- ---

                                81  82  83  84  85  86  87  88  89  90  91  92  93  94  95  96  97  98  99 100
                                        S   G   T   D   F   T   L   T   I   S   S   L   Q   P   E   D   V   A
X63398,IGKV1-27*01,A20         ... ... TCT GGG ACA GAT TTC ACT CTC ACC ATC AGC AGC CTG CAG CCT GAA GAT GTT GCA
X93622,IGKV1-27*01,DPK4        ... ... --- --- --- --- --- --- --- --- --- --- --- --- --- --- --- --- --- ---

                                               _______CDR3-IMGT__________
                               101 102 103 104 105 106 107 108 109 110 111
                                T   Y   Y   C   Q   K   Y   N   S   A   P
X63398,IGKV1-27*01,A20         ACT TAT TAC TGT CAA AAG TAT AAC AGT GCC CCT CC
X93622,IGKV1-27*01,DPK4        --  --- --- --- --- --- --- --- --- --- --- --
```

Framework and complementarity determining regions

FR1-IMGT: 26

FR2-IMGT: 17

FR3-IMGT: 36 (-3 aa: 73, 81, 82)

CDR1-IMGT: 6

CDR2-IMGT: 3

CDR3-IMGT: 7

Collier de Perles for human IGKV1-27*01

Accession number: IMGT X63398 EMBL/GenBank/DDBJ: X63398

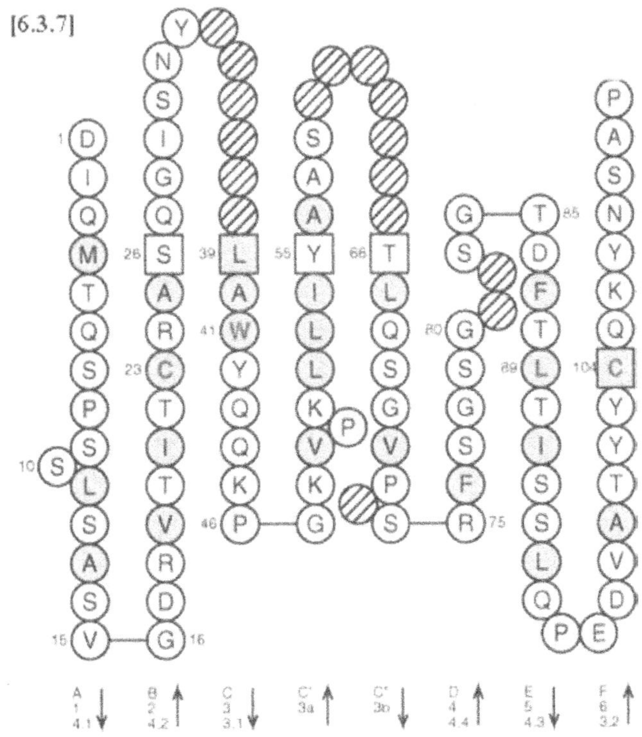

Genome database accession numbers

GDB:9953450 LocusLink: 28935

IGKV1-33 and IGKV1D-33

Nomenclature

IGKV1-33: Immunoglobulin kappa variable 1-33.
IGKV1D-33: Immunoglobulin kappa variable 1D-33.

Definition and functionality

IGKV1-33 and IGKV1D-33 are two of the 15–16 functional genes of the IGKV1 subgroup which comprises 29 mapped genes in the IGK locus and 14–15 orphons.

Gene location

IGKV1-33 is in the proximal cluster, and IGKV1D-33 is in the distal cluster of the IGK locus on chromosome 2 at 2p11.2.

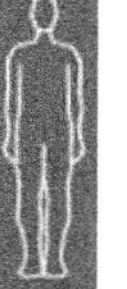

Nucleotide and amino acid sequences for human IGKV1-33 and IGKV1D-33

```
                                        1   2   3   4   5   6   7   8   9  10  11  12  13  14  15  16  17  18  19  20
                                        D   I   Q   M   T   Q   S   P   S   S   L   S   A   S   V   G   D   R   V   T
M64856,IGKV1-33*01,O18            [26] GAC ATC CAG ATG ACC CAG TCT CCA TCC TCC CTG TCT GCA TCT GTA GGA GAC AGA GTC ACC
M64857,IGKV1-33*01,O18a           [26] --- --- --- --- --- --- --- --- --- --- --- --- --- --- --- --- --- --- --- ---
M64855,IGKV1D-33*01,O8            [26] --- --- --- --- --- --- --- --- --- --- --- --- --- --- --- --- --- --- --- ---
X93620,IGKV1-33*01/1D-33*01,DPK1  [6]  --- --- --- --- --- --- --- --- --- --- --- --- --- --- --- --- --- --- --- ---

                                                               ___________________CDR1-IMGT___________________
                                       21  22  23  24  25  26  27  28  29  30  31  32  33  34  35  36  37  38  39  40
                                        I   T   C   Q   A   S   Q   D   I   S   N   Y                           L   N
M64856,IGKV1-33*01,O18                 ATC ACT TGC CAG GCG AGT CAG GAC ATT AGC AAC TAT ... ... ... ... ... ... TTA AAT
M64857,IGKV1-33*01,O18a                --- --- --- --- --- --- --- --- --- --- --- --- ... ... ... ... ... ... --- ---
M64855,IGKV1D-33*01,O8                 --- --- --- --- --- --- --- --- --- --- --- --- ... ... ... ... ... ... --- ---
X93620,IGKV1-33*01/1D-33*01,DPK1       --- --- --- --- --- --- --- --- --- --- --- --- ... ... ... ... ... ... --- ---

                                                                                                   ______________CDR2-
                                       41  42  43  44  45  46  47  48  49  50  51  52  53  54  55  56  57  58  59  60
                                        W   Y   Q   Q   K   P   G   K   A   P   K   L   L   I   Y   D   A   S
M64856,IGKV1-33*01,O18                 TGG TAT CAG CAG AAA CCA GGG AAA GCC CCT AAG CTC CTG ATC TAC GAT GCA TCC ... ...
M64857,IGKV1-33*01,O18a                --- --- --- --- --- --- --- --- --- --- --- --- --- --- --- --- --- --- ... ...
M64855,IGKV1D-33*01,O8                 --- --- --- --- --- --- --- --- --- --- --- --- --- --- --- --- --- --- ... ...
X93620,IGKV1-33*01/1D-33*01,DPK1       --- --- --- --- --- --- --- --- --- --- --- --- --- --- --- --- --- --- ... ...

                                       IMGT_______________
                                       61  62  63  64  65  66  67  68  69  70  71  72  73  74  75  76  77  78  79  80
                                                            N   L   E   T   G   V   P       S   R   F   S   G   S   G
M64856,IGKV1-33*01,O18                 ... ... ... ... ... AAT TTG GAA ACA GGG GTC CCA ... TCA AGG TTC AGT GGA AGT GGA
M64857,IGKV1-33*01,O18a                ... ... ... ... ... --- --- --- --- --- --- --- ... --- --- --- --- --- --- ---
M64855,IGKV1D-33*01,O8                 ... ... ... ... ... --- --- --- --- --- --- --- ... --- --- --- --- --- --- ---
X93620,IGKV1-33*01/1D-33*01,DPK1       ... ... ... ... ... --- --- --- --- --- --- --- ... --- --- --- --- --- --- ---

                                       81  82  83  84  85  86  87  88  89  90  91  92  93  94  95  96  97  98  99  100
                                                S   G   T   D   F   T   F   T   I   S   S   L   Q   P   E   D   I   A
M64856,IGKV1-33*01,O18                 ... ... TCT GGG ACA GAT TTT ACT TTC ACC ATC AGC AGC CTG CAG CCT GAA GAT ATT GCA
M64857,IGKV1-33*01,O18a                ... ... --- --- --- --- --- --- --- --- --- --- --- --- --- --- --- --- --- ---
M64855,IGKV1D-33*01,O8                 ... ... --- --- --- --- --- --- --- --- --- --- --- --- --- --- --- --- --- ---
X93620,IGKV1-33*01/1D-33*01,DPK1       ... ... --- --- --- --- --- --- --- --- --- --- --- --- --- --- --- --- --- ---

                                                       __________CDR3-IMGT___________
                                       101 102 103 104 105 106 107 108 109 110 111
                                        T   Y   Y   C   Q   Q   Y   D   N   L   P
M64856,IGKV1-33*01,O18                 ACA TAT TAC TGT CAA CAG TAT GAT AAT CTC CCT CC
M64857,IGKV1-33*01,O18a                --- --- --- --- --- --- --- --- --- --- --- --
M64855,IGKV1D-33*01,O8                 --- --- --- --- --- --- --- --- --- --- --- --
X93620,IGKV1-33*01/1D-33*01,DPK1       --- --- --- --- --- --- --- --- --- --- --- --
```

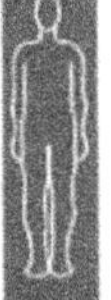

Framework and complementarity determining regions

FR1-IMGT: 26
FR2-IMGT: 17
FR3-IMGT: 36 (-3 aa: 73, 81, 82)

CDR1-IMGT: 6
CDR2-IMGT: 3
CDR3-IMGT: 7

Collier de Perles for human IGKV1-33*01 and IGKV1D-33*01

IGKV1-33*01
Accession number: IMGT M64856 EMBL/GenBank/DDBJ: M64856
IGKV1D-33*01
Accession number: IMGT M64855 EMBL/GenBank/DDBJ: M64855

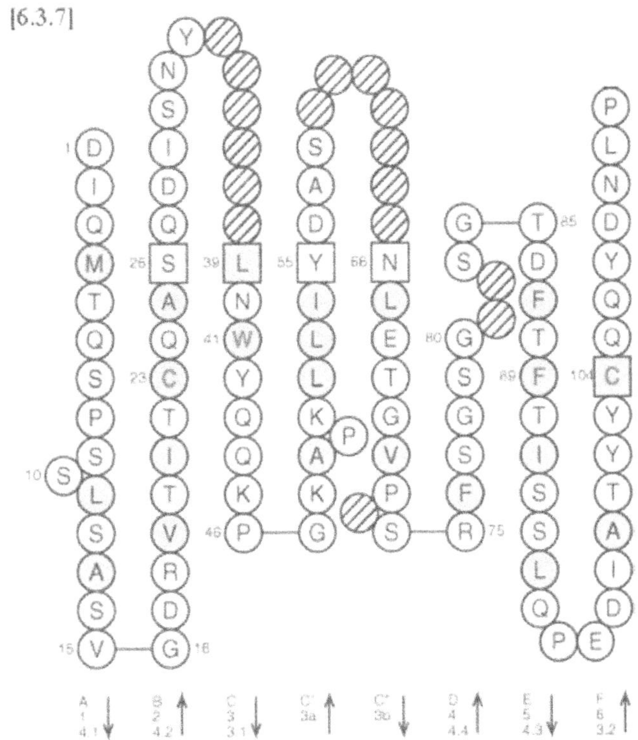

Genome database accession numbers

IGKV1-33:	GDB:9953454	LocusLink: 28933
IGKV1D-33:	GDB:9953528	LocusLink: 28896

IGKV1-37 and IGKV1D-37

Nomenclature

IGKV1-37: Immunoglobulin kappa variable 1-37.
IGKV1D-37: Immunoglobulin kappa variable 1D-37.

Definition and functionality

IGKV1-37 and IGKV1D-37 are two of the five ORF genes of the IGKV1 subgroup, in the IGK locus. The IGKV1 subgroup comprises 29 mapped genes (of which 15–16 are functional) in the IGK locus and 14–15 orphons. IGKV1-37 and IGKV1D-37 are ORF due to the 2nd_CYS (position 104 in FR3-IMGT) being replaced by a Glycine (ggt).

Gene location

IGKV1-37 is in the proximal cluster, and IGKV1D-37 is in the distal cluster of the IGK locus on chromosome 2 at 2p11.2.

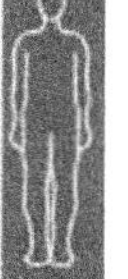

Nucleotide and amino acid sequences for human IGKV1-37 and IGKV1D-37

```
                                          1   2   3   4   5   6   7   8   9  10  11  12  13  14  15  16  17  18  19  20
                                         D   I   Q   L   T   Q   S   P   S   S   L   S   A   S   V   G   D   R   V   T
X59316,IGKV1-37*01,O14             [18] GAC ATC CAG TTG ACC CAG TCT CCA TCC TCC CTG TCT GCA TCT GTA GGA GAC AGA GTC ACC
X71893,IGKV1D-37*01,O4             [24] --- --- --- --- --- --- --- --- --- --- --- --- --- --- --- --- --- --- --- ---
X70466,IGKV1-37*01/1D-37*01,O4/14a [17] --- --- --- --- --- --- --- --- --- --- --- --- --- --- --- --- --- --- --- ---
X93629,IGKV1-37*01/1D-37*01,DPK11   [6] --- --- --- --- --- --- --- --- --- --- --- --- --- --- --- --- --- --- --- ---

                                                                ____________CDR1-IMGT______________
                                         21  22  23  24  25  26  27  28  29  30  31  32  33  34  35  36  37  38  39  40
                                         I   T   C   R   V   S   Q   G   I   S   S   Y                           L   N
X59316,IGKV1-37*01,O14                  ATC ACT TGC CGG GTG AGT CAG GGC ATT AGC AGT TAT ... ... ... ... ... ... TTA AAT
X71893,IGKV1D-37*01,O4                  --- --- --- --- --- --- --- --- --- --- --- --- ... ... ... ... ... ... --- ---
X70466,IGKV1-37*01/1D-37*01,O4/14a      --- --- --- --- --- --- --- --- --- --- --- --- ... ... ... ... ... ... --- ---
X93629,IGKV1-37*01/1D-37*01,DPK11       --- --- --- --- --- --- --- --- --- --- --- --- ... ... ... ... ... ... --- ---

                                                                                                    ___________CDR2-
                                         41  42  43  44  45  46  47  48  49  50  51  52  53  54  55  56  57  58  59  60
                                         W   Y   R   Q   K   P   G   K   V   P   K   L   L   I   Y   S   A   S
X59316,IGKV1-37*01,O14                  TGG TAT CGG CAG AAA CCA GGG AAA GTT CCT AAG CTC CTG ATC TAT AGT GCA TCC ... ...
X71893,IGKV1D-37*01,O4                  --- --- --- --- --- --- --- --- --- --- --- --- --- --- --- --- --- --- ... ...
X70466,IGKV1-37*01/1D-37*01,O4/14a      --- --- --- --- --- --- --- --- --- --- --- --- --- --- --- --- --- --- ... ...
X93629,IGKV1-37*01/1D-37*01,DPK11       --- --- --- --- --- --- --- --- --- --- --- --- --- --- --- --- --- --- ... ...

                                        IMGT_______________
                                         61  62  63  64  65  66  67  68  69  70  71  72  73  74  75  76  77  78  79  80
                                                             N   L   Q   S   G   V   P       S   R   F   S   G   S   G
X59316,IGKV1-37*01,O14                  ... ... ... ... ... AAT TTG CAA TCT GGA GTC CCA ... TCT CGG TTC AGT GGC AGT GGA
X71893,IGKV1D-37*01,O4                  ... ... ... ... ... --- --- --- --- --- --- --- ... --- --- --- --- --- --- ---
X70466,IGKV1-37*01/1D-37*01,O4/14a      ... ... ... ... ... --- --- --- --- --- --- --- ... --- --- --- --- --- --- ---
X93629,IGKV1-37*01/1D-37*01,DPK11       ... ... ... ... ... --- --- --- --- --- --- --- ... --- --- --- --- --- --- ---

                                         81  82  83  84  85  86  87  88  89  90  91  92  93  94  95  96  97  98  99 100
                                                 S   G   T   D   F   T   L   T   I   S   S   L   Q   P   E   D   V   A
X59316,IGKV1-37*01,O14                  ... ... TCT GGG ACA GAT TTC ACT CTC ACT ATC AGC AGC CTG CAG CCT GAA GAT GTT GCA
X71893,IGKV1D-37*01,O4                  ... ... --- --- --- --- --- --- --- --- --- --- --- --- --- --- --- --- --- ---
X70466,IGKV1-37*01/1D-37*01,O4/14a      ... ... --- --- --- --- --- --- --- --- --- --- --- --- --- --- --- --- --- ---
X93629,IGKV1-37*01/1D-37*01,DPK11       ... ... --- --- --- --- --- --- --- --- --- --- --- --- --- --- --- --- --- ---

                                                        _________CDR3-IMGT___________
                                        101 102 103 104 105 106 107 108 109 110 111
                                         T   Y   Y   G   Q   R   T   Y   N   A   P
X59316,IGKV1-37*01,O14                  ACT TAT TAC GGT CAA CGG ACT TAC AAT GCC CCT CC
X71893,IGKV1D-37*01,O4                  --- --- --- --- --- --- --- --- --- --- --- --
X70466,IGKV1-37*01/1D-37*01,O4/14a      --- --- --- --- --- --- --- --- --- --- ---
X93629,IGKV1-37*01/1D-37*01,DPK11       --- --- --- --- --- --- --- --- --- --- --- --
```

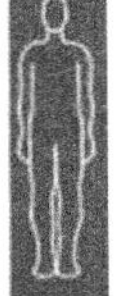

Framework and complementarity determining regions

FR1-IMGT: 26	CDR1-IMGT: 6
FR2-IMGT: 17	CDR2-IMGT: 3
FR3-IMGT: 36 (-3 aa: 73, 81, 82)	CDR3-IMGT: 7

Collier de Perles for human IGKV1-37*01 and IGKV1D-37*01

IGKV1-37*01
Accession number: IMGT X59316 EMBL/GenBank/DDBJ: X59316
IGKV1D-37*01
Accession number: IMGT X71893 EMBL/GenBank/DDBJ: X71893

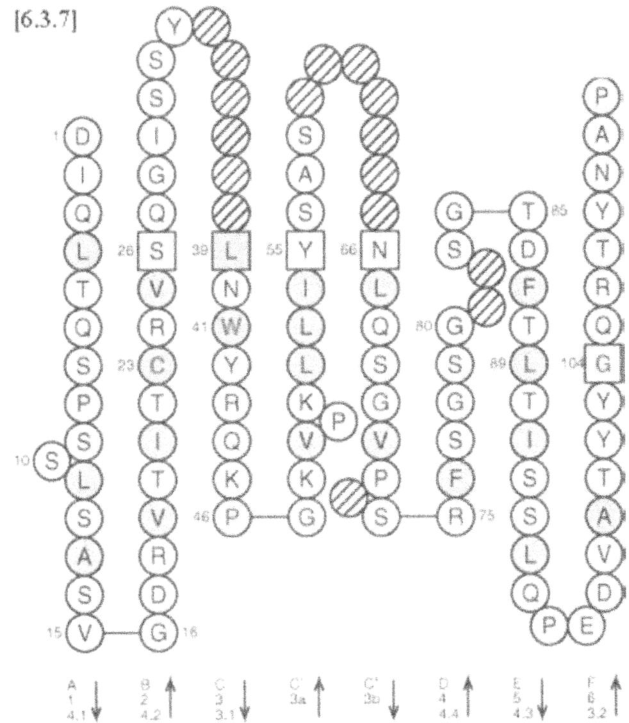

Genome database accession numbers

IGKV1-37:	GDB:9953458	LocusLink: 28931
IGKV1D-37:	GDB:9953532	LocusLink: 28894

IGKV1-39 and IGKV1D-39

Nomenclature

IGKV1-39: Immunoglobulin kappa variable 1-39.
IGKV1D-39: Immunoglobulin kappa variable 1D-39.

Definition and functionality

IGKV1-39 is a functional gene (allele *01) or a pseudogene (allele *02) and IGKV1D-39 a functional gene of the IGKV1 subgroup, in the IGK locus. The IGKV1 subgroup comprises 29 mapped genes (of which 15–16 are functional) in the IGK locus and 14–15 orphons.
IGKV1-39*02 is a pseudogene due to a STOP-CODON in L-PART1[24].

Gene location

IGKV1-39 is in the proximal cluster, and IGKV1D-39 is in the distal cluster of the IGK locus on chromosome 2 at 2p11.2.

Nucleotide and amino acid sequences for human IGKV1-39 and IGKV1D-39

```
                                        1   2   3   4   5   6   7   8   9  10  11  12  13  14  15  16  17  18  19  20
                                        D   I   Q   M   T   Q   S   P   S   S   L   S   A   S   V   G   D   R   V   T
X59315,IGKV1-39*01,O12            [18] GAC ATC CAG ATG ACC CAG TCT CCA TCC TCC CTG TCT GCA TCT GTA GGA GAC AGA GTC ACC
                                                                        F
X59318,IGKV1-39*02,O12a           [18] --- --- --- --- --- --- --- --- --- -T- --- --- --- --- --- --- --- --- --- ---
X59312,IGKV1D-39*01,O2            [18] --- --- --- --- --- --- --- --- --- --- --- --- --- --- --- --- --- --- --- ---
X93627,IGKV1-39*01/1D-39*01,DPK9   [6] --- --- --- --- --- --- --- --- --- --- --- --- --- --- --- --- --- --- --- ---

                                                               ______________CDR1-IMGT______________
                                       21  22  23  24  25  26  27  28  29  30  31  32  33  34  35  36  37  38  39  40
                                        I   T   C   R   A   S   Q   S   I   S   S   Y                           L   N
X59315,IGKV1-39*01,O12                 ATC ACT TGC CGG GCA AGT CAG AGC ATT AGC AGC TAT ... ... ... ... ... ... TTA AAT
X59318,IGKV1-39*02,O12a                --- --- --- --- --- --- --- --- --- --- --- --- ... ... ... ... ... ... --- ---
X59312,IGKV1D-39*01,O2                 --- --- --- --- --- --- --- --- --- --- --- --- ... ... ... ... ... ... --- ---
X93627,IGKV1-39*01/1D-39*01,DPK9       --- --- --- --- --- --- --- --- --- --- --- --- ... ... ... ... ... ... --- ---

                                                                                                ____________CDR2-
                                       41  42  43  44  45  46  47  48  49  50  51  52  53  54  55  56  57  58  59  60
                                        W   Y   Q   Q   K   P   G   K   A   P   K   L   L   I   Y   A   A   S
X59315,IGKV1-39*01,O12                 TGG TAT CAG CAG AAA CCA GGG AAA GCC CCT AAG CTC CTG ATC TAT GCT GCA TCC ... ...
X59318,IGKV1-39*02,O12a                --- --- --- --- --- --- --- --- --- --- --- --- --- --- --- --- --- --- ... ...
X59312,IGKV1D-39*01,O2                 --- --- --- --- --- --- --- --- --- --- --- --- --- --- --- --- --- --- ... ...
X93627,IGKV1-39*01/1D-39*01,DPK9       --- --- --- --- --- --- --- --- --- --- --- --- --- --- --- --- --- --- ... ...

                                       IMGT______________
                                       61  62  63  64  65  66  67  68  69  70  71  72  73  74  75  76  77  78  79  80
                                                            S   L   Q   S   G   V   P       S   R   F   S   G   S   G
X59315,IGKV1-39*01,O12                 ... ... ... ... ... AGT TTG CAA AGT GGG GTC CCA ... TCA AGG TTC AGT GGC AGT GGA
X59318,IGKV1-39*02,O12a                ... ... ... ... ... --- --- --- --- --- --- --- ... --- --- --- --- --- --- ---
X59312,IGKV1D-39*01,O2                 ... ... ... ... ... --- --- --- --- --- --- --- ... --- --- --- --- --- --- ---
X93627,IGKV1-39*01/1D-39*01,DPK9       ... ... ... ... ... --- --- --- --- --- --- --- ... --- --- --- --- --- --- ---

                                       81  82  83  84  85  86  87  88  89  90  91  92  93  94  95  96  97  98  99 100
                                                S   G   T   D   F   T   L   T   I   S   S   L   Q   P   E   D   F   A
X59315,IGKV1-39*01,O12                 ... ... TCT GGG ACA GAT TTC ACT CTC ACC ATC AGC AGT CTG CAA CCT GAA GAT TTT GCA
X59318,IGKV1-39*02,O12a                ... ... --- --- --- --- --- --- --- --- --- --- --- --- --- --- --- --- --- ---
X59312,IGKV1D-39*01,O2                 ... ... --- --- --- --- --- --- --- --- --- --- --- --- --- --- --- --- --- ---
X93627,IGKV1-39*01/1D-39*01,DPK9       ... ... --- --- --- --- --- --- --- --- --- --- --- --- --- --- --- --- --- ---

                                                       _________CDR3-IMGT___________
                                       101 102 103 104 105 106 107 108 109 110 111
                                        T   Y   Y   C   Q   Q   S   Y   S   T   P
X59315,IGKV1-39*01,O12                 ACT TAC TAC TGT CAA CAG AGT TAC AGT ACC CCT CC
                                                            C   G
X59318,IGKV1-39*02,O12a                --- --T --- --- --G TGT G-- --- --- --A --- --
X59312,IGKV1D-39*01,O2                 --- --- --- --- --- --- --- --- --- --- --- --
X93627,IGKV1-39*01/1D-39*01,DPK9       --- --- --- --- --- --- --- --- --- --- --- --
```

Framework and complementarity determining regions

FR1-IMGT: 26 | CDR1-IMGT: 6
FR2-IMGT: 17 | CDR2-IMGT: 3
FR3-IMGT: 36 (-3 aa: 73, 81, 82) | CDR3-IMGT: 7

Collier de Perles for human IGKV1-39*01 and IGKV1D-39*01

IGKV1-39*01
Accession number: IMGT X59315 EMBL/GenBank/DDBJ: X59315
IGKV1D-39*01
Accession number: IMGT X59312 EMBL/GenBank/DDBJ: X59312

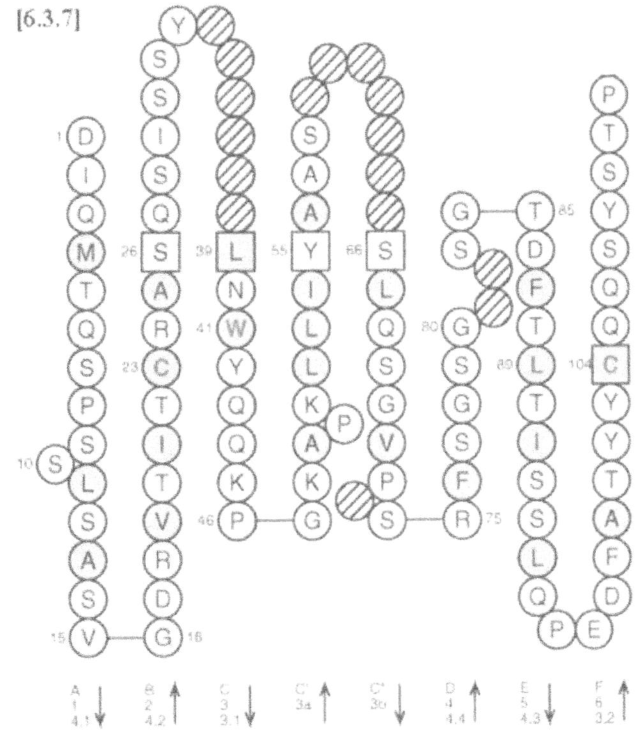

Genome database accession numbers

IGKV1-39:	GDB:9953460	LocusLink: 28930
IGKV1D-39:	GDB:9953534	LocusLink: 28893

IGKV1D-8

Nomenclature

IGKV1D-8: Immunoglobulin kappa variable 1D-8.

Definition and functionality

IGKV1D-8 is one of the 15–16 functional genes of the IGKV1 subgroup which comprises 29 mapped genes in the IGK locus and 14–15 orphons.

Gene location

IGKV1D-8 is in the distal cluster of the IGK locus on chromosome 2 at 2p11.2.

Nucleotide and amino acid sequences for human IGKV1D-8

```
                                       1   2   3   4   5   6   7   8   9  10  11  12  13  14  15  16  17  18  19  20
                                       V   I   W   M   T   Q   S   P   S   L   L   S   A   S   T   G   D   R   V   T
Z00008,IGKV1D-8*01,L24         [12]   GTC ATC TGG ATG ACC CAG TCT CCA TCC TTA CTC TCT GCA TCT ACA GGA GAC AGA GTC ACC
X72819,IGKV1D-8*01,L24a        [10]   ---                 --- --- --- --- --- --- --- ---                     --- ---
X93628,IGKV1D-8*01,DPK10        [6]   --- --- ---             --- --- --- --- --- --- --- --- ---             --- ---

                                                                         CDR1-IMGT
                                      21  22  23  24  25  26  27  28  29  30  31  32  33  34  35  36  37  38  39  40
                                       I   S   C   R   M   S   Q   G   I   S   S   Y                           L   A
Z00008,IGKV1D-8*01,L24                ATC AGT TGT CGG ATG AGT CAG GGC ATT AGC AGT TAT ... ... ... ... ... ... TTA GCC
X72819,IGKV1D-8*01,L24a                   --- --- --- --- --- --- --- ---         ... ... ... ... ... ... ... --
X93628,IGKV1D-8*01,DPK10               -- --- --- --- --- --- --- --- --          ... ... ... ... ... ... ... --

                                                                                                           CDR2-
                                      41  42  43  44  45  46  47  48  49  50  51  52  53  54  55  56  57  58  59  60
                                       W   Y   Q   Q   K   P   G   K   A   P   E   L   L   I   Y   A   A   S
Z00008,IGKV1D-8*01,L24                TGG TAT CAG CAA AAA CCA GGG AAA GCC CCT GAG CTC CTG ATC TAT GCT GCA TCC ... ...
X72819,IGKV1D-8*01,L24a                       --- --- --- --- --- ---                     --- --- --- --- ... ...
X93628,IGKV1D-8*01,DPK10                      --- --- --- --- --- ---     ---             --- --- --- --- ... ...

                                      IMGT
                                      61  62  63  64  65  66  67  68  69  70  71  72  73  74  75  76  77  78  79  80
                                                           T   L   Q   S   G   V   P       S   R   F   S   G   S   G
Z00008,IGKV1D-8*01,L24                ... ... ... ... ... ACT TTG CAA AGT GGG GTC CCA ... TCA AGG TTC AGT GGC AGT GGA
X72819,IGKV1D-8*01,L24a               ... ... ... ... ...                             ... --- --- --- --- ---
X93628,IGKV1D-8*01,DPK10              ... ... ... ... ...      --- --- --- --- --- ... --- --- ---

                                      81  82  83  84  85  86  87  88  89  90  91  92  93  94  95  96  97  98  99 100
                                               S   G   T   D   F   T   L   T   I   S   C   L   Q   S   E   D   F   A
Z00008,IGKV1D-8*01,L24                ... ... TCT GGG ACA GAT TTC ACT CTC ACC ATC AGT TGC CTG CAG TCT GAA GAT TTT GCA
X72819,IGKV1D-8*01,L24a               ... ... --- --- --- --- ---                             --- --- --- --- --- ---
X93628,IGKV1D-8*01,DPK10              ... ... ---     ---                                     --- --- --- --- --- ---

                                                              CDR3-IMGT
                                      101 102 103 104 105 106 107 108 109 110 111
                                       T   Y   Y   C   Q   Q   Y   Y   S   F   P
Z00008,IGKV1D-8*01,L24                ACT TAT TAC TGT CAA CAG TAT TAT AGT TTC CCT CC
X72819,IGKV1D-8*01,L24a               ---         --- --- --  --- --- --- --- --- --
X93628,IGKV1D-8*01,DPK10              --                  --  --- --- --- --- --- --
```

Framework and complementarity determining regions

FR1-IMGT: 26
FR2-IMGT: 17
FR3-IMGT: 36 (-3 aa: 73, 81, 82)

CDR1-IMGT: 6
CDR2-IMGT: 3
CDR3-IMGT: 7

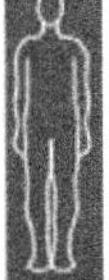

Collier de Perles for human IGKV1D-8*01

Accession number: IMGT Z00008 EMBL/GenBank/DDBJ: Z00008

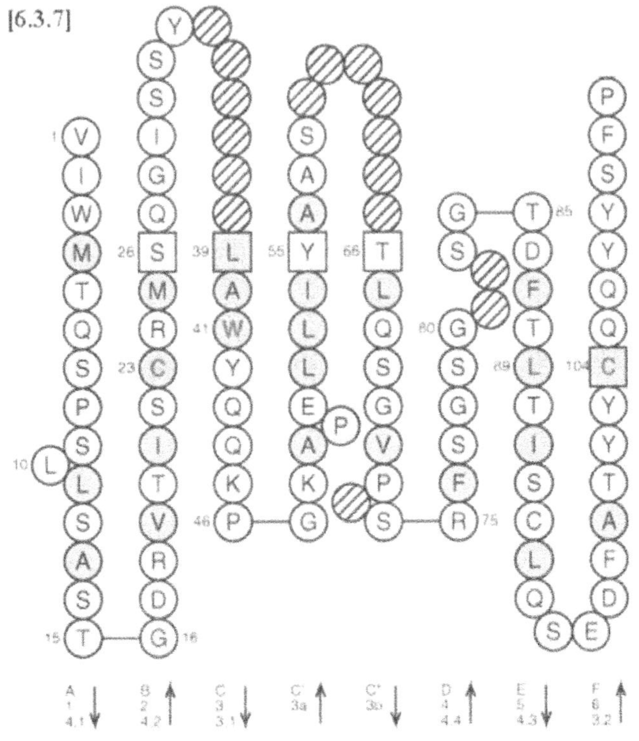

Genome database accession numbers

GDB:9953512 LocusLink: 28904

IGKV1D-16

Nomenclature

IGKV1D-16: Immunoglobulin kappa variable 1D-16.

Definition and functionality

IGKV1D-16 is one of the 15–16 functional genes of the IGKV1 subgroup which comprises 29 mapped genes in the IGK locus and 14–15 orphons.

Gene location

IGKV1D-16 is in the distal cluster of the IGK locus on chromosome 2 at 2p11.2.

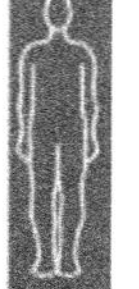

Nucleotide and amino acid sequences for human IGKV1D-16

```
                                   1   2   3   4   5   6   7   8   9  10  11  12  13  14  15  16  17  18  19  20
                                   D   I   Q   M   T   Q   S   P   S   S   L   S   A   S   V   G   D   R   V   T
K01323,IGKV1D-16*01,L15      [3]  GAC ATC CAG ATG ACC CAG TCT CCA TCC TCA CTG TCT GCA TCT GTA GGA GAC AGA GTC ACC
X92329,IGKV1D-16*01,HK134    [3]  --- --- --- --- --- --- --- --- --- --- --- --- --- --- --- --- --- --- --- .
X93625,IGKV1D-16*01,DPK7     [6]  --- --- --- --- --- --- --- --- --- --- --- --- --- --- --- --- --- --- --- ---
V00558,IGKV1D-16*02,L15a     [2]  --- --- --- --- --- --- --- --- --- --- --- --- --- --- --- --- --- --- --- ---
K01322,IGKV1D-16*02,HK146    [3]  --- --- --- --- --- --- --- --- --- --- --- --- --- --- --- --- --- --- --- ---
K01324,IGKV1D-16*02,HK189    [3]  --- --- --- --- --- --- --- --- --- --- --- --- --- --- --- --- --- --- --- ---

                                                          ____________________CDR1-IMGT________________
                                  21  22  23  24  25  26  27  28  29  30  31  32  33  34  35  36  37  38  39  40
                                   I   T   C   R   A   S   Q   G   I   S   S   W                           L   A
K01323,IGKV1D-16*01,L15           ATC ACT TGT CGG GCG AGT CAG GGT ATT AGC AGC TGG ... ... ... ... ... ... TTA GCC
X92329,IGKV1D-16*01,HK134         --- --- --- --- --- --- --- --- --- --- --- --- ... ... ... ... ... ... --- ---
X93625,IGKV1D-16*01,DPK7          --- --- --- --- --- --- --- --- --- --- --- --- ... ... ... ... ... ... --- ---
                                                       R
V00558,IGKV1D-16*02,L15a          --- --- --- --- --- --G --- --- --- --- --- --- ... ... ... ... ... ... --- ---
                                                       R
K01322,IGKV1D-16*02,HK146         --- --- --- --- --- --G --- --- --- --- --- --- ... ... ... ... ... ... --- ---
                                                       R
K01324,IGKV1D-16*02,HK189         --- --- --- --- --- --G --- --- --- --- --- --- ... ... ... ... ... ... --- ---

                                                                                              __________CDR2
                                  41  42  43  44  45  46  47  48  49  50  51  52  53  54  55  56  57  58  59  60
                                   W   Y   Q   Q   K   P   E   K   A   P   K   S   L   I   Y   A   A   S
K01323,IGKV1D-16*01,L15           TGG TAT CAG CAG AAA CCA GAG AAA GCC CCT AAG TCC CTG ATC TAT GCT GCA TCC ... ...
X92329,IGKV1D-16*01,HK134         --- --- --- --- --- --- --- --- --- --- --- --- --- --- --- --- --- --- ... ...
X93625,IGKV1D-16*01,DPK7          --- --- --- --- --- --- --- --- --- --- --- --- --- --- --- --- --- --- ... ...
V00558,IGKV1D-16*02,L15a          --- --- --- --- --- --- --- --- --- --- --- --- --- --- --- --- --- --- ... ...
K01322,IGKV1D-16*02,HK146         --- --- --- --- --- --- --- --- --- --- --- --- --- --- --- --- --- --- ... ...
K01324,IGKV1D-16*02,HK189         --- --- --- --- --- --- --- --- --- --- --- --- --- --- --- --- --- --- ... ...

                                  IMGT_______________
                                  61  62  63  64  65  66  67  68  69  70  71  72  73  74  75  76  77  78  79  80
                                                       S   L   Q   S   G   V   P       S   R   F   S   G   S   G
K01323,IGKV1D-16*01,L15           ... ... ... ... ... AGT TTG CAA AGT GGG GTC CCA ... TCA AGG TTC AGC GGC AGT GGA
X92329,IGKV1D-16*01,HK134         ... ... ... ... ... --- --- --- --- --- --- --- ... --- --- --- --- --- --- ---
X93625,IGKV1D-16*01,DPK7          ... ... ... ... ... --- --- --- --- --- --- --- ... --- --- --- --- --- --- ---
V00558,IGKV1D-16*02,L15a          ... ... ... ... ... --- --- --- --- --- --- --- ... --- --- --- --- --- --- ---
K01322,IGKV1D-16*02,HK146         ... ... ... ... ... --- --- --- --- --- --- --- ... --- --- --- --- --- --- ---
K01324,IGKV1D-16*02,HK189         ... ... ... ... ... --- --- --- --- --- --- --- ... --- --- --- --- --- --- ---

                                  81  82  83  84  85  86  87  88  89  90  91  92  93  94  95  96  97  98  99 100
                                           S   G   T   D   F   T   L   T   I   S   S   L   Q   P   E   D   F   A
K01323,IGKV1D-16*01,L15           ... ... TCT GGG ACA GAT TTC ACT CTC ACC ATC AGC AGC CTG CAG CCT GAA GAT TTT GCA
X92329,IGKV1D-16*01,HK134         ... ... --- --- --- --- --- --- --- --- --- --- --- --- --- --- --- --- --- ---
X93625,IGKV1D-16*01,DPK7          ... ... --- --- --- --- --- --- --- --- --- --- --- --- --- --- --- --- --- ---
V00558,IGKV1D-16*02,L15a          ... ... --- --- --- --- --- --- --- --- --- --- --- --- --- --- --- --- --- ---
K01322,IGKV1D-16*02,HK146         ... ... --- --- --- --- --- --- --- --- --- --- --- --- --- --- --- --- --- ---
K01324,IGKV1D-16*02,HK189         ... ... --- --- --- --- --- --- --- --- --- --- --- --- --- --- --- --- --- ---

                                                  __________CDR3-IMGT__________
                                  101 102 103 104 105 106 107 108 109 110 111
                                   T   Y   Y   C   Q   Q   Y   N   S   Y   P
K01323,IGKV1D-16*01,L15           ACT TAT TAC TGC CAA CAG TAT AAT AGT TAC CCT CC
X92329,IGKV1D-16*01,HK134         --- --- --- --- --- --- --- --- --- --- --- --
X93625,IGKV1D-16*01,DPK7          --- --- --- --- --- --- --- --- --- --- --- --
V00558,IGKV1D-16*02,L15a          --- --- --- --- --- --- --- --- --- --- --- --
K01322,IGKV1D-16*02,HK146         --- --- --- --- --- --- --- --- --- --- --- --
K01324,IGKV1D-16*02,HK189         --- --- --- --- --- --- --- --- --- --- --- --
```

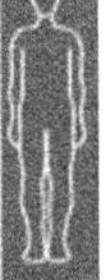

Framework and complementarity determining regions

FR1-IMGT: 26
FR2-IMGT: 17
FR3-IMGT: 36 (-3 aa: 73, 81, 82)
CDR1-IMGT: 6
CDR2-IMGT: 3
CDR3-IMGT: 7

Collier de Perles for human IGKV1D-16*01

Accession number: IMGT K01323 EMBL/GenBank/DDBJ: K01323

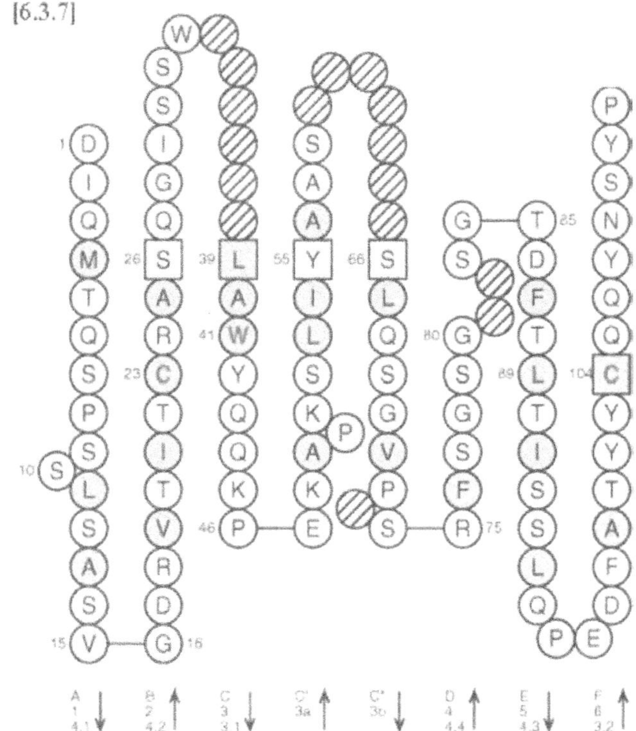

Genome database accession numbers

GDB:9953518 LocusLink: 28901

Nomenclature

IGKV1D-17: Immunoglobulin kappa variable 1D-17.

Definition and functionality

IGKV1D-17 is one of the 15–16 functional genes of the IGKV1 subgroup which comprises 29 mapped genes in the IGK locus and 14–15 orphons.

Gene location

IGKV1D-17 is in the distal cluster of the IGK locus on chromosome 2 at 2p11.2.

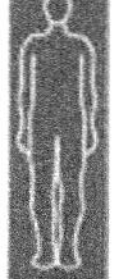

Nucleotide and amino acid sequences for human IGKV1D-17

```
                                     1   2   3   4   5   6   7   8   9  10  11  12  13  14  15  16  17  18  19  20
                                     N   I   Q   M   T   Q   S   P   S   A   M   S   A   S   V   G   D   R   V   T
X63392,IGKV1D-17*01,L14       [15]  AAC ATC CAG ATG ACC CAG TCT CCA TCT GCC ATG TCT GCA TCT GTA GGA GAC AGA GTC ACC

Z27498,IGKV1D-17*01,DPK2       [6]  --- --- --- --- --- --- --- --- --- --- --- --- --- --- --- --- --- --- --- ---

                                                             _________________CDR1-IMGT_________________
                                    21  22  23  24  25  26  27  28  29  30  31  32  33  34  35  36  37  38  39  40
                                     I   T   C   R   A   R   Q   G   I   S   N   Y                           L   A
X63392,IGKV1D-17*01,L14             ATC ACT TGT CGG GCG AGG CAG GGC ATT AGC AAT TAT ... ... ... ... ... ... TTA GCC

Z27498,IGKV1D-17*01,DPK2            --- --- --- --- --- --- --- --- --- --- --- --- ... ... ... ... ... ... --- ---

                                                                                                ___________CDR2-
                                    41  42  43  44  45  46  47  48  49  50  51  52  53  54  55  56  57  58  59  60
                                     W   F   Q   Q   K   P   G   K   V   P   K   H   L   I   Y   A   A   S
X63392,IGKV1D-17*01,L14             TGG TTT CAG CAG AAA CCA GGG AAA GTC CCT AAG CAC CTG ATC TAT GCT GCA TCC ... ...

Z27498,IGKV1D-17*01,DPK2            --- --- --- --- --- --- --- --- --- --- --- --- --- --- --- --- --- --- ... ...

                                    IMGT________________
                                    61  62  63  64  65  66  67  68  69  70  71  72  73  74  75  76  77  78  79  80
                                                         S   L   Q   S   G   V   P       S   R   F   S   G   S   G
X63392,IGKV1D-17*01,L14             ... ... ... ... ... AGT TTG CAA AGT GGG GTC CCA ... TCA AGG TTC AGC GGC AGT GGA

Z27498,IGKV1D-17*01,DPK2            ... ... ... ... ... --- --- --- --- --- --- --- ... --- --- --- --- --- --- ---

                                    81  82  83  84  85  86  87  88  89  90  91  92  93  94  95  96  97  98  99 100
                                             S   G   T   E   F   T   L   T   I   S   S   L   Q   P   E   D   F   A
X63392,IGKV1D-17*01,L14             ... ... TCT GGG ACA GAA TTC ACT CTC ACA ATC AGC AGC CTG CAG CCT GAA GAT TTT GCA

Z27498,IGKV1D-17*01,DPK2            ... ... --- --- --- --- --- --- --- --- --- --- --- --- --- --- --- --- --- ---

                                                     __________CDR3-IMGT___________
                                    101 102 103 104 105 106 107 108 109 110 111
                                     T   Y   Y   C   L   Q   H   N   S   Y   P
X63392,IGKV1D-17*01,L14             ACT TAT TAC TGT CTA CAG CAT AAT AGT TAC CCT CC

Z27498,IGKV1D-17*01,DPK2            --- --- --- --- --- --- --- --- --- --- --- --
```

Framework and complementarity determining regions

FR1-IMGT: 26
FR2-IMGT: 17
FR3-IMGT: 36 (-3 aa: 73, 81, 82)

CDR1-IMGT: 6
CDR2-IMGT: 3
CDR3-IMGT: 7

Collier de Perles for human IGKV1D-17*01

Accession number: IMGT X63392 EMBL/GenBank/DDBJ: X63392

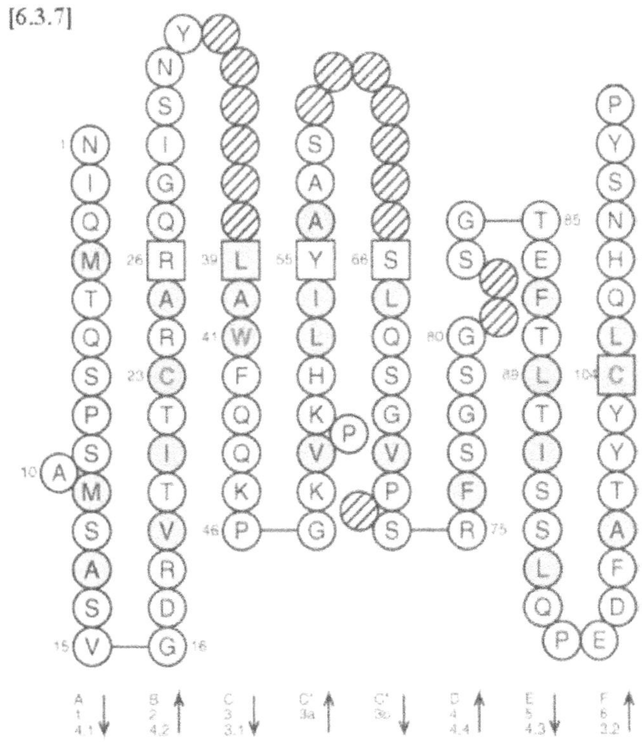

Genome database accession numbers

GDB:9953520 LocusLink: 28900

IGKV1D-42

Nomenclature

IGKV1D-42: Immunoglobulin kappa variable 1D-42.

Definition and functionality

IGKV1D-42 is one of the five ORF genes of the IGKV1 subgroup, in the IGK locus.The IGKV1 subgroup comprises 29 mapped genes (of which 15–16 are functional) in the IGK locus and 14–15 orphons.
IGKV1D-42 is an ORF due to a nucleotide mutation in the core sequence of the DECAMER, and to an altered V-HEPTAMER: cacaggg instead of cacagtg[24].

Gene location

IGKV1D-42 is in the distal cluster of the IGK locus on chromosome 2 at 2p11.2.

Nucleotide and amino acid sequences for human IGKV1D-42

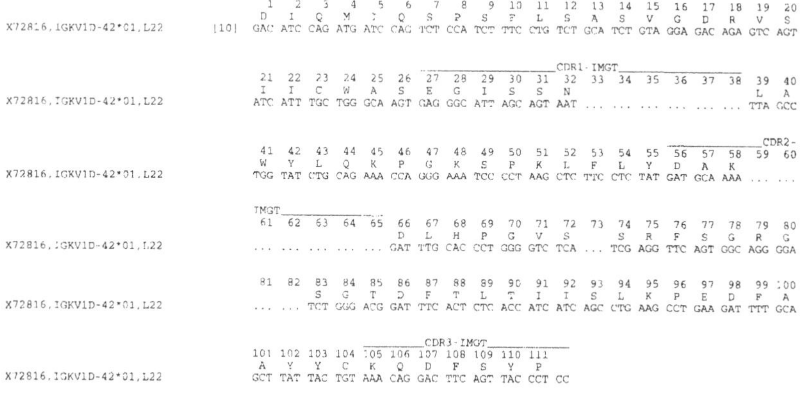

Framework and complementarity determining regions

FR1-IMGT: 26
FR2-IMGT: 17
FR3-IMGT: 36 (-3 aa: 73, 81, 82)

CDR1-IMGT: 6
CDR2-IMGT: 3
CDR3-IMGT: 7

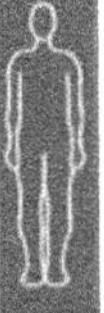

Collier de Perles for human IGKV1D-42*01

Accession number: IMGT X72816 EMBL/GenBank/DDBJ: X72816

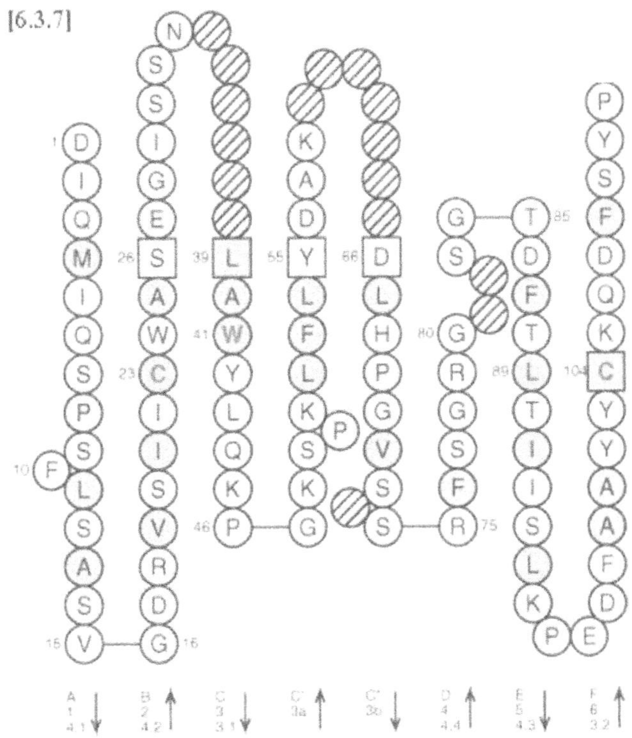

Genome database accession numbers

GDB:9953536 LocusLink: 28892

IGKV1D-43

Nomenclature

IGKV1D-43: Immunoglobulin kappa variable 1D-43.

Definition and functionality

IGKV1D-43 is one of the 15–16 functional genes of the IGKV1 subgroup which comprises 29 mapped genes in the IGK locus and 14–15 orphons.

Gene location

IGKV1D-43 is in the distal cluster of the IGK locus on chromosome 2 at 2p11.2.

Nucleotide and amino acid sequences for human IGKV1D-43

```
                                  1   2   3   4   5   6   7   8   9  10  11  12  13  14  15  16  17  18  19  20
                                  A   I   R   M   T   Q   S   P   F   S   L   S   A   S   V   G   D   R   V   T
X72817,IGKV1D-43*01,L23     (10) GCC ATC CGG ATG ACC CAG TCT CCA TTC TCC CTG TCT GCA TCT GTA GGA GAC AGA GTC ACC
X72818,IGKV1D-43*01,L23a    (10) --- --- --- --- --- --- --- --- --- --- --- --- --- --- --- --- --- --- --- ---

                                                           _______CDR1-IMGT________________
                                 21  22  23  24  25  26  27  28  29  30  31  32  33  34  35  36  37  38  39  40
                                  I   T   C   W   A   S   Q   G   I   S   S   Y                           L   A
X72817,IGKV1D-43*01,L23          ATC ACT TGC TGG GCC AGT CAG GGC ATT AGC AGT TAT ... ... ... ... ... ... TTA GCC
X72818,IGKV1D-43*01,L23a         --- --- --- --- --- --- --- --- --- --- --- --- ... ... ... ... ... ... --- ---

                                                                                             __________CDR2-
                                 41  42  43  44  45  46  47  48  49  50  51  52  53  54  55  56  57  58  59  60
                                  W   Y   Q   Q   K   P   A   K   A   P   K   L   F   I   Y   Y   A   S
X72817,IGKV1D-43*01,L23          TGG TAT CAG CAA AAA CCA GCA AAA GCC CCT AAG CTC TTC ATC TAT TAT GCA TCC ... ...
X72818,IGKV1D-43*01,L23a         --- --- --- --- --- --- --- --- --- --- --- --- --- --- --- --- --- --- ... ...

                                 IMGT_______________
                                 61  62  63  64  65  66  67  68  69  70  71  72  73  74  75  76  77  78  79  80
                                                      S   L   Q   S   G   V   P       S   R   F   S   G   S   G
X72817,IGKV1D-43*01,L23          ... ... ... ... ... AGT TTG CAA AGT GGG GTC CCA ... TCA AGG TTC AGC GGC AGT GGA
X72818,IGKV1D-43*01,L23a         ... ... ... ... ... --- --- --- --- --- --- --- ... --- --- --- --- --- --- ---

                                 81  82  83  84  85  86  87  88  89  90  91  92  93  94  95  96  97  98  99 100
                                          S   G   T   D   Y   T   L   T   I   S   S   L   Q   P   E   D   F   A
X72817,IGKV1D-43*01,L23          ... ... TCT GGG ACG GAT TAC ACT CTC ACC ATC AGC AGC CTG CAG CCT GAA GAT TTT GCA
X72818,IGKV1D-43*01,L23a         ... ... --- --- --- --- --- --- --- --- --- --- --- --- --- --- --- --- --- ---

                                                     ______CDR3-IMGT_______________
                                101 102 103 104 105 106 107 108 109 110 111
                                  T   Y   Y   C   Q   Q   Y   Y   S   T   P
X72817,IGKV1D-43*01,L23          ACT TAT TAC TGT CAA CAG TAT TAT AGT ACC CCT CC
X72818,IGKV1D-43*01,L23a         --- --- --- --- --- --- --- --- --- --- --- --
```

Framework and complementarity determining regions

FR1-IMGT: 26

FR2-IMGT: 17

FR3-IMGT: 36 (-3 aa: 73, 81, 82)

CDR1-IMGT: 6

CDR2-IMGT: 3

CDR3-IMGT: 7

Collier de Perles for human IGKV1D-43*01

Accession number: IMGT X72817 EMBL/GenBank/DDBJ: X72817

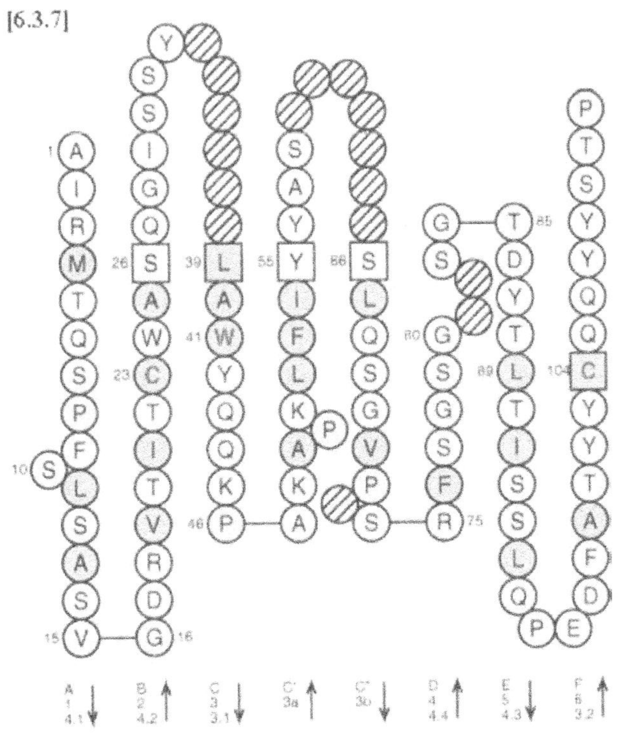

Genome database accession numbers

GDB:9953538 LocusLink: 28891

Nomenclature

IGKV2-24: Immunoglobulin kappa variable 2-24.

Definition and functionality

IGKV2-24 is one of the 8–10 functional genes of the IGKV2 subgroup which comprises 27 mapped genes in the IGK locus and nine orphons.

Gene location

IGKV2-24 is in the proximal cluster of the IGK locus on chromosome 2 at 2p11.2.

Nucleotide and amino acid sequences for human IGKV2-24

```
                                     1   2   3   4   5   6   7   8   9  10  11  12  13  14  15  16  17  18  19  20
                                     D   I   V   M   T   Q   T   P   L   S   S   P   V   T   L   G   Q   P   A   S
X12684,IGKV2-24*01,A23        [29]  GAT ATT GTG ATG ACC CAG ACT CCA CTC TCC TCA CCT GTC ACC CTT GGA CAG CCG GCC TCC
X93633,IGKV2-24*01,DPK16      [6]   --- --- --- --- --- --- --- --- --- --- --- --- --- --- --- --- --- --- --- ---

                                                        ______________CDR1-IMGT______________
                                    21  22  23  24  25  26  27  28  29  30  31  32  33  34  35  36  37  38  39  40
                                     I   S   C   R   S   S   Q   S   L   V   H   S   D   G   N   T   Y       L   S
X12684,IGKV2-24*01,A23              ATC TCC TGC AGG TCT AGT CAA AGC CTC GTA CAC AGT GAT GGA AAC ACC TAC ... TTG AGT
X93633,IGKV2-24*01,DPK16            --- --- --- --- --- --- --- --- --- --- --- --- --- --- --- --- --- ... --- ---

                                                                                            ___________CDR2-
                                    41  42  43  44  45  46  47  48  49  50  51  52  53  54  55  56  57  58  59  60
                                     W   L   Q   Q   R   P   G   Q   P   P   R   L   L   I   Y   K   I   S
X12684,IGKV2-24*01,A23              TGG CTT CAG CAG AGG CCA GGC CAG CCT CCA AGA CTC CTA ATT TAT AAG ATT TCT ... ...
X93633,IGKV2-24*01,DPK16            --- --- --- --- --- --- --- --- --- --- --- --- --- --- --- --- --- --- ... ...

                                    IMGT________________
                                    61  62  63  64  65  66  67  68  69  70  71  72  73  74  75  76  77  78  79  80
                                                         N   R   F   S   G   V   P       D   R   F   S   G   S   G
X12684,IGKV2-24*01,A23              ... ... ... ... ... AAC CGG TTC TCT GGG GTC CCA ... GAC AGA TTC AGT GGC AGT GGG
X93633,IGKV2-24*01,DPK16            ... ... ... ... ... --- --- --- --- --- --- --- ... --- --- --- --- --- --- ---

                                    81  82  83  84  85  86  87  88  89  90  91  92  93  94  95  96  97  98  99 100
                                             A   G   T   D   F   T   L   K   I   S   R   V   E   A   E   D   V   G
X12684,IGKV2-24*01,A23              ... ... GCA GGG ACA GAT TTC ACA CTG AAA ATC AGC AGG GTG GAA GCT GAG GAT GTC GGG
X93633,IGKV2-24*01,DPK16            ... ... --- --- --- --- --- --- --- --- --- --- --- --- --- --- --- --- --- ---

                                                    __________CDR3-IMGT__________
                                   101 102 103 104 105 106 107 108 109 110 111
                                     V   Y   Y   C   M   Q   A   T   Q   F   P
X12684,IGKV2-24*01,A23              GTT TAT TAC TGC ATG CAA GCT ACA CAA TTT CCT CA
X93633,IGKV2-24*01,DPK16            --- --- --- --- --- --- --- --- --- --- --- --
```

Framework and complementarity determining regions

FR1-IMGT: 26
FR2-IMGT: 17
FR3-IMGT: 36 (-3 aa: 73, 81, 82)

CDR1-IMGT: 11
CDR2-IMGT: 3
CDR3-IMGT: 7

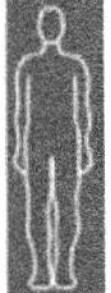

Collier de Perles for human IGKV2-24*01

Accession number: IMGT X12684 EMBL/GenBank/DDBJ: X12684

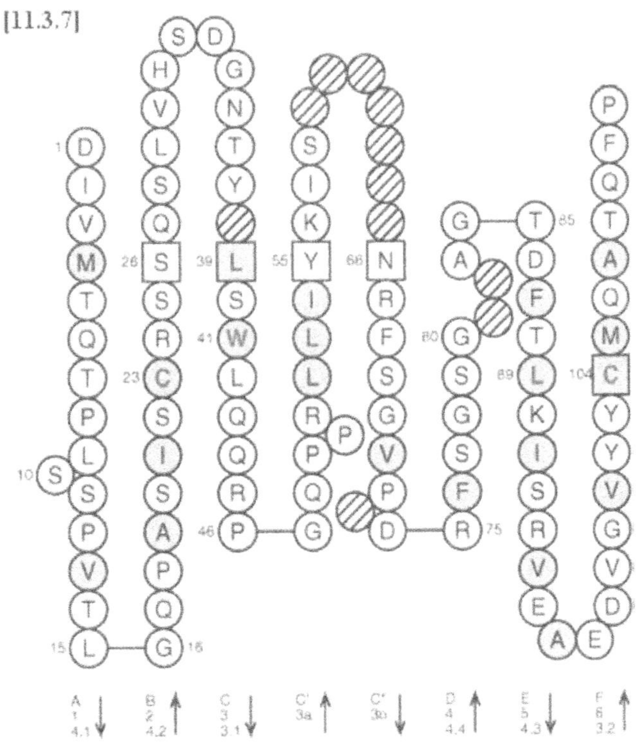

Genome database accession numbers

GDB:9953474 LocusLink: 28923

IGKV2-28 and IGKV2D-28

Nomenclature

IGKV2-28: Immunoglobulin kappa variable 2-28.
IGKV2D-28: Immunoglobulin kappa variable 2D-28.

Definition and functionality

IGKV2-28 and IGKV2D-28 are two of the 8–10 functional genes of the IGKV2 subgroup which comprises 27 mapped genes in the IGK locus and nine orphons.

Gene location

IGKV2-28 is in the proximal cluster, and IGKV2D-28 is in the distal cluster of the IGK locus on chromosome 2 at 2p11.2.

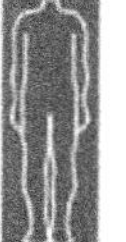

Nucleotide and amino acid sequences for human IGKV2-28 and IGKV2D-28

```
                                          1   2   3   4   5   6   7   8   9  10  11  12  13  14  15  16  17  18  19  20
                                          D   I   V   M   T   Q   S   P   L   S   L   P   V   T   P   G   E   P   A   S
X63397,IGKV2-28*01,A19             [15] GAT ATT GTG ATG ACT CAG TCT CCA CTC TCC CTG CCC GTC ACC CCT GGA GAG CCG GCC TCC
X12691,IGKV2D-28*01,A3             [29] --- --- --- --- --- --- --- --- --- --- --- --- --- --- --- --- --- --- --- ---
Z46616,IGKV2D-28*01,RR25gl          [8] --- --- --- --- --- --- --- --- --- --- --- --- --- --- --- --- --- --- --- ---
X93632,IGKV2-28*01/2D-28*01,DPK15   [6] --- --- --- --- --- --- --- --- --- --- --- --- --- --- --- --- --- --- --- ---

                                                                _______________CDR1-IMGT_______________
                                         21  22  23  24  25  26  27  28  29  30  31  32  33  34  35  36  37  38  39  40
                                          I   S   C   R   S   S   Q   S   L   L   H   S   N   G   Y   N   Y       L   D
X63397,IGKV2-28*01,A19                  ATC TCC TGC AGG TCT AGT CAG AGC CTC CTG CAT AGT AAT GGA TAC AAC TAT ... TTG GAT
X12691,IGKV2D-28*01,A3                  --- --- --- --- --- --- --- --- --- --- --- --- --- --- --- --- --- ... --- ---
Z46616,IGKV2D-28*01,RR25gl              --- --- --- --- --- --- --- --- --- --- --- --- --- --- --- --- --- ... --- ---
X93632,IGKV2-28*01/2D-28*01,DPK15       --- --- --- --- --- --- --- --- --- --- --- --- --- --- --- --- --- ... --- ---

                                                                                                    ___________CDR2-
                                         41  42  43  44  45  46  47  48  49  50  51  52  53  54  55  56  57  58  59  60
                                          W   Y   L   Q   K   P   G   Q   S   P   Q   L   L   I   Y   L   G   S
X63397,IGKV2-28*01,A19                  TGG TAC CTG CAG AAG CCA GGG CAG TCT CCA CAG CTC CTG ATC TAT TTG GGT TCT ... ...
X12691,IGKV2D-28*01,A3                  --- --- --- --- --- --- --- --- --- --- --- --- --- --- --- --- --- --- ... ...
Z46616,IGKV2D-28*01,RR25gl              --- --- --- --- --- --- --- --- --- --- --- --- --- --- --- --- --- --- ... ...
X93632,IGKV2-28*01/2D-28*01,DPK15       --- --- --- --- --- --- --- --- --- --- --- --- --- --- --- --- --- --- ... ...

                                        IMGT_______________
                                         61  62  63  64  65  66  67  68  69  70  71  72  73  74  75  76  77  78  79  80
                                                              N   R   A   S   G   V   P       D   R   F   S   G   S   G
X63397,IGKV2-28*01,A19                  ... ... ... ... ... AAT CGG GCC TCC GGG GTC CCT ... GAC AGG TTC AGT GGC AGT GGA
X12691,IGKV2D-28*01,A3                  ... ... ... ... ... --- --- --- --- --- --- --- ... --- --- --- --- --- --- ---
Z46616,IGKV2D-28*01,RR25gl              ... ... ... ... ... --- --- --- --- --- --- --- ... --- --- --- --- --- --- ---
X93632,IGKV2-28*01/2D-28*01,DPK15       ... ... ... ... ... --- --- --- --- --- --- --- ... --- --- --- --- --- --- ---

                                         81  82  83  84  85  86  87  88  89  90  91  92  93  94  95  96  97  98  99 100
                                                  S   G   T   D   F   T   L   K   I   S   R   V   E   A   E   D   V   G
X63397,IGKV2-28*01,A19                  ... ... TCA GGC ACA GAT TTT ACA CTG AAA ATC AGC AGA GTG GAG GCT GAG GAT GTT GGG
X12691,IGKV2D-28*01,A3                  ... ... --- --- --- --- --- --- --- --- --- --- --- --- --- --- --- --- --- ---
Z46616,IGKV2D-28*01,RR25gl              ... ... --- --- --- --- --- --- --- --- --- --- --- --- --- --- --- --- --- ---
X93632,IGKV2-28*01/2D-28*01,DPK15       ... ... --- --- --- --- --- --- --- --- --- --- --- --- --- --- --- --- --- ---

                                                        ___________CDR3-IMGT___________
                                        101 102 103 104 105 106 107 108 109 110 111
                                          V   Y   Y   C   M   Q   A   L   Q   T   P
X63397,IGKV2-28*01,A19                  GTT TAT TAC TGC ATG CAA GCT CTA CAA ACT CCT CC
X12691,IGKV2D-28*01,A3                  --- --- --- --- --- --- --- --- --- --- --- --
Z46616,IGKV2D-28*01,RR25gl              --- --- --- --- --- --- --- --- --- --- ---
X93632,IGKV2-28*01/2D-28*01,DPK15       --- --- --- --- --- --- --- --- --- --- --- --
```

Framework and complementarity determining regions

FR1-IMGT: 26
FR2-IMGT: 17
FR3-IMGT: 36 (-3 aa: 73, 81, 82)

CDR1-IMGT: 11
CDR2-IMGT: 3
CDR3-IMGT: 7

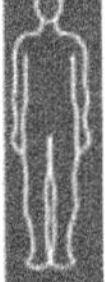

Collier de Perles for human IGKV2-28*01 and IGKV2D-28*01

IGKV2-28*01
Accession number: IMGT X63397 EMBL/GenBank/DDBJ: X63397
IGKV2D-28*01
Accession number: IMGT X12691 EMBL/GenBank/DDBJ: X12691

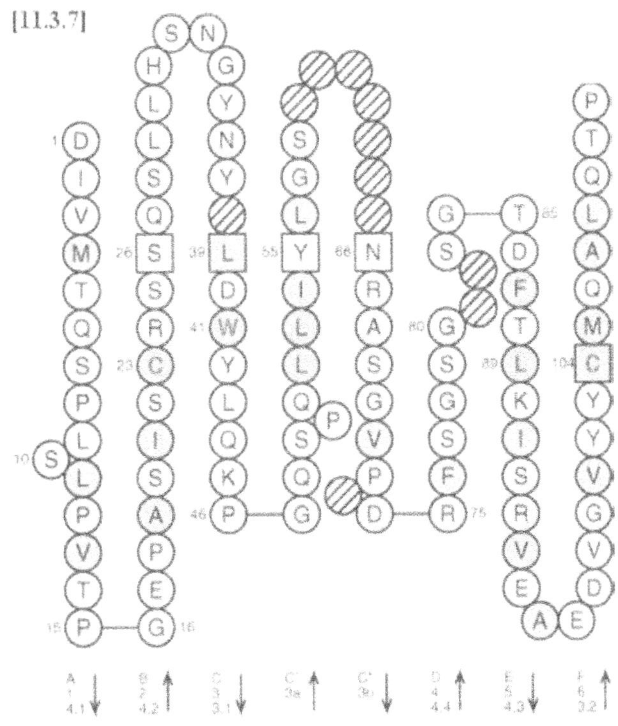

Genome database accession numbers

IGKV2-28:	GDB:9953478	LocusLink: 28921
IGKV2D-28:	GDB:9953554	LocusLink: 28883

Nomenclature

IGKV2-29: Immunoglobulin kappa variable 2-29.

Definition and functionality

IGKV2-29 is a pseudogene (allele *01) or a functional gene (allele *02) of the IGKV2 subgroup. The IGKV2 subgroup comprises 27 mapped genes (of which 8–10 are functional) in the IGK locus and nine orphons.
The IGKV2-29*01 is a pseudogene due to a STOP-CODON at position 104 in the FR3-IMGT[24].

Gene location

IGKV2-29 is in the proximal cluster of the IGK locus on chromosome 2 at 2p11.2.

Nucleotide and amino acid sequences for human IGKV2-29

```
                                    1   2   3   4   5   6   7   8   9  10  11  12  13  14  15  16  17  18  19  20
                                    D   I   V   M   T   Q   T   P   L   S   L   S   V   T   P   G   Q   P   A   S
X63396,IGKV2-29*01,A18/A18a  (15)  GAT ATT GTG ATC ACC CAG ACT CCA CTC TCT CTG TCC GTC ACC CCT GGA CAG CCG GCC TCC
X93644,IGKV2-29*01,DPK28     (6)   --  -   --- --- --- --- --- --- --- --- --- --- --- --- --- --- --- --- --- ---
U41645,IGKV2-29*02,A18b      (1)   -   --- --- --- --- --- --- --- --- --- --- --- --- --- --- --- --- --- --- ---

                                                           ____________CDR1-IMGT_______________
                                   21  22  23  24  25  26  27  28  29  30  31  32  33  34  35  36  37  38  39  40
                                    I   S   C   K   S   S   Q   S   L   L   H   S   D   G   K   T   Y       L   Y
X63396,IGKV2-29*01,A18/A18a        ATC TCC TGC AAG TCT AGT CAG AGC CTC CTG CAT AGT GAT GGA AAG ACC TAT ... TTG TAT
X93644,IGKV2 29*01,DPK28           --- --- --- --- -   --- --- --- --      --- --- --- --- --- --- --- ... --- -
U41645,IGKV2 29*02,A18b            --- --- --- --- --- --- --- --- --- --- -   --- --- --- --- --- --- ... --- ---

                                                                                           ____________CDR2
                                   41  42  43  44  45  46  47  48  49  50  51  52  53  54  55  56  57  58  59  60
                                    W   Y   L   Q   K   P   G   Q   S   P   Q   L   L   I   Y   E   V   S
X63396,IGKV2 29*01,A18/A18a        TGG TAC CTG CAG AAG CCA GGC CAG TCT CCA CAG CTC CTG ATC TAT GAA GTT TCC ... ...
X93644,IGKV2-29*01,DPK28           -   --- --- --- --- --- --- --- --- --- --- --- --- --- --- --- --- --- ... ...
U41645,IGKV2-29*02,A18b            --- --- --- --- --- --- --- --- -   --- --- --- --A --- --- --- --- --- ... ...

IMGT______________
                                   61  62  63  64  65  66  67  68  69  70  71  72  73  74  75  76  77  78  79  80
                                                        S   R   F   S   G   V   P       D   R   F   S   G   S   G
X63396,IGKV2-29*01,A18/A18a        ... ... ... ... ... AGC CGG TTC TCT GGA GTG CCA ... GAT AGG TTC AGT GGC AGC GGG
X93644,IGKV2-29*01,DPK28           ... ... ... ... ... --- --- --- --- --- --- --- ... --- --- --- --- --- ---
U41645,IGKV2-29*02,A18b            ... ... ... ... ... --- --- --- --- --- --- --- ... --- --- --- --- --- --- ---

                                   81  82  83  84  85  86  87  88  89  90  91  92  93  94  95  96  97  98  99 100
                                            S   G   T   D   F   T   L   K   I   S   R   V   E   A   E   D   V   G
X63396,IGKV2 29*01,A18/A18a        ... ... TCA GGG ACA GAT TTC ACA CTG AAA ATC AGC CGG GTG GAG GCT GAG GAT GTT GGG
X93644,IGKV2-29*01,DPK28           ... ... --- --- --- --- --- --- --- --- --- --- --- --- --- --- --- --- --- --
U41645,IGKV2-29*02,A18b            ... ... --  --- --- --- --- --- --- --- --- --- --- --- --- --- --- --- ---

                                                      ________CDR3-IMGT__________
                                  101 102 103 104 105 106 107 108 109 110 111
                                    V   Y   Y   *   M   Q   G   I   H   L   P
X63396,IGKV2-29*01,A18/A18a        GTT TAT TAC TGA ATG CAA GGT ATA CAC CTT CCT CC
X93644,IGKV2-29*01,DPK28           --- --- --- --- --- --- --- --- --- --- --- --
                                                C
U41645,IGKV2-29*02,A18b            --- --- --- --C ---     --- --- --- --- --- --
```

Framework and complementarity determining regions

FR1-IMGT: 26
FR2-IMGT: 17
FR3-IMGT: 36 (-3 aa: 73, 81, 82)

CDR1-IMGT: 11
CDR2-IMGT: 3
CDR3-IMGT: 7

Collier de Perles for human IGKV2-29*01

Accession number: IMGT X63396 EMBL/GenBank/DDBJ: X63396

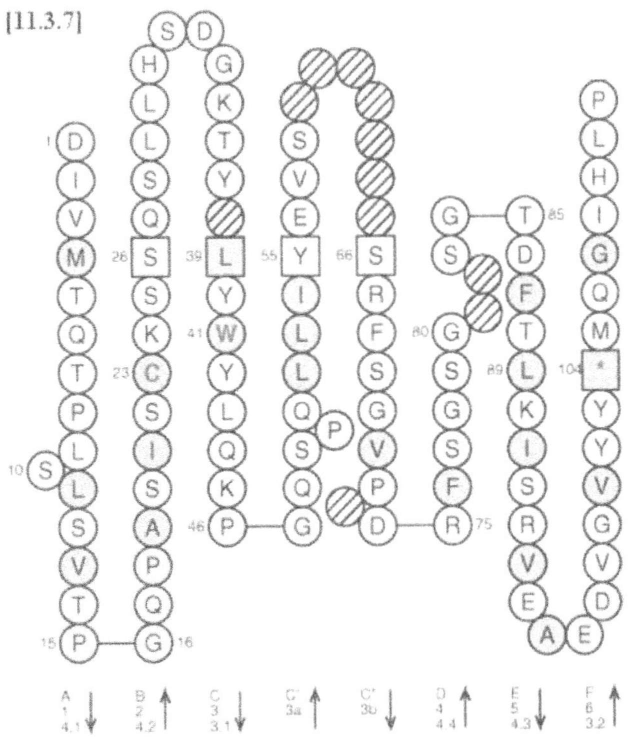

Genome database accession numbers

GDB:9953480 LocusLink: 28920

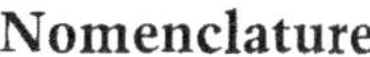

Nomenclature

IGKV2-30: Immunoglobulin kappa variable 2-30.

Definition and functionality

IGKV2-30 is one of the 8–10 functional genes of the IGKV2 subgroup which comprises 27 mapped genes in the IGK locus and nine orphons.

Gene location

IGKV2-30 is in the proximal cluster of the IGK locus on chromosome 2 at 2p11.2.

Nucleotide and amino acid sequences for human IGKV2-30

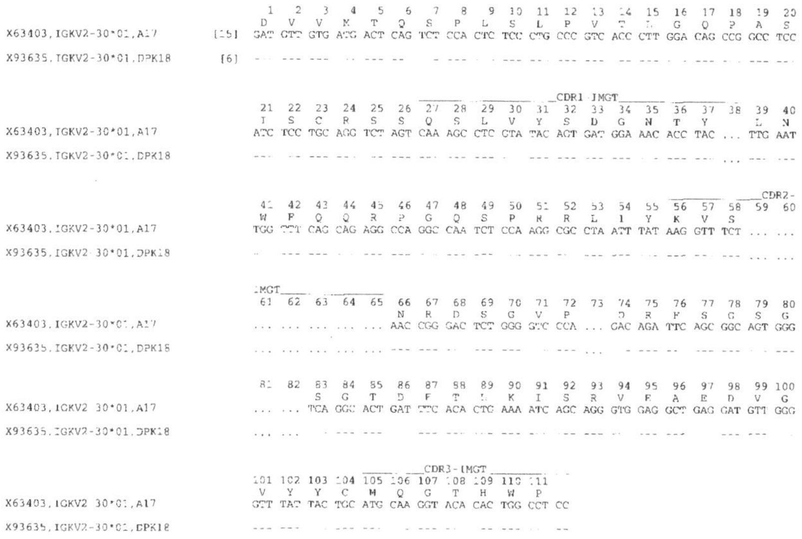

Framework and complementarity determining regions

FR1-IMGT: 26
FR2-IMGT: 17
FR3-IMGT: 36 (-3 aa: 73, 81, 82)

CDR1-IMGT: 11
CDR2-IMGT: 3
CDR3-IMGT: 7

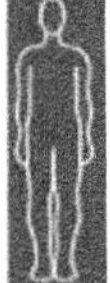

Collier de Perles for human IGKV2-30*01

Accession number: IMGT X63403 EMBL/GenBank/DDBJ: X63403

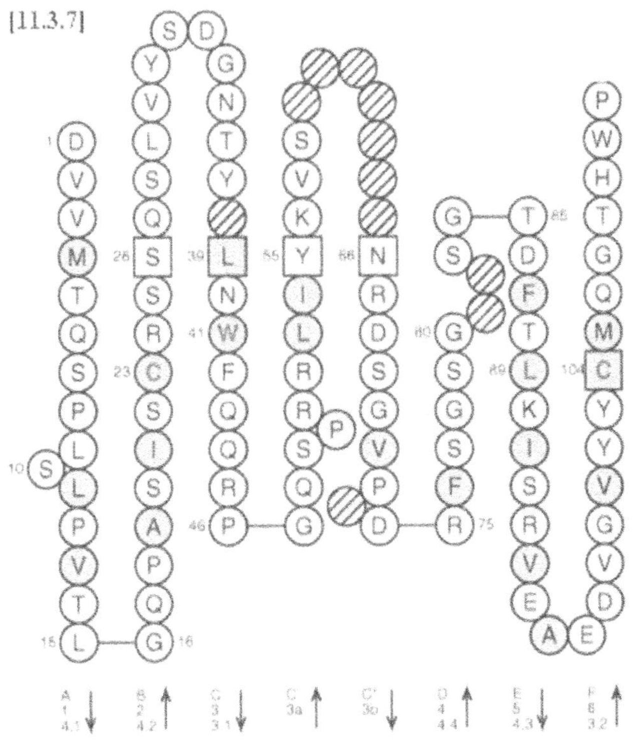

Genome database accession numbers

GDB:9953482 LocusLink: 28919

IGKV2-40 and IGKV2D-40

Nomenclature

IGKV2-40: Immunoglobulin kappa variable 2-40.
IGKV2D-40: Immunoglobulin kappa variable 2D-40.

Definition and functionality

IGKV2-40 and IGKV2D-40 are two of the 8–10 functional genes of the IGKV2 subgroup which comprises 27 mapped genes in the IGK locus and nine orphons.

Gene location

IGKV2-40 is in the proximal cluster, and IGKV2D-40 is in the distal cluster of the IGK locus on chromosome 2 at 2p11.2.

Nucleotide and amino acid sequences for human IGKV2-40 and IGKV2D-40

```
                                          1   2   3   4   5   6   7   8   9  10  11  12  13  14  15  16  17  18  19  20
                                          D   I   V   M   T   Q   T   P   L   S   L   P   V   T   P   G   E   P   A   S
X59314,IGKV2-40*01,O11             [18]  GAT ATT GTG ATG ACC CAG ACT CCA CTC TCC CTG CCC GTC ACC CCT GGA GAG CCG GCC TCC
X59317,IGKV2-40*02,O11a            [18]                                                                          --- ---
X59311,IGKV2D-40*01,O1             [18]  --- --- --- --- --- --- --- --- --- --- --- --- --- --- --- --- --- --- --- ---
X93631,IGKV2-40*01/2D-40*01,DPK13  [6]   --- --- --- --- --- --- --- --- --- --- --- --- --- --- --- --- --- --- --- ---

                                                                   _________________CDR1-IMGT_______________
                                         21  22  23  24  25  26  27  28  29  30  31  32  33  34  35  36  37  38  39  40
                                          I   S   C   R   S   S   Q   S   L   L   D   S   D   D   G   N   T   Y   L   D
X59314,IGKV2-40*01,O11                   ATC TCC TGC AGG TCT AGT CAG AGC CTC TTG GAT AGT GAT GAT GGA AAC ACC TAT TTG GAC
X59317,IGKV2-40*02,O11a                  --- --- --- --- --- --- --- --- --- --- --- --- --- --- --- --- --- --- --- --T
X59311,IGKV2D-40*01,O1                   --- --- --- --- --- --- --- --- --- --- --- --- --- --- --- --- --- --- --- ---
X93631,IGKV2-40*01/2D-40*01,DPK13        --- --- --- --- --- --- --- --- --- --- --- --- --- --- --- --- --- --- --- ---

                                                                                                  __________CDR2-
                                         41  42  43  44  45  46  47  48  49  50  51  52  53  54  55  56  57  58  59  60
                                          W   Y   L   Q   K   P   G   Q   S   P   Q   L   L   I   Y   T   L   S
X59314,IGKV2-40*01,O11                   TGG TAC CTG CAG AAG CCA GGG CAG TCT CCA CAG CTC CTG ATC TAT ACG CTT TCC ... ...
                                          C
X59317,IGKV2-40*02,O11a                  --T --- --- --- --- --- --- --- --- --- --- --- --- --- --- --- --- --- ... ...
X59311,IGKV2D-40*01,O1                   --- --- --- --- --- --- --- --- --- --- --- --- --- --- --- --- --- --- ... ...
X93631,IGKV2-40*01/2D-40*01,DPK13        --- --- --- --- --- --- --- --- --- --- --- --- --- --- --- --- --- --- ... ...

                                         IMGT_______________
                                         61  62  63  64  65  66  67  68  69  70  71  72  73  74  75  76  77  78  79  80
                                                              Y   R   A   S   G   V   P       D   R   F   S   G   S   G
X59314,IGKV2-40*01,O11                   ... ... ... ... ... TAT CGG GCC TCT GGA GTC CCA ... GAC AGG TTC AGT GGC AGT GGG
                                                                                                              D
X59317,IGKV2-40*02,O11a                  ... ... ... ... ... --- --- --- --- --- --- --- ... --- --- --- --- -A- --- ---
X59311,IGKV2D-40*01,O1                   ... ... ... ... ... --- --- --- --- --- --- --- ... --- --- --- --- --- --- ---
X93631,IGKV2-40*01/2D-40*01,DPK13        ... ... ... ... ... --- --- --- --- --- --- --- ... --- --- --- --- --- --- ---

                                         81  82  83  84  85  86  87  88  89  90  91  92  93  94  95  96  97  98  99 100
                                                  S   G   T   D   F   T   L   K   I   S   R   V   E   A   E   D   V   G
X59314,IGKV2-40*01,O11                   ... ... TCA GGC ACT GAT TTC ACA CTG AAA ATC AGC AGG GTG GAG GCT GAG GAT GTT GGA
X59317,IGKV2-40*02,O11a                  ... ... --- --- --- --- --- --- --- --- --- --- --- --- --- --- --- --- --- ---
X59311,IGKV2D-40*01,O1                   ... ... --- --- --- --- --- --- --- --- --- --- --- --- --- --- --- --- --- ---
X93631,IGKV2-40*01/2D-40*01,DPK13        ... ... --- --- --- --- --- --- --- --- --- --- --- --- --- --- --- --- --- ---

                                                             ______CDR3-IMGT_________
                                        101 102 103 104 105 106 107 108 109 110 111
                                          V   Y   Y   C   M   Q   R   I   E   F   P
X59314,IGKV2-40*01,O11                   GTT TAT TAC TGC ATG CAA CGT ATA GAG TTT CCT TC
X59317,IGKV2-40*02,O11a                  --- --- --- --- --- --- --- --- --- --- --- --
X59311,IGKV2D-40*01,O1                   --- --- --- --- --- --- --- --- --- --- --- --
X93631,IGKV2-40*01/2D-40*01,DPK13        --- --- --- --- --- --- --- --- --- --- --- --
```

Framework and complementarity determining regions

FR1-IMGT: 26
FR2-IMGT: 17
FR3-IMGT: 36 (-3 aa: 73, 81, 82)

CDR1-IMGT: 12
CDR2-IMGT: 3
CDR3-IMGT: 7

Collier de Perles for human IGKV2-40*01 and IGKV2D-40*01

IGKV2-40*01
Accession number: IMGT X59314 EMBL/GenBank/DDBJ: X59314
IGKV2D-40*01
Accession number: IMGT X59311 EMBL/GenBank/DDBJ: X59311

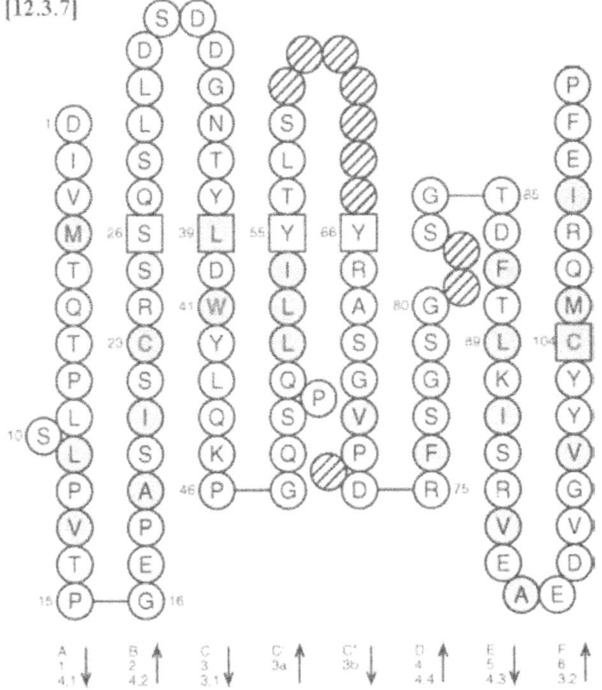

Genome database accession numbers

IGKV2-40:	GDB:9953488	LocusLink: 28916
IGKV2D-40:	GDB:9953564	LocusLink: 28878

Nomenclature

IGKV2D-24: Immunoglobulin kappa variable 2D-24.

Definition and functionality

IGKV2D-24 is one of the 8–10 functional genes of the IGKV2 subgroup which comprises 27 mapped genes and the IGK locus and nine orphons.

Gene location

IGKV2D-24 is in the distal cluster of the IGK locus on chromosome 2 at 2p11.2.

Nucleotide and amino acid sequences for human IGKV2D-24

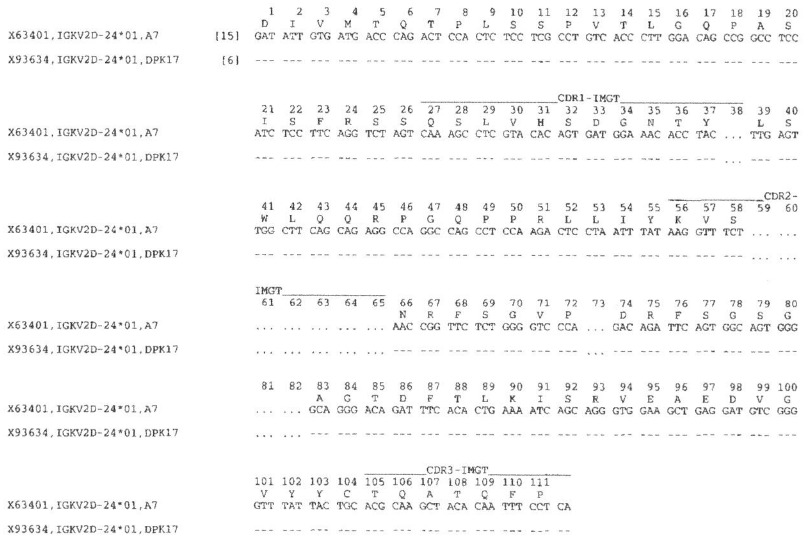

Framework and complementarity determining regions

FR1-IMGT: 26
FR2-IMGT: 17
FR3-IMGT: 36 (-3 aa: 73, 81, 82)

CDR1-IMGT: 11
CDR2-IMGT: 3
CDR3-IMGT: 7

Collier de Perles for human IGKV2D-24*01

Accession number: IMGT X63401 EMBL/GenBank/DDBJ: X63401

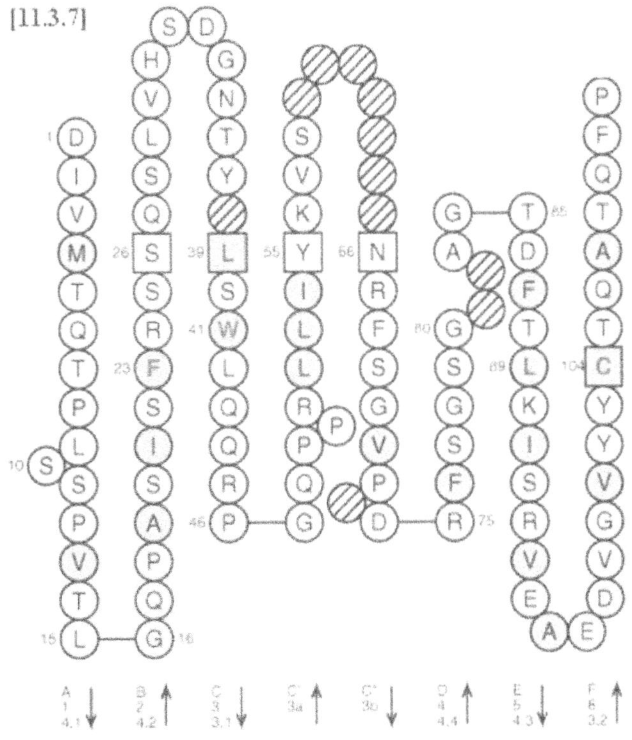

Genome database accession numbers

GDB:9953550 LocusLink: 28885

Nomenclature

IGKV2D-29: Immunoglobulin kappa variable 2D-29.

Definition and functionality

IGKV2D-29 is a functional gene (allele *01, allele *02*A1) or an ORF (allele *02*A2) of the IGKV2 subgroup which comprises 27 mapped genes (of which 8–10 are functional) in the IGK locus and nine orphons.
The IGKV2D-29*01 is functional. The IGKV2D-29*02 is functional (*02*A1) or an ORF (*02*A2), depending on the V-HEPTAMER sequence to which it is associated. Indeed, IGKV2D-29*02*A1 is associated with a canonical V-HEPTAMER, whereas IGKV2D-29*02*A2 is an ORF due to a non-canonical V-HEPTAMER (cacagag instead of cacagtg).

Gene location

IGKV2D-29 is in the distal cluster of the IGK locus on chromosome 2 at 2p11.2.

Nucleotide and amino acid sequences for human IGKV2D-29

```
                                     1   2   3   4   5   6   7   8   9  10  11  12  13  14  15  16  17  18  19  20
                                     D   I   V   M   T   Q   T   P   L   S   L   S   V   T   P   G   Q   P   A   S
M31952,IGKV2D-29*01,A2/A2a    [25]  GAT ATT GTG ATG ACC CAG ACT CCA CTC TCT CTG TCC GTC ACC CCT GGA CAG CCG GCC TCC
X93630,IGKV2D-29*01,DPK12     [6]   --- --- --- --- --- --- --- --- --- --- --- --- --- --- --- --- --- --- --- --
U41644,IGKV2D-29*02*A1,A2c    [1]   --- --- --- --- --- --- --- --- --- --- --- --- --- --- --- --- --- --- --- ---
U41643,IGKV2D-29*02*A2,A2b    [1]   --- --- --- --- --- --- --- --- --- --- --- --- --- --- --- --- --- --- --- ---

                                                            ____________CDR1-IMGT______________
                                    21  22  23  24  25  26  27  28  29  30  31  32  33  34  35  36  37  38  39  40
                                     I   S   C   K   S   S   Q   S   L   L   H   S   D   G   K   T   Y       L   Y
M31952,IGKV2D-29*01,A2/A2a          ATC TCC TGC AAG TCT AGT CAG AGC CTC CTG CAT AGT GAT GGA AAG ACC TAT ... TTG TAT
X93630,IGKV2D-29*01,DPK12           --- --- --- --- --- --- --- --- --- --- --- --- --- --- --- --- --- ... --- ---
U41644,IGKV2D-29*02*A1,A2c          --- --- --- --- --- --- --- --- --- --- --- --- --- --- --- --- --- ... --- ---
U41643,IGKV2D-29*02*A2,A2b          --- --- --- --- --- --- --- --- --- --- --- --- --- --- --- --- --- ... --- ---

                                                                                            ___________CDR2-
                                    41  42  43  44  45  46  47  48  49  50  51  52  53  54  55  56  57  58  59  60
                                     W   Y   L   Q   K   P   G   Q   P   P   Q   L   L   I   Y   E   V   S
M31952,IGKV2D-29*01,A2/A2a          TGG TAC CTG CAG AAG CCA GGC CAG CCT CCA CAG CTC CTG ATC TAT GAA GTT TCC ... ...
X93630,IGKV2D-29*01,DPK12           --- --- --- --- --- --- --- --- --- --- --- --- --- --- --- --- --- --- ... ...
                                                                     S
U41644,IGKV2D-29*02*A1,A2c          --- --- --- --- --- --- --- --- T-- --- --- --- --- --- --- --- --- --- ... ...
                                                                     S
U41643,IGKV2D-29*02*A2,A2b          --- --- --- --- --- --- --- --- T-- --- --- --- --- --- --- --- --- --- ... ...

                                    IMGT_______________
                                    61  62  63  64  65  66  67  68  69  70  71  72  73  74  75  76  77  78  79  80
                                                         N   R   F   S   G   V   P       D   R   F   S   G   S   G
M31952,IGKV2D-29*01,A2/A2a          ... ... ... ... ... AAC CGG TTC TCT GGA GTG CCA ... GAT AGG TTC AGT GGC AGC GGG
X93630,IGKV2D-29*01,DPK12           ... ... ... ... ... --- --- --- --- --- --- --- ... --- --- --- --- --- --- ---
U41644,IGKV2D-29*02*A1,A2c          ... ... ... ... ... --- --- --- --- --- --- --- ... --- --- --- --- --- --- ---
U41643,IGKV2D-29*02*A2,A2b          ... ... ... ... ... --- --- --- --- --- --- --- ... --- --- --- --- --- --- ---

                                    81  82  83  84  85  86  87  88  89  90  91  92  93  94  95  96  97  98  99 100
                                             S   G   T   D   F   T   L   K   I   S   R   V   E   A   E   D   V   G
M31952,IGKV2D-29*01,A2/A2a          ... ... TCA GGG ACA GAT TTC ACA CTG AAA ATC AGC CGG GTG GAG GCT GAG GAT GTT GGG
X93630,IGKV2D-29*01,DPK12           ... ... --- --- --- --- --- --- --- --- --- --- --- --- --- --- --- --- --- ---
U41644,IGKV2D-29*02*A1,A2c          ... ... --- --- --- --- --- --- --- --- --- --- --- --- --- --- --- --- --- ---
U41643,IGKV2D-29*02*A2,A2b          ... ... --- --- --- --- --- --- --- --- --- --- --- --- --- --- --- --- --- ---

                                                    _______CDR3-IMGT__________
                                    101 102 103 104 105 106 107 108 109 110 111
                                     V   Y   Y   C   M   Q   S   I   Q   L   P
M31952,IGKV2D-29*01,A2/A2a          GTT TAT TAC TGC ATG CAA AGT ATA CAG CTT CCT CC
X93630,IGKV2D-29*01,DPK12           --- --- --- --- --- --- --- --- --- --- --- --
U41644,IGKV2D-29*02*A1,A2c          --- --- --- --- --- --- --- --- --- --- --- --
U41643,IGKV2D-29*02*A2,A2b          --- --- --- --- --- --- --- --- --- --- --- --
```

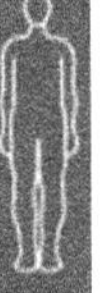

Framework and complementarity determining regions

FR1-IMGT: 26
FR2-IMGT: 17
FR3-IMGT: 36 (-3 aa: 73, 81, 82)

CDR1-IMGT: 11
CDR2-IMGT: 3
CDR3-IMGT: 7

Collier de Perles for human IGKV2D-29*01

Accession number: IMGT M31952 EMBL/GenBank/DDBJ: M31952

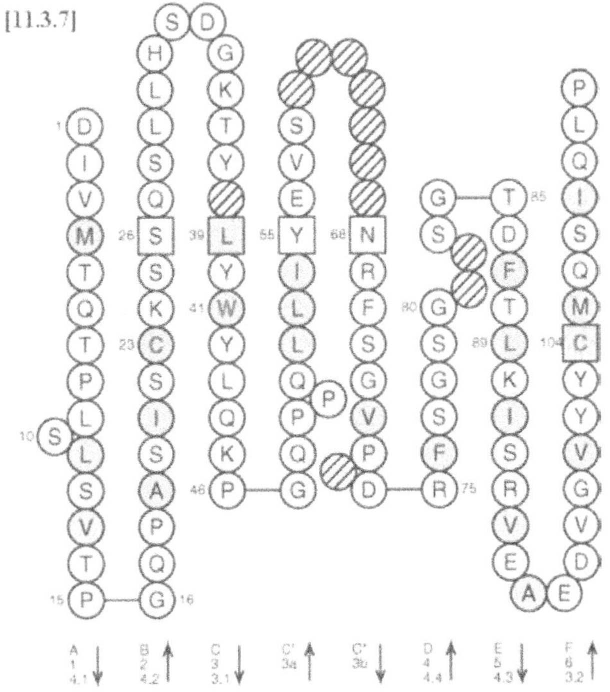

Genome database accession numbers

GDB:9953556 LocusLink: 28882

IGKV2D-30

Nomenclature

IGKV2D-30: Immunoglobulin kappa variable 2D-30.

Definition and functionality

IGKV2D-30 is one of the 8–10 functional genes of the IGKV2 subgroup which comprises 27 mapped genes in the IGK locus and nine orphons.

Gene location

IGKV2D-30 is in the distal cluster of the IGK locus on chromosome 2 at 2p11.2.

Nucleotide and amino acid sequences for human IGKV2D-30

```
                                  1   2   3   4   5   6   7   8   9  10  11  12  13  14  15  16  17  18  19  20
                                  D   V   V   M   T   Q   S   P   L   S   L   P   V   T   L   G   Q   P   A   S
X63402,IGKV2D-30*01,A1    [15]  GAT GTT GTG ATG ACT CAG TCT CCA CTC TCC CTG CCC GTC ACC CTT GGA CAG CCG GCC TCC
X93636,IGKV2D-30*01,DPK19  [6]  --- --- --- --- --- --- --- --- --- --- --- --- --- --- --- --- --- --- --- ---

                                                        ___________________CDR1-IMGT___________________
                                 21  22  23  24  25  26  27  28  29  30  31  32  33  34  35  36  37  38  39  40
                                  I   S   C   R   S   S   Q   S   L   V   Y   S   D   G   N   T   Y       L   N
X63402,IGKV2D-30*01,A1          ATC TCC TGC AGG TCT AGT CAA AGC CTC GTA TAC AGT GAT GGA AAC ACC TAC ... TTG AAT
X93636,IGKV2D-30*01,DPK19       --- --- --- --- --- --- --- --- --- --- --- --- --- --- --- --- --- ... --- ---

                                                                                            ______________CDR2-
                                 41  42  43  44  45  46  47  48  49  50  51  52  53  54  55  56  57  58  59  60
                                  W   F   Q   Q   R   P   G   Q   S   P   R   R   L   I   Y   K   V   S
X63402,IGKV2D-30*01,A1          TGG TTT CAG CAG AGG CCA GGC CAA TCT CCA AGG CGC CTA ATT TAT AAG GTT TCT ... ...
X93636,IGKV2D-30*01,DPK19       --- --- --- --- --- --- --- --- --- --- --- --- --- --- --- --- --- --- ... ...

                                IMGT_______________
                                 61  62  63  64  65  66  67  68  69  70  71  72  73  74  75  76  77  78  79  80
                                                      N   W   D   S   G   V   P       D   R   F   S   G   S   G
X63402,IGKV2D-30*01,A1          ... ... ... ... ... AAC TGG GAC TCT GGG GTC CCA ... GAC AGA TTC AGC GGC AGT GGG
X93636,IGKV2D-30*01,DPK19       ... ... ... ... ... --- --- --- --- --- --- --- ... --- --- --- --- --- --- ---

                                 81  82  83  84  85  86  87  88  89  90  91  92  93  94  95  96  97  98  99 100
                                          S   G   T   D   F   T   L   K   I   S   R   V   E   A   E   D   V   G
X63402,IGKV2D-30*01,A1          ... ... TCA GGC ACT GAT TTC ACA CTG AAA ATC AGC AGG GTG GAG GCT GAG GAT GTT GGG
X93636,IGKV2D-30*01,DPK19       ... ... --- --- --- --- --- --- --- --- --- --- --- --- --- --- --- --- --- ---

                                                __________CDR3-IMGT___________
                                101 102 103 104 105 106 107 108 109 110 111
                                  V   Y   Y   C   M   Q   G   T   H   W   P
X63402,IGKV2D-30*01,A1          GTT TAT TAC TGC ATG CAA GGT ACA CAC TGG CCT CC
X93636,IGKV2D-30*01,DPK19       --- --- --- --- --- --- --- --- --- --- --- --
```

Framework and complementarity determining regions

FR1-IMGT: 26
FR2-IMGT: 17
FR3-IMGT: 36 (-3 aa: 73, 81, 82)

CDR1-IMGT: 11
CDR2-IMGT: 3
CDR3-IMGT: 7

Collier de Perles for human IGKV2D-30*01

Accession number: IMGT X63402 EMBL/GenBank/DDBJ: X63402

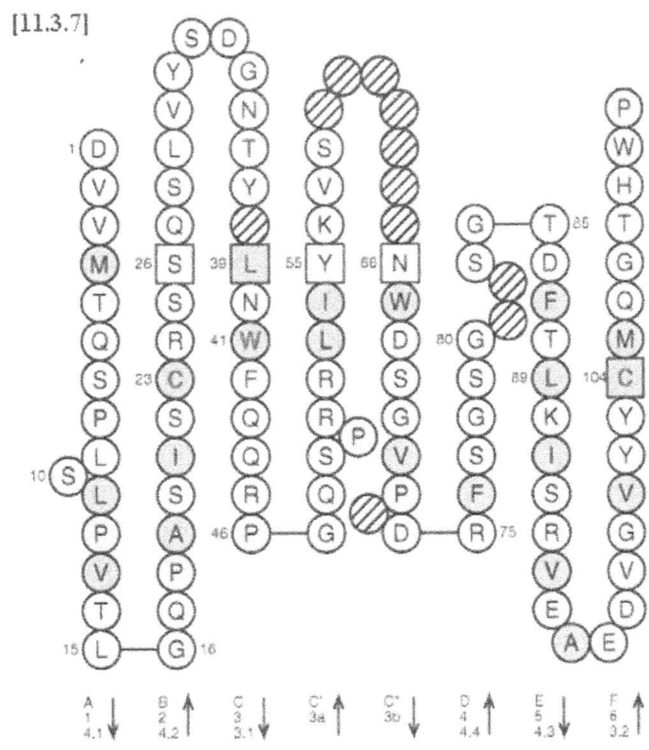

Genome database accession numbers

GDB:9953558 LocusLink: 28881

IGKV3-7

Nomenclature

IGKV3-7: Immunoglobulin kappa variable 3-7.

Definition and functionality

IGKV3-7 is the only ORF gene of the IGKV3 subgroup which comprises 14 mapped genes (of which six–seven are functional) in the IGK locus and three–four orphons.
The IGKV3-7 is an ORF due to the mutation in the ACCEPTOR_SPLICE (nggnn instead of nagnn)[24].

Gene location

IGKV3-7 is in the proximal cluster of the IGK locus on chromosome 2 at 2p11.2.

Nucleotide and amino acid sequences for human IGKV3-7

```
                                   1   2   3   4   5   6   7   8   9  10  11  12  13  14  15  16  17  18  19  20
                                   E   I   V   M   T   Q   S   P   P   T   L   S   L   S   P   G   E   R   V   T
X02725,IGKV3-7*01,L10      [19]  GAA ATT GTA ATG ACA CAG TCT CCA CCC ACC CTG TCT TTG TCT CCA GGG GAA AGA GTC ACC

X72812,IGKV3-7*02,L10a     [10]  --- --- --- --- --- --- --- --- --- --- --- --- --- --- --- --- --- --- --- ---

K02769,IGKV3-7*03,Vh       [19]  --- --- --- --- --- --- --- --- --- --- --- --- --- --- --- --- --- --- --- ---

                                                         ________________CDR1-IMGT______________________
                                  21  22  23  24  25  26  27  28  29  30  31  32  33  34  35  36  37  38  39  40
                                   L   S   C   R   A   S   Q   S   V   S   S   S   Y                       L   T
X02725,IGKV3-7*01,L10            CTC TCC TGC AGG GCC AGT CAG AGT GTT AGC AGC AGC TAC ... ... ... ... ... TTA ACC
                                                                                                           S
X72812,IGKV3-7*02,L10a           --- --- --- --- --- --- --- --- --- --- --- --- --- ... ... ... ... ... --- T--

K02769,IGKV3-7*03,Vh             --- --- --- --- --- --- --- --- --- --- --- --- --- ... ... ... ... ... --- ---

                                                                                             ___________CDR2-
                                  41  42  43  44  45  46  47  48  49  50  51  52  53  54  55  56  57  58  59  60
                                   W   Y   Q   Q   K   P   G   Q   A   P   R   L   L   I   Y   G   A   S
X02725,IGKV3-7*01,L10            TGG TAT CAG CAG AAA CCT GGC CAG GCG CCC AGG CTC CTC ATC TAT GGT GCA TCC ... ...

X72812,IGKV3-7*02,L10a           --- --C --- --- --- --- --G --- --T --- --- --- --- --- --- --- --- --- ... ...

K02769,IGKV3-7*03,Vh             --- --- --- --- --- --- --- --- --- --- --- --- --- --- --- --- --- --- ... ...

                                 IMGT_______________
                                  61  62  63  64  65  66  67  68  69  70  71  72  73  74  75  76  77  78  79  80
                                                       T   R   A   T   S   I   P       A   R   F   S   G   S   G
X02725,IGKV3-7*01,L10            ... ... ... ... ... ACC AGG GCC ACT AGC ATC CCA ... GCC AGG TTC AGT GGC AGT GGG
                                                                       G
X72812,IGKV3-7*02,L10a           ... ... ... ... ... --- --- --- --- G-- --- --- ... --- --- --- --- --- --- ---

K02769,IGKV3-7*03,Vh             ... ... ... ... ... --- --- --- --- --- --- --- ... --- --- --- --- --- --- ---

                                  81  82  83  84  85  86  87  88  89  90  91  92  93  94  95  96  97  98  99 100
                                           S   G   T   D   F   T   L   T   I   S   S   L   Q   P   E   D   F   A
X02725,IGKV3-7*01,L10            ... ... TCT GGG ACA GAC TTC ACT CTC ACC ATC AGC AGC CTG CAG CCT GAA GAT TTT GCA

X72812,IGKV3-7*02,L10a           ... ... --- --- --- --- --- --- --- --- --- --- --- --- --- --- --- --- --- ---
                                                   R
K02769,IGKV3-7*03,Vh             ... ... --- --- -G- --- --- --- --- --- --- --- --- --- --- --- --- --- --- ---

                                                 __________CDR3-IMGT_____________
                                 101 102 103 104 105 106 107 108 109 110 111
                                   V   Y   Y   C   Q   Q   D   H   N   L   P
X02725,IGKV3-7*01,L10            GTT TAT TAC TGT CAG CAG GAT CAT AAC TTA CCT CC
                                                               Y
X72812,IGKV3-7*02,L10a           --- --- --- --- --- --- --- T-- --- --- --- --

K02769,IGKV3-7*03,Vh             --- --- --- --- --- --- --- --- --- --- --- --
```

Framework and complementarity determining regions

FR1-IMGT: 26
FR2-IMGT: 17
FR3-IMGT: 36 (-3 aa: 73, 81, 82)

CDR1-IMGT: 7
CDR2-IMGT: 3
CDR3-IMGT: 7

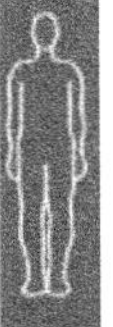

Collier de Perles for human IGKV3-7*01

Accession number: IMGT X02725 EMBL/GenBank/DDBJ: X02725

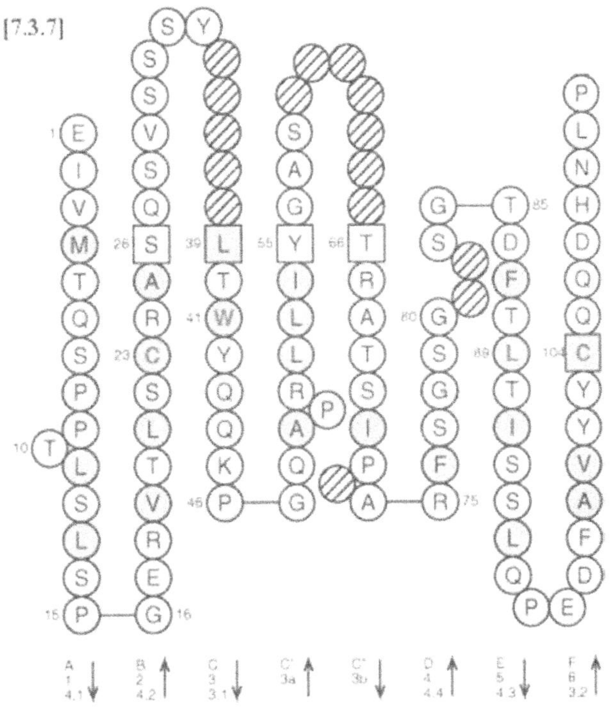

Genome database accession numbers

GDB:9953490 LocusLink: 28915

IGKV3-11

Nomenclature

IGKV3-11: Immunoglobulin kappa variable 3-11.

Definition and functionality

IGKV3-11 is one of the six–seven functional genes of the IGKV3 subgroup which comprises 14 mapped genes in the IGK locus and three–four orphons.

Gene location

IGKV3-11 is in the proximal cluster of the IGK locus on chromosome 2 at 2p11.2.

Nucleotide and amino acid sequences for human IGKV3-11

```
                                     1   2   3   4   5   6   7   8   9  10  11  12  13  14  15  16  17  18  19  20
                                     E   I   V   L   T   Q   S   P   A   T   L   S   L   S   P   G   E   R   A   T
X01668,IGKV3-11*01,L6          [19] GAA ATT GTG TTG ACA CAG TCT CCA GCC ACC CTG TCT TTG TCT CCA GGG GAA AGA GCC ACC

X92342,IGKV3-11*01,L6a         [13] --- --- --- --- --- --- --- --- --- --- --- --- --- --- --- --- --- --- --- ---

L37726,IGKV3-11*01,13K20/13K23 [11] --- --- --- --- --- --- --- --- --- --- --- --- --- --- --- --- --- --- --- ---

K02768,IGKV3-11*02,Vg          [19] --- --- --- --- --- --- --- --- --- --- --- --- --- --- --- --- --- --- --- ---

                                                             ________________CDR1-IMGT_______________
                                    21  22  23  24  25  26  27  28  29  30  31  32  33  34  35  36  37  38  39  40
                                     L   S   C   R   A   S   Q   S   V   S   S   Y                           L   A
X01668,IGKV3-11*01,L6               CTC TCC TGC AGG GCC AGT CAG AGT GTT AGC AGC TAC ... ... ... ... ... ... TTA GCC

X92342,IGKV3-11*01,L6a              --- --- --- --- --- --- --- --- --- --- --- --- ... ... ... ... ... ... --- ---

L37726,IGKV3-11*01,13K20/13K23      --- --- --- --- --- --- --- --- --- --- --- --- ... ... ... ... ... ... --- ---

K02768,IGKV3-11*02,Vg               --- --- --- --- --- --- --- --- --- --- --- --- ... ... ... ... ... ... --- ---

                                                                                           ____________CDR2-
                                    41  42  43  44  45  46  47  48  49  50  51  52  53  54  55  56  57  58  59  60
                                     W   Y   Q   Q   K   P   G   Q   A   P   R   L   L   I   Y   D   A   S
X01668,IGKV3-11*01,L6               TGG TAC CAA CAG AAA CCT GGC CAG GCT CCC AGG CTC CTC ATC TAT GAT GCA TCC ... ...

X92342,IGKV3-11*01,L6a              --- --- --- --- --- --- --- --- --- --- --- --- --- --- --- --- --- --- ... ...

L37726,IGKV3-11*01,13K20/13K23      --- --- --- --- --- --- --- --- --- --- --- --- --- --- --- --- --- --- ... ...

K02768,IGKV3-11*02,Vg               --- --- --- --- --- --- --- --- --- --- --- --- --- --- --- --- --- --- ... ...

                                    IMGT_______________
                                    61  62  63  64  65  66  67  68  69  70  71  72  73  74  75  76  77  78  79  80
                                                         N   R   A   T   G   I   P       A   R   F   S   G   S   G
X01668,IGKV3-11*01,L6               ... ... ... ... ... AAC AGG GCC ACT GGC ATC CCA ... GCC AGG TTC AGT GGC AGT GGG

X92342,IGKV3-11*01,L6a              ... ... ... ... ... --- --- --- --- --- --- --- ... --- --- --- --- --- --- ---

L37726,IGKV3-11*01,13K20/13K23      ... ... ... ... ... --- --- --- --- --- --- --- ... --- --- --- --- --- --- ---

K02768,IGKV3-11*02,Vg               ... ... ... ... ... --- --- --- --- --- --- --- ... --- --- --- --- --- --- ---

                                    81  82  83  84  85  86  87  88  89  90  91  92  93  94  95  96  97  98  99 100
                                             S   G   T   D   F   T   L   T   I   S   S   L   E   P   E   D   F   A
X01668,IGKV3-11*01,L6               ... ... TCT GGG ACA GAC TTC ACT CTC ACC ATC AGC AGC CTA GAG CCT GAA GAT TTT GCA

X92342,IGKV3-11*01,L6a              ... ... --- --- --- --- --- --- --- --- --- --- --- --- --- --- --- --- --- ---

L37726,IGKV3-11*01,13K20/13K23      ... ... --- --- --- --- --- --- --- --- --- --- --- --- --- --- --- --- --- ---
                                                     R
K02768,IGKV3-11*02,Vg               ... ... --- --- -G- --- --- --- --- --- --- --- --- --- --- --- --- --- --- ---

                                                    __________CDR3-IMGT____________
                                    101 102 103 104 105 106 107 108 109 110 111
                                     V   Y   Y   C   Q   Q   R   S   N   W   P
X01668,IGKV3-11*01,L6               GTT TAT TAC TGT CAG CAG CGT AGC AAC TGG CCT CC

X92342,IGKV3-11*01,L6a              --- --- --- --- --- --- --- --- --- --- --- --

L37726,IGKV3-11*01,13K20/13K23      --- --- --- --- --- -

K02768,IGKV3-11*02,Vg               --- --- --- --- --- --- --- --- --- --- --- --
```

Framework and complementarity determining regions

FR1-IMGT: 26
FR2-IMGT: 17
FR3-IMGT: 36 (-3 aa: 73, 81, 82)

CDR1-IMGT: 6
CDR2-IMGT: 3
CDR3-IMGT: 7

Collier de Perles for human IGKV3-11*01

Accession number: IMGT X01668 EMBL/GenBank/DDBJ: X01668

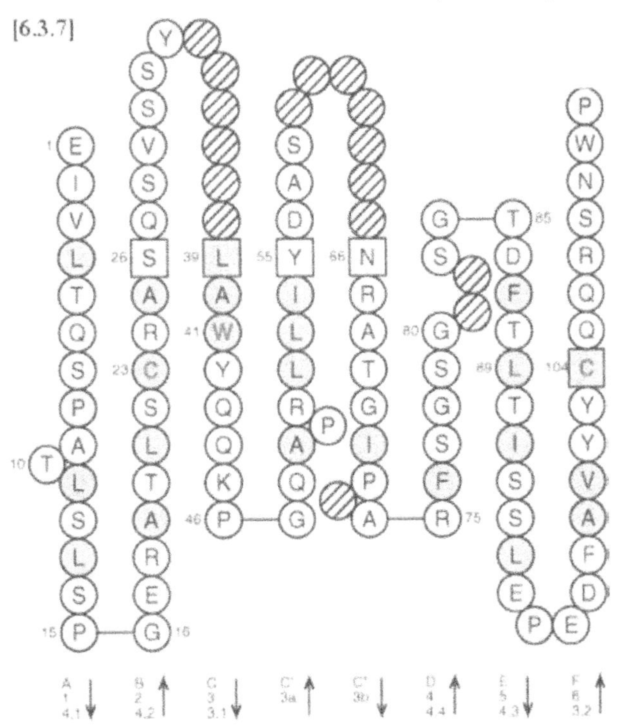

Genome database accession numbers

GDB:9953492 LocusLink: 28914

Nomenclature

IGKV3-15: Immunoglobulin kappa variable 3-15.

Definition and functionality

IGKV3-15 is one of the six–seven functional genes of the IGKV3 subgroup which comprises 14 mapped genes in the IGK locus and three–four orphons.

Gene location

IGKV3-15 is in the proximal cluster of the IGK locus on chromosome 2 at 2p11.2.

Nucleotide and amino acid sequences for human IGKV3-15

```
                                     1   2   3   4   5   6   7   8   9  10  11  12  13  14  15  16  17  18  19  20
                                     E   I   V   M   T   Q   S   P   A   T   L   S   V   S   P   G   E   R   A   T
M23090,IGKV3-15*01,L2        [16]   GAA ATA GTG ATG ACG CAG TCT CCA GCC ACC CTG TCT GTG TCT CCA GGG GAA AGA GCC ACC
X93638,IGKV3-15*01,DPK21     [6]    --- --- --- --- --- --- --- --- --- --- --- --- --- --- --- --- --- --- --- ---

                                                             _____________CDR1-IMGT__________________
                                    21  22  23  24  25  26  27  28  29  30  31  32  33  34  35  36  37  38  39  40
                                     L   S   C   R   A   S   Q   S   V   S   S   N                           L   A
M23090,IGKV3-15*01,L2               CTC TCC TGC AGG GCC AGT CAG AGT GTT AGC AGC AAC ... ... ... ... ... ... TTA GCC
X93638,IGKV3-15*01,DPK21            --- --- --- --- --- --- --- --- --- --- --- --- ... ... ... ... ... ... --- ---

                                                                                                _____________CDR2-
                                    41  42  43  44  45  46  47  48  49  50  51  52  53  54  55  56  57  58  59  60
                                     W   Y   Q   Q   K   P   G   Q   A   P   R   L   L   I   Y   G   A   S
M23090,IGKV3-15*01,L2               TGG TAC CAG CAG AAA CCT GGC CAG GCT CCC AGG CTC CTC ATC TAT GGT GCA TCC ... ...
X93638,IGKV3-15*01,DPK21            --- --- --- --- --- --- --- --- --- --- --- --- --- --- --- --- --- --- ... ...

                                    IMGT________________
                                    61  62  63  64  65  66  67  68  69  70  71  72  73  74  75  76  77  78  79  80
                                                         T   R   A   T   G   I   P       A   R   F   S   G   S   C
M23090,IGKV3-15*01,L2               ... ... ... ... ... ACC AGG GCC ACT GGT ATC CCA ... GCC AGG TTC AGT GGC AGT GGG
X93638,IGKV3-15*01,DPK21            ... ... ... ... ... --- --- --- --- --- --- --- ... --- --- --- --- --- --- ---

                                    81  82  83  84  85  86  87  88  89  90  91  92  93  94  95  96  97  98  99 100
                                             S   G   T   E   F   T   L   T   I   S   S   L   Q   S   E   D   F   A
M23090,IGKV3-15*01,L2               ... ... TCT GGG ACA GAG TTC ACT CTC ACC ATC AGC AGC CTG CAG TCT GAA GAT TTT GCA
X93638,IGKV3-15*01,DPK21            ... ... --- --- --- --- --- --- --- --- --- --- --- --- --- --- --- --- --- ---

                                                 __________CDR3-IMGT_____________
                                    101 102 103 104 105 106 107 108 109 110 111
                                     V   Y   Y   C   Q   Q   Y   N   N   W   P
M23090,IGKV3-15*01,L2               GTT TAT TAC TGT CAG CAG TAT AAT AAC TGG CCT CC
X93638,IGKV3-15*01,DPK21            --- --- --- --- --- --- --- --- --- --- --- --
```

Framework and complementarity determining regions

FR1-IMGT: 26
FR2-IMGT: 17
FR3-IMGT: 36 (-3 aa: 73, 81, 82)

CDR1-IMGT: 6
CDR2-IMGT: 3
CDR3-IMGT: 7

Collier de Perles for human IGKV3-15*01

Accession number: IMGT M23090 EMBL/GenBank/DDBJ: M23090

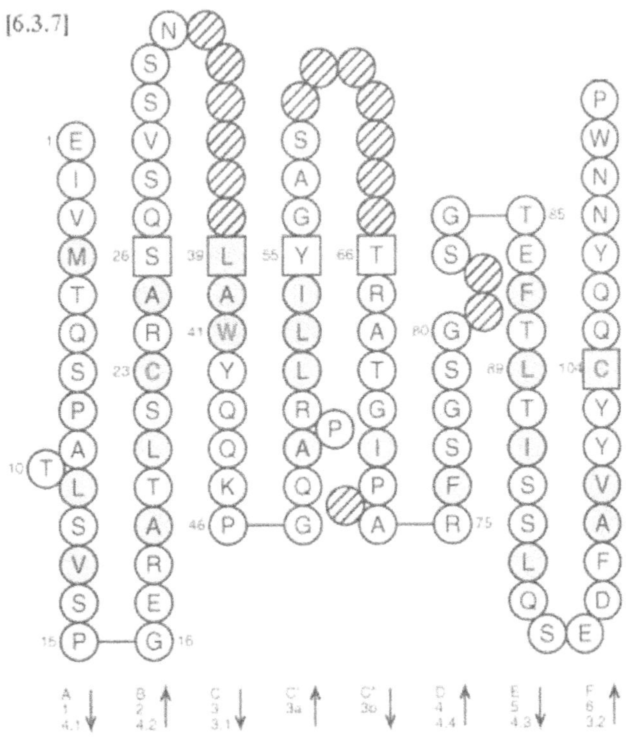

Genome database accession numbers

GDB:9953494 LocusLink: 28913

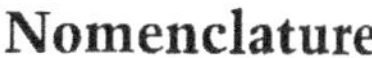

IGKV3-20

Nomenclature

IGKV3-20: Immunoglobulin kappa variable 3-20.

Definition and functionality

IGKV3-20 is one of the six–seven functional genes of the IGKV3 subgroup which comprises 14 mapped genes in the IGK locus and three–four orphons.

Gene location

IGKV3-20 is in the proximal cluster of the IGK locus on chromosome 2 at 2p11.2.

Nucleotide and amino acid sequences for human IGKV3-20

```
                                           1   2   3   4   5   6   7   8   9  10  11  12  13  14  15  16  17  18  19  20
                                           E   I   V   L   T   Q   S   P   G   T   L   S   L   S   P   G   E   R   A   T
X12686,IGKV3-20*01,A27               [29] GAA ATT CTG TTG ACG CAG TCT CCA GGC ACC CTG TCT TTG TCT CCA GGG GAA AGA GCC ACC
M15038,IGKV3-20*01,A27a              [22] --- --- --- --- --- --- --- --- --- --- --- --- --- --- --- --- --- --- --- ---
M15039,IGKV3-20*01,humkv321           [4]
X93639,IGKV3-20*01,DPK22              [6] --- --- --- --- --- --- --- --- --- --- --- --- --- --- --- --- --- --- --- ---
                                                                               P   C   L   C   L
X56593,IGKV3-20*01,Tou-kv325(1)      [31] --- --- --- --- --- --- --- --- --- C-- TGT CT- -GT CTC --- --- --- --- --- ---
                                                                           A
L37729,IGKV3-20*02,13K18             [11] --- --- --- --- --A --- --- --- -C- --- --- --- --- --- --- --- --- --- --- ---

                                                                  _______CDR1 IMGT_____________________
                                          21  22  23  24  25  26  27  28  29  30  31  32  33  34  35  36  37  38  39  40
                                           L   S   C   R   A   S   Q   S   V   S   S   S   Y                       L   A
X12686,IGKV3-20*01,A27                    CTC TCC TGC AGG GCC AGT CAG AGT GTT AGC AGC AGC TAC ... ... ... ... ... TTA GCC
M15038,IGKV3-20*01,A27a                   --- --- --- --- --- --- --- --- --- --- --- --- --- ... ... ... ... ... --- ---
M15039,IGKV3-20*01,humkv321
X93639,IGKV3-20*01,DPK22                  --- --- --- --- --- --- --- --- --- --- --- --- --- ... ... ... ... ... --- ---
X56593,IGKV3-20*01,Tou-kv325              --- --- --- --- --- --- --- --- --- --- --- --- --- ... ... ... ... ... --- ---
L37729,IGKV3-20*02,13K18                  --- --- --- --- --- --- --- --- --- --- --- --- --- ... ... ... ... ... --- ---

                                                                                                   ____________CDR2-
                                          41  42  43  44  45  46  47  48  49  50  51  52  53  54  55  56  57  58  59  60
                                           W   Y   Q   Q   K   P   G   Q   A   P   R   L   L   I   Y   G   A   S
X12686,IGKV3-20*01,A27                    TGG TAC CAG CAG AAA CCT GGC CAG GCT CCC AGG CTC CTC ATC TAT GGT GCA TCC ... ...
M15038,IGKV3-20*01,A27a                   --- --- --- --- --- --- --- --- --- --- --- --- --- --- --- --- --- --- ... ...
M15039,IGKV3-20*01,humkv321                                           --- --- --- --- --- --- --- --- --- --- --- ... ...
X93639,IGKV3-20*01,DPK22                  --- --- --- --- --- --- --- --- --- --- --- --- --- --- --- --- --- --- ... ...
X56593,IGKV3-20*01,Tou-kv325              --- --- --- --- --- --- --- --- --- --- --- --- --- --- --- --- --- --- ... ...
L37729,IGKV3-20*02,13K18                  --- --- --- --- --- --- --- --- --- --- --- --- --- --- --- --- --- --- ... ...

                                          IMGT_______________
                                          61  62  63  64  65  66  67  68  69  70  71  72  73  74  75  76  77  78  79  80
                                                               S   R   A   T   G   I   P       D   R   F   S   G   S   G
X12686,IGKV3-20*01,A27                    ... ... ... ... ... AGC AGG GCC ACT GGC ATC CCA ... GAC AGG TTC AGT GGC AGT GGG
M15038,IGKV3-20*01,A27a                   ... ... ... ... ... --- --- --- --- --- --- --- ... --- --- --- --- --- --- ---
M15039,IGKV3-20*01,humkv321               ... ... ... ... ... --- --- --- --- --- --- --- ... --- --- --- --- --- --- ---
X93639,IGKV3-20*01,DPK22                  ... ... ... ... ... --- --- --- --- --- --- --- ... --- --- --- --- --- --- ---
X56593,IGKV3-20*01,Tou-kv325              ... ... ... ... ... --- --- --- --- --- --- --- ... --- --- --- --- --- --- ---
                                                                                               A
L37729,IGKV3-20*02,13K18                  ... ... ... ... ... --- --- --- --- --- --- --- ... -CA --- --- --- --- --- ---

                                          81  82  83  84  85  86  87  88  89  90  91  92  93  94  95  96  97  98  99 100
                                                   S   G   T   D   F   T   L   T   I   S   R   L   E   P   E   D   F   A
X12686,IGKV3-20*01,A27                    ... ... TCT GGG ACA GAC TTC ACT CTC ACC ATC AGC AGA CTG GAG CCT GAA GAT TTT GCA
M15038,IGKV3-20*01,A27a                   ... ... --- --- --- --- --- --- --- --- --- --- --- --- --- --- --- --- --- ---
M15039,IGKV3-20*01,humkv321               ... ... --- --- --- --- --- --- --- --- --- --- --- --- --- --- --- --- --- ---
X93639,IGKV3-20*01,DPK22                  ... ... --- --- --- --- --- --- --- --- --- --- --- --- --- --- --- --- --- ---
X56593,IGKV3-20*01,Tou-kv325              ... ... --- --- --- --- --- --- --- --- --- --- --- --- --- --- --- --- --- ---
L37729,IGKV3-20*02,13K18                  ... ... --- --- --- --- --- --- --- --- --- --- --- --- --- --- --- --- --- ---
```

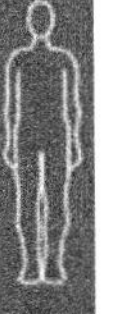

```
                                            ____________CDR3-IMGT____________
                               101 102 103 104 105 106 107 108 109 110 111
                                V   Y   Y   C   Q   Q   Y   G   S   S   P
X12686,IGKV3-20*01,A27         GTG TAT TAC TGT CAG CAG TAT GGT AGC TCA CCT CC
M15038,IGKV3-20*01,A27a        --- --- --- --- --- --- --- --- --- --- --- --
M15039,IGKV3-20*01,humkv321    --- --- --- --- --- --- --- --- --- --- --- --
X93639,IGKV3-20*01,DPK22       --- --- --- --- --- --- --- --- --- --- --- --
X56593,IGKV3-20*01,Tou-kv325   --- --- --- --- --- --- --- --- --- --- --
L37729,IGKV3-20*02,13K18       --T --- --- --- --- -
```

Note:

(1) DELETION of 1 nt (a) between nt 27 and 28, and INSERTION of 1 nt (c) between nt 42 and 43 in X56593 EMBL flat file are probably sequencing errors.

Framework and complementarity determining regions

FR1-IMGT: 26
FR2-IMGT: 17
FR3-IMGT: 36 (-3 aa: 73, 81, 82)

CDR1-IMGT: 7
CDR2-IMGT: 3
CDR3-IMGT: 7

Collier de Perles for human IGKV3-20*01

Accession number: IMGT X12686 EMBL/GenBank/DDBJ: X12686

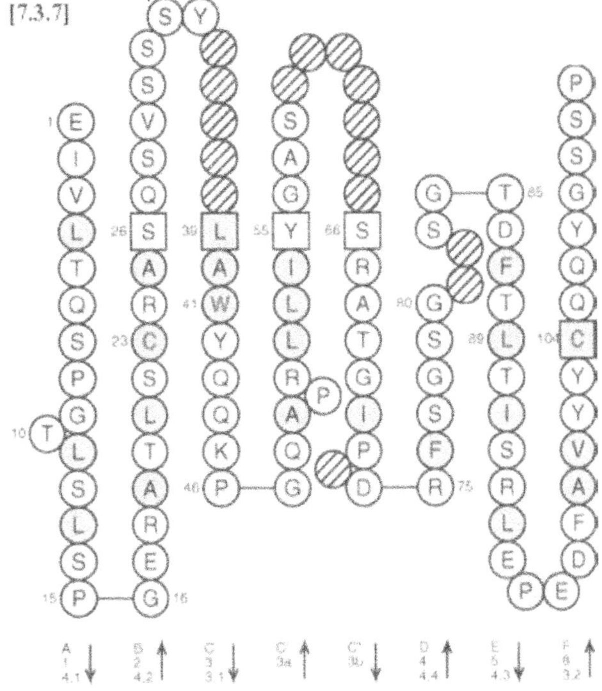

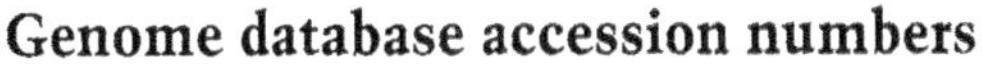

Genome database accession numbers

GDB:9953496 LocusLink: 28912

Nomenclature

IGKV3D-7: Immunoglobulin kappa variable 3D-7.

Definition and functionality

IGKV3D-7 is one of the six–seven functional genes of the IGKV3 subgroup which comprises 14 mapped genes in the IGK locus and three–four orphons.

Gene location

IGKV3D-7 is in the distal cluster of the IGK locus on chromosome 2 at 2p11.2.

Nucleotide and amino acid sequences for human IGKV3D-7

```
                                   1   2   3   4   5   6   7   8   9  10  11  12  13  14  15  16  17  18  19  20
                                   E   I   V   M   T   Q   S   P   A   T   L   S   L   S   P   G   E   R   A   T
X72820,IGKV3D-7*01,L25      [10]  GAA ATT GTA ATG ACA CAG TCT CCA GCC ACC CTG TCT TTG TCT CCA GGG GAA AGA GCC ACC
Z27500,IGKV3D-7*01,DPK23    [6]   --- --- --- --- --- --- --- --- --- --- --- --- --- --- --- --- --- --- --- ---

                                                          ____________________CDR1-IMGT____________________
                                  21  22  23  24  25  26  27  28  29  30  31  32  33  34  35  36  37  38  39  40
                                   L   S   C   R   A   S   Q   S   V   S   S   S   Y                       L   S
X72820,IGKV3D-7*01,L25            CTC TCC TGC AGG GCC AGT CAG AGT GTT AGC AGC AGC TAC ... ... ... ... ... TTA TCC
Z27500,IGKV3D-7*01,DPK23          --- --- --- --- --- --- --- --- --- --- --- --- --- ... ... ... ... ... --- ---

                                                                                              ____________CDR2-
                                  41  42  43  44  45  46  47  48  49  50  51  52  53  54  55  56  57  58  59  60
                                   W   Y   Q   Q   K   P   G   Q   A   P   R   L   L   I   Y   G   A   S
X72820,IGKV3D-7*01,L25            TGG TAC CAG CAG AAA CCT GGG CAG GCT CCC AGG CTC CTC ATC TAT GGT GCA TCC ... ...
Z27500,IGKV3D-7*01,DPK23          --- --- ---     --- --- --- --- --- --- --- --- --- --- --- --- --- --- ... ...

                                  IMGT_______________
                                  61  62  63  64  65  66  67  68  69  70  71  72  73  74  75  76  77  78  79  80
                                                       T   R   A   T   G   I   P       A   R   F   S   G   S   G
X72820,IGKV3D-7*01,L25            ... ... ... ... ... ACC AGG GCC ACT GGC ATC CCA ... GCC AGG TTC AGT GGC AGT GGG
Z27500,IGKV3D-7*01,DPK23          ... ... ... ... ... --- --- --- --- --- --- --- ... --- --- --- --- --- --- ---

                                  81  82  83  84  85  86  87  88  89  90  91  92  93  94  95  96  97  98  99 100
                                           S   G   T   D   F   T   L   T   I   S   S   L   Q   P   E   D   F   A
X72820,IGKV3D-7*01,L25            ... ... TCT GGG ACA GAC TTC ACT CTC ACC ATC AGC AGC CTG CAG CCT GAA GAT TTT GCA
Z27500,IGKV3D-7*01,DPK23          ... ... --- --- --- --- --- --- --- --- --- --- --- --- --- --- --- --- --- ---

                                                  _________CDR3-IMGT___________
                                  101 102 103 104 105 106 107 108 109 110 111
                                   V   Y   Y   C   Q   Q   D   Y   N   L   P
X72820,IGKV3D-7*01,L25            GTT TAT TAC TGT CAG CAG GAT TAT AAC TTA CCT CC
Z27500,IGKV3D-7*01,DPK23          --- --- --- --- --- --- --- --- --- --- --- --
```

Framework and complementarity determining regions

FR1-IMGT: 26	CDR1-IMGT: 7
FR2-IMGT: 17	CDR2-IMGT: 3
FR3-IMGT: 36 (-3 aa: 73, 81, 82)	CDR3-IMGT: 7

Collier de Perles for human IGKV3D-7*01

Accession number: IMGT X72820 EMBL/GenBank/DDBJ: X72820

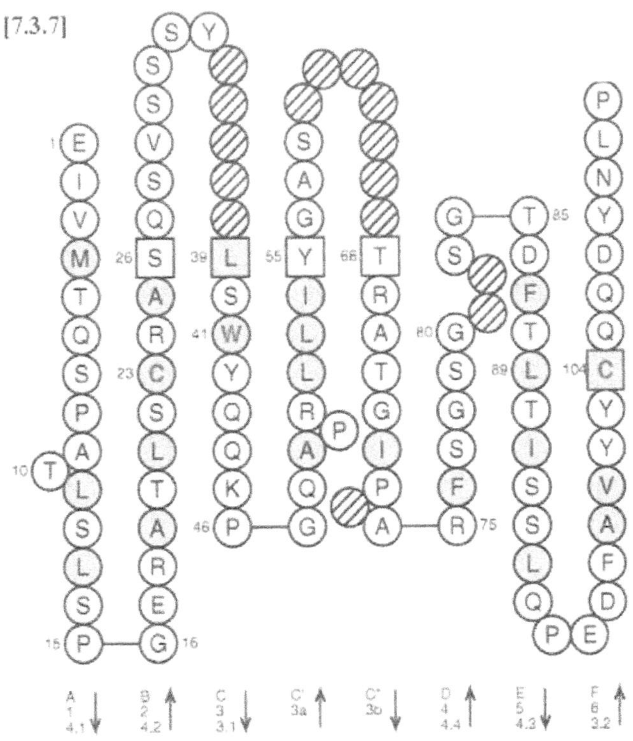

Genome database accession numbers

GDB:9953566 LocusLink: 28877

Nomenclature

IGKV3D-11: Immunoglobulin kappa variable 3D-11.

Definition and functionality

IGKV3D-11 is one of the six–seven functional genes of the IGKV3 subgroup which comprises 14 mapped genes in the IGK locus and three–four orphons.

Gene location

IGKV3D-11 is in the distal cluster of the IGK locus on chromosome 2 at 2p11.2.

Nucleotide and amino acid sequences for human IGKV3D-11

```
                                          1   2   3   4   5   6   7   8   9  10  11  12  13  14  15  16  17  18  19  20
                                          E   I   V   L   T   Q   S   P   A   T   L   S   L   S   P   G   E   R   A   T
X17264,IGKV3D-11*01,L20             [21] GAA ATT GTG TTG ACA CAG TCT CCA GCC ACC CTG TCT TTG TCT CCA GGG GAA AGA GCC ACC

L37725,IGKV3D-11*01,13K11/13K12     [11] --- --- --- --- --- --- --- --- --- --- --- --- --- --- --- --- --- --- --- ---

                                                              ________________CDR1-IMGT__________________
                                         21  22  23  24  25  26  27  28  29  30  31  32  33  34  35  36  37  38  39  40
                                          L   S   C   R   A   S   Q   G   V   S   S   Y                           L   A
X17264,IGKV3D-11*01,L20                  CTC TCC TGC AGG GCC AGT CAG GGT GTT AGC AGC TAC ... ... ... ... ... ... TTA GCC

L37725,IGKV3D-11*01,13K11/13K12          --- --- --- --- --- --- --- --- --- --- --- --- ... ... ... ... ... ... --- ---

                                                                                                  ___________CDR2-
                                         41  42  43  44  45  46  47  48  49  50  51  52  53  54  55  56  57  58  59  60
                                          W   Y   Q   Q   K   P   G   Q   A   P   R   L   L   I   Y   D   A   S
X17264,IGKV3D-11*01,L20                  TGG TAC CAG CAG AAA CCT GGC CAG GCT CCC AGG CTC CTC ATC TAT GAT GCA TCC ... ...

L37725,IGKV3D-11*01,13K11/13K12          --- --- --- --- --- --- --- --- --- --- --- --- --- --- --- --- --- --- ... ...

                                         IMGT_______________
                                         61  62  63  64  65  66  67  68  69  70  71  72  73  74  75  76  77  78  79  80
                                                              N   R   A   T   G   I   P       A   R   F   S   G   S   G
X17264,IGKV3D-11*01,L20                  ... ... ... ... ... AAC AGG GCC ACT GGC ATC CCA ... GCC AGG TTC AGT GGC AGT GGG

L37725,IGKV3D-11*01,13K11/13K12          ... ... ... ... ... --- --- --- --- --- --- --- ... --- --- --- --- --- --- ---

                                         81  82  83  84  85  86  87  88  89  90  91  92  93  94  95  96  97  98  99 100
                                                  P   G   T   D   F   T   L   T   I   S   S   L   E   P   E   D   F   A
X17264,IGKV3D-11*01,L20                  ... ... CCT GGG ACA GAC TTC ACT CTC ACC ATC AGC AGC CTA GAG CCT GAA GAT TTT GCA

L37725,IGKV3D-11*01,13K11/13K12          ... ... --- --- --- --- --- --- --- --- --- --- --- --- --- --- --- --- --- ---

                                                         __________CDR3-IMGT____________
                                         101 102 103 104 105 106 107 108 109 110 111
                                          V   Y   Y   C   Q   Q   R   S   N   W   H
X17264,IGKV3D-11*01,L20                  GTT TAT TAC TGT CAG CAG CGT AGC AAC TGG CAT CC

L37725,IGKV3D-11*01,13K11/13K12          --- --- --- --- --- -
```

Framework and complementarity determining regions

FR1-IMGT: 26
FR2-IMGT: 17
FR3-IMGT: 36 (-3 aa: 73, 81, 82)

CDR1-IMGT: 6
CDR2-IMGT: 3
CDR3-IMGT: 7

Collier de Perles for human IGKV3D-11*01

Accession number: IMGT X17264 EMBL/GenBank/DDBJ: X17264

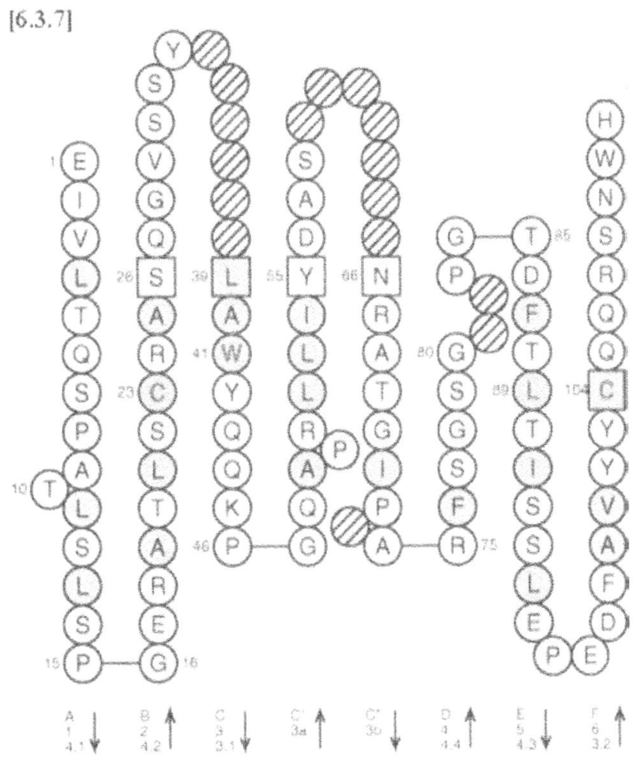

Genome database accession numbers

GDB:9953568 LocusLink: 28876

IGKV3D-15

Nomenclature

IGKV3D-15: Immunoglobulin kappa variable 3D-15.

Definition and functionality

IGKV3D-15 is functional (allele *01) or a pseudogene (allele *02) of the IGKV3 subgroup. This subgroup comprises 14 mapped genes (of which six–seven are functional) in the IGK locus and three–four orphons.
The IGKV3D-15*02 is a pseudogene due to a STOP-CODON at position 110 in the CDR3-IMGT.

Gene location

IGKV3D-15 is in the distal cluster of the IGK locus on chromosome 2 at 2p11.2.

Nucleotide and amino acid sequences for human IGKV3D-15

```
                                     1   2   3   4   5   6   7   8   9   10  11  12  13  14  15  16  17  18  19  20
                                     E   I   V   M   T   Q   S   P   A   T   L   S   V   S   P   G   E   R   A   T
X72815,IGKV3D-15*01,L16       [10]  GAA ATA GTG ATG ACG CAG TCT CCA GCC ACC CTG TCT GTG TCT CCA GGG GAA AGA GCC ACC
M23089,IGKV3D-15*01,L16a      [16]  --- --- --- --- --- --- --- --- --- --- --- --- --- --- --- --- --- --- --- ---
M23088,IGKV3D-15*01,L16b      [16]  --- --- --- --- --- --- --- --- --- --- --- --- --- --- --- --- --- --- --- ---
Y00640,IGKV3D-15*01,humkv31es [5]   --- --- --- --- --- --- --- --- --- --- --- --- --- --- --- --- --- --- --- ---
                                                    M
M23091,IGKV3D-15*02,L16c      [16]  --- --- --- --- T-- --- --- --- --- --- --- --- --- --- --- --- --- --- --- ---

                                                        ______________CDR1-IMGT_________________________
                                     21  22  23  24  25  26  27  28  29  30  31  32  33  34  35  36  37  38  39  40
                                     L   S   C   R   A   S   Q   S   V   S   S   N                           L   A
X72815,IGKV3D-15*01,L16             CTC TCC TGC AGG GCC AGT CAG AGT GTT AGC AGC AAC ... ... ... ... ... ... TTA GCC
M23089,IGKV3D-15*01,L16a            --- --- --- --- --- --- --- --- --- --- --- --- ... ... ... ... ... ... --- ---
M23088,IGKV3D-15*01,L16b            --- --- --- --- --- --- --- --- --- --- --- --- ... ... ... ... ... ... --- ---
Y00640,IGKV3D-15*01,humkv31es       --- --- --- --- --- --- --- --- --- --- --- --- ... ... ... ... ... ... --- ---
M23091,IGKV3D-15*02,L16c            --- --- --- --- --- --- --- --- --- --- --- --- ... ... ... ... ... ... --- ---

                                                                                    ________________CDR2-
                                     41  42  43  44  45  46  47  48  49  50  51  52  53  54  55  56  57  58  59  60
                                     W   Y   Q   Q   K   P   G   Q   A   P   R   L   L   I   Y   G   A   S
X72815,IGKV3D-15*01,L16             TGG TAC CAG CAG AAA CCT GGC CAG GCT CCC AGG CTC CTC ATC TAT GGT GCA TCC ... ...
M23089,IGKV3D-15*01,L16a            --- --- --- --- --- --- --- --- --- --- --- --- --- --- --- --- --- --- ... ...
M23088,IGKV3D-15*01,L16b            --- --- --- --- --- --- --- --- --- --- --- --- --- --- --- --- --- --- ... ...
Y00640,IGKV3D-15*01,humkv31es       --- --- --- --- --- --- --- --- --- --- --- --- --- --- --- --- --- --- ... ...
M23091,IGKV3D-15*02,L16c            --- --- --- --- --- --- --- --- --- --- --- --- --- --- --- --- --- --- ... ...

                                    IMGT________________
                                     61  62  63  64  65  66  67  68  69  70  71  72  73  74  75  76  77  78  79  80
                                                         T   R   A   T   G   I   P       A   R   F   S   G   S   G
X72815,IGKV3D-15*01,L16             ... ... ... ... ... ACC AGG GCC ACT GGC ATC CCA ... GCC AGG TTC AGT GGC AGT GGG
M23089,IGKV3D-15*01,L16a            ... ... ... ... ... --- --- --- --- --- --- --- ... --- --- --- --- --- --- ---
M23088,IGKV3D-15*01,L16b            ... ... ... ... ... --- --- --- --- --- --- --- ... --- --- --- --- --- --- ---
Y00640,IGKV3D-15*01,humkv31es       ... ... ... ... ... --- --- --- --- --- --- --- ... --- --- --- --- --- --- ---
M23091,IGKV3D-15*02,L16c            ... ... ... ... ... --- --- --- --- --- --- --- ... --- --- --- --- --- --- ---

                                     81  82  83  84  85  86  87  88  89  90  91  92  93  94  95  96  97  98  99 100
                                             S   G   T   E   F   T   L   T   I   S   S   L   Q   S   E   D   F   A
X72815,IGKV3D-15*01,L16             ... ... TCT GGG ACA GAG TTC ACT CTC ACC ATC AGC AGC CTG CAG TCT GAA GAT TTT GCA
M23089,IGKV3D-15*01,L16a            ... ... --- --- --- --- --- --- --- --- --- --- --- --- --- --- --- --- --- ---
M23088,IGKV3D-15*01,L16b            ... ... --- --- --- --- --- --- --- --- --- --- --- --- --- --- --- --- --- ---
Y00640,IGKV3D-15*01,humkv31es       ... ... --- --- --- --- --- --- --- --- --- --- --- --- --- --- --- --- --- ---
M23091,IGKV3D-15*02,L16c            ... ... --- --- --- --- --- --- --- --- --- --- --- --- --- --- --- --- --- ---

                                                    _______________CDR3-IMGT_______________
                                    101 102 103 104 105 106 107 108 109 110 111
                                     V   Y   Y   C   Q   Q   Y   N   N   W   P
X72815,IGKV3D-15*01,L16             GTT TAT TAC TGT CAG CAG TAT AAT AAC TGG CCT CC
M23089,IGKV3D-15*01,L16a            --- --- --- --- --- --- --- --- --- --- --- --
M23088,IGKV3D-15*01,L16b            --- --- --- --- --- --- --- --- --- --- --- --
Y00640,IGKV3D-15*01,humkv31es       --- --- --- --- --- --- --- --- --- --- --- --
                                                                          *
M23091,IGKV3D-15*02,L16c            --- --- --- --- --- --- --- --- --- --A --- --
```

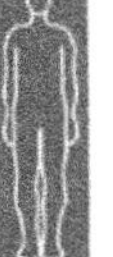

Framework and complementarity determining regions

FR1-IMGT: 26
FR2-IMGT: 17
FR3-IMGT: 36 (-3 aa: 73, 81, 82)

CDR1-IMGT: 6
CDR2-IMGT: 3
CDR3-IMGT: 7

Collier de Perles for human IGKV3D-15*01

Accession number: IMGT X72815 EMBL/GenBank/DDBJ: X72815

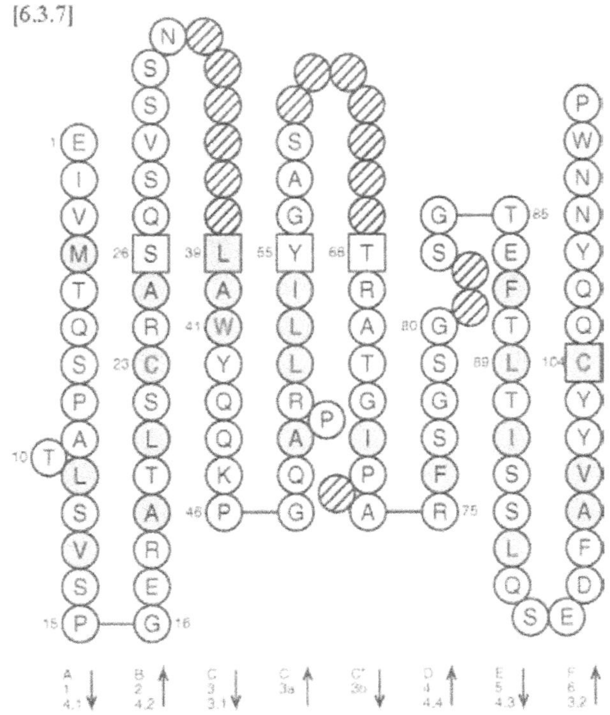

Genome database accession numbers

GDB:9953570 LocusLink: 28875

IGKV3D-20

Nomenclature

IGKV3D-20: Immunoglobulin kappa variable 3D-20.

Definition and functionality

IGKV3D-20 is one of the six–seven functional genes of the IGKV3 subgroup which comprises 14 mapped genes in the IGK locus and three–four orphons.

Gene location

IGKV3D-20 is in the distal cluster of the IGK locus on chromosome 2 at 2p11.2.

Nucleotide and amino acid sequences for human IGKV3D-20

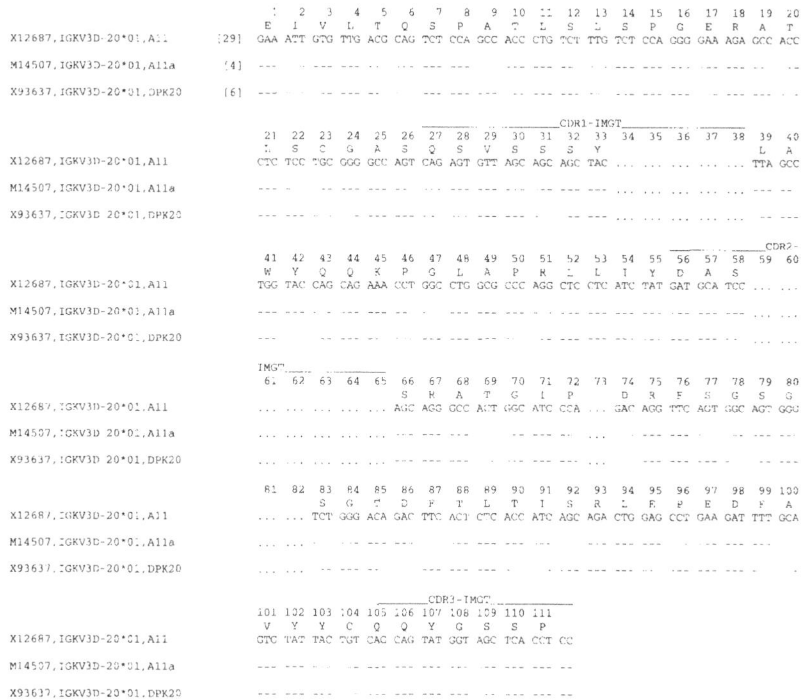

```
                                     1   2   3   4   5   6   7   8   9  10  11  12  13  14  15  16  17  18  19  20
                                     E   I   V   L   T   Q   S   P   A   T   L   S   L   S   P   G   E   R   A   T
X12687,IGKV3D-20*01,A11      [29]  GAA ATT GTG TTG ACG CAG TCT CCA GCC ACC CTG TCT TTG TCT CCA GGG GAA AGA GCC ACC
M14507,IGKV3D-20*01,A11a      [4]  --- --- --- --- --- --- --- --- --- --- --- --- --- --- --- --- --- --- --- ---
X93637,IGKV3D-20*01,DPK20     [6]  --- --- --- --- --- --- --- --- --- --- --- --- --- --- --- --- --- --- --- ---

                                                             ______________CDR1-IMGT______________
                                    21  22  23  24  25  26  27  28  29  30  31  32  33  34  35  36  37  38  39  40
                                     L   S   C   G   A   S   Q   S   V   S   S   S   Y                       L   A
X12687,IGKV3D-20*01,A11            CTC TCC TGC GGG GCC AGT CAG AGT GTT AGC AGC AGC TAC ... ... ... ... ... TTA GCC
M14507,IGKV3D-20*01,A11a           --- --- --- --- --- --- --- --- --- --- --- --- --- ... ... ... ... ... --- ---
X93637,IGKV3D-20*01,DPK20          --- --- --- --- --- --- --- --- --- --- --- --- --- ... ... ... ... ... --- ---

                                                                                             _________CDR2-
                                    41  42  43  44  45  46  47  48  49  50  51  52  53  54  55  56  57  58  59  60
                                     W   Y   Q   Q   K   P   G   L   A   P   R   L   L   I   Y   D   A   S
X12687,IGKV3D-20*01,A11            TGG TAC CAG CAG AAA CCT GGC CTG GCG CCC AGG CTC CTC ATC TAT GAT GCA TCC ... ...
M14507,IGKV3D-20*01,A11a           --- --- --- --- --- --- --- --- --- --- --- --- --- --- --- --- --- --- ... ...
X93637,IGKV3D-20*01,DPK20          --- --- --- --- --- --- --- --- --- --- --- --- --- --- --- --- --- --- ... ...

                                   IMGT_______________
                                    61  62  63  64  65  66  67  68  69  70  71  72  73  74  75  76  77  78  79  80
                                                         S   R   A   T   G   I   P       D   R   F   S   G   S   G
X12687,IGKV3D-20*01,A11            ... ... ... ... ... AGC AGG GCC ACT GGC ATC CCA ... GAC AGG TTC AGT GGC AGT GGG
M14507,IGKV3D-20*01,A11a           ... ... ... ... ... --- --- --- --- --- --- --- ... --- --- --- --- --- --- ---
X93637,IGKV3D-20*01,DPK20          ... ... ... ... ... --- --- --- --- --- --- --- ... --- --- --- --- --- --- ---

                                    81  82  83  84  85  86  87  88  89  90  91  92  93  94  95  96  97  98  99 100
                                             S   G   T   D   F   T   L   T   I   S   R   L   E   P   E   D   F   A
X12687,IGKV3D-20*01,A11            ... ... TCT GGG ACA GAC TTC ACT CTC ACC ATC AGC AGA CTG GAG CCT GAA GAT TTT GCA
M14507,IGKV3D-20*01,A11a           ... ... --- --- --- --- --- --- --- --- --- --- --- --- --- --- --- --- --- ---
X93637,IGKV3D-20*01,DPK20          ... ... --- --- --- --- --- --- --- --- --- --- --- --- --- --- --- --- --- ---

                                                 _________CDR3-IMGT__________
                                   101 102 103 104 105 106 107 108 109 110 111
                                     V   Y   Y   C   Q   Q   Y   G   S   S   P
X12687,IGKV3D-20*01,A11            GTC TAT TAC TGT CAC CAG TAT GGT AGC TCA CCT CC
M14507,IGKV3D-20*01,A11a           --- --- --- --- --- --- --- --- --- --- --- --
X93637,IGKV3D-20*01,DPK20          --- --- --- --- --- --- --- --- --- --- --- --
```

Framework and complementarity determining regions

FR1-IMGT: 26	CDR1-IMGT: 7
FR2-IMGT: 17	CDR2-IMGT: 3
FR3-IMGT: 36 (-3 aa: 73, 81, 82)	CDR3-IMGT: 7

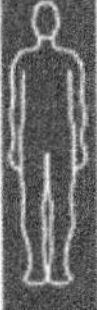

Collier de Perles for human IGKV3D-20*01

Accession number: IMGT X12687 EMBL/GenBank/DDBJ: X12687

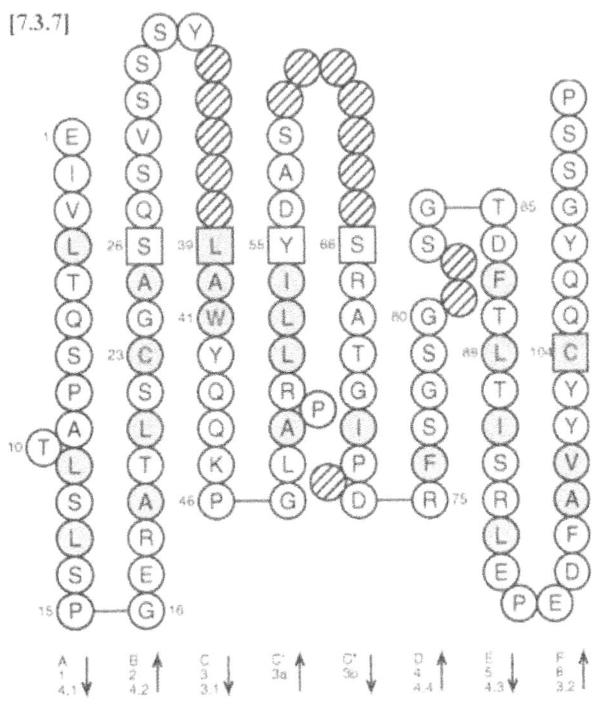

Genome database accession numbers

GDB:9953572 LocusLink: 28874

Nomenclature

IGKV4-1: Immunoglobulin kappa variable 4-1.

Definition and functionality

IGKV4-1 is the unique functional gene of the IGKV4 subgroup which only comprises one mapped gene in the IGK locus.

Gene location

IGKV4-1 is in the proximal cluster of the IGK locus on chromosome 2 at 2p11.2.

Nucleotide and amino acid sequences for human IGKV4-1

```
                                   1   2   3   4   5   6   7   8   9   10  11  12  13  14  15  16  17  18  19  20
                                   D   I   V   M   T   Q   S   P   D   S   L   A   V   S   L   G   E   R   A   T
Z00023,IGKV4-1*01,B3         [14]  GAC ATC GTG ATG ACC CAG TCT CCA GAC TCC CTG GCT GTG TCT CTG GGC GAG AGG GCC ACC
X93640,IGKV4-1*01,DPK24      [6]   --- --- --- --- --- --- --- --- --- --- --- --- --- --- --- --- --- --- --- ---
S56916,IGKV4-1*01,VkIV-GL    [30]  --- --- --- --- --- --- --- --- --- --- --- --- --- --- --- --- --- --- --- ---
Z46615,IGKV4-1*01,FSA10gl    [8]   --- --- --- --- --- --- --- --- --- --- --- --- --- --- --- --- --- --- --- ---
L33853,IGKV4-1*01            [7]   --- --- --- --- --- --- --- --- --- --- --- --- --- --- --- --- --- --- --- ---

                                                           _____________CDR1-IMGT_______________
                                   21  22  23  24  25  26  27  28  29  30  31  32  33  34  35  36  37  38  39  40
                                   I   N   C   K   S   S   Q   S   V   L   Y   S   S   N   N   K   N   Y   L   A
Z00023,IGKV4-1*01,B3               ATC AAC TGC AAG TCC AGC CAG AGT GTT TTA TAC AGC TCC AAC AAT AAG AAC TAC TTA GCT
X93640,IGKV4-1*01,DPK24            --- --- --- --- --- --- --- --- --- --- --- --- --- --- --- --- --- --- --- ---
S56916,IGKV4-1*01,VkIV-GL          --- --- --- --- --- --- --- --- --- --- --- --- --- --- --- --- --- --- --- ---
Z46615,IGKV4-1*01,FSA10gl          --- --- --- --- --- --- --- --- --- --- --- --- --- --- --- --- --- --- --- ---
L33853,IGKV4-1*01                  --- --- --- --- --- --- --- --- --- --- --- --- --- --- --- --- --- --- --- ---

                                                                                   ____________________CDR2-
                                   41  42  43  44  45  46  47  48  49  50  51  52  53  54  55  56  57  58  59  60
                                   W   Y   Q   Q   K   P   G   Q   P   P   K   L   L   I   Y   W   A   S
Z00023,IGKV4-1*01,B3               TGG TAC CAG CAG AAA CCA GGA CAG CCT CCT AAG CTG CTC ATT TAC TGG GCA TCT ... ...
X93640,IGKV4-1*01,DPK24            --- --- --- --- --- --- --- --- --- --- --- --- --- --- --- --- --- --- ... ...
S56916,IGKV4-1*01,VkIV-GL          --- --- --- --- --- --- --- --- --- --- --- --- --- --- --- --- --- --- ... ...
Z46615,IGKV4-1*01,FSA10gl          --- --- --- --- --- --- --- --- --- --- --- --- --- --- --- --- --- --- ... ...
L33853,IGKV4-1*01                  --- --- --- --- --- --- --- --- --- --- --- --- --- --- --- --- --- --- ... ...

                                   IMGT_______________
                                   61  62  63  64  65  66  67  68  69  70  71  72  73  74  75  76  77  78  79  80
                                                       T   R   E   S   G   V   P       D   R   F   S   G   S   G
Z00023,IGKV4-1*01,B3               ... ... ... ... ... ACC CGG GAA TCC GGG GTC CCT ... GAC CGA TTC AGT GGC AGC GGG
X93640,IGKV4-1*01,DPK24            ... ... ... ... ... --- --- --- --- --- --- --- ... --- --- --- --- --- --- ---
S56916,IGKV4-1*01,VkIV-GL          ... ... ... ... ... --- --- --- --- --- --- --- ... --- --- --- --- --- --- ---
Z46615,IGKV4-1*01,FSA10gl          ... ... ... ... ... --- --- --- --- --- --- --- ... --- --- --- --- --- --- ---
L33853,IGKV4-1*01                  ... ... ... ... ... --- --- --- --- --- --- --- ... --- --- --- --- --- --- ---

                                   81  82  83  84  85  86  87  88  89  90  91  92  93  94  95  96  97  98  99  100
                                           S   G   T   D   F   T   L   T   I   S   S   L   Q   A   E   D   V   A
Z00023,IGKV4-1*01,B3               ... ... TCT GGG ACA GAT TTC ACT CTC ACC ATC AGC AGC CTG CAG GCT GAA GAT GTG GCA
X93640,IGKV4-1*01,DPK24            ... ... --- --- --- --- --- --- --- --- --- --- --- --- --- --- --- --- --- ---
S56916,IGKV4-1*01,VkIV-GL          ... ... --- --- --- --- --- --- --- --- --- --- --- --- --- --- --- --- --- ---
Z46615,IGKV4-1*01,FSA10gl          ... ... --- --- --- --- --- --- --- --- --- --- --- --- --- --- --- --- --- ---
L33853,IGKV4-1*01                  ... ... --- --- --- --- --- --- --- --- --- --- --- --- --- --- --- --- --- ---

                                                       _________CDR3-IMGT___________
                                   101 102 103 104 105 106 107 108 109 110 111
                                   V   Y   Y   C   Q   Q   Y   Y   S   T   P
Z00023,IGKV4-1*01,B3               GTT TAT TAC TGT CAG CAA TAT TAT AGT ACT CCT CC
X93640,IGKV4-1*01,DPK24            --- --- --- --- --- --- --- --- --- --- --- --
S56916,IGKV4-1*01,VkIV-GL          --- --- --- --- --- --- --- --- --- --- --- --
Z46615,IGKV4-1*01,FSA10gl          --- --- --- --- --- --- --- --- --- --- ---
L33853,IGKV4-1*01                  --- --- --- --- --- --- --- --- --- --- ---
```

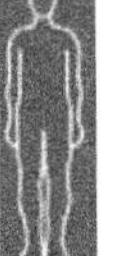

Framework and complementarity determining regions

FR1-IMGT: 26
FR2-IMGT: 17
FR3-IMGT: 36 (-3 aa: 73, 81, 82)

CDR1-IMGT: 12
CDR2-IMGT: 3
CDR3-IMGT: 7

Collier de Perles for human IGKV4-1*01

Accession number: IMGT Z00023 EMBL/GenBank/DDBJ: Z00023

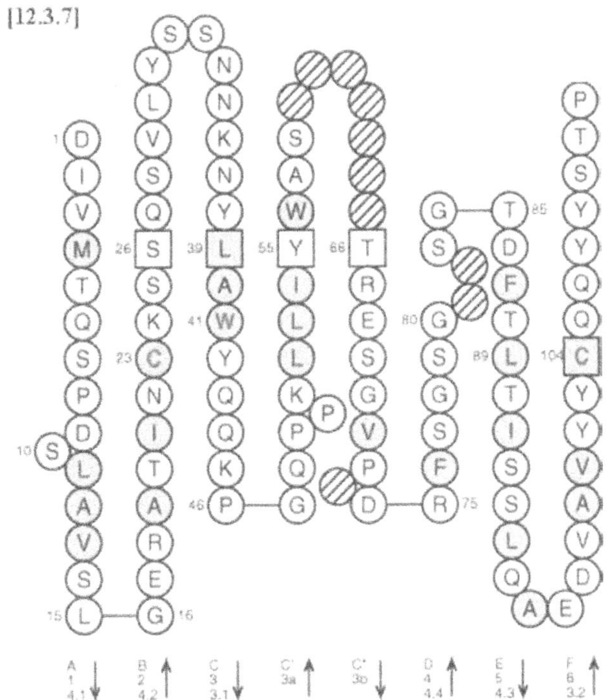

Genome database accession numbers

GDB:9953504 LocusLink: 28908

Nomenclature

IGKV5-2: Immunoglobulin kappa variable 5-2.

Definition and functionality

IGKV5-2 is the unique functional gene of the IGKV5 subgroup which only comprises one mapped gene in the IGK locus.

Gene location

IGKV5-2 is in the proximal cluster of the IGK locus on chromosome 2 at 2p11.2.

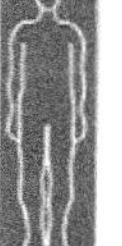

Nucleotide and amino acid sequences for human IGKV5-2

```
                                 1   2   3   4   5   6   7   8   9  10  11  12  13  14  15  16  17  18  19  20
                               E   T   T   L   T   Q   S   P   A   F   M   S   A   T   P   G   D   K   V   N
X02485,IGKV5-2*01,B2      [27] GAA ACG ACA CTC ACG CAG TCT CCA GCA TTC ATG TCA GCG ACT CCA GGA GAC AAA GTC AAC

                                                       ___________________CDR1-IMGT___________________
                                21  22  23  24  25  26  27  28  29  30  31  32  33  34  35  36  37  38  39  40
                               I   S   C   K   A   S   Q   D   I   D   D   D                           M   N
X02485,IGKV5-2*01,B2           ATC TCC TGC AAA GCC AGC CAA GAC ATT GAT GAT GAT ... ... ... ... ... ... ATG AAC

                                                                                           _______________CDR2-
                                41  42  43  44  45  46  47  48  49  50  51  52  53  54  55  56  57  58  59  60
                               W   Y   Q   Q   K   P   G   E   A   A   I   F   I   I   Q   E   A   T
X02485,IGKV5-2*01,B2           TGG TAC CAA CAG AAA CCA GGA GAA GCT GCT ATT TTC ATT ATT CAA GAA GCT ACT ... ...

                               IMGT________________
                                61  62  63  64  65  66  67  68  69  70  71  72  73  74  75  76  77  78  79  80
                                                   T   L   V   P   G   I   P       P   R   F   S   G   S   G
X02485,IGKV5-2*01,B2           ... ... ... ... ... ACT CTC GTT CCT GGA ATC CCA ... CCT CGA TTC AGT GGC AGC GGG

                                81  82  83  84  85  86  87  88  89  90  91  92  93  94  95  96  97  98  99 100
                                       Y   G   T   D   F   T   L   T   I   N   N   I   E   S   E   D   A   A
X02485,IGKV5-2*01,B2           ... ... TAT GGA ACA GAT TTT ACC CTC ACA ATT AAT AAC ATA GAA TCT GAG GAT GCT GCA

                                               ___________CDR3-IMGT____________
                               101 102 103 104 105 106 107 108 109 110 111
                               Y   Y   F   C   L   Q   H   D   N   F   P
X02485,IGKV5-2*01,B2           TAT TAC TTC TGT CTA CAA CAT GAT AAT TTC CCT CT
```

Framework and complementarity determining regions

FR1-IMGT: 26
FR2-IMGT: 17
FR3-IMGT: 36 (-3 aa: 73, 81, 82)

CDR1-IMGT: 6
CDR2-IMGT: 3
CDR3-IMGT: 7

Collier de Perles for human IGKV5-2*01

Accession number: IMGT X02485 EMBL/GenBank/DDBJ: X02485

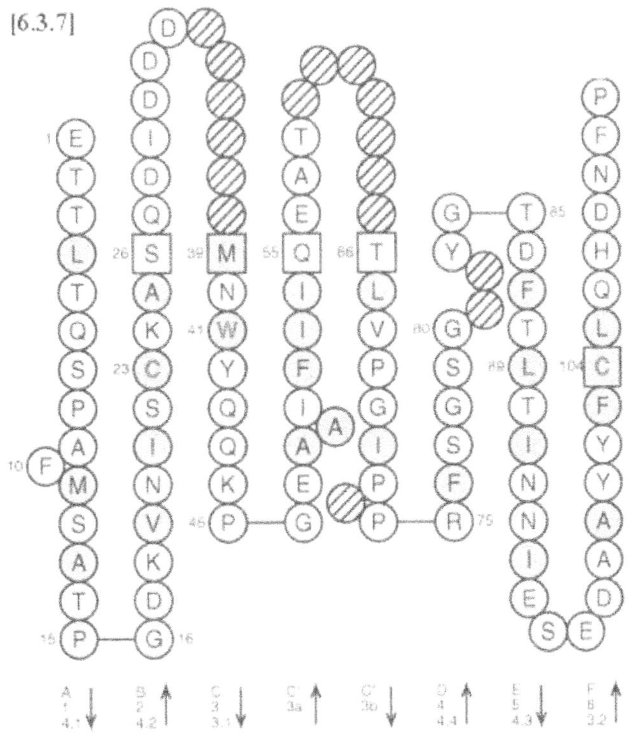

Genome database accession numbers

GDB:9953506 LocusLink: 28907

IGKV6-21 and IGKV6D-21

Nomenclature

IGKV6-21: Immunoglobulin kappa variable 6-21.
IGKV6D-21: Immunoglobulin kappa variable 6D-61.

Definition and functionality

IGKV6-21 and IGKV6D-21 are two of the three ORF genes of the IGKV6 subgroup, the third one being IGKV6D-41. The IGKV6 subgroup only comprises these three mapped genes in the IGK locus.
IGKV6-21 and IGKV6D-21 genes are ORF due to a non-canonical V-HEPTAMER: cactgtg instead of cacagtg[24].

Gene location

IGKV6-21 is in the proximal cluster, and IGKV6D-21 is in the distal cluster of the IGK locus on chromosome 2 at 2p11.2.

Nucleotide and amino acid sequences for human IGKV6-21 and IGKV6D-21

```
                                         1   2   3   4   5   6   7   8   9  10  11  12  13  14  15  16  17  18  19  20
                                         E   I   V   L   T   Q   S   P   D   F   Q   S   V   T   P   K   E   K   V   T
X63399,IGKV6-21*01,A26              [15] GAA ATT GTG CTG ACT CAG TCT CCA GAC TTT CAG TCT GTG ACT CCA AAG GAG AAA GTC ACC
X12683,IGKV6D-21*01,A10             [28] --- --- --- --- --- --- --- --- --- --- --- --- --- --- --- --- --- --- --- ---
M27750,IGKV6D-21*01,A10             [28] --- --- --- --- --- --- --- --- --- --- --- --- --- --- --- --- --- --- --- ---
X93642,IGKV6-21*01/6D-21*01,DPK26   [6]  --- --- --- --- --- --- --- --- --- --- --- --- --- --- --- --- --- --- --- ---

                                                                 _______________CDR1-IMGT_______________
                                        21  22  23  24  25  26  27  28  29  30  31  32  33  34  35  36  37  38  39  40
                                         I   T   C   R   A   S   Q   S   I   G   S   S                           L   H
X63399,IGKV6-21*01,A26                  ATC ACC TGC CGG GCC AGT CAG AGC ATT GGT AGT AGC ... ... ... ... ... ... TTA CAC
X12683,IGKV6D-21*01,A10                 --- --- --- --- --- --- --- --- --- --- --- --- ... ... ... ... ... ... --- ---
M27750,IGKV6D-21*01,A10                 --- --- --- --- --- --- --- --- --- --- --- --- ... ... ... ... ... ... --- ---
X93642,IGKV6-21*01/6D-21*01,DPK26       --- --- --- --- --- --- --- --- --- --- --- --- ... ... ... ... ... ... --- ---

                                                                                                   ___________CDR2
                                        41  42  43  44  45  46  47  48  49  50  51  52  53  54  55  56  57  58  59  60
                                         W   Y   Q   Q   K   P   D   Q   S   P   K   L   L   I   K   Y   A   S
X63399,IGKV6-21*01,A26                  TGG TAC CAG CAG AAA CCA GAT CAG TCT CCA AAG CTC CTC ATC AAG TAT GCT TCC ... ...
X12683,IGKV6D-21*01,A10                 --- --- --- --- --- --- --- --- --- --- --- --- --- --- --- --- --- --- ... ...
M27750,IGKV6D-21*01,A10                 --- --- --- --- --- --- --- --- --- --- --- --- --- --- --- --- --- --- ... ...
X93642,IGKV6-21*01/6D-21*01,DPK26       --- --- --- --- --- --- --- --- --- --- --- --- --- --- --- --- --- --- ... ...

                                        IMGT_______________
                                        61  62  63  64  65  66  67  68  69  70  71  72  73  74  75  76  77  78  79  80
                                                             Q   S   F   S   G   V   P       S   R   F   S   G   S   G
X63399,IGKV6-21*01,A26                  ... ... ... ... ... CAG TCC TTC TCA GGG GTC CCC ... TCG AGG TTC AGT GGC AGT GGA
X12683,IGKV6D-21*01,A10                 ... ... ... ... ... --- --- --- --- --- --- --- ... --- --- --- --- --- --- ---
M27750,IGKV6D-21*01,A10                 ... ... ... ... ... --- --- --- --- --- --- --- ... --- --- --- --- --- --- ---
X93642,IGKV6-21*01/6D-21*01,DPK26       ... ... ... ... ... --- --- --- --- --- --- --- ... --- --- --- --- --- --- ---

                                        81  82  83  84  85  86  87  88  89  90  91  92  93  94  95  96  97  98  99 100
                                                 S   G   T   D   F   T   L   T   I   N   S   L   E   A   E   D   A   A
X63399,IGKV6-21*01,A26                  ... ... TCT GGG ACA GAT TTC ACC CTC ACC ATC AAT AGC CTG GAA GCT GAA GAT GCT GCA
X12683,IGKV6D-21*01,A10                 ... ... --- --- --- --- --- --- --- --- --- --- --- --- --- --- --- --- --- ---
M27750,IGKV6D-21*01,A10                 ... ... --- --- --- --- --- --- --- --- --- --- --- --- --- --- --- --- --- ---
X93642,IGKV6-21*01/6D-21*01,DPK26       ... ... --- --- --- --- --- --- --- --- --- --- --- --- --- --- --- --- --- ---

                                                            _______CDR3-IMGT_______
                                       101 102 103 104 105 106 107 108 109 110 111
                                         T   Y   Y   C   H   Q   S   S   S   L   P
X63399,IGKV6-21*01,A26                  ACG TAT TAC TGT CAT CAG AGT AGT AGT TTA CCT CA
X12683,IGKV6D-21*01,A10                 --- --- --- --- --- --- --- --- --- --- --- --
M27750,IGKV6D-21*01,A10                 --- --- --- --- --- --- --- --- --- --- --- --
X93642,IGKV6-21*01/6D-21*01,DPK26       --- --- --- --- --- --- --- --- --- --- --- --
```

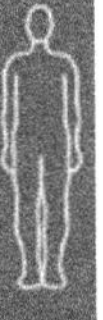

Framework and complementarity determining regions

FR1-IMGT: 26
FR2-IMGT: 17
FR3-IMGT: 36 (-3 aa: 73, 81, 82)

CDR1-IMGT: 6
CDR2-IMGT: 3
CDR3-IMGT: 7

Collier de Perles for human IGKV6-21*01 and IGKV6D-21*01

IGKV6-21*01
Accession code: IMGT X63399 EMBL/GenBank/DDBJ: X63399
IGKV6D-21*01
Accession code: IMGT X12683 EMBL/GenBank/DDBJ: X12683

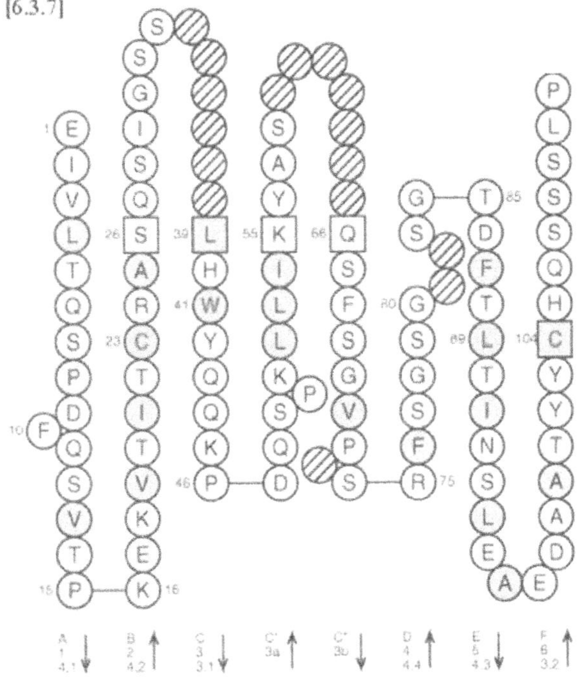

Genome database accession numbers

IGKV6-21:	GDB:9953508	LocusLink: 28906
IGKV6D-21:	GDB:9953580	LocusLink: 28870

IGKV6D-41

Nomenclature

IGKV6D-41: Immunoglobulin kappa variable 6D-41.

Definition and functionality

IGKV6D-41 is one of the three ORF genes of the IGKV6 subgroup, the two others being IGKV6-21 and IGKV6D-21. The IGKV6 subgroup only comprises these three mapped genes in the IGK locus.
The IGKV6D-41 gene is an ORF due to a non-canonical V-HEPTAMER: cactgtg instead of cacagtg[24].

Gene location

IGKV6D-41 is in the distal cluster of the IGK locus on chromosome 2 at 2p11.2.

Nucleotide and amino acid sequences for human IGKV6D-41

```
                                   1   2   3   4   5   6   7   8   9  10  11  12  13  14  15  16  17  18  19  20
                                   D   V   V   M   T   Q   S   P   A   F   L   S   V   T   P   G   E   K   V   T
X12688,IGKV6D-41*01,A14     [28]  GAT GTT GTG ATG ACA CAG TCT CCA GCT TTC CTC TCT GTG ACT CCA GGG GAG AAA GTC ACC
M27751,IGKV6D-41*01,A14     [28]  --- --- --- --                                  --- --- --- --- --- --- --- --- - -
X93641,IGKV6D-41*01,DPK25   [6]                   --- --- --- --- --- --- --- --- --- --- --- --- --- --- --- ---

                                                          ________________CDR1 IMGT________________
                                  21  22  23  24  25  26  27  28  29  30  31  32  33  34  35  36  37  38  39  40
                                   I   T   C   Q   A   S   E   G   I   G   N   Y                           L   Y
X12688,IGKV6D-41*01,A14           ATC ACC TGC CAG GCC AGT GAA GGC ATT GGC AAC TAC ... ... ... ... ... ... TTA TAC
M27751,IGKV6D-41*01,A14           --- --- --- --- --- --- --- --          --- --- --- ... ... ... ... ... ... --- ---
X93641,IGKV6D-41*01,DPK25          -- --- --- --- --- --- --- --- ---             ... ... ... ... ... ...

                                                                                                ________ CDR2-
                                  41  42  43  44  45  46  47  48  49  50  51  52  53  54  55  56  57  58  59  60
                                   W   Y   Q   Q   K   P   D   Q   A   P   K   L   L   I   K   Y   A   S
X12688,IGKV6D-41*01,A14           TGG TAC CAG CAG AAA CCA GAT CAA GCC CCA AAG CTC CTC ATC AAG TAT GCT TCC ... ...
M27751,IGKV6D-41*01,A14            -- --- --- --- --- --- --- --- --- --- --- --- --- --- --- --- ---  -  ... ...
X93641,IGKV6D-41*01,DPK25         --- --- --- --- ---                     --- --- --- --- --- --- --- --- ... ...

IMGT______________
                                  61  62  63  64  65  66  67  68  69  70  71  72  73  74  75  76  77  78  79  80
                                                       Q   S   I   S   G   V   P       S   R   F   S   G   S   G
X12688,IGKV6D-41*01,A14           ... ... ... ... ... CAG TCC ATC TCA GGG GTC CCC ... TCG AGG TTC AGT GGC AGT GGA
M27751,IGKV6D-41*01,A14           ... ... ... ... ... --- --- --- --- ---         ...      --  --  --- --- --- ---
X93641,IGKV6D-41*01,DPK25         ... ... ... ... ... --- --- --- --          --- ... --- --- --- --- --- --- ---

                                  81  82  83  84  85  86  87  88  89  90  91  92  93  94  95  96  97  98  99 100
                                           S   G   T   D   F   T   F   T   I   S   S   L   E   A   E   D   A   A
X12688,IGKV6D-41*01,A14           ... ... TCT GGG ACA GAT TTC ACC TTT ACC ATC AGT AGC CTG GAA GCT GAA GAT GCT GCA
M27751,IGKV6D-41*01,A14           ... ... - - --- --- --- --- --- --- --- --- ---              - - --- --- --- ---
X93641,IGKV6D-41*01,DPK25         ... ... --- --- --- --- --- --- --- --              --- --- --- --- --- --- ---

                                                  _________CDR3-IMGT__________
                                 101 102 103 104 105 106 107 108 109 110 111
                                   T   Y   Y   C   Q   Q   G   N   K   H   P
X12688,IGKV6D-41*01,A14           ACA TAT TAC TGT CAG CAG GGC AAT AAG CAC CCT CA
M27751,IGKV6D-41*01,A14            -- --- --- --- --- --- --- --- --- ---
X93641,IGKV6D-41*01,DPK25         --- --- --      --- --- --- --- --- --- --- --
```

Framework and complementarity determining regions

FR1-IMGT: 26	CDR1-IMGT: 6
FR2-IMGT: 17	CDR2-IMGT: 3
FR3-IMGT: 36 (-3 aa: 73, 81, 82)	CDR3-IMGT: 7

Collier de Perles for human IGKV6D-41*01

Accession number: IMGT X12688 EMBL/GenBank/DDBJ: X12688

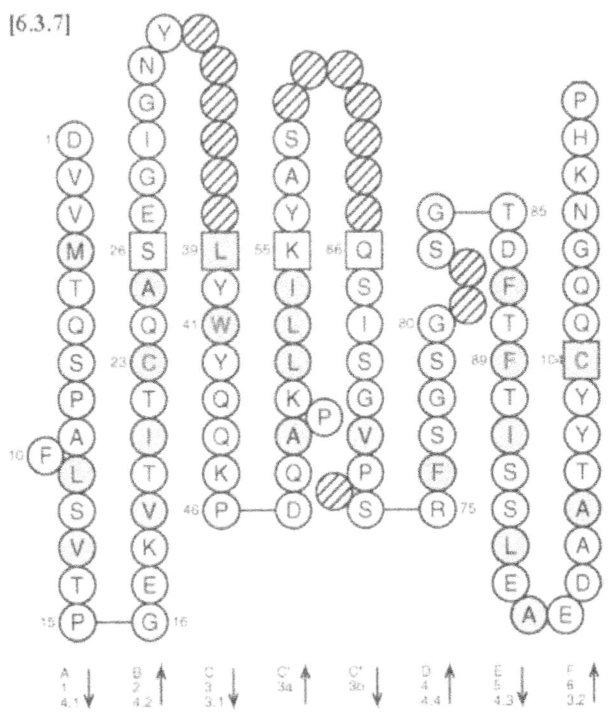

Genome database accession numbers

GDB:9953582 LocusLink: 28869

Protein display of the human IGK V-REGIONs

Only the *01 allele of each functional or ORF V-REGION is shown, except for IGKV2-29 for which the *02 allele is displayed, the IGKV2-29*01 allele being a pseudogene. IGKV genes are listed, for each subgroup, according to their position from 3′ to 5′ in the locus.

IGKV gene	FR1-IMGT (1-26)	CDR1-IMGT (27-38)	FR2-IMGT (39-55)	CDR2-IMGT (56-65)	FR3-IMGT (66-104)	CDR3-IMGT (105-115)
	1 10 20	30	40 50	60	70 80 90 100	110
Z00001,IGKV1-5	DIQMTQSPSTLSASVGDRVTITCRAS	QSISSW......	LAWYQQKPGKAPKLLIY	DAS.......	SLESGVP.SRFSGSG..SGTEFTLTISSLQPDDFATYYC	QQYNSYS....
M64858,IGKV1-6	AIQMTQSPSSLSASVGDRVTITCRAS	QGIRND......	LGWYQQKPGKAPKLLIY	AAS.......	SLQSGVP.SRFSGSG..SGTDFTLTISSLQPEDFATYYC	LQDYNYP....
Z00014,IGKV1-8	AIRMTQSPSSFSASTGDRVTITCRAS	QGISSY......	LAWYQQKPGKAPKLLIY	AAS.......	TLQSGVP.SRFSGSG..SGTDFTLTISCLQSEDFATYYC	QQYYSYP....
Z00008,IGKV1D-8	VIWMTQSPSLLSASTGDRVTISCRMS	QGISSY......	LAWYQQKPGKAPELLIY	AAS.......	TLQSGVP.SRFSGSG..SGTDFTLTISCLQSEDFATYYC	QQYYSFP....
Z00013,IGKV1-9	DIQLTQSPSFLSASVGDRVTITCRAS	QGISSY......	LAWYQQKPGKAPKLLIY	AAS.......	TLQSGVP.SRFSGSG..SGTEFTLTISSLQPEDFATYYC	QQLNSYP....
V01577,IGKV1-12	DIQMTQSPSSVSASVGDRVTITCRAS	QGISSW......	LAWYQQKPGKAPKLLIY	AAS.......	SLQSGVP.SRFSGSG..SGTDFTLTISSLQPEDFATYYC	QQANSFP....
X17263,IGKV1D-12	DIQMTQSPSSVSASVGDRVTITCRAS	QGISSW......	LAWYQQKPGKAPKLLIY	AAS.......	SLQSGVP.SRFSGSG..SGTDFTLTISSLQPEDFATYYC	QQANSFP....
X17262,IGKV1D-13	AIQLTQSPSSLSASVGDRVTITCRAS	QGISSA......	LAWYQQKPGKAPKLLIY	DAS.......	SLESGVP.SRFSGSG..SGTDFTLTISSLQPEDFATYYC	QQFNNYP....
J00248,IGKV1-16	DIQMTQSPSSLSASVGDRVTITCRAS	QGISNY......	LAWFQQKPGKAPKSLIY	AAS.......	SLQSGVP.SRFSGSG..SGTDFTLTISSLQPEDFATYYC	QQYNSYP....
K01323,IGKV1D-16	DIQMTQSPSSLSASVGDRVTITCRAS	QGISSW......	LAWYQQKPEKAPKSLIY	AAS.......	SLQSGVP.SRFSGSG..SGTDFTLTISSLQPEDFATYYC	QQYNSYP....
X72808,IGKV1-17	DIQMTQSPSSLSASVGDRVTITCRAS	QGIRND......	LGWYQQKPGKAPKRLIY	AAS.......	SLQSGVP.SRFSGSG..SGTEFTLTISSLQPEDFATYYC	LQHNSYP....
X63392,IGKV1D-17	NIQMTQSPSAMSASVGDRVTITCRAR	QGISNY......	LAWFQQKPGKVPKHLIY	AAS.......	SLQSGVP.SRFSGSG..SGTEFTLTISSLQPEDFATYYC	LQHNSYP....
X63398,IGKV1-27	DIQMTQSPSSLSASVGDRVTITCRAS	QGISNY......	LAWYQQKPGKVPKLLIY	AAS.......	TLQSGVP.SRFSGSG..SGTDFTLTISSLQPEDVATYYC	QKYNSAP....
M64856,IGKV1-33	DIQMTQSPSSLSASVGDRVTITCQAS	QDISNY......	LNWYQQKPGKAPKLLIY	DAS.......	NLETGVP.SRFSGSG..SGTDFTFTISSLQPEDIATYYC	QQYDNLP....
M64855,IGKV1D-33	DIQMTQSPSSLSASVGDRVTITCQAS	QDISNY......	LNWYQQKPGKAPKLLIY	DAS.......	NLETGVP.SRFSGSG..SGTDFTFTISSLQPEDIATYYC	QQYDNLP....
X59316,IGKV1-37	DIQLTQSPSSLSASVGDRVTITCRVS	QGISSY......	LNWYRQKPGKVPKLLIY	SAS.......	NLQSGVP.SRFSGSG..SGTDFTLTISSLQPEDVATYYG	QRTYNAP....
X71893,IGKV1D-37	DIQLTQSPSSLSASVGDRVTITCRVS	QGISSY......	LNWYRQKPGKVPKLLIY	SAS.......	NLQSGVP.SRFSGSG..SGTDFTLTISSLQPEDVATYYG	QRTYNAP....
X59315,IGKV1-39	DIQMTQSPSSLSASVGDRVTITCRAS	QSISSY......	LNWYQQKPGKAPKLLIY	AAS.......	SLQSGVP.SRFSGSG..SGTDFTLTISSLQPEDFATYYC	QQSYSTP....
X59312,IGKV1D-39	DIQMTQSPSSLSASVGDRVTITCRAS	QSISSY......	LNWYQQKPGKAPKLLIY	AAS.......	SLQSGVP.SRFSGSG..SGTDFTLTISSLQPEDFATYYC	QQSYSTP....
X72816,IGKV1D-42	DIQMIQSPSFLSASVGDRVSIICWAS	EGISSN......	LAWYLQKPGKSPKLFLY	DAK.......	DLHPGVS.SRFSGRG..SGTDFTLTIISLKPEDFAAYYC	KQDFSYP....
X72817,IGKV1D-43	AIRMTQSPFSLSASVGDRVTITCWAS	QGISSY......	LAWYQQKPAKAPKLFIY	YAS.......	SLQSGVP.SRFSGSG..SGTDYTLTISSLQPEDFATYYC	QQYYSTP....
X12684,IGKV2-24	DIVMTQTPLSSPVTLGQPASISCRSS	QSLVHSDGNTY.	LSWLQQRPGQPPRLLIY	KIS.......	NRFSGVP.DRFSGSG..AGTDFTLKISRVEAEDVGVYYC	MQATQFP....
X63401,IGKV2D-24	DIVMTQTPLSSPVTLGQPASISFRSS	QSLVHSDGNTY.	LSWLQQRPGQPPRLLIY	KVS.......	NRFSGVP.DRFSGSG..AGTDFTLKISRVEAEDVGVYYC	TQATQFP....
X63397,IGKV2-28	DIVMTQSPLSLPVTPGEPASISCRSS	QSLLHSNGYNY.	LDWYLQKPGQSPQLLIY	LGS.......	NRASGVP.DRFSGSG..SGTDFTLKISRVEAEDVGVYYC	MQALQTP....
X12691,IGKV2D-28	DIVMTQSPLSLPVTPGEPASISCRSS	QSLLHSNGYNY.	LDWYLQKPGQSPQLLIY	LGS.......	NRASGVP.DRFSGSG..SGTDFTLKISRVEAEDVGVYYC	MQALQTP....
U41645,IGKV2-29*02	DIVMTQTPLSLSVTPGQPASISCKSS	QSLLHSDGKTY.	LYWYLQKPGQSPQLLIY	EVS.......	SRFSGVP.DRFSGSG..SGTDFTLKISRVEAEDVGVYYC	MQGIHLP....
M31952,IGKV2D-29	DIVMTQTPLSLSVTPGQPASISCKSS	QSLLHSDGKTY.	LYWYLQKPGQPPQLLIY	EVS.......	NRFSGVP.DRFSGSG..SGTDFTLKISRVEAEDVGVYYC	MQSIQLP....
X63403,IGKV2-30	DVVMTQSPLSLPVTLGQPASISCRSS	QSLVYSDGNTY.	LNWFQQRPGQSPRRLIY	KVS.......	NRDSGVP.DRFSGSG..SGTDFTLKISRVEAEDVGVYYC	MQGTHWP....
X63402,IGKV2D-30	DVVMTQSPLSLPVTLGQPASISCRSS	QSLVYSDGNTY.	LNWFQQRPGQSPRRLIY	KVS.......	NWDSGVP.DRFSGSG..SGTDFTLKISRVEAEDVGVYYC	MQGTHWP....

IGKV gene	FR1-IMGT (1-26)	CDR1-IMGT (27-38)	FR2-IMGT (39-55)	CDR2-IMGT (56-65)	FR3-IMGT (66-104)	CDR3-IMGT (105-115)
	1 10 20	30	40 50	60	70 80 90 100	110
	.........|.........|......	...|........	.|.........|.....	....|.....	....|.........|.........|.........|....	.....|.....
X59314,IGKV2-40	DIVMTQTPLSLPVTPGEPASISCRSS	QSLLDSDDGNTY	LDWYLQKPGQSPQLLIY	TLS.......	YRASGVP.DRFSGSG..SGTDFTLKISRVEAEDVGVYYC	MQRIEFP....
X59311,IGKV2D-40	DIVMTQTPLSLPVTPGEPASISCRSS	QSLLDSDDGNTY	LDWYLQKPGQSPQLLIY	TLS.......	YRASGVP.DRFSGSG..SGTDFTLKISRVEAEDVGVYYC	MQRIEFP....
X02725,IGKV3-7	EIVMTQSPPTLSLSPGERVTLSCRAS	QSVSSSY.....	LTWYQQKPGQAPRLLIY	GAS.......	TRATSIP.ARFSGSG..SGTDFTLTISSLQPEDFAVYYC	QQDHNLP....
X72820,IGKV3D-7	EIVMTQSPATLSLSPGERATLSCRAS	QSVSSSY.....	LSWYQQKPGQAPRLLIY	GAS.......	TRATGIP.ARFSGSG..SGTDFTLTISSLQPEDFAVYYC	QQDYNLP....
X01668,IGKV3-11	EIVLTQSPATLSLSPGERATLSCRAS	QSVSSY......	LAWYQQKPGQAPRLLIY	DAS.......	NRATGIP.ARFSGSG..SGTDFTLTISSLEPEDFAVYYC	QQRSNWP....
X17264,IGKV3D-11	EIVLTQSPATLSLSPGERATLSCRAS	QGVSSY......	LAWYQQKPGQAPRLLIY	DAS.......	NRATGIP.ARFSGSG..PGTDFTLTISSLEPEDFAVYYC	QQRSNWH....
M23090,IGKV3-15	EIVMTQSPATLSVSPGERATLSCRAS	QSVSSN......	LAWYQQKPGQAPRLLIY	GAS.......	TRATGIP.ARFSGSG..SGTEFTLTISSLQSEDFAVYYC	QQYNNWP....
X72815,IGKV3D-15	EIVMTQSPATLSVSPGERATLSCRAS	QSVSSN......	LAWYQQKPGQAPRLLIY	GAS.......	TRATGIP.ARFSGSG..SGTEFTLTISSLQSEDFAVYYC	QQYNNWP....
X12686,IGKV3-20	EIVLTQSPGTLSLSPGERATLSCRAS	QSVSSSY.....	LAWYQQKPGQAPRLLIY	GAS.......	SRATGIP.DRFSGSG..SGTDFTLTISRLEPEDFAVYYC	QQYGSSP....
X12687,IGKV3D-20	EIVLTQSPATLSLSPGERATLSCGAS	QSVSSSY.....	LAWYQQKPGLAPRLLIY	DAS.......	SRATGIP.DRFSGSG..SGTDFTLTISRLEPEDFAVYYC	QQYGSSP....
Z00023,IGKV4-1	DIVMTQSPDSLAVSLGERATINCKSS	QSVLYSSNNKNY	LAWYQQKPGQPPKLLIY	WAS.......	TRESGVP.DRFSGSG..SGTDFTLTISSLQAEDVAVYYC	QQYYSTP....
X02485,IGKV5-2	ETTLTQSPAFMSATPGDKVNISCKAS	QDIDDD......	MNWYQQKPGEAAIFIIQ	EAT.......	TLVPGIP.PRFSGSG..YGTDFTLTINNIESEDAAYYFC	LQHDNFP....
X63399,IGKV6-21	EIVLTQSPDFQSVTPKEKVTITCRAS	QSIGSS......	LHWYQQKPDQSPKLLIK	YAS.......	QSFSGVP.SRFSGSG..SGTDFTLTINSLEAEDAATYYC	HQSSSLP....
X12683,IGKV6D-21	EIVLTQSPDFQSVTPKEKVTITCRAS	QSIGSS......	LHWYQQKPDQSPKLLIK	YAS.......	QSFSGVP.SRFSGSG..SGTDFTLTINSLEAEDAATYYC	HQSSSLP....
X12688,IGKV6D-41	DVVMTQSPAFLSVTPGEKVTITCQAS	EGIGNY......	LYWYQQKPDQAPKLLIK	YAS.......	QSISGVP.SRFSGSG..SGTDFTFTISSLEAEDAATYYC	QQGNKHP....

Recombination signals

Only the recombination signals of the *01 allele of each functional or ORF IGK V-REGION is shown. Non-conserved nucleotides taken into account for the ORF functionality definition are shown in bold and italic.

IGKV gene name	V recombination signal		
	V-HEPTAMER	(bp)	V-NONAMER
IGKV1-5*01	CACAGTG	12	ACATAAACC
IGKV1-6*01	CACAGTG	12	ACAGAAACC
IGKV1-8*01 (ORF)	CACAGTG	12	ACAAAAACC
IGKV1-9*01	CACAGTG	12	ACATAAACC
IGKV1-12*01	CACAGTG	12	ACATAAACC
IGKV1-16*01	CACAGTG	12	ACATAAACC
IGKV1-17*01	CACAGTG	12	ACATAAACC
IGKV1-27*01	CACTGTG	12	ACATAAACC
IGKV1-33*01	CACAGTG	12	ACATAAATC
IGKV1-37*01 (ORF)	CACAGTG	12	ACATAAACC
IGKV1-39*01	CACAGTG	12	ACATAAACC
IGKV1D-8*01	CACAGTG	12	ACAAAAACC
IGKV1D-12*01	CACAGTG	12	ACATAAACC
IGKV1D-13*01 (ORF)	CA***T***AGTG	12	ACATAAACC
IGKV1D-16*01	CACAGTG	12	ACATAAACC
IGKV1D-17*01	CACAGTG	12	ACATAAACC
IGKV1D-33*01	CACAGTG	12	ACATAAATC
IGKV1D-37*01 (ORF)	CACAGTG	12	ACATAAACC
IGKV1D-39*01	CACAGTG	12	ACATAAACC
IGKV1D-42*01 (ORF)	CACAG***G***G	12	ACATAAGCC
IGKV1D-43*01	CACAGTG	12	ACAAAAACC
IGKV2-24*01	CACAGTG	12	ACAAAAACC
IGKV2-28*01	CACAGTG	12	ACAGAAACC
IGKV2-29*01	CACAGTG	12	ACAGAAACC
IGKV2-30*01	CACAGTG	12	ACAAAAACC
IGKV2-40*01	CACAGTG	12	ACAGAAACC
IGKV2D-24*01	CACAGTG	12	ACAAAAACC
IGKV2D-28*01	CACAGTG	12	ACAGAAACC
IGKV2D-29*01	CACAGTG	12	ACAGAAACC
IGKV2D-30*01	CACAGTG	12	ACAAAAACC
IGKV2D-40*01	CACAGTG	12	ACAGAAACC
IGKV3-7*01 (ORF)	CACAGTG	12	ACAAAAACC
IGKV3-11*01	CACAGTG	12	ACAAAAACC
IGKV3-15*01	CACAGTG	12	ACAAAAACC
IGKV3-20*01	CACAGTG	12	ACAAAAACC
IGKV3D-7*01	CACAGTG	12	ACAAAAACC
IGKV3D-11*01	CACAGTG	12	ACAAAAACC
IGKV3D-15*01	CACAGTG	12	ACAAAAACC
IGKV3D-20*01	CACAGTG	12	ACAAAAACC
IGKV4-1*01	CACAGTG	12	ACACAAACC
IGKV5-2*01	CACAGTG	12	ACAAAAACC
IGKV6-21*01 (ORF)	CAC***T***GTG	12	ACAAAAACT
IGKV6D-21*01 (ORF)	CAC***T***GTG	12	ACAAAAACT
IGKV6D-41*01 (ORF)	CAC***T***GTG	12	ACAAAAATT

IGKV references

1. Atkinson, M.J. et al. (1996) Immunogenetics 44, 115–120.
2. Bentley, D.L. and Rabbitts, T.H. (1980) Nature 288, 730–733.
3. Bentley, D.L. and Rabbitts, T.H. (1983) Cell 32, 181–189.
4. Chen, P.P. et al. (1986) Proc. Natl Acad. Sci. USA, 83, 8318–8322.
5. Chen, P.P. et al. (1987) J. Exp. Med. 166, 1900–1905.
6. Cox, J.P.L. et al. (1994) Eur. J. Immunol. 24, 827–836.
7. Fang, Q. et al. (1995) Clin. Immunol. Immunopathol. 75, 159–167.
8. Giachino, C. et al. (1995) J. Exp. Med. 181, 1245–1250.
9. Harada, T. et al. (1994) J. Immunol. 153, 4806–4815.
10. Huber, C. et al. (1993) Eur. J. Immunol. 23, 2868–2875.
11. Ichiyoshi, Y. et al. (1995) J. Immunol. 154, 226–238.
12. Jaenichen, H.R. et al. (1984) Nucleic Acids Res. 12, 5249–5263.
13. Kimberly, V.D. et al. (1991) J. Clin. Invest. 87, 1603–1613.
14. Klobeck, H.G. et al. (1985) Nucleic Acids Res. 13, 6515–6529.
15. Lautner-Rieske, A. et al. (1992) Eur. J. Immunol. 22, 1023–1029.
16. Liu, M.F. et al. (1989) J. Immunol. 142, 688–694.
17. Manheimer-Lory, A. et al. (1991) J. Exp. Med. 174, 1639–1652.
18. Pargent, W. et al. (1991) Eur. J. Immunol. 21, 1821–1827.
19. Pech, M. and Zachau, H.G. (1984) Nucleic Acids Res. 12, 9229–9236.
20. Pech, M. et al. (1984) J. Mol. Biol. 176, 189–204.
21. Pech, M. et al. (1985) J. Mol. Biol. 183, 291–299.
22. Radoux, V. et al. (1986) J. Exp. Med. 164, 2119–2124.
23. Schäble, K.F. and Zachau, H.G. (1993) Biol. Chem. Hoppe-Seyler 374, 1001–1022.
24. Schäble, K.F. et al. (1994) Biol. Chem. Hoppe-Seyler 375, 189–199.
25. Scott, M.G. et al. (1989) J. Immunol. 143, 4110–4116.
26. Scott, M.G. et al. (1991) J. Immunol. 147, 4007–4013.
27. Stavnezer, J. et al. (1985) Nucleic Acids Res. 13, 3495–3514.
28. Straubinger, B. et al. (1988) Biol. Chem. Hoppe-Seyler 369, 601–607.
29. Straubinger, B. et al. (1988) J. Mol. Biol. 199, 23–34.
30. Timmers, E. et al. (1993) Eur. J. Immunol. 23, 619–624.
31. Van Es, J.H. et al. (1991) J. Exp. Med. 173, 461–470.
32. Youngblood, K. et al. (1994) J. Clin. Invest. 93, 852–861.

Part 1

IGLC

Nomenclature

IGLC1: Immunoglobulin lambda constant 1.

Definition and functionality

IGLC1 is one of the five functional genes of the IGLC group which comprises 7–11 mapped genes, depending on the haplotypes.

Gene location

IGLC1 is in the IGL locus on chromosome 22 at 22q11.2.

Isotypes

The IGLC1 gene encodes Mcg⁺ Ke⁺ Oz⁻ lambda chains.
The Mcg isotype marker is characterized by Asparagine 6, Threonine 8 and Lysine 57. The Ke isotype marker is characterized by Glycine 46. Oz⁻ corresponds to Arginine 83.

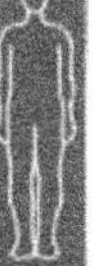

Nucleotide and amino acid sequences for human IGLC1

The nucleotide between parentheses at the beginning of exons comes from a DONOR_SPLICE (n from ngt).

The Cysteines involved in the intrachain disulphide bridges are shown with their number and letter **C** in bold. The Cysteines involved in the H-L interchain disulphide bridges are shown with only the letter **C** in bold. References are shown in square brackets.

```
                                    1   2   3   4   5   6   7   8   9  10  11  12  13  14  15  16  17  18  19  20
                                            P   K   A   N   P   T   V   T   L   F   P   P   S   S   E   E   L   Q
J00252  ,IGLC1*01,Mcg   [9]                CCC AAG GCC AAC CCC ACG GTC ACT CTG TTC CCG CCC TCC TCT GAG GAG CTC CAA
                                    G   Q
X51755  ,IGLC1*02       [16]   (G)GT  CAG --- --- --- --- --- --T --- --- --- --- --- --- --- --- --- --- --- ---
                                    G   Q
D87023  ,IGLC1*02       [20]   (G)GT  CAG --- --- --- --- --- --T --- --- --- --- --- --- --- --- --- --- --- ---

                                   21  22  23  24  25  26  27  28  29  30  31  32  33  34  35  36  37  38  39  40
                                    A   N   K   A   T   L   V   C   L   I   S   D   F   Y   P   G   A   V   T   V
J00252  ,IGLC1*01,Mcg              GCC AAC AAG GCC ACA CTA GTG TGT CTG ATC AGT GAC TTC TAC CCG GGA GCT GTG ACA GTG
X51755  ,IGLC1*02                  --- --- --- --- --- --- --- --- --- --- --- --- --- --- --- --- --- --- --- ---
D87023  ,IGLC1*02                  --- --- --- --- --- --- --- --- --- --- --- --- --- --- --- --- --- --- --- ---

                                   41  42  43  44  45  46  47  48  49  50  51  52  53  54  55  56  57  58  59  60
                                    A   W   K   A   D   G   S   P   V   K   A   G   V   E   T   T   K   P   S   K
J00252  ,IGLC1*01,Mcg              GCT TGG AAG GCA GAT GGC AGC CCC GTC AAG GCG GGA GTG GAG ACG ACC AAA CCC TCC AAA
X51755  ,IGLC1*02                  --C --- --- --- --- --- --- --- --- --- --- --- --- --- --C --- --- --- --- ---
D87023  ,IGLC1*02                  --C --- --- --- --- --- --- --- --- --- --- --- --- --- --C --- --- --- --- ---

                                   61  62  63  64  65  66  67  68  69  70  71  72  73  74  75  76  77  78  79  80
                                    Q   S   N   N   K   Y   A   A   S   S   Y   L   S   L   T   P   E   Q   W   K
J00252  ,IGLC1*01,Mcg              CAG AGC AAC AAC AAG TAC GCG GCC AGC AGC TAC CTG AGC CTG ACG CCC GAG CAG TGG AAG
X51755  ,IGLC1*02                  --- --- --- --- --- --- --- --- --- --- --- --- --- --- --- --- --- --- --- ---
D87023  ,IGLC1*02                  --- --- --- --- --- --- --- --- --- --- --- --- --- --- --- --- --- --- --- ---

                                   81  82  83  84  85  86  87  88  89  90  91  92  93  94  95  96  97  98  99 100
                                    S   H   R   S   Y   S   C   Q   V   T   H   E   G   S   T   V   E   K   T   V
J00252  ,IGLC1*01,Mcg              TCC CAC AGA AGC TAC AGC TGC CAG GTC ACG CAT GAA GGG AGC ACC GTG GAG AAG ACA GTG
X51755  ,IGLC1*02                  --- --- --- --- --- --- --- --- --- --- --- --- --- --- --- --- --- --- --- ---
D87023  ,IGLC1*02                  --- --- --- --- --- --- --- --- --- --- --- --- --- --- --- --- --- --- --- ---

                                  101 102 103 104 105 106
                                    A   P   T   E   C   S  *
J00252  ,IGLC1*01,Mcg              GCC CCT ACA GAA TGT TCA
X51755  ,IGLC1*02                  --- --- --- --- --- ---
D87023  ,IGLC1*02                  --- --- --- --- --- ---
```

Genome database accession numbers

GDB:120690 LocusLink: 3537

IGLC2

Nomenclature

IGLC2: Immunoglobulin lambda constant 2.

Definition and functionality

IGLC2 is one of the five functional genes of the IGLC group which comprises 7–11 mapped genes, depending on the haplotypes.

Gene location

IGLC2 is in the IGL locus on chromosome 22 at 22q11.2.

Isotypes

The IGLC2 gene encodes Mcg$^-$ Ke$^-$ Oz$^-$ lambda chains.
Mcg$^-$ corresponds to Alanine 6, Serine 8 and Threonine 57. The Ke$^-$ corresponds to Serine 46. Oz$^-$ corresponds to Arginine 83.

Nucleotide and amino acid sequences for human IGLC2

The nucleotide between parentheses at the beginning of exons comes from a DONOR_SPLICE (n from ngt).

The Cysteines involved in the intrachain disulphide bridges are shown with their number and letter **C** in bold. The Cysteines involved in the H-L interchain disulphide bridges are shown with only the letter **C** in bold.

```
                                         1   2   3   4   5   6   7   8   9  10  11  12  13  14  15  16  17  18  19  20
                                         G   Q   P   K   A   A   P   S   V   T   L   F   P   P   S   S   E   E   L   Q
J00253  ,IGLC2*01,Ke- Oz-  [9]   (G)GT CAG CCC AAG GCT GCC CCC TCG GTC ACT CTG TTC CCG CCC TCC TCT GAG GAG CTT CAA
X06875  ,IGLC2*02          [15]  (-)-- --- --- --- --- --- --- --- --- --- --- --- --- --- --- --- --- --- --- ---
X51754  ,IGLC2*02          [16]  (-)-- --- --- --- --- --- --- --- --- --- --- --- --- --- --- --- --- --- --- ---
X51755  ,IGLC2*02          [16]  (-)-- --- --- --- --- --- --- --- --- --- --- --- --- --- --- --- --- --- --- ---
D87023  ,IGLC2*02          [20]  (-)-- --- --- --- --- --- --- --- --- --- --- --- --- --- --- --- --- --- --- ---

                                   21  22  23  24  25  26  27  28  29  30  31  32  33  34  35  36  37  38  39  40
                                    A   N   K   A   T   L   V   C   L   I   S   D   F   Y   P   G   A   V   T   V
J00253  ,IGLC2*01,Ke- Oz-          GCC AAC AAG GCC ACA CTG GTG TGT CTC ATA AGT GAC TTC TAC CCG GGA GCC GTG ACA GTG
X06875  ,IGLC2*02                  --- --- --- --- --- --- --- --- --- --- --- --- --- --- --- --- --- --- --- ---
X51754  ,IGLC2*02                  --- --- --- --- --- --- --- --- --- --- --- --- --- --- --- --- --- --- --- ---
X51755  ,IGLC2*02                  --- --- --- --- --- --- --- --- --- --- --- --- --- --- --- --- --- --- --- ---
D87023  ,IGLC2*02                  --- --- --- --- --- --- --- --- --- --- --- --- --- --- --- --- --- --- --- ---

                                   41  42  43  44  45  46  47  48  49  50  51  52  53  54  55  56  57  58  59  60
                                    A   W   K   A   D   S   S   P   V   K   A   G   V   E   T   T   T   P   S   K
J00253  ,IGLC2*01,Ke- Oz-          GCT TGG AAA GCA GAT AGC AGC CCC GTC AAG GCG GGA GTG GAG ACC ACC ACA CCC TCC AAA
X06875  ,IGLC2*02                  --C --- --G --- --- --- --- --- --- --- --- --- --- --- --- --- --- --- --- ---
X51754  ,IGLC2*02                  --C --- --G --- --- --- --- --- --- --- --- --- --- --- --- --- --- --- --- ---
X51755  ,IGLC2*02                  --C --- --G --- --- --- --- --- --- --- --- --- --- --- --- --- --- --- --- ---
D87023  ,IGLC2*02                  --C --- --G --- --- --- --- --- --- --- --- --- --- --- --- --- --- --- --- ---

                                   61  62  63  64  65  66  67  68  69  70  71  72  73  74  75  76  77  78  79  80
                                    Q   S   N   N   K   Y   A   A   S   S   Y   L   S   L   T   P   E   Q   W   K
J00253  ,IGLC2*01,Ke- Oz-          CAA AGC AAC AAC AAG TAC GCG GCC AGC AGC TAT CTG AGC CTG ACG CCT GAG CAG TGG AAG
X06875  ,IGLC2*02                  --- --- --- --- --- --- --- --- --- --- --- --- --- --- --- --- --- --- --- ---
X51754  ,IGLC2*02                  --- --- --- --- --- --- --- --- --- --- --- --- --- --- --- --- --- --- --- ---
X51755  ,IGLC2*02                  --- --- --- --- --- --- --- --- --- --- --- --- --- --- --- --- --- --- --- ---
D87023  ,IGLC2*02                  --- --- --- --- --- --- --- --- --- --- --- --- --- --- --- --- --- --- --- ---

                                   81  82  83  84  85  86  87  88  89  90  91  92  93  94  95  96  97  98  99 100
                                    S   H   R   S   Y   S   C   Q   V   T   H   E   G   S   T   V   E   K   T   V
J00253  ,IGLC2*01,Ke- Oz-          TCC CAC AGA AGC TAC AGC TGC CAG GTC ACG CAT GAA GGG AGC ACC GTG GAG AAG ACA GTG
X06875  ,IGLC2*02                  --- --- --- --- --- --- --- --- --- --- --- --- --- --- --- --- --- --- --- ---
X51754  ,IGLC2*02                  --- --- --- --- --- --- --- --- --- --- --- --- --- --- --- --- --- --- --- ---
X51755  ,IGLC2*02                  --- --- --- --- --- --- --- --- --- --- --- --- --- --- --- --- --- --- --- ---
D87023  ,IGLC2*02                  --- --- --- --- --- --- --- --- --- --- --- --- --- --- --- --- --- --- --- ---
```

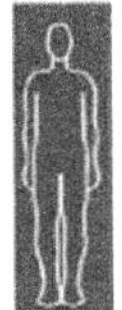

```
                                101 102 103 104 105 106
                                 A   P   T   E   C   S  •
J00253  ,IGLC2*01,Ke- Oz-       GCC CCT ACA GAA TGT TCA

X06875  ,IGLC2*02               --- --- --- --- --- ---

X51754  ,IGLC2*02               --- --- --- --- --- ---

X51755  ,IGLC2*02               --- --- --- --- --- ---

D87023  ,IGLC2*02               --- --- --- --- --- ---
```

Genome database accession numbers

GDB:120691 LocusLink: 3538

IGLC3

Nomenclature

IGLC3: Immunoglobulin lambda constant 3.

Definition and functionality

IGLC3 is one of the five functional genes of the IGLC group which comprises 7–11 mapped genes, depending on the haplotypes.

Gene location

IGLC3 is in the IGL locus on chromosome 22 at 22q11.2.

Isotypes

The IGLC3 gene encodes Mcg$^-$ Ke$^-$ Oz$^+$ (alleles IGLC3*01, *02 and *03) or Mcg$^-$ Ke$^-$ Oz$^-$ (allele IGLC3*04) lambda chains.
The Oz isotype marker is characterized by Lysine 83 whereas Oz$^-$ corresponds to Arginine 83. Mcg$^-$ correponds to Alanine 6, Serine 8 and Threonine 57 and Ke$^-$ corresponds to Serine 46.

Nucleotide and amino acid sequences for human IGLC3

The nucleotide between parentheses at the beginning of exons comes from a DONOR_SPLICE (n from ngt).

The Cysteines involved in the intrachain disulphide bridges are shown with their number and letter **C** in bold. The Cysteines involved in the H-L interchain disulphide bridges are shown with only the letter **C** in bold.

```
                                     1  2   3   4   5   6   7   8   9  10  11  12  13  14  15  16  17  18  19  20
                                            P   K   A   A   P   S   V   T   L   F   P   P   S   S   E   E   L   Q
J00254  ,IGLC3*01,Ke-Oz+    [9]            CCC AAG GCT GCC CCC TCG GTC ACT CTG TTC CCA CCC TCC TCT GAG GAG CTT CAA
                                    G   Q
K01326  ,IGLC3*02           [10] (G)GT CAG --- --- --- --- --- --- --- --- --- --- --- --- --- --- --- --- --- ---
                                    G   Q
X06876  ,IGLC3*03           [15] (G)GT CAG --- --- --- --- --- --- --- --- --- --- --- --- --- --- --- --- --- ---
                                    G   Q
D87017  ,IGLC3*04           [20] (G)GT CAG --- --- --- --- --- --- --- --- --- --- --G --- --- --- --- --- --- ---
                                    G   Q
D87023  ,IGLC3*04           [20] (G)GT CAG --- --- --- --- --- --- --- --- --- --- --G --- --- --- --- --- --- ---

                                 21  22  23  24  25  26  27  28  29  30  31  32  33  34  35  36  37  38  39  40
                                  A   N   K   A   T   L   V   C   L   I   S   D   F   Y   P   G   A   V   T   V
J00254  ,IGLC3*01,Ke-Oz+         GCC AAC AAG GCC ACA CTG GTG TGT CTC ATA AGT GAC TTC TAC CCG GGA GCC GTG ACA GTT
                                                                                                  P
K01326  ,IGLC3*02                --- --- --- --- --- --- --- --- --- --- --- --- --- --- --- --G C-A --- --- ---
X06876  ,IGLC3*03                --- --- --- --- --- --- --- --- --- --- --- --- --- --- --- --- --- --- --- --G
D87017  ,IGLC3*04                --- --- --- --- --- --- --- --- --- --- --- --- --- --- --- --- --- --- --- --G
D87023  ,IGLC3*04                --- --- --- --- --- --- --- --- --- --- --- --- --- --- --- --- --- --- --- --G

                                 41  42  43  44  45  46  47  48  49  50  51  52  53  54  55  56  57  58  59  60
                                  A   W   K   A   D   S   S   P   V   K   A   G   V   E   T   T   T   P   S   K
J00254  ,IGLC3*01,Ke-Oz+         GCC TGG AAG GCA GAT AGC AGC CCC GTC AAG GCG GGG GTG GAG ACC ACC ACA CCC TCC AAA
K01326  ,IGLC3*02                --- --- --- --- --- --- --- --- --- --- --- --- --- --- --- --- --- --- --- ---
X06876  ,IGLC3*03                --- --- --- --- --- --- --- --- --- --- --- --A --- --- --- --- --- --- --- ---
D87017  ,IGLC3*04                --- --- --- --- --- --- --- --- --- --- --- --A --- --- --- --- --- --- --- ---
D87023  ,IGLC3*04                --- --- --- --- --- --- --- --- --- --- --- --A --- --- --- --- --- --- --- ---

                                 61  62  63  64  65  66  67  68  69  70  71  72  73  74  75  76  77  78  79  80
                                  Q   S   N   N   K   Y   A   A   S   S   Y   L   S   L   T   P   E   Q   W   K
J00254  ,IGLC3*01,Ke-Oz+         CAA AGC AAC AAC AAG TAC GCG GCC AGC AGC TAC CTG AGC CTG ACG CCT GAG CAG TGG AAG
K01326  ,IGLC3*02                --- --- --- --- --- --- --- --- --- --- --- --- --- --- --- --- --- --- --- ---
X06876  ,IGLC3*03                --- --- --- --- --- --- --- --- --- --- --- --- --- --- --- --- --- --- --- ---
D87017  ,IGLC3*04                --- --- --- --- --- --- --- --- --- --- --- --- --- --- --- --- --- --- --- ---
D87023  ,IGLC3*04                --- --- --- --- --- --- --- --- --- --- --- --- --- --- --- --- --- --- --- ---
```

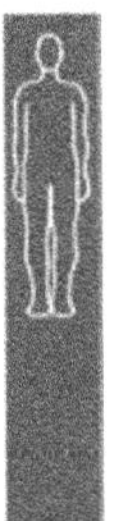

```
                            81  82  83  84  85  86  87  88  89  90  91  92  93  94  95  96  97  98  99 100
                             S   H   K   S   Y   S   C   Q   V   T   H   E   G   S   T   V   E   K   T   V
J00254  ,IGLC3*01,Ke-Oz+   TCC CAC AAA AGC TAC AGC TGC CAG GTC ACG CAT GAA GGG AGC ACC GTG GAG AAG ACA GTT

K01326  ,IGLC3*02          --- --- --- --- --- --- --- --- --- --- --- --- --- --- --- --- --- --- --- --G

X06876  ,IGLC3*03          --- --- --- --- --- --- --- --- --- --- --- --- --- --- --- --- --- --- --- --G
                                     R
D87017  ,IGLC3*04          --- --- -G- --- --- --- --- --- --- --- --- --- --- --- --- --- --- --- --- --G
                                     R
D87023  ,IGLC3*04          --- --- -G- --- --- --- --- --- --- --- --- --- --- --- --- --- --- --- --- --G


                           101 102 103 104 105 106
                             A   P   T   E   C   S   *
J00254  ,IGLC3*01,Ke-Oz+   GCC CCT ACG GAA TGT TCA

K01326  ,IGLC3*02          --- --- --- --- --- ---

X06876  ,IGLC3*03          --- --- --A --- --- ---

D87017  ,IGLC3*04          --- --- --A --- --- ---

D87023  ,IGLC3*04          --- --- --A --- --- ---
```

Genome database accession numbers

GDB:120692 LocusLink: 3539

Nomenclature

IGLC6: Immunoglobulin lambda constant 6.

Definition and functionality

IGLC6 is a functional gene (allele *01) or a pseudogene (alleles *02, *03, *04, and *05) of the IGLC group. The IGLC group comprises 7–11 genes, depending on the haplotypes. In the 7-gene haplotype, four–five genes are functional. The additional IGLC genes in the 8-, 9-, 10- and 11-gene haplotypes have not yet been sequenced. The alleles *02, *03, *04 and *05 are pseudogenes due to a 4 nt (agct) INSERTION (note the agct duplication) leading to a frameshift. The allele *02 has an additional 1 nt INSERTION.

Gene location

IGLC6 is in the IGL locus on chromosome 22 at 22q11.2.

Isotypes

The IGLC6 gene encodes Mcg⁻ Ke⁺ Oz⁻ lambda chains.
Mcg⁻ isotype corresponds to Alanine 6, Serine 8 and Threonine 57. The Ke isotype marker is characterized by Glycine 46. Oz⁻ corresponds to Arginine 83.

Nucleotide and amino acid sequences for human IGLC6

The nucleotide between parentheses at the beginning of exons comes from a DONOR_SPLICE (n from ngt).

The Cysteines involved in the intrachain disulphide bridges are shown with their number and letter **C** in bold. The Cysteines involved in the H-L interchain disulphide bridges are shown with only the letter **C** in bold.

```
                                           1   2   3   4   5   6   7   8   9  10  11  12  13  14  15  16  17  18  19  20
                                           G   Q   P   K   A   A   P   S   V   T   L   F   P   P   S   S   E   E   L   Q
J03011  ,IGLC6*01,Ke-Oz-  [13]  (G)GT CAG CCC AAG GCT GCC CCA TCG GTC ACT CTG TTC CCG CCC TCC TCT GAG GAG CTT CAA
X51755  ,IGLC6*02         [16]  (-)-- --- --- --- --- --- --C --- --- --- --- --- --- --- --- --- --- --- --- ---
M61769  ,IGLC6*03         [17]  (-)-- --- --- --- --- --- --- --- --- --- --- --- --- --- --- --- --- --- --- ---
X57808  ,IGLC6*04         [18]  (-)-- --- --- --- --- --- --- --- --- --- --- --- --- --- --- --- --- --- --- ---
D87017  ,IGLC6*05         [20]  (-)-- --- --- --- --- --- --C --- --- --- --- --- --- --- --- --- --- --- --- ---

                                  21  22  23  24  25  26  27  28  29  30  31  32  33  34  35  36  37  38  39  40
                                   A   N   K   A   T   L   V   C   L   I   S   D   F   Y   P   G   A   V   K   V
J03011  ,IGLC6*01,Ke+Oz         GCC AAC AAG GCC ACA CTG GTG TGC CTG ATC AGT GAC TTC TAC CCG GGA GCT GTG AAA GTG
X51755  ,IGLC6*02               --- --- --- --- --- --- --- --- --- --- --- --- --- --- --- --- --- --A --- ---
M61769  ,IGLC6*03               --- --- --- --- --- --- --- --- --- --- --- --- --- --- --- --- --- --- --- ---
X57808  ,IGLC6*04               --- --- --- --- --- --- --- --- --- --- --- --- --- --- --- --- --- --- --- ---
D87017  ,IGLC6*05               --- --- --- --- --- --- --- --- --- --- --- --- --- --- --- --- --- --- --- ---

                                  41  42  43  44  45  46  47  48  49  50  51  52  53  54  55  56  57  58  59  60
                                   A   W   K   A   D   G   S   P   V   N   T   G   V   E   T   T   T   P   S   K
J03011  ,IGLC6*01,Ke+Oz-        GCC TGG AAG GCA GAT GGC AGC CCC GTC AAC ACG GGA GTG GAG ACC ACC ACA CCC TCC AAA
                                                                         A
X51755  ,IGLC6*02               --- --- --- --- --- --- --- --- --- --- G-- --- --- --- --- --- --- --- --- ---
                                                                         A
M61769  ,IGLC6*03               --- --- --- --- --- --- --- --- --- --- G-- --- --- --- --- --- --- --- --- ---
X57808  ,IGLC6*04               --- --- --- --- --- --- --- --- --- --- --- --- --- --- --- --- --- --- --- ---
                                                                         A
D87017  ,IGLC6*05               --- --- --- --- --- --- --- --- --- --- G-- --- --- --- --- --- --- --- --- ---
```

```
                                 61  62  63  64  65  66  67  68  69      70  71  72  73  74  75  76  77  78  79  80
                                  Q   S   N   N   K   Y   A   A   S       S   Y   L   S   L   T   P   E   Q   W   K
J03011  ,IGLC6*01,Ke+Oz-        CAG AGC AAC AAC AAG TAC GCG GCC AGC     AGC TAC CTG AGC CTG ACG CCT GAG CAG TGG AAG
                                                                      S #
X51755  ,IGLC6*02               --- --- --- --- --- --- --- --- ---AGCT--- --- --- --- --- --- --- --- --- --- ---
                                                                      S #
M61769  ,IGLC6*03               --- --- --- --- --- --- --- --- ---AGCT--- --- --- --- --- --- --- --- --- --- ---
                                                                      S #
X57808  ,IGLC6*04               --- --- --- --- --- --- --- --- ---AGCT--- --- --- --- --- --- --- --- --- --- ---
                                                                      S #
D87017  ,IGLC6*05               --- --- --- --- --- --- --- --- ---AGCT--- --- --- --- --- --- --- --- --- --- ---

                                 81  82  83  84  85  86  87  88  89  90  91  92  93  94  95  96  97  98  99 100
                                  S   H   R   S   Y   S   C   Q   V   T   H   E   G   S   T   V   E   K   T   V
J03011  ,IGLC6*01,Ke+Oz-        TCC CAC AGA AGC TAC AGC TGC CAG GTC ACG CAT GAA GGG AGC ACC GTG GAG AAG ACA GTG
X51755  ,IGLC6*02               --- --- --- --- --- --T --- --- --- --- --- --- --- --- --- --- --- --- --- ---
M61769  ,IGLC6*03               --- --- --- --- --- --T --- --- --- --- --- --- --- --- --- --- --- --- --- ---
X57808  ,IGLC6*04               --- --- --- --- --- --T --- --- --- --- --- --- --- --- --- --- --- --- --- ---
D87017  ,IGLC6*05               --- --- --- --- --- --T --- --- --- --- --- --- --- --- --- --- --- --- --- ---

                                101 102 103 104 105 106
                                  A   P   A   E   C   S  *
J03011  ,IGLC6*01,Ke+Oz         GCC CCT GCA GAA TGT TCA
                                           #
X51755  ,IGLC6*02               --- --- ---G--- --C --T
M61769  ,IGLC6*03               --- --- --- --- --C --T
X57808  ,IGLC6*04               --- --- --- --- --C --T
D87017  ,IGLC6*05               --- --- --- --- --C --T
```

#: frameshift

Genome database accession numbers

GDB:120524　　　LocusLink: 3542

Nomenclature

IGLC7: Immunoglobulin lambda constant 7.

Definition and functionality

IGLC7 is one of the five functional genes of the IGLC group which comprises 7–11 mapped genes, depending on the haplotypes.

Gene location

IGLC7 is in the IGL locus on chromosome 22 at 22q11.2.

Isotypes

The IGLC7 gene encodes characteristic amino acids of the Ke^{+} Oz^{-} and two of the three characteristic amino acids of Mcg^{-} isotype (Mcg^{-} : Alanine 6 and Serine 8). However, it encodes a Lysine at position 57 whereas all Mcg^{-} myeloma proteins described to date contain a Threonine at this position[4].

Nucleotide and amino acid sequences for human IGLC7

The nucleotide between parentheses at the beginning of exons comes from a DONOR_SPLICE (n from ngt).

The Cysteines involved in the intrachain disulphide bridges are shown with their number and letter **C** in bold. The Cysteines involved in the H-L interchain disulphide bridges are shown with only the letter **C** in bold.

```
                                1   2   3   4   5   6   7   8   9   10  11  12  13  14  15  16  17  18  19  20
                                G   Q   P   K   A   A   P   S   V   T   L   F   P   P   S   S   E   E   L   Q
X51755  ,IGLC7*01   [16]   (G)GT CAG CCC AAG GCT GCC CCC TCG GTC ACT CTG TTC CCA CCC TCC TCT GAG GAG CTT CAA
X57808  ,IGLC7*01   [18]   (-)-- --- --- --- --- --- --- --- --- --- --- --- --- --- --- --- --- --- --- ---
D87017  ,IGLC7*01   [20]   (-)-- --- --- --- --- --- --- --- --- --- --- --- --- --- --- --- --- --- --- ---
M61771  ,IGLC7*02   [17]   (-)-- --- --- --- --- --- --A --- --- --- --- --- --- --- --- --- --- --- --- ---

                             21  22  23  24  25  26  27  28  29  30  31  32  33  34  35  36  37  38  39  40
                             A   N   K   A   T   L   V   C   L   V   S   D   F   Y   P   G   A   V   T   V
X51755  ,IGLC7*01           GCC AAC AAG GCC ACA CTG GTG TGT CTC GTA AGT GAC TTC TAC CCG GGA GCC GTG ACA GTG
X57808  ,IGLC7*01           --- --- --- --- --- --- --- --- --- --- --- --- --- --- --- --- --- --- --- ---
D87017  ,IGLC7*01           --- --- --- --- --- --- --- --- --- --- --- --- --- --- --- --- --- --- --- ---
M61771  ,IGLC7*02           --- --- --- --- --- --- --- --- --- --- --- --- --- --- --- --- --- --- --- ---

                             41  42  43  44  45  46  47  48  49  50  51  52  53  54  55  56  57  58  59  60
                             A   W   K   A   D   G   S   P   V   K   V   G   V   E   T   T   K   P   S   K
X51755  ,IGLC7*01           GCC TGG AAG GCA GAT GGC AGC CCC GTC AAG GTG GGA GTG GAG ACC ACC AAA CCC TCC AAA
X57808  ,IGLC7*01           --- --- --- --- --- --- --- --- --- --- --- --- --- --- --- --- --- --- --- ---
D87017  ,IGLC7*01           --- --- --- --- --- --- --- --- --- --- --- --- --- --- --- --- --- --- --- ---
M61771  ,IGLC7*02           --- --- --- --- --- --- --- --- --- --- --- --- --- --- --- --- --- --- --- ---

                             61  62  63  64  65  66  67  68  69  70  71  72  73  74  75  76  77  78  79  80
                             Q   S   N   N   K   Y   A   A   S   S   Y   L   S   L   T   P   E   Q   W   K
X51755  ,IGLC7*01           CAA AGC AAC AAC AAG TAT GCG GCC AGC AGC TAC CTG AGC CTG ACG CCC GAG CAG TGG AAG
X57808  ,IGLC7*01           --- --- --- --- --- --- --- --- --- --- --- --- --- --- --- --- --- --- --- ---
D87017  ,IGLC7*01           --- --- --- --- --- --- --- --- --- --- --- --- --- --- --- --- --- --- --- ---
M61771  ,IGLC7*02           --- --- --- --- --- --- --- --- --- --- --- --- --- --- --- --- --- --- --- ---
```

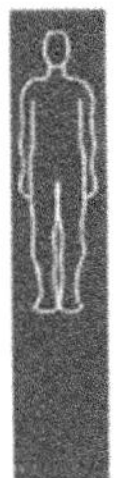

		81 S	82 H	83 R	84 S	85 Y	86 S	87 C	88 R	89 V	90 T	91 H	92 E	93 G	94 S	95 T	96 V	97 E	98 K	99 T	100 V
X51755	,IGLC7*01	TCC	CAC	AGA	AGC	TAC	AGC	TGC	CGG	GTC	ACG	CAT	GAA	GGG	AGC	ACC	GTG	GAG	AAG	ACA	GTG
X57808	,IGLC7*01	---	---	---	---	---	---	---	---	---	---	---	---	---	---	---	---	---	---	---	---
D87017	,IGLC7*01	---	---	---	---	---	---	---	---	---	---	---	---	---	---	---	---	---	---	---	---
M61771	,IGLC7*02	---	---	---	---	---	---	---	---	---	---	---	---	---	---	---	---	---	---	---	---

		101 A	102 P	103 A	104 E	105 C	106 S	*
X51755	,IGLC7*01	GCC	CCT	GCA	GAA	TGC	TCT	
X57808	,IGLC7*01	---	---	---	---	---	---	
D87017	,IGLC7*01	---	---	---	---	---	---	
M61771	,IGLC7*02	---	---	---	---	---	---	

Genome database accession numbers

GDB:9953636 LocusLink: 28834

[1] Apella, E. and Ein, D. (1967) Proc. Natl Acad. Sci. USA 57, 1449–1454.
[2] Ein, D. and Fahey, J.L. (1967) Science 156, 957–950.
[3] Ein, D. (1968) Proc. Natl Acad. Sci. USA 60, 982–985.
[4] Fett, J.W. and Deutsh, H.F. (1975) Immunochemistry 12, 643–652.
[5] Ponstingl, H. et al. (1968) Hoppe-Seyler's Z. Physiol. Chem. 349, 867–871.
[6] Hess, M. et al. (1971) Nature New Biol. 234, 58–61.
[7] Fett, J.W. and Deutsh, H.F. (1974) Biochemistry 13, 4102–4114.
[8] Fett, J.W. and Deutsh, H.F. (1975) Immunochemistry 12, 643–652.
[9] Hieter, P. A. et al. (1981) Nature 294, 536–540.
[10] Hollis, G. F. et al. (1982) Nature 296, 321–325.
[11] Taub, R. A. et al. (1983) Nature 304, 172–174.
[12] Frangione, B. et al. (1985) Proc. Acad. Natl Sci. USA 82, 3415–3419.
[13] Dariavach, P. et al. (1987) Proc. Natl Acad. Sci. USA, 84, 9074–9078.
[14] Ghanem, N. et al. (1988) Exp. Clin. Immunogenet. 5, 186–195.
[15] Udey, J. A. and Blomberg, B. (1988) Nucleic Acids Res. 16, 2959–2969.
[16] Vasicek, T.J. and Leder, P. (1990) J. Exp. Med. 172, 609–620.
[17] Bauer, T.R. et al. (1991) J. Immunol. 146, 2813–2820.
[18] Combriato, G. and Klobeck, H.G. (1991) Eur. J. Immunol. 21, 1513–1522.
[19] Kay, P.H. et al. (1992) Immunogenetics 35, 341–343.
[20] Kawasaki, K. et al. (1997) Genome Res. 7, 250–261.

Protein display

The protein display of IGLC genes is shown on page 94.

Part 2

IGLJ

IGLJ group

Nomenclature

Immunoglobulin lambda joining group.

Definition and functionality

The human IGLJ group comprises seven mapped genes, in the 7-IGLC gene haplotype. Four IGLJ genes are functional (IGLJ1, IGLJ2, IGLJ3 and IGLJ7), and three are ORF (IGLJ4, IGLJ5 and IGLJ6). The IGLJ4 and IGLJ5 have a non-canonical DONOR_SPLICE. Moreover, the IGLJ4 and IGLJ5 have a non-conserved J-HEPTAMER and precede the IGLC4 and IGLC5 pseudogenes. The IGLC6 has a conserved J-HEPTAMER and it is not excluded that it can be used in a productive lambda chain when it precedes the rare IGLC6 functional allele.

The potential additional IGLJ genes from the 8- to 11- IGLC gene haplotypes have not yet been characterized and sequenced.

Gene location

The human IGLJ genes are located in the IGL locus on chromosome 22 at 22q11.2. Each IGLJ gene is localized upstream from the corresponding IGLC gene, in the 7- IGLC gene haplotype (the only haplotype in which all seven IGLJ and IGLC genes have been sequenced so far).

Nucleotide and amino acid sequences for the human functional IGLJ genes

The conserved <u>**FGXG**</u> motif, characteristic of the IGL (and IGK) J-REGION is underlined. References are shown in square brackets.

```
IGLJ1
Immunoglobulin lambda joining 1

                           Y   V   F   G   T   G   T   K   V   T   V   L
  X04457  ,IGLJ1*01   [1]  T TAT GTC TTC GGA ACT GGG ACC AAG GTC ACC GTC CTA G

IGLJ2
Immunoglobulin lambda joining 2

                           V   V   F   G   G   G   T   K   L   T   V   L
  M15641  ,IGLJ2*01   [3]  T GTG GTA TTC GGC GGA GGG ACC AAG CTG ACC GTC CTA G

IGLJ3
Immunoglobulin lambda joining 3

                           V   V   F   G   G   G   T   K   L   T   V   L
  M15642  ,IGLJ3*01   [3]  T GTG GTA TTC GGC GGA GGG ACC AAG CTG ACC GTC CTA G
                           W
  D87023  ,IGLJ3*02   [5]  - TG- --G --- --- --- --- --- --- --- --- --- -

IGLJ4
Immunoglobulin lambda joining 4

                           F   V   F   G   G   G   T   Q   L   I   I   L
  X51755  ,IGLJ4*01   [4]  T TTT GTA TTT GGT GGA GGA ACC CAG CTC ATC ATT TTA G

IGLJ5
Immunoglobulin lambda joining 5

                           W   V   F   G   E   G   T   E   L   T   V   L
  X51755  ,IGLJ5*01   [4]  C TGG GTG TTT GGT GAG GGG ACC GAG CTC ACC GTC CTA G
  D87017  ,IGLJ5*02   [5]  - --- --- --- --- --- --- --G --- --- --- --- -

IGLJ6
Immunoglobulin lambda joining 6

                           N   V   F   G   S   G   T   K   V   T   V   L
  M18338  ,IGLJ6*01   [2]  T AAT GTG TTC GGC AGT GGC ACC AAG GTG ACC GTC CTC G

IGLJ7
Immunoglobulin lambda joining 7

                           A   V   F   G   G   G   T   Q   L   T   V   L
  X51755  ,IGLJ7*01   [4]  T GCT GTG TTC GGA GGA GGC ACC CAG CTG ACC GTC CTC G
                                                                 A
  D87017  ,IGLJ7*02   [5]  - --- --- --- --- --- --- --- --- --- C-- --- -
```

Recombination signals

Non-conserved nucleotides taken into account for the ORF functionality definition are shown in bold and italic.

J recombination signal			IGLJ gene and allele name
J-NONAMER	(bp)	J-HEPTAMER	
GGTTTTGGT	12	CACTGTG	IGLJ1*01
GGTTTTTGT	12	CACAGTG	IGLJ2*01
GGTTTTTGT	12	CACAGTG	IGLJ3*01
GGTTTTTGT	12	CACAGTG	IGLJ3*02
AGTTTTTGT	12	CACCG***CA***	IGLJ4*01 (ORF)
GGGTTTTGT	12	CACAG***CA***	IGLJ5*01 (ORF)
GGTTTTTGT	12	CACAG***CA***	IGLJ5*02 (ORF)
GGTTTGTGT	12	CACAGTG	IGLJ6*01 (ORF)
GGTTTGTGT	12	CACTGTG	IGLJ7*01
GGTTTGTGT	12	CACTGTG	IGLJ7*02

References

[1] Chang, H. et al. (1986) J. Exp. Med. 163, 425–435.
[2] Dariavach, P. et al. (1987) Proc. Natl Acad. Sci. USA 84, 9074–9078.
[3] Udey, J.A. and Blomberg, B. (1987) Immunogenetics 25, 63–70.
[4] Vasicek, T.J. and Leder, P.J. (1990) J. Exp. Med. 172, 609–620.
[5] Kawasaki, K. et al. (1997) Genome Res. 7, 250–261.

Part 3

IGLV

IGLV1-36

Nomenclature

IGLV1-36: Immunoglobulin lambda variable 1-36.

Definition and functionality

IGLV1-36 is one of the five functional genes of the IGLV1 subgroup which comprises eight mapped genes.

Gene location

IGLV1-36 is located in the cluster B of the IGL locus on chromosome 22 at 22q11.2.

Nucleotide and amino acid sequences for human IGLV1-36

References are shown in square brackets.

```
                                       1   2   3   4   5   6   7   8   9  10  11  12  13  14  15  16  17  18  19  20
                                       Q   S   V   L   T   Q   P   P   S       V   S   E   A   P   R   Q   R   V   T
Z73653,IGLV1-36*01,1a.11.2      [27]  CAG TCT GTG CTG ACT CAG CCA CCC TCG ... GTG TCT GAA GCC CCC AGG CAG AGG GTC ACC
D87009,IGLV1-36*01,V1-11        [19]  --- --- --- --- --- --- --- --- --- ... --- --- --- --- --- --- --- --- --- ---
D87010,IGLV1-36*01,V1-11        [19]  --- --- --- --- --- --- --- --- --- ... --- --- --- --- --- --- --- --- --- ---
Z22187,IGLV1-36*01,DPL1         [26]  --- --- --- --- --- --- --- --- --- ... --- --- --- --- --- --- --- --- --- ---
U03900,IGLV1-36*01,lv1c2        [12]  --- --- --- --- --- --- --- --- --- ... --- --- --- --- --- --- --- --- --- ---
U03901,IGLV1-36*01,lv1c2c       [12]  --- --- --- --- --- --- --- --- --- ... --- --- --- --- --- --- --- --- --- ---

                                                               ____________CDR1-IMGT______________
                                      21  22  23  24  25  26  27  28  29  30  31  32  33  34  35  36  37  38  39  40
                                       I   S   C   S   G   S   S   S   N   I   G   N   N   A                   V   N
Z73653,IGLV1-36*01,1a.11.2            ATC TCC TGT TCT GGA AGC AGC TCC AAC ATC GGA AAT AAT GCT ... ... ... ... GTA AAC
D87009,IGLV1-36*01,V1-11              --- --- --- --- --- --- --- --- --- --- --- --- --- --- ... ... ... ... --- ---
D87010,IGLV1-36*01,V1-11              --- --- --- --- --- --- --- --- --- --- --- --- --- --- ... ... ... ... --- ---
Z22187,IGLV1-36*01,DPL1               --- --- --- --- --- --- --- --- --- --- --- --- --- --- ... ... ... ... --- ---
U03900,IGLV1-36*01,lv1c2              --- --- --- --- --- --- --- --- --- --- --- --- --- --- ... ... ... ... --- ---
U03901,IGLV1-36*01,lv1c2c             --- --- --- --- --- --- --- --- --- --- --- --- --- --- ... ... ... ... --- ---

                                                                                               ___________ CDR2-
                                      41  42  43  44  45  46  47  48  49  50  51  52  53  54  55  56  57  58  59  60
                                       W   Y   Q   Q   L   P   G   K   A   P   K   L   L   I   Y   Y   D   D
Z73653,IGLV1-36*01,1a.11.2            TGG TAC CAG CAG CTC CCA GGA AAG GCT CCC AAA CTC CTC ATC TAT TAT GAT GAT ... ...
D87009,IGLV1-36*01,V1-11              --- --- --- --- --- --- --- --- --- --- --- --- --- --- --- --- --- --- ... ...
D87010,IGLV1-36*01,V1-11              --- --- --- --- --- --- --- --- --- --- --- --- --- --- --- --- --- --- ... ...
Z22187,IGLV1-36*01,DPL1               --- --- --- --- --- --- --- --- --- --- --- --- --- --- --- --- --- --- ... ...
U03900,IGLV1-36*01,lv1c2              --- --- --- --- --- --- --- --- --- --- --- --- --- --- --- --- --- --- ... ...
U03901,IGLV1-36*01,lv1c2c             --- --- --- --- --- --- --- --- --- --- --- --- --- --- --- --- --- --- ... ...

                                      IMGT________________
                                      61  62  63  64  65  66  67  68  69  70  71  72  73  74  75  76  77  78  79  80
                                                           L   L   P   S   G   V   S       D   R   F   S   G   S   K
Z73653,IGLV1-36*01,1a.11.2            ... ... ... ... ... CTG CTG CCC TCA GGG GTC TCT ... GAC CGA TTC TCT GGC TCC AAG
D87009,IGLV1-36*01,V1-11              ... ... ... ... ... --- --- --- --- --- --- --- ... --- --- --- --- --- --- ---
D87010,IGLV1-36*01,V1-11              ... ... ... ... ... --- --- --- --- --- --- --- ... --- --- --- --- --- --- ---
Z22187,IGLV1-36*01,DPL1               ... ... ... ... ... --- --- --- --- --- --- --- ... --- --- --- --- --- --- ---
U03900,IGLV1-36*01,lv1c2              ... ... ... ... ... --- --- --- --- --- --- --- ... --- --- --- --- --- --- ---
U03901,IGLV1-36*01,lv1c2c             ... ... ... ... ... --- --- --- --- --- --- --- ... --- --- --- --- --- --- ---

                                      81  82  83  84  85  86  87  88  89  90  91  92  93  94  95  96  97  98  99 100
                                               S   G   T   S   A   S   L   A   I   S   G   L   Q   S   E   D   E   A
Z73653,IGLV1-36*01,1a.11.2            ... ... TCT GGC ACC TCA GCC TCC CTG GCC ATC AGT GGG CTC CAG TCT GAG GAT GAG GCT
D87009,IGLV1-36*01,V1-11              ... ... --- --- --- --- --- --- --- --- --- --- --- --- --- --- --- --- --- ---
D87010,IGLV1-36*01,V1-11              ... ... --- --- --- --- --- --- --- --- --- --- --- --- --- --- --- --- --- ---
Z22187,IGLV1-36*01,DPL1               ... ... --- --- --- --- --- --- --- --- --- --- --- --- --- --- --- --- --- ---
U03900,IGLV1-36*01,lv1c2              ... ... --- --- --- --- --- --- --- --- --- --- --- --- --- --- --- --- --- ---
U03901,IGLV1-36*01,lv1c2c             ... ... --- --- --- --- --- --- --- --- --- --- --- --- --- --- --- --- --- ---
```

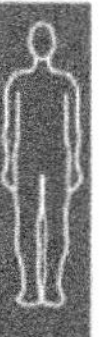

```
                                                             ___________CDR3-IMGT____________
                               101 102 103 104 105 106 107 108 109 110 111 112 113
                                D   Y   Y   C   A   A   W   D   D   S   L   N   G
Z73653,IGLV1-36*01,1a.11.2     GAT TAT TAC TGT GCA GCA TGG GAT GAC AGC CTG AAT GGT CC
D87009,IGLV1-36*01,V1-11       --- --- --- --- --- --- --- --- --- --- --- --- --- --
D87010,IGLV1-36*01,V1-11       --- --- --- --- --- --- --- --- --- --- --- --- --- --
Z22187,IGLV1-36*01,DPL1        --- --- --- --- --- --- --- --- --- --- --- --- ---
U03900,IGLV1-36*01,lv1c2       --- --- --- --- --- --- --- --- --- --- --- --- --- --
U03901,IGLV1-36*01,lv1c2c      --- --- --- --- --- --- --- --- --- --- --- --- --- --
```

Framework and complementarity determining regions

FR1-IMGT: 25 (-1 aa: 10)
FR2-IMGT: 17
FR3-IMGT: 36 (-3 aa: 73,81,82)

CDR1-IMGT: 8
CDR2-IMGT: 3
CDR3-IMGT: 9

Collier de Perles for human IGLV1-36*01

Accession number: IMGT Z73653

EMBL/GenBank/DDBJ: Z73653

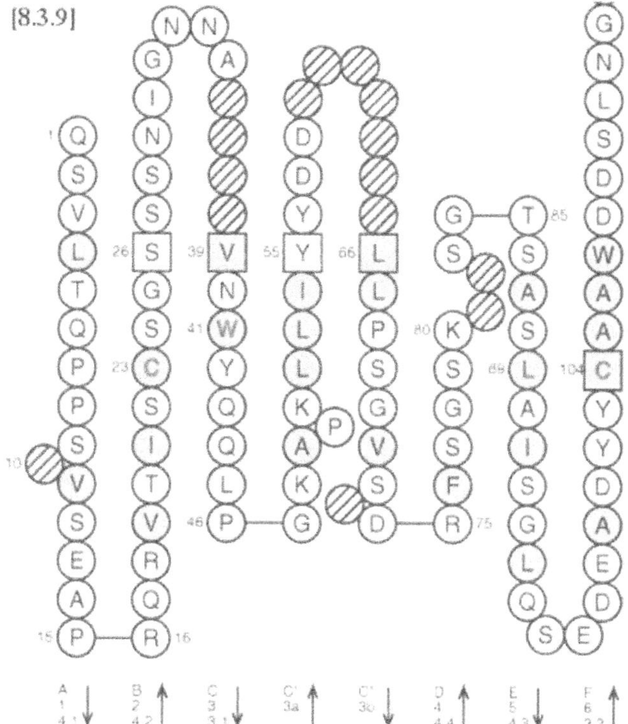

Genome database accession numbers

GDB:9953652

LocusLink: 28826

IGLV1-40

Nomenclature

IGLV1-40: Immunoglobulin lambda variable 1-40.

Definition and functionality

IGLV1-40 is one of the five functional genes of the IGLV1 subgroup which comprises eight mapped genes.

Gene location

IGLV1-40 is located in the cluster B of the IGL locus on chromosome 22 at 22q11.2.

Nucleotide and amino acid sequences for human IGLV1-40

```
                                         1   2   3   4   5   6   7   8   9  10  11  12  13  14  15  16  17  18  19  20
                                         Q   S   V   L   T   Q   P   P   S       V   S   G   A   P   G   Q   R   V   T
M94116,IGLV1-40*01,Humlv1042       [9]  CAG TCT GTG CTG ACG CAG CCG CCC TCA ... GTG TCT GGG GCC CCA GGG CAG AGG GTC ACC

Z73656,IGLV1-40*01,1e-10-2        [27]  --- --- --- --- --- --- --- --- --- ... --- --- --- --- --- --- --- --- --- ---

D87010,IGLV1-40*01,V1-13          [19]  --- --- --- --- --- --- --- --- --- ... --- --- --- --- --- --- --- --- --- ---

Z22194,IGLV1-40*01,DPL8           [26]  --- --- --- --- --- --- --- --- --- ... --- --- --- --- --- --- --- --- --- ---
                                                     V
X53936,IGLV1-40*02,Vlambda1.2      [4]  --- --- --C G-- --- --- --- --- --- ... --- --- --- --- --- --- --- --- --- ---
                                                     V
Z22193,IGLV1-40*02,DPL7           [26]  --- --- --C G-- --- --- --- --- --- ... --- --- --- --- --- --- --- --- --- ---
                                                     V
Z22192,IGLV1-40*03,DPL6           [26]  --- --- --C G-- --- --- --- --- --- ... --- --- --- --- --- --- --- --- --- ---

                                                                ____________________CDR1-IMGT___________________
                                        21  22  23  24  25  26  27  28  29  30  31  32  33  34  35  36  37  38  39  40
                                         I   S   C   T   G   S   S   S   N   I   G   A   G   Y   D               V   H
M94116,IGLV1-40*01,Humlv1042            ATC TCC TGC ACT GGG AGC AGC TCC AAC ATC GGG GCA GGT TAT GAT ... ... ... GTA CAC

Z73656,IGLV1-40*01,1e-10-2              --- --- --- --- --- --- --- --- --- --- --- --- --- --- --- ... ... ... --- ---

D87010,IGLV1-40*01,V1-13                --- --- --- --- --- --- --- --- --- --- --- --- --- --- --- ... ... ... --- ---

Z22194,IGLV1-40*01,DPL8                 --- --- --- --- --- --- --- --- --- --- --- --- --- --- --- ... ... ... --- ---

X53936,IGLV1-40*02,Vlambda1.2           --- --- --- --- --- --- --- --- --- --- --- --- --- --- --- ... ... ... --- ---

Z22193,IGLV1-40*02,DPL7                 --- --- --- --- --- --- --- --- --- --- --- --- --- --- --- ... ... ... --- ---

Z22192,IGLV1-40*03,DPL6                 --- --- --- --- --- --- --- --- --- --- --- --- --- --- --- ... ... ... --- ---

                                                                                                    ____________________CDR2-
                                        41  42  43  44  45  46  47  48  49  50  51  52  53  54  55  56  57  58  59  60
                                         W   Y   Q   Q   L   P   G   T   A   P   K   L   L   I   Y   G   N   S
M94116,IGLV1-40*01,Humlv1042            TGG TAC CAG CAG CTT CCA GGA ACA GCC CCC AAA CTC CTC ATC TAT GGT AAC AGC ... ...

Z73656,IGLV1-40*01,1e-10-2              --- --- --- --- --- --- --- --- --- --- --- --- --- --- --- --- --- --- ... ...

D87010,IGLV1-40*01,V1-13                --- --- --- --- --- --- --- --- --- --- --- --- --- --- --- --- --- --- ... ...

Z22194,IGLV1-40*01,DPL8                 --- --- --- --- --- --- --- --- --- --- --- --- --- --- --- --- --- --- ... ...

X53936,IGLV1-40*02,Vlambda1.2           --- --- --- --- --- --- --- --- --- --- --- --- --- --- --- --- --- --- ... ...

Z22193,IGLV1-40*02,DPL7                 --- --- --- --- --- --- --- --- --- --- --- --- --- --- --- --- --- --- ... ...

Z22192,IGLV1-40*03,DPL6                 --- --- --- --- --- --- --- --- --- --- --- --- --- --- --- --- --- --- ... ...

                                        IMGT________________
                                        61  62  63  64  65  66  67  68  69  70  71  72  73  74  75  76  77  78  79  80
                                                             N   R   P   S   G   V   P       D   R   F   S   G   S   K
M94116,IGLV1-40*01,Humlv1042            ... ... ... ... ... AAT CGG CCC TCA GGG GTC CCT ... GAC CGA TTC TCT GGC TCC AAG

Z73656,IGLV1-40*01,1e-10-2              ... ... ... ... ... --- --- --- --- --- --- --- ... --- --- --- --- --- --- ---

D87010,IGLV1-40*01,V1-13                ... ... ... ... ... --- --- --- --- --- --- --- ... --- --- --- --- --- --- ---

Z22194,IGLV1-40*01,DPL8                 ... ... ... ... ... --- --- --- --- --- --- --- ... --- --- --- --- --- --- ---

X53936,IGLV1-40*02,Vlambda1.2           ... ... ... ... ... --- --- --- --- --- --- --- ... --- --- --- --- --- --- ---

Z22193,IGLV1-40*02,DPL7                 ... ... ... ... ... --- --- --- --- --- --- --- ... --- --- --- --- --- --- ---

Z22192,IGLV1-40*03,DPL6                 ... ... ... ... ... --- --- --- --- --- --- --- ... --- --- --- --- --- --- ---

                                        81  82  83  84  85  86  87  88  89  90  91  92  93  94  95  96  97  98  99 100
                                                 S   G   T   S   A   S   L   A   I   T   G   L   Q   A   E   D   E   A
M94116,IGLV1-40*01,Humlv1042            ... ... TCT GGC ACC TCA GCC TCC CTG GCC ATC ACT GGG CTC CAG GCT GAG GAT GAG GCT

Z73656,IGLV1-40*01,1e-10-2              ... ... --- --- --- --- --- --- --- --- --- --- --- --- --- --- --- --- --- ---

D87010,IGLV1-40*01,V1-13                ... ... --- --- --- --- --- --- --- --- --- --- --- --- --- --- --- --- --- ---

Z22194,IGLV1-40*01,DPL8                 ... ... --- --- --- --- --- --- --- --- --- --- --- --- --- --- --- --- --- ---

X53936,IGLV1-40*02,Vlambda1.2           ... ... --- --- --- --- --- --- --- --- --- --- --- --- --- --- --- --- --- ---

Z22193,IGLV1-40*02,DPL7                 ... ... --- --- --- --- --- --- --- --- --- --- --- --- --- --- --- --- --- ---
                                                         A
Z22192,IGLV1-40*03,DPL6                 ... ... --- --- G-- --- --- --- --- --- --- --- --- --- --- --- --- --- --- ---
```

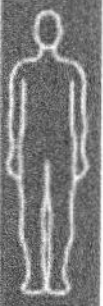

```
                                                              CDR3-IMGT
                                     101 102 103 104 105 106 107 108 109 110 111 112 113
                                      D   Y   Y   C   Q   S   Y   D   S   S   L   S   G
M94116,IGLV1-40*01,Humlv1042         GAT TAT TAC TGC CAG TCC TAT GAC AGC AGC CTG AGT GGT TC
Z73656,IGLV1-40*01,1e-10-2           --  --- . . ---   - --- -   --- ---  -- ---   .  --- --
D87010,IGLV1-40*01,V1-13                 --- .   --- --   -- --- ... ---   . --- -   --- --
Z22194,IGLV1-40*01,DPL8              --  --  --- ---   - --- -   --- --   -- --   -  ---
X53936,IGLV1-40*02,Vlambda1.2        --   -- ---      --- --  -- ---     --- --  --- ---
Z22193,IGLV1-40*02,DPL7              --  --- -  --- ---   . --- -   --- --   -- ---
Z22192,IGLV1-40*03,DPL6                  --- .  --- --   -- ---   . --- . .  --- ---  - -
```

Framework and complementarity determining regions

FR1-IMGT: 25 (-1 aa: 10)
FR2-IMGT: 17
FR3-IMGT: 36 (-3 aa: 73,81,82)

CDR1-IMGT: 9
CDR2-IMGT: 3
CDR3-IMGT: 9

Collier de Perles for human IGLV1-40*01

Accession number: IMGT M94116 EMBL/Genbank/DDBJ: M94116

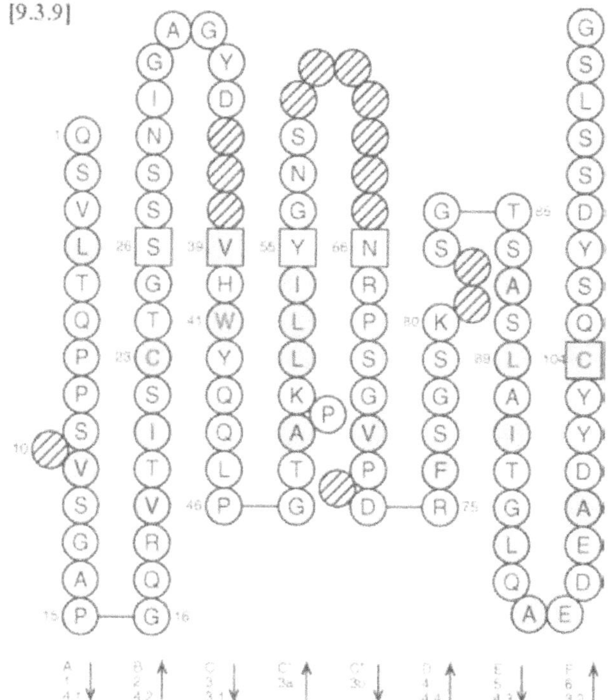

Genome database accession numbers

GDB:9953654 LocusLink: 28825

IGLV1-41

Nomenclature

IGLV1-41: Immunoglobulin lambda variable 1-41.

Definition and functionality

IGLV1-41 is an ORF (allele *01) or a pseudogene (allele *02). IGLV1-41 belongs to the IGLV1 subgroup which comprises eight mapped genes, of which five are functional.
IGLV1-41*01 is an ORF due to a non-canonical ACCEPTOR_SPLICE: nacnn instead of nagnn.
IGLV1-41*02 is a pseudogene due to a STOP-CODON at position 99 in the FR3-IMGT.

Gene location

IGLV1-41 is located in the cluster B of the IGL locus on chromosome 22 at 22q11.2.

Nucleotide and amino acid sequences for human IGLV1-41

```
                                       1   2   3   4   5   6   7   8   9  10  11  12  13  14  15  16  17  18  19  20
                                      Q   S   V   L   T   Q   P   P   S       V   S   A   A   P   G   Q   K   V   T
M94118,IGLV1-41*01,Humlv1041    [1]  CAG TCT GTG TTG ACG CAG CCG CCT TCA ... GTG TCT GCG GCC CCA GGA CAG AAG GTC ACC
X14615,IGLV1-41*01,Vlambda1.1   [9]  --- --- --- --- --- --- --- --- --- ... --- --- --- --- --- --- --- --- --- ---
Z73655,IGLV1-41*01,1d.8.3      [27]  --- --- --- --- --- --- --- --- --- ... --- --- --- --- --- --- --- --- --- ---
Z22212,IGLV1-41*01,DPL4        [26]  --- --- --- --- --- --- --- --- --- ... --- --- --- --- --- --- --- --- --- ---
D87010,IGLV1-41*02,V1-14P      [19]  --- --- --- --- --- --- --- --- --- ... --- --- --- --- --- --- --- --- --- ---

                                                             ______________CDR1-IMGT_______________
                                      21  22  23  24  25  26  27  28  29  30  31  32  33  34  35  36  37  38  39  40
                                      I   S   C   S   G   S   S   S   D   M   G   N   Y   A                   V   S
M94118,IGLV1-41*01,Humlv1041         ATC TCC TGC TCT GGA AGC AGC TCC GAC ATG GGG AAT TAT GCG ... ... ... ... GTA TCC
X14615,IGLV1-41*01,Vlambda1.1        --- --- --- --- --- --- --- --- --- --- --- --- --- --- ... ... ... ... --- ---
Z73655,IGLV1-41*01,1d.8.3            --- --- --- --- --- --- --- --- --- --- --- --- --- --- ... ... ... ... --- ---
Z22212,IGLV1-41*01,DPL4              --- --- --- --- --- --- --- --- --- --- --- --- --- --- ... ... ... ... --- ---
D87010,IGLV1-41*02,V1-14P            --- --- --- --- --- --- --- --- --- --- --- --- --- --- ... ... ... ... --- ---

                                                                                             ______________ CDR2-
                                      41  42  43  44  45  46  47  48  49  50  51  52  53  54  55  56  57  58  59  60
                                      W   Y   Q   Q   L   P   G   T   A   P   K   L   L   I   Y   E   N   N
M94118,IGLV1-41*01,Humlv1041         TGG TAC CAG CAG CTC CCA GGA ACA GCC CCC AAA CTC CTC ATC TAT GAA AAT AAT ... ...
X14615,IGLV1-41*01,Vlambda1.1        --- --- --- --- --- --- --- --- --- --- --- --- --- --- --- --- --- --- ... ...
Z73655,IGLV1-41*01,1d.8.3            --- --- --- --- --- --- --- --- --- --- --- --- --- --- --- --- --- --- ... ...
Z22212,IGLV1-41*01,DPL4              --- --- --- --- --- --- --- --- --- --- --- --- --- --- --- --- --- --- ... ...
D87010,IGLV1-41*02,V1-14P            --- --- --- --- --- --- --- --- --- --- --- --- --- --- --- --- --- --- ... ...

                                     IMGT_______________
                                      61  62  63  64  65  66  67  68  69  70  71  72  73  74  75  76  77  78  79  80
                                                          K   R   P   S   G   I   P       D   R   F   S   G   S   K
M94118,IGLV1-41*01,Humlv1041         ... ... ... ... ... AAG CGA CCC TCA GGG ATT CCT ... GAC CGA TTC TCT GGC TCC AAG
X14615,IGLV1-41*01,Vlambda1.1        ... ... ... ... ... --- --- --- --- --- --- --- ... --- --- --- --- --- --- ---
Z73655,IGLV1-41*01,1d.8.3            ... ... ... ... ... --- --- --- --- --- --- --- ... --- --- --- --- --- --- ---
Z22212,IGLV1-41*01,DPL4              ... ... ... ... ... --- --- --- --- --- --- --- ... --- --- --- --- --- --- ---
D87010,IGLV1-41*02,V1-14P            ... ... ... ... ... --- --- --- --- --- --- --- ... --- --- --- --- --- --- ---

                                      81  82  83  84  85  86  87  88  89  90  91  92  93  94  95  96  97  98  99 100
                                              S   G   T   S   A   T   L   G   I   T   G   L   W   P   E   D   E   A
M94118,IGLV1-41*01,Humlv1041         ... ... TCT GGC ACC TCA GCC ACC CTG GGC ATC ACT GGC CTC TGG CCT GAG GAC GAG GCC
X14615,IGLV1-41*01,Vlambda1.1        ... ... --- --- --- --- --- --- --- --- --- --- --- --- --- --- --- --- --- ---
Z73655,IGLV1-41*01,1d.8.3            ... ... --- --- --- --- --- --- --- --- --- --- --- --- --- --- --- --- --- ---
Z22212,IGLV1-41*01,DPL4              ... ... --- --- --- --- --- --- --- --- --- --- --- --- --- --- --- --- --- ---
                                                                                                          *
D87010,IGLV1-41*02,V1-14P            ... ... --- --- --- --- --- --- --- --- --- --- --- --- --- --- --- --- T-- ---
```

```
                                                        _CDR3-IMGT___________
                              101 102 103 104 105 106 107 108 109 110 111 112 113
                               D   Y   Y   C   L   A   W   D   T   S   P   R   A
M94118,IGLV1-41*01,Humlv1041  GAT TAT TAC TGC TTA GCA TGG GAT ACC AGC CCG AGA GCT TG
X14615,IGLV1-41*01,Vlambda1.1 --- --- --- --- --- --- --- --- --- --- --- --- --- --
Z73655,IGLV1-41*01,1d.8.3     --- --- --- --- --- --- --- --- --- --- --- --- --- -
Z22212,IGLV1-41*01,DPL4       --- --- --- --- --- --- --- --- --- --- --- --- ---
                                                                      L
D87010,IGLV1-41*02,V1-14P     --- --- --- --- --- --- --- --- --- --- -T- --- --- --
```

Framework and complementarity determining regions

FR1-IMGT: 25 (-1 aa: 10)
FR2-IMGT: 17
FR3-IMGT: 36 (-3 aa: 73,81,82)

CDR1-IMGT: 8
CDR2-IMGT: 3
CDR3-IMGT: 9

Collier de Perles for human IGLV1-41*01

Accession number: IMGT M94118 EMBL/GenBank/DDBJ: M94118

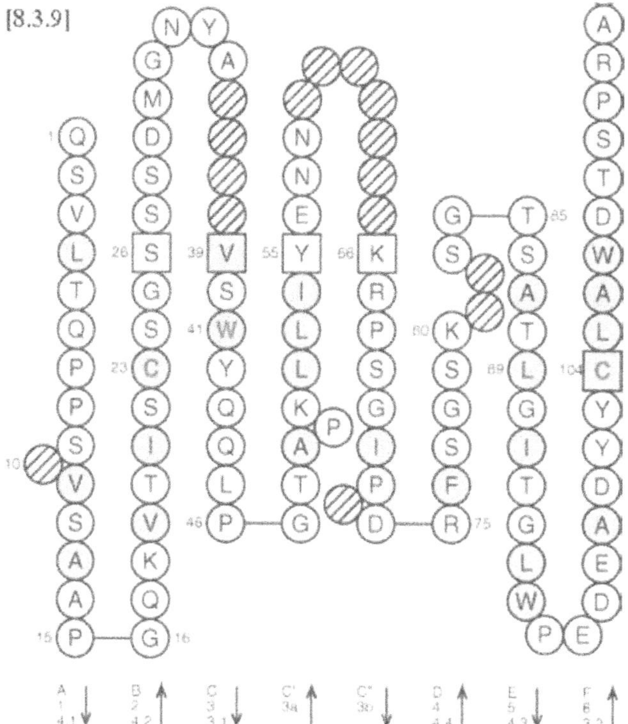

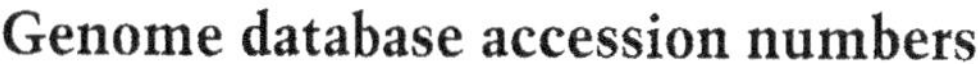

Genome database accession numbers

GDB:9953656 LocusLink: 28824

IGLV1-44

Nomenclature

IGLV1-44: Immunoglobulin lambda variable 1-44.

Definition and functionality

IGLV1-44 is one of the five functional genes of the IGLV1 subgroup which comprises eight mapped genes.

Gene location

IGLV1-44 is located in the cluster B of the IGL locus on chromosome 22 at 22q11.2.

Nucleotide and amino acid sequences for human IGLV1-44

```
                                     1   2   3   4   5   6   7   8   9  10  11  12  13  14  15  16  17  18  19  20
                                     Q   S   V   L   T   Q   P   P   S       A   S   G   T   P   G   Q   R   V   T
Z73654,IGLV1-44*01,1c.10.2    [27]  CAG TCT GTG CTG ACT CAG CCA CCC TCA ... GCG TCT GGG ACC CCC GGG CAG AGG GTC ACC
D86999,IGLV1-44*01,V1-16      [19]  --- --- --- --- --- --- --- --- --- ... --- --- --- --- --- --- --- --- --- ---
Z22188,IGLV1-44*01,DPL2       [26]  --- --- --- --- --- --- --- --- --- ... --- --- --- --- --- --- --- --- --- ---
X59707,IGLV1-44*01,Humlv1L1   [9]   --- --- --- --- --- --- --- --- --- ... --- --- --- --- --- --- --- --- --- ---
U03902,IGL1-44*01,lv1L1a      [12]  --- --- --- --- --- --- --- --- --- ... --- --- --- --- --- --- --- --- --- ---

                                                                        CDR1-IMGT
                                    21  22  23  24  25  26  27  28  29  30  31  32  33  34  35  36  37  38  39  40
                                     I   S   C   S   G   S   S   S   N   I   G   S   N   T                   V   N
Z73654,IGLV1-44*01,1c.10.2          ATC TCT TGT TCT GGA AGC AGC TCC AAC ATC GGA AGT AAT ACT ... ... ... ... GTA AAC
D86999,IGLV1-44*01,V1-16            --- --- --- --- --- --- --- --- --- --- --- --- --- --- ... ... ... ... --- ---
Z22188,IGLV1-44*01,DPL2             --- --- --- --- --- --- --- --- --- --- --- --- --- --- ... ... ... ... --- ---
X59707,IGLV1-44*01,Humlv1L1         --- --- --- --- --- --- --- --- --- --- --- --- --- --- ... ... ... ... --- ---
U03902,IGL1-44*01,lv1L1a            --- --- --- --- --- --- --- --- --- --- --- --- --- --- ... ... ... ... --- ---

                                                                                                         CDR2
                                    41  42  43  44  45  46  47  48  49  50  51  52  53  54  55  56  57  58  59  60
                                     W   Y   Q   Q   L   P   G   T   A   P   K   L   L   I   Y   S   N   N
Z73654,IGLV1-44*01,1c.10.2          TGG TAC CAG CAG CTC CCA GGA ACG GCC CCC AAA CTC CTC ATC TAT AGT AAT AAT ... ...
D86999,IGLV1-44*01,V1-16            --- --- --- --- --- --- --- --- --- --- --- --- --- --- --- --- --- --- ... ...
Z22188,IGLV1-44*01,DPL2             --- --- --- --- --- --- --- --- --- --- --- --- --- --- --- --- --- --- ... ...
X59707,IGLV1-44*01,Humlv1L1         --- --- --- --- --- --- --- --- --- --- --- --- --- --- --- --- --- --- ... ...
U03902,IGL1-44*01,lv1L1a            --- --- --- --- --- --- --- --- --- --- --- --- --- --- --- --- --- --- ... ...

                                    IMGT
                                    61  62  63  64  65  66  67  68  69  70  71  72  73  74  75  76  77  78  79  80
                                                         Q   R   P   S   G   V   P       D   R   F   S   G   S   K
Z73654,IGLV1-44*01,1c.10.2          ... ... ... ... ... CAG CGG CCC TCA GGG GTC CCT ... GAC CGA TTC TCT GGC TCC AAG
D86999,IGLV1-44*01,V1-16            ... ... ... ... ... --- --- --- --- --- --- --- ... --- --- --- --- --- --- ---
Z22188,IGLV1-44*01,DPL2             ... ... ... ... ... --- --- --- --- --- --- --- ... --- --- --- --- --- --- ---
X59707,IGLV1-44*01,Humlv1L1         ... ... ... ... ... --- --- --- --- --- --- --- ... --- --- --- --- --- --- ---
U03902,IGL1-44*01,lv1L1a            ... ... ... ... ... --- --- --- --- --- --- --- ... --- --- --- --- --- --- ---

                                    81  82  83  84  85  86  87  88  89  90  91  92  93  94  95  96  97  98  99 100
                                             S   G   T   S   A   S   L   A   I   S   G   L   Q   S   E   D   E   A
Z73654,IGLV1-44*01,1c.10.2          ... ... TCT GGC ACC TCA GCC TCC CTG GCC ATC AGT GGG CTC CAG TCT GAG GAT GAG GCT
D86999,IGLV1-44*01,V1-16            ... ... --- --- --- --- --- --- --- --- --- --- --- --- --- --- --- --- --- ---
Z22188,IGLV1-44*01,DPL2             ... ... --- --- --- --- --- --- --- --- --- --- --- --- --- --- --- --- --- ---
X59707,IGLV1-44*01,Humlv1L1         ... ... --- --- --- --- --- --- --- --- --- --- --- --- --- --- --- --- --- ---
U03902,IGL1-44*01,lv1L1a            ... ... --- --- --- --- --- --- --- --- --- --- --- --- --- --- --- --- --- ---

                                                     _____________CDR3-IMGT___________
                                    101 102 103 104 105 106 107 108 109 110 111 112 113
                                     D   Y   Y   C   A   A   W   D   D   S   L   N   G
Z73654,IGLV1-44*01,1c.10.2          GAT TAT TAC TGT GCA GCA TGG GAT GAC AGC CTG AAT GGT CC
D86999,IGLV1-44*01,V1-16            --- --- --- --- --- --- --- --- --- --- --- --- --- --
Z22188,IGLV1-44*01,DPL2             --- --- --- --- --- --- --- --- --- --- --- --- ---
X59707,IGLV1-44*01,Humlv1L1         --- --- --- --- --- --- --- --- --- --- --- --- --- --
U03902,IGL1-44*01,lv1L1a            --- --- --- --- --- --- --- --- --- --- --- --- --- --
```

Framework and complementarity determining regions

FR1-IMGT: 25 (-1 aa: 10)
FR2-IMGT: 17
FR3-IMGT: 36 (-3 aa: 73,81,82)

CDR1-IMGT: 8
CDR2-IMGT: 3
CDR3-IMGT: 9

Collier de Perles for human IGLV1-44*01

Accession number: IMGT Z73654

EMBL/GenBank/DDBJ: Z73654

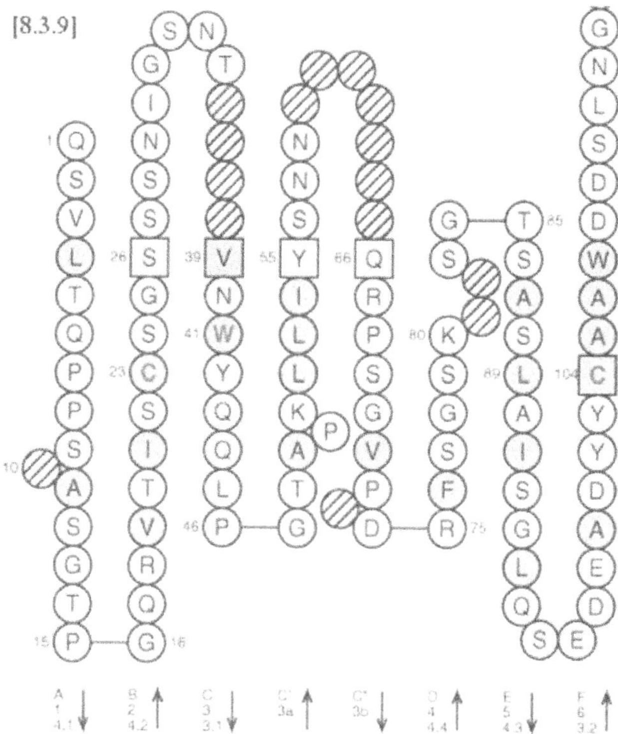

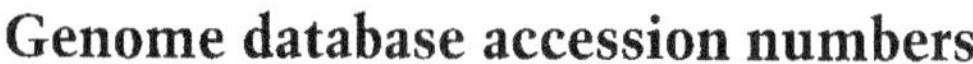

Genome database accession numbers

GDB:9953658

LocusLink: 28823

Nomenclature

IGLV1-47: Immunoglobulin lambda variable 1-47.

Definition and functionality

IGLV1-47 is one of the five functional genes of the IGLV1 subgroup which comprises eight mapped genes.

Gene location

IGLV1-47 is located in the cluster B of the IGL locus on chromosome 22 at 22q11.2.

Nucleotide and amino acid sequences for human IGLV1-47

```
                                       1   2   3   4   5   6   7   8   9  10  11  12  13  14  15  16  17  18  19  20
                                       Q   S   V   L   T   Q   P   P   S       A   S   G   T   P   G   Q   R   V   T
Z73663,IGLV1-47*01,1g.400B5     [27]  CAG TCT GTG CTG ACT CAG CCA CCC TCA ... GCG TCT GGG ACC CCC GGG CAG AGG GTC ACC
Z22189,IGLV1-47*01,DPL3         [26]  --- --- --- --- --- --- --- --- --- ... --- --- --- --- --- --- --- --- --- ---
M94114,IGLV1-47*01,Humlv122      [9]  --- --- --- --- --- --- --- --- --- ... --- --- --- --- --- --- --- --- --- ---
D87016,IGLV1-47*02,V1-17        [19]  --- --- --- --- --- --- --- --- --- ... --- --- --- --- --- --- --- --- --- ---

                                                              ___________________CDR1-IMGT____________________
                                      21  22  23  24  25  26  27  28  29  30  31  32  33  34  35  36  37  38  39  40
                                       I   S   C   S   G   S   S   S   N   I   G   S   N   Y                   V   Y
Z73663,IGLV1-47*01,1g.400B5           ATC TCT TGT TCT GGA AGC AGC TCC AAC ATC GGA AGT AAT TAT ... ... ... ... GTA TAC
Z22189,IGLV1-47*01,DPL3               --- --- --- --- --- --- --- --- --- --- --- --- --- --- ... ... ... ... --- ---
M94114,IGLV1-47*01,Humlv122           --- --- --- --- --- --- --- --- --- --- --- --- --- --- ... ... ... ... --- --
D87016,IGLV1-47*02,V1-17              --- --- --- --- --- --- --- --- --- --- --- --- --- --- ... ... ... ... --- ---

                                                                                          _____________ CDR2-
                                      41  42  43  44  45  46  47  48  49  50  51  52  53  54  55  56  57  58  59  60
                                       W   Y   Q   Q   L   P   G   T   A   P   K   L   L   I   Y   R   N   N
Z73663,IGLV1-47*01,1g.400B5           TGC TAC CAG CAG CTC CCA GGA ACG GCC CCC AAA CTC CTC ATC TAT AGG AAT AAT ... ...
Z22189,IGLV1-47*01,DPL3               --- --- --- --- --- --- --- --- --- --- --- --- --- --- --- --- --- --- ... ...
M94114,IGLV1-47*01,Humlv122           --- --- --- --- --- --- --- --- --- --- --- --- --- --- --- --- --- --- ... ...
                                                                                               S
D87016,IGLV1-47*02,V1-17              --- --- --- --- --- --- --- --- --- --- --- --- --- --- --- --T --- --- ... ...

                                      IMGT________________
                                      61  62  63  64  65  66  67  68  69  70  71  72  73  74  75  76  77  78  79  80
                                                           Q   R   P   S   G   V   P       D   R   F   S   G   S   K
Z73663,IGLV1-47*01,1g.400B5           ... ... ... ... ... CAG CGG CCC TCA GGG GTC CCT ... GAC CGA TTC TCT GGC TCC AAG
Z22189,IGLV1-47*01,DPL3               ... ... ... ... ... --- --- --- --- --- --- --- ... --- --- --- --- --- --- ---
M94114,IGLV1-47*01,Humlv122           ... ... ... ... ... --- --- --- --- --- --- --- ... --- --- --- --- --- --- ---
D87016,IGLV1-47*02,V1-17              ... ... ... ... ... --- --- --- --- --- --- --- ... --- --- --- --- --- --- ---

                                      81  82  83  84  85  86  87  88  89  90  91  92  93  94  95  96  97  98  99 100
                                               S   G   T   S   A   S   L   A   I   S   G   L   R   S   E   D   E   A
Z73663,IGLV1-47*01,1g.400B5           ... ... TCT GGC ACC TCA GCC TCC CTG GCC ATC AGT GGG CTC CGG TCC GAG GAT GAG GCT
Z22189,IGLV1-47*01,DPL3               ... ... --- --- --- --- --- --- --- --- --- --- --- --- --- --- --- --- --- ---
M94114,IGLV1-47*01,Humlv122           ... ... --- --- --- --- --- --- --- --- --- --- --- --- --- --- --- --- --- ---
D87016,IGLV1-47*02,V1-17              ... ... --- --- --- --- --- --- --- --- --- --- --- --- --- --- --- --- --- ---

                                                      _____________CDR3-IMGT______________
                                      101 102 103 104 105 106 107 108 109 110 111 112 113
                                       D   Y   Y   C   A   A   W   D   D   S   L   S   G
Z73663,IGLV1-47*01,1g.400B5           GAT TAT TAC TGT GCA GCA TGG GAT GAC AGC CTG AGT GGT CC
Z22189,IGLV1-47*01,DPL3               --- --- --- --- --- --- --- --- --- --- --- --- ---
M94114,IGLV1-47*01,Humlv122           --- --- --- --- --- --- --- --- --- --- --- --- --- --
D87016,IGLV1-47*02,V1-17              --- --- --- --- --- --- --- --- --- --- --- --- --- --
```

Framework and complementarity determining regions

FR1-IMGT: 25 (-1 aa: 10)
FR2-IMGT: 17
FR3-IMGT: 36 (-3 aa: 73,81,82)

CDR1-IMGT: 8
CDR2-IMGT: 3
CDR3-IMGT: 9

Collier de Perles for human IGLV1-47*01

Accession number: IMGT Z73663 EMBL/GenBank/DDBJ: Z73663

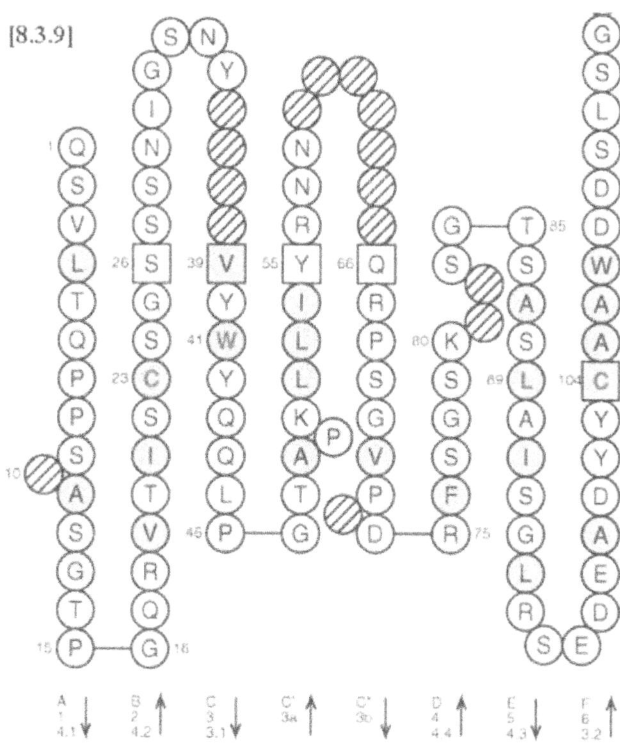

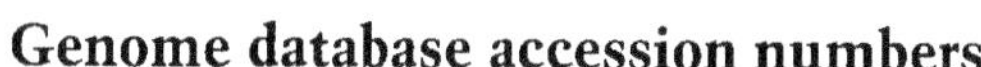

Genome database accession numbers

GDB:9953660 LocusLink: 28822

Nomenclature

IGLV1-50: Immunoglobulin lambda variable 1-50.

Definition and functionality

IGLV1-50 is an ORF due to a non-conserved OCTAMER sequence: ctgcacaa instead of atttgcat, a non-conserved TATA_BOX and an unusual V-NONAMER sequence: agaagaacc instead of acaaaaacc. IGLV1-50 belongs to the IGLV1 subgroup which comprises eight mapped genes, of which five are functional.

Gene location

IGLV1-50 is located in the cluster B of the IGL locus on chromosome 22 at 22q11.2.

Nucleotide and amino acid sequences for human IGLV1-50

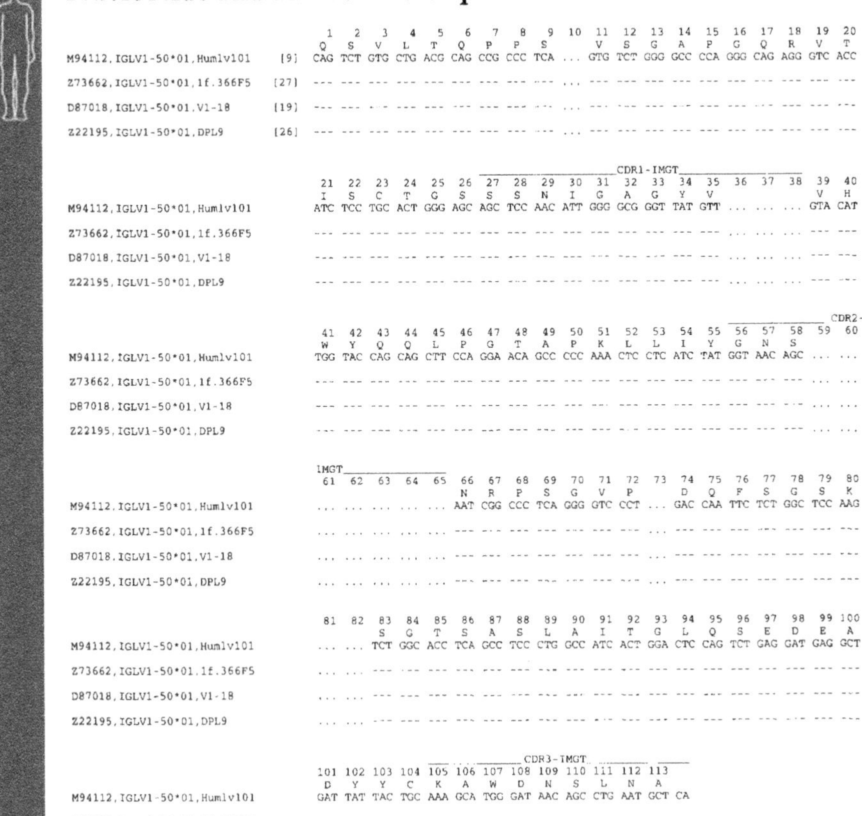

Framework and complementarity determining regions

FR1-IMGT: 25 (-1 aa: 10)
FR2-IMGT: 17
FR3-IMGT: 36 (-3 aa: 73,81,82)

CDR1-IMGT: 9
CDR2-IMGT: 3
CDR3-IMGT: 9

Collier de Perles for human IGLV1-50*01

Accession number: IMGT M94112 EMBL/GenBank/DDBJ: M94112

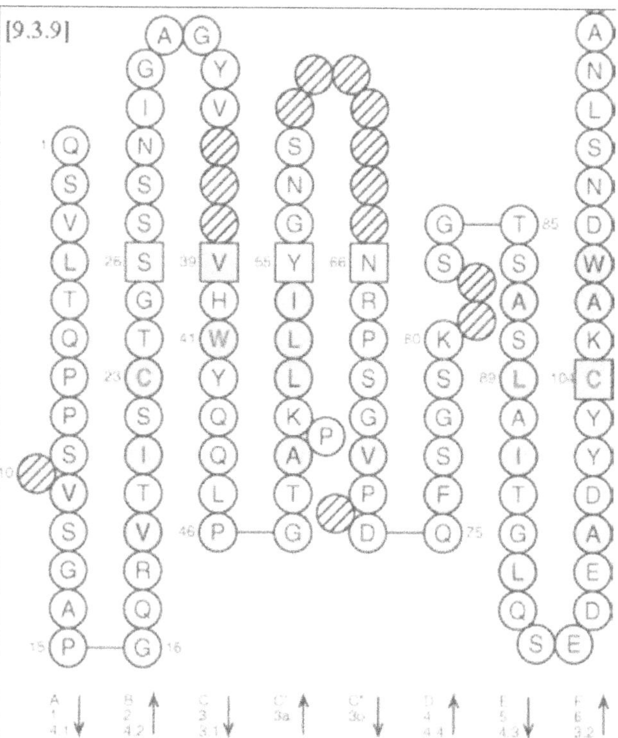

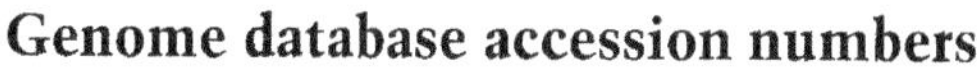

Genome database accession numbers

GDB:9953662 LocusLink: 28821

Nomenclature

IGLV1-51: Immunoglobulin lambda variable 1-51.

Definition and functionality

IGLV1-51 is one of the five functional genes of the IGLV1 subgroup which comprises eight mapped genes.

Gene location

IGLV1-51 is located in the cluster B of the IGL locus on chromosome 22 at 22q11.2.

Nucleotide and amino acid sequences for human IGLV1-51

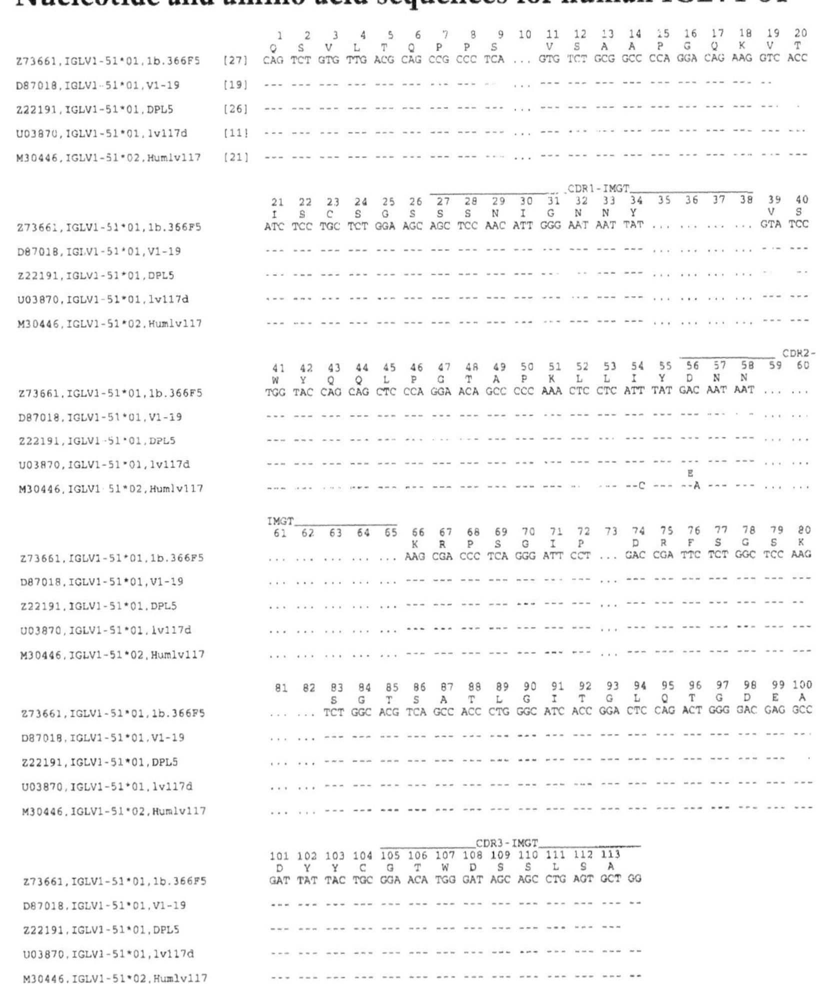

```
                                     1   2   3   4   5   6   7   8   9  10  11  12  13  14  15  16  17  18  19  20
                                     Q   S   V   L   T   Q   P   P   S       V   S   A   A   P   G   Q   K   V   T
Z73661,IGLV1-51*01,1b.366F5   [27]  CAG TCT GTG TTG ACG CAG CCG CCC TCA ... GTG TCT GCG GCC CCA GGA CAG AAG GTC ACC
D87018,IGLV1-51*01,V1-19      [19]  --- --- --- --- --- --- --- --- --- ... --- --- --- --- --- --- --- --- --
Z22191,IGLV1-51*01,DPL5       [26]  --- --- --- --- --- --- --- --- --- ... --- --- --- --- --- --- --- --- ---
U03870,IGLV1-51*01,lv117d     [11]  --- --- --- --- --- --- --- --- --- ... --- --- --- --- --- --- --- --- --- ---
M30446,IGLV1-51*02,Humlv117   [21]  --- --- --- --- --- --- --- --- --- ... --- --- --- --- --- --- --- --- --- ---

                                                            _________________CDR1-IMGT________________
                                    21  22  23  24  25  26  27  28  29  30  31  32  33  34  35  36  37  38  39  40
                                     I   S   C   S   G   S   S   S   N   I   G   N   N   Y                   V   S
Z73661,IGLV1-51*01,1b.366F5         ATC TCC TGC TCT GGA AGC AGC TCC AAC ATT GGG AAT AAT TAT ... ... ... ... GTA TCC
D87018,IGLV1-51*01,V1-19            --- --- --- --- --- --- --- --- --- --- --- --- --- --- ... ... ... ... --- ---
Z22191,IGLV1-51*01,DPL5             --- --- --- --- --- --- --- --- --- --- --- --- --- --- ... ... ... ... --- ---
U03870,IGLV1-51*01,lv117d           --- --- --- --- --- --- --- --- --- --- --- --- --- --- ... ... ... ... --- ---
M30446,IGLV1-51*02,Humlv117         --- --- --- --- --- --- --- --- --- --- --- --- --- --- ... ... ... ... --- ---

                                                                                        ____________ CDR2-
                                    41  42  43  44  45  46  47  48  49  50  51  52  53  54  55  56  57  58  59  60
                                     W   Y   Q   Q   L   P   G   T   A   P   K   L   L   I   Y   D   N   N
Z73661,IGLV1-51*01,1b.366F5         TGG TAC CAG CAG CTC CCA GGA ACA GCC CCC AAA CTC CTC ATT TAT GAC AAT AAT ... ...
D87018,IGLV1-51*01,V1-19            --- --- --- --- --- --- --- --- --- --- --- --- --- --- --- --- --- --- ... ...
Z22191,IGLV1-51*01,DPL5             --- --- --- --- --- --- --- --- --- --- --- --- --- --- --- --- --- --- ... ...
U03870,IGLV1-51*01,lv117d           --- --- --- --- --- --- --- --- --- --- --- --- --- --- --- --- --- --- ... ...
                                                                                             E
M30446,IGLV1-51*02,Humlv117         --- --- --- --- --- --- --- --- --- --- --- --- --- --C --- --A --- --- ... ...

                                    IMGT_______________
                                    61  62  63  64  65  66  67  68  69  70  71  72  73  74  75  76  77  78  79  80
                                                         K   R   P   S   G   I   P       D   R   F   S   G   S   K
Z73661,IGLV1-51*01,1b.366F5         ... ... ... ... ... AAG CGA CCC TCA GGG ATT CCT ... GAC CGA TTC TCT GGC TCC AAG
D87018,IGLV1-51*01,V1-19            ... ... ... ... ... --- --- --- --- --- --- --- ... --- --- --- --- --- --- ---
Z22191,IGLV1-51*01,DPL5             ... ... ... ... ... --- --- --- --- --- --- --- ... --- --- --- --- --- --- --
U03870,IGLV1-51*01,lv117d           ... ... ... ... ... --- --- --- --- --- --- --- ... --- --- --- --- --- --- ---
M30446,IGLV1-51*02,Humlv117         ... ... ... ... ... --- --- --- --- --- --- --- ... --- --- --- --- --- --- ---

                                    81  82  83  84  85  86  87  88  89  90  91  92  93  94  95  96  97  98  99 100
                                             S   G   T   S   A   T   L   G   I   T   G   L   Q   T   G   D   E   A
Z73661,IGLV1-51*01,1b.366F5         ... ... TCT GGC ACG TCA GCC ACC CTG GGC ATC ACC GGA CTC CAG ACT GGG GAC GAG GCC
D87018,IGLV1-51*01,V1-19            ... ... --- --- --- --- --- --- --- --- --- --- --- --- --- --- --- --- --- ---
Z22191,IGLV1-51*01,DPL5             ... ... --- --- --- --- --- --- --- --- --- --- --- --- --- --- --- ---
U03870,IGLV1-51*01,lv117d           ... ... --- --- --- --- --- --- --- --- --- --- --- --- --- --- --- --- --- ---
M30446,IGLV1-51*02,Humlv117         ... ... --- --- --- --- --- --- --- --- --- --- --- --- --- --- --- --- --- ---

                                                        ____________CDR3-IMGT__________
                                    101 102 103 104 105 106 107 108 109 110 111 112 113
                                     D   Y   Y   C   G   T   W   D   S   S   L   S   A
Z73661,IGLV1-51*01,1b.366F5         GAT TAT TAC TGC GGA ACA TGG GAT AGC AGC CTG AGT GCT GG
D87018,IGLV1-51*01,V1-19            --- --- --- --- --- --- --- --- --- --- --- --- --- --
Z22191,IGLV1-51*01,DPL5             --- --- --- --- --- --- --- --- --- --- --- --- ---
U03870,IGLV1-51*01,lv117d           --- --- --- --- --- --- --- --- --- --- --- --- --- --
M30446,IGLV1-51*02,Humlv117         --- --- --- --- --- --- --- --- --- --- --- --- --- --
```

Framework and complementarity determining regions

FR1-IMGT: 25 (-1 aa: 10)
FR2-IMGT: 17
FR3-IMGT: 36 (-3 aa: 73,81,82)

CDR1-IMGT: 8
CDR2-IMGT: 3
CDR3-IMGT: 9

Collier de Perles for human IGLV1-51*01

Accession number: IMGT Z73661

EMBL/GenBank/DDBJ: Z73661

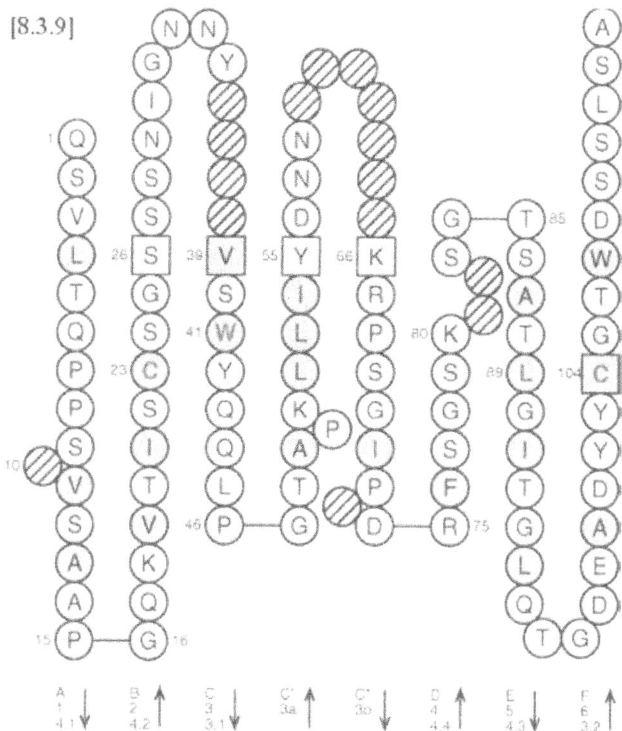

Genome database accession numbers

GDB:9953664

LocusLink: 28820

IGLV2-8

Nomenclature

IGLV2-8: Immunoglobulin lambda variable 2-8.

Definition and functionality

IGLV2-8 is one of the five functional genes of the IGLV2 subgroup which comprises nine mapped genes.

Gene location

IGLV2-8 is located in the cluster A of the IGL locus on chromosome 22 at 22q11.2.

Nucleotide and amino acid sequences for human IGLV2-8

```
                                   1   2   3   4   5   6   7   8   9  10  11  12  13  14  15  16  17  18  19  20
                                   Q   S   A   L   T   Q   P   P   S       A   S   G   S   P   G   Q   S   V   T
X97462,IGLV2-8*01,2c.118D9  [27]  CAG TCT GCC CTG ACT CAG CCT CCC TCC ... GCG TCC GGG TCT CCT GGA CAG TCA GTC ACC

D87021,IGLV2-8*01,V1-2      [16]  --- --- --- --- --- --- --- --- --- ... --- --- --- --- --- --- --- --- --- ---

Y12417,IGLV2-8*01,EKL2046    [3]                              --- --- ... --- --- --- --- --- --- --- --- --- ---
                                                                  R
L27695,IGLV2-8*02,1v2046    [18]  --- --- --- --- --- --- --- --- --- ... --- --- A-- --- --- --- --- --- --- ---

Y12418,IGLV2-8*03,RXL2046    [3]                              --- --- ... --- --- --- --- --- --- --- --- --- ---


                                                         ______CDR1-IMGT______
                                  21  22  23  24  25  26  27  28  29  30  31  32  33  34  35  36  37  38  39  40
                                   I   S   C   T   G   T   S   S   D   V   G   G   Y   N   Y               V   S
X97462,IGLV2-8*01,2c.118D9        ATC TCC TGC ACT GGA ACC AGC AGT GAC GTT GGT GGT TAT AAC TAT ... ... ... GTC TCC

D87021,IGLV2-8*01,V1-2            --- --- --- --- --- --- --- --- --- --- --- --- --- --- --- ... ... ... --- ---

Y12417,IGLV2-8*01,EKL2046         --- --- --- --- --- --- --- --- --- --- --- --- --- --- --- ... ... ... --- ---

L27695,IGLV2-8*02,1v2046          --- --- --- --- --- --- --- --- --- --- --- --- --- --- --- ... ... ... --- ---

Y12418,IGLV2-8*03,RXL2046         --- --- --- --- --- --- --- --- --- --- --- --- --- --- --- ... ... ... --- ---


                                                                                              ____________ CDR2
                                  41  42  43  44  45  46  47  48  49  50  51  52  53  54  55  56  57  58  59  60
                                   W   Y   Q   Q   H   P   G   K   A   P   K   L   M   I   Y   E   V   S
X97462,IGLV2-8*01,2c.118D9        TGG TAC CAA CAG CAC CCA GGC AAA GCC CCC AAA CTC ATG ATT TAT GAG GTC AGT ... ...

D87021,IGLV2-8*01,V1-2            --- --- --- --- --- --- --- --- --- --- --- --- --- --- --- --- --- --- ... ...

Y12417,IGLV2-8*01,EKL2046         --- --- --- --- --- --- --- --- --- --- --- --- --- --- --- --- --- --- ... ...

L27695,IGLV2-8*02,1v2046          --- --- --- --- --- --- --- --- --- --- --- --- --- --- --- --- --- --- ... ...

Y12418,IGLV2-8*03,RXL2046         --- --- --- --- --- --- --- --- --- --- --- --- --- --- --- --- --- --- ... ...


                                  IMGT________________
                                  61  62  63  64  65  66  67  68  69  70  71  72  73  74  75  76  77  78  79  80
                                                       K   R   P   S   G   V   P       D   R   F   S   G   S   K
X97462,IGLV2-8*01,2c.118D9        ... ... ... ... ... AAG CGG CCC TCA GGG GTC CCT ... GAT CGC TTC TCT GGC TCC AAG

D87021,IGLV2-8*01,V1-2            ... ... ... ... ... --- --- --- --- --- --- --- ... --- --- --- --- --- --- ---

Y12417,IGLV2-8*01,EKL2046         ... ... ... ... ... --- --- --- --- --- --- --- ... --- --- --- --- --- --- ---

L27695,IGLV2-8*02,1v2046          ... ... ... ... ... --- --- --- --- --- --- --- ... --- --- --- --- --- --- ---
                                                                                                   F
Y12418,IGLV2-8*03,RXL2046         ... ... ... ... ... --- --- --- --- --- --- --- ... --- --- --- -T- --- --- ---


                                  81  82  83  84  85  86  87  88  89  90  91  92  93  94  95  96  97  98  99  100
                                           S   G   N   T   A   S   L   T   V   S   G   L   Q   A   E   D   E   A
X97462,IGLV2-8*01,2c.118D9        ... ... TCT GGC AAC ACG GCC TCC CTG ACC GTC TCT GGG CTC CAG GCT GAG GAT GAG GCT

D87021,IGLV2-8*01,V1-2            ... ... --- --- --- --- --- --- --- --- --- --- --- --- --- --- --- --- --- ---

Y12417,IGLV2-8*01,EKL2046         ... ... --- --- --- --- --- --- --- --- --- --- --- --- --- --- --- --- --- ---

L27695,IGLV2-8*02,1v2046          ... ... --- --- --- --- --- --- --- --- --- --- --- --- --- --- --- --- --- ---

Y12418,IGLV2-8*03,RXL2046         ... ... --- --- --- --- --- --- --- --- --- --- --- --- --- --- --- --- --- ---


                                                  ________CDR3-IMGT________
                                  101 102 103 104 105 106 107 108 109 110 111 112 113
                                   D   Y   Y   C   S   S   Y   A   G   S   N   N   F
X97462,IGLV2-8*01,2c.118D9        GAT TAT TAC TGC AGC TCA TAT GCA GGC AGC AAC AAT TTC

D87021,IGLV2-8*01,V1-2            --- --- --- ---             --- --- --- --- --- ---

Y12417,IGLV2-8*01,EKL2046         --- --- --- --- --- ---     --- --- --- --- --- ---

L27695,IGLV2-8*02,1v2046          --- --- --- --- --- --- --- --- ---         --- ---

Y12418,IGLV2-8*03,RXL2046         --- --- ---                 --- --- --- --- --- ---
```

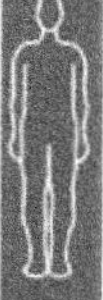

Framework and complementarity determining regions

FR1-IMGT: 25 (-1 aa: 10)
FR2-IMGT: 17
FR3-IMGT: 36 (-3 aa: 73,81,82)

CDR1-IMGT: 9
CDR2-IMGT: 3
CDR3-IMGT: 9

Collier de Perles for human IGLV2-8*01

Accession number: IMGT X97462

EMBL/GenBank/DDBJ: X97462

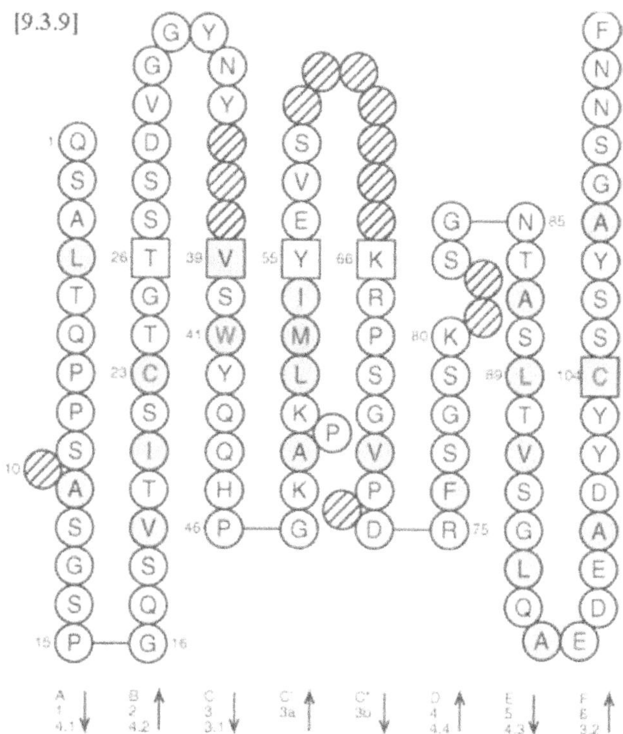

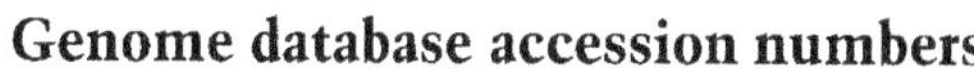

Genome database accession numbers

GDB:9953670 LocusLink: 28817

Nomenclature

IGLV2-11: Immunoglobulin lambda variable 2-11.

Definition and functionality

IGLV2-11 is one of the five functional genes of the IGLV2 subgroup which comprises nine mapped genes.

Gene location

IGLV2-11 is located in the cluster A of the IGL locus on chromosome 22 at 22q11.2.

Nucleotide and amino acid sequences for human IGLV2-11

```
                                    1   2   3   4   5   6   7   8   9   10  11  12  13  14  15  16  17  18  19  20
                                    Q   S   A   L   T   Q   P   R   S       V   S   G   S   P   G   Q   S   V   T
Z73657,IGLV2-11*01,2e.2.2     [27] CAG TCT GCC CTG ACT CAG CCT CGC TCA ... GTG TCC GGG TCT CCT GGA CAG TCA GTC ACC
D86998,IGLV2-11*01,V1-3       [19] --- --- --- --- --- --- --- --- --- ... --- --- --- --- --- --- --- --- --- ---
Y12414,IGLV2-11*01,EKL12       [3]                                 - --- ... --- --- --- --- --- --- --- --- --- ---
Z22198,IGLV2-11*02,DPL12      [26] --- --- --- --- --- ---     --- --- ... --- --- --- --- --- --- --- --- --- ---
Y12415,IGLV2-11*03,RXL12       [3]                                 - --- ... --- --- --- --- --- --- --- --- --- ---

                                                        ________________CDR1-IMGT________________
                                    21  22  23  24  25  26  27  28  29  30  31  32  33  34  35  36  37  38  39  40
                                    I   S   C   T   G   T   S   S   D   V   G   G   Y   N   Y               V   S
Z73657,IGLV2-11*01,2e.2.2          ATC TCC TGC ACT GGA ACC AGC AGT GAT GTT GGT GGT TAT AAC TAT ... ... ... GTC TCC
D86998,IGLV2-11*01,V1-3            --- --- --   -- --- --- --- --- --- --- --- --- --- --- --- ... ... ... ---   -
Y12414,IGLV2-11*01,EKL12           --- --- --- --- --- --- --- --- --- --- --- --- --- --- --- ... ... ... --- ---
Z22198,IGLV2-11*02,DPL12           --- --- --- --- --- --- --- --- --- --- --- --G --- --- --- ... ... ... --- ---
Y12415,IGLV2-11*03,RXL12           --- --- --- --- --- --- --- --- --- --- --- --- --- --- --- ... ... ... --- ---

                                                                                       __________ CDR2-
                                    41  42  43  44  45  46  47  48  49  50  51  52  53  54  55  56  57  58  59  60
                                    W   Y   Q   Q   H   P   G   K   A   P   K   L   M   I   Y   D   V   S
Z73657,IGLV2-11*01,2e.2.2          TGG TAC CAA CAG CAC CCA GGC AAA GCC CCC AAA CTC ATG ATT TAT GAT GTC AGT ... ...
D86998,IGLV2-11*01,V1-3            --  --- --- --- --- --- --- --- --- --- --- --- --- --- --- --- --- --- ... ...
Y12414,IGLV2-11*01,EKL12           --- --- --- --- --- --- --- --- --- --- --- --- --- --- --- --- --- --- ... ...
Z22198,IGLV2-11*02,DPL12           --- --- --- --- --- --- --- --- --- --- --- --- --- --- --- --- --- --- ... ...
Y12415,IGLV2-11*03,RXL12           --- --- --- --A --- --- --- --- --- --- --- --- --- --- --- --- --- --- ... ...

                                   IMGT_______________
                                    61  62  63  64  65  66  67  68  69  70  71  72  73  74  75  76  77  78  79  80
                                                        K   R   P   S   G   V   P       D   R   F   S   G   S   K
Z73657,IGLV2-11*01,2e.2.2          ... ... ... ... ... AAG CGG CCC TCA GGG GTC CCT ... GAT CGC TTC TCT GGC TCC AAG
D86998,IGLV2-11*01,V1-3            ... ... ... ... ... --- --- --- --- --- --- --- ... --- --- --- --- --- --- ---
Y12414,IGLV2-11*01,EKL12           ... ... ... ... ... --- --- --- --- --- --- --- ... --- --- --- --- --- --- ---
Z22198,IGLV2-11*02,DPL12           ... ... ... ... ... --- --- --- --- --- --- --- ... --- --- --- --- --- --- ---
Y12415,IGLV2-11*03,RXL12           ... ... ... ... ... --- --- --- --- --- --- --- ... --- --- --- --- --- --- ---

                                    81  82  83  84  85  86  87  88  89  90  91  92  93  94  95  96  97  98  99  100
                                            S   G   N   T   A   S   L   T   I   S   G   L   Q   A   E   D   E   A
Z73657,IGLV2-11*01,2e.2.2          ... ... TCT GGC AAC ACG GCC TCC CTG ACC ATC TCT GGG CTC CAG GCT GAG GAT GAG GCT
D86998,IGLV2-11*01,V1-3            ... ... --- --- --- --- --- --- --- --- --- --- --- --- --- --- --- --- --- ---
Y12414,IGLV2-11*01,EKL12           ... ... --- --- --- --- --- --- --- --- --- --- --- --- --- --- --- --- --- ---
Z22198,IGLV2-11*02,DPL12           ... ... --- --- --- --- --- --- --- --- --- --- --- --- --- --- --- --- --- ---
Y12415,IGLV2-11*03,RXL12           ... ... --- --- --- --- --- --- --- --- --- --- --- --- --- --- --- --- --- ---

                                                    __________CDR3-IMGT__________
                                   101 102 103 104 105 106 107 108 109 110 111 112 113
                                    D   Y   Y   C   C   S   Y   A   G   S   Y   T   F
Z73657,IGLV2-11*01,2e.2.2          GAT TAT TAC TGC TGC TCA TAT GCA GGC AGC TAC ACT TTC
D86998,IGLV2-11*01,V1-3            --- --- --- --- --- --- --- --- --- --- --- --- ---
Y12414,IGLV2-11*01,EKL12           --- --- --- --- --- --- --- --- --- --- --- --- ---
Z22198,IGLV2-11*02,DPL12           --- --- --- --- --- --- --- --- --- --- --- --- ---
Y12415,IGLV2-11*03,RXL12           --- --- --- --- --- --- --- --- --- --- --- --- ---
```

Framework and complementarity determining regions

FR1-IMGT: 25 (-1 aa: 10)
FR2-IMGT: 17
FR3-IMGT: 36 (-3 aa: 73,81,82)

CDR1-IMGT: 9
CDR2-IMGT: 3
CDR3-IMGT: 9

Collier de Perles for human IGLV2-11*01

Accession number: IMGT Z73657

EMBL/GenBank/DDBJ: Z73657

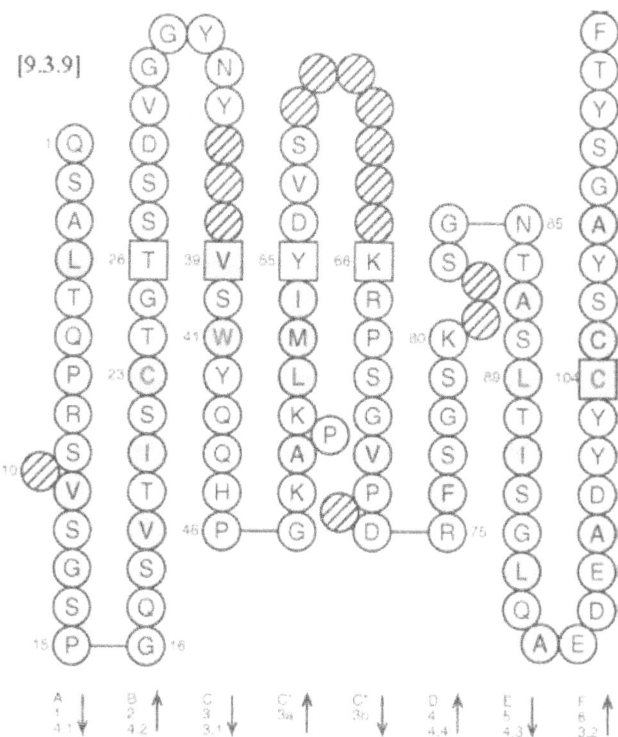

Genome database accession numbers

GDB:9953674

LocusLink: 28816

IGLV2-14

Nomenclature

IGLV2-14: Immunoglobulin lambda variable 2-14.

Definition and functionality

IGLV2-14 is one of the five functional genes of the IGLV2 subgroup which comprises nine mapped genes.

Gene location

IGLV2-14 is located in the cluster A of the IGL locus on chromosome 22 at 22q11.2.

Nucleotide and amino acid sequences for human IGLV2-14

```
                                          1   2   3   4   5   6   7   8   9  10  11  12  13  14  15  16  17  18  19  20
                                          Q   S   A   L   T   Q   P   A   S       V   S   G   S   P   G   Q   S   I   T
Z73664,IGLV2-14*01,2a2.272A12     [27]  CAG TCT GCC CTG ACT CAG CCT GCC TCC ... GTG TCT GGG TCT CCT GGA CAG TCG ATC ACC
D87015,IGLV2-14*01,V1-4           [19]  --- --- --- --- --- --- --- --- --- ... --- --- --- --- --- --- --- --- --- ---
Z22197,IGLV2-14*01,DPL11          [26]  --- --- --- --- --- --- --- --- --- ... --- --- --- --- --- --- --- --- --- ---
L27693,IGLV2-14*01,1v215.23       [18]  --- --- --- --- --- --  --- --- --- ... --- --- --- --- --- --- --- --- --- ---
L27822,IGLV2-14*02,1v2018         [18]  --- --- --- --- --- --- --- --- --- ... --- --- --- --- --- --- --- --- --- ---
Y12412,IGLV2-14*03,EKL11           [3]                                -  --- ... --- --- --- --- --- --- --- --- --- ---
Y12413,IGLV2-14*04,RXL11           [3]                                -  --- ... --- --- --- --- --- --- --- --- --- ---

                                                                  _____________CDR1-IMGT______________
                                         21  22  23  24  25  26  27  28  29  30  31  32  33  34  35  36  37  38  39  40
                                          I   S   C   T   G   T   S   S   D   V   G   G   Y   N   Y               V   S
Z73664,IGLV2-14*01,2a2.272A12           ATC TCC TGC ACT GGA ACC AGC AGT GAC GTT GGT GGT TAT AAC TAT ... ... ... GTC TCC
D87015,IGLV2-14*01,V1-4                 --- --- --- --- --- --- --- --- --- --- --- --- --- --- --- ... ... ... --- ---
Z22197,IGLV2-14*01,DPL11                --- --- --- --- --- --- --- --- --- --- --- --- --- --- --- ... ... ... --- ---
L27693,IGLV2-14*01,1v215.23             --- --- --- --- --- --- --- --  --- --- --- --- --- --- --- ... ... ... --- ---
                                                                                  S           L
L27822,IGLV2-14*02,1v2018               --- --- --- --- --- --- --- --- --T --- --G A-- --- --- CT- ... ... ... --- ---
Y12412,IGLV2-14*03,EKL11                --- --- --- --- --- --- --- --- --- --- --- --- --- --- --- ... ... ... --- ---
Y12413,IGLV2-14*04,RXL11                --- --- --- --- --- --- --- --- --- --- --- --- --- --- --- ... ... ... --- ---

                                                                                              ____________  CDR2-
                                         41  42  43  44  45  46  47  48  49  50  51  52  53  54  55  56  57  58  59  60
                                          W   Y   Q   Q   H   P   G   K   A   P   K   L   M   I   Y   E   V   S
Z73664,IGLV2-14*01,2a2.272A12           TGG TAC CAA CAG CAC CCA GGC AAA GCC CCC AAA CTC ATG ATT TAT GAG GTC AGT ... ...
D87015,IGLV2-14*01,V1-4                 --- --- --- --- --- --- --- --- --- --- --- --- --- --- --- --- --- --- ... ...
Z22197,IGLV2-14*01,DPL11                --- --- --- --- --- --- --- --- --- --- --- --- --- --- --- --- --- --- ... ...
L27693,IGLV2-14*01,1v215.23             --- --- --- --- --- --- --- --- --- --- --- --- --- --- --- --- --- --- ... ...
                                                                                                      G
L27822,IGLV2-14*02,1v2018               --- --- --- --- --- --- --- --- --- --- --- --- --- --- --- --- -G- --- ... ...
                                                                                                  D
Y12412,IGLV2-14*03,EKL11                --- --- --- --A --- --- --- --- --- --- --- --- --- --- --- --T --- --- ... ...
                                                                                                  D
Y12413,IGLV2-14*04,RXL11                --- --- --- --- --- --- --- --- --- --- --- --- --- --- --- --T --- --- ... ...

                                        IMGT________________
                                         61  62  63  64  65  66  67  68  69  70  71  72  73  74  75  76  77  78  79  80
                                                              N   R   P   S   G   V   S       N   R   F   S   G   S   K
Z73664,IGLV2-14*01,2a2.272A12           ... ... ... ... ... AAT CGG CCC TCA GGG GTT TCT ... AAT CGC TTC TCT GGC TCC AAG
D87015,IGLV2-14*01,V1-4                 ... ... ... ... ... --- --- --- --- --- --- --- ... --- --- --- --- --- --- ---
Z22197,IGLV2-14*01,DPL11                ... ... ... ... ... --- --- --- --- --- --- --- ... --- --- --- --- --- --- ---
L27693,IGLV2-14*01,1v215.23             ... ... ... ... ... --- --- --- --- --- --- --- ... --- --- --- --- --- --- ---
                                                              K
L27822,IGLV2-14*02,1v2018               ... ... ... ... ... --G --- --- --- --- --- --- ... --- --- --- --- --- --- ---
Y12412,IGLV2-14*03,EKL11                ... ... ... ... ... --- --- --- --- --- --- --- ... --- --- --- --- --- --- ---
Y12413,IGLV2-14*04,RXL11                ... ... ... ... ... --- --- --- --- --- --- --- ... --- --- --- --- --- --- ---

                                         81  82  83  84  85  86  87  88  89  90  91  92  93  94  95  96  97  98  99 100
                                                  S   G   N   T   A   S   L   T   I   S   G   L   Q   A   E   D   E   A
Z73664,IGLV2-14*01,2a2.272A12           ... ... TCT GGC AAC ACG GCC TCC CTG ACC ATC TCT GGG CTC CAG GCT GAG GAC GAG GCT
D87015,IGLV2-14*01,V1-4                 ... ... --- --- --- --- --- --- --- --- --- --- --- --- --- --- --- --- --- ---
Z22197,IGLV2-14*01,DPL11                ... ... --- --- --- --- --- --- --- --- --- --- --- --- --- --- --- --- --- ---
L27693,IGLV2-14*01,1v215.23             ... ... --- --- --- --- --- --- --- --- --- --- --- --- --- --- --- --- --- ---
L27822,IGLV2-14*02,1v2018               ... ... --- --- --- --- --- --- --- --- --- --- --- --- --- --- --- --- --- ---
Y12412,IGLV2-14*03,EKL11                ... ... --- --- --- --- --- --- --- --- --- --- --- --- --- --- --- --- --- ---
Y12413,GLV2 14*04,RXL11                 ... ... --- --- --- --- --- --- --- --- --- --- --- --- --- --- --- --- --- ---
```

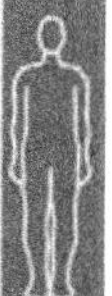

	101	102	103	104	105	106	107	108	109	110	111	112	113
					CDR3-IMGT								
	D	Y	Y	C	S	S	Y	T	S	S	S	T	L
Z73664,IGLV2-14*01,2a2.272A12	GAT	TAT	TAC	TGC	AGC	TCA	TAT	ACA	AGC	AGC	AGC	ACT	CTC
D87015,IGLV2-14*01,V1-4	---	---	---	---	---	---	---	---	---	---	---	--	
Z22197,IGLV2-14*01,DPL11	---	---	---	---	---	---	---	---	---	--			--
L27693,IGLV2-14*01,1v215.23		--	---	---	---	---	---	---	- -				--
L27822,IGLV2-14*02,1v2018	---	---	---	---	---	---	---	---	---	---	---	---	
Y12412,IGLV2-14*03,EKL11	---	---	---	---	---	---	---	---			---	---	---
Y12413,IGLV2-14*04,RXL11	---	---	---	---	---	---	---	---	---	-		---	---

Framework and complementarity determining regions

FR1-IMGT: 25 (-1 aa: 10)
FR2-IMGT: 17
FR3-IMGT: 36 (-3 aa: 73,81,82)

CDR1-IMGT: 9
CDR2-IMGT: 3
CDR3-IMGT: 9

Collier de Perles for human IGLV2-14*01

Accession number: IMGT Z73664

EMBL/GenBank/DDBJ: Z73664

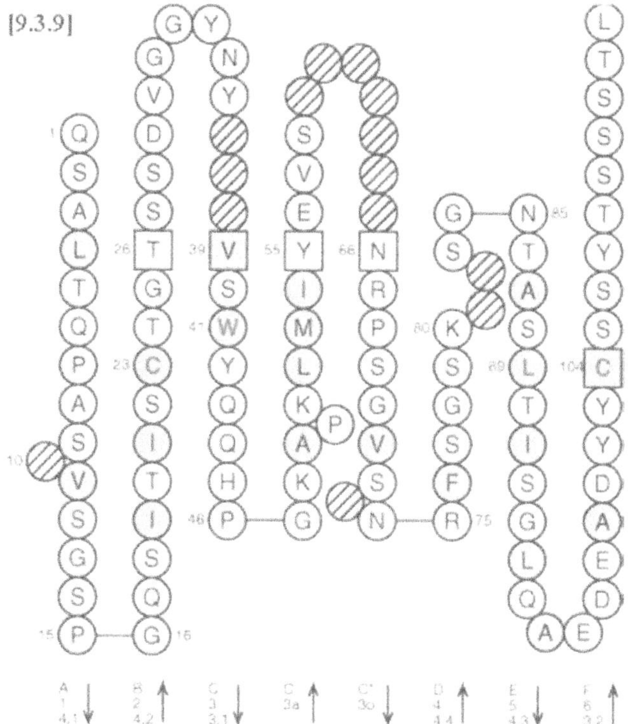

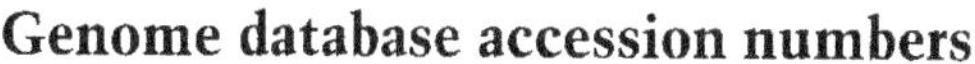

Genome database accession numbers

GDB:9953676

LocusLink: 28815

Nomenclature

IGLV2-18: Immunoglobulin lambda variable 2-18.

Definition and functionality

IGLV2-18 is one of the five functional genes of the IGLV2 subgroup which comprises nine mapped genes.

Gene location

IGLV2-18 is located in the cluster A of the IGL locus on chromosome 22 at 22q11.2.

Nucleotide and amino acid sequences for human IGLV2-18

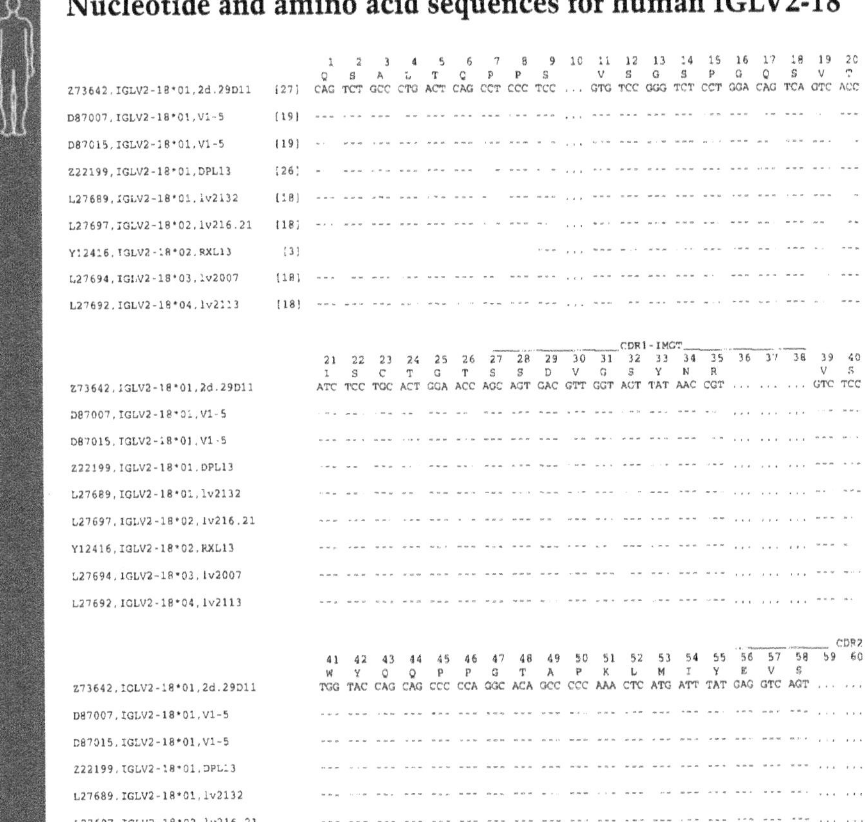

		1	2	3	4	5	6	7	8	9	10	11	12	13	14	15	16	17	18	19	20
		Q	S	A	L	T	Q	P	P	S		V	S	G	S	P	G	Q	S	V	T
Z73642,IGLV2-18*01,2d.29D11	[27]	CAG	TCT	GCC	CTG	ACT	CAG	CCT	CCC	TCC	...	GTG	TCC	GGG	TCT	CCT	GGA	CAG	TCA	GTC	ACC
D87007,IGLV2-18*01,V1-5	[19]	---	---	---	---	---	---	---	---	---	...	---	---	---	---	---	---	---	---	---	---
D87015,IGLV2-18*01,V1-5	[19]	---	---	---	---	---	---	---	---	---	...	---	---	---	---	---	---	---	---	---	---
Z22199,IGLV2-18*01,DPL13	[26]	---	---	---	---	---	---	---	---	---	...	---	---	---	---	---	---	---	---	---	---
L27689,IGLV2-18*01,lv2132	[18]	---	---	---	---	---	---	---	---	---	...	---	---	---	---	---	---	---	---	---	---
L27697,IGLV2-18*02,lv216.21	[18]	---	---	---	---	---	---	---	---	---	...	---	---	---	---	---	---	---	---	---	---
Y12416,IGLV2-18*02,RXL13	[3]									---	...	---	---	---	---	---	---	---	---	---	---
L27694,IGLV2-18*03,lv2007	[18]	---	---	---	---	---	---	---	---	---	...	---	---	---	---	---	---	---	---	---	---
L27692,IGLV2-18*04,lv2113	[18]	---	---	---	---	---	---	---	---	---	...	---	---	---	---	---	---	---	---	---	---

								CDR1-IMGT												
	21	22	23	24	25	26	27	28	29	30	31	32	33	34	35	36	37	38	39	40
	I	S	C	T	G	T	S	S	D	V	G	S	Y	N	R				V	S
Z73642,IGLV2-18*01,2d.29D11	ATC	TCC	TGC	ACT	GGA	ACC	AGC	AGT	GAC	GTT	GGT	AGT	TAT	AAC	CGT	...	...	...	GTC	TCC
D87007,IGLV2-18*01,V1-5	---	---	---	---	---	---	---	---	---	---	---	---	---	---	---	...	...	...	---	---
D87015,IGLV2-18*01,V1-5	---	---	---	---	---	---	---	---	---	---	---	---	---	---	---	...	...	...	---	---
Z22199,IGLV2-18*01,DPL13	---	---	---	---	---	---	---	---	---	---	---	---	---	---	---	...	...	...	---	---
L27689,IGLV2-18*01,lv2132	---	---	---	---	---	---	---	---	---	---	---	---	---	---	---	...	...	...	---	---
L27697,IGLV2-18*02,lv216.21	---	---	---	---	---	---	---	---	---	---	---	---	---	---	---	...	...	...	---	---
Y12416,IGLV2-18*02,RXL13	---	---	---	---	---	---	---	---	---	---	---	---	---	---	---	...	...	...	---	---
L27694,IGLV2-18*03,lv2007	---	---	---	---	---	---	---	---	---	---	---	---	---	---	---	...	...	...	---	---
L27692,IGLV2-18*04,lv2113	---	---	---	---	---	---	---	---	---	---	---	---	---	---	---	...	...	...	---	---

																				CDR2
	41	42	43	44	45	46	47	48	49	50	51	52	53	54	55	56	57	58	59	60
	W	Y	Q	Q	P	P	G	T	A	P	K	L	M	I	Y	E	V	S		
Z73642,IGLV2-18*01,2d.29D11	TGG	TAC	CAG	CAG	CCC	CCA	GGC	ACA	GCC	CCC	AAA	CTC	ATG	ATT	TAT	GAG	GTC	AGT	...	...
D87007,IGLV2-18*01,V1-5	---	---	---	---	---	---	---	---	---	---	---	---	---	---	---	---	---	---	...	...
D87015,IGLV2-18*01,V1-5	---	---	---	---	---	---	---	---	---	---	---	---	---	---	---	---	---	---	...	...
Z22199,IGLV2-18*01,DPL13	---	---	---	---	---	---	---	---	---	---	---	---	---	---	---	---	---	---	...	...
L27689,IGLV2-18*01,lv2132	---	---	---	---	---	---	---	---	---	---	---	---	---	---	---	---	---	---	...	...
L27697,IGLV2-18*02,lv216.21	---	---	---	---	---	---	---	---	---	---	---	---	---	---	---	---	---	---	...	...
Y12416,IGLV2-18*02,RXL13	---	---	---	---	---	---	---	---	---	---	---	---	---	---	---	---	---	---	...	...
L27694,IGLV2-18*03,lv2007	---	---	---	---	---	---	---	---	---	---	---	---	---	---	---	---	---	---	...	...
L27692,IGLV2-18*04,lv2113	---	---	---	---	---	---	---	---	---	---	---	---	---	---	---	---	---	---	...	...

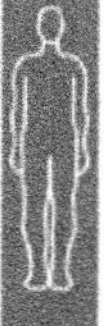

```
                               IMGT_____________ . ...
                               61  62  63  64  65  66  67  68  69  70  71  72  73  74  75  76  77  78  79  80
                                                   N   R   P   S   G   V   P       D   R   F   S   G   S   K
Z73642,IGLV2-18*01,2d.29D11    ... ... ... ... ... AAT CGG CCC TCA GGG GTC CCT ... GAT CGC TTC TCT GGG TCC AAG
D87007,IGLV2-18*01,V1-5        ... ... ... ... ... --- --- --- ---         --- --- ... --- ---     --- --- ---
D87015,IGLV2-18*01,V1-5        ... ... ... ... ...     --- ---     ---         --- ... --- --- --- --- --- ---
Z22199,IGLV2-18*01,DPL13       ... ... ... ... ... --- --- --- --- --- --- --- ... --- --- --- --- --- --- ---
L27689,IGLV2-18*01,1v2132      ... ... ... ... ...     --- --- --- ---         ... ---     ---     --- --- ---
L27697,IGLV2-18*02,1v216.21    ... ... ... ... ... --- --- --- --- --- --- --- ... --- --- --- --- --- --- ---
Y12416,IGLV2-18*02,RXL13       ... ... ... ... ...                 ---     --- ... --- --- --- --- ---
L27694,IGLV2-18*03,1v2007      ... ... ... ... ... --- --- ---     ---     --- --- ... --- --- --- --- --- ---
                                                                                           S
L27692,IGLV2-18*04,1v2113      ... ... ... ... ... --- --- --- --- --- --- --- ... --- --- -C- --- --- ...

                               81  82  83  84  85  86  87  88  89  90  91  92  93  94  95  96  97  98  99  100
                                       S   G   N   T   A   S   L   T   I   S   G   L   Q   A   E   D   E   A
Z73642,IGLV2-18*01,2d.29D11    ... ... TCT GGC AAC ACG GCC TCC CTG ACC ATC TCT GGG CTC CAG GCT GAG GAC GAG GCT
D87007,IGLV2-18*01,V1-5        ... ... --- -   --- --- --- --- --- --- --- --- --- --- --- --- --- --- --- ---
D87015,IGLV2-18*01,V1-5        ... ... --- -   --- --- --- --- --- --- --- --- --- --- --- --- --- --- --- ---
Z22199,IGLV2-18*01,DPL13       ... ... --- --- --- --- --- --- --- --- --- --- --- --- --- --- --- --- --- ---
L27689,IGLV2-18*01,1v2132      ... ... --- -   --- --- --- --- --- --- --- --- --- --- --- --- --- --- --- ---
L27697,IGLV2-18*02,1v216.21    ... ... --- --- ---                     ---     --- ---     --- --- --- --- ---
Y12416,IGLV2-18*02,RXL13       ... ... --- --- ---                 --- ---     --- --- --- --- --- ---
                                                                       T
L27694,IGLV2-18*03,1v2007      ... ... --- --- --- ---             --- -C- --- --- ---     --- --- ---
L27692,IGLV2-18*04,1v2113      ... ... --C -   --- --- --- --- --- --- --- --- --- --- --- --- --- --- --- ---

                                                   ____________CDR3 IMGT____________
                               101 102 103 104 105 106 107 108 109 110 111 112 113
                               D   Y   Y   C   S   L   Y   T   S   S   S   T   F
Z73642,IGLV2-18*01,2d.29D11    GAT TAT TAC TGC AGC TTA TAT ACA AGC AGC AGC ACT TTC
D87007,IGLV2-18*01,V1-5        --- ---         --- --- --- --- --- --- --- --- ---
D87015,IGLV2-18*01,V1-5        --- --- --- -   --- --- --- --- --- --- --- --- ---
Z22199,IGLV2-18*01,DPL13       --- --- -       --- --- --- --- --- --- --- --- ---
L27689,IGLV2-18*01,1v2132      ---     --- --- --- --- -   -   -   -   --- -   --
                                                   S
L27697,IGLV2-18*02,1v216.21    --- ---         --- -C- --- --- --- --- --- --- ---
                                                   S
Y12416,IGLV2-18*02,RXL13       --- --- --- ---  -  C-- --- ---     --- ---     ---
                                                   S
L27694,IGLV2-18*03,1v2007      --- --- --- --- --  C-- --- ---     --- ---     ---
                                                   S
L27692,IGLV2-18*04,1v2113      --- --- --- --- --  C-- --- --- --- --  --      --
```

Framework and complementarity determining regions

FR1-IMGT: 25 (-1 aa: 10)
FR2-IMGT: 17
FR3-IMGT: 36 (-3 aa: 73,81,82)

CDR1-IMGT: 9
CDR2-IMGT: 3
CDR3-IMGT: 9

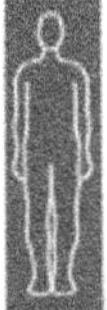

Collier de Perles for human IGHV2-18*01

Accession number: IMGT Z73642 EMBL/GenBank/DDBJ: Z73642

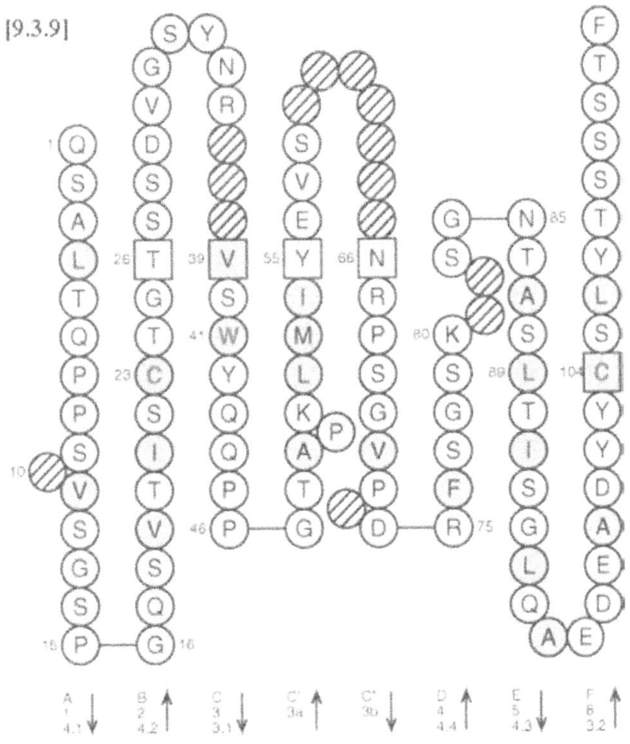

Genome database accession numbers

GDB:9953679 LocusLink: 28814

IGLV2-23

Nomenclature

IGLV2-23: Immunoglobulin lambda variable 2-23.

Definition and functionality

IGLV2-23 is one of the five functional genes of the IGLV2 subgroup which comprises nine mapped genes.

Gene location

IGLV2-23 is located in the cluster A of the IGL locus on chromosome 22 at 22q11.2.

Nucleotide and amino acid sequences for human IGLV2-23

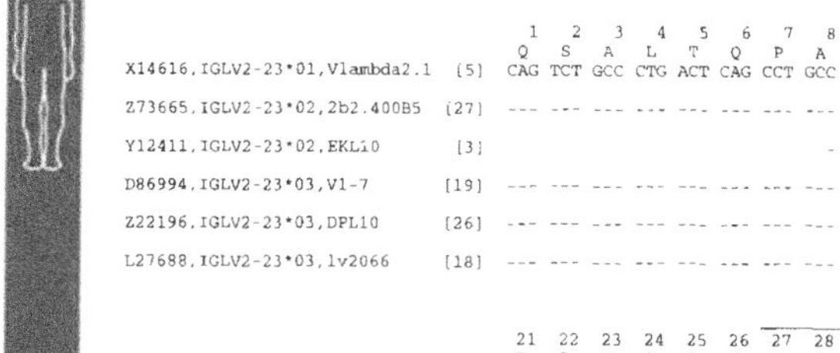

```
                                        1   2   3   4   5   6   7   8   9   10  11  12  13  14  15  16  17  18  19  20
                                        Q   S   A   L   T   Q   P   A   S       V   S   G   S   P   G   Q   S   I   T
X14616,IGLV2-23*01,Vlambda2.1   [5]    CAG TCT GCC CTG ACT CAG CCT GCC TCC ... GTG TCT GGG TCT CCT GGA CAG TCG ATC ACC
Z73665,IGLV2-23*02,2b2.400B5    [27]   --- --- --- --- --- --- --- --- --- ... --- --- --- --- --- --- --- --- --- ---
Y12411,IGLV2-23*02,EKL10        [3]                                  -  --- ... --- --- --- --- --- --- --- --- --- ---
D86994,IGLV2-23*03,V1-7         [19]   --- --- --- --- --- --- --- --- --- ... --- --- --- --- --- --- --- --- --- ---
Z22196,IGLV2-23*03,DPL10        [26]   --- --- --- --- --- --- --- --- --- ... --- --- --- --- --- --- --- --- --- ---
L27688,IGLV2-23*03,1v2066       [18]   --- --- --- --- --- --- --- --- --- ... --- --- --- --- --- --- --- --- --- ---

                                                               ______________CDR1-IMGT________________
                                        21  22  23  24  25  26  27  28  29  30  31  32  33  34  35  36  37  38  39  40
                                        I   S   C   T   G   T   S   S   D   V   G   S   Y   N   L               V   S
X14616,IGLV2-23*01,Vlambda2.1          ATC TCC TGC ACT GGA ACC AGC AGT GAT GTT GGG AGT TAT AAC CTT ... ... ... GTC TCC
Z73665,IGLV2-23*02,2b2.400B5           --- --- --- --- --- --- --- --- --- --- --- --- --- --- --- ... ... ... --- ---
Y12411,IGLV2-23*02,EKL10               --- --- --- --- --- --- --- --- --- --- --- --- --- --- --- ... ... ... --- ---
D86994,IGLV2-23*03,V1-7                --- --- --- --- --- --- --- --- --- --- --- --- --- --- --- ... ... ... --- ---
Z22196,IGLV2-23*03,DPL10               --- --- --- --- --- --- --- --- --- --- --- --- --- --- --- ... ... ... --- ---
L27688,IGLV2-23*03,1v2066              --- --- --- --- --- --- --- --- --- --- --- --- --- --- --- ... ... ... --- ---

                                                                                            ____________ CDR2-
                                        41  42  43  44  45  46  47  48  49  50  51  52  53  54  55  56  57  58  59  60
                                        W   Y   Q   Q   H   P   G   K   A   P   K   L   M   I   Y   E   G   S
X14616,IGLV2-23*01,Vlambda2.1          TGG TAC CAA CAG CAC CCA GGC AAA GCC CCC AAA CTC ATG ATT TAT GAG GGC AGT ... ...
                                                                                                    V
Z73665,IGLV2-23*02,2b2.400B5           --- --- --- --- --- --- --- --- --- --- --- --- --- --- --- --- -T- --- ... ...
                                                                                                    V
Y12411,IGLV2-23*02,EKL10               --- --- --- --- --- --- --- --- --- --- --- --- --- --- --- --- -T- --- ... ...
D86994,IGLV2-23*03,V1-7                --- --- --- --- --- --- --- --- --- --- --- --- --- --- --- --- --- --- ... ...
Z22196,IGLV2-23*03,DPL10               --- --- --- --- --- --- --- --- --- --- --- --- --- --- --- --- --- --- ... ...
L27688,IGLV2-23*03,1v2066              --- --- --- --- --- --- --- --- --- --- --- --- --- --- --- --- --- --- ... ...

                                       IMGT_______________
                                        61  62  63  64  65  66  67  68  69  70  71  72  73  74  75  76  77  78  79  80
                                                            K   R   P   S   G   V   S       N   R   F   S   G   S   K
X14616,IGLV2-23*01,Vlambda2.1          ... ... ... ... ... AAG CGG CCC TCA GGG GTT TCT ... AAT CGC TTC TCT GGC TCC AAG
Z73665,IGLV2-23*02,2b2.400B5           ... ... ... ... ... --- --- --- --- --- --- --- ... --- --- --- --- --- --- ---
Y12411,IGLV2-23*02,EKL10               ... ... ... ... ... --- --- --- --- --- --- --- ... --- --- --- --- --- --- ---
D86994,IGLV2-23*03,V1-7                ... ... ... ... ... --- --- --- --- --- --- --- ... --- --- --- --- --- --- ---
Z22196,IGLV2-23*03,DPL10               ... ... ... ... ... --- --- --- --- --- --- --- ... --- --- --- --- --- --- ---
L27688,IGLV2-23*03,1v2066              ... ... ... ... ... --- --- --- --- --- --- --- ... --- --- --- --- --- --- ---

                                        81  82  83  84  85  86  87  88  89  90  91  92  93  94  95  96  97  98  99  100
                                                S   G   N   T   A   S   L   T   I   S   G   L   Q   A   E   D   E   A
X14616,IGLV2-23*01,Vlambda2.1          ... ... TCT GGC AAC ACG GCC TCC CTG ACA ATC TCT GGG CTC CAG GCT GAG GAC GAG GCT
Z73665,IGLV2-23*02,2b2.400B5           ... ... --- --- --- --- --- --- --- --- --- --- --- --- --- --- --- --- --- ---
Y12411,IGLV2-23*02,EKL10               ... ... --- --- --- --- --- --- --- --- --- --- --- --- --- --- --- --- --- ---
D86994,IGLV2-23*03,V1-7                ... ... --- --- --- --- --- --- --- --- --- --- --- --- --- --- --- --- --- ---
Z22196,IGLV2-23*03,DPL10               ... ... --- --- --- --- --- --- --- --- --- --- --- --- --- --- --- --- --- ---
L27688,IGLV2-23*03,1v2066              ... ... --- --- --- --- --- --- --- --- --- --- --- --- --- --- --- --- --- ---
```

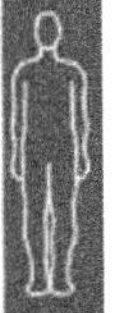

```
                                                    ______CDR3-IMGT____________
                                   101 102 103 104 105 106 107 108 109 110 111 112 113
                                    D   Y   Y   C   C   S   Y   A   G   S   S   T   L
X14616,IGLV2-23*01,Vlambda2.i      GAT TAT TAC TGC TGC TCA TAT GCA GGT AGT AGC ACT TTA C
                                                                                    F
Z73665,IGLV2-23*02,2b2.400B5       --- --- --- --- --- --- --- --- --- --- --- --- --C
                                                                                    F
Y12411,IGLV2-23*02,EKL10           --- --- --- --- --- --- --- --- --- --- --- --- --C
                                                                                    F
D86994,IGLV2-23*03,V1-7            --- --- --- --- --- --- --- --- --- --- --- --- --C
                                                                                    F
Z22196,IGLV2-23*03,DPL10           --- --- --- --- --- --- --- --- --- --- --- --- --C
                                                                                    F
L27688,IGLV2-23*03,lv2066          --- --- --- --- --- --- --- --- --- --- --- --- --C
```

Framework and complementarity determining regions

FR1-IMGT: 25 (-1 aa: 10)
FR2-IMGT: 17
FR3-IMGT: 36 (-3 aa: 73,81,82)

CDR1-IMGT: 9
CDR2-IMGT: 3
CDR3-IMGT: 9

Collier de Perles for human IGLV2-23*01

Accession number: IMGT X14616

EMBL/GenBank/DDBJ: X14616

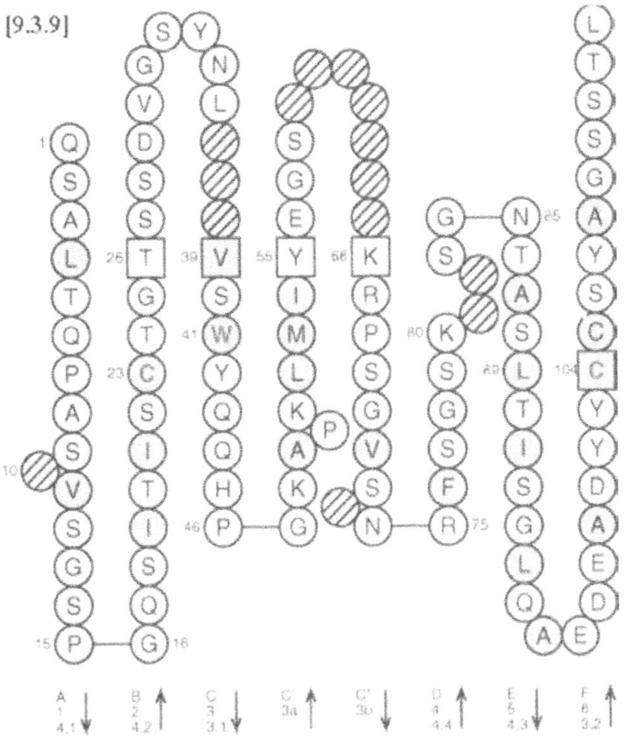

Genome database accession numbers

GDB:9953681 LocusLink: 28813

Nomenclature

IGLV2-33: Immunoglobulin lambda variable 2-33.

Definition and functionality

IGLV2-33 is an ORF due to an unusual V-HEPTAMER sequence: catagtg instead of cacagtg. IGLV2-33 belongs to the IGLV2 subgroup which comprises nine mapped genes, of which five are functional.

Gene location

IGLV2-33 is located in the cluster A of the IGL locus on chromosome 22 at 22q11.2.

Nucleotide and amino acid sequences for human IGLV2-33

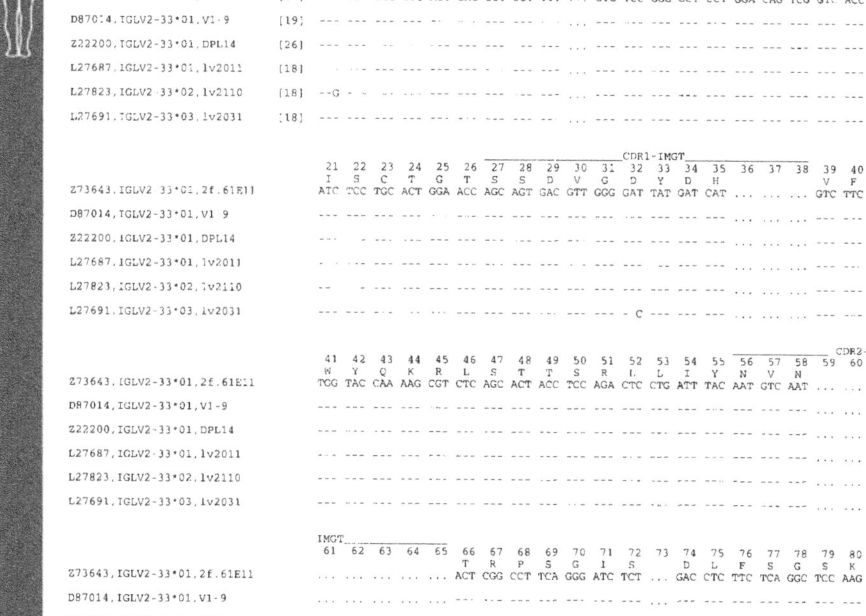

	101	102	103	104	105	106	107	108	109	110	111	112	113
					CDR3-IMGT								
	N	Y	H	C	S	L	Y	S	S	S	Y	T	F
Z73643,IGLV2-33*01,2f.61E11	AAT	TAT	CAC	TGC	AGC	TTA	TAT	TCA	AGT	AGT	TAC	ACT	TTC
D87014,IGLV2-33*01,V1-9	---	---	---	---	---	---	---	---	---	---	---	---	---
Z22200,IGLV2-33*01,DPL14	---	---	---	---	---	---	---	---	---	---	---	---	---
L27687,IGLV2-33*01,lv2011	---	---	---	---	---	---	---	---	---	---	---	---	---
L27823,IGLV2-33*02,lv2110	---	---	---	---	---	---	---	---	---	---	---	---	---
L27691,IGLV2-33*03,lv2031	---	---	---	---	---	---	---	---	---	---	---	---	---

Framework and complementarity determining regions

FR1-IMGT: 25 (-1 aa: 10)
FR2-IMGT: 17
FR3-IMGT: 36 (-3 aa: 73,81,82)

CDR1-IMGT: 9
CDR2-IMGT: 3
CDR3-IMGT: 9

Collier de Perles for human IGLV2-33*01

Accession number: IMGT Z73643 EMBL/Genbank/DDBJ: Z73643

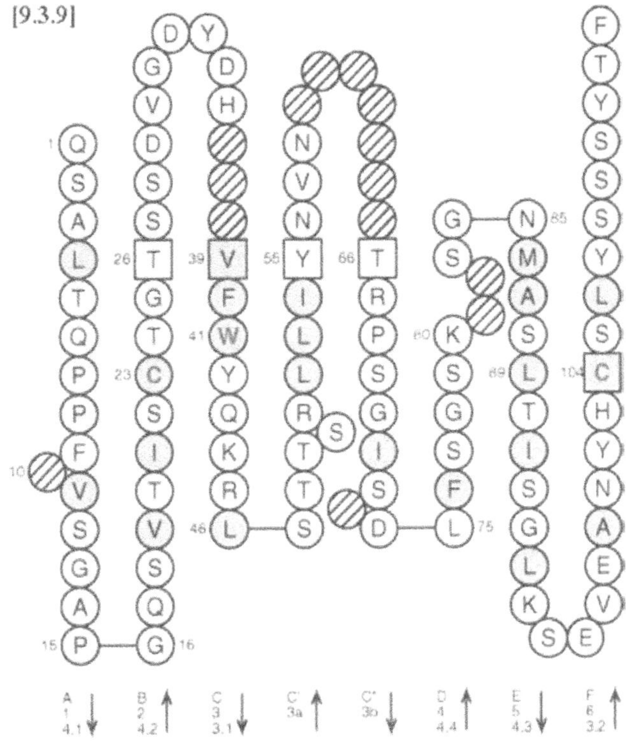

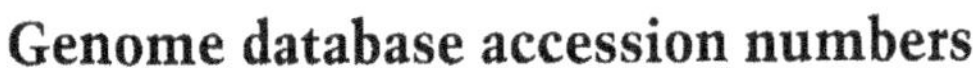

Genome database accession numbers

GDB:9953685 LocusLink: 28811

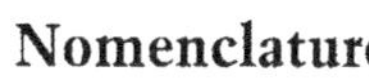

IGLV3-1

Nomenclature

IGLV3-1: Immunoglobulin lambda variable 3-1.

Definition and functionality

IGLV3-1 is one of the 8–10 functional genes of the IGLV3 subgroup which comprises 23 mapped genes.

Gene location

IGLV3-1 is located in the cluster A of the IGL locus on chromosome 22 at 22q11.2.

Nucleotide and amino acid sequences for human IGLV3-1

```
                                         1   2   3   4   5   6   7   8   9  10  11  12  13  14  15  16  17  18  19  20
                                        S   Y   E   L   T   Q   P   P   S       V   S   V   S   P   G   Q   T   A   S
X57826,IGLV3-1*01,VLIII.1         [8]  TCC TAT GAG CTG ACT CAG CCA CCC TCA ... GTG TCC GTG TCC CCA GGA CAG ACA GCC AGC
Z73647,IGLV3-1*01,3r.9C5          [27] --- --- --- --- --- --- --- --- --- ... --- --- --- --- --- --- --- --- --- ---
D87023,IGLV3-1*01,V2-1            [19] --- --- --- --- --- --- --- --- --- ... --- --- --- --- --- --- --- --- --- ---
Z22208,IGLV3-1*01,DPL23           [26] --- --- --- --- --- --- --- --- --- ... --- --- --- --- --- --- --- --- --- ---
L26403,IGLV3-1*01,BHGL6-13        [17] --- --- --- --- --- --- --- --- --- ... --- --- --- --- --- --- --- --- --- ---

                                                               ____________CDR1-IMGT_______________________
                                        21  22  23  24  25  26  27  28  29  30  31  32  33  34  35  36  37  38  39  40
                                        I   T   C   S   G   D   K   L   G   D   K   Y                           A   C
X57826,IGLV3-1*01,VLIII.1              ATC ACC TGC TCT GGA GAT AAA TTG GGG GAT AAA TAT ... ... ... ... ... ... GCT TGC
Z73647,IGLV3-1*01,3r.9C5               --- --- --- --- --- --- --- --- --- --- --- --- ... ... ... ... ... ... --- ---
D87023,IGLV3-1*01,V2-1                 --- --- --- --- --- --- --- --- --- --- --- --- ... ... ... ... ... ... --- ---
Z22208,IGLV3-1*01,DPL23                --- --- --- --- --- --- --- --- --- --- --- --- ... ... ... ... ... ... --- ---
L26403,IGLV3-1*01,BHGL6-13             --- --- --- --- --- --- --- --- --- --- --- --- ... ... ... ... ... ... --- ---

                                                                                         ______________CDR2-
                                        41  42  43  44  45  46  47  48  49  50  51  52  53  54  55  56  57  58  59  60
                                        W   Y   Q   Q   K   P   G   Q   S   P   V   L   V   I   Y   Q   D   S
X57826,IGLV3-1*01,VLIII.1              TGG TAT CAG CAG AAG CCA GGC CAG TCC CCT GTG CTG GTC ATC TAT CAA GAT AGC ... ...
Z73647,IGLV3-1*01,3r.9C5               --- --- --- --- --- --- --- --- --- --- --- --- --- --- --- --- --- --- ... ...
D87023,IGLV3-1*01,V2-1                 --- --- --- --- --- --- --- --- --- --- --- --- --- --- --- --- --- --- ... ...
Z22208,IGLV3-1*01,DPL23                --- --- --- --- --- --- --- --- --- --- --- --- --- --- --- --- --- --- ... ...
L26403,IGLV3-1*01,BHGL6-13             --- --- --- --- --- --- --- --- --- --- --- --- --- --- --- --- --- --- ... ...

                                       IMGT_______________
                                        61  62  63  64  65  66  67  68  69  70  71  72  73  74  75  76  77  78  79  80
                                                            K   R   P   S   G   I   P       E   R   F   S   G   S   N
X57826,IGLV3-1*01,VLIII.1              ... ... ... ... ... AAG CGG CCC TCA GGG ATC CCT ... GAG CGA TTC TCT GGC TCC AAC
Z73647,IGLV3-1*01,3r.9C5               ... ... ... ... ... --- --- --- --- --- --- --- ... --- --- --- --- --- --- ---
D87023,IGLV3-1*01,V2-1                 ... ... ... ... ... --- --- --- --- --- --- --- ... --- --- --- --- --- --- ---
Z22208,IGLV3-1*01,DPL23                ... ... ... ... ... --- --- --- --- --- --- --- ... --- --- --- --- --- --- ---
L26403,IGLV3-1*01,BHGL6-13             ... ... ... ... ... --- --- --- --- --- --- --- ... --- --- --- --- --- --- ---

                                        81  82  83  84  85  86  87  88  89  90  91  92  93  94  95  96  97  98  99 100
                                                S   G   N   T   A   T   L   T   I   S   G   T   Q   A   M   D   E   A
X57826,IGLV3-1*01,VLIII.1              ... ... TCT GGG AAC ACA GCC ACT CTG ACC ATC AGC GGG ACC CAG GCT ATG GAT GAG GCT
Z73647,IGLV3-1*01,3r.9C5               ... ... --- --- --- --- --- --- --- --- --- --- --- --- --- --- --- --- --- ---
D87023,IGLV3-1*01,V2-1                 ... ... --- --- --- --- --- --- --- --- --- --- --- --- --- --- --- --- --- ---
Z22208,IGLV3-1*01,DPL23                ... ... --- --- --- --- --- --- --- --- --- --- --- --- --- --- --- --- --- ---
L26403,IGLV3-1*01,BHGL6-13             ... ... --- --- --- --- --- --- --- --- --- --- --- --- --- --- --- --- ---

                                                        ___________CDR3-IMGT___________
                                       101 102 103 104 105 106 107 108 109 110 111 112 113
                                        D   Y   Y   C   Q   A   W   D   S   S   T   A
X57826,IGLV3-1*01,VLIII.1              GAC TAT TAC TGT CAG GCG TGG GAC AGC AGC ACT GCA
Z73647,IGLV3-1*01,3r.9C5               --- --- --- --- --- --- --- --- --- --- --- ---
D87023,IGLV3-1*01,V2-1                 --- --- --- --- --- --- --- --- --- --- --- ---
Z22208,IGLV3-1*01,DPL23                --- --- --- --- --- --- --- --- --- --- --- ---
                                                                                        H
L26403,IGLV3-1*01,BHGL6-13             --- --- --- --- --- --- --- --- --- --- --- --- CAC AG
```

Framework and complementarity determining regions

FR1-IMGT: 25 (-1 aa: 10)
FR2-IMGT: 17
FR3-IMGT: 36 (-3 aa: 73,81,82)

CDR1-IMGT: 6
CDR2-IMGT: 3
CDR3-IMGT: 8

Collier de Perles for human IGLV3-1*01

Accession number: IMGT X57826

EMBL/GenBank/DDBJ: X57826

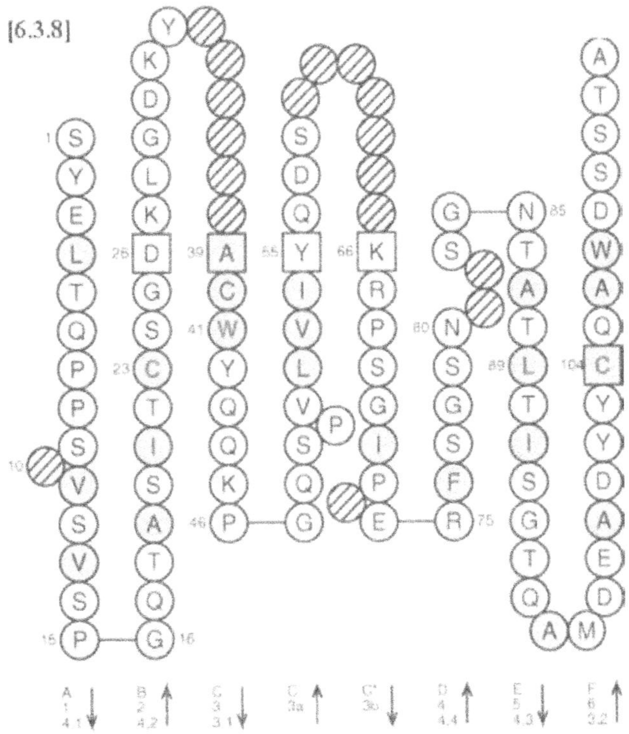

Genome database accession numbers

GDB:9953689

LocusLink: 28809

IGLV3-9

Nomenclature

IGLV3-9: Immunoglobulin lambda variable 3-9.

Definition and functionality

IGLV3-9 is a functional gene (alleles *01 and *02) or a pseudogene (allele *03). IGLV3-9 belongs to the IGLV3 subgroup which comprises 23 mapped genes, of which 8–10 are functional.
IGLV3-9*03 is a pseudogene due to a 1 nt DELETION in codon 32 leading to a frameshift in CDR1-IMGT.

Gene location

IGLV3-9 is located in the cluster A of the IGL locus on chromosome 22 at 22q11.2.

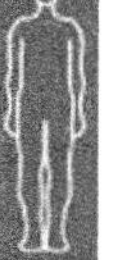

Nucleotide and amino acid sequences for human IGLV3-9

```
                                    1   2   3   4   5   6   7   8   9   10  11  12  13  14  15  16  17  18  19  20
                                    S   Y   E   L   T   Q   P   L   S       V   S   V   A   L   G   Q   T   A   R
X97473,IGLV3-9*01,3j.118D9     [27] TCC TAT GAG CTG ACT CAG CCA CTC TCA ... GTG TCA GTG GCC CTG GGA CAG ACG GCC AGG
D87021,IGLV3-9*01,V2-6         [19] --- --- --- --- --- --- --- --- --- ... --- --- --- --- --- --- --- --- --- ---
                                                                                                    A
X74288,IGLV3-9*02,IGLV3S6      [15] --- --- --- --- --- --- --- --- --- ... --- --- --- --- --- --- --- G-- --- ---
                                                                                                    A
X51754,IGLV3-9*03,psiVlambda1  [25] --- --- --- --- --- --- --- --- --- ... --- --- --- --- --- --- --- G-- --- ---

                                                        ___________________CDR1-IMGT___________________
                                    21  22  23  24  25  26  27  28  29  30  31  32  33  34  35  36  37  38  39  40
                                    I   T   C   G   G   N   N   I   G   S   K   N                           V   H
X97473,IGLV3-9*01,3j.118D9          ATT ACC TGT GGG GGA AAC AAC ATT GGA AGT AAA AAT ... ... ... ... ... ... GTG CAC
D87021,IGLV3-9*01,V2-6              --- --- --- --- --- --- --- --- --- --- --- --- ... ... ... ... ... ... --- ---
                                                                L       Y       S
X74288,IGLV3-9*02,IGLV3S6           --- --- --- --- --- --- --- C-- --- TA- --- -G- ... ... ... ... ... ... --- ---
                                                                L       Y       #
X51754,IGLV3-9*03,psiVlambda1       --- --- --- --- --- --- --- C-- --- TA- --- -.. ... ... ... ... ... ... --- ---

                                                                                                ___________ CDR2
                                    41  42  43  44  45  46  47  48  49  50  51  52  53  54  55  56  57  58  59  60
                                    W   Y   Q   Q   K   P   G   Q   A   P   V   L   V   I   Y   R   D   S
X97473,IGLV3-9*01,3j.118D9          TGG TAC CAG CAG AAG CCA GGC CAG GCC CCT GTG CTG GTC ATC TAT AGG GAT AGC ... ...
D87021,IGLV3-9*01,V2-6              --- --- --- --- --- --- --- --- --- --- --- --- --- --- --- --- --- --- ... ...
                                                                                                     N
X74288,IGLV3-9*02,IGLV3S6           --- --- --- --- --- --- --- --- --- --- --- --- --- --- --- --- --- -A- ... ...
                                                                                                     N
X51754,IGLV3-9*03,psiVlambda1       --- --- --- --- --- --- --- --- --- --- --- --- --- --- --- --- --- -A- ... ...

                                   IMGT_______________
                                    61  62  63  64  65  66  67  68  69  70  71  72  73  74  75  76  77  78  79  80
                                                        N   R   P   S   G   I   P       E   R   F   S   G   S   N
X97473,IGLV3-9*01,3j.118D9          ... ... ... ... ... AAC CGG CCC TCT GGG ATC CCT ... GAG CGA TTC TCT GGC TCC AAC
D87021,IGLV3-9*01,V2-6              ... ... ... ... ... --- --- --- --- --- --- --- ... --- --- --- --- --- --- ---
X74288,IGLV3-9*02,IGLV3S6           ... ... ... ... ... --- --- --- --- --- --- --- ... --- --- --- --- --- --- ---
X51754,IGLV3-9*03,psiVlambda1       ... ... ... ... ... --- --- --- --- --- --- --- ... --- --- --- --- --- --- ---

                                    81  82  83  84  85  86  87  88  89  90  91  92  93  94  95  96  97  98  99 100
                                            S   G   N   T   A   T   L   T   I   S   R   A   Q   A   G   D   E   A
X97473,IGLV3-9*01,3j.118D9          ... ... TCG GGG AAC ACG GCC ACC CTG ACC ATC AGC AGA GCC CAA GCC GGG GAT GAG GCT
D87021,IGLV3-9*01,V2-6              ... ... --- --- --- --- --- --- --- --- --- --- --- --- --- --- --- --- --- ---
X74288,IGLV3-9*02,IGLV3S6           ... ... --- --- --- --- --- --- --- --- --- --- --- --- --- --- --- --- --- ---
X51754,IGLV3-9*03,psiVlambda1       ... ... --- --- --- --- --- --- --- --- --- --- --- --- --- --- --- --- --- ---

                                                    _____________CDR3-IMGT_____________
                                   101 102 103 104 105 106 107 108 109 110 111 112 113
                                    D   Y   Y   C   Q   V   W   D   S   S   T   A
X97473,IGLV3-9*01,3j.118D9         GAC TAT TAC TGT CAG GTG TGG GAC AGC AGC ACT GCA
D87021,IGLV3-9*01,V2-6             --- --- --- --- --- --- --- --- --- --- --- ---
                                                                                H
X74288,IGLV3-9*02,IGLV3S6          --- --- --- --- --- --- --- --- --- --- --- --- CAC CC
X51754,IGLV3-9*03,psiVlambda1      --- --- --- --- --- --- --- --- --- --- --- ---
```

Framework and complementarity determining regions

FR1-IMGT: 25 (-1 aa: 10)
FR2-IMGT: 17
FR3-IMGT: 36 (-3 aa: 73,81,82)

CDR1-IMGT: 6
CDR2-IMGT: 3
CDR3-IMGT: 8

Collier de Perles for human IGLV3-9*01

Accession number: IMGT X97473 EMBL/GenBank/DDBJ: X97473

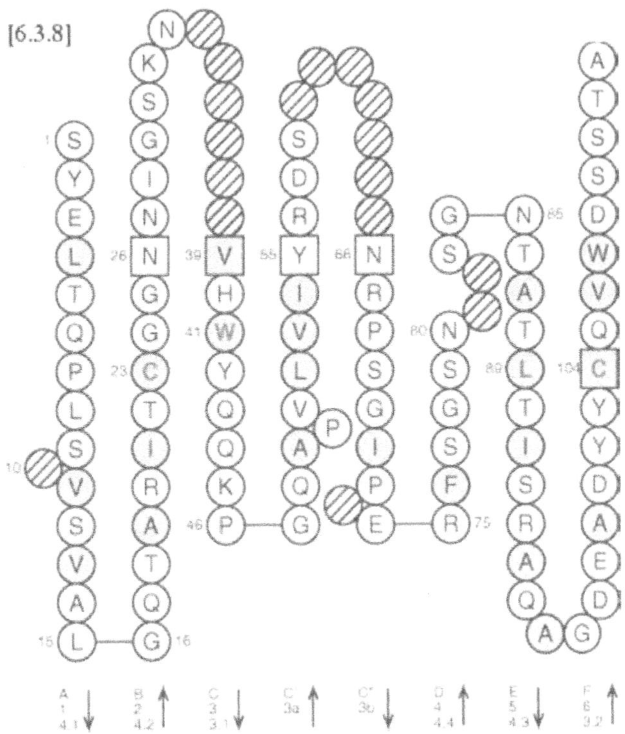

Genome database accession numbers

GDB:9953699 LocusLink: 28804

IGLV3-10

Nomenclature

IGLV3-10: Immunoglobulin lambda variable 3-10.

Definition and functionality

IGLV3-10 is one of the 8–10 functional genes of the IGLV3 subgroup which comprises 23 mapped genes.

Gene location

IGLV3-10 is located in the cluster A of the IGL locus on chromosome 22 at 22q11.2.

Nucleotide and amino acid sequences for human IGLV3-10

```
                                   1   2   3   4   5   6   7   8   9  10  11  12  13  14  15  16  17  18  19  20
                                   S   Y   E   L   T   Q   P   P   S       V   S   V   S   P   G   Q   T   A   R
X97464,IGLV3-10*01,3p.81A4  [27]  TCC TAT GAG CTG ACA CAG CCA CCC TCG ... GTG TCA GTG TCC CCA GGA CAA ACG GCC AGG
D87021,IGLV3-10*01,V2-7     [19]  --- --- --- --- --- --- --- --- --- ... --- --- --- --- --- --- --- --- --- ---
L29166,IGLV3-10*02,IGGLL295 [14]  --- --- --- --- --- --- --- --- --- ... --- --- --- --- --- --- --- --- --- ---

                                                          _____________CDR1-IMGT_____________________
                                  21  22  23  24  25  26  27  28  29  30  31  32  33  34  35  36  37  38  39  40
                                   I   T   C   S   G   D   A   L   P   K   K   Y                           A   Y
X97464,IGLV3-10*01,3p.81A4        ATC ACC TGC TCT GGA GAT GCA TTG CCA AAA AAA TAT ... ... ... ... ... ... GCT TAT
D87021,IGLV3-10*01,V2-7           --- --- --- --- --- --- --- --- --- --- --- --- ... ... ... ... ... ... --- ---
L29166,IGLV3-10*02,IGGLL295       --- --- --- --- --- --- --- --- --- --- --- --- ... ... ... ... ... ... --- ---

                                                                                              ___________ CDR2-
                                  41  42  43  44  45  46  47  48  49  50  51  52  53  54  55  56  57  58  59  60
                                   W   Y   Q   Q   K   S   G   Q   A   P   V   L   V   I   Y   E   D   S
X97464,IGLV3-10*01,3p.81A4        TGG TAC CAG CAG AAG TCA GGC CAG GCC CCT GTG CTG GTC ATC TAT GAG GAC AGC ... ...
D87021,IGLV3-10*01,V2-7           --- --- --- --- --- --- --- --- --- --- --- --- --- --- --- --- --- --- ... ...
                                                                                               K
L29166,IGLV3-10*02,IGGLL295       --- --- --- --- --- --- --- --- --- --- --- --- --- --- --- A-- --- --- ... ...

                                  IMGT_______________
                                  61  62  63  64  65  66  67  68  69  70  71  72  73  74  75  76  77  78  79  80
                                                       K   R   P   S   G   I   P       E   R   F   S   G   S   S
X97464,IGLV3-10*01,3p.81A4        ... ... ... ... ... AAA CGA CCC TCC GGG ATC CCT ... GAG AGA TTC TCT GGC TCC AGC
D87021,IGLV3-10*01,V2-7           ... ... ... ... ... --- --- --- --- --- --- --- ... --- --- --- --- --- --- ---
L29166,IGLV3-10*02,IGGLL295       ... ... ... ... ... --- --- --- --A --- --- --- ... --- --- --- --- --- --- ---

                                  81  82  83  84  85  86  87  88  89  90  91  92  93  94  95  96  97  98  99 100
                                           S   G   T   M   A   T   L   T   I   S   G   A   Q   V   E   D   E   A
X97464,IGLV3-10*01,3p.81A4        ... ... TCA GGG ACA ATG GCC ACC TTG ACT ATC AGT GGG GCC CAG GTG GAG GAT GAA GCT
D87021,IGLV3-10*01,V2-7           ... ... --- --- --- --- --- --- --- --- --- --- --- --- --- --- --- --- --- ---
                                                                                                           D
L29166,IGLV3-10*02,IGGLL295       ... ... --- --- --- --- --- --- --- --C --- --- --- --- --- --- --- --- --- -A-

                                                  ____________CDR3-IMGT______________
                                  101 102 103 104 105 106 107 108 109 110 111 112 113
                                   D   Y   Y   C   Y   S   T   D   S   S   G   N   H
X97464,IGLV3-10*01,3p.81A4        GAC TAC TAC TGT TAC TCA ACA GAC AGC AGT GGT AAT CAT AG
D87021,IGLV3-10*01,V2-7           --- --- --- --- --- --- --- --- --- --- --- --- --- --
                                                           A       Y
L29166,IGLV3-10*02,IGGLL295       --- --- --- --- --- --- G-- --- TA- --- --- ---
```

Framework and complementarity determining regions

FR1-IMGT: 25 (-1 aa: 10)
FR2-IMGT: 17
FR3-IMGT: 36 (-3 aa: 73,81,82)

CDR1-IMGT: 6
CDR2-IMGT: 3
CDR3-IMGT: 9

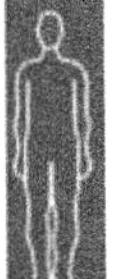

Collier de Perles for human IGLV3-10*01

Accession number: IMGT X97464 EMBL/GenBank/DDBJ: X97464

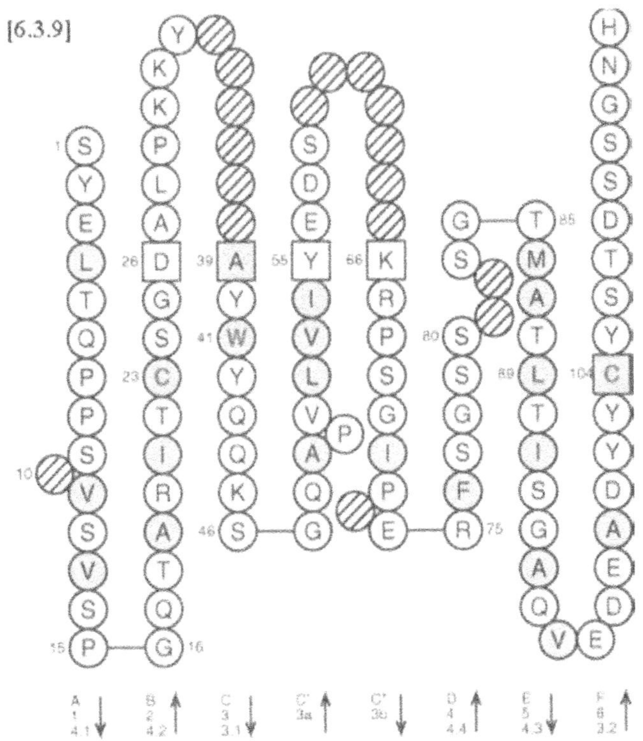

Genome database accession numbers

GDB:9953701 LocusLink: 28803

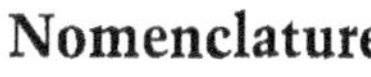

IGLV3-12

Nomenclature

IGLV3-12: Immunoglobulin lambda variable 3-12.

Definition and functionality

IGLV3-12 is one of the 8–10 functional genes of the IGLV3 subgroup which comprises 23 mapped genes.

Gene location

IGLV3-12 is located in the cluster A of the IGL locus on chromosome 22 at 22q11.2.

Nucleotide and amino acid sequences for human IGLV3-12

```
                                     1   2   3   4   5   6   7   8   9  10  11  12  13  14  15  16  17  18  19  20
                                     S   Y   E   L   T   Q   P   H   S       V   S   V   A   T   A   Q   M   A   R
Z73658,IGLV3-12*01,3i.2.2     [27]  TCC TAT GAG CTG ACT CAG CCA CAC TCA ... GTG TCA GTG GCC ACA GCA CAG ATG GCC AGG
D86998,IGLV3-12*02,V2-8       [19]  --- --- --- --- --- --- --- --- --- ... --- --- --- --- --- --- --- --- --- ---

                                                             ________CDR1-IMGT______________________
                                    21  22  23  24  25  26  27  28  29  30  31  32  33  34  35  36  37  38  39  40
                                     I   T   C   G   G   N   N   I   G   S   K   A                           V   H
Z73658,IGLV3-12*01,3i.2.2           ATC ACC TGT GGG GGA AAC AAC ATT GGA AGT AAA GCT ... ... ... ... ... ... GTG CAC
D86998,IGLV3-12*02,V2-8             --- --- --- --- --- --- --- --- --- --- --- --- ... ... ... ... ... ... --- ---

                                                                                                 ___________ CDR2-
                                    41  42  43  44  45  46  47  48  49  50  51  52  53  54  55  56  57  58  59  60
                                     W   Y   Q   Q   K   P   G   Q   D   P   V   L   V   I   Y   S   D   S
Z73658,IGLV3-12*01,3i.2.2           TGG TAC CAG CAA AAG CCA GGC CAG GAC CCT GTG CTG GTC ATC TAT AGC GAT AGC ... ...
D86998,IGLV3-12*02,V2-8             --- --- --- --- --- --- --- --- --- --- --- --- --- --- --- --- --- --- ... ...

                                    IMGT_______________
                                    61  62  63  64  65  66  67  68  69  70  71  72  73  74  75  76  77  78  79  80
                                                         N   R   P   S   G   I   P       E   R   F   S   G   S   N
Z73658,IGLV3-12*01,3i.2.2           ... ... ... ... ... AAC CGG CCC TCA GGG ATC CCT ... GAG CGA TTC TCT GGC TCC AAC
D86998,IGLV3-12*02,V2-8             ... ... ... ... ... --- --- --- --- --- --- --- ... --- --- --- --- --- --- ---

                                    81  82  83  84  85  86  87  88  89  90  91  92  93  94  95  96  97  98  99 100
                                             P   G   N   T   T   T   L   T   I   S   R   I   E   A   G   D   E   A
Z73658,IGLV3-12*01,3i.2.2           ... ... CCA GGG AAC ACC ACC ACC CTA ACC ATC AGC AGG ATC GAG GCT GGG GAT GAG GCT
                                                             A
D86998,IGLV3-12*02,V2-8             ... ... --- --- --- --- G-- --- --- --- --- --- --- --- --- --- --- --- --- ---

                                                    _____________CDR3-IMGT____________
                                   101 102 103 104 105 106 107 108 109 110 111 112 113
                                     D   Y   Y   C   Q   V   W   D   S   S   S   D   H
Z73658,IGLV3-12*01,3i.2.2           GAC TAT TAC TGT CAG GTG TGG GAC AGT AGT AGT GAT CAT CC
D86998,IGLV3-12*02,V2-8             --- --- --- --- --- --- --- --- --- --- --- --- --- --
```

Framework and complementarity determining regions

FR1-IMGT: 25 (-1 aa: 10)
FR2-IMGT: 17
FR3-IMGT: 36 (-3 aa: 73,81,82)

CDR1-IMGT: 6
CDR2-IMGT: 3
CDR3-IMGT: 9

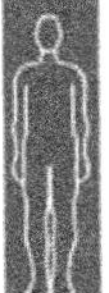

Collier de Perles for human IGLV3-12*01

Accession number: IMGT Z73658 EMBL/GenBank/DDBJ: Z73658

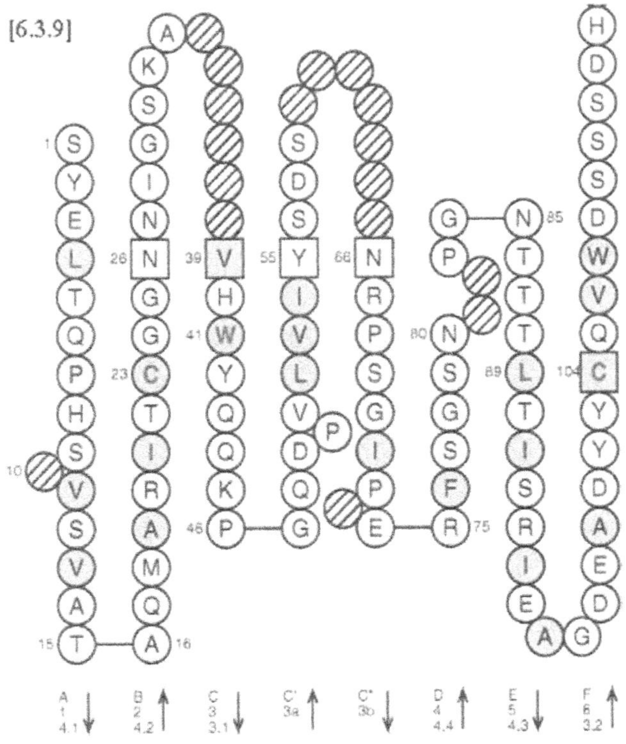

Genome database accession numbers

GDB:9953703 LocusLink: 28802

Nomenclature

IGLV3-16: Immunoglobulin lambda variable 3-16.

Definition and functionality

IGLV3-16 is one of the 8–10 functional genes of the IGLV3 subgroup which comprises 23 mapped genes.

Gene location

IGLV3-16 is located in the cluster A of the IGL locus on chromosome 22 at 22q11.2.

Nucleotide and amino acid sequences for human IGLV3-16

```
                                          1   2   3   4   5   6   7   8   9  10  11  12  13  14  15  16  17  18  19  20
                                          S   Y   E   L   T   Q   P   P   S       V   S   V   S   L   G   Q   M   A   R
X97471,IGLV3-16*01,3a.119B4        [27]  TCC TAT GAG CTG ACA CAG CCA CCC TCG ... GTG TCA GTG TCC CTA GGA CAG ATG GCC AGG

D87015,IGLV3-16*01,V2-11           [19]  --- --- --- --- --- --- --- --- --- ... --- --- --- --- --- --- --- --- --- ---

                                                                 ____________CDR1-IMGT________________
                                         21  22  23  24  25  26  27  28  29  30  31  32  33  34  35  36  37  38  39  40
                                          I   T   C   S   G   E   A   L   P   K   K   Y                           A   Y
X97471,IGLV3-16*01,3a.119B4              ATC ACC TGC TCT GGA GAA GCA TTG CCA AAA AAA TAT ... ... ... ... ... ... GCT TAT

D87015,IGLV3-16*01,V2-11                 --- --- --- --- --- --- --- --- --- --- --- --- ... ... ... ... ... ... --- ---

                                                                                                      __________ CDR2-
                                         41  42  43  44  45  46  47  48  49  50  51  52  53  54  55  56  57  58  59  60
                                          W   Y   Q   Q   K   P   G   Q   F   P   V   L   V   I   Y   K   D   S
X97471,IGLV3-16*01,3a.119B4              TGG TAC CAG CAG AAG CCA GGC CAG TTC CCT GTG CTG GTG ATA TAT AAA GAC AGC ... ...

D87015,IGLV3-16*01,V2-11                 --- --- --- --- --- --- --- --- --- --- --- --- --- --- --- --- --- --- ... ...

                                         IMGT_______________
                                         61  62  63  64  65  66  67  68  69  70  71  72  73  74  75  76  77  78  79  80
                                                              E   R   P   S   G   I   P       E   R   F   S   G   S   S
X97471,IGLV3-16*01,3a.119B4              ... ... ... ... ... GAG AGG CCC TCA GGG ATC CCT ... GAG CGA TTC TCT GGC TCC AGC

D87015,IGLV3-16*01,V2-11                 ... ... ... ... ... --- --- --- --- --- --- --- ... --- --- --- --- --- --- ---

                                         81  82  83  84  85  86  87  88  89  90  91  92  93  94  95  96  97  98  99 100
                                                  S   G   T   I   V   T   L   T   I   S   G   V   Q   A   E   D   E   A
X97471,IGLV3-16*01,3a.119B4              ... ... TCA GGG ACA ATA GTC ACA TTG ACC ATC AGT GGA GTC CAG GCA GAA GAC GAG GCT

D87015,IGLV3-16*01,V2-11                 ... ... --- --- --- --- --- --- --- --- --- --- --- --- --- --- --- --- --- ---

                                                         ___________CDR3-IMGT_______________
                                        101 102 103 104 105 106 107 108 109 110 111 112 113 114
                                          D   Y   Y   C   L   S   A   D   S   S   G   T   Y
X97471,IGLV3-16*01,3a.119B4              GAC TAT TAC TGT CTA TCA GCA GAC AGC AGT GGT ACT TAT CC

D87015,IGLV3-16*01,V2-11                 --- --- --- --- --- --- --- --- --- --- --- --- --- --
```

Framework and complementarity determining regions

FR1-IMGT: 25 (-1 aa: 10)
FR2-IMGT: 17
FR3-IMGT: 36 (-3 aa: 73,81,82)

CDR1-IMGT: 6
CDR2-IMGT: 3
CDR3-IMGT: 9

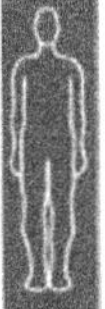

Collier de Perles for human IGLV3-16*01

Accession number: IMGT X97471 EMBL/GenBank/DDBJ: X97471

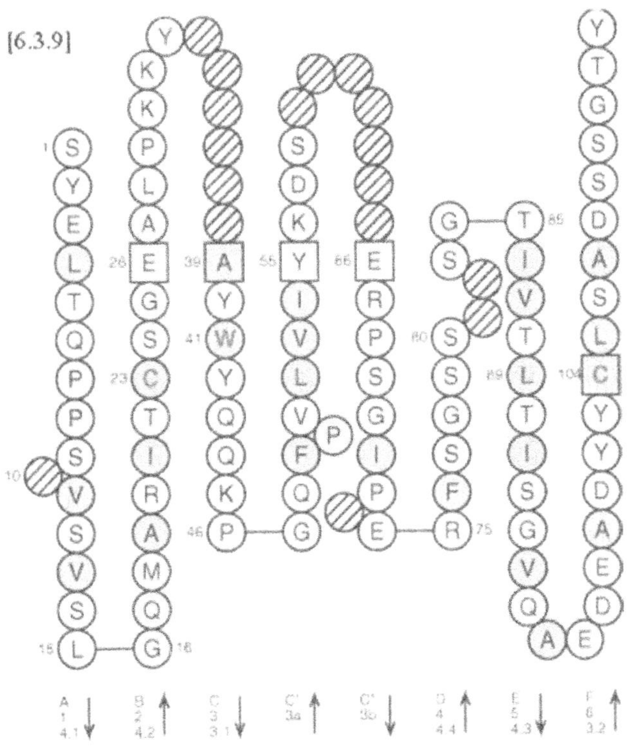

Genome database accession numbers

GDB:9953709 LocusLink: 28799

Nomenclature

IGLV3-19: Immunoglobulin lambda variable 3-19.

Definition and functionality

IGLV3-19 is one of the 8–10 functional genes of the IGLV3 subgroup which comprises 23 mapped genes.

Gene location

IGLV3-19 is located in the cluster A of the IGL locus on chromosome 22 at 22q11.2.

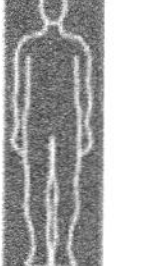

Nucleotide and amino acid sequences for human IGLV3-19

```
                                         1   2   3   4   5   6   7   8   9   10  11  12  13  14  15  16  17  18  19  20
                                         S   S   E   L   T   Q   D   P   A       V   S   V   A   L   G   Q   T   V   R
X56178,IGLV3-19*01,Vlambda3.1   [16]    TCT TCT GAG CTG ACT CAG GAC CCT GCT ... GTG TCT GTG GCC TTG GGA CAG ACA GTC AGG
D87007,IGLV3-19*01,V2-13        [19]    --- --- --- --- --- --- --- --- --- ... --- --- --- --- --- --- --- --- --- ---
Z22202,IGLV3-19*01,DPL16        [26]    --- --- --- --- --- --- --- --- --- ... --- --- --- --- --- --- --- --- --- ---
M94113,IGLV3-19*01,Humlv418     [10]    --- --- --- --- --- --- --- --- --- ... --- --- --- --- --- --- --- --- --- ---
L35919,IGLV3-19*01,IGLV4BR      [13]    --- --- --- --- --- --- --- --- --- ... --- --- --- --- --- --- --- --- --- ---

                                                                ___________CDR1-IMGT___________________
                                         21  22  23  24  25  26  27  28  29  30  31  32  33  34  35  36  37  38  39  40
                                         I   T   C   Q   G   D   S   L   R   S   Y   Y                           A   S
X56178,IGLV3-19*01,Vlambda3.1           ATC ACA TGC CAA GGA GAC AGC CTC AGA AGC TAT TAT ... ... ... ... ... ... GCA AGC
D87007,IGLV3-19*01,V2-13                --- --- --- --- --- --- --- --- --- --- --- --- ... ... ... ... ... ... --- ---
Z22202,IGLV3-19*01,DPL16                --- --- --- --- --- --- --- --- --- --- --- --- ... ... ... ... ... ... --- ---
M94113,IGLV3-19*01,Humlv418             --- --- --- --- --- --- --- --- --- --- --- --- ... ... ... ... ... ... --- ---
L35919,IGLV3-19*01,IGLV4BR              --- --- --- --- --- --- --- --- --- --- --- --- ... ... ... ... ... ... --- ---

                                                                                                    ______________CDR2-
                                         41  42  43  44  45  46  47  48  49  50  51  52  53  54  55  56  57  58  59  60
                                         W   Y   Q   Q   K   P   G   Q   A   P   V   L   V   I   Y   G   K   N
X56178,IGLV3-19*01,Vlambda3.1           TGG TAC CAG CAG AAG CCA GGA CAG GCC CCT GTA CTT GTC ATC TAT GGT AAA AAC ... ...
D87007,IGLV3-19*01,V2-13                --- --- --- --- --- --- --- --- --- --- --- --- --- --- --- --- --- --- ... ...
Z22202,IGLV3-19*01,DPL16                --- --- --- --- --- --- --- --- --- --- --- --- --- --- --- --- --- --- ... ...
M94113,IGLV3-19*01,Humlv418             --- --- --- --- --- --- --- --- --- --- --- --- --- --- --- --- --- --- ... ...
L35919,IGLV3-19*01,IGLV4BR              --- --- --- --- --- --- --- --- --- --- --- --- --- --- --- --- --- --- ... ...

                                        IMGT______________
                                         61  62  63  64  65  66  67  68  69  70  71  72  73  74  75  76  77  78  79  80
                                                             N   R   P   S   G   I   P       D   R   F   S   G   S   S
X56178,IGLV3-19*01,Vlambda3.1           ... ... ... ... ... AAC CGG CCC TCA GGG ATC CCA ... GAC CGA TTC TCT GGC TCC AGC
D87007,IGLV3-19*01,V2-13                ... ... ... ... ... --- --- --- --- --- --- --- ... --- --- --- --- --- --- ---
Z22202,IGLV3-19*01,DPL16                ... ... ... ... ... --- --- --- --- --- --- --- ... --- --- --- --- --- --- ---
M94113,IGLV3-19*01,Humlv418             ... ... ... ... ... --- --- --- --- --- --- --- ... --- --- --- --- --- --- ---
L35919,IGLV3-19*01,IGLV4BR              ... ... ... ... ... --- --- --- --- --- --- --- ... --- --- --- --- --- --- ---

                                         81  82  83  84  85  86  87  88  89  90  91  92  93  94  95  96  97  98  99  100
                                                 S   G   N   T   A   S   L   T   I   T   G   A   Q   A   E   D   E   A
X56178,IGLV3-19*01,Vlambda3.1           ... ... TCA GGA AAC ACA GCT TCC TTG ACC ATC ACT GGG GCT CAG GCG GAA GAT GAG GCT
D87007,IGLV3-19*01,V2-13                ... ... --- --- --- --- --- --- --- --- --- --- --- --- --- --- --- --- --- ---
Z22202,IGLV3-19*01,DPL16                ... ... --- --- --- --- --- --- --- --- --- --- --- --- --- --- --- --- --- ---
M94113,IGLV3-19*01,Humlv418             ... ... --- --- --- --- --- --- --- --- --- --- --- --- --- --- --- --- --- ---
L35919,IGLV3-19*01,IGLV4BR              ... ... --- --- --- --- --- --- --- --- --- --- --- --- --- --- --- --- --- ---

                                                             ___________CDR3-IMGT___________
                                         101 102 103 104 105 106 107 108 109 110 111 112 113
                                         D   Y   Y   C   N   S   R   D   S   S   G   N   H
X56178,IGLV3-19*01,Vlambda3.1           GAC TAT TAC TGT AAC TCC CGG GAC AGC AGT GGT AAC CAT CT
D87007,IGLV3-19*01,V2-13                --- --- --- --- --- --- --- --- --- --- --- --- --- --
Z22202,IGLV3-19*01,DPL16                --- --- --- --- --- --- --- --- --- --- --- --- ---
M94113,IGLV3-19*01,Humlv418             --- --- --- --- --- --- --- --- --- --- --- --- --- --
L35919,IGLV3-19*01,IGLV4BR              --- --- --- --- --- --- --- --- --- --- --- --- --- --
```

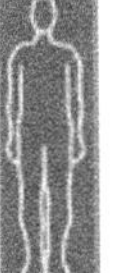

Framework and complementarity determining regions

FR1-IMGT: 25 (-1 aa: 10)
FR2-IMGT: 17
FR3-IMGT: 36 (-3 aa: 73,81,82)

CDR1-IMGT: 6
CDR2-IMGT: 3
CDR3-IMGT: 9

Collier de Perles for human IGLV3-19*01

Accession number: IMGT X56178   EMBL/GenBank/DDBJ: X56178

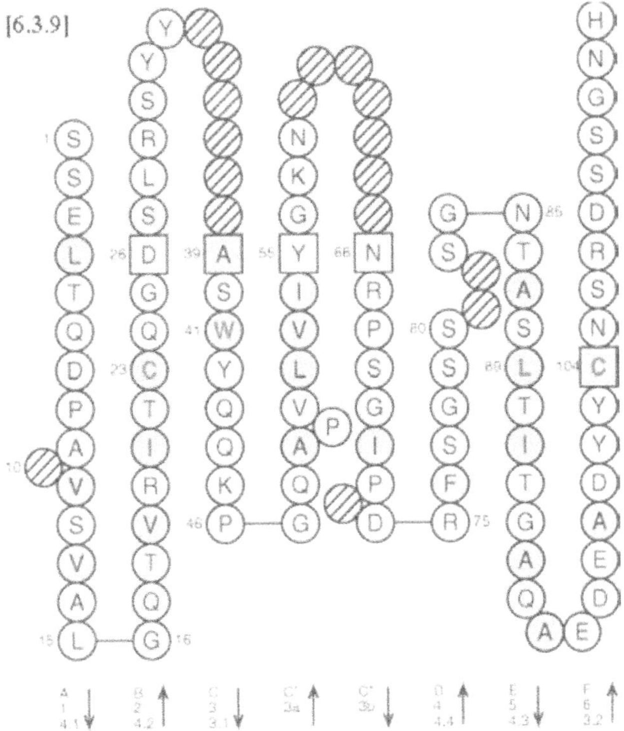

Genome database accession numbers

GDB:9953713   LocusLink: 28797

IGLV3-21

Nomenclature

IGLV3-21: Immunoglobulin lambda variable 3-21.

Definition and functionality

IGLV3-21 is one of the 8–10 functional genes of the IGLV3 subgroup which comprises 23 mapped genes.

Gene location

IGLV3-21 is located in the cluster A of the IGL locus on chromosome 22 at 22q11.2.

Nucleotide and amino acid sequences for human IGLV3-21

```
                                    1   2   3   4   5   6   7   8   9  10  11  12  13  14  15  16  17  18  19  20
                                  S   Y   V   L   T   Q   P   P   S       V   S   V   A   P   G   K   T   A   R
X71966,IGLV3-21*01,IGLV3S2   [15] TCC TAT GTG CTG ACT CAG CCA CCC TCA ... GTG TCA GTG GCC CCA GGA AAG ACG GCC AGG
                                                                                                  Q
D87007,IGLV3-21*02,V2-14     [19] --- --- --- --- --- --- --- --- --G ... --- --- --- --- --- --- C-- --- --- ---

M94115,IGLV3-21*03,Humlv318  [10] --- --- --- --- --- --- --- --- --G ... --- --- --- --- --- --- --- --- --- ---


                                                          ____________________CDR1-IMGT___________________
                                   21  22  23  24  25  26  27  28  29  30  31  32  33  34  35  36  37  38  39  40
                                  I   T   C   G   G   N   N   I   G   S   K   S                           V   H
X71966,IGLV3-21*01,IGLV3S2        ATT ACC TGT GGG GGA AAC AAC ATT GGA AGT AAA AGT ... ... ... ... ... ... GTG CAC

D87007,IGLV3-21*02,V2-14          --- --- --- --- --- --- --- --- --- --- --- --- ... ... ... ... ... ... --- ---

M94115,IGLV3-21*03,Humlv318       --- --- --- --- --- --- --- --- --- --- --- --- ... ... ... ... ... ... --- ---


                                                                                      ________________ CDR2-
                                   41  42  43  44  45  46  47  48  49  50  51  52  53  54  55  56  57  58  59  60
                                  W   Y   Q   Q   K   P   G   Q   A   P   V   L   V   I   Y   Y   D   S
X71966,IGLV3-21*01,IGLV3S2        TGG TAC CAG CAG AAG CCA GGC CAG GCC CCT GTG CTG GTC ATC TAT TAT GAT AGC ... ...
                                                                                      V       D
D87007,IGLV3-21*02,V2-14          --- --- --- --- --- --- --- --- --- --- --- --- --- G-- --- G-- --- --- ... ...
                                                                                      V       D
M94115,IGLV3-21*03,Humlv318       --- --- --- --- --- --- --- --- --- --- --- --- --- G-- --- G-- --- --- ... ...


                                  IMGT________________
                                   61  62  63  64  65  66  67  68  69  70  71  72  73  74  75  76  77  78  79  80
                                                      D   R   P   S   G   I   P       K   R   F   S   G   S   N
X71966,IGLV3-21*01,IGLV3S2        ... ... ... ... ... GAC CGG CCC TCA GGG ATC CCT ... GAG CGA TTC TCT GGC TCC AAC

D87007,IGLV3-21*02,V2-14          ... ... ... ... ... --- --- --- --- --- --- --- ... --- --- --- --- --- --- ---

M94115,IGLV3-21*03,Humlv318       ... ... ... ... ... --- --- --- --- --- --- --- ... --- --- --- --- --- --- ---


                                   81  82  83  84  85  86  87  88  89  90  91  92  93  94  95  96  97  98  99 100
                                          S   G   N   T   A   T   L   T   I   S   R   V   E   A   G   D   E   A
X71966,IGLV3-21*01,IGLV3S2        ... ... TCT GGG AAC ACG GCC ACC CTG ACC ATC AGC AGG GTC GAA GCC GGG GAT GAG GCC

D87007,IGLV3-21*02,V2-14          ... ... --- --- --- --- --- --- --- --- --- --- --- --- --- --- --- --- --- ---

M94115,IGLV3-21*03,Humlv318       ... ... --- --- --- --- --- --- --- --- --- --- --- --- --- --- --- --- --- ---


                                                  ______________CDR3-IMGT_______________
                                  101 102 103 104 105 106 107 108 109 110 111 112 113
                                  D   Y   Y   C   Q   V   W   D   S   S   S   D   H
X71966,IGLV3-21*01,IGLV3S2        GAC TAT TAC TGT CAG GTG TGG GAC AGT AGT AGT GAT CAT CC

D87007,IGLV3-21*02,V2-14          --- --- --- --- --- --- --- --T --- --- --- --- --- --

M94115,IGLV3-21*03,Humlv318       --- --- --- --- --- --- --- --T --- --- --- --- --- --
```

Framework and complementarity determining regions

FR1-IMGT: 25 (-1 aa: 10)
FR2-IMGT: 17
FR3-IMGT: 36 (-3 aa: 73,81,82)

CDR1-IMGT: 6
CDR2-IMGT: 3
CDR3-IMGT: 9

Collier de Perles for human IGLV3-21*01

Accession number: IMGT X71966 EMBL/GenBank/DDBJ: X71966

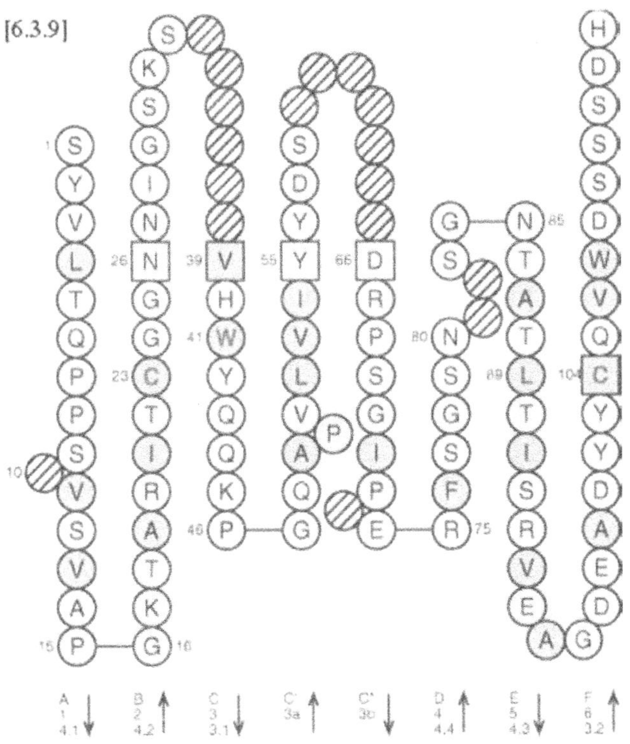

Genome database accession numbers

GDB:9953715 LocusLink: 28796

IGLV3-22

Nomenclature

IGLV3-22: Immunoglobulin lambda variable 3-22.

Definition and functionality

IGLV3-22 is a functional gene (allele *01) or a pseudogene (allele *02). IGLV3-22 belongs to the IGLV3 subgroup which comprises 23 mapped genes, of which 8–10 are functional.

IGLV3-22*02 is a pseudogene due to a 1 nt DELETION in codon 47, and to a 5 nt INSERTION between positions 50 and 51 in FR2-IMGT, leading to frameshifts.

Gene location

IGLV3-22 is located in the cluster A of the IGL locus on chromosome 22 at 22q11.2.

Nucleotide and amino acid sequences for human IGLV3-22

```
                                      1   2   3   4   5   6   7   8   9   10  11  12  13  14  15  16  17  18  19  20
                                      S   Y   E   L   T   Q   L   P   S       V   S   V   S   P   G   Q   T   A   R
Z73666,IGLV3-22*01,3e.272A12   [27]  TCC TAT GAG CTG ACA CAG CTA CCC TCG ... GTG TCA GTG TCC CCA GGA CAG ACA GCC AGG
D87007,IGLV3-22*01,V2-15       [19]  --- --- --- --- --- --- --- --- --- ... --- --- --- --- --- --- --- --- --- ---
                                                                                                        K
X71967,IGLV3-22*02,IGLV3S3P    [15]  --- --- --- --- --- --- --- --- --- ... --- --- --- --- --- --- --- -A- --- ---

                                                              _______________CDR1-IMGT_______________
                                      21  22  23  24  25  26  27  28  29  30  31  32  33  34  35  36  37  38  39  40
                                      I   T   C   S   G   D   V   L   G   E   N   Y                           A   D
Z73666,IGLV3-22*01,3e.272A12         ATC ACC TGC TCT GGA GAT GTA CTG GGG GAA AAT TAT ... ... ... ... ... ... GCT GAC
D87007,IGLV3-22*01,V2-15             --- --- --- --- --- --- --- --- --- --- --- --- ... ... ... ... ... ... --- ---
                                                                          K
X71967,IGLV3-22*02,IGLV3S3P          --- --- --- --- --- --- --- --- --- A-- --- --- ... ... ... ... ... ... --- ---

                                                                                               ___________CDR2-
                                      41  42  43  44  45  46  47  48  49  50       51  52  53  54  55  56  57  58  59  60
                                      W   Y   Q   Q   K   P   G   Q   A   P        E   L   V   I   Y   E   D   S
Z73666,IGLV3-22*01,3e.272A12         TGG TAC CAG CAG AAG CCA GGC CAG GCC CCT      GAG TTG GTG ATA TAC GAA GAT AGT ... ...
D87007,IGLV3-22*01,V2-15             --- --- --- --- --- --- --- --- --- ---      --- --- --- --- --- --- --- --- ... ...
                                                              #               * Y #
X71967,IGLV3-22*02,IGLV3S3P          --- --- --- --- --- --- .-- --- --- TGATATAC--- --- --- --- --- --- --- --- ... ...

                                     IMGT________________
                                      61  62  63  64  65  66  67  68  69  70  71  72  73  74  75  76  77  78  79  80
                                                          E   R   Y   P   G   I   P       E   R   F   S   G   S   T
Z73666,IGLV3-22*01,3e.272A12         ... ... ... ... ... GAG CGG TAC CCT GGA ATC CCT ... GAA CGA TTC TCT GGG TCC ACC
D87007,IGLV3-22*01,V2-15             ... ... ... ... ... --- --- --- --- --- --- --- ... --- --- --- --- --- --- ---
X71967,IGLV3-22*02,IGLV3S3P          ... ... ... ... ... --- --- --- --- --- --- --- ... --- --- --- --- --- --- ---

                                      81  82  83  84  85  86  87  88  89  90  91  92  93  94  95  96  97  98  99  100
                                              S   G   N   T   T   T   L   T   I   S   R   V   L   T   E   D   E   A
Z73666,IGLV3-22*01,3e.272A12         ... ... TCA GGG AAC ACG ACC ACC CTG ACC ATC AGC AGG GTC CTG ACC GAA GAC GAG GCT
D87007,IGLV3-22*01,V2-15             ... ... --- --- --- --- --- --- --- --- --- --- --- --- --- --- --- --- --- ---
X71967,IGLV3-22*02,IGLV3S3P          ... ... --- --- --- --- --- --- --- --- --- --- --- --- --- --- --- --- --- ---

                                                      ___________CDR3-IMGT____________
                                      101 102 103 104 105 106 107 108 109 110 111
                                      D   Y   Y   C   L   S   G   D   E   D   N
Z73666,IGLV3-22*01,3e.272A12         GAC TAT TAC TGT TTG TCT GGG GAT GAG GAC AAT CC
D87007,IGLV3-22*01,V2-15             --- --- --- --- --- --- --- --- --- --- --- --
                                                                  N
X71967,IGLV3-22*02,IGLV3S3P          --- --- --- --- --- --- --- A-- --- --T --- --
```

Framework and complementarity determining regions

FR1-IMGT: 25 (-1 aa: 10)
FR2-IMGT: 17
FR3-IMGT: 36 (-3 aa: 73,81,82)

CDR1-IMGT: 6
CDR2-IMGT: 3
CDR3-IMGT: 7

Collier de Perles for human IGLV3-22*01

Accession number: IMGT Z73666 EMBL/GenBank/DDBJ: Z73666

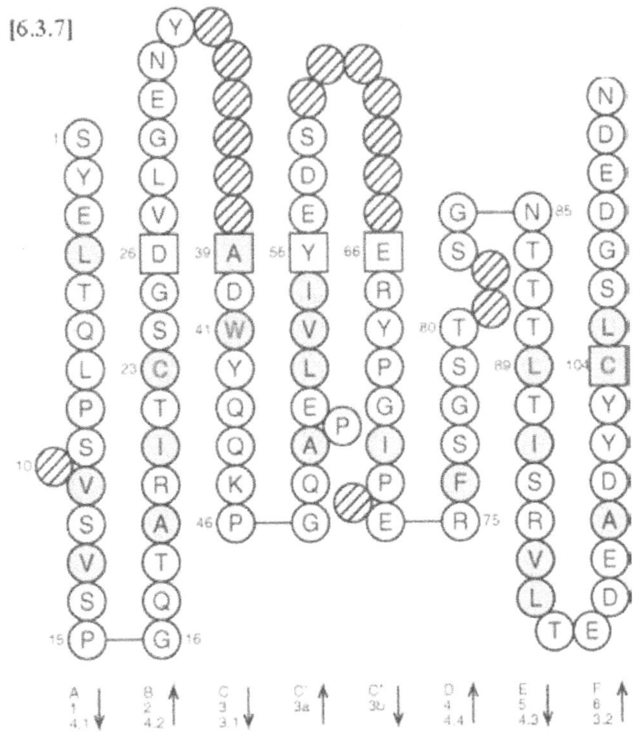

Genome database accession numbers

GDB:9953717 LocusLink: 28795

Nomenclature

IGLV3-25: Immunoglobulin lambda variable 3-25.

Definition and functionality

IGLV3-25 is one of the 8–10 functional genes of the IGLV3 subgroup which comprises 23 mapped genes.

Gene location

IGLV3-25 is located in the cluster A of the IGL locus on chromosome 22 at 22q11.2.

Nucleotide and amino acid sequences for human IGLV3-25

```
                                         1   2   3   4   5   6   7   8   9  10  11  12  13  14  15  16  17  18  19  20
                                         S   Y   E   L   M   Q   P   P   S       V   S   V   S   P   G   Q   T   A   R
X97474,IGLV3-25*01,3m.102D1      [27]   TCC TAT GAG CTG ATG CAG CCA CCC TCG ... GTG TCA GTG TCC CCA GGA CAG ACG GCC AGG
                                                     T
D86994,IGLV3-25*02,V2-17         [19]       --- --- --- -CA --- ---     - -     ... --- --- --  --- --- --- --- --- --- ---
                                                     T
L29165,IGLV3-25*03,IGGLL150      [14]       --- --- --- -CA - -             ...       -       - -     --- --  --- --  ---

                                                                                   CDR1-IMGT
                                        21  22  23  24  25  26  27  28  29  30  31  32  33  34  35  36  37  38  39  40
                                         I   T   C   S   G   D   A   L   P   K   Q   Y                           A   Y
X97474,IGLV3-25*01,3m.102D1             ATC ACC TGC TCT GGA GAT GCA TTG CCA AAG CAA TAT ... ... ... ... ... ... GCT TAT
D86994,IGLV3-25*02,V2-17                --               .   . --- --- --- --- --- --- --- ... ... ... ... ... ... --  -
L29165,IGLV3-25*03,IGGLL150             --- --  --  --- --- --- --- --- --- --- --- --- ... ... ... ... ... ... --- ---

                                                                                                        CDR2-
                                        41  42  43  44  45  46  47  48  49  50  51  52  53  54  55  56  57  58  59  60
                                         W   Y   Q   Q   K   P   G   Q   A   P   V   L   V   I   Y   K   D   S
X97474,IGLV3-25*01,3m.102D1             TGG TAC CAG CAG AAG CCA GGC CAG GCC CCT GTG CTG GTC ATA TAT AAA GAC AGT ... ...
D86994,IGLV3-25*02,V2-17                --- --- --- --- --- ---  -       -                                       ... ...
L29165,IGLV3-25*03,IGGLL150             - - --- --- --- --- --- --- --- --- --- --- --- --- --- --- --- --- --- ... ...

                                        IMGT
                                        61  62  63  64  65  66  67  68  69  70  71  72  73  74  75  76  77  78  79  80
                                                             E   R   P   S   G   I   P       E   R   F   S   G   S   S
X97474,IGLV3-25*01,3m.102D1             ... ... ... ... ... GAG AGG CCC TCA GGG ATC CCT ... GAG CGA TTC TCT GGC TCC AGC
D86994,IGLV3-25*02,V2-17                ... ... ... ... ...                             ...         --- --- --- --- ---
L29165,IGLV3-25*03,IGGLL150             ... ... ... ... ...                 - -   - --- ... ---   - --- --- --- --- ---

                                        81  82  83  84  85  86  87  88  89  90  91  92  93  94  95  96  97  98  99 100
                                                 S   G   T   T   V   T   L   T   I   S   G   V   Q   A   E   D   E   A
X97474,IGLV3-25*01,3m.102D1             ... ... TCA GGG ACA ACA GTC ACG TTG ACC ATC AGT GGA GTC CAG GCA GAA GAT GAG GCT
D86994,IGLV3-25*02,V2-17                ... ... --- --- --- --- --- --- --- --- --- --- --- --- --- --- --- --- --- ---
L29165,IGLV3-25*03,IGGLL150             ... ... --- --- --- --- --- --- --- --- --- --- --- --- --- --- --- --C --- ---

                                                           CDR3-IMGT
                                        101 102 103 104 105 106 107 108 109 110 111 112 113
                                         D   Y   Y   C   Q   S   A   D   S   S   G   T   Y
X97474,IGLV3-25*01,3m.102D1             GAC TAT TAC TGT CAA TCA GCA GAC AGC AGT GGT ACT TAT CC
D86994,IGLV3-25*02,V2-17                    ..          .           - - - --- --- --- --- --
L29165,IGLV3-25*03,IGGLL150                     .   - --    - --  --- --- --- ---
```

Framework and complementarity determining regions

FR1-IMGT: 25 (-1 aa: 10)
FR2-IMGT: 17
FR3-IMGT: 36 (-3 aa: 73,81,82)

CDR1-IMGT: 6
CDR2-IMGT: 3
CDR3-IMGT: 9

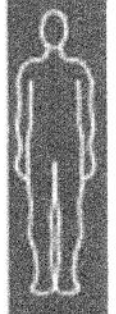

Collier de Perles for human IGLV3-25*01

Accession number: IMGT X97474 EMBL/GenBank/DDBJ: X97474

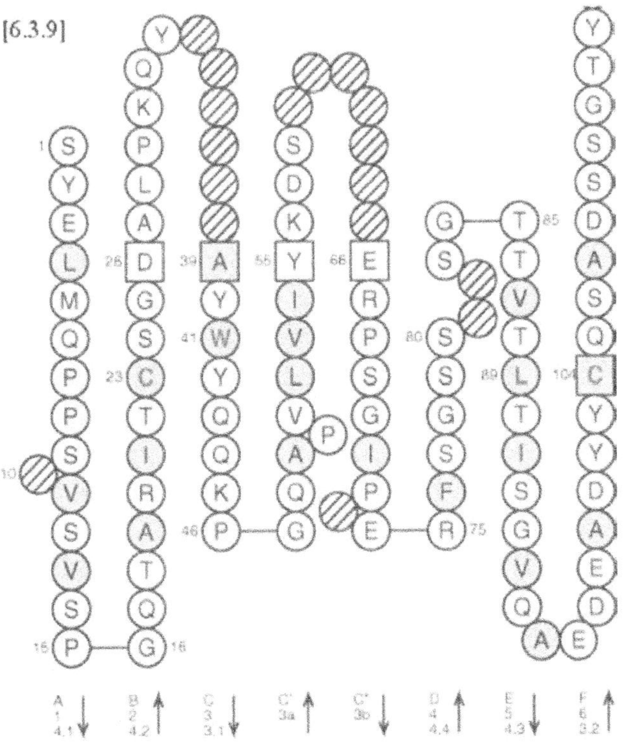

Genome database accession numbers

GDB:9953721 LocusLink: 28793

Nomenclature

IGLV3-27: Immunoglobulin lambda variable 3-27.

Definition and functionality

IGLV3-27 is one of the 8–10 functional genes of the IGLV3 subgroup which comprises 23 mapped genes.

Gene location

IGLV3-27 is located in the cluster A of the IGL locus on chromosome 22 at 22q11.2.

Nucleotide and amino acid sequences for human IGLV3-27

```
                                   1   2   3   4   5   6   7   8   9  10  11  12  13  14  15  16  17  18  19  20
                                   S   Y   E   L   T   Q   P   S   S       V   S   V   S   P   G   Q   T   A   R
D86994,IGLV3-27*01,V2-19    [19]  TCC TAT GAG CTG ACA CAG CCA TCC TCA ... GTG TCA GTG TCT CCG GGA CAG ACA GCC AGG

                                                          ____________CDR1-IMGT____________________
                                  21  22  23  24  25  26  27  28  29  30  31  32  33  34  35  36  37  38  39  40
                                   T   T   C   S   G   D   V   L   A   K   K   Y                           A   R
D86994,IGLV3-27*01,V2-19          ATC ACC TGC TCA GGA GAT GTA CTG GCA AAA AAA TAT ... ... ... ... ... ... GCT CGG

                                                                                             ____________CDR2-
                                  41  42  43  44  45  46  47  48  49  50  51  52  53  54  55  56  57  58  59  60
                                   W   F   Q   Q   K   P   G   Q   A   P   V   L   V   I   Y   K   D   S
D86994,IGLV3-27*01,V2-19          TGG TTC CAG CAG AAG CCA GGC CAG GCC CCT GTG CTG GTG ATT TAT AAA GAC AGT ... ...

                                  IMGT_______________
                                  61  62  63  64  65  66  67  68  69  70  71  72  73  74  75  76  77  78  79  80
                                                       E   R   P   S   G   I   P       E   R   F   S   G   S   S
D86994,IGLV3-27*01,V2-19          ... ... ... ... ... GAG CGG CCC TCA GGG ATC CCT ... GAG CGA TTC TCC GGC TCC AGC

                                  81  82  83  84  85  86  87  88  89  90  91  92  93  94  95  96  97  98  99 100
                                           S   G   T   T   V   T   L   T   I   S   G   A   Q   V   E   D   E   A
D86994,IGLV3-27*01,V2-19          ... ... TCA GGG ACC ACA GTC ACC TTG ACC ATC AGC GGG GCC CAG GTT GAG GAT GAG GCT

                                                  ________CDR3-IMGT_________
                                  101 102 103 104 105 106 107 108 109 110 111
                                   D   Y   Y   C   Y   S   A   A   D   N   N
D86994,IGLV3-27*01,V2-19          GAC TAT TAC TGT TAC TCT GCG GCT GAC AAC AAT CT
```

Framework and complementarity determining regions

FR1-IMGT: 25 (-1 aa: 10)
FR2-IMGT: 17
FR3-IMGT: 36 (-3 aa: 73,81,82)

CDR1-IMGT: 6
CDR2-IMGT: 3
CDR3-IMGT: 7

Collier de Perles for human IGLV3-27*01

Accession number: IMGT D86994 EMBL/Genbank/DDBJ: D86994

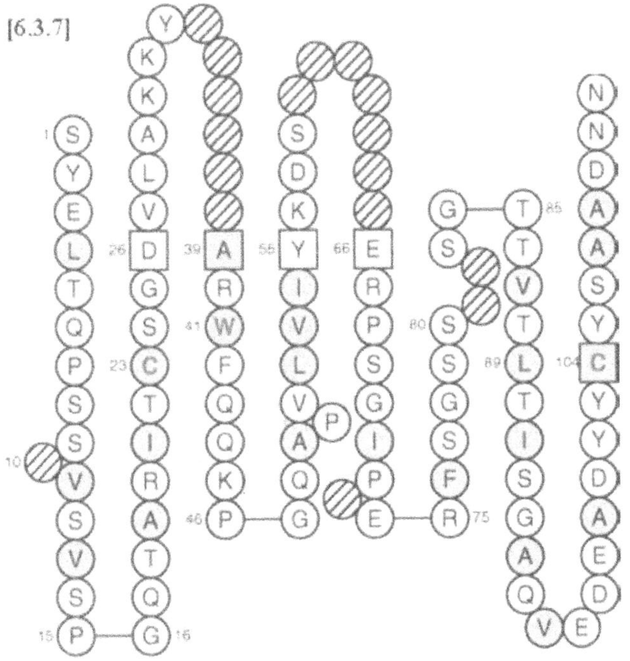

Genome database accession numbers

GDB:9953725 LocusLink: 28791

IGLV3-32

Nomenclature

IGLV3-32: Immunoglobulin lambda variable 3-32.

Definition and functionality

IGLV3-32 is an ORF due to a non-canonical ACCEPTOR_SPLICE: nggnn instead of nagnn, and to a Tyrosine (tat) instead of the 2nd_CYS at position 104. IGLV3-32 belongs to the IGLV3 subgroup which comprises 23 mapped genes, of which 8–10 are functional.

Gene location

IGLV3-32 is located in the cluster A of the IGL locus on chromosome 22 at 22q11.2.

Nucleotide and amino acid sequences for human IGLV3-32

```
                                     1   2   3   4   5   6   7   8   9  10  11  12  13  14  15  16  17  18  19  20
                                     S   S   G   P   T   Q   V   P   A       V   S   V   A   L   G   Q   M   A   R
Z73645,IGLV3-32*01,3il.61E11   [27] TCC TCT GGG CCA ACT CAG GTG CCT GCA ... GTG TCT GTG GCC TTG GGA CAA ATG GCC AGG

D87014,IGLV3-32*01,V2-23P      [19] --- --- --- --- --- --- --- --- --- ... --- --- --- --- --- --- --- --- --- ---

                                                            ____________________CDR1-IMGT___________________
                                    21  22  23  24  25  26  27  28  29  30  31  32  33  34  35  36  37  38  39  40
                                     I   T   C   Q   G   D   S   M   E   G   S   Y                           E   H
Z73645,IGLV3-32*01,3il.61E11        ATC ACC TGC CAG GGA GAC AGC ATG GAA GGC TCT TAT ... ... ... ... ... ... GAA CAC

D87014,IGLV3-32*01,V2-23P           --- --- --- --- --- --- --- --- --- --- --- --- ... ... ... ... ... ... --- ---

                                                                                                ____________CDR2-
                                    41  42  43  44  45  46  47  48  49  50  51  52  53  54  55  56  57  58  59  60
                                     W   Y   Q   Q   K   P   G   Q   A   P   V   L   V   I   Y   D   S   S
Z73645,IGLV3-32*01,3il.61E11        TGG TAC CAG CAG AAG CCA GGC CAG GCC CCC GTG CTG GTC ATC TAT GAT AGC AGT ... ...

D87014,IGLV3-32*01,V2-23P           --- --- --- --- --- --- --- --- --- --- --- --- --- --- --- --- --- --- ... ...

                                    IMGT________________
                                    61  62  63  64  65  66  67  68  69  70  71  72  73  74  75  76  77  78  79  80
                                                         D   R   P   S   R   I   P       E   R   F   S   G   S   K
Z73645,IGLV3-32*01,3il.61E11        ... ... ... ... ... GAC CGG CCC TCA AGG ATC CCT ... GAG CGA TTC TCT GGC TCC AAA

D87014,IGLV3-32*01,V2-23P           ... ... ... ... ... --- --- --- --- --- --- --- ... --- --- --- --- --- --- ---

                                    81  82  83  84  85  86  87  88  89  90  91  92  93  94  95  96  97  98  99  100
                                             S   G   N   T   T   T   L   T   I   T   G   A   Q   A   E   D   F   A
Z73645,IGLV3-32*01,3il.61E11        ... ... TCA GGC AAC ACA ACC ACC CTG ACC ATC ACT GGG GCC CAG GCT GAG GAT GAG GCT

D87014,IGLV3-32*01,V2-23P           ... ... --- --- --- --- --- --- --- --- --- --- --- --- --- --- --- --- --- ---

                                                    ______________CDR3-IMGT______________
                                    101 102 103 104 105 106 107 108 109 110 111 112 113
                                     D   Y   Y   Y   Q   L   I   D   N   H   A
Z73645,IGLV3-32*01,3il.61E11        GAT TAT TAC TAT CAG TTG ATA GAC AAC CAT GCT AC
                                                                                  T   Q
D87014,IGLV3-32*01,V2-23P           --- --- --- --- --- --- --- --- --- --- --- --T CAA CT
```

Framework and complementarity determining regions

FR1-IMGT: 25 (-1 aa: 10)
FR2-IMGT: 17
FR3-IMGT: 36 (-3 aa: 73,81,82)

CDR1-IMGT: 6
CDR2-IMGT: 3
CDR3-IMGT: 7

Collier de Perles for human IGLV3-32*01

Accession number: IMGT Z73645 EMBL/GenBank/DDBJ: Z73645

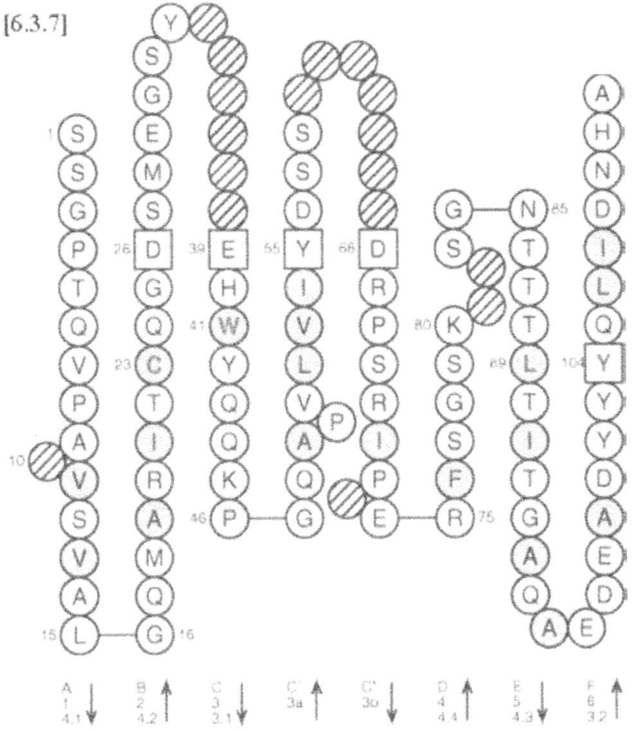

Genome database accession numbers

GDB:9953733 LocusLink: 28787

IGLV4-3

Nomenclature

IGLV4-3: Immunoglobulin lambda variable 4-3.

Definition and functionality

IGLV4-3 is considered as one of the three functional genes of the IGLV4 subgroup which comprises three mapped genes, despite the presence of a STOP-CODON at position 116 in the CDR3-IMGT. Indeed, this STOP-CODON at the 3′ end of the V-REGION can disappear during the V-J rearrangement by deletion and/or mutation at the V-J junction.

Gene location

IGLV4-3 is located in the cluster A of the IGL locus on chromosome 22 at 22q11.2.

Nucleotide and amino acid sequences for human IGLV4-3

```
                                       1   2   3   4   5   6   7   8   9  10  11  12  13  14  15  16  17  18  19  20
                                       L   P   V   L   T   Q   P   P   S       A   S   A   L   L   G   A   S   I   K
X57828,IGLV4-3*01,VlambdaN.2   [8]   CTG CCT GTG CTG ACT CAG CCC CCG TCT ... GCA TCT GCC TTG CTG GGA GCC TCG ATC AAG
Z73652,IGLV4-3*01,4c127E5     [27]   --- --- --- --- --- --- --- --- --- ... --- --- --- --- --- --- --- --- --- ---
D87024,IGLV4-3*01,V5-1        [19]   --- --- --- --- --- --- --- --- --- ... --- --- --- --- --- --- --- --- --- ---
Z22211,IGLV4-3*01,DPL24       [26]   --- --- --- --- --- --- --- --- --- ... --- --- --- --- --- --- --- --- --- ---

                                                          _______________CDR1-IMGT_______________
                                      21  22  23  24  25  26  27  28  29  30  31  32  33  34  35  36  37  38  39  40
                                       L   T   C   T   L   S   S   E   H   S   T   Y   T                       I   E
X57828,IGLV4-3*01,VlambdaN.2         CTC ACC TGC ACC CTA AGC AGT GAG CAC AGC ACC TAC ACC ... ... ... ... ... ATC GAA
Z73652,IGLV4-3*01,4c127E5            --- --- --- --- --- --- --- --- --- --- --- --- --- ... ... ... ... ... ---
D87024,IGLV4-3*01,V5-1               --- --- --- --- --- --- --- ---     --- --- --- --- ... ... ... ... ... --- ---
Z22211,IGLV4-3*01,DPL24                  --- --- --- --- --- --- --- --- --- --- --- --- ... ... ... ... ... --- ---

                                                                                         ____________________CDR2-
                                      41  42  43  44  45  46  47  48  49  50  51  52  53  54  55  56  57  58  59  60
                                       W   Y   Q   Q   R   P   G   R   S   P   Q   Y   I   M   K   V   K   S   D   G
X57828,IGLV4-3*01,VlambdaN.2         TGG TAT CAA CAG AGA CCA GGG AGG TCC CCC CAG TAT ATA ATG AAG GTT AAG AGT GAT GGC
Z73652,IGLV4-3*01,4c127E5            --- --- --- --- --- --- --- --- --- --- ---                 --- --- --- --- --- ---
D87024,IGLV4-3*01,V5-1               --- --- --- --- --- ---     --- --- --- --- --- --- --- --- --- --- --- --- ---
Z22211,IGLV4-3*01,DPL24              --- --- --- --- --- --- --- --- --- --- --- --- --- --- --- --- --- --- --- ---

                                     IMGT_______________
                                      61  62  63  64  65  66  67  68  69  70  71  72  73  74  75  76  77  78  79  80
                                       S   H               S   K   G   D   G   I   P       D   R   F   M   G   S   S
X57828,IGLV4-3*01,VlambdaN.2         AGC CAC ... ... ... AGC AAG GGG GAC GGG ATC CCC ... GAT CGC TTC ATG GGC TCC AGT
Z73652,IGLV4-3*01,4c127E5            --- --- ... ... ... --- --- --- --- --- --- --- ... --- --- --- --- --- --- ---
D87024,IGLV4-3*01,V5-1               --- --- ... ... ... --- --- ---             --- ... --- --- --- --- --- --- ---
Z22211,IGLV4-3*01,DPL24              --- --- ... ... ... --- --- --- --- --- --- --- ... --- --- --- --- --- --- ---

                                      81  82  83  84  85  86  87  88  89  90  91  92  93  94  95  96  97  98  99 100
                                               S   G   A   D   R   Y   L   T   F   S   N   L   Q   S   D   D   E   A
X57828,IGLV4-3*01,VlambdaN.2         ... ... TCT GGG GCT GAC CGC TAC CTC ACC TTC TCC AAC CTC CAG TCT GAC GAT GAG GCT
Z73652,IGLV4-3*01,4c127E5            ... ... --- --- --- --- --- --- --- --- --- --- --- --- --- --- --- --- --- ---
D87024,IGLV4-3*01,V5-1               ... ... --- --- --- --- --- --- --- --- --- --- --- --- --- --- --- --- --- ---
Z22211,IGLV4-3*01,DPL24              ... ... ---     --- --- --- --- --- --- --- --- ---                 --- --- ---

                                                          ______________CDR3-IMGT______________________
                                     101 102 103 104 105 106 107 108 109 110 111 112 113 114 115 116
                                       E   Y   H   C   G   E   S   H   T   I   D   G   Q   V   G   *
X57828,IGLV4-3*01,VlambdaN.2         GAG TAT CAC TGT GGA GAG AGC CAC ACG ATT GAT GGC CAA GTC GGT TGA GC
Z73652,IGLV4-3*01,4c127E5            --- --- --- --- --- --- --- --- --- --- ---     --- --- --- --- --
D87024,IGLV4-3*01,V5-1               --- --- --- --- --- --- --- --- --- --- --- --- --- --- --- --- --
Z22211,IGLV4-3*01,DPL24              --- --- --- --- --- --- --- --- --- --- --- --- --- --- ---
```

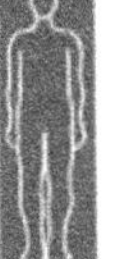

Framework and complementarity determining regions

FR1-IMGT: 25 (-1 aa: 10)
FR2-IMGT: 17
FR3-IMGT: 36 (-3 aa: 73,81,82)

CDR1-IMGT: 7
CDR2-IMGT: 7
CDR3-IMGT: 11

Collier de Perles for human IGLV4-3*01

Accession number: IMGT X57828

EMBL/GenBank/DDBJ: X57828

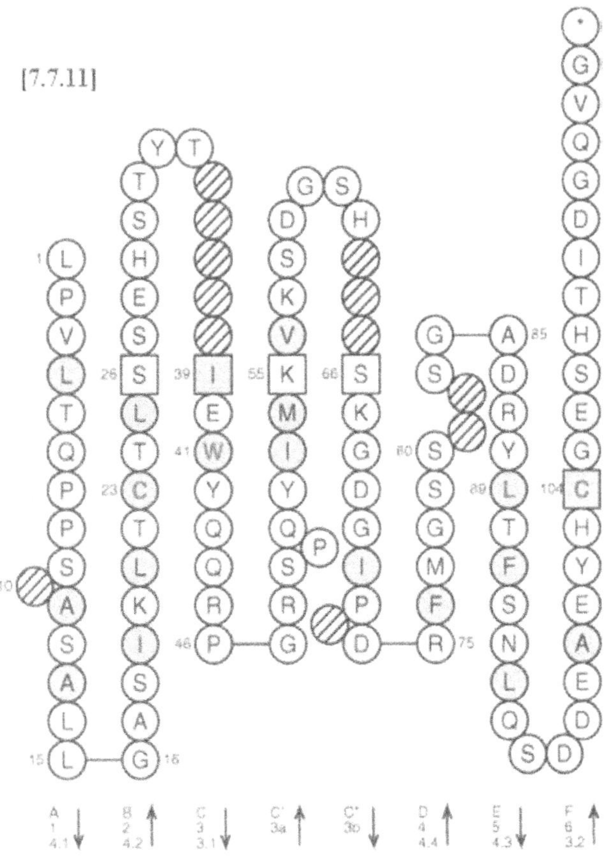

Genome database accession numbers

GDB:9953735

LocusLink: 28786

IGLV4-60

Nomenclature

IGLV4-60: Immunoglobulin lambda variable 4-60.

Definition and functionality

IGLV4-60 is one of the three functional genes of the IGLV4 subgroup which comprises three mapped genes.

Gene location

IGLV4-60 is located in the cluster C of the IGL locus on chromosome 22 at 22q11.2.

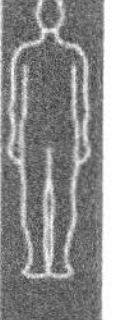

Nucleotide and amino acid sequences for human IGLV4-60

```
                                    1   2   3   4   5   6   7   8   9  10  11  12  13  14  15  16  17  18  19  20
                                    Q   P   V   L   T   Q   S   S   S       A   S   A   S   L   G   S   S   V   K
Z73667,IGLV4-60*01,4a.366F5  [27]  CAG CCT GTG CTG ACT CAA TCA TCC TCT ... GCC TCT GCT TCC CTG GGA TCC TCG GTC AAG

D87000,IGLV4-60*02,V5-4      [19]  --- --- --- --- --- --- --- --- --- ... --- --- --- --- --- --- --- --- --- ---


                                                   ___________________CDR1-IMGT___________________
                                   21  22  23  24  25  26  27  28  29  30  31  32  33  34  35  36  37  38  39  40
                                    L   T   C   T   L   S   S   G   H   S   S   Y   I                       I   A
Z73667,IGLV4-60*01,4a.366F5        CTC ACC TGC ACT CTG AGC AGT GGG CAC AGT AGC TAC ATC ... ... ... ... ... ATC GCA

D87000,IGLV4-60*02,V5-4            --- --- --- --- --- --- --- --- --- --- --- --- --- ... ... ... ... ... --- ---


                                                                                           _____________ CDR2-
                                   41  42  43  44  45  46  47  48  49  50  51  52  53  54  55  56  57  58  59  60
                                    W   H   Q   Q   Q   P   G   K   A   P   R   Y   L   M   K   L   E   G   S   G
Z73667,IGLV4-60*01,4a.366F5        TGG CAT CAG CAG CAG CCA GGG AAG GCC CCT CGG TAC TTG ATG AAC CTT GAA GGT AGT GGA

D87000,IGLV4-60*02,V5-4            --- --- --- --- --- --- --- --- --- --- --- --- --- --- --- --- --- --- --- ---


                                   IMGT_______________
                                   61  62  63  64  65  66  67  68  69  70  71  72  73  74  75  76  77  78  79  80
                                    S   Y               N   K   G   S   G   V   P       D   R   F   S   G   S   S
Z73667,IGLV4-60*01,4a.366F5        AGC TAC ... ... ... AAC AAG GGG AGC GGA GTT CCT ... GAT CGC TTC TCA GGC TCC AGC

D87000,IGLV4-60*02,V5-4            --- --- ... ... ... --- --- --- --- --- --- --- ... --- --- --- --- --- --- ---


                                   81  82  83  84  85  86  87  88  89  90  91  92  93  94  95  96  97  98  99 100
                                            S   G   A   D   R   Y   L   T   I   S   N   L   Q   L   E   D   E   A
Z73667,IGLV4-60*01,4a.366F5        ... ... TCT GGG GCT GAC CGC TAC CTC ACC ATC TCC AAC CTC CAG TTA GAG GAT GAG GCT
                                                                                                F
D87000,IGLV4-60*02,V5-4            ... ... --- --- --- --- --- --- --- --- --- --- --- --- --- --T --- --- --- ---


                                                   _________CDR3-IMGT__________
                                   101 102 103 104 105 106 107 108 109 110 111
                                    D   Y   Y   C   E   T   W   D   S   N   T
Z73667,IGLV4-60*01,4a.366F5        GAT TAT TAC TGT GAG ACC TGG GAC AGT AAC ACT

D87000,IGLV4-60*02,V5-4            --- --- --- --- --- --- --- --- --- --- --- CA
```

Framework and complementarity determining regions

FR1-IMGT: 25 (-1 aa: 10)
FR2-IMGT: 17
FR3-IMGT: 36 (-3 aa: 73,81,82)

CDR1-IMGT: 7
CDR2-IMGT: 7
CDR3-IMGT: 7

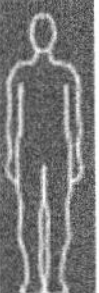

Collier de Perles for human IGLV4-60*01

Accession number: IMGT Z73667 EMBL/GenBank/DDBJ: Z73667

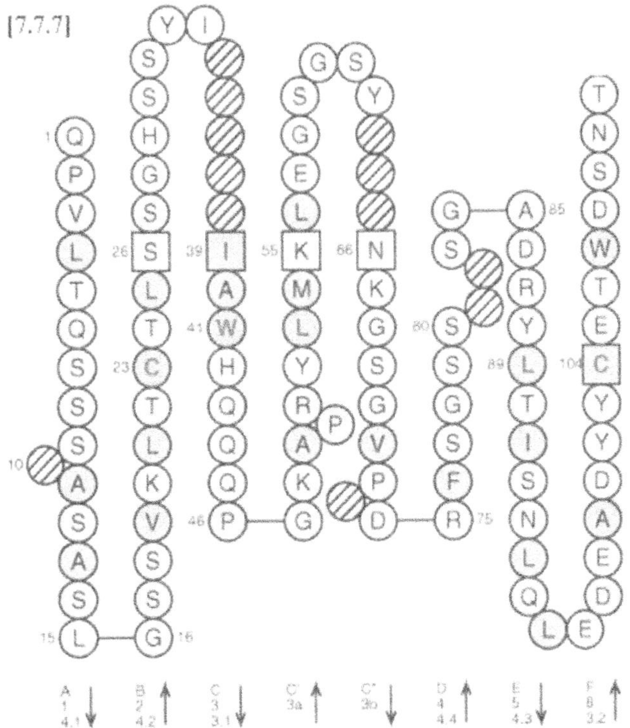

Genome database accession numbers

GDB:9953737 LocusLink: 28785

IGLV4-69

Nomenclature

IGLV4-69: Immunoglobulin lambda variable 4-69.

Definition and functionality

IGLV4-69 is one of the three functional genes of the IGLV4 subgroup which comprises three mapped genes.

Gene location

IGLV4-69 is located in the cluster C of the IGL locus on chromosome 22 at 22q11.2.

Nucleotide and amino acid sequences for human IGLV4-69

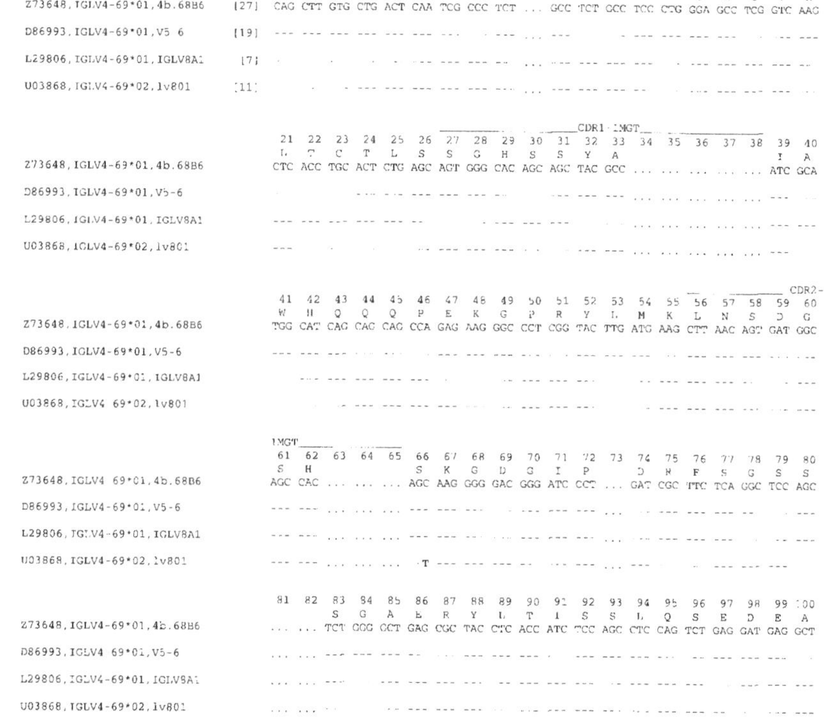

```
                                           1   2   3   4   5   6   7   8   9   10  11  12  13  14  15  16  17  18  19  20
                                           Q   L   V   L   T   Q   S   P   S       A   S   A   S   L   G   A   S   V   K
Z73648,IGLV4-69*01,4b.68B6        [27]    CAG CTT GTG CTG ACT CAA TCG CCC TCT ... GCC TCT GCC TCC CTG GGA GCC TCG GTC AAG
D86993,IGLV4-69*01,V5-6           [19]
L29806,IGLV4-69*01,IGLV8A1        [7]
U03868,IGLV4-69*02,1v801          [11]

                                                                      CDR1-IMGT
                                           21  22  23  24  25  26  27  28  29  30  31  32  33  34  35  36  37  38  39  40
                                           L   T   C   T   L   S   S   G   H   S   S   Y   A                       I   A
Z73648,IGLV4-69*01,4b.68B6                CTC ACC TGC ACT CTG AGC AGT GGG CAC AGC AGC TAC GCC ... ... ... ... ... ATC GCA
D86993,IGLV4-69*01,V5-6
L29806,IGLV4-69*01,IGLV8A1
U03868,IGLV4-69*02,1v801

                                                                                                                   CDR2-
                                           41  42  43  44  45  46  47  48  49  50  51  52  53  54  55  56  57  58  59  60
                                           W   H   Q   Q   Q   P   E   K   G   P   R   Y   L   M   K   L   N   S   D   G
Z73648,IGLV4-69*01,4b.68B6                TGG CAT CAG CAG CAG CCA GAG AAG GGC CCT CGG TAC TTG ATG AAG CTT AAC AGT GAT GGC
D86993,IGLV4-69*01,V5-6
L29806,IGLV4-69*01,IGLV8A1
U03868,IGLV4-69*02,1v801

                                          IMGT
                                           61  62  63  64  65  66  67  68  69  70  71  72  73  74  75  76  77  78  79  80
                                           S   H               S   K   G   D   G   I   P       D   R   F   S   G   S   S
Z73648,IGLV4-69*01,4b.68B6                AGC CAC ... ... ... AGC AAG GGG GAC GGG ATC CCT ... GAT CGC TTC TCA GGC TCC AGC
D86993,IGLV4-69*01,V5-6
L29806,IGLV4-69*01,IGLV8A1
U03868,IGLV4-69*02,1v801

                                           81  82  83  84  85  86  87  88  89  90  91  92  93  94  95  96  97  98  99  100
                                                   S   G   A   E   R   Y   L   T   I   S   S   L   Q   S   E   D   E   A
Z73648,IGLV4-69*01,4b.68B6                ... ... TCT GGG GCT GAG CGC TAC CTC ACC ATC TCC AGC CTC CAG TCT GAG GAT GAG GCT
D86993,IGLV4-69*01,V5-6
L29806,IGLV4-69*01,IGLV8A1
U03868,IGLV4-69*02,1v801

                                                            CDR3-IMGT
                                           101 102 103 104 105 106 107 108 109 110 111
                                           D   Y   Y   C   Q   T   W   G   T   G   I
Z73648,IGLV4-69*01,4b.68B6                GAC TAT TAC TGT CAG ACC TGG GGC ACT GGC ATT CA
D86993,IGLV4-69*01,V5-6
L29806,IGLV4-69*01,IGLV8A1
U03868,IGLV4-69*02,1v801
```

Framework and complementarity determining regions

FR1-IMGT: 25 (-1 aa: 10) | CDR1-IMGT: 7
FR2-IMGT: 17 | CDR2-IMGT: 7
FR3-IMGT: 36 (-3 aa: 73,81,82) | CDR3-IMGT: 7

Collier de Perles for human IGLV4-69*01

Accession number: IMGT Z73648 EMBL/GenBank/DDBJ: Z73648

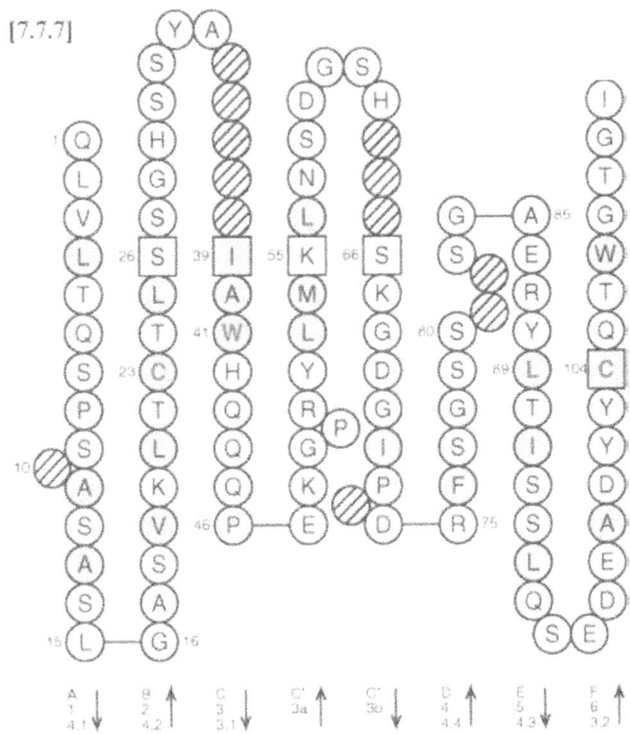

Genome database accession numbers

GDB:9953739 LocusLink: 28784

Nomenclature

IGLV5-37: Immunoglobulin lambda variable 5-37.

Definition and functionality

IGLV5-37 is one of three or four functional genes of the IGLV5 subgroup which comprises four or five mapped genes, depending on the haplotypes.

Gene location

IGLV5-37 is located in the cluster B of the IGL locus on chromosome 22 at 22q11.2.

Nucleotide and amino acid sequences for human IGLV5-37

```
                                     1   2   3   4   5   6   7   8   9   10  11  12  13  14  15  16  17  18  19  20
                                     Q   P   V   L   T   Q   P   P   S       S   S   A   S   P   G   E   S   A   R
Z73672,IGLV5-37*01,5e.366F5   [27]  CAG CCT GTG CTG ACT CAG CCA CCT TCC ... TCC TCC GCA TCT CCT GGA GAA TCC GCC AGA
D87009,IGLV5-37*01,V4-1       [19]  --- --- --- --- --- --- --- --- --- ... --- --- --- --- --- --- --- --- --- ---
D87010,IGLV5-37*01,V4-1       [19]  --- --- --- --- --- --- --- --- --- ... --- --- --- --- --- --- --- --- --- ---

                                                             _______________ CDR1-IMGT ____________________
                                     21  22  23  24  25  26  27  28  29  30  31  32  33  34  35  36  37  38  39  40
                                     L   T   C   T   L   P   S   D   I   N   V   G   S   Y   N               I   Y
Z73672,IGLV5-37*01,5e.366F5         CTC ACC TGC ACC TTG CCC AGT GAC ATC AAT GTT GGT AGC TAC AAC ... ... ... ATA TAC
D87009,IGLV5-37*01,V4-1             --- --- --- --- --- --- --- --- --- --- --- --- --- --- --- ... ... ... --- ---
D87010,IGLV5-37*01,V4-1             --- --- --- --- --- --- --- --- --- --- --- --- --- --- --- ... ... ... --- ---

                                                                                             ____________ CDR2-
                                     41  42  43  44  45  46  47  48  49  50  51  52  53  54  55  56  57  58  59  60
                                     W   Y   Q   Q   K   P   G   S   P   P   R   Y   L   L   Y   Y   Y   S   D   S
Z73672,IGLV5-37*01,5e.366F5         TGG TAC CAG CAG AAG CCA GGG AGC CCT CCC AGG TAT CTC CTG TAC TAC TAC TCA GAC TCA
D87009,IGLV5-37*01,V4-1             --- --- --- --- --- --- --- --- --- --- --- --- --- --- --- --- --- --- --- ---
D87010,IGLV5-37*01,V4-1             --- --- --- --- --- --- --- --- --- --- --- --- --- --- --- --- --- --- --- ---

                                    IMGT _______
                                     61  62  63  64  65  66  67  68  69  70  71  72  73  74  75  76  77  78  79  80
                                     D   K               G   Q   G   S   G   V   P       S   R   F   S   G   S   K
Z73672,IGLV5-37*01,5e.366F5         GAT AAG ... ... ... GGC CAG GGC TCT GGA GTC CCC ... AGC CGC TTC TCT GGA TCC AAA
D87009,IGLV5-37*01,V4-1             --- --- ... ... ... --- --- --- --- --- --- --- ... --- --- --- --- --- --- ---
D87010,IGLV5-37*01,V4-1             --- --- ... ... ... --- --- --- --- --- --- --- ... --- --- --- --- --- --- ---

                                     81  82  83  84  85  86  87  88  89  90  91  92  93  94  95  96  97  98  99  100
                                     D   A   S   A   N   T   G   I   L   L   I   S   G   L   Q   S   E   D   E   A
Z73672,IGLV5-37*01,5e.366F5         GAT GCT TCA GCC AAT ACA GGG ATT TTA CTC ATC TCC GGG CTC CAG TCT GAG GAT GAG GCT
D87009,IGLV5-37*01,V4-1             --- --- --- --- --- --- --- --- --- --- --- --- --- --- --- --- --- --- --- ---
D87010,IGLV5-37*01,V4-1             --- --- --- --- --- --- --- --- --- --- --- --- --- --- --- --- --- --- --- ---

                                                     __________ CDR3-IMGT ____________
                                    101 102 103 104 105 106 107 108 109 110 111 112
                                     D   Y   Y   C   M   I   W   P   S   N   A   S
Z73672,IGLV5-37*01,5e.366F5         GAC TAT TAC TGT ATG ATT TGG CCA AGC AAT GCT TCT
D87009,IGLV5-37*01,V4-1             --- --- --- --- --- --- --- --- ---         --- ---
D87010,IGLV5-37*01,V4-1             --- ---     --- --- --- --- ---             --- ---
```

Framework and complementarity determining regions

FR1-IMGT: 25 (-1 aa: 10)
FR2-IMGT: 17
FR3-IMGT: 38 (-1 aa: 73)

CDR1-IMGT: 9
CDR2-IMGT: 7
CDR3-IMGT: 8

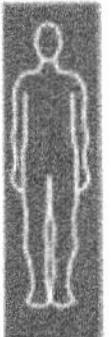

Collier de Perles for human IGLV5-37*01

Accession number: IMGT Z73672 EMBL/GenBank/DDBJ: Z73672

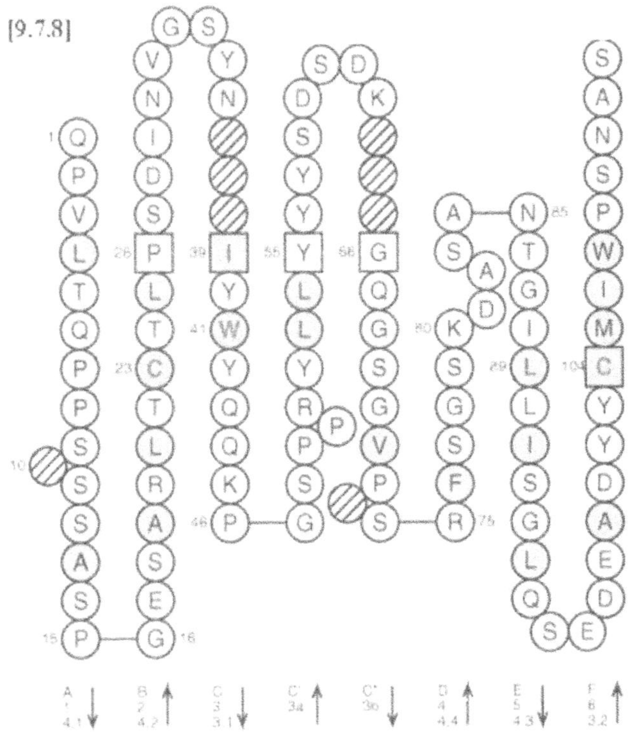

Genome database accession numbers

GDB:9953741 LocusLink: 28783

Nomenclature

IGLV5-39: Immunoglobulin lambda variable 5-39.

Definition and functionality

IGLV5-39 is a functional gene, which may, or may not, be present due to a polymorphism by insertion/deletion. IGLV5-39 belongs to the IGLV5 subgroup which comprises four or five mapped genes, depending on the haplotypes, of which three or four are functional.

Gene location

IGLV5-39, in the haplotypes where it is present, is located in the cluster B of the IGL locus on chromosome 22 at 22q11.2.

Nucleotide and amino acid sequences for human IGLV5-39

```
                                      1   2   3   4   5   6   7   8   9  10  11  12  13  14  15  16  17  18  19  20
                                      Q   P   V   L   T   Q   P   T   S       L   S   A   S   P   G   A   S   A   R
Z73668,IGLV5-39*01,5a.366F5   [27]  CAG CCT GTG CTG ACT CAG CCA ACC TCC ... CTC TCA GCA TCT CCT GGA GCA TCA GCC AGA

                                                              _______       CDR1 IMGT     ________ __
                                     21  22  23  24  25  26  27  28  29  30  31  32  33  34  35  36  37  38  39  40
                                      F   T   C   T   L   R   S   G   I   N   V   G   T   Y   R               I   Y
Z73668,IGLV5-39*01,5a.366F5         TTC ACC TGC ACC TTG CGC AGT GGC ATC AAT GTT GGT ACC TAC AGG ... ... ... ATA TAC

                                                                                                     _______  CDR2
                                     41  42  43  44  45  46  47  48  49  50  51  52  53  54  55  56  57  58  59  60
                                      W   Y   Q   Q   K   P   G   S   L   P   R   Y   L   L   R   Y   K   S   D   S
Z73668,IGLV5-39*01,5a.366F5         TGG TAC CAG CAG AAG CCA GGC AGT CTT CCC CGG TAT CTC CTG AGG TAC AAA TCA GAC TCA

                                    IMGT__ . . ._______
                                     61  62  63  64  65  66  67  68  69  70  71  72  73  74  75  76  77  78  79  80
                                      D   K               Q   Q   G   S   G   V   P       S   R   F   S   G   S   K
Z73668,IGLV5-39*01,5a.366F5         GAT AAG ... ... ... CAG CAG GGC TCT GGA GTC CCC ... AGC CGC TTC TCT GGA TCC AAA

                                     81  82  83  84  85  86  87  88  89  90  91  92  93  94  95  96  97  98  99 100
                                      D   A   S   T   N   A   G   L   L   L   I   S   G   L   Q   S   E   D   E   A
Z73668,IGLV5-39*01,5a.366F5         GAT GCT TCA ACC AAT GCA GGC CTT TTA CTC ATC TCT GGG CTC CAG TCT GAA GAT GAG GCT

                                                     ___     ____CDR3-IMGT________
                                    101 102 103 104 105 106 107 108 109 110 111 112
                                      D   Y   Y   C   A   I   W   Y   S   S   T   S
Z73668,IGLV5-39*01,5a.366F5         GAC TAT TAC TGT GCC ATT TGG TAC AGC AGC ACT TCT
```

Framework and complementarity determining regions

FR1-IMGT: 25 (-1 aa: 10)
FR2-IMGT: 17
FR3-IMGT: 38 (-1 aa: 73)

CDR1-IMGT: 9
CDR2-IMGT: 7
CDR3-IMGT: 8

Collier de Perles for human IGLV5-39*01

Accession number: IMGT Z73668 EMBL/GenBank/DDBJ: Z73668

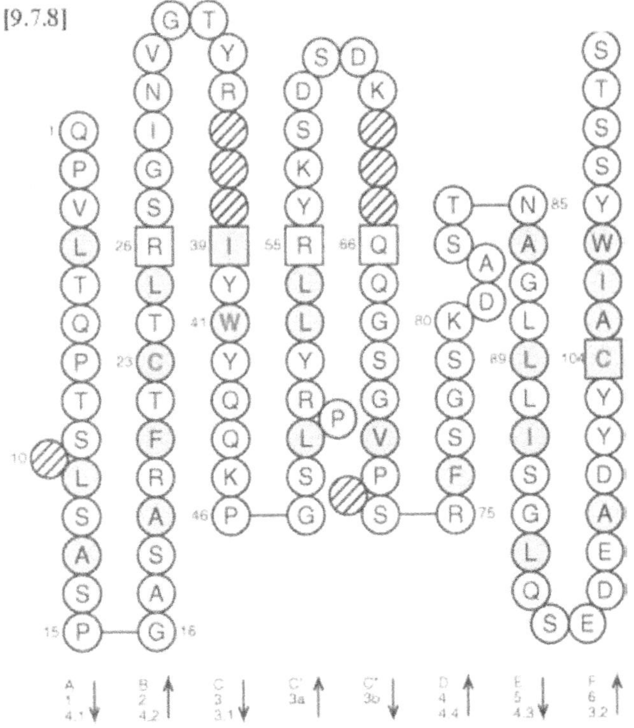

Genome database accession numbers

GDB:9953743 LocusLink: 28782

Nomenclature

IGLV5-45: Immunoglobulin lambda variable 5-45.

Definition and functionality

IGLV5-45 is one of the three or four functional genes of the IGLV5 subgroup which comprises four or five mapped genes, depending on the haplotypes.

Gene location

IGLV5-45 is located in the cluster B of the IGL locus on chromosome 22 at 22q11.2.

Nucleotide and amino acid sequences for human IGLV5-45

```
                                          1   2   3   4   5   6   7   8   9  10  11  12  13  14  15  16  17  18  19  20
                                          Q   A   V   L   T   Q   P   A   S       L   S   A   S   P   G   A   S   A   S
Z73670,IGLV5-45*01,5c.366F5     [27]    CAG GCT GTG CTG ACT CAG CCG GCT TCC ... CTC TCT GCA TCT CCT GGA GCA TCA GCC AGT
                                                                      S
Z73671,IGLV5-45*02,5c.400B5     [27]    --- --- --- --- --- --- --- T-- --- ... --- --- --- --- --- --- --- --- --- ---
                                                                      S
U93494,IGLV5-45*02,IGLV5-1      [22]    --- --- --- --- --- --- --- T-- --- ... --- --- --- --- --- --- --- --- --- ---
                                                                      S
D86999,IGLV5-45*03,V4-2         [19]    --- --- --- --- --- --- --- T-- --- ... --- --- --- --- --- --- --- --- --- ---

                                                                  _______________CDR1 IMGT__________________
                                         21  22  23  24  25  26  27  28  29  30  31  32  33  34  35  36  37  38  39  40
                                          L   T   C   T   L   R   S   G   I   N   V   G   T   Y   R               I   Y
Z73670,IGLV5-45*01,5c.366F5             CTC ACC TGC ACC TTG CGC AGT GGC ATC AAT GTT GGT ACC TAC AGG ... ... ... ATA TAC
Z73671,IGLV5-45*02,5c.400B5             --- --- --- --- --A --- --- --- --- --- --- --- --- --- --- ... ... ... --- ---
U93494,IGLV5-45*02,IGLV5-1              --- --- --- --- --A --- --- --- --- --- --- --- --- --- --- ... ... ... --- ---
D86999,IGLV5-45*03,V4-2                 --- --- --- --- --- --- --- --- --- --- --- --- --- --- --- ... ... ... --- ---

                                                                                                  ____________CDR2-
                                         41  42  43  44  45  46  47  48  49  50  51  52  53  54  55  56  57  58  59  60
                                          W   Y   Q   Q   K   P   G   S   P   P   Q   Y   L   L   R   Y   K   S   D   S
Z73670,IGLV5-45*01,5c.366F5             TGG TAC CAG CAG AAG CCA GGG AGT CCT CCC CAG TAT CTC CTG AGG TAC AAA TCA GAC TCA
Z73671,IGLV5-45*02,5c.400B5             --- --- --- --- --- --- --- --- --- --- --- --- --- --- --- --- --- --- --- ---
U93494,IGLV5-45*02,IGLV5-1              --- --- --- --- --- --- --- --- --- --- --- --- --- --- --- --- --- --- --- ---
D86999,IGLV5-45*03,V4-2                 --- --- --- --- --- --- --- --- --- --- --- --- --- --- --- --- --- --- --- ---

                                        IMGT _____________
                                         61  62  63  64  65  66  67  68  69  70  71  72  73  74  75  76  77  78  79  80
                                          D   K               Q   Q   G   S   G   V   P       S   R   F   S   G   S   K
Z73670,IGLV5-45*01,5c.366F5             GAT AAG ... ... ... CAG CAG GGC TCT GGA GTC CCC ... AGC CGC TTC TCT GGA TCC AAA
Z73671,IGLV5-45*02,5c.400B5             --- --- ... ... ... --- --- --- --- --- --- --- ... --- --- --- --- --- --- ---
U93494,IGLV5-45*02,IGLV5-1              --- --- ... ... ... --- --- --- --- --- --- --- ... --- --- --- --- --- --- ---
D86999,IGLV5-45*03,V4-2                 --- --- ... ... ... --- --- --- --- --- --- --- ... --- --- --- --- --- --- ---

                                         81  82  83  84  85  86  87  88  89  90  91  92  93  94  95  96  97  98  99 100
                                          D   A   S   A   N   A   G   I   L   L   I   S   G   L   Q   S   E   D   E   A
Z73670,IGLV5-45*01,5c.366F5             GAT GCT TCG GCC AAT GCA GGG ATT TTA CTC ATC TCT GGG CTC CAG TCT GAG GAT GAG GCT
Z73671,IGLV5-45*02,5c.400B5             --- --- --- --- --- --- --- --- --- --- --- --- --- --- --- --- --- --- --- ---
U93494,IGLV5-45*02,IGLV5-1              --- --- --- --- --- --- --- --- --- --- --- --- --- --- --- --- --- --- --- ---
D86999,IGLV5-45*03,V4-2                 --- --- --- --- --- --- --- --- --- --- --- --- --- --- --- --- --- --- --- ---

                                                         _________CDR3-IMGT__________
                                        101 102 103 104 105 106 107 108 109 110 111 112
                                          D   Y   Y   C   M   I   W   H   S   S   A   S
Z73670,IGLV5-45*01,5c.366F5             GAC TAT TAC TGT ATG ATT TGG CAC AGC AGC GCT TCT
Z73671,IGLV5-45*02,5c.400B5             --- --- --- --- --- --- --- --- --- --- --- ---
U93494,IGLV5-45*02,IGLV5-1              --- --- --- --- --- --- --- --- --- --- --- ---
D86999,IGLV5-45*03,V4-2                 --- --- --- --- --- --- --- --- --- --- --- ---
```

Framework and complementarity determining regions

FR1-IMGT: 25 (-1 aa: 10)
FR2-IMGT: 17
FR3-IMGT: 38 (-1 aa: 73)

CDR1-IMGT: 9
CDR2-IMGT: 7
CDR3-IMGT: 8

Collier de Perles for human IGLV5-45*01

Accession number: IMGT Z73670 EMBL/GenBank/DDBJ: Z73670

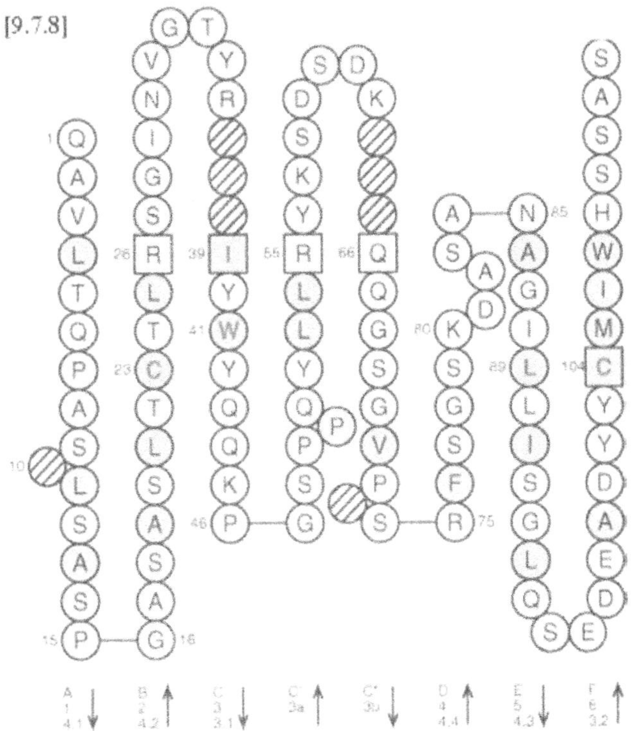

Genome database accession numbers

GDB:9953745 LocusLink: 28781

Nomenclature

IGLV5-48: Immunoglobulin lambda variable 5-48.

Definition and functionality

IGLV5-48 is an ORF due to an unusual V-HEPTAMER sequence: tacagtg instead of cacagtg. IGLV5-48 belongs to the IGLV5 subgroup which comprises four or five mapped genes, depending on the haplotypes, of which three or four are functional.

Gene location

IGLV5-48 is located in the cluster B of the IGL locus on chromosome 22 at 22q11.2.

Nucleotide and amino acid sequences for human IGLV5-48

```
                                     1   2   3   4   5   6   7   8   9   10  11  12  13  14  15  16  17  18  19  20
                                     Q   P   V   L   T   Q   P   T   S       L   S   A   S   P   G   A   S   A   R
Z73649,IGLV5-48*01,5d.75A1    [27]  CAG CCT GTG CTG ACT CAG CCA ACT TCC ... CTC TCA GCA TCT CCT GGA GCA TCA GCC AGA
D87016,IGLV5-48*01,V4-3       [19]  --- --                  --- --- --- --- ... ---  - -                 --- --- --- ---

                                                                         _______CDR1-IMGT_______________
                                     21  22  23  24  25  26  27  28  29  30  31  32  33  34  35  36  37  38  39  40
                                     L   T   C   T   L   R   S   G   I   N   L   G   S   Y   R               I   F
Z73649,IGLV5-48*01,5d.75A1          CTC ACC TGC ACC TTG CGC AGT GGC ATC AAT CTT GGT AGC TAC AGG ... ... ... ATA TTC
D87016,IGLV5-48*01,V4-3             --- --              --- --- --- --- --- --- --- --- --- - - ... ... ... - - ---

                                                                                              _______________CDR2-
                                     41  42  43  44  45  46  47  48  49  50  51  52  53  54  55  56  57  58  59  60
                                     W   Y   Q   Q   K   P   E   S   P   P   R   Y   L   L   S   Y   Y   S   D   S
Z73649,IGLV5-48*01,5d.75A1          TGG TAC CAG CAG AAG CCA GAG AGC CCT CCC CGG TAT CTC CTG AGC TAC TAC TCA GAC TCA
D87016,IGLV5-48*01,V4-3             -               - - --- --- --- --                         --- --- --- --- ---

                                    IMGT
                                     61  62  63  64  65  66  67  68  69  70  71  72  73  74  75  76  77  78  79  80
                                     S   K               H   Q   G   S   G   V   P       S   R   F   S   G   S   K
Z73649,IGLV5-48*01,5d.75A1          AGT AAG ... ... ... CAT CAG GGC TCT GGA GTC CCC ... AGC CGC TTC TCT GGA TCC AAA
D87016,IGLV5-48*01,V4-3             --- --- ... ... ...             --- --- --- --- ... --- --- --- --

                                     81  82  83  84  85  86  87  88  89  90  91  92  93  94  95  96  97  98  99  100
                                     D   A   S   S   N   A   G   I   L   V   I   S   G   L   Q   S   E   D   E   A
Z73649,IGLV5-48*01,5d.75A1          GAT GCT TCC AGC AAT GCA GGG ATT TTA GTC ATC TCT GGG CTC CAG TCT GAG GAT GAG GCT
D87016,IGLV5-48*01,V4-3             --- --- --- --- ---  - -                           - - --- --- --- --- ---

                                                             ___CDR3-IMGT_____________
                                     101 102 103 104 105 106 107 108 109 110 111 112
                                     D   Y   Y   C   M   T   W   H   S   S   A   S
Z73649,IGLV5-48*01,5d.75A1          GAC TAT TAC TGT ATG ATT TGG CAC AGC AGT GCT TCT
D87016,IGLV5-48*01,V4-3              -- --- --- --- --- --- - - -           - ---
```

Framework and complementarity determining regions

FR1-IMGT: 25 (-1 aa: 10) CDR1-IMGT: 9

FR2-IMGT: 17 CDR2-IMGT: 7

FR3-IMGT: 38 (-1 aa: 73) CDR3-IMGT: 8

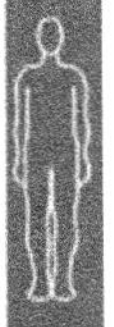

Collier de Perles for human IGLV5-48*01

Accession number: IMGT Z73649 EMBL/GenBank/DDBJ: Z73649

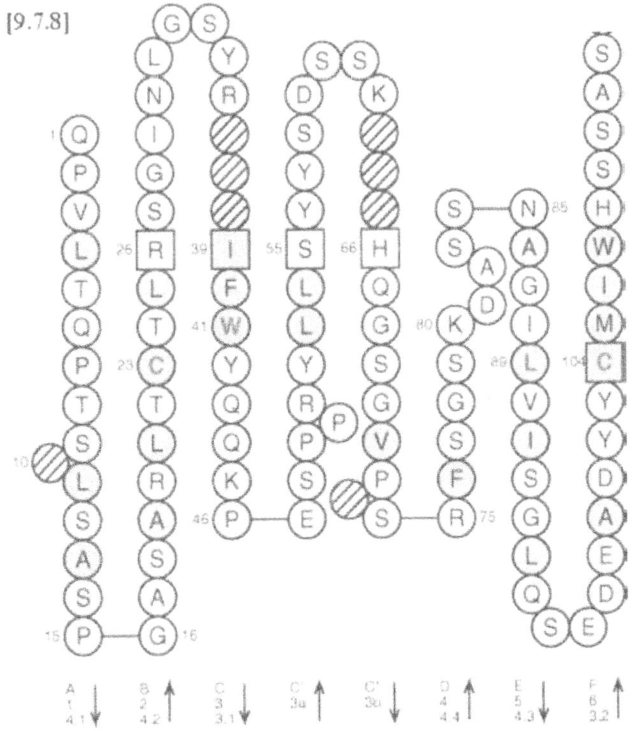

Genome database accession numbers

GDB:9953747 LocusLink: 28780

IGLV5-52

Nomenclature

IGLV5-52: Immunoglobulin lambda variable 5-52.

Definition and functionality

IGLV5-52 is one of the three or four functional genes of the IGLV5 subgroup which comprises four or five mapped genes, depending on the haplotypes.

Gene location

IGLV5-52 is located in the cluster B of the IGL locus on chromosome 22 at 22q11.2.

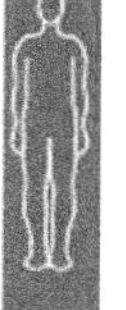

Nucleotide and amino acid sequences for human IGLV5-52

```
                                        1   2   3   4   5   6   7   8   9   10  11  12  13  14  15  16  17  18  19  20
                                        Q   P   V   L   T   Q   P   S   S       H   S   A   S   S   G   A   S   V   R
Z73669,IGLV5-52*01,5b.366F5      [27]  CAG CCT GTG CTG ACT CAG CCA TCT TCC ... CAT TCT GCA TCT TCT GGA GCA TCA GTC AGA
D87018,IGLV5-52*01,V4-4          [19]  --- --   - --- --      --- --- --- ... --      -   --- --- ---     --- --- ---

                                                                            ________CDR1-IMGT______________
                                        21  22  23  24  25  26  27  28  29  30  31  32  33  34  35  36  37  38  39  40
                                        L   T   C   M   L   S   S   G   F   S   V   G   D   F   W               I   R
Z73669,IGLV5-52*01,5b.366F5            CTC ACC TGC ATG CTG AGC AGT GGC TTC AGT GTT GGG GAC TTC TGG ... ... ... ATA AGG
D87018,IGLV5-52*01,V4-4                --- ---     --- --- - -     -- --- --- --- ---     --- --- ... ... ... -- ---

                                                                                                ____________ CDR2-
                                        41  42  43  44  45  46  47  48  49  50  51  52  53  54  55  56  57  58  59  60
                                        W   Y   Q   Q   K   P   G   N   P   P   R   Y   L   L   Y   Y   H   S   D   S
Z73669,IGLV5-52*01,5b.366F5            TGG TAC CAA CAA AAG CCA GGG AAC CCT CCC CGG TAT CTC CTG TAC TAC CAC TCA GAC TCC
D87018,IGLV5-52*01,V4-4                --- --- -   -- --- ---     --- --- --- --- ---       --- --- --- --      ---

                                       IMGT_________   ___
                                        61  62  63  64  65  66  67  68  69  70  71  72  73  74  75  76  77  78  79  80
                                        N   K                   G   Q   G   S   G   V   P       S   R   F   S   G   S   N
Z73669,IGLV5-52*01,5b.366F5            AAT AAG ... ... ... GGC CAA GGC TCT GGA GTT CCC ... AGC CGC TTC TCT GGA TCC AAC
D87018,IGLV5-52*01,V4-4                --- --  ... ... ... --- --- --          --- --- ... --      -- --- --- --- ---

                                        81  82  83  84  85  86  87  88  89  90  91  92  93  94  95  96  97  98  99  100
                                        D   A   S   A   N   A   G   I   L   R   I   S   G   L   Q   P   E   D   E   A
Z73669,IGLV5-52*01,5b.366F5            GAT GCA TCA GCC AAT GCA GGG ATT CTG CGT ATC TCT GGG CTC CAG CCT GAG GAT GAG GCT
D87018,IGLV5-52*01,V4-4                --- ---         -- --- --- --          --- --- --- -           --- --- ---

                                                       ____________CDR3-IMGT____________
                                        101 102 103 104 105 106 107 108 109 110 111 112 113
                                        D   Y   Y   C   G   T   W   H   S   N   S   K   T
Z73669,IGLV5-52*01,5b.366F5            GAC TAT TAC TGT GGT ACA TGG CAC AGC AAC TCT AAG ACT CA
D87018,IGLV5-52*01,V4-4                --- ---         - --- --- ---       --- --- --- --
```

Framework and complementarity determining regions

FR1-IMGT: 25 (-1 aa: 10)
FR2-IMGT: 17
FR3-IMGT: 38 (-1 aa: 73)

CDR1-IMGT: 9
CDR2-IMGT: 7
CDR3-IMGT: 9

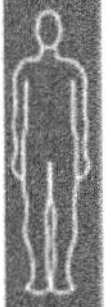

Collier de Perles for human IGLV5-52*01

Accession number: IMGT Z73669 EMBL/GenBank/DDBJ: Z73669

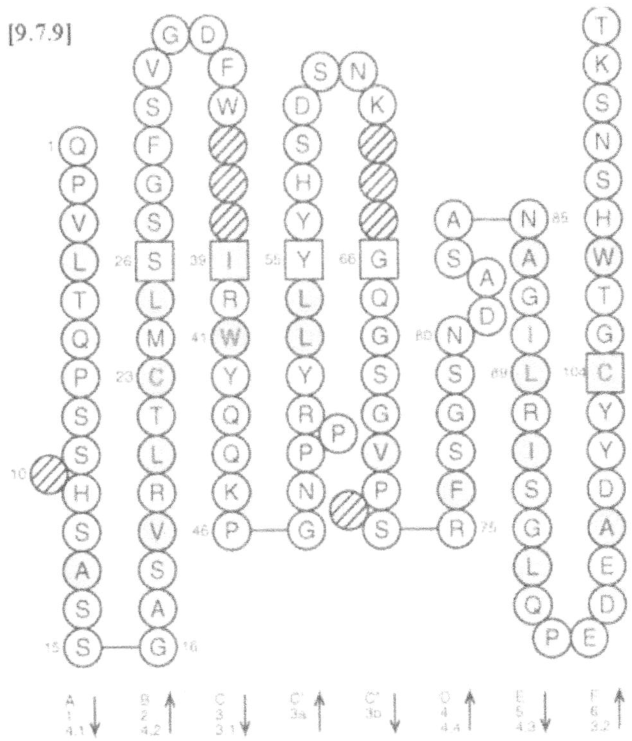

Genome database accession numbers

GDB:9953749 LocusLink: 28779

Nomenclature

IGLV6-57: Immunoglobulin lambda variable 6-57.

Definition and functionality

IGLV6-57 is the unique functional gene of the IGLV6 subgroup which only comprises this mapped gene.

Gene location

IGLV6-57 is located in the cluster C of the IGL locus on chromosome 22 at 22q11.2.

Nucleotide and amino acid sequences for human IGLV6-57

```
                                      1   2   3   4   5   6   7   8   9   10  11  12  13  14  15  16  17  18  19  20
                                      N   F   M   L   T   Q   P   H   S       V   S   E   S   P   G   K   T   V   T
Z73673,IGLV6-57*01,6a.366F5     [27]  AAT TTT ATG CTG ACT CAG CCC CAC TCT ... GTG TCG GAG TCT CCG GGG AAG ACG GTA ACC
D86996,IGLV6-57*01,V1-22        [19]  --- --- --- --- --- --- --- --- --- ... --- --- --- --- --- --- --- --- --- ---
M87320,IGLV6-57*01,IGLV6S1       [6]      --- --- --- --- --- --- --- --- ... --- --- --- --- --- --- --- --- --- ---
X92337,IGLV6-57*01,LV6SW-G       [6]  --- --- --- --- --- --- --- --- --- ... --- --- --- --- --- --- --- --- --- ---
X92338,IGLV6-57*01,VlambdaVI-3.6 [34] --- --- --- --- --- --- --- --- --- ... --- --- --- --- --- --- --- --- --- ---

                                                                  ________CDR1-IMGT__
                                      21  22  23  24  25  26  27  28  29  30  31  32  33  34  35  36  37  38  39  40
                                      I   S   C   T   R   S   S   G   S   I   A   S   N   Y                   V   Q
Z73673,IGLV6-57*01,6a.366F5           ATC TCC TGC ACC CGC AGC AGT GGC AGC ATT GCC AGC AAC TAT ... ... ... ... GTG CAG
D86996,IGLV6-57*01,V1-22              --- --- --- --- --- --- --- --- --- --- --- --- --- --- ... ... ... ... --- ---
M87320,IGLV6-57*01,IGLV6S1            --- --- --- --- --- --- --- --- --- --- --- --- --- --- ... ... ... ... --- ---
X92337,IGLV6-57*01,LV6SW-G            --- --- --- --- --- --- --- --- --- --- --- --- --- --- ... ... ... ... --- ---
X92338,IGLV6-57*01,VlambdaVI-3.6      --- --- --- --- --- --- --- --- --- --- --- --- --- --- ... ... ... ... --- ---

                                                                                                  ___________CDR2-
                                      41  42  43  44  45  46  47  48  49  50  51  52  53  54  55  56  57  58  59  60
                                      W   Y   Q   Q   R   P   G   S   S   P   T   T   V   I   Y   E   D   N
Z73673,IGLV6-57*01,6a.366F5           TGG TAC CAG CAG CGC CCG GGC AGT TCC CCC ACC ACT GTG ATC TAT GAG GAT AAC ... ...
D86996,IGLV6-57*01,V1-22              --- --- --- --- --- --- --- --- --- --- --- --- --- --- --- --- --- --- ... ...
M87320,IGLV6-57*01,IGLV6S1            --- --- --- --- --- --- --- --- --- --- --- --- --- --- --- --- --- --- ... ...
X92337,IGLV6-57*01,LV6SW-G            --- --- --- --- --- --- --- --- --- --- --- --- --- --- --- --- --- --- ... ...
X92338,IGLV6-57*01,VlambdaVI-3.6      --- --- --- --- --- --- --- --- --- --- --- --- --- --- --- --- --- --- ... ...

                                      IMGT____
                                      61  62  63  64  65  66  67  68  69  70  71  72  73  74  75  76  77  78  79  80
                                                          Q   R   P   S   G   V   P       D   R   F   S   G   S   I
Z73673,IGLV6-57*01,6a.366F5           ... ... ... ... ... CAA AGA CCC TCT GGG GTC CCT ... GAT CGG TTC TCT GGC TCC ATC
D86996,IGLV6-57*01,V1-22              ... ... ... ... ... --- --- --- --- --- --- --- ... --- --- --- --- --- --- ---
M87320,IGLV6-57*01,IGLV6S1            ... ... ... ... ... --- --- --- --- --- --- --- ... --- --- --- --- --- --- ---
X92337,IGLV6-57*01,LV6SW-G            ... ... ... ... ... --- --- --- --- --- --- --- ... --- --- --- --- --- --- ---
X92338,IGLV6-57*01,VlambdaVI-3.6      ... ... ... ... ... --- --- --- --- --- --- --- ... --- --- --- --- --- --- ---

                                      81  82  83  84  85  86  87  88  89  90  91  92  93  94  95  96  97  98  99  100
                                      D   S   S   S   N   S   A   S   L   T   I   S   G   L   K   T   E   D   E   A
Z73673,IGLV6-57*01,6a.366F5           GAC AGC TCC TCC AAC TCT GCC TCC CTC ACC ATC TCT GGA CTG AAG ACT GAG GAC GAG GCT
D86996,IGLV6-57*01,V1-22              --- --- --- --- --- --- --- --- --- --- --- --- --- --- --- --- --- --- --- ---
M87320,IGLV6-57*01,IGLV6S1            --- --- --- --- --- --- --- --- --- --- --- --- --- --- --- --- --- --- --- ---
X92337,IGLV6-57*01,LV6SW-G            --- --- --- --- --- --- --- --- --- --- --- --- --- --- --- --- --- --- --- ---
X92338,IGLV6-57*01,VlambdaVI-3.6      --- --- --- --- --- --- --- --- --- --- --- --- --- --- --- --- --- --- --- ---

                                                     ____CDR3-IMGT______
                                      101 102 103 104 105 106 107 108 109 110 111
                                      D   Y   Y   C   Q   S   Y   D   S   S   N
Z73673,IGLV6-57*01,6a.366F5           GAC TAC TAC TGT CAG TCT TAT GAT AGC AGC AAT CA
D86996,IGLV6-57*01,V1-22              --- --- --- --- --- --- --- --- --- --- --- --
M87320,IGLV6-57*01,IGLV6S1
X92337,IGLV6-57*01,LV6SW-G            --- --- --- --- --- --- --- --- --- --- --- --
X92338,IGLV6-57*01,VlambdaVI-3.6      --- --- --- --- --- --- --- --- --- --- --- --
```

Framework and complementarity determining regions

FR1-IMGT: 25 (-1 aa: 10)
FR2-IMGT: 17
FR3-IMGT: 38 (-1 aa: 73)

CDR1-IMGT: 8
CDR2-IMGT: 3
CDR3-IMGT: 7

Collier de Perles for human IGLV6-57*01

Accession number: IMGT Z73673 EMBL/GenBank/DDBJ: Z73673

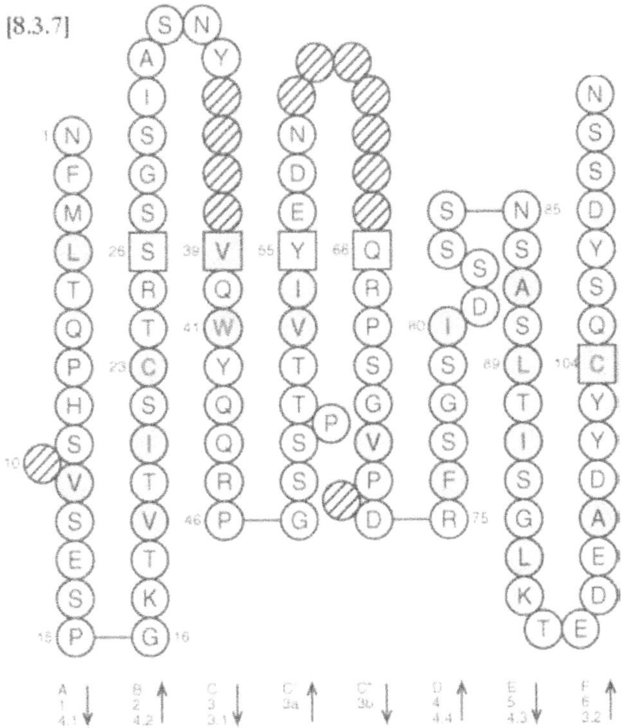

Genome database accession numbers

GDB:9953751 LocusLink: 28778

Nomenclature

IGLV7-43: Immunoglobulin lambda variable 7-43.

Definition and functionality

IGLV7-43 is one of the one–two functional genes of the IGLV7 subgroup which comprises three mapped genes.

Gene location

IGLV7-43 is located in the cluster B of the IGL locus on chromosome 22 at 22q11.2.

Nucleotide and amino acid sequences for human IGLV7-43

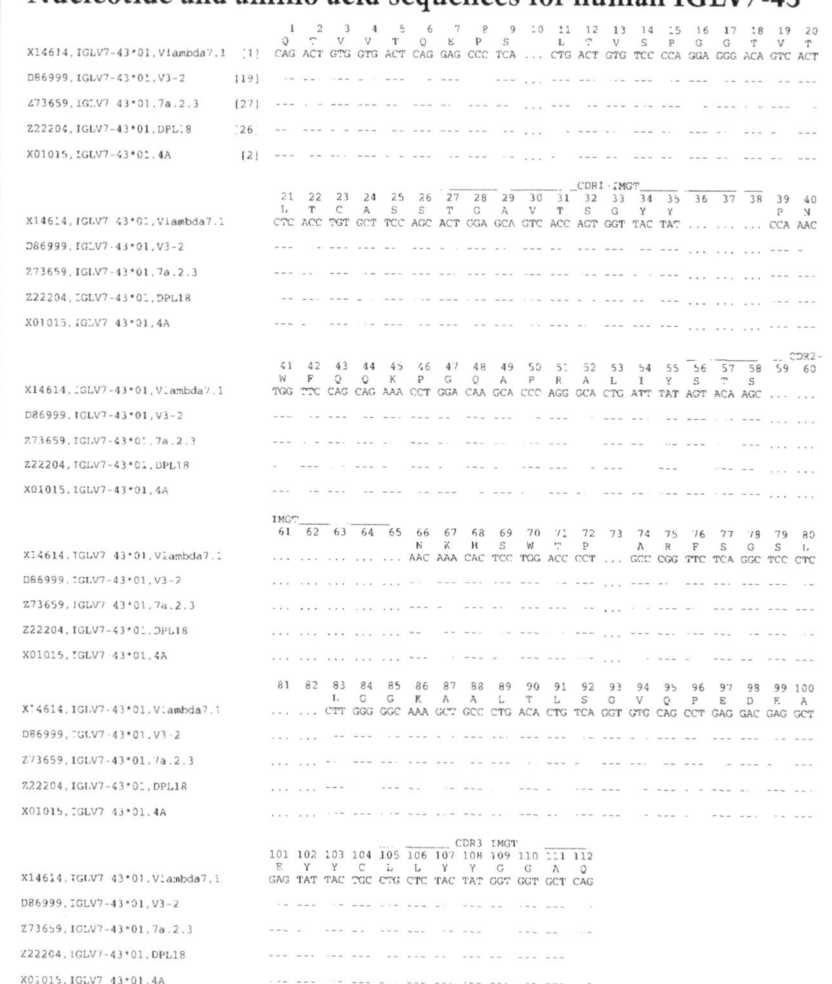

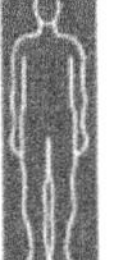

Framework and complementarity determining regions

FR1-IMGT: 25 (-1 aa: 10)
FR2-IMGT: 17
FR3-IMGT: 36 (-3 aa: 73,81,82)

CDR1-IMGT: 9
CDR2-IMGT: 3
CDR3-IMGT: 8

Collier de Perles for human IGLV7-43*01

Accession number: IMGT X14614   EMBL/GenBank/DDBJ: X14614

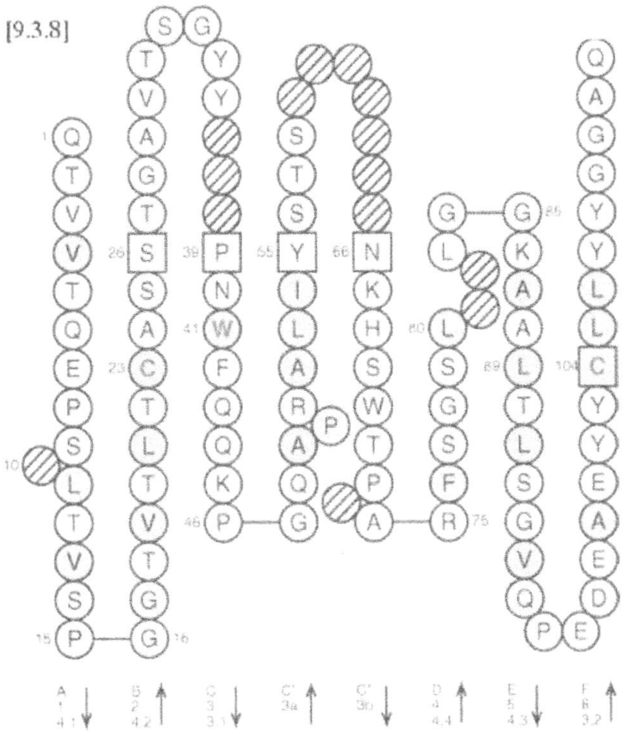

Genome database accession numbers

GDB:9953755   LocusLink: 28776

Nomenclature

IGLV7-46: Immunoglobulin lambda variable 7-46.

Definition and functionality

IGLV7-46 is a functional gene (alleles *01 and *02) or a pseudogene (allele *03). IGLV7-46 belongs to the IGLV7 subgroup which comprises three mapped genes, of which one–two are functional.
IGLV7-46*03 is a pseudogene due to a 1 nt DELETION in codon 91 leading to a frameshift in FR3-IMGT.

Gene location

IGLV7-46 is located in the cluster B of the IGL locus on chromosome 22 at 22q11.2.

Nucleotide and amino acid sequences for human IGLV7-46

```
                                          1   2   3   4   5   6   7   8   9  10  11  12  13  14  15  16  17  18  19  20
                                          Q   A   V   V   T   Q   E   P   S       L   T   V   S   P   G   G   T   V   T
Z73674,IGLV7-46*01,7b.400B5       [27]  CAG GCT GTG GTG ACT CAG GAG CCC TCA ... CTG ACT GTG TCC CCA GGA GGG ACA GTC ACT
Z22205,IGLV7-46*01,DPL19          [26]  --- --- --- --- --- --- --- --- --- ... --- --- --- --- --- --- --- --- --- ---
D86999,IGLV7-46*02,V3-3           [19]  --- --- --- --- --- --- --- --- --- ... --- --- --- --- --- --- --- --- --- ---
Z22210,IGLV7-46*03,DPL20          [26]  --- --- --- --- --- --- --- --- --- ... --- --- --- --- --- --- --- --- --- ---

                                                                  _______________CDR1-IMGT_______________
                                         21  22  23  24  25  26  27  28  29  30  31  32  33  34  35  36  37  38  39  40
                                          L   T   C   G   S   S   T   G   A   V   T   S   G   H   Y               P   Y
Z73674,IGLV7-46*01,7b.400B5             CTC ACC TGT GGC TCC AGC ACT GGA GCT GTC ACC AGT GGT CAT TAT ... ... ... CCC TAC
Z22205,IGLV7-46*01,DPL19                --- --- --- --- --- --- --- --- --- --- --- --- --- --- --- ... ... ... --- ---
D86999,IGLV7-46*02,V3-3                 --- --- --- --- --- --- --- --- --- --- --- --- --- --- --- ... ... ... --- ---
Z22210,IGLV7-46*03,DPL20                --- --- --- --- --- --- --- --- --- --- --- --- --- --- --- ... ... ... --- ---

                                                                                                      ____________CDR2-
                                         41  42  43  44  45  46  47  48  49  50  51  52  53  54  55  56  57  58  59  60
                                          W   F   Q   Q   K   P   G   Q   A   P   R   T   L   I   Y   D   T   S
Z73674,IGLV7-46*01,7b.400B5             TGG TTC CAG CAG AAG CCT GGC CAA GCC CCC AGG ACA CTG ATT TAT GAT ACA AGC ... ...
Z22205,IGLV7-46*01,DPL19                --- --- --- --- --- --- --- --- --- --- --- --- --- --- --- --- --- --- ... ...
D86999,IGLV7-46*02,V3-3                 --- --- --- --- --- --- --- --- --- --- --- --- --- --- --- --- --- --- ... ...
Z22210,IGLV7-46*03,DPL20                --- --- --- --- --- --- --- --- --- --- --- --- --- --- --- --- --- --- ... ...

                                        IMGT________________
                                         61  62  63  64  65  66  67  68  69  70  71  72  73  74  75  76  77  78  79  80
                                                              N   K   H   S   W   T   P       A   R   F   S   G   S   L
Z73674,IGLV7-46*01,7b.400B5             ... ... ... ... ... AAC AAA CAC TCC TGG ACA CCT ... GCC CGG TTC TCA GGC TCC CTC
Z22205,IGLV7-46*01,DPL19                ... ... ... ... ... --- --- --- --- --- --- --- ... --- --- --- --- --- --- ---
D86999,IGLV7-46*02,V3-3                 ... ... ... ... ... --- --- --- --- --- --- --- ... --- --- --- --- --- --- ---
Z22210,IGLV7-46*03,DPL20                ... ... ... ... ... --- --- --- --- --- --- --- ... --- --- --- --- --- --- ---

                                         81  82  83  84  85  86  87  88  89  90  91  92  93  94  95  96  97  98  99 100
                                                  L   G   G   K   A   A   L   T   L   S   G   A   Q   P   E   D   E   A
Z73674,IGLV7-46*01,7b.400B5             ... ... CTT GGG GGC AAA GCT GCC CTG ACC CTT TCG GGT GCG CAG CCT GAG GAT GAG GCT
Z22205,IGLV7-46*01,DPL19                ... ... --- --- --- --- --- --- --- --- --- --- --- --- --- --- --- --- --- ---
                                                                                      L
D86999,IGLV7-46*02,V3-3                 ... ... --- --- --- --- --- --- --- --- --- -T- --- --- --- --- --- --- --- ---
                                                                                  #
Z22210,IGLV7-46*03,DPL20                ... ... --- --- --- --- --- --- --- --- .-- --- --- --- --- --- --- --- --- ---

                                                         _____________CDR3-IMGT________
                                        101 102 103 104 105 106 107 108 109 110 111 112
                                          E   Y   Y   C   L   L   S   Y   S   G   A   R
Z73674,IGLV7-46*01,7b.400B5             GAG TAT TAC TGC TTG CTC TCC TAT AGT GGT GCT CGG
Z22205,IGLV7-46*01,DPL19
D86999,IGLV7-46*02,V3-3                 --- --- --- --- --- --- --- --- --- --- --- ---
Z22210,IGLV7-46*03,DPL20                --- --- --- --- --- --- --- --- --- --- --- --
```

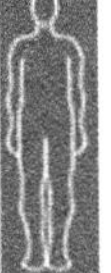

Framework and complementarity determining regions

FR1-IMGT: 25 (-1 aa: 10)
FR2-IMGT: 17
FR3-IMGT: 36 (-3 aa: 73,81,82)

CDR1-IMGT: 9
CDR2-IMGT: 3
CDR3-IMGT: 8

Collier de Perles for human IGLV7-46*01

Accession number: IMGT Z73674 EMBL/GenBank/DDBJ: Z73674

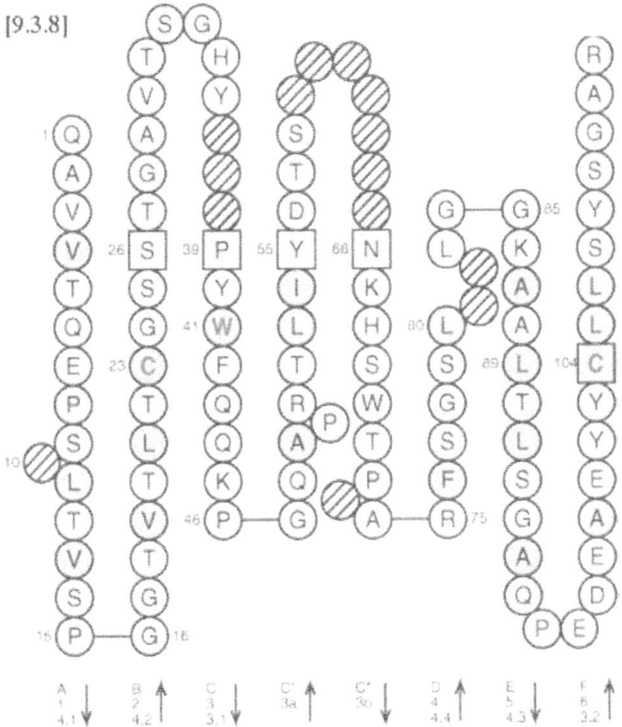

Genome database accession numbers

GDB:9953757 LocusLink: 28775

Nomenclature

IGLV8-61: Immunoglobulin lambda variable 8-61.

Definition and functionality

IGLV8-61 is the unique functional gene of the IGLV8 subgroup which only comprises this mapped gene and two orphons.

Gene location

IGLV8-61 is located in the cluster C of the IGL locus on chromosome 22 at 22q11.2.

Nucleotide and amino acid sequences for human IGLV8-61

```
                                        1   2   3   4   5   6   7   8   9  10  11  12  13  14  15  16  17  18  19  20
                                        Q   T   V   V   T   Q   E   P   S       F   S   V   S   P   G   G   T   V   T
Z73650,IGLV8-61*01,8a.88E1     [27]  CAG ACT GTG GTG ACC CAG GAG CCA TCG ... TTC TCA GTG TCC CCT GGA GGG ACA GTC ACA
D87022,IGLV8-61*01,V3-4        [19]  --- --- --- --- --- --- --- --- --- ... --- --- --- --- --- --- --- --- --- ---
Z22206,IGLV8-61*01,DPL21       [26]  --- --- --- --- --- --- --- --- --- ... --- --- --- --- --- --- --- --- --- ---
S39395,IGLV8-61*01,VL8         [28]  --- --- --- --- --- --- --- --- --- ... --- --- --- --- --- --- --- --- --- ---
U03639,IGLV8-61*01,FL7         [20]  --- --- --- --- --- --- --- --- --- ... --- --- --- --- --- --- --- --- --- ---
U03635,IGLV8-61*01,TL7         [20]  --- --- --- --- --- --- --- --- --- ... --- --- --- --- --- --- --- --- --- ---
U03637,IGLV8-61*02,BL7         [20]  --- --- --- --- --- --- --- --- --- ... --- --- --- --- --- --- --- --- --- ---

                                                             _____________CDR1-IMGT_________________
                                       21  22  23  24  25  26  27  28  29  30  31  32  33  34  35  36  37  38  39  40
                                        L   T   C   G   L   S   S   G   S   V   S   T   S   Y   Y               P   S
Z73650,IGLV8-61*01,8a.88E1           CTC ACT TGT GGC TTG AGC TCT GGC TCA GTC TCT ACT AGT TAC TAC ... ... ... CCC AGC
D87022,IGLV8-61*01,V3-4              --- --- --- --- --- --- --- --- --- --- --- --- --- --- --- ... ... ... --- ---
Z22206,IGLV8-61*01,DPL21             --- --- --- --- --- --- --- --- --- --- --- --- --- --- --- ... ... ... --- ---
S39395,IGLV8-61*01,VL8               --- --- --- --- --- --- --- --- --- --- --- --- --- --- --- ... ... ... --- ---
U03639,IGLV8-61*01,FL7               --- --- --- --- --- --- --- --- --- --- --- --- --- --- --- ... ... ... --- ---
U03635,IGLV8-61*01,TL7               --- --- --- --- --- --- --- --- --- --- --- --- --- --- --- ... ... ... --- ---
U03637,IGLV8-61*02,BL7               --- --- --- --- --- --- --- --- --- --- --- --- --- --- --- ... ... ... --- ---

                                                                                                ____________CDR2-
                                       41  42  43  44  45  46  47  48  49  50  51  52  53  54  55  56  57  58  59  60
                                        W   Y   Q   Q   T   P   G   Q   A   P   R   T   L   I   Y   S   T   N
Z73650,IGLV8-61*01,8a.88E1           TGG TAC CAG CAG ACC CCA GGC CAG GCT CCA CGC ACG CTC ATC TAC AGC ACA AAC ... ...
D87022,IGLV8-61*01,V3-4              --- --- --- --- --- --- --- --- --- --- --- --- --- --- --- --- --- --- ... ...
Z22206,IGLV8-61*01,DPL21             --- --- --- --- --- --- --- --- --- --- --- --- --- --- --- --- --- --- ... ...
S39395,IGLV8-61*01,VL8               --- --- --- --- --- --- --- --- --- --- --- --- --- --- --- --- --- --- ... ...
U03639,IGLV8-61*01,FL7               --- --- --- --- --- --- --- --- --- --- --- --- --- --- --- --- --- --- ... ...
U03635,IGLV8-61*01,TL7               --- --- --- --- --- --- --- --- --- --- --- --- --- --- --- --- --- --- ... ...
U03637,IGLV8-61*02,BL7               --- --- --- --- --- --- --- --- --- --- --- --- --- --- --- --- --- --- ... ...

                                     IMGT_______________
                                       61  62  63  64  65  66  67  68  69  70  71  72  73  74  75  76  77  78  79  80
                                                            T   R   S   S   G   V   P       D   R   F   S   G   S   I
Z73650,IGLV8-61*01,8a.88E1           ... ... ... ... ... ACT CGC TCT TCT GGG GTC CCT ... GAT CGC TTC TCT GGC TCC ATC
D87022,IGLV8-61*01,V3-4              ... ... ... ... ... --- --- --- --- --- --- --- ... --- --- --- --- --- --- ---
Z22206,IGLV8-61*01,DPL21             ... ... ... ... ... --- --- --- --- --- --- --- ... --- --- --- --- --- --- ---
S39395,IGLV8-61*01,VL8               ... ... ... ... ... --- --- --- --- --- --- --- ... --- --- --- --- --- --- ---
U03639,IGLV8-61*01,FL7               ... ... ... ... ... --- --- --- --- --- --- --- ... --- --- --- --- --- --- ---
U03635,IGLV8-61*01,TL7               ... ... ... ... ... --- --- --- --- --- --- --- ... --- --- --- --- --- --- ---
                                                                                            C
U03637,IGLV8-61*02,BL7               ... ... ... ... ... --- --- --- --- --- --- --- ... --- T-- --- --- --- --- ---

                                       81  82  83  84  85  86  87  88  89  90  91  92  93  94  95  96  97  98  99 100
                                                L   G   N   K   A   A   L   T   I   T   G   A   Q   A   D   D   E   S
Z73650,IGLV8-61*01,8a.88E1           ... ... CTT GGG AAC AAA GCT GCC CTC ACC ATC ACG GGG GCC CAG GCA GAT GAT GAA TCT
D87022,IGLV8-61*01,V3-4              ... ... --- --- --- --- --- --- --- --- --- --- --- --- --- --- --- --- --- ---
Z22206,IGLV8-61*01,DPL21             ... ... --- --- --- --- --- --- --- --- --- --- --- --- --- --- --- --- --- ---
S39395,IGLV8-61*01,VL8               ... ... --- --- --- --- --- --- --- --- --- --- --- --- --- --- --- --- --- ---
U03639,IGLV8-61*01,FL7               ... ... --- --- --- --- --- --- --- --- --- --- --- --- --- --- --- --- --- ---
U03635,IGLV8-61*01,TL7               ... ... --- --- --- --- --- --- --- --- --- --- --- --- --- --- --- --- --- ---
U03637,IGLV8-61*02,BL7               ... ... --- --- --- --- --- --- --- --- --- --- --- --- --- --- --- --- --- ---
```

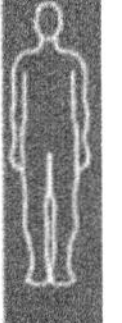

```
                                                 ______CDR3 IMGT_______________
                              101 102 103 104 105 106 107 108 109 110 111 112 113
                               D   Y   Y   C   V   L   Y   M   G   S   G   I
Z73650,IGLV8 61*01,8a.88E1    GAT TAT TAC TGT GTG CTG TAT ATG GGT ACT GGC ATT TC
D87022,IGLV8-61*01,V3-4       ---          -- --- ---     --- --      --- --
Z22206,IGLV8-61*01,DPL21          --- --- ---             --- --    - --- -   ---
S39395,IGLV8-61*01,VL8        ---          --- --- ---     --- --      --- -
U03639,IGLV8 61*01,FL7        --- ---          --- --         ---  -      - ---  -
U03635,IGLV8-61*01,TL7        ---          --- --- --      --- -       --- -
U03637,IGLV8-61*02,BL7                --- --- --- -   --- ---      -  - ---     --
```

Framework and complementarity determining regions

FR1-IMGT: 25 (-1 aa: 10)
FR2-IMGT: 17
FR3-IMGT: 36 (-3 aa: 73,81,82)

CDR1-IMGT: 9
CDR2-IMGT: 3
CDR3-IMGT: 8

Collier de Perles for human IGLV8-61*01

Accession number: IMGT Z73650 EMBL/GenBank/DDBJ: Z73650

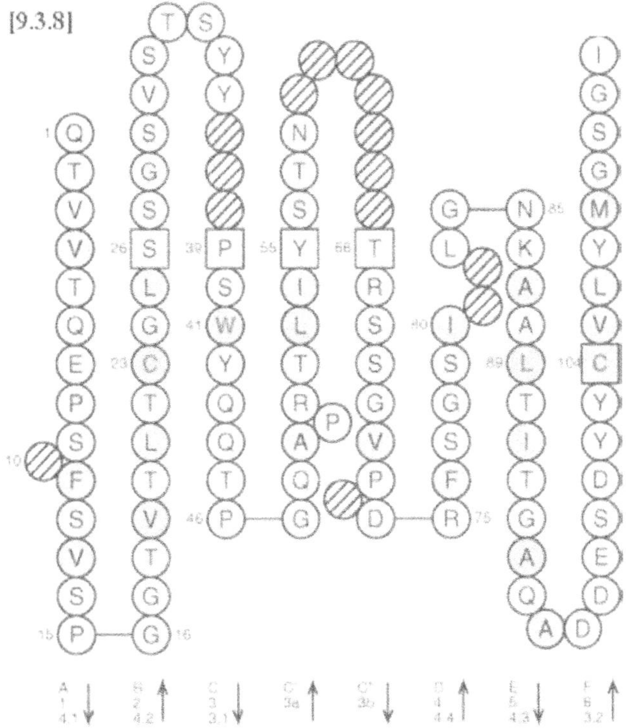

Genome database accession numbers

GDB:9953759 LocusLink: 28774

Nomenclature

IGLV9-49: Immunoglobulin lambda variable 9-49.

Definition and functionality

IGLV9-49 is considered as the unique functional gene of the IGLV9 subgroup which only comprises this mapped gene, despite the presence of a STOP-CODON at position 116 in the CDR3-IMGT. Indeed, this STOP-CODON at the 3′ end of the V-REGION can disappear during the V-J rearrangement by deletion and/or mutation at the V-J junction.

Gene location

IGLV9-49 is located in the cluster B of the IGL locus on chromosome 22 at 22q11.2.

Nucleotide and amino acid sequences for human IGLV9-49

```
                                     1   2   3   4   5   6   7   8   9  10  11  12  13  14  15  16  17  18  19  20
                                     Q   P   V   L   T   Q   P   P   S       A   S   A   S   L   G   A   S   V   T
Z73675,IGLV9-49*01,9a.366F5   [27]  CAG CCT GTG CTG ACT CAG CCA CCT TCT ... GCA TCA GCC TCC CTG GGA GCC TCG GTC ACA
Z22207,IGLV9-49*01,DPL22      [16]  --- --- --- --- --- --- --- --- --- ... --- --- --- --- --- --- --- --- --- ---
D87016,IGLV9-49*02,V5-2       [19]  --- --- --- --- --- --- --- --- --- ... --- --- --- --- --- --- --- --- --- ---
U03869,IGLV9-49*03,1v901e     [11]  --- --- --- --- --- --- --- --- --- ... --- --- --- --- --- --- --- --- --- ---

                                                            _______________CDR1-IMGT______________
                                    21  22  23  24  25  26  27  28  29  30  31  32  33  34  35  36  37  38  39  40
                                     L   T   C   T   L   S   S   G   Y   S   N   Y   K                       V   D
Z73675,IGLV9-49*01,9a.366F5         CTC ACC TGC ACC CTG AGC AGC GGC TAC AGT AAT TAT AAA ... ... ... ... ... GTG GAC
Z22207,IGLV9-49*01,DPL22            --- --- --- --- --- --- --- --- --- --- --- --- --- ... ... ... ... ... --- ---
D87016,IGLV9-49*02,V5-2             --- --- --- --- --- --- --- --- --- --- --- --- --- ... ... ... ... ...
U03869,IGLV9-49*03,1v901e                   --- --- --- --- --- --- --- --- --- --- --- ... ... ... ... ... --- ---

                                                                                        ________________CDR2-
                                    41  42  43  44  45  46  47  48  49  50  51  52  53  54  55  56  57  58  59  60
                                     W   Y   Q   Q   R   P   G   K   G   P   R   F   V   M   R   V   G   T   G   G
Z73675,IGLV9-49*01,9a.366F5         TGG TAC CAG CAG AGA CCA GGG AAG GGC CCC CGG TTT GTG ATG CGA GTG GGC ACT GGT GGG
Z22207,IGLV9-49*01,DPL22            --- --- --- --- --- --- --- --- --- --- --- --- --- --- --- --- ---
D87016,IGLV9-49*02,V5-2             --- --- --- --- --- --- --- --- --- --- --- --- --- --- --- --- --- ---
U03869,IGLV9-49*03,1v901e                               --- --- --- --- --A --- --- --- --- --- --- --- --- ---

                                    IMGT_______________
                                    61  62  63  64  65  66  67  68  69  70  71  72  73  74  75  76  77  78  79  80
                                     I   V   G           S   K   G   D   G   I   P       D   R   F   S   V   L   G
Z73675,IGLV9-49*01,9a.366F5         ATT GTG GGA ... ... TCC AAG GGG GAT GGC ATC CCT ... GAT CGC TTC TCA GTC TTG GGC
Z22207,IGLV9-49*01,DPL22            --- --- --- ... ... --- --- --- --- --- --- --- ... --- --- --- --- --- --- ---
D87016,IGLV9-49*02,V5-2             --- --- --- ... ... --- --- --- --- --- --- --- ... --- --- --- --- --- --- ---
U03869,IGLV9-49*03,1v901e           --- --- --- ... ... --- --- --- --- --- --- --- ... --- --- --- --- --- --- ---

                                    81  82  83  84  85  86  87  88  89  90  91  92  93  94  95  96  97  98  99 100
                                             S   G   L   N   R   Y   L   T   I   K   N   I   Q   E   E   D   E   S
Z73675,IGLV9-49*01,9a.366F5         ... ... TCA GGC CTG AAT CGG TAC CTG ACC ATC AAG AAC ATC CAG GAA GAG GAT GAG AGT
Z22207,IGLV9-49*01,DPL22            ... ... --- --- --- --- --- --- --- --- --- --- --- --- --- --- --- --- --- ---
D87016,IGLV9-49*02,V5-2             ... ... --- --- --- --- --- --- --- --- --- --- --- --- --- --- -A- --- ---
U03869,IGLV9-49*03,1v901e           ... ... --- --- --- --- --- --- --- --- --- --- --- --- --- --- --- --- --- ---

                                                  ____________________CDR3-IMGT______________________
                                   101 102 103 104 105 106 107 108 109 110 111 112 113 114 115 116
                                     D   Y   H   C   G   A   D   H   G   S   G   S   N   F   V   *
Z73675,IGLV9-49*01,9a.366F5         GAC TAC CAC TGT GGG GCA GAC CAT GGC AGT GGG AGC AAC TTC GTG TAA CC
Z22207,IGLV9-49*01,DPL22                --- --- --- --- --- --- --- --- --- --- --- --- --- ---
D87016,IGLV9-49*02,V5-2             --- --- --- --- --- --- --- --- --- --- --- --- --- --- --- ---
U03869,IGLV9-49*03,1v901e                   --- --- --- --- --- --- --- --- --- --- --- --- --- --
```

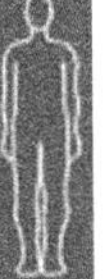

Framework and complementarity determining regions

FR1-IMGT: 25 (-1 aa: 10)
FR2-IMGT: 17
FR3-IMGT: 36 (-3 aa: 73,81,82)

CDR1-IMGT: 7
CDR2-IMGT: 8
CDR3-IMGT: 11

Collier de Perles for human IGLV9-49*01

Accession number: IMGT Z73675

EMBL/GenBank/DDBJ: Z73675

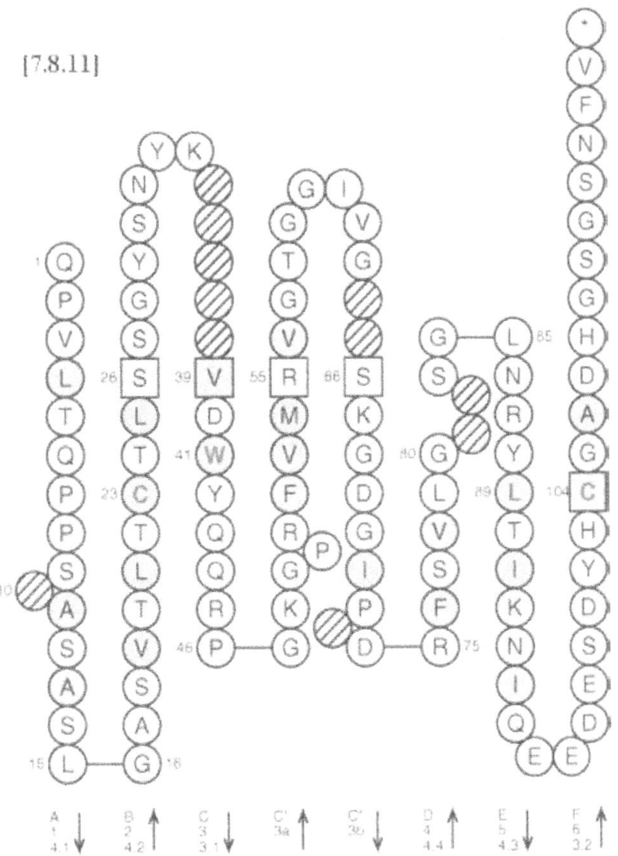

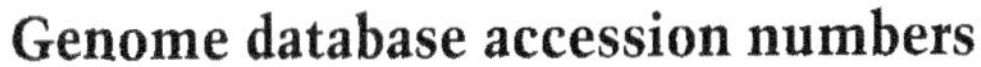

Genome database accession numbers

GDB:9953761

LocusLink: 28773

IGLV10-54

Nomenclature

IGLV10-54: Immunoglobulin lambda variable 10-54.

Definition and functionality

IGLV10-54 is the unique functional gene of the IGLV10 subgroup which comprises two mapped genes (this functional gene and one pseudogene).

Gene location

IGLV10-54 is located in the cluster C of the IGL locus on chromosome 22 at 22q11.2.

Nucleotide and amino acid sequences for human IGLV10-54

```
                                        1   2   3   4   5   6   7   8   9  10  11  12  13  14  15  16  17  18  19  20
                                        Q   A   G   L   T   Q   P   P   S       V   S   K   G   L   R   Q   T   A   T
Z73676,IGLV10-54*01,10a.872F9      [27] CAG GCA GGG CTG ACT CAG CCA CCC TCG ... GTG TCC AAG GGC TTG AGA CAG ACC GCC ACA

D86996,IGLV10-54*02,V1-20          [19] --- --- --- --- --- --- --- --- --- ... --- --- --- --- --- --- --- --- --- ---

S70116,IGLV10-54*03,gVlambdaX-4.4  [23] --- --- --- --- --- --- --- --- --- ... -C- --- --- --- --- --- --- --- --- ---


                                                                ________________CDR1-IMGT________________
                                       21  22  23  24  25  26  27  28  29  30  31  32  33  34  35  36  37  38  39  40
                                        L   T   C   T   G   N   S   N   N   V   G   N   Q   G                   A   A
Z73676,IGLV10-54*01,10a.872F9           CTC ACC TGC ACT GGG AAC AGC AAC AAT GTT GGC AAC CAA GGA ... ... ... ... GCA GCT
                                                                        I
D86996,IGLV10-54*02,V1-20               --- --- --- --- --- --- --- --- -T- --- --- --- --- --- ... ... ... ... --- ---

S70116,IGLV10-54*03,gVlambdaX-4.4       --- --- --- --- --- --- --- --- --- --- --- --- --- --- ... ... ... ... --- ---


                                                                                                 __________ CDR2-
                                       41  42  43  44  45  46  47  48  49  50  51  52  53  54  55  56  57  58  59  60
                                        W   L   Q   Q   H   Q   G   H   P   P   K   L   L   S   Y   R   N   N
Z73676,IGLV10-54*01,10a.872F9           TGC CTG CAG CAG CAC CAG GGC CAC CCT CCC AAA CTC CTA TCC TAC AGG AAT AAC ... ...

D86996,IGLV10-54*02,V1-20               --- --- --- --- --- --- --- --- --- --- --- --- --- --- --- --- --- --- ... ...
                                            P   E
S70116,IGLV10-54*03,gVlambdaX-4.4       --- -CT G-- --- --- --- --- --- --- --- --- --- --- --- --- --- --- --- ... ...


                                        IMGT________________
                                       61  62  63  64  65  66  67  68  69  70  71  72  73  74  75  76  77  78  79  80
                                                            N   R   P   S   G   I   S       E   R   L   S   A   S   R
Z73676,IGLV10-54*01,10a.872F9           ... ... ... ... ... AAC CGG CCC TCA GGG ATC TCA ... GAG AGA TTA TCT GCA TCC AGG
                                                                                                F
D86996,IGLV10-54*02,V1-20               ... ... ... ... ... --- --- --- --- --- --- --- ... --- --- --C --- --- --- ---

S70116,IGLV10-54*03,gVlambdaX-4.4       ... ... ... ... ... --- --- --- --- --- --- --- ... --- --- --- --- --- --- ---


                                       81  82  83  84  85  86  87  88  89  90  91  92  93  94  95  96  97  98  99 100
                                                S   G   N   T   A   S   L   T   I   T   G   L   Q   P   E   D   E   A
Z73676,IGLV10-54*01,10a.872F9           ... ... TCA GGA AAC ACA GCC TCC CTG ACC ATT ACT GGA CTC CAG CCT GAG GAC GAG GCT

D86996,IGLV10-54*02,V1-20               ... ... --- --- --- --- --- --- --- --- --- --- --- --- --- --- --- --- --- ---

S70116,IGLV10-54*03,gVlambdaX-4.4       ... ... --- --- --- --- --- --- --- --- --- --- --- --- --- --- --- --- --- ---


                                                        ____________CDR3-IMGT____________
                                       101 102 103 104 105 106 107 108 109 110 111 112 113
                                        D   Y   Y   C   S   A   W   D   S   S   L   S   A
Z73676,IGLV10-54*01,10a.872F9           GAC TAT TAC TGC TCA GCA TGG GAC AGC AGC CTC AGT GCT CA
                                                                L
D86996,IGLV10-54*02,V1-20               --- --- --- --- --- --- -T- --- --- --- --- --- --- --

S70116,IGLV10-54*03,gVlambdaX-4.4       --- --- --- --- --- --- --- --- --- --- --- --- --- --
```

Framework and complementarity determining regions

FR1-IMGT: 25 (-1 aa: 10)
FR2-IMGT: 17
FR3-IMGT: 36 (-3 aa: 73,81,82)

CDR1-IMGT: 8
CDR2-IMGT: 3
CDR3-IMGT: 9

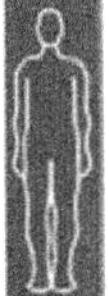

Collier de Perles for human IGLV10-54*01

Accession number: IMGT Z73676 EMBL/GenBank/DDBJ: Z73676

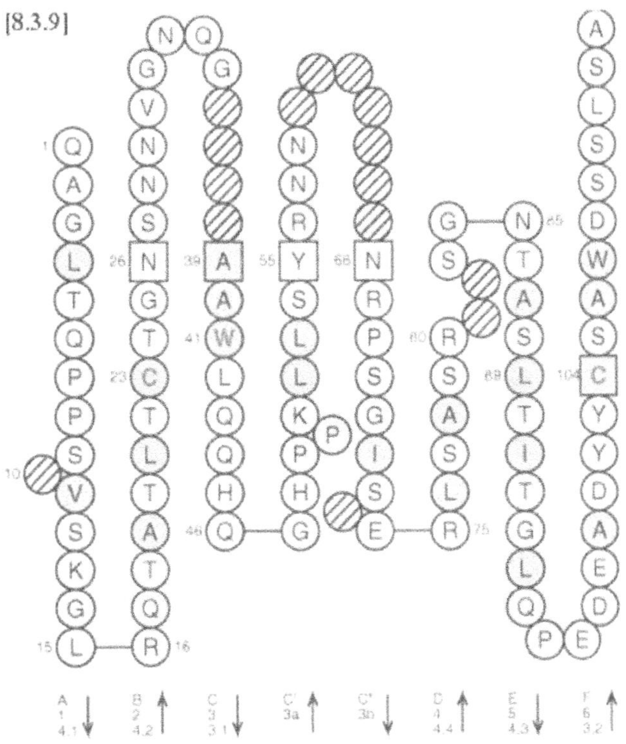

Genome database accession numbers

GDB:9953763 LocusLink: 28772

Nomenclature

IGLV11-55: Immunoglobulin lambda variable 11-55.

Definition and functionality

IGLV11-55 is an ORF due to an unusual V-SPACER length: 19 nt instead of 23 nt. IGLV11-55 belongs to the IGLV11 subgroup which only comprises this mapped gene.

Gene location

IGLV11-55 is located in the cluster C of the IGL locus on chromosome 22 at 22q11.2.

Nucleotide and amino acid sequences for human IGLV11-55

```
                                      1   2   3   4   5   6   7   8   9  10  11  12  13  14  15  16  17  18  19  20
                                      R   P   V   L   T   Q   P   P   S       L   S   A   S   P   G   A   T   A   R
D86996,IGLV11-55*01,V4-6       [19]  CGG CCC GTG CTG ACT CAG CCG CCC TCT ... CTG TCT GCA TCC CCG GGA GCA ACA GCC AGA

                                                              _______________CDR1-IMGT______________
                                     21  22  23  24  25  26  27  28  29  30  31  32  33  34  35  36  37  38  39  40
                                      L   P   C   T   L   S   S   D   L   S   V   G   G   K   N               M   F
D86996,IGLV11-55*01,V4-6             CTC CCC TGC ACC CTG AGC AGT GAC CTC AGT GTT GGT GGT AAA AAC ... ... ... ATG TTC

                                                                                                ____________ CDR2-
                                     41  42  43  44  45  46  47  48  49  50  51  52  53  54  55  56  57  58  59  60
                                      W   Y   Q   Q   K   P   G   S   S   P   R   L   F   L   Y   H   Y   S   D   S
D86996,IGLV11-55*01,V4-6             TGG TAC CAG CAG AAG CCA GGG AGC TCT CCC AGG TTA TTC CTG TAT CAC TAC TCA GAC TCA

                                     IMGT_______________
                                     61  62  63  64  65  66  67  68  69  70  71  72  73  74  75  76  77  78  79  80
                                      D   K               Q   L   G   P   G   V   P       S   R   V   S   G   S   K
D86996,IGLV11-55*01,V4-6             GAC AAG ... ... ... CAG CTG GGA CCT GGG GTC CCC ... AGT CGA GTC TCT GGC TCC AAG

                                     81  82  83  84  85  86  87  88  89  90  91  92  93  94  95  96  97  98  99 100
                                      E   T   S   S   N   T   A   F   L   L   I   S   G   L   Q   P   E   D   E   A
D86996,IGLV11-55*01,V4-6             GAG ACC TCA AGT AAC ACA GCG TTT TTG CTC ATC TCT GGG CTC CAG CCT GAG GAC GAG GCC

                                                     __________CDR3-IMGT__________
                                     101 102 103 104 105 106 107 108 109 110 111 112
                                      D   Y   Y   C   Q   V   Y   E   S   S   A   N
D86996,IGLV11-55*01,V4-6             GAT TAT TAC TGC CAG GTG TAC GAA AGT AGT GCT AAT
```

Framework and complementarity determining regions

FR1-IMGT: 25 (-1 aa: 10)
FR2-IMGT: 17
FR3-IMGT: 38 (-1 aa: 73)

CDR1-IMGT: 9
CDR2-IMGT: 7
CDR3-IMGT: 8

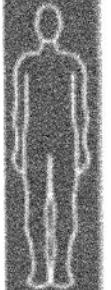

Collier de Perles for human IGLV11-55*01

Accession number: IMGT D86996 EMBL/GenBank/DDBJ: D86996

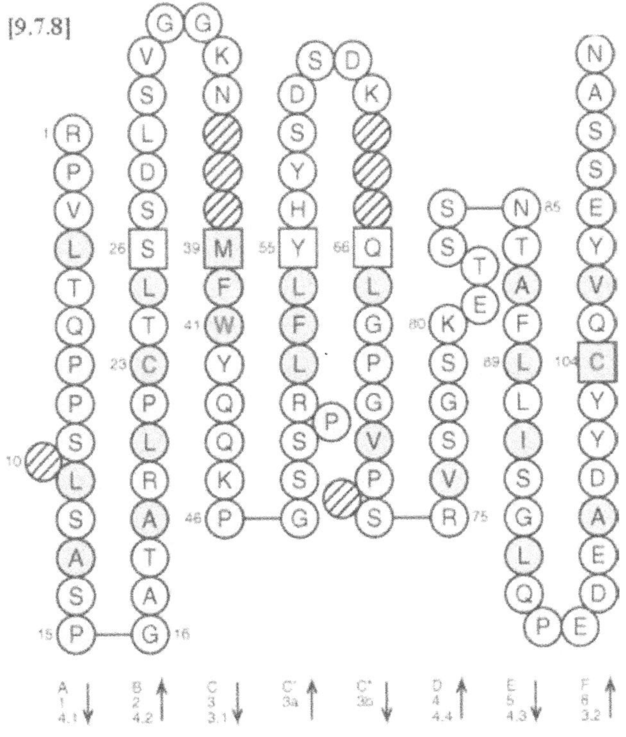

Genome database accession numbers

GDB:9953767 LocusLink: 28770

Protein display of the human IGL V-REGIONs

Only the *01 allele of each functional or ORF V-REGION is shown. IGLV genes are listed, for each subgroup, according to their position from 3′ to 5′ in the locus.

IGLV gene	FR1-IMGT (1-26)	CDR1-IMGT (27-38)	FR2-IMGT (39-55)	CDR2-IMGT (56-65)	FR3-IMGT (66-104)	CDR3-IMGT (105-115)
	1 10 20	30	40 50	60	70 80 90 100	110
Z73653,IGLV1-36	QSVLTQPPS.VSEAPRQRVTISCSGS	SSNIGNNA....	VNWYQQLPGKAPKLLIY	YDD.......	LLPSGVS.DRFSGSK..SGTSASLAISGLQSEDEADYYC	AAWDDSLNG...
M94116,IGLV1-40	QSVLTQPPS.VSGAPGQRVTISCTGS	SSNIGAGYD...	VHWYQQLPGTAPKLLIY	GNS.......	NRPSGVP.DRFSGSK..SGTSASLAITGLQAEDEADYYC	QSYDSSLSG...
M94118,IGLV1-41	QSVLTQPPS.VSAAPGQKVTISCSGS	SSDMGNYA....	VSWYQQLPGTAPKLLIY	ENN.......	KRPSGIP.DRFSGSK..SGTSATLGITGLWPEDEADYYC	LAWDTSPRA...
Z73654,IGLV1-44	QSVLTQPPS.ASGTPGQRVTISCSGS	SSNIGSNT....	VNWYQQLPGTAPKLLIY	SNN.......	QRPSGVP.DRFSGSK..SGTSASLAISGLQSEDEADYYC	AAWDDSLNG...
Z73663,IGLV1-47	QSVLTQPPS.ASGTPGQRVTISCSGS	SSNIGSNY....	VYWYQQLPGTAPKLLIY	RNN.......	QRPSGVP.DRFSGSK..SGTSASLAISGLRSEDEADYYC	AAWDDSLSG...
M94112,IGLV1-50	QSVLTQPPS.VSGAPGQRVTISCTGS	SSNIGAGYV...	VHWYQQLPGTAPKLLIY	GNS.......	NRPSGVP.DQFSGSK..SGTSASLAITGLQSEDEADYYC	KAWDNSLNA...
Z73661,IGLV1-51	QSVLTQPPS.VSAAPGQKVTISCSGS	SSNIGNNY....	VSWYQQLPGTAPKLLIY	DNN.......	KRPSGIP.DRFSGSK..SGTSATLGITGLQTGDEADYYC	GTWDSSLSA...
X97462,IGLV2-8	QSALTQPPS.ASGSPGQSVTISCTGT	SSDVGGYNY...	VSWYQQHPGKAPKLMIY	EVS.......	KRPSGVP.DRFSGSK..SGNTASLTVSGLQAEDEADYYC	SSYAGSNNF...
Z73657,IGLV2-11	QSALTQPRS.VSGSPGQSVTISCTGT	SSDVGGYNY...	VSWYQQHPGKAPKLMIY	DVS.......	KRPSGVP.DRFSGSK..SGNTASLTISGLQAEDEADYYC	CSYAGSYTF...
Z73664,IGLV2-14	QSALTQPAS.VSGSPGQSITISCTGT	SSDVGGYNY...	VSWYQQHPGKAPKLMIY	EVS.......	NRPSGVS.NRFSGSK..SGNTASLTISGLQAEDEADYYC	SSYTSSSTL...
Z73642,IGLV2-18	QSALTQPPS.VSGSPGQSVTISCTGT	SSDVGSYNR...	VSWYQQPPGTAPKLMIY	EVS.......	NRPSGVP.DRFSGSK..SGNTASLTISGLQAEDEADYYC	SLYTSSSTF...
X14616,IGLV2-23	QSALTQPAS.VSGSPGQSITISCTGT	SSDVGSYNL...	VSWYQQHPGKAPKLMIY	EGS.......	KRPSGVS.NRFSGSK..SGNTASLTISGLQAEDEADYYC	CSYAGSSTL...
Z73643,IGLV2-33	QSALTQPPF.VSGAPGQSVTISCTGT	SSDVGDYDH...	VFWYQKRLSTTSRLLIY	NVN.......	TRPSGIS.DLFSGSK..SGNMASLTISGLKSEVEANYHC	SLYSSSYTF...
X57826,IGLV3-1	SYELTQPPS.VSVSPGQTASITCSGD	KLGDKY......	ACWYQQKPGQSPVLVIY	QDS.......	KRPSGIP.ERFSGSN..SGNTATLTISGTQAMDEADYYC	QAWDSSTA....
X97473,IGLV3-9	SYELTQPLS.VSVALGQTARITCGGN	NIGSKN......	VHWYQQKPGQAPVLVIY	RDS.......	NRPSGIP.ERFSGSN..SGNTATLTISRAQAGDEADYYC	QVWDSSTA....
X97464,IGLV3-10	SYELTQPPS.VSVSPGQTARITCSGD	ALPKKY......	AYWYQQKSGQAPVLVIY	EDS.......	KRPSGIP.ERFSGSS..SGTMATLTISGAQVEDEADYYC	YSTDSSGNH...
Z73658,IGLV3-12	SYELTQPHS.VSVATAQMARITCGGN	NIGSKA......	VHWYQQKPGQDPVLVIY	SDS.......	NRPSGIP.ERFSGSN..PGNTTTLTISRIEAGDEADYYC	QVWDSSSDH...
X97471,IGLV3-16	SYELTQPPS.VSVSLGQMARITCSGE	ALPKKY......	AYWYQQKPGQFPVLVIY	KDS.......	ERPSGIP.ERFSGSS..SGTIVTLTISGVQAEDEADYYC	LSADSSGTY...
X56178,IGLV3-19	SSELTQDPA.VSVALGQTVRITCQGD	SLRSYY......	ASWYQQKPGQAPVLVIY	GKN.......	NRPSGIP.DRFSGSS..SGNTASLTITGAQAEDEADYYC	NSRDSSGNH...
X71966,IGLV3-21	SYVLTQPPS.VSVAPGKTARITCGGN	NIGSKS......	VHWYQQKPGQAPVLVIY	YDS.......	DRPSGIP.ERFSGSN..SGNTATLTISRVEAGDEADYYC	QVWDSSSDH...
Z73666,IGLV3-22	SYELTQLPS.VSVSPGQTARITCSGD	VLGENY......	ADWYQQKPGQAPELVIY	EDS.......	ERYPGIP.ERFSGST..SGNTTTLTISRVLTEDEADYYC	LSGDEDN.....
X97474,IGLV3-25	SYELMQPPS.VSVSPGQTARITCSGD	ALPKQY......	AYWYQQKPGQAPVLVIY	KDS.......	ERPSGIP.ERFSGSS..SGTTVTLTISGVQAEDEADYYC	QSADSSGTY...
D86994,IGLV3-27	SYELTQPSS.VSVSPGQTARITCSGD	VLAKKY......	ARWFQQKPGQAPVLVIY	KDS.......	ERPSGIP.ERFSGSS..SGTTVTLTISGAQVEDEADYYC	YSAADNN.....
Z73645,IGLV3-32	SSGPTQVPA.VSVALGQMARITCQGD	SMEGSY......	EHWYQQKPGQAPVLVIY	DSS.......	DRPSRIP.ERFSGSK..SGNTTTLTITGAQAEDEADYYY	QLIDNHA.....

IGLV gene	FR1-IMGT (1-26)	CDR1-IMGT (27-38)	FR2-IMGT (39-55)	CDR2-IMGT (56-65)	FR3-IMGT (66-104)	CDR3-IMGT (105-115)
	1 10 20	30	40 50	60	70 80 90 100	110
X57828,IGLV4-3	LPVLTQPPS.ASALLGASIKLTCTLS	SEHSTYT.....	IEWYQQRPGRSPQYIMK	VKSDGSH...	SKGDGIP.DRFMGSS..SGADRYLTFSNLQSDDEAEYHC	GESHTIDGQVG*
Z73667,IGLV4-60	QPVLTQSSS.ASASLGSSVKLTCTLS	SGHSSYI.....	IAWHQQQPGKAPRYLMK	LEGSGSY...	NKGSGVP.DRFSGSS..SGADRYLTISNLQLEDEADYYC	ETWDSNT.....
Z73648,IGLV4-69	QLVLTQSPS.ASASLGASVKLTCTLS	SGHSSYA.....	IAWHQQQPEKGPRYLMK	LNSDGSH...	SKGDGIP.DRFSGSS..SGAERYLTISSLQSEDEADYYC	QTWGTGI.....
Z73672,IGLV5-37	QPVLTQPPS.SSASPGESARLTCTLP	SDINVGSYN...	IYWYQQKPGSPPRYLLY	YYSDSDK...	GQGSGVP.SRFSGSKDASANTGILLISGLQSEDEADYYC	MIWPSNAS....
Z73668,IGLV5-39	QPVLTQPTS.LSASPGASARFTCTLR	SGINVGTYR...	IYWYQQKPGSLPRYLLR	YKSDSDK...	QQGSGVP.SRFSGSKDASTNAGLLLISGLQSEDEADYYC	AIWYSSTS....
Z73670,IGLV5-45	QAVLTQPAS.LSASPGASASLTCTLR	SGINVGTYR...	IYWYQQKPGSPPQYLLR	YKSDSDK...	QQGSGVP.SRFSGSKDASANAGILLISGLQSEDEADYYC	MIWHSSAS....
Z73649,IGLV5-48	QPVLTQPTS.LSASPGASARLTCTLR	SGINLGSYR...	IFWYQQKPESPPRYLLS	YYSDSSK...	HQGSGVP.SRFSGSKDASSNAGILVISGLQSEDEADYYC	MIWHSSAS....
Z73669,IGLV5-52	QPVLTQPSS.HSASSGASVRLTCMLS	SGFSVGDFW...	IRWYQQKPGNPPRYLLY	YHSDSNK...	GQGSGVP.SRFSGSNDASANAGILRISGLQPEDEADYYC	GTWHSNSKT...
Z73673,IGLV6-57	NFMLTQPHS.VSESPGKTVTISCTRS	SGSIASNY....	VQWYQQRPGSSPTTVIY	EDN.......	QRPSGVP.DRFSGSIDSSSNSASLTISGLKTEDEADYYC	QSYDSSN.....
X14614,IGLV7-43	QTVVTQEPS.LTVSPGGTVTLTCASS	TGAVTSGYY...	PNWFQQKPGQAPRALIY	STS.......	NKHSWTP.ARFSGSL..LGGKAALTLSGVQPEDEAEYYC	LLYYGGAQ....
Z73674,IGLV7-46	QAVVTQEPS.LTVSPGGTVTLTCGSS	TGAVTSGHY...	PYWFQQKPGQAPRTLIY	DTS.......	NKHSWTP.ARFSGSL..LGGKAALTLSGAQPEDEAEYYC	LLSYSGAR....
Z73650,IGLV8-61	QTVVTQEPS.FSVSPGGTVTLTCGLS	SGSVSTSYY...	PSWYQQTPGQAPRTLIY	STN.......	TRSSGVP.DRFSGSI..LGNKAALTITGAQADDESDYYC	VLYMGSGI....
Z73675,IGLV9-49	QPVLTQPPS.ASASLGASVTLTCTLS	SGYSNYK.....	VDWYQQRPGKGPRFVMR	VGTGGIVG..	SKGDGIP.DRFSVLG..SGLNRYLTIKNIQEEDESDYHC	GADHGSGSNFV*
Z73676,IGLV10-54	QAGLTQPPS.VSKGLRQTATLTCTGN	SNNVGNQG....	AAWLQQHQGHPPKLLSY	RNN.......	NRPSGIS.ERLSASR..SGNTASLTITGLQPEDEADYYC	SAWDSSLSA...
D86996,IGLV11-55	RPVLTQPPS LSASPGATARLPCTLS	SDLSVGGKN...	MFWYQQKPGSSPRLFLY	HYSDSDK...	QLGPGVP.SRVSGSKETSSNTAFLLISGLQPEDEADYYC	QVYESSAN....

Recombination signals

Only the recombination signals of the allele *01 of each functional or ORF IGL V-REGION are shown.
Non-conserved nucleotides taken into account for the ORF functionality definition are shown in bold and italic.

IGLV	V recombination signal		
gene name	V-HEPTAMER	(bp)	V-NONAMER
IGLV1-36*01	CACAGTG	23	ACAAGAACC
IGLV1-40*01	CACAGTG	23	ACAAGAACC
IGLV1-41*01 (ORF)	CACAGTG	23	ACAAGAACC
IGLV1-44*01	CACAGTG	23	ACAAGAACC
IGLV1-47*01	CACAGTG	23	ACAAGAACC
IGLV1-50*01 (ORF)	CACAGTG	23	A***G***AAGAACC
IGLV1-51*01	CACAGTG	23	ACAAGAACC
IGLV2-8*01	CACAGTG	23	ACCAAAACC
IGLV2-11*01	CACAGTG	23	ACCAAAACC
IGLV2-14*01	CACAGTG	23	ACCAAAACC
IGLV2-18*01	CACAGAG	23	ACCAAAACC
IGLV2-23*01	CACAGTG	23	ACCAAAACC
IGLV2-33*01 (ORF)	CA***T***AGTG	23	ACCAAAACC
IGLV3-1*01	CACAGTG	23	ACAGAAACC
IGLV3-9*01	CACAGTG	23	ACACAAACC
IGLV3-10*01	CACAGTG	23	ACACAAACC
IGLV3-12*01	CACGGTG	23	ACACAAACA
IGLV3-16*01	CACAGTG	23	ACATAAACC
IGLV3-19*01	CACAGTG	23	ACAGAAACC
IGLV3-21*01	CACGGTG	23	ACAAAAACA
IGLV3-22*01	CTCAGTG	23	ACACAAACT
IGLV3-25*01	CACAGTG	23	ACATAAACC
IGLV3-27*01	CACAGTG	23	ACACAAACC
IGLV3-32*01 (ORF)	CACAGTG	23	ACACAAACC
IGLV4-3*01	CACAGTG	23	CCAAAACTC
IGLV4-60*01	CACAGTG	23	ACAAAATCC
IGLV4-69*01	CACAGTG	23	ACAGAAACC
IGLV5-37*01	CACAGTG	23	ACAAAAACC
IGLV5-39*01 (ORF)	nd	nd	nd
IGLV5-45*01	CACAGTG	23	ACAAAAACC
IGLV5-48*01 (ORF)	***T***ACAGTG	23	ACAAAAACT
IGLV5-52*01	CACAGTG	23	ACAAAACCC
IGLV6-57*01	CACAGTG	23	ACAGAAACT
IGLV7-43*01	CACAGTG	23	ACATAAACC
IGLV7-46*01	CACAGTG	23	ACATAAACC
IGLV8-61*01	CACAGTG	23	ACTAAAACC
IGLV9-49*01	CACAGTG	23	ACAAAAACC
IGLV10-54*01	CACAGTG	23	ATAAAAACT
IGLV11-55*01 (ORF)	CACAGTG	19	ACAAAAACC

nd: not defined

IGLV references

[1] Alexandre, D. et al. (1989) Nucleic Acids Res. 17, 3975.
[2] Anderson, M.L.M. et al. (1984) Nucleic Acids Res. 12, 6647–6661.
[3] Berek, C. et al. (1997) Immunogenetics 46, 533–534.
[4] Bernard, O. et al. (1990) Nucleic. Acids Res. 18, 7139.
[5] Brockly, F. et al. (1989) Nucleic Acids Res. 17, 3976.
[6] Ch'ang, L.-Y. et al. (1994) Mol. Immunol. 31, 531–536.
[7] Ch'ang, L.-Y. et al. (1995) Mol. Immunol. 32, 49–55.
[8] Combriato, G. and Klobeck, H.-G. (1991) Eur. J. Immunol. 21, 1513–1522.
[9] Daley, M.D. et al. (1992) Mol. Immunol. 29, 1031–1042.
[10] Daley, M.D. et al. (1992) Mol. Immunol. 29, 1515–1518.
[11] Deftos, M. et al. (1994) Scand. J. Immunol. 39, 95–103.
[12] Deftos, M. et al. (1994) J. Clin. Invest. 93, 2545–2553.
[13] Eulitz, M. et al. (1995) J. Immunol. 154, 3256–3265.
[14] Fang, Q. et al. (1994) J. Exp. Med. 179, 1445–1456.
[15] Frippiat, J.-P. and Lefranc, M.-P. (1994) Mol. Immunol. 31, 657–670.
[16] Frippiat, J.-P. et al. (1990) Nucleic Acids Res. 18, 7134.
[17] Harmer, I.J. et al. (1995) Arthritis Rheum. 38, 1068–1076.
[18] Irigoyen, M. et al. (1994) J. Clin. Invest. 94, 532–538.
[19] Kawasaki, K. et al. (1997) Genome Res. 7, 250–261.
[20] Lee, G. et al. (1994) J. Neuroimmunol. 51, 45–52.
[21] Siminovitch, K.A. et al. (1989) J. Clin. Invest. 84, 1675–1678.
[22] Solomon, A. et al. (1997) Mol. Immunol. 6, 463–470.
[23] Stiernholm, N.B.J. et al. (1994) J. Immunol. 152, 4969–4975.
[24] Stiernholm, N.B.J. and Berinstein, N.L. (1995) J. Immunol. 154, 1748–1761.
[25] Vasicek, T.J. and Leder, P. (1990) J. Exp. Med. 172, 609–620.
[26] Williams, S.C. and Winter, G. (1993) Eur. J. Immunol. 23, 1456–1461.
[27] Williams, et al. (1996) J. Mol. Biol. 264, 220–232.
[28] Winkler, T.H. et al. (1992) Eur. J. Immunol. 22, 1719–1728.

Index

Printed and bound by CPI Group (UK) Ltd, Croydon, CR0 4YY
03/10/2024
01040421-0005